*P. B. Kirti, PhD*
*Editor*

# Handbook of New Technologies for Genetic Improvement of Legumes

*Pre-publication*
*REVIEWS,*
*COMMENTARIES,*
*EVALUATIONS . . .*

"This is a timely publication that provides exhaustive and insightful commentary on the recent successful work on genetic improvement of many economically important legumes by distinguished experts. It should be of interest and use to anyone interested in legume biology and improvement."

**Indra K. Vasil**
*Associate Director, Genetics Institute, Graduate Research Professor Emeritus, University of Florida*

"This book provides a wide range of practical methods for legume cell culture and transformation. In addition, many chapters contain comprehensive accounts of the most recent advances in fundamental and applied aspects of legume biotechnology. Perhaps the most impressive feature of this monumental work is that it manages to incorporate the contributions from leading researchers engaged in different geographical regions and working with a wide variety of food, forage, oilseed, and pulse crops. I highly recommend this book as a most comprehensive and meaningful compilation of legume transformation literature, which is difficult to capture in a book format. In sum, the book should be useful for both young and experienced researchers who wish to evaluate what has been, and what remains to be, accomplished in this fast-growing field."

**Praveen K. Saxena, PhD**
*Professor, Department of Plant Agriculture, University of Guelph*

# HANDBOOK OF NEW TECHNOLOGIES FOR GENETIC IMPROVEMENT OF LEGUMES

# HANDBOOK OF NEW TECHNOLOGIES FOR GENETIC IMPROVEMENT OF LEGUMES

P. B. KIRTI, PHD

EDITOR

CRC Press
Taylor & Francis Group
Boca Raton London New York

CRC Press is an imprint of the
Taylor & Francis Group, an **informa** business

CRC Press
Taylor & Francis Group
6000 Broken Sound Parkway NW, Suite 300
Boca Raton, FL 33487-2742

First issued in hardback 2020



ISBN-13: 978-1-56022-308-5 (hbk)



**Library of Congress Cataloging-in-Publication Data**

Handbook of new technologies for genetic improvement of legumes / P. B. Kirti, editor.
p. cm.
ISBN: 978-1-56022-308-5 (hard : alk. paper)
ISBN: 978-1-56022-309-2 (soft : alk. paper)
1. Legumesó Genetic engineeringó Handbooks, manuals, etc. 2. Forage plantsó Genetic engineeringó Handbooks, manuals, etc. 3. Transgenic plantsó Handbooks, manuals, etc. I. Kirti, P. B.

SB203.H36 2007
633.3í042336 dc22 2007034357

**Visit the Taylor & Francis Web site at**
**http://www.taylorandfrancis.com**

**and the CRC Press Web site at**
**http://www.crcpress.com**

# CONTENTS

**Contributors** xix

**Foreword** xxv

*V. L. Chopra*

**Preface** xxvii

**Chapter 1. Genetic Engineering for Forage Quality in Alfalfa** 1

*Champa Sengupta-Gopalan*
*Suman Bagga*
*Jose Luis Ortega*
*Carol Potenza*

Introduction 1
Promoters for Genetic Engineering in Alfalfa 2
Nutritional Improvement of Alfalfa 3
Improving Digestibility of Alfalfa Forage 8
Improving Efficiency of Protein Utilization 11

**Chapter 2. Genetic Transformation in *Trifolium* and *Medicago*** 19

*Stephen Panter*
*Aidyn Mouradov*
*Emma Ludlow*
*Marcel Labandera*
*Michael Emmerling*
*German Spangenberg*

Introduction 19
Materials 21
*Agrobacterium*-Mediated Transformation Method 22
Isogenic Transformation of White Clover 25
Conclusion 27

**Chapter 3. *Agrobacterium*-Mediated Genetic Transformation of Alfalfa** **29**

*Michele Bellucci*
*Francesca De Marchis*
*Sergio Arcioni*

Introduction 29
Somatic Embryogenesis and Plant Regeneration 31
Materials 31
Methods 34
Conclusion 37

**Chapter 4. *Agrobacterium*-Mediated Transformation of *Medicago truncatula*** **45**

*M. Chabaud*
*A. Boisson-Dernier*
*E.-P. Journet*
*F. de Carvalho-Niebel*
*D. G. Barker*

Introduction 45
*Agrobacterium tumefaciens*-Mediated Transformation and In Vitro Regeneration of Transgenic Plants 46
*Agrobacterium tumefaciens*-Mediated Transformation and Regeneration via Organogenesis 55
*Agrobacterium rhizogenes*-Mediated Transformation 57
Key Parameters for *Agrobacterium rhizogenes*-Mediated Transformation of *M. truncatula* 58
Conclusion and Applications of *A. rhizogenes*-Mediated Transformation 62
Final Comments 63

**Chapter 5. Biotechnological Approaches to Soybean Improvement** **69**

*Qungang Qi*
*Henry E. Valentin*

Introduction 69
Summary of Field Trial Applications of Transgenic Soybean for the Period 1987-2005 70
Improvement of Soybean Seed Quality Traits 70

Enhancing Seed Tryptophan Levels 73
Engineering Seed Oil Composition and Content 76
Transgenic Approaches for Enhancement of Vitamin E 77
Practical Considerations 83
Conclusion and Future Prospects 84

**Chapter 6. *Agrobacterium*-Mediated Genetic Transformation of Soybean** **89**

*Randy D. Dinkins*
*Glenn B. Collins*

Introduction 89
*Agrobacterium*-Soybean Interaction 90
Culture Protocols 91
Meristematic Cell Proliferation and Plant Regeneration 91
Somatic Embryogenesis Protocols 94
Meristematic Cell Proliferation versus Somatic Embryogenesis Protocols 96
Conclusion 97

**Chapter 7. Genetic Transformation of Soybean Using Particle Bombardment and SAAT Approaches** **103**

*John J. Finer*
*Kathryn M. Larkin*

Soybean Transformation 103
Embryogenic Cultures of Soybean as a Target for Gene Transfer 104
Description of Soybean Embryogenic Cultures 106
Particle Bombardment 107
A Recommended Method for Soybean Transformation via Particle Bombardment of Embryogenic Tissues 109
*Agrobacterium*-Mediated Transformation 111
Sonication-Assisted *Agrobacterium*-Mediated Transformation 113
SAAT Optimization Strategies 113

**Chapter 8. Genetic Transformation as a Tool for Improvement of Pigeon Pea, *Cajanus Cajan* (L.) Millsp.** **125**

*G. Lakshmi Sita*
*P. Venkatachalam*

Introduction 125
Major Constraints for the Crop 125
Scope for Biotechnological Improvement 126
Different Approaches to Crop Improvement 127
In Vitro Approaches 128
Plant Regeneration via Organogenesis 128
Plant Regeneration via Somatic Embryogenesis and Protoplast Culture 134
Genetic Transformation 135
Future Prospects 141
Conclusion 142

**Chapter 9. Genetic Transformation of Chickpea (*Cicer arietinum* L.) Using Cotyledonary Node Explants** **147**

*Indraneel Sanyal*
*D. V. Amla*

Introduction 147
Materials and Procedures 148
Results 152
Discussion 155

**Chapter 10. Genetic Transformation of *Phaseolus* and *Vigna* Species** **159**

*Nancy Terryn*
*Marc van Montagu*

Introduction 159
State of the Art of Genetic Transformation in Cowpea and *Phaseolus* 160
Some Problems of Bean and Cowpea That Could Be Solved by Genetic Engineering 162
New Avenues for Genetic Transformation 164

**Chapter 11. Genetic Engineering in Azuki Bean** **169**
*Masao Ishimoto*

Introduction 169
Genetic and Molecular Features of Azuki Bean 169
Plant Regeneration from Explants and Protoplasts 170
*Agrobacterium*-Mediated Transformation Using Epicotyl Explants 171
A Transformation Protocol for Azuki Bean 172
Conclusion 173

**Chapter 12. Somaclonal Variation and Genetic Transformation in *Lathyrus sativus*** **177**
*S. L. Mehta*
*I. M. Santha*

Introduction 177
Exploitation of Somaclonal Variation 178
Biochemical and Molecular Characterization 180
Transformation of Lathyrus sativus 181
Somatic Embryogenesis 183
Future Prospects 184

**Chapter 13. *Agrobacterium*-Mediated Transformation of Pea (*Pisum sativum* L.)** **187**
*Anna Nadolska-Orczyk*

Introduction 187
Toward *Agrobacterium*-Mediated Transformation 188
Methods of *Agrobacterium*-Mediated Pea Transformation 190
Transformation Efficiency, Expression, and Inheritance 194
Genetically Improved Peas 195
Results 196
Protocols 197

**Chapter 14. Genetic Transformation of Pea by Microprojectile Bombardment** **203**
*Z. Molnar*

Introduction 203
Gene Transfer to Plants Using Biolistics 203

Factors Affecting Transformation by Microprojectile Bombardment 204
Genetic Transformation in Pea 205
Protocol for Pea Transformation by Microprojectile Bombardment 206
Results 208

**Chapter 15. *In Planta* Transformation of *Pisum sativum* L. Using Microinjection and Electroporation** **217**

*Paul F. Lurquin*

Introduction 217
Methodology, Analytical Procedures, and Results 219
Response of PEMV Transgenics to Viral Challenge 222
Conclusion 222

**Chapter 16. *Agrobacterium*-Mediated Genetic Transformation of Peanut** **227**

*P. Bhatnagar-Mathur*
*V. Anjaiah*
*P. B. Kirti*
*Kiran K. Sharma*

Introduction 227
Tissue Culture Methods 229
Genetic Transformation 233
Direct Gene Transfer 239
Genetically Engineered Plants for Peanut Improvement 241
Conclusion 244

**Chapter 17. Cotyledonary Node and Embryo Axes as Explants in Legume Transformation with Special Reference to Peanut** **253**

*T. Swathi Anuradha*
*S. K. Jami*
*M. R. Beena*
*P. B. Kirti*

Introduction 253
Genetic Transformation of Grain Legumes Using Cotyledonary Node as an Explant 254

Embryo Axes 257
Genetic Transformation of Peanut Using Cotyledonary Node Explants 264

**Chapter 18. A Novel *In Planta* Approach to Gene Transfer for Legumes** **273**

*K. Sankara Rao*
*V. K. Rohini*

Introduction 273
Targets for Genetic Modification of Legumes 274
Transformation of Legumes 274
Embryo Transformation: An *In Planta* Method for Large Seed Legumes 278
Conclusion 282

**Chapter 19. *Agrobacterium*-Mediated Genetic Transformation of Faba Bean** **287**

*M. Hanafy*
*H.-J. Jacobsen*
*P. Böttinger*
*T. Pickardt*

Introduction 287
Genetic Transformation 288
Methodology 289
De Novo Regeneration Protocol 291
Embryo Axis Transformation System 292
Results 294
De Novo Regeneration Versus Embryo Axis Transformation Protocols 297

**Chapter 20. Genetic Transformation of *Lotus* Species** **301**

*Francesco Damiani*
*Francesco Paolocci*
*Sergio Arcioni*

Introduction 301
The Genus *Lotus* 302
In Vitro Manipulation of *Lotus* Species 302
Genetic Transformation of *Lotus* Species 303

Application of Genetic Transformation in *Lotus* Species 306
Protocols 308
Conclusion 310

**Chapter 21. Genetic Transformation of Tree Legume, *Leucaena leucocephala*** **317**
*Smita Rastogi*
*U. N. Dwivedi*

Introduction 317
In Vitro Regeneration of *Leucaena* 317
Transformation Protocol for *Leucaena leucocephala* 319
Results 323

**Chapter 22. Genetic Transformation of Tree Legume, *Robinia*** **331**
*Tomohiro Igasaki*
*Kenji Shinohara*

Introduction 331
Procedure for Transformation of *R. pseudoacacia* 331
Identification of Suitable Selection Marker, *A. tumefaciens* Strain and Plant Tissue for Transformation of *R. pseudoacacia* 333
Characterization of Transgenic Plants 334
General Considerations 336

**Chapter 23. *Medicago truncatula* As a Model System in Plant Molecular Biology** **339**
*Julia Frugoli*

Model System 339
The Availability of Powerful Tools of Traditional and Molecular Genetics 340
The Research Community 346
Acknowledgments 348

**Chapter 24. Reverse Genetic Approaches in *Medicago truncatula*** **353**

*Lysiane Brocard*
*Isabelle d'Erfurth*
*Adam Kondorosi*
*Pascal Ratet*

Reverse Genetics in Plants 353
Development of Molecular Tools for Model Legumes 354
Reverse Genetics Tools in Legumes 355
Analysis of the *Tnt1*-Induced Mutants 362
Conclusion 362

**Chapter 25. Fluorescence *In Situ* Hybridization on *Medicago truncatula* Chromosomes** **371**

*Olga Kulikova*
*Hans de Jong*
*Ton Bisseling*

Development of FISH Technology in *M. truncatula* 372
General Technical Aspects of FISH on *M. truncatula* Chromosomes 375

**Chapter 26. Molecular and Genetic Analyses of Transgenic Plants** **385**

*S. R. Bhat*
*R. Srinivasan*

Introduction 385
Causes of Variation Among Transgenics 385
Methods of Analysis of Transgenic Plants 386

**Chapter 27. Tagging Regulatory Elements in Plants** **403**

*Raju Datla*
*Vivijan Babic*
*P. B. Kirti*
*Gopalan Selvaraj*

Introduction 403
Plant Promoters 404
In Vivo Promoter Tagging for Plants 405

Isolation of Promoters from Tagged Lines 408
Perspectives 409

**Chapter 28. cDNA-AFLP Transcriptome Analysis in Legumes** **413**

*Márcia Margis-Pinheiro*
*Jean Luis Simões-Araújo*

Differentially Expressed Genes 413
cDNA-AFLP Systems Used for Gene Discovery 416
cDNA-AFLP Protocols 419
Conclusion 423

**Chapter 29. Suppression Subtractive Hybridization to Identify Novel Genes in Legumes** **427**

*Katrien Schroeyers*
*Sofie Goormachtig*
*Marcelle Holsters*

Introduction 427
The SSH Technique 429
Applications 433
Experimental Setup 439
Large-Scale Primary Confirmation 444
Small-Scale Secondary Confirmation 447
SSH and Legumes 449
Conclusion 451

**Chapter 30. Quantitative PCR in the Analysis of Transgenic Plants** **459**

*Hiroyuki Koyama*
*Tomonori Kihara*
*Eiji Takita*
*Munechika Honda*

Introduction 459
Quantitative PCR Procedures 460
Determination of the Copy Number in Transgenic Plants Using Quantitative PCR 462
Quantitative RT-PCR for Determination of Transcript Levels 463

**Chapter 31. Gene Expression Analysis with DNA Array Technology in Legumes** **467**

*Masao Watanabe*
*Makoto Endo*
*Hirokazu Hakozaki*
*Hiromi Masuko*
*Jong-In Park*
*Makiko Kawagishi-Kobayashi*
*Tohru Tsuchiya*

Introduction 467
Analysis of Expressed Sequence Tags and Determination of Genome Sequence 468
Gene Expression Analysis with Micro- and Macroarrays 469
Conclusion 472

**Chapter 32. Advent of Diverse DNA Markers to Decipher Genome Sequence Polymorphism** **477**

*Deepmala Sehgal*
*Vishnu Bhat*
*Soom Nath Raina*

Introduction 477
Non-PCR-Based Markers 478
PCR-Based Markers 478
Microsatellite-Based Markers 482
Expressed Sequence Tag (EST)-Based Markers 483
Retrotransposon/Miniature-Inverted Repeat Transposable Element (MITE)-Based Markers 485
Microsatellite-and Minisatellite-Based Markers 488
Other Markers 490

**Chapter 33. Applicability of DNA Markers for Genome Diagnostics of Grain Legumes** **497**

*Deepmala Sehgal*
*Vishnu Bhat*
*Soom Nath Raina*

Introduction 497
DNA Markers in Grain Legumes 498
Conclusions 533

**Chapter 34. Genetic Maps and Molecular Markers in the Genus *Medicago*** **559**

*Bernadette Julier*
*Thierry Huguet*

Introduction 559
Markers in *Medicago* Spp. 560
Genetic Maps in *Medicago truncatula* 562
Genetic Maps in *Medicago sativa* at the Diploid Level 565
Genetic Maps at the Tetraploid Level 566
Synteny 568
Prospects 569

**Chapter 35. Management of Abiotic Stresses in Grain Legumes Through Manipulation of Genes for Compatible Solutes** **577**

*P. Sharmila*
*Firoz Anwar*
*K. R. Sharma*
*P. Pardha Saradhi*

Introduction 577
Strategies for Enhancing Abiotic Stress Tolerance 579
Compatible Solutes 580
Epilogue 594

**Chapter 36. Legume-Nematode Interactions with Reference to Model Legumes** **605**

*Dasharath P. Lohar*

Introduction 605
Development of Legume Models in Plant-Nematode Interactions 606
Resistance to Root Parasitic Nematodes in Model Legumes 609
Advances in Understanding Legume-Nematode Interactions Using Model Legumes 611
Conclusion 613

**Chapter 37. Defensins in Legumes** **617**

*Jack H. Wong*
*T. B. Ng*

Introduction 617
Definition and Structure of Defensin 618
Mechanisms of Action of Defensins 620
$M(IP)_2C$ as Target for Dahlia Plant Defensin DmAMP1 620
Fungal Glucosyl Ceramides as Target for the Radish Plant Defensin RsAFP2 621
L-Type Calcium Channel as Target for Alfalfa Defensin MsDef1 622
Applications of Plant Defensins 622
Expression of Plant Defensins in Transgenic Plants 622
Isolation of Legume Defensins 623

**Index** **629**

## ABOUT THE EDITOR

**P. B. Kirti, PhD,** is Professor in the Department of Plant Sciences at the University of Hyderabad, India. He has extensive experience in genetic manipulation in mustard and chickpea and in development of cytoplasmic male sterility-fertility restorer systems based on the use of cell fusion hybrids. Dr. Kirti's current research interests revolve around the genetic manipulation of peanut and pigeonpea for pest and disease resistance as well as transposon insertion mutagenesis. He has been a Fellow at the Alexander von Humboldt Foundation in Germany and FAO Visiting Fellow at the Ohio State Biotechnology Center and is a currently a Fellow of the National Academy of Agricultural Scientists, an Elected Member of the Plant Tissue Association of India, and a Visiting Principal Fellow at the University of Melbourne.

# Contributors

**D. V. Amla,** Plant Transgenic Laboratory, National Botanical Research Institute, Lucknow, India.

**V. Anjaiah,** Genetic Transformation Laboratory, International Crops Research Institute for Semi Arid Tropics, Patancheru, Hyderabad, India.

**T. Swathi Anuradha,** Department of Plant Sciences, University of Hyderabad, Hyderabad, India.

**Firoz Anwar,** Department of Environmental Biology, University of Delhi, Delhi, India.

**Sergio Arcioni,** Institute of Plant Genetics, Research Division of Perugia, CNR, Perugia, Italy.

**Vivijan Babic,** Plant Biotechnology Institute, National Research Council of Canada, Saskatoon, Saskatchewan, Canada.

**Suman Bagga,** Department of Plant and Environmental Sciences, New Mexico State University, Las Cruces, New Mexico.

**D. G. Barker,** Laboratory of Plant-Microbe Interactions, UMR INRA-CNRS, Castanet Tolosan Cedex, France.

**M. R. Beena,** Department of Plant Sciences, University of Hyderabad, Hyderabad, India.

**Michele Bellucci,** Institute of Plant Genetics, Research Division of Perugia, CNR, Perugia, Italy.

**S. R. Bhat,** National Research Centre for Plant Biotechnology, Indian Agricultural Research Institute, New Delhi, India.

**Vishnu Bhat,** Department of Botany, University of Delhi, Delhi, India.

**P. Bhatnagar-Mathur,** Genetic Transformation Laboratory, International Crops Research Institute for Semi Arid Tropics, Patancheru, Hyderabad, India.

**Ton Bisseling,** Wageningen University, Laboratory of Molecular Biology, Wageningen, Netherlands.

**A. Boisson-Dernier,** Division of Biological Sciences, University of California, San Diego, La Jolla, California.

**P. Böttinger,** Institute for Applied Genetics, Department of Biology/Chemistry/Pharmacy of the Free University of Berlin, Germany.

**Lysiane Brocard,** Institut des Sciences du Végétal, CNRS, Cedex, France.

**M. Chabaud,** Laboratory of Plant-Microbe Interactions, UMR INRA-CNRS, Castanet Tolosan Cedex, France.

**Glenn B. Collins,** Department of Agronomy, University of Kentucky, Lexington, Kentucky.

**Francesco Damiani,** Institute of Plant Genetics, Research Division of Perugia, CNR, Perugia, Italy.

**Raju Datla,** Plant Biotechnology Institute, National Research Council of Canada, Saskatoon, Saskatchewan, Canada.

**F. de Carvalho-Niebel,** Laboratory of Plant-Microbe Interactions, UMR INRA-CNRS, Castanet Tolosan Cedex, France.

**Hans de Jong,** Wageningen University, Laboratory of Genetics, Wageningen, Netherlands.

**Francesca De Marchis,** Institute of Plant Genetics, Research Division of Perugia, CNR, Perugia, Italy.

**Isabelle d'Erfurth,** Institut des Sciences du Végétal, CNRS, Cedex, France.

**Randy D. Dinkins,** USDA-ARS Animal-Forage Production Research Unit, University of Kentucky, Kentucky.

**U. N. Dwivedi,** Department of Biochemistry, Lucknow University, Lucknow, India.

**Michael Emmerling,** QIAGEN Pty Ltd., Doncaster, Victoria, Australia.

**Makoto Endo,** Graduate School of Life Sciences, Tohoku University, Sendai, Japan; Laboratory of Biotechnology, National Institute of Crop Science Research Center, Mie, Japan; Promotion of Basic Research Activities for Innovative Biosciences, Bio-oriented Technology Research Advancement Institution, Minato-ku, Tokyo, Japan.

**John J. Finer,** Department of Horticulture and Crop Science, OARDC, The Ohio State University, Wooster, Ohio.

**Julia Frugoli,** Department of Genetics, Biochemistry and Life Science Studies, Clemson University, Clemson, South Carolina.

**Sofie Goormachtig,** Department of Plant Systems Biology, University of Ghent, Ghent, Belgium.

**Hirokazu Hakozaki,** Graduate School of Life Sciences, Tohoku University, Sendai, Japan.

**M. Hanafy,** Plant Biotechnology Department, National Research Centre, Cairo, Egypt.

**Munechika Honda,** Gifu Presectural Research Institute for Bioengineering, Minokamo, Gifu, Japan.

**Marcelle Holsters,** Department of Plant Systems Biology, University of Ghent, Ghent, Belgium.

**Thierry Huguet,** Laboratoire Biotechnologie et Amélioration des Plantes, Pôle de Biotechnologie Végétale, INP-ENSAT, Castanet Tolosan Cedex, France.

**Tomohiro Igasaki,** Department of Molecular and Cell Biology, Forestry and Forest Products Research Institute, Tsukuba, Ibaraki, Japan.

**Masao Ishimoto,** National Agricultural Research Center for Hokkaido Region, Sapporo, Hokkaido 062-8555, Japan.

**H.-J. Jacobsen,** LG Molekulargenetik, Hannover, Germany.

**S. K. Jami,** Department of Plant Sciences, University of Hyderabad, Hyderabad, India.

**E.-P. Journet,** Laboratory of Plant-Microbe Interactions, UMR INRA-CNRS, Castanet Tolosan Cedex, France.

**Bernadette Julier,** INRA, Unité de Génétique et d'Amélioration des Plantes Fourragères, Lusignan, France.

**Makiko Kawagishi-Kobayashi,** Laboratory of Biotechnology, National Institute of Crop Science Research Center, Mie, Japan; Promotion of Basic Research Activities for Innovative Biosciences, Bio-oriented Technology Research Advancement Institution, Minato-ku, Tokyo, Japan.

**Tomonori Kihara,** Forestry Research Institute, Oji Paper Company, Kameyama, Mie, Japan.

**Adam Kondorosi,** Institut des Sciences du Végétal, CNRS, Cedex, France; Institute of Genetics, BRC, Hungarian Academy of Sciences, Szeged, Hungary.

**Hiroyuki Koyama,** Laboratory of Plant Cell Technology, Faculty of Applied Biological Sciences, Gifu University, Yanagido, Gifu, Japan.

**Olga Kulikova,** Wageningen University, Laboratory of Molecular Biology, Wageningen, Netherlands.

**Marcel Labandera,** Montevideo, Uruguay.

**G. Lakshmi Sita,** Department of Microbiology and Cell Biology, Indian Institute of Science, Bangalore, India.

**Kathryn M. Larkin,** Plant Sciences, University of Arizona, Tucson, Arizona.

**Dasharath P. Lohar,** Biological Sciences, University of Minnesota, St. Paul, Minnesota.

**Emma Ludlow,** Victorian AgriBiosciences Centre, Melbourne, Victoria, Australia.

**Paul F. Lurquin,** School of Molecular Biosciences, Washington State University, Pullman Washington.

**Márcia Margis-Pinheiro,** Laboratório de Genética Vegetal, Departamento de Genética Universidade Federal do Rio Grande, do Sul, Brazil.

**Hiromi Masuko,** Faculty of Agriculture, Iwate University, Morioka, Japan.

**S. L. Mehta,** Division of Biochemistry, Indian Agricultural Research Institute, New Delhi, India.

**Z. Molnar,** Department of Plant Physiology and Plant Biotechnology, Faculty of Food and Agriculture Sciences, University of West Hungary, Mosonmagyorovar, Hungary.

**Aidyn Mouradov,** Victorian AgriBiosciences Centre, Melbourne, Victoria, Australia.

**Anna Nadolska-Orczyk,** Plant Transformation and Cell Engineering Lab, Plant Breeding and Acclimatization Institute, Radzikow, Poland.

**T. B. Ng,** Department of Biochemistry, Faculty of Medicine, Chinese University of Hong Kong, Shatin, New Territories, Hong Kong, China.

**Jose Luis Ortega,** Department of Plant and Environmental Sciences, New Mexico State University, Las Cruces, New Mexico.

**Stephen Panter,** Victorian AgriBiosciences Centre, Melbourne, Victoria, Australia.

**Francesco Paolocci,** Institute of Plant Genetics, Research Division of Perugia, CNR, Perugia, Italy.

**P. Pardha Saradhi,** Department of Environmental Biology, University of Delhi, Delhi, India.

**Jong-In Park,** Graduate School of Life Sciences, Tohoku University, Sendai, Japan.

**T. Pickardt,** Institute for Applied Genetics, Department of Biology/Chemistry/Pharmacy of the Free University of Berlin, Germany.

**Carol Potenza,** Department of Plant and Environmental Sciences, New Mexico State University, Las Cruces, New Mexico.

**Qungang Qi,** Monsanto Company, Chesterfield, Missouri.

**Soom Nath Raina,** Department of Botany, University of Delhi, Delhi, India.

**Smita Rastogi,** Department of Biochemistry, Lucknow University, Lucknow, India.

**Pascal Ratet,** Institut des Sciences du Végétal, CNRS, Cedex, France.

**V. K. Rohini,** Department of Crop Physiology, University of Agricultural Sciences, Bangalore, India.

**K. Sankara Rao,** Department of Biochemistry, Indian Institute of Science, Bangalore, India.

**Indraneel Sanyal,** Plant Transgenic Laboratory, National Botanical Research Institute, Lucknow, India.

**I. M. Santha,** Division of Biochemistry, Indian Agriculture Research Institute, New Delhi, India.

**Katrien Schroeyers,** Department of Plant Systems Biology, University of Ghent, Ghent, Belgium.

**Deepmala Sehgal,** Department of Botany, University of Delhi, Delhi, India.

**Gopalan Selvaraj,** Plant Biotechnology Institute, National Research Council of Canada, Saskatoon, Saskatchewan, Canada.

**Champa Sengupta-Gopalan,** Department of Plant and Environmental Sciences, New Mexico State University, Las Cruces, New Mexico.

**Kiran K. Sharma,** Genetic Transformation Laboratory, International Crops Research Institute for Semi Arid Tropics, Patancheru, Hyderabad, India.

**K. R. Sharma,** Department of Botany, Hindu College, University of Delhi, Delhi, India.

**P. Sharmila,** Department of Environmental Biology, University of Delhi, Delhi, India.

**Kenji Shinohara,** Department of Molecular and Cell Biology, Forestry and Forest Products Research Institute, Tsukuba, Ibaraki, Japan.

**Jean Luis Simões-Araújo,** Laboratório de Genética e Bioquímica, Seropédica, Brazil.

**German Spangenberg,** Victorian AgriBiosciences Centre, Primary Industries Research Victoria, Department of Primary Industries, La Trobe R&D Park, Victoria, Australia.

**R. Srinivasan,** National Research Centre on Plant Biotechnology, Indian Agricultural Research Institute, New Delhi, India.

**Eiji Takita,** Department of Plant Genome Biotechnology, Kazusa DNA Research Institute, Kisarazu, Chiba, Japan.

**Nancy Terryn,** Institute of Plant Biotechnology for Developing Countries, Department of Molecular Genetics, Ghent University, Ghent, Belgium.

**Tohru Tsuchiya,** Life Science Research Center, Mie University, Japan.

**Henry E. Valentin,** Monsanto Company, Chesterfield, Missouri.

**Marc van Montagu,** Institute of Plant Biotechnology for Developing Countries, Department of Molecular Genetics, Ghent University, Ghent, Belgium.

**P. Venkatachalam,** Biotechnology Division, Rubber Research Institute of India, Kottayam, India.

**Masao Watanabe,** Graduate School of Life Sciences, Tohoku University, Sendai, Japan; 21st Century Center of Excellence Program, Iwate University, Morioka, Japan.

**Jack H. Wong,** Department of Biochemistry, Faculty of Medicine, Chinese University of Hong Kong, Shatin, New Territories, Hong Kong, China.

# Foreword

Legumes occupy a very important place in agriculture, following only cereals. Because of their protein-rich nature, the grain legumes became the mainstay of vegetarian food across the world, particularly in India. Apart from this, nonvegetarians around the world are slowly changing their food choices because of the risks associated with animal-based diets. Hence, there is a need to rectify deficiencies in legume proteins like insufficient availability of essential amino acids and to improve the quality and quantity of proteins, making them more suitable for human consumption. Besides consumption by humans, forage legumes are important as cattle feed. Similar to the value addition of making them suitable as a major vegetarian diet component, there is an urgent need to improve the forage legumes in terms of protein quality and quantity, reduced lignin content, and so on, making them easily digestible by cattle. Besides these very important economic aspects, legumes are a source of timber, high-quality wood, paper pulp, oil, and so on. In addition, legumes have the unique quality of fixing atmospheric nitrogen, making their cultivation advantageous in enriching the soil organically. Hence, this group of plants deserves appropriate attention from scientists, and the present treatment, *Handbook of New Technologies for the Genetic Improvement of Legumes,* is highly justified.

As the title indicates, this book serves as a reference for researchers interested in legume genetic manipulation, besides offering protocols and other practical aspects that are useful to young workers who want to enter the field of legume biotechnology. The contributors to this compendium includes a battery of experts drawn from all over the world, making it really valuable.

It is heartening to note that the volume has been edited by my long-term associate, Dr. P. B. Kirti, who worked with me on *Brassica* genetic manipulation at the National Research Centre for Plant Biotechnology, Indian Agricultural Research Institute, New Delhi, and is at present practicing legume genetic manipulation at the University of Hyderabad.

I wish him all good luck in this endeavor.

V. L. Chopra
Member, Planning Commission
Government of India
New Delhi

# Preface

Legumes, belonging to the family Fabaceae, follow only the cereals in importance. Apart from being sources of nutritious protein, species belonging to this group have applications of economic importance. For example, the peanut has the distinction of being a source of oil as well as protein; its seed is consumed even without processing and is a favorite snack for humans. No other crop has this distinction, even though the soybean has the distinction of being a rich source of protein and oil as well. The other applications include forage and feed in the species of *Medicago* and *Trifolium,* highly preferred vegetable oils in soybean and peanut, paper pulp in *Leucaena leucocephala,* firewood in many species, hardwood for construction purposes, mulching material, improving soil fertility, and so on.

As experimental material, the importance of legumes need not be exaggerated. A member of this legume family, the pea, offered itself as the study material to the father of genetics, Mendel, who proposed the laws of inheritance, which subsequently became the basis for modern genetics. In terms of molecular biology too, legumes did not lag behind the model system for molecular biology studies, *Arabidopsis thaliana,* which lacks an important biological activity, the interaction with soil rhizobia that results in atmospheric nitrogen fixation. With suitable characteristics like short generation time, relatively small genome size (though not as small as *Arabidopsis*), suitable plant morphology, low chromosome number, availability of easy regeneration and genetic transformation system, and the capacity to fix atmospheric nitrogen through the interaction with rhizobium, *Medicago truncatula* started posing a threat to the supremacy of *Arabidopsis*. Intense efforts are being made to sequence the genome of this species, and already as many expressed sequence tags from this species have been deposited in the GenBanks as for *Arabidopsis*. Along with these features, *M. truncatula* is a favorable material for plant-rhizobial interaction. It is the dream of scientists to transfer the trait of atmospheric nitrogen fixation from the legumes to other economically important crops like cereals, making their cultivation independent of chemical fertilizers. Simultaneously, other legume species like *Lotus japonicus* are also trying to grab attention from *Arabidopsis* as models for molecular biology studies. Julia Frugoli has reviewed this aspect and the importance of *Medicago truncatula* in this volume, followed by a contribution from P. Ratet, and A. Kondorosi, and others on reverse genetic approaches that are applicable to this model legume species. M. Chabaud has dealt with the transformation technology and the various possibilities in this material.

Most of the legumes are cultivated under adverse conditions, and the cultivation is generally rain fed. In such a situation, the crop is dependent on the vagaries of nature. Various stresses in the form of insect predators, disease-causing pathogens, and environmental stresses including drought, extended rainfall, salinity, chilling, and so on cause considerable damage to the crop. Some environmental stresses also make the plant susceptible to diseases, resulting in serious loss of productivity. When adequate genetic variability is available for the crop, different stresses can be tackled through suitable crop improvement programs. Unfortunately, many legumes suffer from a serious lack of adequate genetic variability. This can be tackled by recourse to genetic engineering and introducing suitable alien genes into crop cultivars. However, this is incumbent on the availability of suitable genetic transformation and regeneration technology for legumes.

Traditionally, legumes have been branded as recalcitrant to in vitro regeneration, making plant genetic engineering difficult for this set of species. Very few legumes are amenable to regeneration through the callus phase. However, the successful deployment of cotyledonary node and embryo axes explants in genetic transformation experiments has hastened the process of genetic engineering in legumes. To date, the cotyledonary node–mediated transformation has been the preferred method of genetic manipulation in legumes. Of late, there is an increasing involvement of these two explant systems in genetic manipulation in legumes.

Champa Sengupta-Gopalan in Chapter 1 has summarized the current trends in the genetic manipulation of forage legumes, whereas Qi and Valentin (Chapter 5) discuss the recent trends in nutritional enhancement of soybeans. The species belonging to the legume family treated in this compendium are *Medicago* and *Trifolium* species (Chapter 2), soybean (Chapters 6 and 7), pigeon pea (Chapter 8), peanut (Chapters 16 and 17), chickpea (Chapter 9), Lathyrus (Chapter 12), pea (Chapters 13, 14, and 15), *Phaseolus* and *Vigna* (Chapter 10), *Lotus* species (Chapter 20), azuki bean (Chapter 11), broad bean (Chapter 19), *Leucaena* (Chapter 21), and *Robinia* (Chapter 22). As there are efforts from various laboratories across the world aimed at simplifying the transformation procedure for different species, Chapter 18 summarizes *in planta* genetic transformation methodology with reference to legumes. An attempt has been made in various chapters to give suitable protocols and hints.

Along with the tissue culture–plant regeneration technology for some selected legumes, some of the recent developments in the field of molecular biology and genomics as applicable to legumes are also summarized in this volume.

The analysis of transgenic plants is more technically demanding than their production. It is a fundamental principle in plant biotechnology that transgenic plants carrying single copies of introduced genes with optimum expression should be developed to make the technology successful. Chapter 26 details the strategies for analyzing transgenic plants in detail. Real-time polymerase chain reaction (RT-PCR) is fast becoming a standard technique in quantifying gene expression. Chapter 30 provides detailed information on the use of quantitative real-time-PCR in the analysis of transgenic plants. Chapters 28, 29, and 31 discuss techniques based on complementary cDNA for characterizing differential gene expression and cloning novel genes.

Plant biotechnology is often limited by the lack of promoters for proper tissue-specific expression of foreign genes in transgenic plants and only a few promoters are available for deployment in plants. Hence, the isolation and characterization of novel promoters should be given utmost priority. Chapter 27 explains the strategies in tagging regulatory elements for successful biotechnological applications.

Genetic maps occupy a very important place in modern genetics and dense molecular genetic maps with very closely placed DNA markers have often been used in cloning genes of agricultural importance. The importance of synteny between closely related species is being appreciated. It essentially means that detailed molecular genetic maps available for one species can be utilized in related species for which such information is not readily available. Chapters 32 and 34 give extensive information about the various options available for developing molecular markers and compiling them into molecular genetic maps. Chapter 25 summarizes the fluorescence *in situ* hybridization techniques for characterizing chromosomes in legumes, the information from which ultimately is utilized in placing genes and gene blocks on chromosomes.

Legumes are a rich source of proteins, oil, and other essential biochemicals. Hence, they are prone to attack by various invaders. To protect themselves from the impending attack, they develop proteins of immense agricultural value. Chapter 37 summarizes the recent information on defensins in legumes. Chapter 36 describes recent trends in legume-*rhizobium* interaction. Chapter 35 gives elaborate information on manipulating abiotic stress tolerance in legumes.

With these chapters, it is hoped that the handbook becomes a handy reference with protocols for research workers in the field of legume genetic manipulation and young workers, who are getting initiated into this area.

I thank all the contributors for their valuable contributions and for making this project a success by sparing their time. I thank my associate, Ms. M. R. Beena, for helping me in formatting the volume suitably.

# Chapter 1

# Genetic Engineering for Forage Quality in Alfalfa

Champa Sengupta-Gopalan
Suman Bagga
Jose Luis Ortega
Carol Potenza

## *INTRODUCTION*

Alfalfa is a very challenging crop to breed because of its complicated genetics and self-breeding restrictions. Therefore, development of new cultivars is usually done as a population. Strategies have included breeding for resistance to pests and diseases, resistance to abiotic stresses like drought, salinity, and heat, and breeding for increased biomass, protein content, and forage quality. Successes have been variable, but quality-breeding programs are successfully introducing important genetic traits into local populations that are beginning to address some of these problems.

The advent of molecular biology and genetic engineering has only begun to have an impact on alfalfa usage. While there are many reports in the literature about transgenic alfalfa with many different kinds of transgenes, the focus of this chapter is improvement of forage quality. Genetic engineering approaches require a good transformation system, desirable target genes based on basic understanding of the biochemical pathways leading to the desired trait, and appropriate promoters for fine-tuning the expression pattern of the transgenes. Since the transformation of alfalfa is dealt with elsewhere in this book, the focus here is on the genetic engineering strategies used to target nutritional quality, digestibility, and efficiency of protein utilization.

## PROMOTERS FOR GENETIC ENGINEERING IN ALFALFA

As the application of genetically engineered plants has widened over the past decades, so has the need to develop methods to fine-tune the expression of integrated transgenes (Potenza et al., 2004). The development of a broad spectrum of promoters has increased the successful application of genetic engineering technologies in both research and industrial/agricultural settings. The use of a variety of promoters, when integrated properly into both common and uncommon modular vector constructs, enhances the goals of any genetic engineering project.

Alfalfa has both benefited and been penalized in promoter development. For instance, the virally derived cauliflower mosaic virus (CaMV) 35S promoter (Odell et al., 1985) has been used with great success in proof-of-concept research that used the promoter's high constitutive expression to drive the integrated transgene (examples include Ortega et al., 2001; Tesfaye et al., 2001; Samis et al., 2002). The 35S promoter has an almost universal capacity to express transgenes at high levels in most tested plants using host nuclear RNA polymerase, without the necessity of any trans-acting viral gene products. However, the 35S promoter has also been shown to drive expression at lower levels in alfalfa than when used in other, nonleguminous plants (Bagga et al., 2005). Because of the success of the 35S promoter, other viral promoters have also been developed for use in alfalfa, including the cassava vein mosaic virus promoter (CsVMV), which showed up to 24-fold greater activity than the CaMV 35S promoter in the leaves of alfalfa (Samac et al., 2004).

Alfalfa has been penalized in promoter development because of the difficulty in funding such projects, and because of its complex genetics/breeding and length of transformation time. Even so, both root-specific and nodule-specific promoters have been developed (Gregerson et al., 1994; Fang and Hirsch, 1998; Winicov et al., 2004). However, much of the agricultural value of alfalfa lies in its green tissues (leaves and stems). Light-inducible and green-tissue-specific promoters have been isolated and tested in both research and industrial settings. Alfalfa light-inducible promoters that have been isolated include small subunit *rbcS* (Khoudi et al., 1997; Aleman, 2001), chlorophyll a/b binding protein (Aleman, 2001), and rubisco activase (Aleman, 2001). Thus far, those isolated promoters lack the strength to drive transgenes to higher levels than those of the viral constitutive promoters (Aleman, 2001; Moore, 2003; Samac et al., 2004).

While the use of promoters from other plant species has been successful in alfalfa (examples include Brears et al., 1991; Cordoba et al., 2003), the specificity of *trans*-acting factors to promoter *cis* elements might complicate the expression dynamics of promoters used across species. The development of *Medicago truncatula* as a model legume, the extensive monetary support given to the study of this species, and its genetic closeness to alfalfa can be exploited in the development of alfalfa promoters. This is especially true when studying gene expression patterns based on *M. truncatula* transcriptomics. By comparing *M. truncatula* microarray expression patterns under a number of conditions, there might be increased success for mining alfalfa promoters with similar spatial and temporal expression patterns.

## *NUTRITIONAL IMPROVEMENT OF ALFALFA*

Human and monogastric animals cannot synthesize 10 out of 20 essential amino acids and, therefore, must obtain them in their diet. Alfalfa supplied as forage is an important source of protein for animal feeding and is considered the most important forage crop in the world (Hanson et al., 1988). However, it is deficient in sulfur-containing amino acids (Kaldy et al., 1979) and it has been shown that wool growth in sheep, milk production in dairy animals, and meat production are limited by the availability of S-amino acids (Rogers et al., 1979; Barry, 1981; Pickering and Reis, 1993). Rumen fermentation contributes partly to the S-amino acid deficiency, since rumen microflora degrade the feed protein and in some cases resynthesize proteins with a lower nutrient value. Efforts to use conventional breeding and cell selection techniques to improve the S-amino acid content of alfalfa have met with little success (Reish et al., 1981); such efforts are now being made to use a genetic engineering approach. Although most of the concerted efforts with regard to nutritional improvement by genetic engineering have focused on seed proteins (Hoffman et al., 1988; De Clercq et al., 1990; Guerche et al., 1990; Altenbach et al., 1992; Molvig et al., 1997), some efforts have been made to improve amino acid balance in the leaves, with particular attention to forage legumes. The production of transgenic forage legumes expressing genes encoding different rumen bypass proteins rich in S-amino acid, such as chicken ovalbumin, pea albumin, Brazil nut 2S albumin, and sunflower seed albumin have been reported. However, the accumulation of the proteins in the leaves was very low in all cases, probably because the proteins were targeted to the protease-rich vacuoles. To protect transgenic proteins from being degraded in the vacuoles, efforts were also made to retain the protein in the endoplasmic reticulum (ER) by engineering the ER retention signal (KDEL) in the carboxy terminal end of the protein. While the modified protein accumulated to reasonably high levels in subterranean clover, the accumulation was very low in alfalfa (Khan et al., 1996).

More recently, efforts have been made to engineer plants with the genes for the ER-targeted methionine (Met)-rich zein proteins from corn (Bagga et al., 1995, 2004; Bagga, Rodriguez, et al., 1997). The β-zein (15 kD) and the two δ-zeins (10 kD and 18 kD) are Met-rich proteins with a Met content of 22 and 27 percent, respectively. The genes encoding these proteins were engineered behind the CaMV 35S promoter and introduced into alfalfa and other plants. These transformants showed high-level accumulation of the zein proteins, and the proteins were found to accumulate in unique ER-derived protein bodies (Bagga et al., 1995; Bagga, Rodriguez, et al., 1997; Hinchliffe and Kemp, 2002). Preliminary studies have also indicated that the zein proteins are highly stable in the rumen of cows (Bagga et al., 2004). Thus, the β- and δ-zein proteins appear to be ideal candidates for use in improving the Met content of alfalfa forage: rich in sulfur-containing amino acids, resistant to ruminant bacteria, and stable in the vegetative tissues of transformed plants.

To increase zein accumulation in the vegetative tissues, multiple zein genes were introduced by retransformation and by sexual crosses into the same plant, and analysis of these transformants showed a zein gene dose-dependent increase in the accumulation of bound Met (Bagga et al., 2004). This would imply that the plant cells are able to handle high-level accumulation of the zein protein in spite of the fact that the protein accumulates in the ER. To further increase the level of accumulation of the zein protein, efforts were made to express them in the chloroplasts. However, unlike many other proteins that have been shown to be very stable in the chloroplasts, the zeins did not accumulate in the chloroplasts (Bellucci et al., 2005).

Comparison of the accumulation pattern of the two different zein proteins showed that the β-zeins accumulated to higher levels compared to the δ-zeins (Bagga et al., 2004). To make an impact on the S-amino acid content, the zeins have to accumulate in the leaves at appreciably high levels and since the δ-zeins have higher Met content, it will be more desirable to accumulate the δ-zeins than the β-zeins. In an attempt to increase the accumulation of the δ-zein, efforts were made to introduce multiple copies of the δ-zein or combinations of δ- and β-zein. Analysis of the transformants with the different combinations of transgenes showed that coexpression of β-zein and δ-zein genes yielded a fivefold increase in the accumulation of δ-zein proteins and that the two zein proteins were colocalized in the same protein bodies (Bagga, Temple, et al., 1997; Hinchliffe and Kemp, 2002). The data suggested that the interaction between β-zein and δ-zein had a stabilizing effect on the δ-zein. To keep the two zein genes in the same linkage group, the β- and δ-zein genes have been engineered behind two different promoters in the same T-DNA construct and introduced into alfalfa. These transformants have been analyzed and, as seen in Figure 1.1, the level of accumulation of the δ-zein is highly enhanced in plants containing this dual gene construct when compared to transformants with just the δ-zein or the δ-/β-zein coexpressors produced by sexual cross.

While coexpression of the β- and δ-zein genes increased the level of δ-zein fivefold, the accumulation level is still not enough to make an impact on the amino acid balance in alfalfa. The inability of the zein proteins to accumulate in alfalfa at a higher level could reflect instability of the zein pro-

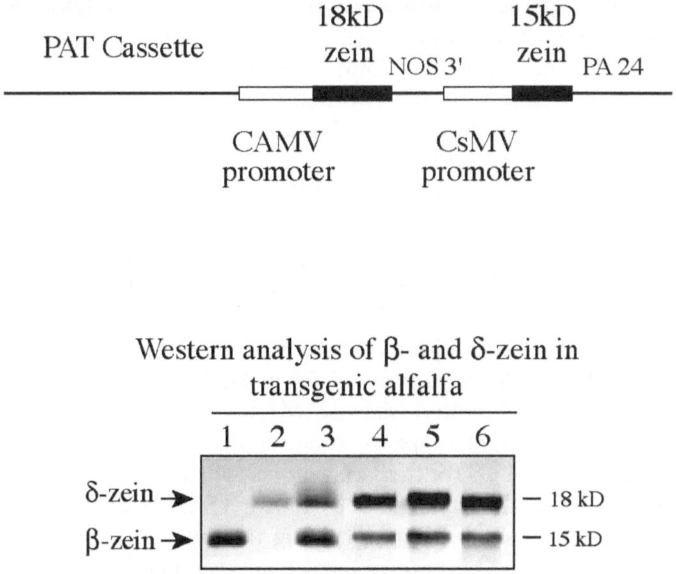


FIGURE 1.1. Coexpression of the δ- and β-zein in transgenic alfalfa leaves. Top: Representation of the δ-/β-zein dual gene construct used in alfalfa transformation. The δ-zein is driven by the cauliflower mosaic virus promoter (CaMV) and uses the NOS 3' terminator. The β-zein is driven by the cassava vein mosaic virus promoter (CsVMV) and uses the PA 24 terminator. Plants are selected using phosphinothricin acetyl transferase (PAT cassette), which confers resistance to the herbicide glufosinate. Bottom: Western analysis: EtOH-soluble fractions (equivalent to 100 μg of phosphate buffered saline (PBS) soluble fraction) from the alfalfa leaves were subjected to sodium dodecyl sulfate-polyacrylamide gel elctrophoresis (SDS-PAGE), followed by immunoblot analysis using a mixture of δ-zein and β-zein antibodies. Lanes: (1) Alfalfa transformant expressing only β-zein; (2) alfalfa transformant expressing only δ-zein; (3) alfalfa transformants coexpressing β-zein and δ-zein produced by sexual cross; (4-6) independent alfalfa transformants expressing the δ-/β-zein dual gene construct.

tein or inefficient translation of the zein transcripts due to a limited supply of Met tRNA in alfalfa. The first possibility has been ruled out based on incubation experiments done with alfalfa leaf extracts. To check if limited supply of free Met is the basis for low-level accumulation of the Met-rich zein proteins, callus tissues from alfalfa transformants (β-zein) were incubated with Met and analyzed for zein accumulation. A threefold increase in zein levels was seen following incubation with free Met (Bagga et al., 2005).

Met belongs to the aspartate (Asp) family of amino acids, which includes asparagine (Asn), lysine (Lys), threonine (Thr), isoleucine (Ile), and Met (Giovanelli et al., 1985, 1988; Giovanelli, 1990). The biosynthetic pathway for the Asp family has two major branch points. The first is at 3-Asp semialdehyde, which separates Lys biosynthesis from that of *O*-phosphohomoserine (OPH). OPH is the branch-point substrate leading to Met and Thr/Ile. The committing step in Met synthesis occurs when the side chain of OPH condenses with the thiol group of cysteine (Cys) to form cystathionine in an irreversible reaction catalyzed by cystathionine γ-synthase (CγS; EC 4.2.99.9; Figure 1.2; Ravanel et al., 1998; Matthews, 1999; Kim and Leustek, 2000). Threonine synthase (TS; EC 4.2.99.2) converts OPH into Thr. The major metabolic fates of Met include its incorporation into protein, adenosylation to form *S*-adenosyl Met (SAM), and methylation to form *S*-methyl Met (SMM).

Met synthesis in plants appears to be regulated at the level of competition between CγS and TS for their common sugbstrate OPH (Bartlem et al., 2000; Gakiere et al., 2000; Zeh et al., 2001). TS is allosterically regulated by SAM (Curien et al., 1998), suggesting that Met synthesis could influence TS activity. The Met pool in most plants studied is tightly regulated. Excess amounts of Met (or related metabolite, SAM) are known to reduce the stability of CγS mRNA in *Lemna pauciscostata* and *Arabidopsis* (Giovanelli et al., 1985; Chiba et al., 1999, 2003). The characterization of an *Arabidopsis* mutant (*mto1*), in which a mutation in the CγS gene resulted in a transcript that is resistant to Met-dependent degradation, suggests that CγS is autoregulated at the posttranscriptional level, presumably via a mechanism involving the *N*-terminal region of the *Arabidopsis* CγS protein (Chiba et al., 1999; Hacham et al., 2002).

The experimental approach undertaken to increase Met levels in alfalfa was to simultaneously increase both the free Met pools and a Met sink in the form of the zein proteins (Bagga et al., 2005). Several lines of evidence indicate that CγS controls the rate of Met synthesis (Gakiere et al., 2000, 2002; Hacham et al., 2002; Kim et al., 2002; Di et al., 2003). Therefore, the *Arabidopsis* CγS (*AtC*γ*S*) gene was introduced into alfalfa driven by the

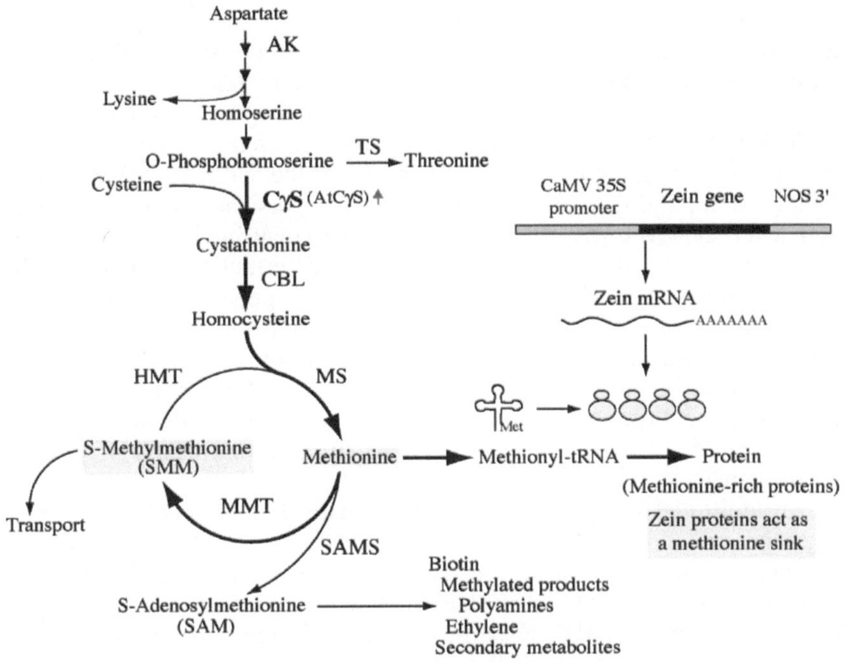


FIGURE 1.2. The sink source model for increasing bound Met in alfalfa. The pathway for Met synthesis and metabolism in plants. Cystathionine γ-synthase (CγS) is the key enzyme regulating the Met biosynthetic pathway. CγS catalyzes the committing step in Met synthesis to make cystathionine, which is converted into homocysteine. Homocysteine is then methylated to form Met. All of these reactions occur in the chloroplast/plastid. The *Arabidopsis* cystathionine-γ-synthase cDNA (AtCγS) driven by the CaMV 35S promoter was introduced into alfalfa to create a source of methionine. Met has multiple fates: SMM (*S*-methyl methionine) synthesis, SAM (*S*-adenosyl methionine) synthesis, or incorporation into protein. Abbreviations: AK, aspartate kinase; TS, threonine synthase; CβL, cystathionine β-lyase; MS, methionine synthase; SAMS, SAM synthase. SAM is the precursor for biotin, ethylene, and polyamines and also acts as a methyl donor. The sink is represented by the synthesis of Met-rich zein proteins requiring the availability of Met-tRNA.

CaMV 35S promoter. The data showed that alfalfa plants expressing *At*CγS in a constitutive manner have an increased free Met pool but an even larger increase in the level of *S*-methyl methionine (SMM), a transport form of Met, compared to control plants. To check if the increased free Met pools resulting from the constitutive expression of the *At*CγS gene could be directed toward the synthesis of the zein proteins, alfalfa plants were co-transformed with the *AtC*γS and β-zein genes driven by the CaMV 35S promoter. These coexpressors showed an increase in the zein level compared to the level in the β-zein transformants. While the coexpressors showed a drop in the free Met level, the SMM level remained unchanged, suggesting that the Met was being used for zein synthesis while the SMM could not be used (Bagga et al., 2005). Even though the coexpressors showed an increase in zein level, the level was still not high enough to make an impact. The question is whether zein levels could be increased further by increasing Met level by diverting SMM back to Met. Another genetic engineering approach to increase free Met pools, currently in progress, is to down-regulate *TS* or *SAM* (*S*-adenosyl Met synthase; Bagga, unpublished results). Metabolic engineering is a tricky venture because of the high degree of networking among the different pathways. Thus, it is not clear how down-regulating SAMs would affect downstream functions and how that would affect plant performance.

An additional approach to increasing Met level in alfalfa plants expressing genes encoding Met-rich proteins is to increase aspartate synthesis, the precursor for the synthesis of the Met. Alfalfa overexpressing the cytosolic form of glutamine synthetase has shown increased levels of Asp/Asn (Moguel-Esponda, 2004). Experiments are in progress to produce alfalfa transformants overexpressing the zeins, *AtC*γS, *and the cytosolic glutamine synthetase genes simultaneously.*

Avraham et al. (2005) expressed the *Arabidopsis* cystathionine γ-synthase (*AtC*γS), the enzyme that controls the synthesis of the first intermediate metabolite in the methionine pathway, in transgenic alfalfa plants. The *AtC*γS cDNA was driven by the *Arabidopsis* rubisco small subunit promoter to obtain strong light induced expression in leaves. In the transgenic lines, the contents of Met, *S*-methyl Met (SMM), and Met incorporated into the water-soluble protein fraction increased up to 32-fold, 19-fold, and 2.2-fold, respectively, compared with that in wild-type plants. Interestingly, the levels of free Cys (the sulphur donor for Met synthesis), glutathione (the Cys storage and transport form), and protein-bound Cys increased up to 2.6-fold, 5.5-fold, and 2.3-fold, respectively in the transgenic plants relative to that in wild-type plants indicating that the transgenic plants expressing *AtC*γS accumulated higher levels of soluble and protein bound Met and Cys.

## *IMPROVING DIGESTIBILITY OF ALFALFA FORAGE*

Forages are the primary source of fiber in ruminant diets. A minimum amount of dietary fiber is required for normal rumen function, animal health, and milkfat content. However, fiber is the least digestible part of most dairy diets and has the lowest energy content. The component that contributes to nondigestibility of forage is lignin. Lignin is an indigestible cell wall fiber. It cross-links with cellulose to increase cell wall strength but decreases the usefulness of cellulose as an extractable energy component of forage. Lignin is a phenolic compound found in most plant secondary cell walls and is the major structural component of secondarily thickened plant cell walls. It is a complex polymer of hydroxylated and methoxylated phenylpropane units. Dicotyledonous angiosperm lignins contain two major monomer species, guaiacyl (G) and syringyl (S) units. Lignin levels increase as the plant matures and there is also change in the lignin composition with age toward a progressively higher S/G ratio (Buxton and Russel, 1988). Both lignin content and the S/G ratio have been negatively correlated with forage digestibility.

Lignin is derived from the phenyl propanoid pathway (Figure 1.3). The formation of the G and S units of lignin is catalyzed by *O*-methyl transferase enzymes. Caffeic acid 3-*O*-methyl transferase (COMT) converts caffeic acid to ferulic acid, which then via 5-hydroxy-ferulic acid leads to the formation of sinapic acid.

Methylation of caffeate moiety is also catalyzed by caffeoyl CoA 3-*O*-methyl transferase (CCOMT); this leads to the synthesis of coniferaldehyde via feruloyl CoA (Guo, Chen, Inoue, et al., 2001). The enzyme cinnamyl alcohol dehydrogenase (CAD) catalyzes the formation of sinapyl alcohol from sinapaldehyde and coniferyl alcohol from coniferylaldehyde. Sinapyl alcohol is the precursor for the S units and the coniferyl alcohol is the precursor for the G units.

Alfalfa has also been transformed with antisense gene constructs of CAD driven by CaMV 35S promoter, in an attempt to down-regulate lignin synthesis (Baucher et al., 1999). These transformants showed no change in lignin quantity but the lignin composition was altered with a lower S/G ratio and a lower S + G yield. While these transformants were not tested for rumen digestibility, they showed increased digestibility over control in *in situ* degradation tests (Baucher et al., 1999). Independent alfalfa transformants down-regulated in COMT and CCOMT were produced by gene-silencing and antisense approaches using the bean phenylalanine ammonia-lyase PAL2 promoter (Guo, Chen, Inoue, et al., 2001). PAL is the first key enzyme in the lignin biosynthetic pathway. Strong down-regulation of COMT resulted in decreased lignin content, a reduction in total G lignin units, and a near total loss of S units in monomeric and dimeric lignin. In contrast, strong down-regulation of CCOMT led to reduced lignin levels and a reduction in G units without reduction in S units. Analysis of rumen digestibility

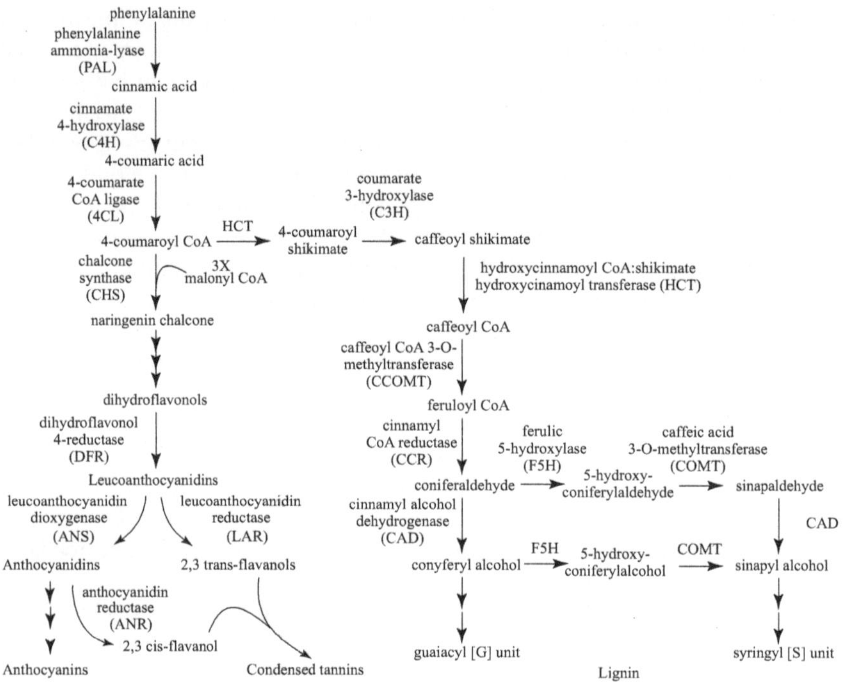


FIGURE 1.3. Schematic representations of the pathways for synthesis of condensed tannins and lignins.

of alfalfa forage in fistulated steers showed that both the COMT and CCOMT down-regulated alfalfa had improved digestibility, the latter showing greater improvement (Guo, Chen, Wheeler, et al., 2001).

In field tests (2002-2004), several of these transgenic plants demonstrated a significant reduction in lignin content and an increase in cell wall digestibility (>10 percent). This improvement can be compared to conventional breeding where over 15 years of selection has resulted in a 2 to 3 percent increase in cell wall digestibility. The effect of low lignin content was accompanied by a depression in crop yields and reduction of long-term survival of some perennial plants. The interactions of genes involved in lignin metabolism with genetic background and environment play an important role in determining the performance of these genetically engineered low-lignin plants (Pederson et al., 2005). However, Reddy et al. (2005) generated transgenic alfalfa lines with a range of differences in lignin content and composition by targeting three specific cytochrome P450 enzymes of the lignin pathway for antisense down-regulation. They have observed a strong negative relationship between lignin content and rumen digestibility, but no relationship between lignin composition and digestibility, in these transgenic lines. They also suggested that the down-regulation of 4-coumarate 3-hydroxylase (C3H) provides the largest improvements in digestibility yet seen in a forage crop.

## *IMPROVING EFFICIENCY OF PROTEIN UTILIZATION*

Condensed tannins (CTs) or the proanthocyanidins have been implicated in the prevention of pasture bloat. Other legumes that contain CTs may have lower protein degradation than alfalfa with its low CT content. Low levels of CT (between 3 and 4 percent) in forage have been shown to increase animal production (Woodward et al., 2000). CTs bind to protein and decrease ruminal protein degradability of forage, but the bonds are reversible at the low pH found in the abomasums, where the protein is released. CTs are dimers or higher oligomers of flavan-3-ol units and are powerful antioxidants (Hagerman et al., 1998).

Proanthocyanidin (PA) biosynthesis branches from the phenylpropanoid pathway after the synthesis of anthocyanidins (Figure 1.3). They share the same upstream biosynthetic pathway as the lignins, and the branch point between the two pathways occurs at the level of coumarate. The two key enzymes that convert the immediate precursors into CTs are leucoanthocyanidin dioxygenase (ANS) and leucoanthocyanidin reductase (LAR). More recently, anthocyanidin reductase (ANR) encoded by the *BANYULS* gene has been shown to convert anthocyanidin into 2,3-*cis*-flavanol (epicatechin), which condenses with 2,3-trans-flavanol (catechin) to form CTs (Xie et al., 2004). While alfalfa foliar tissues accumulate anthocyanins at senescence or locally under certain stress conditions, no known conditions induce PAs in alfalfa. However, PAs do accumulate in seed coat, suggesting that alfalfa does have the genes for the pathway. Since several of these genes have been isolated, it is feasible to express the key members in CT synthesis in the leaves and stem of alfalfa by engineering the genes behind the appropriate leaf/stem promoters.

A second approach to manipulate CT content is at the level of the regulatory genes in the flavonoid pathway. Several of the maize *myc* (*R, B-Peru, Sn, Lc*) and *myb* (*Cl, P*) flavonoid regulatory genes have been tested for their ability to influence the accumulation of flavonoids or PAs when expressed in heterologous plants (Bradley et al., 1998; de Majnik et al., 2000;

Bovy et al., 2002; Ray et al., 2003; Robbins et al., 2003). The expression of the myc protein Lc in alfalfa was accompanied by changes in PA accumulation under high light intensity or cold (Ray et al., 2003). Constitutive overexpression of *Sn* in *Lotus corniculatus* had subtle effects on anthocyanin accumulation, but stimulated PA accumulation in the leaf tissues (Robbins et al., 2003). One of the regulatory genes affecting PA accumulation has been isolated from *Lotus uliginosus* and it encodes for a basic helix-loop-helix myc-like protein (Ray et al., 2003). Other regulatory genes like *TT8* (encoding a bHLH domain transcription factor) and *TT2* (encoding a R2R3 MYB domain protein) have been identified as key determinants for PA synthesis in developing seeds of *Arabidopsis* (Nesi et al., 2000, 2001). More recently, Sharma and Dixon (2005) have shown that *Arabidopsis* transformed with an *Arabidopsis TT2MYB* transcription factor gene leads to the accumulation of CTs.

During the ensiling process, there is a heavy loss of proteins due to proteolytic degradation. For some ensiled forages, such as alfalfa, proteolytic losses are especially high, with degradation of up to about 80 percent of the forage protein. In contrast, red clover has up to 90 percent less proteolysis than alfalfa during ensiling (Papadopoulos and McKersie, 2005). The lower extent of postharvest proteolysis in red clover is related to the presence of soluble polyphenol oxidase (PPO) and *o*-diphenol PPO substrates in the leaves. Both PPO and *o*-diphenol are absent or present in very low amounts in alfalfa (Jones et al., 1995). PPO catalyzes the oxidation of *o*-diphenols to *o*-quinones, which can bind and inactivate endogenous proteases. A cDNA clone for PPO, isolated from red clover and engineered behind the CaMV-35S promoter, was introduced into alfalfa. These transformants showed increased PPO activity and the proteolysis was dramatically reduced in the presence of an *o*-diphenol compared to controls (Sullivan et al., 2004).

The different desirable traits to improve forage quality in alfalfa are being worked on independently. However, to produce the ideal alfalfa plant with all the desirable traits described above, the engineering manipulations will have to be done on the same plant. This would entail stacking the different gene constructs and ensuring that they are driven by different promoters to avoid homology-driven gene silencing (Halpin, 2005).

## NOTE

The enzyme hydroxycinnamoyl CoA: shikimate hydroxycinnamoyl transferase (HCT) catalyzes the reactions both immediately preceding and following the insertion of the 3-hydroxyl group into monolignol precursors. Shadle et al. (2007) have recently generated a number of independent transgenic lines of alfalfa (*Medicago sativa* L.) in which the levels of HCT were reduced through antisense HCT expression under control of the bean PAL2 promoter, which is preferentially expressed in vascular tissue. Reduction of enzyme activity in these lines was from at least 15 to 50 percent. The most severely down-regulated lines exhibited significant stunting, reduction of biomass, and delayed flowering. HCT down-regulation resulted in strongly reduced lignin content and striking changes in lignin monomer composition, with predominant deposition of 4-hydroxyphenyl units in the lignin. Vascular structure was impaired in the most strongly down-regulated lines. Analysis of forage quality parameters showed strong reductions of neutral- and acid-detergent fiber in the down-regulated lines, in parallel with large increases (up to 20 percent) in dry matter forage digestibility. It has been suggested that, although manipulation of lignin biosynthesis can greatly improve forage digestibility, accompanying effects on plant development should be considered carefully.

## REFERENCES

Aleman, L. (2001). Isolation and characterization of leaf-specific promoters from alfalfa (*Medicago sativa*). Master's Thesis, New Mexico State University, Las Cruces, NM.

Altenbach, S.B., C.C. Kua, L.C. Staraci, K.W. Pearson, C. Wainwright, A. Georgescu, and J. Townsend (1992). Accumulation of a Brazil nut albumin in seeds of transgenic canola results in enhanced levels of seed protein methionine. *Plant Molecular Biology* 18:235-245.

Avraham, T., H. Badani, S. Galili, and R. Amir (2005). Enhanced levels of methionine and cysteine in transgenic alfalfa (*Medicago sativa* L.) plants over-expressing the Arabidopsis cystathionine γ-synthase gene. Plant *Biotechnology Journal* 3:71-79.

Bagga, S., H. Adams, J.D. Kemp, and C. Sengupta-Gopalan (1995). Accumulation of the 15 kD zein protein in novel protein bodies in transgenic tobacco. *Plant Physiology* 107:13-23.

Bagga, S., A. Armendaris, N. Klypina, I. Ray, S. Ghoshroy, M. Endress, D. Sutton, J.D. Kemp, and C. Sengupta-Gopalan (2004). Genetic engineering ruminal stable high methionine protein in the foliage of alfalfa. *Plant Science* 166:273-283.

Bagga, S., C. Potenza, J. Ross, M.N. Martin, T. Leustek, and C. Sengupta-Gopalan (2005). A transgene for high methionine protein is posttranscriptionally regulated by methionine. *In Vitro Cellular and Developmental Biology—Plant* 41: 731-741.

Bagga, S., F. Rodriguez, J.D. Kemp, and C. Sengupta-Gopalan (1997). Co-expression of the maize β- and δ-zein genes results in stable accumulation of the δ-zein in ER derived protein bodies formed by the β-zein. *Plant Cell* 9:1683-1696.

Bagga, S., S.J. Temple, J.D. Kemp, and C. Sengupta-Gopalan (1997). A genetic engineering approach for improving protein quality of forage legumes. In *Plant Molecular Biology and Biotechnology*, K.K. Tewari and G.S. Singhal (eds.). New Delhi, India: Narosa Publishing House, pp. 305-319.

Barry, T.N. (1981). Protein metabolism in growing lambs fed on fresh rye grass (*Lolium perenne*) white clover (*Trifolium repens*) pastures ad lib. *British Journal of Nutrition* 46:521-532.
Bartlem, D., I. Lambein, T. Okamoto, A. Itaya, Y. Uda, F. Kijima, Y. Tamaki, E. Nambara, and S. Naito (2000). Mutation in the threonine synthase gene results in an over accumulation of soluble methionine in *Arabidopsis*. *Plant Physiology* 123:101-110.
Baucher, M., M.A. Bernard Vaihe, B. Chabbert, J.M. Besle, C. Opsomer, M. van Montagu, and J. Botterman (1999). Down regulation of cinnamyl alcohol dehydrogenase in transgenic alfalfa (*Medicago sativa* L.) and the effect on lignin composition and digestibility. *Plant Molecular Biology* 39:437-447.
Bellucci, M., F. De Marchis, R. Mannucci, R. Bock, and S. Arcioni (2005). Cytoplasm and chloroplast are not suitable sub-cellular locations for β-zein accumulation in transgenic plants. *Journal of Experimental Botany* 56:1205-1212.
Bovy, A.G., R. de Vos, M. Kemper, E. Schijlen, M.A. Pertejo, S. Muir, G. Collins, S. Robinson, M. Verhoeyen, S. Hughes, C. Santos-Buelga, and A. van Tunen (2002). High-flavonol tomatoes resulting from the heterologous expression of the maize transcription factor genes *LC* and *C1*. *Plant Cell* 14:2509-2526.
Bradley, J.M., K.M. Davies, S.C. Deroles, S.J. Bloor, and D.H. Lewis (1998). The maize *Lc* regulatory gene up-regulates the flavonoid biosynthetic pathway in *Petunia*. *Plant Journal* 13:381-392.
Brears, T., E.L. Walker, and G.M. Coruzzi (1991). A promoter sequence involved in cell-specific expression of the pea glutamine synthetase *GS3A* gene in organs of transgenic tobacco and alfalfa. *Plant Journal* 1:235-244.
Buxton, D.R. and J.R. Russell (1988). Lignin constituents and cell wall digestibility of grass and legume stems. *Crop Science* 28:553-558.
Chiba, Y., M. Ishikawa, F. Kijima, R.H. Tyson, J. Kim, A. Yamamoto, E. Nambara, T. Leustek, R.M. Wallsgrove, and S. Naito (1999). Evidence for autoregulation of cystathionine gamma-synthase mRNA stability in *Arabidopsis*. *Science* 286: 1371-1374.
Chiba, Y., R. Sakurai, M. Yoshino, K. Ominato, M. Ishikawa, H. Onouchi, and S. Naito (2003). *S*-adenosyl-*L*-methionine is an effector in the posttranscriptional autoregulation of the cystathionine γ-synthase gene in *Arabidopsis*. *Proceedings of the National Academy of Sciences USA* 100:10225-10230.
Cordoba, E., S. Shishkova, C.P. Vance, and G. Hernandez (2003). Antisense inhibition of NADH glutamate synthase impairs carbon/nitrogen assimilation in nodules of alfalfa (*Medicago sativa* L.). *Plant Journal* 33:1039-1044.
Curien, G., D. Job, R. Douce, and R. Dumas (1998). Allosteric activation of *Arabidopsis* threonine synthase by *S*-adenosyl methionine. *Biochemistry* 37: 13212-13221.
De Clercq, A., M. Vandewlele, R. De Rycke, J. Van Damme, M. Van Montagu, E. Krebbers, and J. Vandekerckjove (1990). Expression and processing of an *Arabidopsis* 2S albumin in transgenic tobacco. *Plant Physiology* 94:970-979.
De Majnik, J., J.J. Weinman, M.A. Djordjevic, B.G. Rolfe, G.J. Tanner, R.G. Joseph, and P.J. Larkin (2000). Anthocyanin regulatory gene expression in trans-

genic white clover can result in an altered pattern of pigmentation. *Australian Journal of Plant Physiology* 27:659-667.

Di, R., J. Kim, M.N. Martin, T. Leustek, J. Jhoo, C.T. Ho, and N.E. Tumer (2003). Enhancement of the primary flavor compound methional in potato by increasing the level of soluble methionine. *Journal of Agricultural Food Chemistry* 51: 5695-5702.

Fang, Y. and A.M. Hirsh (1998). Studying early nodulin gene ENOD40 expression and induction by nodulation factor and cytokinin in transgenic alfalfa. *Plant Physiology* 116:53-68.

Gakiere, B., L. Denis, M. Droux, and D. Job (2002). Over-expression of cystathionine gamma-synthase in *Arabidopsis thaliana* leads to increased levels of methionine and *S*-methyl methionine. *Plant Physiology and Biochemistry* 40: 119-126.

Gakiere, B., S. Ravanel, M. Droux, R. Douce, and D. Job (2000). Mechanisms to account for maintenance of the soluble methionine pools in transgenic *Arabidopsis* plants expressing antisense cystathionine γ-synthase cDNA. *Comptes Rendus Academie des Sciences* 323:841-851.

Giovanelli, J. (1990). Regulation of cysteine and methionine biosynthesis. In: *Sulfur nutrition and sulfur assimilation in higher plants. Fundamental, environmental and agricultural aspects,* H. Rennenberg, C. Brunold, L.J. De Kok, and I. Stulen (eds.). The Hague: SBP Academic Publishers, pp. 33-48.

Giovanelli, J., S.H. Mudd, and A.H. Datko (1985). *In vivo* regulation of methionine biosynthesis in a higher plant (*Lemna*). *Plant Physiology* 77:450-455.

Giovanelli, J., S.H. Mudd, and A.H. Datko (1988). *In vivo* regulation of threonine and isoleucine biosynthesis in *Lemna paucicostata* Hegelm. 6746. *Plant Physiology* 86:369-377.

Gregerson, R.G., S.S. Miller, M. Petrowski, J.S. Gantt, and C.P. Vance (1994). Genomic structure, expression and evaluation of the alfalfa aspartate amino transferase genes. *Plant Molecular Biology* 25:387-399.

Guerche, P., E.R.P. De Almeida, M.A. Schwarztein, E.S. Gander, E. Krebbers, and G. Pelletier (1990). Expression of the 2S albumin from *Bertholletia excelsa* in *Brassica napus*. *Molecular and General Genetics* 221:305-314.

Guo, D., F. Chen, K. Inoue, J.W. Blount, and R.A. Dixon (2001). Down regulation of caffeic acid 3-*O*-methyltransferase and caffeoyl CoA 3-*O*-methyltransferase in transgenic alfalfa: Impacts on lignin structure and implications for the biosynthesis of G and S lignin. *Plant Cell* 13:73-88.

Guo, D., F. Chen, J. Wheeler, J. Winder, S. Selman, M. Peterson, and R.A. Dixon (2001). Improvement of in-rumen digestibility of alfalfa forage by genetic manipulation of lignin *O*-methyl transferases. *Transgenic Research* 10:457-464.

Hacham, Y., T. Avraham, and R. Amir (2002). The N-terminal region of *Arabidopsis* cystathionine-γ-synthase plays an important regulatory role in methionine metabolism. *Plant Physiology* 128:454-462.

Hagerman, A.E., K.M. Riedl, G.A. Jones, K.N. Sovik, N.T. Ritchard, P.W. Hartzfeld, and T.L. Riechel (1998). High molecular weight plant polyphenolics (tannins) as biological antioxidants. *Journal of Agricultural Food Chemistry* 46: 1887-1892.

Halpin, C. (2005). Gene stacking in transgenic plants—the challenge for 21st century plant biotechnology. *Plant Biotechnology Journal* 2:141-155.

Hanson, A.A., D.K. Barnes, and R.R. Hill, Jr. (1988). Alfalfa and alfalfa improvements. *Agronomy Monograph* 29, Madison, WI: ASA, CSSA, SSSA.

Hinchliffe, D.J. and J.D. Kemp (2002). β-Zein protein bodies sequester and protect the 18kD δ-zein protein from degradation. *Plant Science* 163:741-752.

Hoffman, L.M., D.D. Donaldson, and E.M. Herman (1988). A modified storage protein is synthesized, processed and degraded in the seeds of transgenic plants. *Plant Molecular Biology* 11:717-729.

Jones, B.A., R.D. Hatfield, and R.E. Muck (1995). Screening legume forages for soluble phenols, polyphenol oxidase and extract browning. *Journal of Scientific Food and Agriculture* 67:109-112.

Kaldy, M.S., M.R. Hanna, and S. Smoliak (1979). Amino acid composition of sanfoin forage. *Grass and Forage Science* 34:145-148.

Khan, M.R.I., A. Ceriotti, I. Tabe, A. Aryan, W. McNabb, A. Moore, S. Craig, D. Spencer, and T.J.V. Higgins (1996). Accumulation of a sulfur-rich seed albumin from sunflower in the leaves of transgenic subterranean clover (*Trifolium subterranean* L.). *Transgenic Research* 5:81-88.

Khoudi, H., L.P. Vezina, J. Mercier, Y. Castonguay, G. Allard, and S. Laberge (1997). An alfalfa rubisco small subunit homologue shares *cis*-acting elements with the regulatory sequences of the *RbcS-3A* gene from pea. *Gene* 197:1-2.

Kim, J., M. Lee, R. Chalam, M.N. Martin, T. Leustek, and W. Boerjan (2002). Constitutive overexpression of cystathione γ-synthase in *Arabidopsis* leads to accumulation of soluble methionine and S-methylmethionine. *Plant Physiology* 128:95-107.

Kim, J. and T. Leustek (2000). Repression of cystathionine γ-synthase in *Arabidopsis thaliana* produces partial methionine auxotrophy and developmental abnormalities. *Plant Science* 151:9-18.

Matthews, B.F. (1999). Lysine, threonine and methionine biosynthesis. In *Plant amino acids: Biochemistry and biotechnology*. B.K. Singh (ed.). New York: Marcel Dekker, pp. 205-225.

Moguel-Esponda, S. (2004). The 3′ untranslated region of cytosolic glutamine synthetase gene from soybean (*Glycine max*) mediates stability of its transcript. PhD dissertation, New Mexico State University, Las Cruces, NM.

Molvig, L., L.M. Tabe, B.O. Eggum, A.E. Moore, S. Craig, D. Spencer, and T.J.V. Higgins (1997). Enhanced methionine levels and increased nutritive value of seeds of transgenic lupins (*Lupinus angustifolius* L.) expressing a sunflower seed albumin gene. *Proceedings of the National Academy of Sciences USA* 94:8393-8398.

Moore, R.L. (2003). Functional analysis of alfalfa (*Medicago sativa*) green-tissue specific promoters. Master's thesis, New Mexico State University, Las Cruces, NM.

Nesi, N., I. Debeaujon, C. Jond, G. Pelletier, M. Caboche, and L. Lepiniec (2000). The TT8 gene encodes a basic helix-loop helix domain protein required for expression of *DFR* and *BAN* genes in *Arabidopsis* siliques. *Plant Cell* 12:1863-1878.

Nesi, N., C. Jond, I. Debeaujon, M. Caboche, and L. Lepiniec (2001). The *Arabidopsis* TT2 gene encodes an R2R3 MYB domain protein that acts as a key determinant for proanthocyanidin accumulation in developing seeds. *Plant Cell* 13:2099-2114.

Odell, J.T., F. Nagy, and N.H. Chua (1985). Identification of DNA sequences required for activity of the cauliflower mosaic virus 35S promoter. *Nature* 313: 810-812.

Ortega, J.L., S.J. Temple, and C. Sengupta-Gopalan (2001). Constitutive over expression of cytosolic glutamine synthetase ($GS_1$) gene in transgenic alfalfa demonstrates that $GS_1$ may be regulated at the level of RNA stability and protein turnover. *Plant Physiology* 126:109-121.

Papadopoulos, Y.A. and B.D. McKersie (2005). A comparison of protein degradation during wilting and ensiling of six forage species. *Canadian Journal of Plant Science* 63:903-912.

Pedersen, J.F., K.P. Vogel, and D.L. Funnell (2005). Impact of reduced lignin on plant fitness. *Crop Science* 45:812-819.

Pickering, F.S. and P.J. Reis (1993). Effects of abomasal supplements of methionine on wool growth of grazing sheep. *Australian Journal of Experimental Agriculture* 33:7-12.

Potenza, C., L. Aleman, and C. Sengupta-Gopalan (2004). Targeting transgene expression in research, agriculture and environmental applications: Promoters used in plant transformation. *In Vitro Cellular and Developmental Biology—Plant* 40:1-22.

Ravanel, S., B. Gakiere, D. Job, and R. Douce (1998). Cystathionine γ-synthase from *Arabidopsis thaliana*: Purification and biochemical characterization of the recombinant enzyme overexpressed in *Escherichia coli*. *Biochemical Journal* 331:639-648.

Ray, H., M. Yu, P. Auser, L. Blahut-Beatty, B. McKersie, S. Bowley, N. Westcott, B. Coulman, A. Lloyd, and M.Y. Gruber (2003). Expression of anthocyanins and proanthocyanidins after transformation of alfalfa with maize *Lc*. *Plant Physiology* 132:1448-1463.

Reddy, M.S,, F. Chen, G. Shadle, L. Jackson, H. Aljoi, and R.A. Dixon (2005). Targeted down-regulation of cytochrome P450 enzymes for forage quality improvement in alfalfa (*Medicago sativa* L.). *Proceedings of the National Academy of Sceinces USA* 102:16573-16578.

Reish, B., S.H. Duke, and E.T. Bingham (1981). Selection and characterization of ethionine-resistant alfalfa (*Medicago sativa* L.) cell lines. *Theoretical and Applied Genetics* 59:89-94.

Robbins, M.P., F. Paolocci, J.W. Hughes, V. Turchetti, G. Allison, S. Arcioni, P. Morris, and F. Damiani (2003). *Sn,* a maize bHLH gene, modulates anthocyanin and condensed tannins pathways in *Lotus corniculatus*. *Journal of Experimental Botany* 54:239-248.

Rogers, G., A. Bryant, and L. McLeay (1979). Silage and dairy cow production. Abomasal infusions of casein, methionine and glucose, and milk yield and compositions. *New Zealand Journal of Agricultural Research* 22:533-541.

Samac, D.A., M. Tesfaye, M. Dornbusch, P. Saruul, and S.J. Temple (2004). A comparison of constitutive promoters for expression of transgenes in alfalfa (*Medicago sativa*). *Transgenic Research* 13:349-361.

Samis, K., S. Bowley, and B. McKersie (2002). Pyramiding Mn-superoxide dismutase transgenes to improve persistence and biomass production in alfalfa. *Journal of Experimental Botany* 37:1342-1350.

Shadle, G., F. Chen, M.S. Srinivasa Reddy, L. Jackson, J. Nakashima, and R.A. Dixon (2007). Down-regulation of hydroxycinnamoyl CoA: shikimate hydroxycinnamoyl transferase in transgenic alfalfa affects lignification, development and forage quality. *Phytochemistry* 68:1521-1529.

Sharma, S.B. and R.A. Dixon (2005). Metabolic engineering of proanthocyanidins by ectopic expression of transcription factors in *Arabidopsis thaliana*. *Plant Journal* 44:62-75.

Sullivan, M.L., R.D. Hatfield, S.L. Thoma, and D.A. Samac (2004). Cloning and characterization of red clover polyphenol oxidase cDNAs and expression of active protein in *Escherichia coli* and transgenic alfalfa. *Plant Physiology* 136: 3234-3244.

Tesfaye, M., S.J. Temple, D.L. Allan, C.P. Vance, and D.A. Samac (2001). Over expression of malate dehydrogenase gene expression in transgenic alfalfa enhances organic acid synthesis and confers tolerance to aluminum. *Plant Physiology* 127:1836-1844.

Winicov, I., B. Valliyodan, L. Xue, and J.K. Hoober (2004). The MsPRP2 promoter enables strong heterologous gene expression in a root-specific manner and is enhanced by over expression of *Alfin1*. *Planta* 219:925-935.

Woodward, S.L., M.J. Laboyrie, and E.B.I. Jansen (2000). *Lotus corniculatus* and condensed tannins—effects on milk production by dairy cows. *Asian-Australasian Journal of Animal Sciences* 13:521-525.

Xie, D.Y., S.B. Sharma, and R.A. Dixon (2004). Anthocyanidin reductases from *Medicago truncatula* and *Arabidopsis thaliana*. *Archives in Biochemistry and Biophysics* 422:91-102.

Zeh, M., A.P. Casazza, O. Kreft, U. Roessner, K. Bieberich, L. Willmitzer, R. Hoefgen, and H. Hesse (2001). Antisense inhibition of threonine synthase leads to high methionine content in transgenic potato plants. *Plant Physiology* 127: 792-802.

## Chapter 2

# Genetic Transformation in *Trifolium* and *Medicago*

Stephen Panter
Aidyn Mouradov
Emma Ludlow
Marcel Labandera
Michael Emmerling
German Spangenberg

### *INTRODUCTION*

Legumes contribute to human and animal nutrition worldwide by providing a source of high-quality protein. Legumes can fix atmospheric nitrogen in the soil in symbiosis with the *Rhizobia* class of bacteria, reducing the need for application of nitrogenous fertilizers. Inclusion of legumes as a break crop in crop rotations reduces the need for chemical control of pests and pathogens. Furthermore, legumes produce a wide range of complex natural products, which may be involved in the establishment of symbiosis and protection of plants from abiotic and biotic stresses. These valuable features make legumes an attractive target for functional genomics research. Forage quality traits including digestibility, nutritional value, silage properties, and adaptation to environmental stresses, are primary targets for genetic modification in *Trifolium* and *Medicago* species, such as white clover and alfalfa (or lucerne).

Both white clover (*Trifolium repens*) and alfalfa or lucerne (*Medicago sativa*) are important components of temperate pastures throughout the world. In addition, *Medicago truncatula* is the preferred model legume for functional genomics. This species has a small, diploid genome (approximately $5 \times 10^8$ bp), is self-fertile, has a rapid generation time, and sets a large amount of seed. The numerous ecotypes of *M. truncatula* that have been collected represent an important resource for classical and molecular genetics. *M. truncatula* is also a host to *Sinorhizobium meliloti,* a microsymbiont species whose genome has been fully sequenced.

Over the last decade, genetic modification has been used to complement plant breeding approaches aimed at improving forage quality, yield, and adaptation to stress in legume species. White clover is susceptible to infection by alfalfa mosaic virus (AMV), white clover mosaic virus (WCMV), and clover yellow vein virus (CYVV). Infection with these viruses can reduce potential biomass production in white clover-dominated pastures by up to 30 percent (Campbell and Moyer, 1984; Dudas et al., 1998; Gibson et al., 1981; Latch and Skipp, 1987). While potential sources of virus resistance or tolerance have been described in some legumes (Barnett and Gibson, 1975; Crill et al., 1971), introgression of genes for resistance to AMV and WCMV into white clover dependent on conventional breeding methods has not been successful. This is due to virus strain limitations, lack of durability of natural resistance, and barriers to interspecific sexual and/or somatic hybridization.

Condensed tannins (also known as proanthocyanidins) are polymers or oligomers of flavan-3-ol units derived from the phenylpropanoid pathway. Low levels of condensed tannins (1-3 percent by dry mass) in forage can improve protein uptake by ruminants and reduce the risk of pasture bloat by slowing ruminal degradation of plant proteins and increasing crude protein flow to the intestine (Waghorn et al., 1986; Wang et al., 1994). Modification of the phenylpropanoid pathway in legumes could potentially lead to bloat-safe forage cultivars with an elevated level and improved composition of condensed tannins in leaves.

Large-scale generation of transgenic white clover and *M. truncatula* plants is essential for functional genomic studies in these species and requires robust and efficient methods for genetic transformation and plant regeneration that are largely genotype independent.

Two rapid and simple *in planta* transformation methods have been described for *M. truncatula* (Trieu and Harrison, 1996; Trieu et al., 2000). The first approach is based on vacuum infiltration of flowering plants. This procedure had been developed earlier and was used successfully for the transformation of different *Arabidopsis thaliana* ecotypes. A second approach developed by the same group involved vacuum infiltration of young seedlings with *Agrobacterium.* Although transformation efficiencies ranging from 2.9 to 76 percent were reported, both protocols are difficult to reproduce and have not gained widespread acceptance for legume transformation.

The protocol described in this chapter is based on our long-term experience of successful regeneration and genetic transformation of different

*Trifolium* species, namely, white clover (*Trifolium repens* cultivars Haifa, Huia, Irrigation, and Mink); red clover (*T. pratense* cultivars Astred, Colenso, Cherokee, Quinequeli, Redquin, and Renegade); subterranean clover (*T. subterraneum* subsp. *brachycalycinum* cv. Clare, subsp. *subterraneum* cultivars Denmark and Woogenellup, and subsp. *yanninicum* cultivars Larisa and Trikkala); *T. michelianum; T. isthmocarpum;* and *Medicago* species including alfalfa (*M. sativa*), *M. polymorpha, M. truncatula, M. littoralis,* and *M. tonata*. This protocol is highly reproducible, robust, and genotype independent. Detailed molecular analyses have confirmed the presence and expression of transgenes in three transformed clover species. Stable meiotic transmission of transgenes was demonstrated for selected transgenic clover plants carrying single T-DNA inserts. A further application of this technology is in isogenic transformation, demonstrated here for white clover, which provides an untransformed control plant with a genetic background otherwise identical to the transgenic plant.

## *MATERIALS*

### *Growth Media for* Agrobacterium tumefaciens

- Luria-Bertani (LB) medium, per liter: 10 g Bacto-tryptone (Becton-Dickinson and Co., Sparks), 5 g Bacto yeast extract (Becton-Dickinson and Co.), 10 g NaCl (Merck, Darmstadt, Germany), appropriate selective agents.
- MGL medium, per liter: 5 g Bacto-tryptone (Becton-Dickinson and Co.), 2.5 g Bacto yeast extract (Becton-Dickinson and Co.), 1.15 g *L*-glutamic acid, K-salt (Sigma), 5 g mannitol (Sigma), 250 mg $KH_2PO_4$ (Merck), 100 mg NaCl (Merck), 100 mg $MgSO_4 \cdot 7H_2O$ (Sigma).

### *Plant Tissue Culture*

- Surface sterilizing agents: 70 percent (v/v) ethanol and 25 percent (v/v) commercial Domestos bleach solution (12.5 $g \cdot l^{-1}$ active chlorine) with 0.1 percent (v/v) Tween 20.
- RM73 + acetosyringone (cocultivation media), per liter: 4.4 g Murashige and Skoog (MS) Basal Medium (Sigma), 30 g sucrose (Merck), 8 g Bacto-Agar (Becton-Dickinson and Co.), pH 5.75. This medium is supplemented with 250 $mg \cdot l^{-1}$ of cefotaxime (Sandoz, Vienna, Austria), 5 μM thidiazuron (Sigma), 0.5 μM napthalene acetic acid (Sigma) and 40 $mg \cdot l^{-1}$ acetosyringone (Sigma).

- RM73 + selection (regeneration media with selection), per liter: RM73 containing 250 mg·l$^{-1}$ of cefotaxime (Sandoz) and working concentrations of antibiotics: 50 mg·l$^{-1}$ kanamycin (Sigma), 75 mg·l$^{-1}$ gentamicin (Sigma) or 50 mg·l$^{-1}$ hygromycin (Sigma).
- RIM73 + selection (root-inducing media with selection), per liter: 4.4 g MS Basal Medium (Sigma), 8 g·l$^{-1}$ Bacto-Agar (Becton-Dickinson and Co.), 15 g sucrose (Merck), supplemented with 250 mg·l$^{-1}$ cefotaxime (Sandoz), 1.2 µM indole-3-butyric acid (Sigma) and appropriate antibiotics, pH 5.75.

## AGROBACTERIUM-*MEDIATED TRANSFORMATION METHOD*

### *Preparation of* **Agrobacterium** *Cultures*

- Day 1: Inoculate freshly grown *Agrobacterium* in 20 ml of LB medium containing 40 mg·l$^{-1}$ acetosyringone and the appropriate antibiotic for selection of bacteria (e.g., 50 mg·l$^{-1}$ spectinomycin for pPZP200-derived vectors). Cover the flask with aluminium foil to exclude light. Incubate at 28°C and 250 rpm for 24-48 hours.

### *Seed Sterilization and Imbibition*

- Day 1: Place 1 teaspoon of seeds (approximately 500) into a 280 µm mesh sieve (Saulas, Paisy Cosdon, France) and wash seeds for 5 minutes under running tap water.
- In a laminar flow hood, transfer seeds to a sterile 120 ml tissue culture vessel (Techno-Plas, Adelaide, Australia) with a sterile spatula. Add a magnetic stirrer bar (wiped with 70 percent ethanol) and add approximately 30 ml of 70 percent ethanol to the vessel. Stir gently for 5 minutes. Discard ethanol.
- Add 50 ml of the bleach + Tween 20 surface sterilant to the vessel.
- Stir for 20-30 minutes. The duration of this step may require evaluation for different cultivars or seed batches. The age of the seed may affect imbibition time.
- Discard sodium hypochlorite and rinse the seeds 6-8 times with approximately 80 ml of sterile distilled water.
- Add approximately 30 ml of water and cover the vessel with aluminium foil. Incubate the seeds overnight at 15-18°C.

### *Preparation of* Agrobacterium *Inoculum*

- Day 2: Centrifuge *Agrobacterium* cultures in a bench-top centrifuge (e.g., Beckman Orbital 500) for 10 minutes at 3,000 rpm. Discard supernatant.
- Add 20 ml MGL medium supplemented with acetosyringone (40 $mg \cdot l^{-1}$) to each tube. Resuspend by vortexing.
- Assess bacterial growth and adjust the $OD_{600}$ to 0.4-0.5 with MGL medium plus 40 $mg \cdot l^{-1}$ acetosyringone. Incubate at 28°C with shaking (250 rpm) for 2-4 hours. When the $OD_{600}$ value is 0.7-0.8, the *Agrobacterium* is ready to use for transformation.

### *Dissection of Seeds*

- Day 2: Pour approximately 20 ml MGL medium into a 90 × 90 × 20 mm Petri dish. Place imbibed seeds in another Petri dish under the dissecting microscope. Remove the seed coat and endosperm layer with sterile needles.
- Cut the hypocotyl, making sure to leave approximately 1.5 mm of the cotyledonary petioles attached to the cotyledons. Separate the cotyledons and transfer them to the Petri dish with MGL medium. Continue steps for all cotyledons. If seeds become dry during dissection, add a small quantity (1-2 ml) of autoclaved water. It is important not to let the seeds dry out.

### *Inoculation of Explants and Cocultivation*

- Day 2: Remove the MGL medium from the Petri dish with cotyledonary explants using a pipette and pour or pipette the *Agrobacterium* solution into the dish or wells containing the explants. Lid and seal plates with laboratory Parafilm and cover completely with aluminium foil. Incubate plates for 45 minutes at room temperature with gentle shaking.
- Remove the *Agrobacterium* culture with a pipette. Blot cotyledonary explants on one or two sheets of sterilized filter paper with sterilized paper towel underneath and transfer cotyledonary explants onto plates of RM73 media + acetosyringone with sterile forceps, without letting them dry out.
- Using the forceps, distribute the cotyledonary explants evenly (no more than 100 cotyledonary explants per dish). Incubate in a growth room at 25°C for 3 days (16-hour photoperiod).

### *Selection and Regeneration of Transgenic Plants*

- Day 5: After cocultivation, transfer the cotyledonary explants from the RM73 + acetosyringone plates with sterile forceps into 9 × 2.0 cm Petri dishes containing approximately 20-30 ml of sterile distilled water.
- Wash the cotyledonary explants by gently shaking them in the water-filled Petri dish.
- Repeat the previous step two to three times.
- Transfer the explants individually to 9 × 2.0 cm Petri dishes containing appropriate RM73 selective medium with sterile forceps by inserting the hypocotyl end into the medium (50 explants per plate). Seal the dishes with Parafilm. Incubate at 25°C with a 16-hour photoperiod. For selection: 50 $mg{\cdot}l^{-1}$ kanamycin, 50 $mg{\cdot}l^{-1}$ hygromycin, 75 $mg{\cdot}l^{-1}$ gentamicin (final concentrations).
- Subculture cotyledonary explants every 2 weeks for a period of 6 weeks by transferring them individually onto fresh Petri dishes containing RM73 selective medium, sealing the dishes with laboratory film, and labeling appropriately.

### *Root Induction*

- Week 7: Excise developing transgenic shoots (with a green base) using a sterile scalpel and transfer to sterile 120 ml tissue culture vessels containing rooting media (RIM73) supplemented with appropriate antibiotics, as for selective RM73 media (one plant per vessel). Roots will develop within 8-20 days for white clover. If shoots are too small for transfer to RIM73, they can be dissected out with a sterile scalpel and transferred to another RM73 plate with selection.

### *Transfer to Soil*

- Weeks 10-12: Gently remove plantlets with well-formed root systems from culture vessels and wash medium from the roots with tap water.
- Transfer each plantlet to a 10.16 cm (4 inch) pot containing moist potting mix. Place in a misting bench for 2 weeks and then grow under standard glasshouse conditions, potting up to 15.24 cm (8 inch) pots as necessary.

## *ISOGENIC TRANSFORMATION OF WHITE CLOVER*

White clover is a self-incompatible species in which individual seeds of a given cultivar are genetically distinct. Isogenic transformation, the production of pairs of transgenic and nontransgenic plants in an identical genetic background, can help to distinguish natural variability between different genotypes of a recipient variety, and phenotypes caused by expression of transgenes (Ding et al., 2003). This method is based on separate regeneration of both cotyledonary explants of the same seed. One of them is inoculated with *Agrobacterium* and thus used for genetic transformation while the second isogenic cotyledonary explant is used for regeneration without *Agrobacterium*-mediated transformation (Figure 2.1).

### *Modifications to the Standard Method for Isogenic Transformation*

- After dissecting a seed, place one of the cotyledonary explants in a numbered well of a sterile 96-well cell microtiter plate containing MGL medium for transformation. Draw a numbered grid on the base of an RM73 (regeneration, no antibiotics except cefotaxime) plate and place the untransformed cotyledonary explant of each seed directly on a numbered position on this plate.
- Repeat the process for the remaining seeds, taking care to keep track of pairs of untransformed and transformed cotyledonary explants.
- Transform cotyledonary explants in the microtiter plate. Draw a numbered grid on the base of an RM73 + acetosyringone (cocultivation) plate and transfer transformed cotyledonary explants directly from numbered wells of the 96-well microtiter plate to numbered positions of the cocultivation plate with a sterile 18-21G needle on a 10 ml syringe. Instead of the described procedure, draw numbered grids on the base of an RM73 + selection plate and transfer transformed cotyledonary explants directly from the RM73 + acetosyringone plate to corresponding positions on the RM73 + selection plate.
- Regeneration of untransformed partner cotyledonary explants requires RM73 plates with no antibiotics except for cefotaxime. After seed dissection, a numbered grid needs to be drawn on the base of these plates and cotyledonary explants must be transferred from the corresponding positions of one plate to another. Since growth of the untransformed partner cotyledonary explants occurs faster than the corresponding transformed ones, more transfers are needed for the untransformed partner cotyledonary explants.

- For root induction as detailed earlier, transfer untransformed partner cotyledonary explants to RIM73 with no antibiotics except for cefotaxime.

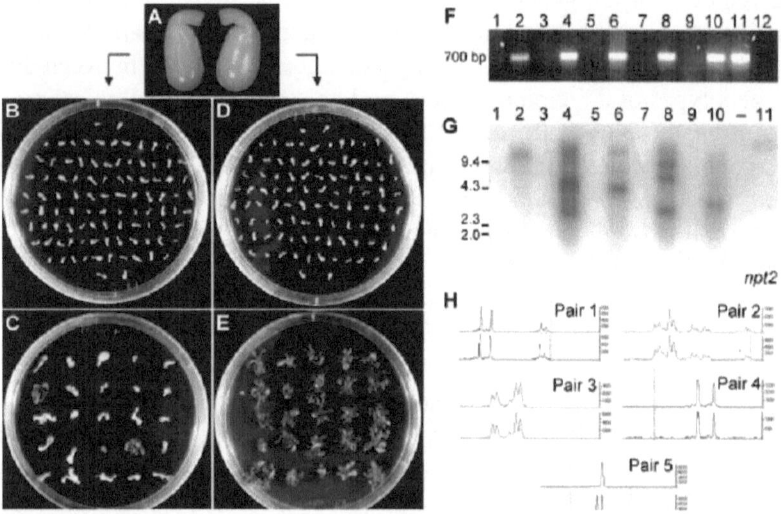


FIGURE 2.1. Development of white clover transgenic and nontransgenic lines. (A) Isogenic pair of white clover cotyledonary explants from a mature seed; (B) 96 treated independent cotyledonary explants from cotyledon pairs arranged in grid pattern after three days of cocultivation with *A. tumefaciens* on RM73 medium; (C) kanamycin-resistant shoots derived from transformed cotyledonary explants on RM73 medium supplemented with 50 $mg{\cdot}l^{-1}$ of kanamycin; (D) 96 nontreated independent cotyledonary explants on antibiotic-free RM73 medium in a grid pattern corresponding in position with their isogenic transformants in plate B; (E) proliferating nontransgenic shoots regenerated from nontreated cotyledonary explants growing on antibiotic-free RM73 medium; (F) PCR analysis (*nptII*) of isogenic transgenic and nontransgenic white clover plants. Lanes 1-10: five pairs of nontransgenic (odd lanes) and transgenic white clover plants (even lanes); lanes 1-2: pair 1; lanes 3-4: pair 2; lanes 5-6: pair 3; lanes 7-8: pair 4; lanes 9-10: pair 5, nonisogenic; lane 11: pBincits (positive control); lane 12: no-DNA negative control. (G) Southern hybridization with *Bam*HI-digested genomic DNA isolated from pairs of nontransgenic and transgenic white clover plants derived from direct regeneration and *A. tumefaciens*-mediated transformation of cotyledonary explants. Lanes 1-11: as in F. (H) SSR polymorphism profile of four isogenic pairs (as in F) of nontransgenic and transgenic white clover plants (pairs 1-4) and one nonisogenic pair derived from unrelated cotyledonary explants (pair 5, negative control); top traces: nontransgenic white clover plants; bottom traces: transgenic white clover plants.

## *CONCLUSION*

The use of cotyledonary explants freshly dissected from imbibed, mature seeds allows for a robust and genotype-independent method for generating transgenic *Trifolium* and *Medicago* plants using *Agrobacterium*. Plant regeneration frequencies, expressed as percentage of explants producing shoots, ranged from 65 to 97 percent for all tested cultivars of *Trifolium* species. Plants were similarly regenerated from excised cotyledonary explants from 16 cultivars of *Medicago sativa* and 6 cultivars of four annual medics, *M. polymorpha, M. truncatula, M. littoralis*, and *M. tonata*. Plant regeneration frequencies for all the tested cultivars of alfalfa ranged from 33 to 85 percent and from 35 to 65 percent for the 6 cultivars of medics. Transformation efficiency, calculated as number of independent transformants confirmed by Southern hybridization analysis per 100 transformed cotyledonary explants, varied between 0.3 percent and 6 percent. The protocol described is efficient, reproducible, robust, and genotype independent. It has been used successfully for the recovery of transgenic plants in different *Trifolium* and *Medicago* species with a range of binary vectors carrying different selectable marker genes (Ding et al., 2003).

## REFERENCES

Barnett, O.W. and P.B. Gibson (1975). Identification and prevalence of white clover viruses and the resistance of *Trifolium* species to these viruses. *Crop Science* 15:32-37.

Campbell, C.L. and J.W. Moyer (1984). Yield responses of 6 white clover clones to virus infection under field condition. *Plant Disease* 68:1033-1035.

Crill, P., E.W. Hanson, and D.J. Hagedorn (1971). Resistance and tolerance to alfalfa mosaic virus in alfalfa. *Phytopathology* 61:371-379.

Ding, Y.-L., G. Aldao-Humble, E. Ludlow, M. Drayton, Y.-H. Lin, J. Nagel, M. Dupal, G.-Q. Zhao, C. Pallaghy, R. Kalla, M. Emmerling, and G. Spangenberg (2003). Efficient plant regeneration and *Agrobacterium*-mediated transformation in *Medicago* and *Trifolium* species. *Plant Science* 165:1419-1427.

Dudas, B., D.R. Woodfield, P. M. Tong, M.F. Nicholls, G.R. Cousins, R. Burgess, D.W.R. White, D.L. Beck, T.J. Lough, and R.L.S. Forster (1998). Estimating the agronomic impact of white clover mosaic virus on white clover performance in the North Island of New Zealand. *New Zealand Journal of Agricultural Research* 41:171-178.

Gibson, P.B., O.W. Barnett, H.D. Skipper, and M.R. McLaughlin (1981). Effects of three viruses on growth of white clover. *Plant Disease* 65:50-51.

Latch, G.C.M. and R.A. Skipp (1987). Diseases. In *White clover,* M.J. Baker and W.M. Williams (eds.). Wallingford-Oxon: CAB International, pp. 421-446.

Trieu, A.T., S.H. Burleigh, I.V. Kardailsky, I.E. Maldonado-Mendoza, W.K. Versaw, L.A. Blaylock, H. Shin, T.J. Chiou, H. Katagi, G.R. Dewbre, et al. (2000). Transformation of *Medicago truncatula* via infiltration of seedlings or flowering plants with *Agrobacterium. Plant Journal* 22:531-541.

Trieu, A.T. and M.J. Harrison (1996). Rapid transformation of *Medicago truncatula*: Regeneration via shoot organogenesis. *Plant Cell Reports* 16:6-11.

Waghorn, G.C., M.J. Ulyatt, A. John, and M.T. Fisher (1986). The effect of condensed tannins on the site of digestion of amino acids and other nutrients in sheep fed on *Lotus corniculatus* L. *British Journal of Nutrition* 57:115-126.

Wang, Y., G.C. Waghorn, T.N. Barry, and I.D. Shelton (1994). The effect of condensed tannins in *Lotus corniculatus* on plasma metabolism of methionine, cysteine and inorganic sulphate by sheep. *British Journal of Nutrition* 72:923-935.

## Chapter 3

# *Agrobacterium*-Mediated Genetic Transformation of Alfalfa

Michele Bellucci
Francesca De Marchis
Sergio Arcioni

## *INTRODUCTION*

Obtaining plants producing recombinant proteins, which opens up the way to many technological applications that include industrial and pharmaceutical use for human and animal nutrition, is now a reality. The fundamental technology to obtain plants accumulating heterologous proteins is genetic transformation. As a fodder crop, alfalfa (*Medicago sativa* L.) is the most important protein source of all legumes. Alfalfa is also important for its high biomass production and versatility. It is mainly used as a hay and silage crop, but it can also be grazed. Moreover, when it is dried, ground, and pelleted, it is sold as a dehydrated feed for many animals. For all these reasons, alfalfa is the main forage crop in temperate agriculture, and the third largest crop in the United States where its direct economic value is about $7.1 billion each year. Alfalfa is considered the queen of forage crops in most of the European countries, too, and it is particularly appreciated in the Mediterranean basin.

Is alfalfa easy to transform? According to Somers et al. (2003), relatively rapid and efficient alfalfa transformation methods have been developed using cocultivation of tissue pieces with *Agrobacterium tumefaciens*. However, a genotype-independent and widely applicable protocol for an efficient plant regeneration and genetic transformation of alfalfa is not yet available. Genetic transformation of alfalfa has been initially inhibited by two major limiting factors, one being related to the method of transformation and the other to poor regeneration from tissue cultures. The first limiting factor lies in the fact that most attempts at transforming alfalfa have been based on infection by *A. tumefaciens*, although other methods for plant transformation such as use of *A. rhizogenes* (Spanò et al., 1987) or particle bombardment (Pereira and Erickson, 1995) have been reported. *A. tumefaciens* exhibits host strain specificity, so that only certain bacterial strains will infect a few alfalfa genotypes. The second obstacle in alfalfa transformation is its own poor regeneration frequency, as reviewed in Samac and Temple (2004). Saunders and Bingham (1972) reported the first successes in regenerating this important fodder crop via somatic embryogenesis, either directly from leaf explants or indirectly from callus cells. However, most alfalfa varieties do not show high levels of plant regeneration (Bingham et al., 1988). According to some authors (e.g., Buising and Tomes, 2003), the combination of these two limiting factors has created a serious bottleneck in achieving alfalfa transformation. Nevertheless, efficient transformation protocols have been developed for alfalfa in the past decade (Atkins and Smith, 1997). A key explant source tissue for alfalfa regeneration and transformation seems to be cotyledons. A robust genotype-independent protocol for regeneration employing cotyledonary explants has been reported for a range of clovers (*Trifolium* species) and medics (*Medicago* species), including alfalfa (Ding et al., 2003). When this procedure was used, plant regeneration frequencies for 16 tested cultivars of alfalfa ranged from 33 to 85 percent. Moreover, improved methods for the transformation and regeneration of alfalfa using immature cotyledons have been patented (Buising and Tomes, 2003).

Several protocols for alfalfa transformation are available in the literature. According to Samac and Temple (2004), these protocols use regeneration systems with a two- or three-step procedure. In the first case, callus formation is stimulated on a basal medium with 2,4-dichlorophenoxyacetic acid (2,4-D) and kinetin followed by transfer of calli to growth regulator-free medium for embryo formation (Desgagnés et al., 1995; Hernández-Fernández and Christie, 1989). The three-step process consists of callus initiation on a medium with a synthetic auxin and kinetin, transfer to a medium with a high 2,4-D content for a short period of time, then transfer to growth regulator-free medium (Brown and Atanassov, 1985; Skokut et al., 1985). In this chapter, some practical aspects of a two-step protocol for alfalfa genetic transformation by *A. tumefaciens* are described. This protocol is adapted from a procedure developed by Austin et al. (1995) after testing several modifications of the basic explant cocultivation system (Horsch et al., 1985), in order to optimize genetic transformation in one specific alfalfa genotype that was derived from a cross between Regen S and Regen Y. Overall transformation frequencies using this method are 25 to 55 percent. Rooted plantlets can be obtained in as short a time as 9 to 11 weeks.

## *SOMATIC EMBRYOGENESIS AND PLANT REGENERATION*

Somatic embryogenesis is largely applied in alfalfa genetic transformation. Alfalfa regeneration is feasible when embryos develop into whole plantlets. While regeneration is genotype independent in some species, this is not the case in alfalfa. Many efforts have been made to understand the role of the genetic background in the somatic embryogenesis of *Medicago* species (reviewed in Arcioni et al., 1990). Several works have shown that *M. falcata* germplasm has contributed significantly to the in vitro regeneration response of *M. sativa* (Meijer and Brown, 1985; Chen et al., 1987). The genotypic variability existing in all the alfalfa varieties allows the identification of regenerable genotypes with proper screening (Mitten et al., 1984; Brown and Atanassov, 1985). To overcome the problems arising out of scarcity of embryogenic genotypes, it would be useful to constitute a synthetic variety with a high regeneration capacity. For this purpose the trait should be under simple genetic control. Studies carried out in diploid (Reisch and Bingham, 1980) and tetraploid cultivars of alfalfa (Hérnandez-Fernández and Christie, 1989; Kielly and Bowley, 1992) showed that the process by which embryos are formed from vegetative cells is under the control of at least two independent genes though with different interactions. One major progress in alfalfa transformation has been made possible with the identification of the highly regenerable variety Regen S, which was developed by recurrent selection for regeneration (Bingham et al., 1975). Following this procedure, the frequency of genotypes that regenerate increased from 12 to 67 percent, and two tetraploid varieties, Regen S and Regen Y, with improved regeneration capacity were released (Bingham, 1991). The loci controlling regeneration and embryo formation in alfalfa have not yet been identified and cloned, in spite of the availability of a randomly amplified polymorphic DNA (RAPD) marker associated with this trait (Yu and Pauls, 1993).

## *MATERIALS*

### *Plant Material*

Alfalfa is an autotetraploid ($2n = 4x = 32$) and prevalently allogamous species, which makes its genetics and breeding difficult and time consuming. Each variety is a highly heterozygous population. Thus, when highly regenerable alfalfa genotype is selected either from regenerable germplasm (usually Regen S or Regen SY) or from an elite breeding population (Matheson et al., 1990), vegetative cuttings of the genotype must be used in order to have enough material for transformation experiments. Alternatively, the genotype of interest can be multiplied by tissue culture of its explants for callus induction and somatic embryo formation. The plantlets obtained from the embryos will be identical to the initial genotype, even though somaclonal variation may occur in some plants as a result of the tissue culture process.

The protocol described here is optimized for transformation of a highly embryogenic plant selected from Regen SY. Clonal plants of this genotype are vegetatively propagated in vitro on MS medium without growth regulators (Murashige and Skoog, 1962) and cultured at 23 ± 1°C with a light intensity of approximately 80 to 100 $\mu E \cdot m^{-2} \cdot sec^{-1}$. For production of explants, plants are transplanted to a soil mix and maintained in a growth room with a 16 hour photoperiod of 300 $\mu E \cdot m^{-2} \cdot sec^{-1}$ and day and night temperatures of 21°C and 19°C respectively. Do not use alfalfa plants that are flowering. The third to sixth fully expanded dark green trifoliate leaves are the best material. Plants are fertilized weekly with a complete fertilizer and watered daily.

## *Culture Media*

Alfalfa needs a special sequence of media and hormonal stimuli to be cultured and regenerated in vitro. Explants can be induced to callus in different basal medium compositions (reviewed in Arcioni et al., 1990), and this process requires the presence of synthetic growth regulators (2,4-D) or naphthalene acetic acid (NAA) in combination with kinetin or benzyl adenine (BA) ranging from 0.5-2 $mg \cdot l^{-1}$ auxin and 0.1-0.5 $mg \cdot l^{-1}$ cytokinin, depending on the alfalfa genotype. In fact, the optimal medium for regeneration has been shown to vary with the genotype and, apart from 2,4-D, which is always required for somatic embryo induction, other variables (e.g., media chemical composition, type of explants) interact to modulate somatic embryo formation and have been reviewed in detail by McKersie and Brown (1997). Several studies have elucidated the role of reduced nitrogen, amino acids, and carbohydrate sources in somatic embryo formation in alfalfa. The level of $NH_4^+$ is critical for embryo formation; and concentrations between 12.5 and 100 mM are indicated in successful alfalfa regeneration (Walker and Sato, 1981). The effect of adding amino acids to alfalfa tissue cultures was investigated by Stuart and Strickland (1984), who showed that addition of proline stimulated a threefold increase in somatic embryogenesis when compared with the normal medium. Other proline analogs such as alanine, glutamine, or serine have also been proven to be effective. A complex mixture of organic substances such as yeast extract and myo-inositol has been reported to stimulate morphogenesis and embryogenesis. In particular, maltose gave the highest embryo yield and improved embryogenesis of alfalfa genotypes, which were recalcitrant to regeneration (Arcioni et al., 1990).

## Agrobacterium *Strain and Selectable Marker*

*Agrobacterium tumefaciens* has a natural capability of transkingdom DNA transfer (Stachel and Zambryski, 1989). It is the etiologic agent of the crown-gall disease, a neoplastic growth that results from the transfer of a transferred DNA (T-DNA) sequence from the bacterial tumor-inducing (Ti) plasmid into the genome of an infected plant (reviewed by Tzfira et al., 2004). This DNA fragment encodes genes for auxin and cytokinin biosynthesis, and these growth regulators promote growth of undifferentiated cells in the crown gall. The transfer process is dependent upon trans-acting virulence proteins encoded by the Ti-plasmid virulence (*vir*) genes, as well as elements known as border sequences, a set of 25-bp direct repeats that define the limits for the T-DNA transfer. These border sequences are the only required *cis* elements that determine the T-DNA region on the Ti plasmid; therefore the tumor-inducing genes can be removed from Ti plasmid vectors (disarmed vectors) and replaced with the DNA of interest for genetic engineering.

*Medicago* species are sensitive to *A. tumefaciens* and alfalfa has been transformed by cocultivation of stem sections with *A. tumefaciens* strains using kanamycin resistance as a selectable marker from late 1980s (Christou, 1992). Transgenic plants could be obtained from 12 to 51 percent of the inoculated stem explants using the disarmed *A. tumefaciens* strain LBA 4404 (Shahin et al., 1986). In the same year, *M. sativa* spp. *multivaria* also was transformed with the wild-type strain A281 (Deak et al., 1986). The first report in which the parameters affecting the frequency of kanamycin-resistant plants obtained were studied indicated that the A281 strain was better than LBA4404 to transform a genotype of *M. sativa* spp. *multivaria* (Chabaud et al., 1988). Under these conditions, *A. tumefaciens* infection of leaf and petiole tissue produced about 70 percent of transformed plantlets. Later studies using commercial alfalfa germplasm demonstrated that alfalfa transformation by *A. tumefaciens* is strongly characterized by a strain-genotype interaction (Du et al., 1994).

Normally, three *A. tumefaciens* strains have been used to transform alfalfa: A281 (Hood et al., 1986), LBA 4404 (Hoekema et al., 1983), and C58 (Zambryski et al., 1983). The *A. tumefaciens* strains utilized to infect alfalfa tissue harbor a binary vector system, which means that the Ti plasmid is divided into two components, a shuttle vector and a helper plasmid. The helper plasmid carries the virulence genes and is permanently placed in the bacterium, whereas the shuttle vector contains T-DNA borders, a broad-host-range bacterial origin of replication, and antibiotic resistance markers. A selectable marker gene, located together with the gene of interest between the T-DNA borders, is required to recover the transformed cells and to regenerate a transgenic plant. Even though kanamycin has been largely utilized as a selection agent in alfalfa transformation, other antibiotic markers, such as hygromycin or the herbicide phosphinothricin have also been used to obtain transgenic alfalfa plants. Using phosphinothricin in the medium reduces the number of nontransformed escapes (D'Halluin et al., 1990). Kanamycin concentrations in the transformation protocols have been reported in the range of 25-100 $mg{\cdot}l^{-1}$, but lower amounts of kanamycin are preferred (25-50 $mgl^{-1}$). Otherwise, the antibiotic may inhibit embryo production (Desgagnés et al., 1995) and root formation (Austin et al., 1995).

## *METHODS*

### Agrobacterium *Culture*

- The strain that gives the best results with our Regen SY genotype is LBA 4404. Streak out *Agrobacterium* from glycerol stock onto YMB plate with appropriate antibiotics. The LBA 4404 strain carries a chromosomal gene for resistance to rifampicin and has a Ti-plasmid (carrying *vir* genes) that encodes streptomycin resistance. Usually, rifampicin 25-50 $mg{\cdot}l^{-1}$ is used for selection of this strain. Rifampicin is soluble in methanol (vortex immediately to dissolve the antibiotic after adding methanol) or dimethyl sulfoxide (DMSO). Do not filter sterilized rifampicin solutions. Culture at 28-30°C for 2-3 days.
- Inoculate *Agrobacterium* from plate culture into 3-5 ml LB in a 15 ml tube with the same antibiotics. Incubate overnight at 28-29°C with shaking. Prepare 6 ml of *Agrobacterium* culture for each inoculation Petri dish.

### *Alfalfa Transformation*

#### *Day 1*

- New-growth leaves (best leaves are dark green and very healthy looking) are removed from plants, floated immediately in cool tap water, and surface sterilized using alcohol and bleach washes (5-10 seconds in 70 percent alcohol, then transfer to 20 percent bleach with 0.05 percent Tween 20 for 1.5 minutes). Rinse at least three times with sterile distilled water.

- Remove leaves from water and place on sterile Petri dish. Remove the leaf edges and cut the remaining portion of the lamina into sections of 0.5 × 0.5 cm. Place leaf pieces into sterile Petri dishes, 25-30 pieces for each dish, containing 24 ml of SHO liquid medium. After all leaves have been cut, add 6 ml of the *Agrobacterium* suspension cells to each Petri dish. Leave leaf pieces in *Agrobacterium* for 15 minutes. In this way, *Agrobacterium* cell density is not adjusted but it should fall between $OD_{600}$ of 0.2-0.4.
- Remove the explants and gently blot on sterile filter paper to remove excess liquid. Explants are placed on B5H medium in 100 × 15 mm plates (Brown and Atanassov, 1985). Wrap plates with gas-permeable tape (Parafilm) and place inside the growth chamber at 30 $\mu E \cdot m^{-2} \cdot sec^{-1}$ (place a piece of cheesecloth over plates at 60-80 $\mu E \cdot m^{-2} \cdot sec^{-1}$) and 16 hour photoperiod. Leaves may be placed with either the abaxial or adaxial side down; it does not seem to matter much in the case of alfalfa. B5H medium must be fresh, less than 24 hours old. Hence, prepare the medium the day before starting cocultivation, pour into plates, and leave at room temperature overnight. It is always advisable to make a control (inoculation without *Agrobacterium*) each time to make sure that leaves have been sterilized sufficiently, that regeneration is normal, and that kanamycin selection is working.

*Day 8*

- One week after inoculation, remove leaf pieces from plates and rinse three to four times in sterile water. Blot briefly on sterile filter paper and place on B5H containing 25 $mg \cdot l^{-1}$ kanamycin and 500 $mg \cdot l^{-1}$ cefotaxime. Maintain plates at 23 ± 1°C, 16 hour photoperiod, and light intensity of 60-80 $\mu E \cdot m^{-2} \cdot sec^{-1}$.

*Days 21-28*

- Transfer explants with developed calli three weeks after inoculation (and embryos, Figure 3.1) to B5H with antibiotics, but without growth regulators to allow for further embryo production and development of existing embryos (growth regulators inhibit maturation of embryos). If explants do not have a lot of callus and are not forming embryos or pro-embryos (small green spots), wait another week before transfer.

FIGURE 3.1. An alfalfa callus producing green embryos on B5H with 25 mg·l$^{-1}$ kanamycin and 500 mg·l$^{-1}$ cefotaxime. (See also color gallery.)

*Days 35-42*

- Pale green or white embryos, which often appear, are likely to be escapes, so use only the darkest green, well-developed embryos. Gently separate embryos from the callus and transfer to modified MS medium (MMS) containing 25 mg·l$^{-1}$ kanamycin and 500 mg·l$^{-1}$ cefotaxime to promote plantlet development. Callus will form on untreated explants in the presence of 25 mg·l$^{-1}$ kanamycin but embryos are never produced. If embryos cannot be separated, transfer the clump and separate them as they develop. Number each explant piece you take embryos from, which can give rise to multiple (~25) embryos. Developing embryos are most likely the result of a single transformation event since somatic embryos probably arise from a single cell. Hence, use only one transformed plant per explant piece to ensure analysis of independent transformation events.

*Days 49+*

- Embryos will form a shoot and sometimes a root (rooting is inhibited by kanamycin). Transfer green plantlets to MMS with 500 mg·l$^{-1}$ cefotaxime in 25 × 150 mm test tubes or Magenta GA-7 vessels for further development. They will normally root within 2-3 weeks. Once a good root system has formed, move plants to soil in containers in a growth chamber (16 hour photoperiod, 300 μE·m$^{-2}$·sec$^{-1}$, and day and night temperatures of 21°C and 19°C). Plants need to have high humidity at first and then can be hardened off.

### *Media*

- YMB medium (1 L): $K_2HPO_4$ 0.5 g; $MgSO_4 \cdot 7H_2O$ 0.2 g; NaCl 0.1 g; mannitol 10 g; yeast extract 0.4 g; pH to 7.5 with NaOH. The liquid medium is supplemented with 15 g of agar, autoclaved, and poured into Petri dishes.
- LB medium (1 L): 10 g bacto-tryptone; 5 g yeast extract, 5 g NaCl; pH to 7.5 with NaOH. Autoclave.
- SHO medium (1 L): Schenk and Hildebrandt (1972) salts and vitamins; sucrose 30 g; 2-(4-morpholino)-ethane sulfonic acid (MES) 0.5 g; pH to 7.5 with NaOH.
- B5H medium (1 L): 100 ml Gamborg's B5 stock solution 10 X; $KNO_3$ 0.5 g; $MgSO_4 \cdot 7H_2O$ 0.25 g; proline 0.5 g; sucrose 30 g; pH to 5.7 with KOH. The liquid medium is supplemented with 8 g of agar and autoclaved. Before pouring into Petri dish, add 30 ml stock amino acids and growth regulators.
- B5 stock solution 10 X: dissolve the medium powder (ICN Biomedicals, code. 2613022) in 1 L of distilled water and adjust pH between 2.0 and 3.0. Store at 4°C.
- B5H amino acids stock (250 ml): L-glutamine 6.65 g; serine 0.83 g; L-glutathione 0.083 g; adenine 0.004 g. Filter-sterilize and store at 4°C.
- B5H growth regulators: 1 $mg \cdot l^{-1}$ 2,4-D, 0.1 $mg \cdot l^{-1}$ kinetin. Prepare 1 $mg \cdot l^{-1}$ stocks, filter-sterilize, and store in freezer in small aliquots.
- MMS medium (1 L): 100 ml Murashige and Skoog medium stock solution 10 X; myoinositol 0.1 g; nicotinic acid 4.5 mg; folic acid 0.5 mg; thiamine·HCl 0.4 mg; biotin 0.05 mg; sucrose 30 g; pH to 5.7. The liquid medium is supplemented with 7 g of agar and autoclaved.
- MMS stock solution 10 X: dissolve the medium powder (ICN Biomedicals, code. 2610022) in 1 L of distilled water and adjust pH between 2.0 and 3.0. Store at 4°C.

## *CONCLUSION*

Genetic transformation has been extensively practiced in alfalfa in an effort to improve the economic value of this species. Examples of useful traits that have been engineered into alfalfa are reviewed in several publications (Rosellini and Veronesi, 2002; Samac and Temple, 2004). They belong to different research areas, such as production of industrial (Ullah et al., 2002) and pharmaceutical molecules (Wigdorovitz et al., 1999; Bellucci et al., 2007), resistance to biotic and abiotic stresses (Tesfaye et al., 2001), the improvement of nutritional qualities (Bellucci et al., 2002; Bagga et al., 2004). Alfalfa proteins suffer from a limited level of sulphur-containing amino acids, and this reduces wool growth in sheep, milk production by dairy animals, and meat quality (Tabe et al., 1995). In order to increase the methionine level in alfalfa, transgenic alfalfa plants have been produced which expressed *Arabidopsis* cystathionine γ-synthase, the enzyme that controls the synthesis of the first intermediate metabolite in the methionine pathway. In the transformed plants, the content of methionine, *S*-methylmethionine, and methionine incorporated into the water-soluble protein fraction increased up to 32-fold, 19-fold, and 2.2-fold, respectively, compared to wild-type plants (Havraham et al., 2004). Topics like the role of polyphenol oxidase (PPO) activity in the inhibition of postharvest proteolysis (Sullivan et al., 2004) have been analyzed. During alfalfa harvesting and the early stages of ensiling, proteolytic losses can reach 44 to 87 percent of forage production. Several observations suggest that PPO inhibits postharvest proteolysis in red clover (*Trifolium pratense*) and this could explain why this species has up to 90 percent less proteolysis than alfalfa during ensiling (Papadopoulos and McKersie, 1983). Red clover PPO cDNAs expressed in alfalfa, which has little PPO activity, were active and reduced proteolysis in the presence of an *o*-diphenol (Sullivan et al., 2004). Furthermore, a comparative study in transgenic alfalfa plants using different constitutive promoters and two marker genes was carried out (Samac et al., 2004). It showed that the cassava vein mosaic virus promoter is useful for high-level transgene expression in alfalfa. Other alfalfa-specific gene expression cassettes have been designed as an alternative to the low efficiency of the constitutive CaMV 35S promoter in alfalfa. These cassettes are based on regulatory sequences from the alfalfa plastocyanin gene (MED-2000 series) or inducible regulatory elements derived from the alfalfa nitrite reductase (NiR) gene (MED-1000 series; D'Aoust et al., 2004).

A transgenic approach to alfalfa has also been used jointly by Monsanto and Forage Genetics International to develop a Roundup Ready alfalfa tolerant to glyphosate. This agreement became operative in 1997 and it is very important because it has led to the first application for regulatory approval of a genetically modified trait in this species (Samac and Temple, 2004). The Roundup Ready alfalfa system was launched in 2006 and this alfalfa variety is for sale and distribution in the United States only. Roundup Ready alfalfa was developed starting from the transformation of a regenerable alfalfa clone selected from a high-yielding elite breeding population. Alfalfa transformation methods optimized for industrial applications have been recently developed by the privately held biopharmaceutical company Medicago Inc. These methods comprise the *Agrobacterium*-mediated transfer and appear to work very efficiently because 98 percent of the regenerated plants are PCR positive for the gene of interest (D'Aoust et al., 2004).

## NOTES

1. Although the Regen SY alfalfa genotype we used shows high levels of plant regeneration in tissue culture, its agronomic performance is disappointing. Hence, some problems may arise when in vitro propagated-transformed plantlets are moved to soil for obtaining leaf material, flowers, and seeds for subsequent analyses. The young plants often grow slowly and are susceptible to virus attacks. For these reasons, it would be better to analyze them for the presence and expression of the transgene directly in vitro before moving the regenerated transgenic plants to the greenhouse. Even a very small quantity of plant material (~100 mg each analysis) is enough for PCR and RNA blot analyses to reveal the presence of the gene of interest in the plant genome and its mRNA, respectively. Once the plants that suit our purpose have been identified, they can be multiplied in vitro with vegetative cuttings and allowed to grow until they reach an adequate leaf and root expansion to be transplanted. Remember that plants coming from sterile culture containers are extremely sensitive to reduced humidity. For better adaptation to the greenhouse conditions, they should be kept under a transparent plastic cover for at least a few days before being exposed to greenhouse humidity conditions. In order to obtain high seed production from transformed plants, without problems of "inbreeding depression," the Regen SY transgenic plants can be crossed (manually or with honeybees) with other alfalfa genotypes that have better agronomic characteristics. Then the progeny have to be screened for the presence of the transgene either indirectly on the seeds germinating under selection conditions (kanamycin in the medium) or by PCR analyses on the plantlets obtained.

2. Weeks et al. (2007) have developed a simple *in planta* method of transformation of alfalfa by vortexing the seedlings, which are cut at the apical node, with a suspension of *Agrobacterium* cells and sterile sand. These seedlings were grown to maturity and progenies of 7 percent of treated plants segregated for the transgene, caffeic acid o-methyltransferase (*Comt*). The T-DNA region of the binary vector did not carry any marker gene, thereby generating marker free alfalfa transagenic plants.

## REFERENCES

Arcioni, S., F. Damiani, M. Pezzotti, and E. Lupotto (1990). Alfalfa lucerne (*Medicago* spp.). In *Biotechnology in agriculture and forestry 10. Legumes and oilseed crops I*, Y.P.S. Bajaj (ed.). Berlin: Springer-Verlag, pp. 197-241.

Atkins, C.A. and P.M.C. Smith (1997). Genetic transformation and regeneration of legumes. In *Biological fixation of nitrogen for ecology and sustainable agriculture*, A. Legocki, H. Bothe, and A. Puhler (eds.). Berlin: Springer-Verlag, pp. 283-304.

Austin, S., E.T. Bingham, D.E. Matthews, M.N. Shahan, J. Will, and R.R. Burgess (1995). Production and field performance of transgenic alfalfa (*Medicago sativa* L.) expressing alpha amylase and manganese-dependent lignin peroxidase. *Euphytica* 85:381-393.

Bagga, S., A. Armendaris, N. Klypina, I. Ray, S. Ghoshroy, M. Endress, D. Sutton, J.D. Kemp, and C. Sengupta-Gopalan (2004). Genetic engineering ruminal stable high methionine protein in the foliage of alfalfa. *Plant Science* 166:273-283.

Bellucci, M., A. Alpini, and S. Arcioni (2002). Zein accumulation in forage species (*Lotus corniculatus* and *Medicago sativa*) and coexpression of the γ-zein:KDEL and γ-zein:KDEL polypeptides in tobacco leaf. *Plant Cell Reports* 20:848-856.

Bellucci, M., F. De Marchis, and S. Arcioni (2007). Zeolin is a recombinant storage protein that can be used to produce value-added proteins in alfalfa (*Medicago sativa* L.). *Plant Cell Tissue and Organ Culture*, 90:85-91.

Bingham, E.T. (1991). Registration of alfalfa hybrid Regen-SY germplasm for tissue culture and transformation research. *Crop Science* 31:1098.

Bingham, E.T., L.V. Hurley, D.M. Kaatz, and J.W. Saunders (1975). Breeding alfalfa which regenerates from callus tissue in culture. *Crop Science* 15:719-721.

Bingham, E.T., T.J. McCoy, and K.A. Walker (1988). Alfalfa tissue culture. In *Alfalfa and alfalfa improvement*. A.A. Hanson, D.K. Barnes, and R.R. Hill (eds.). Madison, WI: American Society of Agronomy, pp. 903-929.

Brown, D.C.W. and A. Atanassov (1985). Role of genetic background in somatic embryogenesis in *Medicago*. *Plant Cell Tissue and Organ Culture* 4:111-122.

Buising, C.M. and D. Tomes (2003). Method of regeneration of *Medicago sativa* and expressing foreign DNA in same. United States Patent 6,566,137. May 20, 2003.

Chabaud, M., J. Passiatore, F. Cannon, and V. Buchanan-Wollaston (1988). Parameters affecting the frequency of kanamycin resistant alfalfa obtained by *Agrobacterium tumefaciens* mediated transformation. *Plant Cell Reports* 7:512-516.

Chen, T.H.H., J. Marowitch, and B.G. Thompson (1987). Genotypic effects on somatic embryogenesis and plant regeneration from callus cultures of alfalfa. *Plant Cell Tissue and Organ Culture* 8:73-81.

Christou, P. (1992). Alfalfa (*Medicago* spp.). In *Genetic engineering and in vitro culture of crop legumes*, P. Christou (ed.). Lancaster, PA: Technomic Publishing, pp. 175-206.

D'Aoust, M.A., P. Lerouge, U. Busse, P. Bilodeau, S. Trépanier, V. Gomord, L. Faye, and L.P. Vzina (2004). Efficient and reliable production of pharmaceuticals in alfalfa. In *Molecular farming: plant-made pharmaceuticals and technical proteins,* R. Fischer and S. Schillberg (eds.). Weinheim: Wiley-VCH Verlag, pp. 1-11.

Deak, M., G.B. Kiss, C. Koncz, and D. Dudits (1986). Transformation of *Medicago* by *Agrobacterium* mediated gene transfer. *Plant Cell Reports* 5:97-100.

Desgagnés, R., S. Laberge, G. Allard, H. Khoudi, Y. Castonguay, J. Lapoint, R. Michaud, and L.-P. Vézina (1995). Genetic transformation of commercial breeding lines of alfalfa (*Medicago sativa*). *Plant Cell Tissue and Organ Culture* 42:129-140.

D'Halluin, K., J. Botterman, and W. De Greef (1990). Engineering of herbicide-resistant alfalfa and evaluation under field conditions. *Crop Science* 30:866-871.

Ding, Y.L., G. Aldao-Humble, E. Ludlow, M. Drayton, Y.H. Lin, J. Nagel, M. Dupal, G. Zhao, C. Pallaghy, R. Kalla et al. (2003). Efficient plant regeneration and *Agrobacterium*-mediated transformation in *Medicago* and *Trifolium* species. *Plant Science* 165:1419-1427.

Du, S., L. Erickson, and S. Bowley (1994). Effect of plant genotype on the transformation of cultivated alfalfa (*Medicago sativa*) by *Agrobacterium tumefaciens*. *Plant Cell Reports* 13:330-334.

Havraham, T., H. Badani, S. Galili, and R. Amir (2004). Enhanced levels of methionine and cysteine in transgenic alfalfa (*Medicago sativa* L.) plants over-expressing the Arabidopsis cystathionine γ-synthase gene. *Plant Biotechnology Journal*, Online publication date: September 21, 2004, doi: 10.1111/j.1467-7652.2004.00102.x.

Hernández-Fernández, M.M. and B.R. Christie (1989). Inheritance of somatic embryogenesis in alfalfa (*Medicago sativa* L.). *Genome* 32:318-321.

Hoekema, A., P.R. Hirsch, P.J.J. Hooykaas, and R.A. Schilperoort (1983). A binary plant vector strategy based on separation of *vir*- and T-region of the *Agrobacterium tumefaciens* Ti-plasmid. *Nature* 303:179-180.

Hood, E.E., G.L. Helmer, R.T. Fraley, and M.D. Chilton (1986). The hypervirulence of *Agrobacterium tumefaciens* A281 is encoded in a region of pTiBo542 outside of T-DNA. *Journal of Bacteriology* 168:1291-1301.

Horsch, R.B., J.E. Fry, N.L. Hofmann, D. Eicholtz, S.G. Rogers, and R.T. Fraley (1985). A simple and general method for transferring genes into plant. *Science* 227:1229-1231.

Kielly, G.A. and S.R. Bowley (1992). Genetic control of somatic embryogenesis in alfalfa. *Genome* 35:474-477.

Matheson, S.L., J. Nowak, and N.L. MacLean (1990). Selection of regenerative genotypes from highly productive cultivars of alfalfa. *Euphytica* 45:105-112.

McKersie, B.D., and D.C.W. Brown (1997). Somatic embryogenesis and artificial seeds. In: *Biotechnology and the improvement of forage legumes*, Biotechnology in Agriculture Series, No. 17, B. D. McKersie and D.C.W. Brown (eds.). Wallingford: CAB International, pp. 111-143.

Meijer, E.G.M. and D.C.W. Brown (1985). Screening of diploid *Medicago sativa* germplasm for somatic embryogenesis. *Plant Cell Reports* 4:285-288.

Mitten, D.H., S.J. Sato, and T.A. Skokut (1984). In vitro regenerative potential of alfalfa germplasm sources. *Crop Science* 24:943-945.

Murashige, T. and F. Skoog (1962). A revised medium for rapid growth and bioassays with tobacco tissue cultures. *Physiologia Plantarum* 15:473-497.

Papadopoulos, Y.A. and B.D. McKersie (1983). A comparison of protein degradation during wilting and ensiling of six forage species. *Canadian Journal of Plant Science* 63:903-912.

Pereira, L.F. and L. Erickson (1995). Stable transformation of alfalfa (*Medicago sativa* L.) by particle bombardment. *Plant Cell Reports* 14:290-293.

Reisch, B. and E. T. Bingham (1980). The genetic control of bud formation from callus cultures of diploid alfalfa. *Plant Science Letters* 20:71-77.

Rosellini, D. and F. Veronesi (2002). Potential of biotechnology for alfalfa. *AgBiotechNet* 4 (April), ABN 085.

Samac, D.A. and S.J. Temple (2004). Development and utilization of transformation in *Medicago* species. In: *Genetic transformation in crops*, D.Z. Skinner and G. Liang (eds.). Binghamton, NY: The Haworth Press.

Samac, D.A., M. Tesfaye, M. Dornbusch, P. Saruul, and S.J. Temple (2004). A comparison of constitutive promoters for expression of transgenes in alfalfa (*Medicago sativa*). *Transgenic Research* 13:349-361.

Saunders, J. W. and E. T. Bingham (1972). Production of alfalfa plants from callus tissue. *Crop Science* 12:804-808.

Schenk, B.U. and A.C. Hildebrandt (1972). Medium and techniques for induction and growth of monocotyledonous and dicotyledonous plant cell culture. *Canadian Journal of Botany* 50:199-204.

Shahin, E.A., A. Spielmann, K. Sukhapinda, R.B. Simpson, and M. Yashar (1986). Transformation of cultivated alfalfa using disarmed *Agrobacterium tumefaciens*. *Crop Science* 26:1235-1239.

Skokut, T.A., J. Manchester, and J. Schaefer (1985). Regeneration in alfalfa tissue culture. *Plant Physiology* 79:579-583.

Somers, D.A., D.A. Samac, and P.M. Olhoft (2003). Recent advances in legume transformation. *Plant Physiology* 131:892-899.

Spanò, L., D. Mariotti, M. Pezzotti, F. Damiani, and S. Arcioni (1987). Hairy root transformation in alfalfa (*Medicago sativa* L.). *Theoretical and Applied Genetics* 73:523-530.

Stachel, S.E. and P.C. Zambryski (1989). Generic trans-kingdom sex? *Nature* 340:190-191.

Stuart, D.A. and K.A. Strickland (1984). Somatic embryogenesis from cell cultures of *Medicago sativa* L. I. The role of the amino acid addition to the regeneration medium. *Plant Science Letters* 34:165-174.

Sullivan, M.L., R.D. Hatfield, S.L. Thoma, and D.A. Samac (2004). Cloning and characterization of red clover polyphenol oxidase cDNAs and expression of active protein in *Escherichia coli* and transgenic alfalfa. *Plant Physiology* 136:1-11.

Tabe, L., T. Wardley-Richardson, A. Ceriotti, A. Aryan, W. McNabb, A. Moore, and T.J. Higgins (1995). A biotechnological approach to improving the nutritive value of alfalfa. *Journal of Animal Science* 73:2752-2759.

Tesfaye, M., S.J. Temple, C.P. Vance, D.L. Allan, and D.A. Samac (2001). Overexpression of malate dehydrogenase in transgenic alfalfa enhances organic acid synthesis and confers tolerance to aluminum. *Plant Physiology* 127:1836-1844.

Tzfira, T., J. Li, B. Lacroix, and V. Citovsky (2004). Agrobacterium T-DNA integration: Molecules and models. *Trends in Genetics* 20:375-383.

Ullah, A.H., K. Sethumadhavan, E.J. Mullaney, T. Ziegelhoffer, S. Austin-Phillips (2002). Cloned and expressed fungal phyA gene in alfalfa produces a stable phytase. *Biochemical and Biophysical Research Communications* 290:1343-1348.

Walker, K.A. and S.J. Sato (1981). Morphogenesis in callus tissue of *Medicago sativa*: The role of ammonium ion in somatic embryogenesis. *Plant Cell Tissue and Organ Culture* 1:109-121.

Weeks, J.T., J. Ye, and C.M. Rommen (2007). Development of an *in planta* method for transformation of alfalfa (*Medicago sativa*). Transgenic Research (EPub-ahead of publication).

Wigdorovitz, A., C. Carrillo, M.J. Dus Santos, K. Trono, A. Peralta, M.C. Gómez, R.D. Ríos, P.M. Franzone, A.M. Sadir, J.M. Escribano, and M.V. Borca (1999). Induction of a protective antibody response to foot and mouth disease virus in mice following oral or parenteral immunization with alfalfa transgenic plants expressing the viral structural protein VP1. *Virology* 255:347-353.

Yu, K. and K.P. Pauls (1993). Identification of a RAPD marker associated with somatic embryogenesis in alfalfa. *Plant Molecular Biology* 22:269-277.

Zambryski, P., H. Joos, C. Genetello, J. Leemans, M. van Montagu, and J. Schell (1983). Ti plasmid vector for the introducing of DNA into plant cells without alteration of their normal regeneration capacity. *EMBO Journal* 2: 2143-2150.

Chapter 4

# *Agrobacterium*-Mediated Transformation of *Medicago truncatula*

M. Chabaud
A. Boisson-Dernier
E.-P. Journet
F. de Carvalho-Niebel
D. G. Barker

## *INTRODUCTION*

The ability to transform *M. truncatula* via *Agrobacterium tumefaciens* was one of the original criteria used to select this legume as a model species (Barker et al., 1990). The *Agrobacterium* T-DNA transfer machinery generates simple, well-characterized insertion profiles consisting of a small number of T-DNA copies (Tinland, 1996). Conventional *A. tumefaciens* transformation generally comprises an initial coculture step involving T-DNA transfer to plant cells, followed by the selection and subsequent regeneration of the transformed cells. Since the limiting step is most often the in vitro regeneration of transformed cells into entire transgenic plants via organogenesis or embryogenesis, the regeneration capacity of a given genotype is of critical importance. In the case of *M. truncatula,* recurrent breeding for highly regenerative genotypes resulted in successful genetic transformation (Thomas et al., 1992). Since then, various improvements and alternative transformation protocols have been published and are presented in this chapter. We first detail the different techniques currently available for genetic transformation of *M. truncatula* using *A. tumefaciens* and subse-

We thank Pascal Ratet for information about the R108 genotype. We are also grateful to Anelia Iantcheva, Pedro Fevereiro, Maria Harrison, and Xin Zhou for information about transformation and regeneration protocols. We thank Springer and the American Phytopathological Society for permission for using published photographs.

quently describe an alternative approach using *A. rhizogenes,* which avoids in vitro plant regeneration. Examples are presented to illustrate the use of these different transformation techniques and we discuss how to choose the most appropriate method for a given experimental objective.

## AGROBACTERIUM TUMEFACIENS-*MEDIATED TRANSFORMATION AND IN VITRO REGENERATION OF TRANSGENIC PLANTS*

Transformation methods using *A. tumefaciens* are essentially dependent on the procedure used for the regeneration of transformed cells. The efficiency of embryogenesis, which exploits the totipotency of plant cells, is highly genotype dependent. In contrast, direct organogenesis based on shoot formation from cotyledonary nodes is a regeneration process that is genotype independent. Before discussing both approaches applied to *M. truncatula,* it should be mentioned that an attempt to avoid labor-intensive regeneration by means of *A. tumefaciens* infiltration of *M. truncatula* floral parts or seedlings has been reported (Trieu et al., 2000), based on the success of this approach for *Arabidopsis thaliana* (Bechtold et al., 1993). Unfortunately, agro-infiltration has not yet been successfully reproduced in other laboratories working with *M. truncatula,*, and for this reason is not presented in this review.

### *Transformation and Regeneration via Somatic Embryogenesis*

Most of the *A. tumefaciens*-mediated transformation protocols available for *M. truncatula* are based on regeneration of transformed cells via embryogenesis. The impact of these parameters on the transformation efficiency is quite variable. However, as already mentioned, the fundamental parameter, that determines transformation success and efficiency is the genotype dependent embryogenic regeneration step.

### **M. truncatula** *Genotypes*

Two main genotypes and their derivatives have been used for transformation and regeneration via embryogenesis: Jemalong and R108. To increase regeneration efficiency for the cultivar Jemalong, a highly embryogenic genotype, named 2HA, was selected through one cycle of in vitro culture and three cycles of recurrent breeding for embryogenesis (Rose et al., 1999; Thomas et al., 1992). Subsequently, Chabaud et al. (1996)

showed that the use of the genotype 2HA improves by at least 10-fold the percentage of explants regenerating into transgenic plants as compared to the original Jemalong A17 genotype. For both A17 and 2HA, the percentage of explants developing transformed calli is over 50 percent, but somatic embryogenesis and subsequent conversion into plantlets are clearly major limiting steps for A17 (Figure 4.1d-e). The genotype M9-10a, a derivative of Jemalong, was obtained from a single regenerated embryo and subsequently maintained in vitro by vegetative propagation (Santos and Fever-

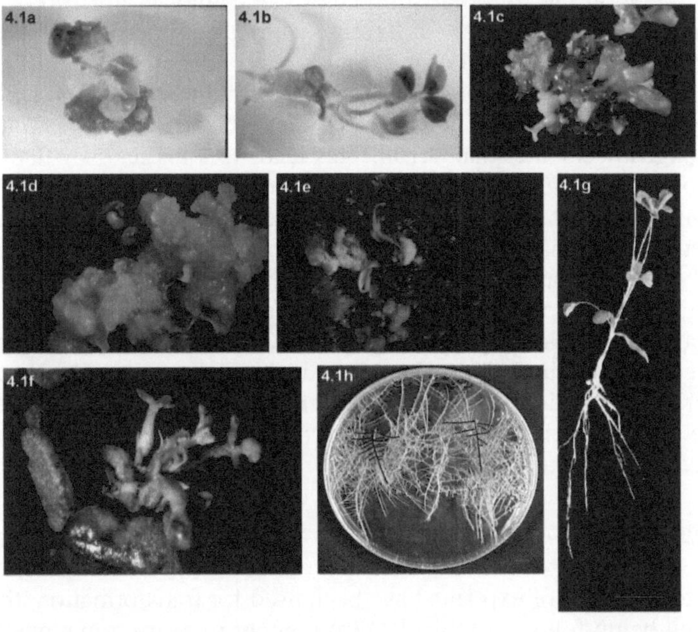


FIGURE 4.1. Transformation and regeneration of *M. truncatula*. a: Green kanamycin-resistant calli formed on a 2HA foliole following transformation. b: Regenerated plantlet. c: Secondary embryogenesis on embryo-development medium. d-e: Comparison of embryogenic calli from different Jemalong lines. Note that embryogenic calli are brown. The A17 line with less than five embryos per callus. e: Callus derived from the highly embryogenic genotype 2HA is covered in globular embryos. f: Shoot formation through organogenesis on a half-cotyledon of Jemalong A17. g: "Composite plant" generated following transformation with *A. rhizogenes*, with Ri T-DNA transformed roots developing from the site of radicle sectioning. Note that transformed roots present a normal phenotype. h: In vitro propagation of excised Ri T-DNA transformed roots. *Source:* Reprinted from Chabaud et al., 1996, with permission. (Bar represents 1 mm in a, b, c, d, e, f, and 1 cm in g, h.) (See also color gallery.)

eiro, 2002). In contrast, a total of five cycles of embryogenesis (Trinh et al., 1998) were used to generate the highly embryogenic genotype R108-1-c3. Although transformation efficiency is not directly comparable in these three improved genotypes (2HA, M9-10a, and R108-1-c3), they all exhibit high frequencies of embryogenesis on transformed calli. The published data suggest that R108-1-c3 has probably the highest embryogenic capacity of the three genotypes.

It should be pointed out that Jemalong and R108 (and their respective derivatives) belong to different subspecies (Le Signor et al., 2005), phenotypically distinct. They have different symbiotic properties and the R108 genotype was originally chosen because of its successful interaction with the *Sinorhizobium meliloti* strain Rm41 (Hoffmann et al., 1997). The two genotypes differ in growth behavior both in vitro and in the greenhouse, and Jemalong flowers later than R108. This leads to a longer growth cycle for Jemalong, but at the same time permits higher seed production. R108 is significantly more sensitive than Jemalong to the pathogenic fungus oïdium, a frequent contaminant in greenhouses. Finally, the genome of R108 is slightly smaller than that of Jemalong (Blondon et al., 1994), and crossing between the genotypes is not straightforward and generally yields chlorotic, but fertile F1 progeny (P. Ratet, CNRS Gif/Yvette, France, personal communication). Since Jemalong is the genotype that has been chosen for mutagenesis and TILLING programs, it is clearly the genotype of choice for projects requiring genetic crossing of transgenic plants with characterized mutants.

### *Type of Explant*

Different types of explants have been used for transformation, the most common being folioles (Table 4.1) from either in vitro-grown plantlets or plants grown in the greenhouse. Using explants from in vitro plantlets has the advantage of avoiding the need to disinfect the tissues prior to transformation, thus reducing the risk of fungal or bacterial contamination of explants. On the other hand, pot-grown-plants can provide large numbers of explants from only a few donor plants. The effect of explant type on embryogenesis (independent of transformation capacity) has been studied. Folioles from Jemalong (genotype A17) were shown to be more embryogenic than petioles and were subsequently used for transformation experiments (Chabaud et al., 1996). Furthermore, Iantcheva et al. (2001) showed that leaf tissues were less polysomatic than petioles, and therefore more likely to yield normal diploid regenerated plants after somatic embryogenesis. In another study, Iantcheva et al. (1999) also compared different regeneration media using explants from 1-month-old plants from the greenhouse and showed that in general hypocotyls and cotyledons were more embryogenic than petioles and nodal explants. Unfortunately, these authors did not evaluate folioles in this study. Kamaté et al. (2000), using young plants of the genotype R108-1-c3 (5 weeks old), showed that embryogenesis of floral parts (especially calyx, petals, carpel, and stamen) was comparable to that of vegetative tissues and better than that of floral branches. Moreover, these authors showed that transformation efficiencies using petal and calyx were higher than using floral branches, correlating with the embryogenic capacity of each tissue. In addition, Crane et al. (2006) obtained high transformation efficiencies using in vitro root explants of R108-1. Finally, Iantcheva et al. (2005) took advantage of secondary embryogenesis frequently observed with *M. truncatula*, and used cotyledon-derived embryo clusters as initial explants for transformation and subsequent direct transgenic plant regeneration.

The age of explants or donor plants is often thought to be correlated with regeneration efficiency (Kamaté et al., 2000; Nolan and Rose, 1998). Using transient GUS assays, we have been able to show that young leaflets are transformed more efficiently than older leaves (M. Chabaud, unpublished results). In conclusion, either folioles from young leaves, root segments, or floral explants (calyx, petals) appear to be the explants of choice to maximize transformation efficiency.

## Agrobacterium tumefaciens *Strains and Cocultivation Conditions*

A variety of *A. tumefaciens* strains have been used successfully for *M. truncatula* transformation (Table 4.1). *A. tumefaciens* strains can be classified according to the opine-metabolism genes of the Ti plasmid of origin: octopine for LBA 4404, C58pGV2260, and C58C1; nopaline for GV3101 (also called C58pMP90); L-succinamopine for the disarmed strains EHA101, EHA105, and AGL1 derived from the hypervirulent strain A281. Hoffman et al. (1997) concluded that C58pGV2260 and A281 have comparable transformation efficiencies using the genotype R108-1. However, these strains were not compared using identical binary vectors and selection protocols. On the other hand, Chabaud et al. (2003) compared four different strains using the same binary vector and showed that the use of the hypervirulent strain AGL1 doubled the transformation efficiency of Jemalong compared to the nonhypervirulent strains LBA 4404, C58pMP90, or C58pGV2260, as measured by the percentage of explants developing kanamycin-resistant calli. In a comparable approach, using the genotype R108-1, Crane et al. (2006) compared four strains with the same binary vector and

TABLE 4.1. Various parameters tested in *A. tumefaciens*-mediated transformation protocols.

| Genotype | Type of explant | *A. tumefaciens* strain | Regeneration media | Selection | Efficiency | Reference |
|---|---|---|---|---|---|---|
| Regeneration via embryogenesis | | | | | | |
| Jemalong 2HA | In vivo folioles | EHA 101 | Callogenesis and embryogenesis: 10 µM NAA; 4µM BAP | Km | No data | Thomas et al., 1992 |
| Jemalong A17 and 2HA | In vitro folioles or petioles | LBA 4404 | Callogenesis: 4.5 µM 2,4-D; 9 µM zeatin embryogenesis: 4.5 µM zeatin | Km | A17: 2% explants developing plants 2HA: 29% explants developing plants | Chabaud et al., 1996 |
| Jemalong 2HA | In vivo folioles | LBA 4404 | Idem Thomas et al., 1992 | Km | 15-19% explants developing embryogenic calli ~1 transgenic plant/embryogenic callus | Wang et al., 1996 |
| Jemalong 2HA | In vitro folioles | LBA 4404<br>C58pMP90<br>C58pGV2260<br>AGL1 | Idem Chabaud et al., 1996 | Km | 25% explants developing transgenic plants within 4-5 months | Chabaud et al., 2003 |
| Jemalong M9-10a | In vitro folioles | EHA 105 | Callogenesis (embryo induction) 0.45 µM 2,4-D; 0.9 µM zeatin | Km | 25-45% explants developing embryos 85% conversion embryos to plants | de Sousa Araujo et al., 2004 |
| R108-1 | Hypocotyls, cotyledons, roots, petioles, folioles | C58pGV2260 A281 | Callogenesis 4.5 µM 2,4-D; 1 µM BAP embryogenesis: no hormones | Km Hyg | No data | Hoffmann et al., 1997 |
| R108-1 (c3) | In vivo folioles | EHA 105 or GV 3101 | Callogenesis and embryogenesis: 22.5 µM 2,4-D; 2.5 µM BAP | Km Hyg | 100% explants developing embryogenic calli 80% conversion embryos to plants | Trinh et al., 1998 |

| | | | | | | |
|---|---|---|---|---|---|---|
| R108-1 (c3) | In vivo folioles and floral parts | EHA 105 | Idem Trinh et al., 1998 | Km | No data for plants | Kamaté et al., 2000 |
| R108-1 (c3) | In vivo folioles | EHA 105 | Idem Trinh et al., 1998 | Km Basta | No data | Scholte et al., 2002 |
| R108-1 | In vitro embryos | LBA 4404 | Secondary embryogenesis: 2.2 μM thidiazuron conversion to plants: 0.2 μM BAP | Km | 65% of embryo clusters developing new resistant embryos<br>Plants obtained within 2 months | Iantcheva et al., 2005 |
| R108-1 (c3) | In vivo folioles | EHA 105 | Callogenesis and embryogenesis: 18 μM 2,4-D; 2.2 μM BAP<br>Shoot development: No hormone | PPT | Up to 50% of explants developing transgenic plants | Cosson et al., 2006 |
| R108-1 | In vitro roots | AGL1<br>C58C1<br>EHA 105<br>LBA 4404 | Callogenesis: 22.5 μM 2,4-D; 2.2 μM BAP,<br>Shoot development: No hormone | PPT | 70% explants giving transgenic calli<br>40% explants giving transgenic shoots within 4 months | Crane et al., 2006 |
| Regeneration via organogenesis | | | | | | |
| Jemalong A17 | Cotyledonary nodes | LBA 4404 | Shoot induction:<br>15 μM BAP, 0.5 μM NAA<br>Shoot development: 0.5 μM NAA, root induction: 1 μM IBA | PPT | 3% explants developing plants | Trieu and Harrison, 1996 |
| Jemalong A17 | Cotyledonary nodes | EHA 105<br>AGL1 | Idem Trieu and Harrison, 1996 | PPT | 15% explants developing shoots<br>~40% rooting | Zhou et al., 2004 |
| Jemalong A17<br>R108-1 | Cotyledonary nodes | AGL1 | Shoot induction and development: 4.4 μM BAP; 0.5 μM NAA, root induction: 1 μM IBA | Km | 5% and 12% of explants developing shoots for Jemalong and R108-1 within 3-4 months | Wright et al., 2006 |

showed that although there was no significant difference in the frequency of resistant calli between the four strains used, the hypervirulent strains AGL1 and EHA 105 led to significantly increased frequencies of explants giving rise to transgenic shoots compared to the strains C58C1 and LBA 4404. In practice, the hypervirulent strains AGL1 for Jemalong (Chabaud et al., 2003) and EHA105 for both R108-1-c3 (Scholte et al., 2002; Crane et al. 2006) and M9-10a (de Sousa Araujo et al., 2004) are now routinely used for *M. truncatula* transformation.

The cocultivation conditions differ between protocols, with cocultivation times of 2 days (Cosson et al., 2006), 3 days (Chabaud et al., 2003; Trinh et al., 1998), and 5 days (de Sousa Araujo et al., 2004). Cocultivation has to be long enough to ensure optimal T-DNA transfer, but if too long can lead to bacterial overgrowth, seriously damaging the explants. Indeed, washing the explants following cocultivation to remove excess bacteria improves transformation efficiency (Chabaud et al., 1996). In some cases a vacuum infiltration step (Trinh et al., 1998) and the addition of acetosyringone (Trinh et al., 1998; de Sousa Araujo et al., 2004) have been included during cocultivation.

### *Media for Regeneration*

In vitro regeneration through somatic embryogenesis involves distinct steps induced sequentially by means of specific culture conditions. The first step is cell dedifferentiation in the presence of auxin and cytokinin leading to the formation of callus (callogenesis, Figure 4.1a). This is followed by embryo induction in the absence of auxin and in either the presence or absence of cytokinin. Embryo conversion to plantlets (Figure 4.1b) generally occurs in the absence of plant growth regulators, whereas rooting, the last step in plant development, can require low levels of auxin and a specific gelling agent (Chabaud et al., 1996; de Sousa Araujo et al., 2004; Kamaté et al., 2000).

The current protocols for regeneration of *M. truncatula* follow this general outline, but it should be underlined that the conversion of embryos into plants is the most inefficient and time-consuming step because secondary embryogenesis frequently occurs instead of normal plant development (Figure 4.1c). As shown in Table 4.1, the concentrations and nature of the various hormones used in different protocols vary considerably, and optimal concentrations depend on the genotype.

Using a direct embryogenesis regeneration protocol in which callogenesis is not required, Iantcheva et al. (1999) have proposed an alternative procedure. In this case, there is only a cytokinin in the first culture medium. This regeneration protocol can be coupled to transformation of embryo clusters to reduce the overall time required for regeneration of transgenic *M. truncatula* (Iantcheva et al., 2005).

### *Selection of Transformed Cells*

So far, resistance to kanamycin (Km; using the *npt*II gene) has generally been used to select transformed cells (Table 4.1). The Km concentrations required for efficient selection depend upon the genotype used: 40 mg·l for R108-1-c3 (Trinh et al., 1998; Cosson et al., 2006) and 50 $mg{\cdot}l^{-1}$ for 2HA (Chabaud et al., 1996). The genotype M9-10a was shown to be more resistant to Km and required higher concentrations (100 $mg{\cdot}l^{-1}$) to completely block embryo development in nontransformed tissues (Duque et al., 2004). Selection pressure can be limited to certain steps of the regeneration procedure, thus allowing a faster development of transformed tissues. For example, Trinh et al. (1998) proposed that Km selection is no longer necessary once the calli have formed. However, it was subsequently shown that this can lead to a large number of so-called escapes, with the formation of nonresistant embryos (d'Erfurth et al., 2003). Finally, Cosson et al. (2006) suggested maintaining Km selection during the first subculture on shoot-inducing medium (SH9) to reduce escapes. Since rooting appears to be particularly sensitive to Km, Chabaud et al. (1996) maintained selection until the conversion of embryos to shoots, but removed the antibiotic during rooting. Under these conditions, 95 percent of regenerated plants are resistant to Km (the remaining 5 percent are escapes or unstable with respect to Km resistance; Chabaud, 1998). Finally, Wang et al. (1996) and de Sousa Araujo et al. (2004) lowered the Km concentration during both embryo development and rooting.

Other agents can also be used for selection of *M. truncatula* transformants. Hygromycin (Hyg-10 $mg{\cdot}l^{-1}$; *hpt* gene) has been shown to be sufficient to select transformed cells of the genotype R108-1 and its derivative R108-1-c3 (Hoffmann et al., 1997; Trinh et al., 1998; Cosson et al., 2006). Selection appears more stringent using Hyg than Km. Finally, Scholte et al. (2002) have reported that selection with 3 $mg{\cdot}l^{-1}$ ammonium glufosinate using the Basta resistance (*bar* gene) with the genotype R108-1-c3 is also very efficient. These alternative selective agents should also be of interest for Jemalong transformation once appropriate concentrations have been determined for this genotype.

### *Transformation Efficiency*

The various means of evaluating transformation efficiencies are based on different steps of the procedure including (1) the percentage of explants developing transformed calli, (2) the percentage of embryogenic transformed calli, and (3) the percentage of embryos converted to plants. In addition, the overall length of the whole procedure is an important parameter in measuring transformation efficiency since regeneration of transformed calli is unfortunately not a synchronous process. Thus, the percentage of explants generating transgenic plants within a defined time turns out to be the most practical measure of transformation efficiency. Finally, it should be added that, when several embryos develop from the same callus, it is frequently the case that they result from the same original transformation event. Since the objective is generally to obtain independent transformants, regenerating only a single plant per callus is therefore a good strategy.

With the genotype R108-1-c3, Trinh et al. (1998) have developed a very efficient transformation protocol with 100 percent of explants giving rise to transformed calli and 80 percent of the embryos converting into plants within 4 to 5 months (P. Ratet, personal communication). In the case of Jemalong 2HA, Chabaud et al. (2003) obtained 90 percent, 75 percent, and 25 percent of explants giving respectively transformed calli, embryos, and plants within 4 to 5 months of culture (increasing culture length increases the number of transformed plants). With the genotype M9-10a, de Sousa Araujo et al. (2004) obtained 25 to 45 percent of explants developing Km-resistant embryos and 85 percent embryo conversion to plants. Even if the published data are not directly comparable for the different genotypes tested, they do suggest that transformation efficiency is directly correlated to the embryogenic capacity of the genotype, and in particular to the number of embryogenesis cycles used for selection (respectively five, three, and one for R108-1-c3, 2HA, and M9-10a).

Detailed technical protocols for Jemalong and R108 are available at the following Web sites respectively: http://medicago.toulouse.inra.fr/Mt/Protocol/Transformation/ and http://www.isv.cnrs-gif.fr/embo01/manuels/ (Module 2, p1-10), as well as the *Medicago truncatula* Handbook Web site (www.noble.org/MedicagoHandbook/).

## AGROBACTERIUM TUMEFACIENS-*MEDIATED TRANSFORMATION AND REGENERATION VIA ORGANOGENESIS*

To avoid time-consuming and genotype-dependent regeneration through embryogenesis, transformation followed by regeneration via organogenesis has also been established for *M. truncatula.* This procedure relies on the direct induction of shoots from cotyledonary nodes using high concentrations of cytokinin (Figure 4.1f). This regeneration process is known to be independent of the regenerative capacity of the genotype and has been successfully developed for other legumes recalcitrant to in vitro regeneration via somatic embryogenesis, such as pea (Schroeder et al., 1993), *Lotus* (Oger et al., 1996), and white clover (Voisey et al., 1994). We present here the different parameters that have been evaluated for this approach.

### *Genotypes, Explants, and Regeneration Conditions*

This transformation technique was first established for Jemalong A17 (Trieu and Harrison, 1996) and subsequently improved by Zhou et al. (2004) for the same genotype and by Wright et al. (2006) for Jemalong A17 and R108-1. Ding et al. (2003) used this regeneration procedure successfully with two *M. truncatula* cultivars, Caliph and Morgul, but unfortunately no data were published concerning transformation with these two genotypes. The explants used in all cases are cotyledonary nodes, which are separated into two parts, so that each has one cotyledon and half of the embryonic axis. The nodes are obtained from young seedlings, grown for either 1 (Ding et al., 2003), 2 (Trieu and Harrison, 1996; Wright et al., 2006) or 4 days (Zhou et al., 2004), from which the radicle and most of the hypocotyl have been removed. In most cases, seeds are germinated in the presence of the cytokinin BAP. To induce adventitious shoots from the embryogenic axis of cotyledonary nodes, Trieu and Harrison (1996) and Zhou et al. (2004) used a high concentration of BAP (15 μM) in conjunction with a low concentration of NAA (0.5 μM). Wright et al. (2006) used a lower concentration of BAP (4.4 μM). These shoots are then rooted in the absence of cytokinin and rooting is promoted by the auxin IAB. Zhou et al. (2004) reported that rooting of the transgenic shoots often fails (~40 percent successful) resulting in an overall loss of transformation efficiency. The authors suggested that increasing the pH from 5.8 to 6.5 in the rooting medium might improve root induction efficiency.

### *Strains and Inoculation Conditions*

Trieu and Harrison (1996) used the strain LBA 4404 and coculture was performed during 5 days in the presence of 15 μM BAP and 100 μM acetosyringone. Zhou et al. (2004) used the hypervirulent strains EHA105 and AGL1 and added 400 $mg \cdot l^{-1}$ L-cysteine and 1 mM DTT to limit tissue browning.

### *Selection of Transformed Cells*

To select transgenic adventitious shoots and avoid escapes and chimeras, selection has to be very efficient in this system. Chimeras can theoretically be obtained through organogenesis since multiple cells are likely to be precursors for the development of new shoots. Trieu and Harrison (1996) and Zhou et al. (2004) both used phosphinothricin (PPT) as the selective agent (respectively 5 and 1.6 $mg \cdot l^{-1}$) in conjunction with the *bar* gene. However, while Trieu and Harrison (1996) maintained selection pressure throughout both the regeneration procedure and rooting, Zhou et al. (2004) delayed applying selection until after 15 days of culture, and then removed it completely during rooting. In contrast to the observations made by Trieu and Harrison (1996) suggesting that Km was not useful for selection since nonresistant shoots can survive even in the presence of 250 $mg \cdot l^{-1}$ Km, Wright et al. (2006) used 50 $mg \cdot l^{-1}$ Km for transgenic shoot selection in their procedure.

### *Transformation Efficiency*

Trieu and Harrison (1996) obtained transgenic plants from 3 percent of explants, with the first transgenic plants developing within 2.5 months. Similarly, Zhou et al. (2004) obtained an average of 6 percent explants yielding transgenic shoots, of which 40 percent developed roots within at least 3 months of culture. Wright et al. (2006) obtained comparable results with 5 and 12 percent of transgenic shoots for Jemalong and R108-1, respectively, within 3 to 4 months of culture. Several shoots often developed from the same explant, and Zhou et al. (2004) showed that they could potentially derive from the same transformation event. When compared to the protocols based on regeneration via embryogenesis, it is clear for the moment that the use of highly embryogenic genotypes is considerably more efficient. Moreover, in the case of Jemalong, there is no obvious advantage in using A17 rather than the highly embryogenic derivative 2HA, since, to our knowledge, it possesses all the phenotypic characteristics of A17.

### *Conclusions and Use of* A. Tumefaciens-*Mediated Transformation*

The development of efficient *M. truncatula* transformation techniques has now opened the way to routine *in planta* analysis of chimeric constructs. For a specific transgene, transgenic plants can be obtained within a few (3-5) months in any laboratory equipped with basic in vitro culture facilities (laminar flow hood and growth chamber). As *A. tumefaciens*-mediated transformation leads to entire hemizygous transgenic plants with Mendelian inheritance of transgenes in most of the cases, stable homozygous transgenic lines can be readily obtained in the $S_2$ generation (second generation following self-pollination). In this way, transgenes of interest (e.g., promoter/gene reporter fusions or overexpression/gene-silencing constructs) can be studied in detail and if necessary crossed into other genetic backgrounds. As discussed below, transgene expression studies of root-specific genes, which do not require S2 material, can also be studied using the simpler and more rapid *A. rhizogenes* approach.

Besides transgene expression experiments, T-DNA can also be used to inactivate plant genes, creating insertional mutants and providing a convenient tag to identify the mutated gene. It has been demonstrated that T-DNA tagging is a potential tool for insertional mutagenesis in *M. truncatula* (Scholte et al., 2002). However, this approach is very labor intensive since each insertional mutant is obtained by individual transformation or regeneration, and each mutant contains only a few T-DNA insertions. An interesting alternative to T-DNA insertional mutagenesis is the use of transposons, and in particular retrotransposons, which can lead to high copy numbers of insertions per plant. In this context, d'Erfurth et al. (2003) have shown that the retrotransposon *Tnt1* from tobacco is a powerful tool for insertional mutagenesis in *M. truncatula* with transposition occurring during in vitro tissue culture. A collaborative European program (within the framework of the FP6 Grain Legumes Integrated Project) is underway aimed at creating a *Tnt1*-based insertional mutant collection for Jemalong.

## AGROBACTERIUM RHIZOGENES-*MEDIATED TRANSFORMATION*

Even though transformation or regeneration via somatic embryogenesis or organogenesis has proved its usefulness for *M. truncatula,*, this procedure remains time consuming. For this reason, an alternative approach for obtaining transgenic roots has been developed using *A. rhizogenes,* which avoids lengthy in vitro plant regeneration (Boisson-Dernier et al., 2001). This procedure generates what is termed a composite plant, comprising nontransformed shoots and leaves attached to a transgenic root system (Figure 4.1g). We have shown that this material is well adapted for studies of root endosymbiosis since the Ri T-DNA-transformed roots of *M. truncatula* (generated using the *A. rhizogenes* strain ARqua-1) have comparable morphology to wild-type roots, and can be both nodulated by *Sinorhizobium meliloti* and colonized by endomycorrhizal fungi (Boisson-Dernier et al., 2001; Harrison et al., 2002). *A. rhizogenes*-mediated transformation takes advantage of the possibility of cotransforming plant cells with more than one T-DNA at the same time. The T-DNA containing the transgene of interest in a disarmed binary vector is cotransformed with the resident Ri T-DNA containing the root locus (*rol*) genes responsible for the development of root tissue. Here we discuss the different parameters that have been evaluated for optimizing the technique of *A. rhizogenes*-mediated transformation of *M. truncatula* (see also Table 4.2).

## *KEY PARAMETERS FOR* AGROBACTERIUM RHIZOGENES-*MEDIATED TRANSFORMATION OF* M. TRUNCATULA

### *Plant Genotypes and* A. rhizogenes *Strains*

In this approach, transformation is, of course, totally independent of the regenerative capacity of the genotype but does depend on the compatibility between the plant host and the *A. rhizogenes* strain required for T-DNA transfer. Over 15 years ago, Thomas et al. (1992) obtained a number of Ri T-DNA-transformed roots after inoculation of Jemalong with *A. rhizogenes*. More recently, Boisson-Dernier et al. (2001) have established an efficient and simple method for cotransformation of two T-DNAs using *A. rhizogenes* strain ARqua-1 and the genotype Jemalong J5 (J5 is very similar to the A17 line; Thoquet et al., 2002). It was also shown that an identical approach could be used to transform both DZA315-16, a natural Algerian variant (Tirichine et al., 2000; Boisson-Dernier et al., 2001) and R108-1-c3 (Limpens et al., 2004; Rodriguez-Llorente et al., 2003), confirming that the procedure is not genotype dependent.

In addition to the ARqua I strain (a streptomycin-derivative of A4T; Quandt et al., 1993) used by Boisson-Dernier et al. (2001), a second pRiA4-containing strain (MSU 440) and the NCPPB 2659 strain (also called K599) have been used respectively by Limpens et al. (2004) and Collier et al. (2005). It should be emphasized that the ARqua I strain was previously shown to elicit a relatively small number of Ri T-DNA transformed roots after inoculation of *Vicia hirsuta,* with growth and morphology comparable to normal roots (Quandt et al., 1993), as it is observed for *M. truncatula*

TABLE 4.2. Various parameters tested for *A. rhizogenes*-mediated transformation.

| Genotype | Material | Growth tempe-rature | A. rhizo-genes strain | Root growth media | Selection | Efficiency | Reference |
|---|---|---|---|---|---|---|---|
| Jemalong J5/A17<br>R108-1<br>DZA 315-16 | 1-day-old seedlings with sectioned radicle | 20°C and 25°C | ARqual | Fahraeus medium + 0.5 mM $NH_4NO_3$ | 25 mg·l$^{-1}$ Km | 33% (at 25°C) or 63% (at 20°C) cotransformed composite plants 95% cotransformed roots under selection | Boisson-Dernier et al., 2001 |
| Jemalong A17 | Idem Boisson-Dernier et al., 2001 | | | Medium poor in phosphate; pH 6.4 | 5 mg·l$^{-1}$ PPT | No data | Harrison et al., 2002<br>Liu et al., 2003 |
| R108-1-c3 | Stab-inoculated etiolated hypocotyls | 25°C | ARqual | Half strength MS with 1% sucrose | No selection-GFP expression | 25 to 30% plants with cotransformed roots | Rodriguez-Llorente et al., 2003 |
| Jemalong A17<br>R108-1-c3 | 5-day-old seedlings with sectioned roots | 21°C | MSU 440 | Fahraeus medium | No selection DsRed expression | 30% cotransformed roots among newly formed roots | Limpens et al., 2004 |
| R108-1 | 2-day-old seedlings, sectioned radicle | No data | ARqual | Modified Fahraeus medium | 5 mg·l$^{-1}$ PPT | 65% cotransformed roots among newly formed roots | Crane et al., 2006 |
| | Apical stem sections | | NCPPB 2659 | Ex vitro culture, deionized water | No selection | 56% plants with cotransformed roots 40% cotransformed roots among newly formed roots | Collier et al., 2005 |

(Boisson-Dernier et al., 2001). Hence, we strongly recommend the use of the low-virulence ARqua 1 strain to ensure that the elicited roots have minimally perturbed phenotypes.

### *Inoculation and Root Growth Conditions*

Boisson-Dernier et al. (2001) sectioned the radicle tip of 1 cm germinated seedlings with a scalpel and inoculated the wound site with *A. rhizogenes.* Cotransformed roots appeared from the site of sectioning after approximately 1 week of coculture onward. Using the same inoculation method, Limpens et al. (2004) used 5-day-old plantlets and transferred the inoculated plants to an emergence medium, richer than the basic Fahraeus medium used previously. Under these conditions, the first cotransformed roots appeared 3 weeks after inoculation. Rodriguez-Llorente et al. (2003) used stab inoculation of elongated hypocotyls from 6 day-old *M. truncatula* plantlets and transformed roots were obtained 2 to 3 weeks later. Taken together, these results suggest that the use of older plantlets delays cotransformed root development. Finally, temperature appears to be a major parameter affecting transformation efficiency, since Boisson-Dernier et al. (2001) showed that higher numbers of transformed roots could be obtained if the coculture is performed at 20°C rather than 25°C.

Plant growth media differ between protocols depending on the subsequent use of the Ri T-DNA transformed roots. In the case of studies involving either *Rhizobium* inoculation or addition of Nod factors, it is sufficient to grow the roots on nitrogen-limited medium (Boisson-Dernier et al., 2001). Liu et al. (2003) used a medium containing low concentrations of phosphate for successful endomycorrhizal fungal infection. For in vivo protein transport or localization studies, Rodriguez-Llorente et al. (2003) used a half-strength MS medium with 1 percent sucrose, a classical plant growth medium. Thus, the nature of the growth medium for coculture and T-DNA transfer does not appear to be of critical importance.

When sucrose is not part of the culture medium, subsequent decontamination of *A. rhizogenes* is not required during plant growth to avoid excessive bacterial development. However, for in vitro propagation of transformed roots on sucrose-containing medium, decontamination with antibiotics such as cefotaxime or augmentin is essential (Boisson-Dernier et al., 2001; Rodriguez-Llorente et al., 2003). Finally, in addition to the diverse methods presented above to produce composite plants in vitro, Collier et al. (2005) have developed a procedure that they have called ex vitro, making use of apical stem sections inoculated directly in basalt-based wool cubes (Table 4.2).

### *Identification and Selection of Transgenic Roots*

It is relatively easy to distinguish Ri T-DNA-transformed roots from nontransformed roots using the protocol developed by Boisson-Dernier et al. (2001) since nontransformed roots generally appear within a few days after sectioning, whereas Ri T-DNA- transformed roots appear only after 1 to 2 weeks of coculture. In addition, the transformed roots are formed at the site of sectioning, whereas nontransformed laterals initiate above the section. Without selection, approximately half of the Ri T-DNA-transformed roots are cotransformed with the T-DNA of interest.

In certain cases, it is advantageous to counterselect nontransformed roots, and Boisson-Dernier et al. (2001) have shown that the addition of Km ($25\ mg \cdot l^{-1}$) to the plant growth medium can be used to directly select cotransformed roots expressing the *npt*II gene. Interestingly, since nontransformed root development is totally blocked in the presence of Km, there is an additional stimulation of cotransformed root development. Thus, while the total percentage of plants developing cotransformed roots is unchanged in the presence of Km, the total number of cotransformed roots per plant is doubled compared to that observed in the absence of selection (Boisson-Dernier et al., 2001). Alternatively, Harrison et al. (2002) selected cotransformant roots using $5\ mg \cdot l^{-1}$ PPT (with the *bar* gene on the T-DNA). In this case, however, contact between the sensitive aerial parts of the composite plant and the PPT must be avoided (Liu et al., 2003). It seems that in certain conditions, the PPT is degraded by the *A. rhizognes* containing the *bar* gene, leading to the suppression of the selection (Chabaud, unpublished results).

Using fluorescent markers such as DsRed, Limpens et al. (2004) have proposed another approach for detecting cotransformed roots without the use of antibiotics or herbicides. Detection of cotransformed roots can thus be performed directly using the fluorescent marker during root development. This system also allows the rapid identification of nonhomogeneously transformed roots (chimeras), a relatively frequent occurrence according to the authors. Such transgenic chimeras appear to be counterselected when direct Km selection is employed, presumably because the development of nontransformed cells is inhibited.

### *Transformation Efficiency*

When inoculation and coculture are performed at 25°C, transformation efficiencies are very similar for the different protocols, with 25 to 33 percent of plants developing transgenic roots (Boisson-Dernier et al., 2001;

Rodriguez-Llorente et al., 2003), whatever the genotype used. However, as stated above, decreasing coculture temperature to 20°C doubles this transformation efficiency to over 50 percent (Boisson-Dernier et al., 2001). Another important parameter is the percentage of cotransformed roots. Using ARqua I, there is an average of four Ri T-DNA-transformed roots per composite plant, of which 50 to 60 percent are cotransformed with the gene of interest in the absence of selection. However, as already stated, the use of a selective agent such as Km raises this percentage to 95 percent (Boisson-Dernier et al., 2001). In the absence of selection, Rodriguez-Llorente et al. (2003) and Collier et al. (2005, using the ex vitro procedure) obtained respectively 80 and 40 percent of cotransformed hairy roots detected by green fluorescent protein (GFP)-fusions, whereas Limpens et al. (2004) identified only 30 percent of newly developed roots as expressing DsRed. The reasons why the percentage of cotransformed roots should be so variable between different protocols is not clear, but may be related to the sensitivity of the detection methods used.

### *Regeneration of Entire Fertile Plants from Hairy Root Fragments*

Since the use of *A. rhizogenes*-mediated transformation leads to generation of composite plants with nontransformed aerial parts, one important limitation of this technique is the impossibility of obtaining transgenic seeds. Crane et al. (2006) have now shown that hairy root explants (genotype R108-1) can be regenerated into entire fertile plants, and that the gene of interested can be segregated from the *rol* genes in the progeny plants.

A detailed technical protocol for *A. rhizogenes* transformation of the genotype Jemalong A17 is available on the following Web site: http://medicago.toulouse.inra.fr/Mt/Protocol/Transformation/ as well as on the *Medicago truncatula* Handbook Web site (www.noble.org/MedicagoHandbook/).

## *CONCLUSION AND APPLICATIONS OF* A. RHIZOGENES-*MEDIATED TRANSFORMATION*

The great advantage of *A. rhizogenes*-mediated transformation leading to composite plants is the rapidity of the technique coupled with its simplicity. A variety of endosymbiosis-specific gene promoters fused to marker genes such as *gusA* (e.g., Boisson-Dernier et al., 2001; Manthey et al., 2004) or *gfp* (Liu et al., 2003), as well as protein fusions to GFP (Rodriguez-Llorente et al., 2003) have been analyzed using this convenient approach. However, even if this has not so far presented difficulties, it should be borne in mind that *rol* gene expression in Ri T-DNA-transformed roots could interfere with transgene expression. In addition to transgene studies, *A. rhizogenes*- transformed roots have also been successfully used for the complementation of plant symbiotic mutants with candidate genes (e.g., Endre et al., 2002). Naturally, this is only possible for mutations where the genetic determinant is root associated. Finally, it has been shown that *A. rhizogenes*-transformed roots can also be used successfully for RNAi-based approaches to study gene function (e.g., Limpens et al., 2004).

An interesting property of Ri T-DNA transformed roots is their amenability to in vitro propagation after excision from the parent composite plant. Root clones can thus be established and maintained in culture for several years (Figure 4.1h). In contrast to *Rhizobium*-dependent nodulation, mycorhization leading to arbuscule formation occurs successfully in excised roots. For example, Chabaud et al. (2002) have shown that such transformed root cultures can be used for transgene analysis during the endomycorrhizal association. Furthermore, observations of the earliest stages of the fungal-plant association are greatly facilitated using such in vitro propagated material (Genre et al., 2005). *M. truncatula* root cultures can also be infected by root-knot nematodes (Boisson-Dernier et al., 2005), thus facilitating the comparative study of genes expressed in both endomycorrhizal and nematode interactions (e.g., Koltai et al., 2001). Finally, transformed root cultures can also provide excellent abundant material for recombinant protein production. As stated earlier, if required, transformed hairy root explants can be regenerated to fertile plants and the transgene of interest subsequently segregated from the *rol* genes (Crane et al., 2006).

## *FINAL COMMENTS*

The optimization of *Agrobacterium*-mediated transformation techniques for *M. truncatula* over recent years now offers a collection of efficient and well-tested protocols, thus contributing to the wider exploitation of this model legume. The use of highly embryogenic genotypes in conjunction with *A. tumefaciens* transformation coupled to regeneration via somatic embryogenesis permits the regeneration of stable, fertile transgenic plants within several months. Alternatively, *A. rhizogenes*-mediated transformation can be used to generate transgenic roots within just a few weeks. With these new tools, in vivo transgene analyses have now become routine, and large-scale functional genomics programs employing RNAi approaches as well as insertional mutagenesis programs using the retrotransposon *Tnt1* have been undertaken. These programs will stimulate current and future research on legumes and thereby contribute to a better understanding of legume biology including symbiotic and pathogenic interactions.

## REFERENCES

Barker, D.G., S. Bianchi, F. Blondon, Y. Dattée, G. Duc, S. Essad, P. Flament, P. Gallusci, G. Génier, P. Guy, et al. (1990). *Medicago truncatula*, a model plant for studying the molecular genetics of the *Rhizobium*-legume symbiosis. *Plant Molecular Biology Reporter* 8:40-49.

Bechtold, N., J. Ellis, and G. Pelletier (1993). *In planta Agrobacterium* mediated gene transfer by infiltration of adult *Arabidopsis thaliana* plants. *Comptes Rendus de l'Academie Sci. Serie III, Sciences de la Vie,* 316.

Blondon, F., D. Marie, S. Brown, and A. Kondorosi (1994). Genome size and base composition in *Medicago sativa* and *M. truncatula* species. *Genome* 37:264-270.

Boisson-Dernier, A., A. Andriank, M. Chabaud, A. Niebel, E. P. Journet, D. Barker, and F. de Carvalho-Neibel (2005). MtENOD11 gene activation during rhizobial infection and mycorrhizal arbuscule development requires a common AT-rich-containing regulatory sequence. *Molecular Plant-Microbe Interactions* 18:1269-1276.

Boisson-Dernier, A., M. Chabaud, F. Garcia, G. Bécard, C. Rosenberg, and D.G. Barker (2001). *Agrobacterium rhizogenes*-transformed roots of *Medicago truncatula* in the study of nitrogen-fixing and endomycorrhizal symbiotic associations. *Molecular Plant-Microbe Interactions* 14: 695-700.

Chabaud, M. (1998). La transformation de *Medicago truncatula* par *Agrobacterium tumefaciens* et la régénération in vitro de plantes transgéniques par embryogenèse somatique: un outil pour l'étude de la symbiose *Rhizobium-Medicago.* PhD thesis, University Paul Sabatier, Toulouse (France). *Spécialité biologie cellulaire et moléculaire végétale,* p. 468.

Chabaud, M., F. de Carvalho Niebel, and D.G. Barker (2003). Efficient transformation of *Medicago truncatula* cv. Jemalong using the hypervirulent *Agrobacterium tumefaciens* strain AGL1. *Plant Cell Reports* 22:46-51.

Chabaud, M., C. Larsonneau, C. Marmouget, and T. Huguet (1996). Transformation of barrel medic (*Medicago truncatula* Gaertn) by *Agrobacterium tumefaciens* and regeneration via somatic embryogenesis of transgenic plants with the MtENOD12 nodulin promoter fused to the *gus* reporter gene. *Plant Cell Reports* 15:305-310.

Chabaud, M., C. Venard, A. Defaux-Petras, G. Becard, and D.G. Barker (2002). Targeted inoculation of *Medicago truncatula* in vitro root cultures reveals *MtENOD11* expression during early stages of infection by arbuscular mycorrhizal fungi. *New Phytologist* 156:265-273.

Collier, R., B. Fuchs, N. Walter, W.K. Litke, and C.G. Taylor (2005). Ex vitro composite plants: An inexpensive, rapid method for root biology. *Plant Journal* 43:449-457.

Cosson, V., P. Durand, I. d'Erfurth, A. Kondorosi, and P. Ratet (2006). *Medicago truncatula* transformation using leaf explants. *Methods in Molecular Biology* 343:115-127.

Crane, C., E. Wright, R.A. Dixon, and Z.-Y. Wang (2006). Transgenic *Medicago truncatula* plants obtained from *Agrobacterium tumefaciens*-transformed roots and *Agrobacterium rhizogenes*-transformed hairy roots. *Planta* 222:1344-1354.

d'Erfurth, I., V. Cosson, A. Eschstruth, H. Lucas, A. Kondorosi, and P. Ratet (2003). Efficient transposition of the *Tnt1* tobacco retrotransposon in the model legume *Medicago truncatula. Plant Journal* 34:95-106.

de Sousa Araujo, S., S.R.L. Duque, D.M.M. dos Santos, and M.P.S. Fevereiro (2004). An efficient transformation method to regenerate a high number of transgenic plants using a new embryogenic line of *Medicago truncatula* cv. Jemalong. *Plant Cell Tissue and Organ Culture* 78:123-131.

Ding, Y.L., G. Aldao Humble, E. Ludlow, M. Drayton, Y.-H. Lin, J. Nagel, M. Dupal, G. Zhao, C. Pallaghy, R. Kalla, et al. (2003). Efficient plant regeneration and *Agrobacterium*-mediated transformation in *Medicago* and *Trifolium* species. *Plant Science* 165:1419-1427.

Duque, A.S.R.L.A., S.d.S. Araujo, D.M.M.F.d. Santos, and M.P.S. Fevereiro (2004). Optimization of a selection scheme using kanamycin to improve transformation of *Medicago truncatula* cv. Jemalong. *Plant Cell Tissue and Organ Culture* 78:277-280.

Endre, G., A. Kereszt, Z. Kevei, S. Mihacea, P. Kalo, and G.B. Kiss (2002). A receptor kinase gene regulating symbiotic nodule development. *Nature* 417:962-966.

Genre, A., M. Chabaud, T. Timmers, P. Bonfante, and D.G. Barker (2005). Arbuscular mycorrhizal fungi elicit a novel intracellular apparatus in *Medicago truncatula* root epidermal cells before infection. *Plant Cell* 17:3489-3499.

Harrison, M.J., G.R. Dewbre, and J. Liu (2002). A phosphate transporter from *Medicago truncatula* involved in the acquisition of phosphate released by arbuscular mycorrhizal fungi. *Plant Cell* 14:2413-2429.

Hoffmann, B., T.H. Trinh, J. Leung, A. Kondorosi, and E. Kondorosi (1997). A new *Medicago truncatula* line with superior in vitro regeneration, transformation, and symbiotic properties isolated through cell culture selection. *Molecular Plant-Microbe Interactions* 10:307-315.

Iantcheva, A., M. Vlahova, and A. Atanassov (2005). Genetic transformation of *Medicago truncatula* using system for direct somatic embryogenesis promoted by TDZ. *Biotechnologie and Biotechnologie Eq.* 19:50-56.

Iantcheva, A., M. Vlahova, E. Bakalova, E. Kondorosi, M.C. Elliott, and A. Atanassov (1999). Regeneration of diploid annual medics via direct somatic embryogenesis promoted by thidiazuron and benzylaminopurine. *Plant Cell Reports* 18:904-910.

Iantcheva, A., M. Vlahova, T.H. Trinh, S.C. Brown, A. Slater, M.C. Elliott, and A. Atanassov (2001). Assessment of polysomaty, embryo formation and regeneration in liquid media for various species of diploid annual *Medicago. Plant Science* 160:621-627.

Kamaté, K., I.D. Rodriguez Llorente, M. Scholte, P. Durand, P. Ratet, E. Kondorosi, A. Kondorosi, and H. Trinh (2000). Transformation of floral organs with GFP in *Medicago truncatula. Plant cell Reports* 19:647-653.

Koltai, H., M. Dhandaydham, C. Opperman, J. Thomas, and D. Bird (2001). Overlapping plant signal transduction pathways induced by a parasitic nematode and a rhizobial endosymbiont. *Molecular Plant-Microbe Interactions* 14:1168-1177.

Le Signor, C., K. Gallardo, J.M. Prosperi, C. Salon, L. Quillien, R. Thompson, and G. Duc (2005). Genetic diversity for seed protein composition in *Medicago truncatula. Plant Genetic Resources* 3:59-71.

Limpens, E., J. Ramos, C. Franken, V. Raz, B. Compaan, H. Franssen, T. Bisseling, and R. Geurts (2004). RNA interference in *Agrobacterium rhizogenes*-transformed roots of *Arabidopsis* and *Medicago truncatula. Journal of Experimental Botany* 55:983-992.

Liu, J.Y., L.A. Blaylock, G. Endre, J. Cho, C.D. Town, K.A. VandenBosch, and M.J. Harrison (2003). Transcript profiling coupled with spatial expression analyses reveals genes involved in distinct developmental stages of an arbuscular mycorrhizal symbiosis. *Plant Cell* 15:2106-2123.

Manthey, K., F. Krajinski, N. Hohnjec, C. Firnhaber, A. Puhler, A.M. Perlick, and H. Kuster (2004). Transcriptome profiling in root nodules and arbuscular mycorrhiza identifies a collection of novel genes induced during *Medicago truncatula* root endosymbioses. *Molecular Plant-Microbe Interactions* 17:1063-1077.

Nolan, K.E. and R.J. Rose (1998). Plant regeneration from cultured *Medicago truncatula* with particular reference to abscisic acid and light treatments. *Australian Journal of Botany* 46:151-160.

Oger, P., A. Petit, and Y. Dessaux (1996). A simple technique for direct transformation and regeneration of the diploid legume species *Lotus japonicus. Plant Science* 116:159-168.

Quandt, H.J., A. Puhler, and I. Broer (1993). Transgenic root nodules of *Vicia hirsuta*: A fast and efficient system for the study of gene expression in indeterminate-type nodules. *Molecular Plant-Microbe Interactions* 6:699-706.

Rodriguez-Llorente, I.D., J. Perez-Hormaeche, M. Dary, M.A. Caviedes, A. Kondorosi, P. Ratet, and A.J. Palomares (2003). Expression of MsPG3-GFP fusions in *Medicago truncatula* hairy roots reveals preferential tip localization of the protein in root hairs. *European Journal of Biochemistry* 270:261-269.

Rose, R.J., K.E. Nolan, and L. Bicego (1999). The development of the highly regenerable seed line Jemalong 2HA for transformation of *Medicago truncatula*-implications for regenerability via somatic embryogenesis. *Journal of Plant Physiology* 155:788-791.

Santos, D. and P. Fevereiro (2002). Loss of DNA methylation affects somatic embryogenesis in *Medicago truncatula. Plant Cell Tissue and Organ Culture* 70:155-161.

Scholte, M., I. d'Erfurth, R. Rippa, S. Mondy, V. Cosson, P. Durand, B. Breda, H. Trinh, I. Rodriguez-Llorente, E. Kondorosi, et al. (2002). T-DNA tagging in the model legume *Medicago truncatula* allows efficient gene discovery. *Molecular Breeding* 10:203-215.

Schroeder, H.E., A.H. Schotz, T. Wardley-Richardson, D. Spencer, and T.J.V. Higgins (1993). Transformation and regeneration of two cultivars of pea (*Pisum sativum* L.). *Plant Physiology* 101:751-757.

Thomas, M.R., R.J. Rose, and K.E. Nolan (1992). Genetic transformation of *Medicago truncatula* using *Agrobacterium* with genetically modified Ri and disarmed Ti plasmids. *Plant Cell Reports* 11:113-117.

Thoquet, P., M. Ghérardi, E.P. Journet, A. Kereszt, J.M. Ané, J.M. Prosperi, and T. Huguet (2002). The molecular genetic linkage map of the model legume *Medicago truncatula*: an essential tool for comparative legume genomics and the isolation of agronomically important genes. *BMC Plant Biology* 2:1.

Tinland, B. (1996). The integration of T-DNA into plant genomes. *Trends in Plant Science* 1:178-184.

Tirichine, L., F. de Billy, and T. Huguet (2000). Mtsym6, a gene conditioning *Sinorhizobium* strain-specific nitrogen fixation in *Medicago truncatula. Plant Physiology* 123:845-852.

Trieu, A.T., S.H. Burleigh, I.V. Kardailsky, I.E. Maldonado-Mendoza, W.K. Versaw, L.A. Blaylock, H. Shin, T.-J. Chiou, H. Katagi, G.R. Dewbre, et al. (2000). Transformation of *Medicago truncatula* via infiltration of seedlings or flowering plants with *Agrobacterium. Plant Journal* 22:531-541.

Trieu, A.T. and M.J. Harrison (1996). Rapid transformation of *Medicago truncatula*: Regeneration via organogenesis. *Plant Cell Reports* 16:6-11.

Trinh, T.H., P. Ratet, E. Kondorosi, P. Durand, F. Kamate, P. Bauer, and A. Kondorosi (1998). Rapid and efficient transformation of diploid *Medicago truncatula* and *Medicago sativa* ssp. *falcata* lines improved in somatic embryogenesis. *Plant Cell Reports* 17:345-355.

Voisey, C.R., D.W.R. White, R.D. Appleby, P.M. Ealing, and A.G. Scott (1994). *Agrobacterium*-mediated transformation of white clover using direct shoot organogenesis. *Plant Cell Reports* 13:309-314.

Wang, J.H., R.J. Rose, and B.I. Donaldson (1996). *Agrobacterium*-mediated transformation and expression of foreign genes in *Medicago truncatula. Australian Journal of Plant Physiology* 23:265-270.

Wright, E., R.A. Dixon, and Z.-Y. Wang (2006). *Medicago truncatula* transformation using cotyledon explants. *Methods in Molecular Biology* 343:129-135.

Zhou, X., M.B. Chandrasekharan, and T.C. Hall (2004). High rooting frequency and functional analysis of GUS and GFP expression in transgenic *Medicago truncatula* A17. *New Phytologist* 162:813-822.

# Chapter 5

# Biotechnological Approaches to Soybean Improvement

Qungang Qi
Henry E. Valentin

## *INTRODUCTION*

Soybean, *Glycine max* (L.), is grown as a commercial crop in over 80 countries and has long been recognized as a valuable component of medicine, food, and animal feed. It is an important crop as it contains both protein and oil resources. Due to its nutritional value, wide availability, low cost, and consistency in composition, soybean meal has displaced most other protein sources and become a vital component of poultry and swine feed. Together, poultry and swine feed account for over two thirds of soybean meal utilization in the United States. Therefore, there have been recent intensive efforts to improve soybean seed quality traits and agronomic characteristics via genetic engineering. This review summarizes the data relating to the commercial growth of biotech soybean, including field-testing applications and patent publications. It describes a variety of biotech soybean traits and methods that have been employed to generate these traits. A major focus is devoted to research and development of seed quality traits (i.e., essential amino acids, protein quality, oil, and tocopherols). Details of the many enabling technologies such as methods for transformation of biotech soybean plants and utilization of regulatory genetic elements (i.e., promoters) are not discussed here.

---

We are grateful to Drs. William D. Rapp, Andre Silvanovich, Lisa Weaver, and Jintai Huang for critical reading of the manuscript.

## *SUMMARY OF FIELD TRIAL APPLICATIONS OF TRANSGENIC SOYBEAN FOR THE PERIOD 1987-2005*

The most complete set of field trials for transgenic crops is compiled by the U.S. Animal and Plant Health Inspection Service (APHIS) and is available at http://www.aphis.usda.gov/brs/database.html. The confidential business information for some applications related to genetic modification of specific genes has been excluded, but much of the relevant information may be obtained through a patent database site such as http://www.patents.uspto.gov (or http://scientific.thomson.com/derwent/).

Analysis of the statistical data collected from 1987 to date reveals several trends. First, the number of U.S. biotech soybean field trials has dramatically increased since the first four trials were performed in 1989. The total number of U.S.-approved biotech soybean trials from 1987 to 2005 reached 1,070 and accounted for approximately 15 percent of the total U.S. field trials of all transgenic plant species during this period (http://www.aphis .usda.gov). Several trials have led to commercial products. Second, there has been a great expansion in the number of new trait genes introduced into transgenic soybean. In the first several years, there was a great emphasis on single gene traits, which were intended to provide herbicide tolerance. For example, data for U.S. biotech soybean trials show that all the trials during the period 1989-1991were related to herbicide tolerance. This figure declined to 40 percent by 2005. During the same period, the number of trials for biotech soybean with a single new trait or multiple traits has dramatically increased. For instance, the trials for soybean seed composition traits such as oil, protein, and essential amino acids increased from 0 percent in 1991 to 43 percent in 2005. Other traits introduced included drought tolerance, cold tolerance, and yield enhancement, as well as insect and fungal resistance.

Lastly, the number of institutes conducting U.S.-approved biotech soybean field trials has increased from 2 during 1987-1990 to more than 40 during 1991-2005, in large part due to the efforts of a small number of companies involved in this area of biotechnology.

## *IMPROVEMENT OF SOYBEAN SEED QUALITY TRAITS*

### *Genetic Engineering for High Tryptophan and Methionine Soybean*

Human and other monogastric animals cannot synthesize the essential amino acids (EAA) isoleucine (Ile), leucine (Leu), lysine (Lys), methionine (Met), cysteine (Cys), tryptophan (Trp), phenylalanine (Phe), threonine

(Thr), and valine (Val). Plants are the primary source of proteins and essential amino acids consumed by humans and livestock. For example, the seeds of legumes contain 20 to 50 percent protein by weight and are an excellent source of dietary protein. However, the amino acid composition of the legume seeds is not balanced for human and livestock nutrition. In maize- and soybean-based animal feed rations, Lys and the sulfur amino acids, Met or Cys are limiting. In soybean-derived feed, Met and Cys are the primary limiting essential amino acids. To enrich maize- and soybean-based animal diets, the rations are routinely supplemented with synthetic or microbe-synthesized Lys and Met resulting in increased costs of these feeds. Other essential amino acids such as Trp and Thr then became limiting in these supplemented feeds. Trp and Thr are very costly to produce, thereby preventing further supplementation of these improved animal feeds. An alternative approach to post-harvest feed supplementation is to increase the content of EAAs in soybean seed through genetic enhancement. Strategies to increase the content of EAA in seeds include genetic selection of high-EAA mutants, plant breeding, and biotechnological approaches. This review focuses on transgenic approaches for increasing seed Met and Trp content.

## *Increasing Seed Methionine Level*

### *Biosynthesis of Methionine and Regulation*

Plants synthesize Met through the trans-sulfuration pathway from Cys (Hesse et al., 2004; Saito, 2000), which diverges from the threonine branch of the aspartate family pathway. *O*-phosphohomoserine (OPHS) is the common precursor for Met and Thr synthesis. OPHS is either directly converted to Thr by threonine synthase (TS) or through three sequential steps (cystathionine γ-synthase [CS], cystathionine-β-lyase [CL], methionine synthase [MS]) to Met. The major metabolic fates of Met include its incorporation into protein, adenosylation to form S-adenosylmethionine (SAM), and methylation to form S-methyl Met (SMM).

Characterization of transgenic and mutant plants with altered CS or TS activities led to the conclusion that Met synthesis in plants is controlled at the branch point between CS and TS (Hesse et al., 2004; Amir et al., 2002). For example, an *Arabidopsis* mutant (*mto2*) deficient in TS activity (Bartlem et al., 2000) and TS-antisense potato plants (Zeh et al., 2001) contained 20- and 240-fold more free Met, respectively, compared to wild-type (WT) plants, while Thr levels were decreased by 6 percent and 45 percent, respectively. The enzymatic activity of plant TS is allosterically activated by SAM, a direct product of Met (Curien et al., 1998), suggesting that carbon flux could be directed into the threonine branch when sufficient Met and hence SAM are available. These results suggest that Met synthesis is regulated by an autogenous mechanism through modeling metabolite flux via the TS and CS branch point, but species-specific regulatory mechanisms may exist as well (Hesse et al., 2004). The regulation of Met synthesis has been reviewed (Hesse et al., 2004; Saito, 2000).

*Approaches to Increase Met Levels in Genetically Modified Plants*

Three major strategies commonly used to enhance the content of Met in genetically modified (GM) plants include (1) increasing free Met content via modulation of the key pathway gene expression, (2) sequence modifications of storage proteins by increasing the number of Met residues, and (3) overexpression of heterologous or endogenous genes encoding Met-rich proteins. In some cases, enhanced free Met can be lost through leaching from plant tissues during boiling and other processing procedures. Therefore, the strategies to improve the Met content in plants frequently combine the free Met approach with expression of Met-rich proteins.

Protein sequence modification and synthetic gene strategies have the flexibility of engineering and designing a gene with increased Met codons. A critical step of this approach is to be able to select a region of the protein that can be modified without changing the overall structure, stability, and other cellular properties of the protein. An example of this is that a part of the variable region in *Arabidopsis* 2S albumin gene (*AT2S1*) was deleted and replaced with a modified sequence containing 11 additional Met codons (Sun, 1999). This chimeric gene was overexpressed and the Met-enriched protein was stably accumulated in the seed of *Arabidopsis, Brassica napus,* and tobacco. This type of modification with the intent to alter seed amino acid composition, while promising, is still in its early stages.

The heterologous or endogenous gene approaches have the advantage of utilizing naturally occurring genes and reducing the risk of protein function loss. Many genes encoding Met-rich proteins have been identified from various plant species. Most of them are seed storage proteins with low molecular weight (i.e., less than 15 kDa; Sun, 1999). For example, seed-specific expression of a sulfur-rich Brazil nut 2S albumin (18 mol percent Met and 8 mol percent Cys) doubled Met content of the transgenic soybean seed (Sun, 1999). However, the Brazil nut protein has been identified as a food allergen (Nordlee et al., 1996). Thus it is not desirable to use this protein to enhance the nutritional value of crop seeds. An edible, sulfur-rich 2S albumin identified from sesame seed could serve as an alternative to the Brazil nut protein. It has been used to elevate Met and Cys content of transgenic rice grains by up to 76 percent and 75 percent, respectively, compared with those of WT rice grains (Lee et al., 2003). Similarly, expression of maize Met-rich 10 kD zein under the regulation of soybean 7S$\alpha$ promoter resulted in a 30-100 percent increase in Met content in canola and soybean seeds (Wandelt, 1992). In summary, these results provide proof of the concept that the Met deficiency problem of crop plants including soybean can be resolved by genetic engineering.

## *ENHANCING SEED TRYPTOPHAN LEVELS*

### *Trp Biosynthesis and Regulation*

The plant Trp biosynthetic pathway is located in plastids, derives its precursors from chorismate, and comprises the same sequence of reactions seen in microorganisms (Siehl, 1999). Anthranilate synthase (AS) catalyzes the first committed reaction. As in most bacteria, all plant AS holoenzymes characterized thus far are tetramers consisting of two $\alpha$- and two $\beta$-subunits, and in most cases function as an $\alpha_2\beta_2$ complex. In some bacteria (*Rhizobium meliloti* and *Agrobacterium tumefaciens;* Weaver et al., 2003) monomeric ASs have been found that comprise a single polypeptide chain that includes the activities of both $\alpha$- and $\beta$-subunits. Both heterotetrameric and monomeric ASs catalyze the formation of anthranilate from glutamine and chorismate. More detailed information about Trp biosynthesis has been reviewed by Siehl (1999).

It has been well documented that AS activity plays a regulatory role in Trp biosynthesis (Siehl, 1999). AS is feedback inhibited by Trp (Li and Last, 1996). For example, rice calli and plants expressing a Trp-insensitive rice AS$\alpha$-subunit gene *OASA1* had up to 180- and 35-fold increased free Trp levels, respectively (Tozawa et al., 2001).

### *Generation of Transgenic High-Trp Soybean*

One way to increase the Trp content in crop seed is to elevate the accumulation of this amino acid in the pool of free amino acids. As described above, AS is the key enzyme that regulates Trp biosynthesis. Thus, most approaches to elevating Trp were directed at transgenic expression of a feedback-insensitive AS. Such AS mutants include natural or genetically engineered chimeric, monomeric, and heterotetrameric enzymes. For example, seed-specific expression of a maize Trp-insensitive AS$\alpha$ gene in soybean resulted in Trp concentrations of 2,000 to 3,000 ppm, which is 10 to 15 times higher than in nontransgenic seeds (Liang et al., 2003). Both greenhouse and field tests through five generations demonstrated that these high-Trp seeds were not substantially different in protein and oil content than WT seeds, implying that there are no penalties for seed oil and protein traits in the AS transgenic seeds. Similarly, transgenic soybean expressing the monomeric *Agrobacterium tumefaciens* AS driven by a seed-specific promoter ($P_{7S\alpha}'$) produced up to 12,000 ppm Trp in seeds, with average Trp levels ranging up to 8,000 ppm, compared to 200 ppm Trp in WT seed (Weaver et al., 2003). These transgenic soybean lines with various levels of Trp are continually field tested for agronomic performance, their profiles of seed metabolites, and the genetic and environmental stability of the trait (Liang et al., 2003). Thus, like other plant systems such as corn and *Arabidopsis,* seed-specific expression of Trp-insensitive AS appears to be an effective strategy for enhanced Trp production in soybean seed.

An alternative approach to improve seed Trp levels is to increase the fixed Trp content of seed proteins through expression of natural or modified Trp-rich proteins in seed. An example of this strategy is that modified soybean β-conglycinin β-subunits having an increased number of Trp or Ile codons were found to fold properly in *E. coli* and accumulated when expressed in transgenic *Arabidopsis* seeds (Rapp et al., 2003).

### *Modification of Seed Protein Composition*

Soybean seed contains 35 to 50 percent protein on a dry weight basis. The majority of soybean seed protein is storage protein. Glycinin (also known as the 11S globulins) and β-conglycinin (also known as the 7S globulins) are two major storage proteins in soybean seed. Together, they comprise 70 to 80 percent of the total seed protein. Due to the health benefits and low cost of soy protein, along with improved isolation techniques resulting in better flavor and increased functionality, soy protein isolates and concentrates are used extensively in a variety of food products in different parts of the world. However, soybean allergies are found in 6 to 8 percent of children and 1 to 2 percent of adults (Burks et al., 1988). Widespread use of soy protein may make its avoidance particularly difficult, especially for babies. In addition, the use of these new products requires novel functional characteristics of the protein products. Therefore, much effort has been devoted to improve the quality of the storage protein with respect to the reduction of immunodominant allergens and to produce novel soy protein products with unique and valuable functional characteristics.

Soybean possesses as many as 15 proteins recognized by IgEs from sensitive individuals (Burks et al., 1988). The immunodominant soybean allergens are the subunit of β-conglycinin and P34 or Gly m Bd 30k, which is a member of the papain superfamily lacking the catalytic cysteine residue that is replaced by a glycine (Ogawa et al., 2000). In a number of IgE binding studies, it has been shown that about two thirds of soybean-sensitive patients have IgE cross-reactivity only with the P34/Gly m Bd 30k subunit (Helm et al., 2000). Gly m Bd 30k is a relatively minor soybean storage protein (less than 1 percent of total seed protein) and therefore regarded as an important target for removal from soybean seed in order to produce hypoallergenic soybeans.

Several approaches could be employed for removing or reducing soybean immunodominant allergens. An attempt to identify mutant line(s) lacking Gly m Bd 30k was made but has not yet succeeded (Ogawa et al., 2000). Protein engineering could also be used to alter amino acid sequences by disrupting allergenic sequences. However, substituting a hypoallergenic variant of a plant with the modified protein still has a high technological threshold and has yet to be achieved (Herman, 2003). The alternative transgenic approach offers the prospect of using gene suppression techniques to reduce or eliminate allergens such as Gly m Bd 30k. For example, Gly m Bd 30k was completely removed from soybean seed using transgene-induced gene silencing of the Gly m Bd 30k (Herman, 2003). The resultant transgenic seeds did not appear to show any other unintended agronomic effects.

Research efforts have been aimed at understanding the functional characteristics of soybean proteins (Hayashi et al., 1998) such as water sorption parameters, viscosity, gelation, and emulsification properties. As a result of direct relation of functional and physiochemical properties of proteins, the structural differences of β-conglycinin and glycinin result in proteins with very different functional characteristics. For example, glycinin forms gels that have greater tensile strain, stress, and shear strength. However, soy protein products produced today are a blend of both glycinin and β-conglycinin and therefore, their functional characteristics are dependent on the blend of glycinin and β-conglycinin's individual characteristics. Breeding efforts to combine genetic mutational techniques have identified and produced soybean lines lacking one or more of the various storage protein subunits (Hayashi et al., 1998). However, a breeding approach for developing agronomically viable soybean varieties having only glycinin or β-conglycinin will be time consuming and costly. In comparison, biotechnological methods have advantages in these aspects, especially with efficient transformation of elite soybean varieties. For example, antisense technology has been used to reduce specific seed storage proteins such as napin (a 2S albumin) and cruciferin (an 11S globulin) in *Brassica napus* (Kohno-Murase et al., 1994). More recently, it has been demonstrated that cosuppression technology can be used to suppress the expression of genes encoding the 7S-globulin class of the seed proteins, resulting in soybean lines with an altered seed storage protein profile (Kinney and Fader, 2004).

## *ENGINEERING SEED OIL COMPOSITION AND CONTENT*

Oil crops are one of the world's most valuable agricultural commodities. In fact, after cereals, oil crops are the second most important source of edible calories for humans. Almost three fourths of global vegetable oil production comes from soybean, oil palm, rapeseed, and sunflower. Plant lipids represent a vast array of chemical structures, which determine the nutritional and industrial properties. The predominant use of plant lipids is as edible oils in the form of triacylglycerols. The specific performance and health attributes of edible oils are determined largely by their fatty acid composition. Typically, oil derived from commercial soybean varieties is composed primarily of palmitic (16:0; 11 percent), stearic (18:0; 4 percent), oleic (18:1; 22 percent), linoleic (18:2; 53 percent), and linolenic (18:3; 7.5 percent) acids. A vegetable oil low in total saturates and high in monounsaturated (e.g., 18:1) would provide significant health benefits to consumers as well as economic benefits to oil processors. The high level of polyunsaturated fatty acids in natural soybean oil renders the oil unstable, and as a result susceptible to the development of disagreeable odors and flavors. Thus, soybean oil with increased oleic acid would be ideal for use as cooking oil. On the other hand, high levels of polyunsaturates can be desirable for specialized uses. For example, linolenic acid has widespread use in domestic and industrial coatings since the double bonds of the fatty acids react rapidly with oxygen to polymerize into a soft and flexible film. Therefore, biotechnology provides many options to tailor oil crops for the production of specialty oils.

In addition to the number of double bonds in fatty acids of oils, functional groups derived from fatty acid double bonds such as hydroxyl and epoxy groups, as well as non-methylene-interrupted double bonds, confer important functional properties to oils used in nonedible applications. Fatty acid chain length also determines specific applications of plant oil in food or nonfood industrial sectors. Such examples are lauric (12:0) oils, which are mainly used in soaps and detergents, and erucic acid (22:1), used primarily in polymers, cosmetics, and pharmaceuticals. Therefore, efforts to genetically modify oil crops such as soybean have been aimed at the optimization of seed oil composition.

The formation of fatty acid double bonds and related functional groups results directly or indirectly from the activity of fatty acid desaturases

(Singh et al., 2005; Kinney et al., 2002). Thus, considerable efforts have been directed toward the manipulation of fatty acid desaturase gene expression. An example of this is the down-regulation of Δ-12 desaturase *FAD2-1* alone, or in combination with down-regulated palmitoyl-thioesterase (*FatB*), resulting in soybean containing 85 percent oleic acid, compared with about 20 percent in WT seed (Buhr et al., 2002). Concomitant with the oleic acid increase was a dramatic reduction of linolenic acid and saturated fatty acids. Other examples of efforts to engineer soybean oil were aimed at enhancing its nutritional quality (Singh et al., 2005). For example, seed-specific expression of a Δ6-desaturase resulted in overproduction of the high-value fatty acids γ-linolenic acid (GLA) and stearidonic acid (SDA; Sato et al., 2004), and more recently the high human nutritional long-chain n-3 polyunsaturated fatty acid eicosapentaenoic acid (EPA) was produced by overexpression of ω3-fatty acid desaturases and elongases in the transgenic somatic soybean embryos (Pereira et al., 2004), demonstrating its potential for use in overproduction of EPA in biotech oilseeds. Similarly, the expression of several desaturases and elongases in somatic embryos resulted in the formation of docosahexaenoic acid (DHA) (Abbadi et al., 2004). Coexpression of the multiple desaturases, elongases, and appropriate acyltransferases might be needed to obtain high-level production of these polyunsaturated fatty acids in transgenic plants.

In addition to fatty acids, oil crops produce many other important lipophilic compounds including vitamins, sterols, flavonoids, and glucosinolates. These compounds have uses either in nutrition or for medicinal purposes. Therefore, they have been targeted for improving quality of soybean seed (Keller et al., 2004; Nes, 1998).

In addition to the above-mentioned efforts to improve oil composition, much effort has also been directed at the identification and modification of genes regulating oil yield in oilseed (Renz et al., 2003; Martini et al., 1999).

## *TRANSGENIC APPROACHES FOR ENHANCEMENT OF VITAMIN E*

Vitamin E is a collective term that refers to the biological activity of a group of eight natural amphipathic compounds, consisting of α-, β-, γ- and δ-tocopherols and tocotrienols (Figure 5.1). Tocotrienols are distinguished from α-, β-, γ-, and δ-tocopherols by the presence of three additional double bonds in the isoprenoid side chain. It is commonly believed that the beneficial effects of tocopherols are derived from their radical scavenging activity in lipophilic environments, resulting in the stabilization of unsaturated fatty acids in membrane lipids.

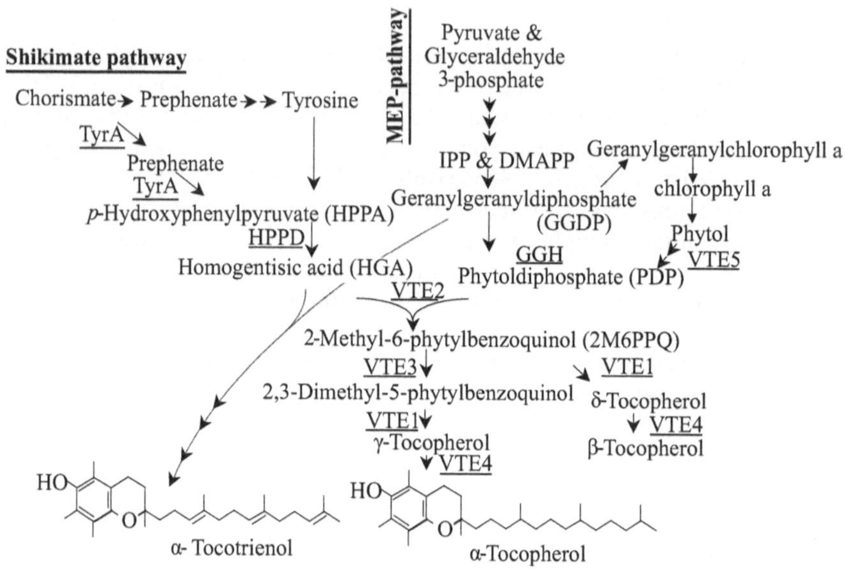


FIGURE 5.1. Schematic drawing of the tocopherol biosynthetic pathway. Abbreviations: DMAPP, dimethylallyldiphosphate; GGH, geranylgeranyl diphosphate reductase; HPPD, *p*-hydroxyphenylpyruvate dioxygenase; IPP, isopentenyldiphosphate; MEP, methylerythritol phosphate; TyrA, bifunctional chorismate mutase and prephenate dehydrogenase; VTE1, tocopherol cyclase; VTE2, homogentisate phytyltransferase; VTE3, 2-methyl-6-phytylbenzoquinol methyltransferase; VTE4, γ-tocopherol methyltransferase; VTE5, phytol kinase.

The vitamin E activity of 1 mg chemically synthesized all racemic α-tocopherol is defined as 1 international unit (IU). Natural isomeric pure RRR-α-tocopherol is 1.5-fold more active than synthetic α-tocopherol (Chow, 2001). The National Institute of Health currently suggests a recommended daily allowance of 15 to 19 mg α-tocopherol, or 22 to 28 IU for humans (http://www.cc.nih.gov/ccc/supplements/vite.html#rda). In addition to human nutrition, large amounts of vitamin E are being used to supplement animal feed to improve the quality and shelf life of meat (Sanders et al., 1997) and for pharmaceutical applications (Chow, 2001). Because of these health benefits associated with vitamin E consumption and the superiority of natural vitamin E with respect to bioactivity, there is considerable interest in engineering plants for vitamin E production. The highest naturally occurring tocopherol concentrations are found in the seed of oilseed crops. In addition, current processing technology is designed to enrich tocopherols during the oil-refining process where they are collected with deodorized distillate. Most commercially sold natural tocopherols originate from soybean. For these reasons, soybean seed represents a prime target for biotechnology-based efforts to increase tocopherol levels.

### *Tocopherol Biosynthesis and Regulation*

Animals and humans depend on plants as a tocopherol source to meet their nutritional needs, as only plants and some photosynthetic bacteria are known to synthesize these compounds. The first committed reaction in tocopherol biosynthesis is the prenylation of homogentisic acid (HGA) with phytyldiphosphate (PDP) or geranylgeranyldiphosphate (GGDP; Figure 5.1). This reaction is catalyzed by homogentisate phytyltransferase, encoded by the *Arabidopsis VITAMIN E 2* (*VTE2*) locus (Savidge et al., 2002), or by the homogentisate geranylgeranyl transferase (HGGT) (Cahoon et al., 2003). The reaction product, 2-methyl-6-phytylbenzoquinol (2M6PBQ), can undergo cyclization catalyzed by tocopherol cyclase (VTE1) to form δ-tocopherol, or it can be methylated by the 2-methyl-6-phytylbenzoquinol methyl transferase (VTE3) to form 2,3-dimethyl-5-phytylbenzoquinol (2,3M5PBQ) followed by a subsequent cyclization to form γ-tocopherol. Delta- and γ-tocopherol can be further methylated by γ-methyltransferase (VTE4), resulting in the formation of β- or α-tocopherol, respectively. Prenylation of HGA with GGDP leads to the formation of tocotrienols with subsequent reaction sequences equivalent to tocopherol formation. More detailed information about tocopherol biosynthesis has been reviewed by Valentin and Qi (2005) and Sattler et al. (2004).

Employing genetic and biochemical approaches, numerous genes encoding tocopherol biosynthetic enzymes from different organisms have been identified (Cahoon et al., 2003; Van Eenennaam et al., 2003; Porfirova et al., 2002; Savidge et al., 2002). The discovery of these genes has facilitated tocopherol pathway engineering. For example, *VTE4* and *VTE3* have been employed to modify tocopherol composition in *Arabidopsis,* canola, and soybean (Van Eenennaam et al., 2003; Shintani and Della Penna, 1998). Overexpression of *Arabidopsis VTE4 (At-VTE4)* in *Arabidopsis* resulted in more than 95 percent conversion of γ-tocopherol to α-tocopherol (Shintani and Della Penna, 1998). This approach resulted in an 80-fold increase in vitamin E activity in *Arabidopsis* seed. However, transgenic expression of these methyltransferases alone did not result in quantitative improvements in total tocopherol content.

Using transgenic approaches, two independent groups have recently demonstrated that VTE2 activity limits total tocopherol synthesis in *Arabidopsis* (Collakova and Della Penna, 2003; Savidge et al., 2002). Interestingly, Cahoon et al. (2003) successfully demonstrated that overexpression of barley *HGGT* resulted in approximately 15- and 5-fold increases in total tocopherols and tocotrienols, with over 75 percent being tocotrienols in *Arabidopsis* leaves and corn kernels, compared to the nontransgenic plants, respectively.

In addition, a limitation in the availability of the HGA precursor has been demonstrated through transgenic expression of *HPPD* and through combined expression of *HPPD* and a prephenate dehydrogenase (TyrA; Qi et al., 2005; Karunanandaa et al., 2005). Although these studies have provided some useful information regarding limiting steps in tocopherol biosynthesis, the regulation and rate-limiting reactions in tocopherol biosynthesis are still not completely understood.

### *Enhanced Vitamin E Content in Oilseed Crops*

The vitamin E activity of α-tocopherol is 2-, 10-, and 33-fold that of β-, γ-, and δ-tocopherols, respectively (Chow, 2001). Therefore, vitamin E activity in oilseed can be increased through enhancement of α-tocopherol content, or through elevation of total tocopherol content. Soybean seed contains a diverse tocopherol composition, consisting of 10 to 20 percent α-tocopherol, 2 to 5 percent β-tocopherol, 60 to 70 percent γ-tocopherol, and 20 to 30 percent δ-tocopherol. Expression of *At-VTE4* converted γ-tocopherol almost completely to α-tocopherol, and δ-tocopherol completely to β-tocopherol (Table 5.1). Overexpression of *At-VTE3* in soybean seed led to the conversion of 94 percent of all tocopherols to γ- and α-tocopherol (Table 5.1). Coexpression of *At-VTE3* and *At-VTE4* in soybean seed resulted in the conversion of over 90 percent of all tocopherols to α-tocopherol, which correlates with, on average, about an 8.0-fold increase in vitamin E activity. These results demonstrate the successful biotechnological optimization of soybean for vitamin E production.

As discussed above, VTE2 activity and availability of HGA and PDP have been identified as limiting factors for tocopherol biosynthesis in plants. Coexpression of *At-HPPD* and *Eh-tyrA* increased mean tocopherol level 3.3-fold relative to WT levels in soybean seed (Table 5.1). Seed-specific coexpression of *At-HPPD, Eh-tyrA,*, and *Synechocystis VTE2 (Syn-VTE2)* increased average total seed tocopherol level 4.2-fold relative to WT levels for soybean. The increase in total seed tocopherols and tocotrienols

TABLE 5.1. Tocopherol profile in transgenic R1 soybean seeds overexpressing tocopherol pathway genes.

| | % of total tocopherol | | | | Total tocopherols | Total tocotrienols | Total tocopherols and tocotrienols |
|---|---|---|---|---|---|---|---|
| | Delta | Beta | Gamma | Alpha | (ng/mg seed) | (ng/mg seed) | (ng/mg seed) |
| WT | 23 | 2.5 | 64.1 | 10.4 | 301 | nd | 301 |
| *VTE3* | 5.2 | 1.2 | 76.7 | 17.3 | 317 | nd | 317 |
| *VTE4* | nd | 26.5 | nd | 73.5 | 308 | nd | 308 |
| *VTE3* and *VTE4* | 1.6 | 2.8 | 5.7 | 89.8 | 296 | nd | 296 |
| WT | 15.7 | 2.6 | 63.3 | 11.9 | 325 | nd | 325 |
| *VTE2* | 18.8 | 2.9 | 61.8 | 15.5 | 430 | 26 | 456 |
| $HPPD_{At}$ and $tryA_{Eh}$ | 27.1 | 3.5 | 48.5 | 18.9 | 340 | 750 | 1,090 |
| *VTE2*, $tryA_{Eh}$ and $HPPD_{At}$ | 31.9 | 3.7 | 47.7 | 17.5 | 347 | 1,025 | 1,372 |

*Source:* Data for this table were summarized from Van Eenennaam et al. (2003) and Karunanandaa et al. (2005). *Note:* nd, below the detection limit (5 ng/mg dry seed).

in soybean was accompanied by the accumulation of substantial tocotrienol levels, accounting for up to 71percent of the total tocopherol pool in soybean seed. Enhanced HGA synthesis in transgenic soybean seeds is evidenced by the fact that soybean seed harboring an *At-HPPD-Eh-tyrA* double gene construct or an *At-HPPD-Eh-tyrA-Syn-VTE2* triple gene construct exhibited an 800-fold increase in HGA levels. In addition, the increased HGA phenotype and tocotrienol accumulation in soybean seeds were accompanied by a dark seed color and a smaller, irregularly shaped seed phenotype.

The substantial accumulation of tocotrienols in the presence of increased HGA suggested a limitation of PDP availability in these transgenic seeds (Figure 5.1). Therefore, it was hypothesized that overexpression of *Arabidopsis* geranylgeranyl diphosphate reductase (*At-GGH*) in seeds harboring the *HPPD-tyrA-VTE2* triple gene expression constructs would increase PDP availability and lead to further increases in tocopherol biosynthesis. The combined seed-specific expression of *At-HPPD, Eh-tyrA, At-VTE2,* and *At-GGH* in soybean resulted in transgenic seed with an average tocopherol increase of more than 10-fold (up to 4,806 ng/mg) relative to WT seeds when single seeds were analyzed. However, the increase in total tocopherols was obtained exclusively through tocotrienol accumulation, indicating that expression of *GGH* in this high-tocopherol soybean seed was not sufficient to redirect the metabolic flux from tocotrienol synthesis to tocopherol formation or that other factors are required in addition to *GGH* expression to effectively access PDP for tocopherol biosynthesis. In addition, the seed phenotypes described above (e.g., dark seed color and poor germination) were maintained in seed harboring the four expression cassettes for *At-HPPD, Eh-tyrA, At-VTE2,* and *At-GGH.*

Characterization of an *Arabidopsis* low-tocopherol mutant has led to the identification of a phytol kinase (Valentin et al., 2005). This *Arabidopsis* mutant accumulated only 20 percent of WT seed tocopherol levels suggesting that the majority of phytol for tocopherol biosynthesis may be provided through a free phytol intermediate (Figure 5.1). This discovery provides an interesting new perspective for the sequences of biochemical reactions leading to tocopherol, chlorophyll, and phylloquinone biosyntheses, though additional studies will be required to define the nature of this enzyme and its impact on phytol metabolism.

In order to further enhance seed vitamin E content through increases in α-tocopherol and total tocopherols, crosses between high (>95 percent) α-tocopherol soybean lines and high total tocopherol soybean lines were performed. The high α-tocopherol soybean harbored transgenic expression cassettes for *VTE3* and *VTE4,* and the high tocopherol soybean lines harbored expression cassettes for *HPPD, tryA,* and *VTE2*. Average seed tocopherol levels in the progeny increased by 1.2-fold from this cross relative to seed exhibiting the high-tocopherol phenotype only. In addition, the average α-tocopherol and α-tocotrienol content in this seed increased to 83.8 percent of total tocopherols resulting in up to 11-fold increased vitamin E activity compared to WT soybean seed. Interestingly, F1 seed germinated at comparable rates to control soybean seed. However, the phenotypes in seed color and shape observed previously in seed harboring *At-HPPD, Eh-tyrA,* and *Syn-VTE2* expression construct were maintained.

In summary, enhanced vitamin E activity in oilseed crops was achieved by combining two biotechnological approaches—optimized seed tocopherol composition in favor of α-tocopherol formation through seed-specific expression of *VTE2* and *VTE4,*, and enhanced total seed tocopherol content through seed-specific expression of *HPPD, TYRA,* and *VTE2*—and a combination of these two approaches obtained through crossing experiments. Further enhancement of seed vitamin E content may be achieved through enhancing the supply of endogenous PDP.

## *PRACTICAL CONSIDERATIONS*

### *Complexity in Introducing Multiple Transgenes into Plants*

Flux control with the goal of end-product overaccumulation through complex biochemical pathways such as tocopherol biosynthesis in plants requires the coordinated regulation or overexpression of multiple transgenes. Expression of multiple transgenes also may be needed to combine multiple crop traits in one plant. A number of approaches have been undertaken to combine or pyramid transgenes in one plant, and these efforts have met with varying degrees of success. These approaches include sexual crossing, retransformation, cotransformation, stacking multiple genes on one vector, the use of linked transgenes, and single chimeric transgenes incorporating multiple protein-coding sequences or gene-suppression cassettes (Halpin et al., 2001).

Each method has its own challenges. For example, the current methods for T-DNA handling quickly become limiting when stacking four or more transgenes into a single binary vector, resulting in reduced transformation efficiency and T-DNA instability in *Agrobacterium*. Second, while a multitude of different gene expression elements have been described in the literature, few have been characterized sufficiently to allow assembly of a set of elements with comparable timing and expression profiles, and even fewer have been characterized to a level that would allow their use for commercial crop transformation. Third, the problem of transgene silencing often occurs, especially if homologous genetic elements have been used in attempts at transgene stacking (Halpin et al., 2001; Finnegan and McElroy, 1994). Therefore, in order to fully exploit the potential for manipulation of plant metabolism, there is a need to improve the efficiency of conventional methods or develop novel methods for the routine introduction or manipulation of multiple genes in plants. Alternatively, transcription factor expression can be manipulated to alter the activity of the set of down stream genes for modifying flux through a metabolic pathway (Broun, 2004). However, the complex mechanism of transcriptional regulation and lack of versatility of this strategy can be an obstacle for obtaining the desired amount and/or balance of metabolites (Broun, 2004; Halpin et al., 2001).

The success of developing a biotech crop with the desired traits largely depends on the appropriate expression of the transgene in plants. Therefore, it is important to select the proper regulatory genetic elements (e.g., promoters) for obtaining the desired transgene expression level in the targeted tissues. For example, expression of *At-VTE2* under the regulation of a seed-specific napin promoter resulted in significant seed tocopherol increases of

1.7-fold in R2 *Arabidopsis* seed, compared to WT seed (Karunanandaa et al., 2005). In contrast, expression of the same gene under the control of a constitutive cauliflower mosaic virus 35S promoter did not significantly increase seed tocopherol levels. Similarly, seed-specific promoters are often used for the modification of other seed quality traits such as oil and protein composition. It should also be considered that other factors such as position effects, gene silencing or suppression, and posttranscriptional or translational effects will affect transgene expression in plants. It is beyond the scope of this chapter to discuss transgene expression issues.

### *Effects on Crop Growth and Yield Performance*

Although many successful examples of improved seed quality traits have been achieved using biotechnological methods, it should be pointed out that some genetically engineered oilseed crops may exhibit pleiotropic effects on agronomic performance. An example of this is the poor germination of transgenic high-HGA soybean seed, as described above. However, $F_1$ seed from crossing experiments between these transgenic lines and high $\alpha$-tocopherol soybean lines displayed normal germination. Furthermore, the integration and expression must have no deleterious pleiotropic effect on plant agronomic and yield performance. Therefore, to be commercially useful, every transgenic event must then be rigorously evaluated for agronomic field performance.

## *CONCLUSION AND FUTURE PROSPECTS*

The research developments described in this chapter document significant progress toward producing a more diversified portfolio of biotech soybean products. The range of soybean material undergoing field testing is now extensive, and there are some promising transgenic lines, which provide a good perspective on the potential next generation of products. Analyses of available data from U.S. field trial applications and patent databases reveal several commercial trends. First, there is a change of emphasis from the single-gene agronomic traits towards stacking of multiple traits. For example, herbicide-tolerant soybean can be stacked with insect resistance or a quality or yield trait for the creation of higher-value products. Second, there have been extensive research activities around developing commercial soybean varieties with improved seed quality traits. There seems little doubt that we can envision the use of modified soybean seed for production of a huge range of products for edible, industrial, and pharmaceutical or nutritional applications. Third, integration of conventional breeding, marker-assisted breeding, and transgenic efforts would accelerate the development of new soybean products. This is especially true for developing new varieties containing the most effective combination of existing characters. It should also be noted that there are challenges to be faced for developing new biotech soybean varieties with more complex agronomic or seed quality traits using multiple gene combinations or single regulatory genes such as transcription factors. Furthermore, biotech crop varieties with modified seed quality traits must possess good agronomic characteristics.

## REFERENCES

Abbadi, A., F. Domergue, J. Bauer, J.A. Napier, R. Welti, U. Zähringer, P. Cirpus, and E. Heinz (2004). Biosynthesis of very-long-chain polyunsaturated fatty acids in transgenic oilseeds: constraints on their accumulation. *Plant Cell* 16:2734-2748.

Amir, R., Y. Hacham, and G. Galili (2002). Cystathionine γ-synthase and threonine synthase operate in concert to regulate carbon flow towards methionine in plants. *Trends in Plant Science* 7:153-156.

Bartlem, D., I. Lambein, T. Okamoto, A. Itaya, Y. Uda, F. Kijima, Y. Tamaki, E. Nambara, and S. Naito (2000). Mutation in the threonine synthase gene results in an over-accumulation of soluble met in *Arabidopsis. Plant Physiology* 123:101-110.

Broun, P. (2004). Transcription factors as tools for metabolic engineering in plants. *Current Opinion in Plant Biology* 7:202-209.

Buhr, T., S. Sato, F. Ebrahim, A. Xing, Y. Zhou, M. Mathiesen, B. Schweiger, A.J. Kinney, P. Staswick, and T. Clemente (2002). Ribozyme termination of RNA transcripts down-regulate seed fatty acid genes in transgenic soybean. *Plant Journal* 30:155-163.

Burks, Jr. A.W., J.R. Brooks, and H.A. Sampson (1988). Allergenicity of major component proteins of soybean determined by enzyme-linked immunosorbent assay (ELISA) and immunoblotting in children with atopic dermatitis and positive soy challenges. *Journal of Allergy and Clinical Immunology* 81:1135-1142.

Cahoon, E.B., S.H. Hall, K.G. Ripp, T.S. Ganzke, W.D. Hitz, and S.J. Coughlan (2003). Metabolic redesign of vitamin E biosynthesis in plants for tocotrienol production and increased antioxidant content. *Nature Bio/Technology* 21:1082-1087.

Chow, C.K. (2001). Vitamin E. In *Handbook of vitamins* 3rd ed., R.B. Rucker, J.W. Suttie, D.B. McCormick, and L.J. Machlin (eds.). New York: Marcel Dekker, pp. 165-197.

Collakova, E. and D. Della Penna (2003). Homogentisate phytyltransferase activity is limiting for tocopherol biosynthesis in *Arabidopsis. Plant Physiology* 131: 632-642.

Curien, G., D. Job, R. Douce, and R. Dumas (1998). Allosteric activation of *Arabidopsis* threonine synthase by S-adenosylmethionine. *Biochemistry* 37: 13212-13221.

Finnegan, J. and D. McElroy (1994). Transgene inactivation: plants fight back. *Nature Bio/Technology* 12:883-888.

Halpin, C., A. Barakate, B.M. Askari, J.C. Abbott, and M.D. Ryan (2001). Enabling technologies for manipulating multiple genes on complex pathways. *Plant Molecular Biology* 47:295-310.

Hayashi, M., K. Harada, T. Fujiwara, and K. Kitamura (1998). Characterization of a 7S globulin-deficient mutant of soybean (*Glycine* max (L.) Merrill). *Molecular and General Genetics* 258:208-214.

Helm, R.M., G. Cockrell, C.M. West, E.M. Herman, H.A. Sampson, G.A. Bannon, and A.W. Burks (2000). Mutational analysis of the IgE-binding epitopes of P34/Gly m1. *Journal of Allergy and Clinical Immunology* 105:378-384.

Herman, E.M. (2003). Genetically modified soybeans and food allergies. *Journal of Experimental Botany* 54:1317-1319.

Hess, H., O. Kraft, S. Maimann, M. Zeh, and R. Hoefgen (2004). Current understanding of the regulation of methionine biosynthesis in plants. *Journal of Experimental Botany* 55:1799-1808.

Karunanandaa, B., Q. Qi, M. Hao, S. Baszis, P.K. Jensen, Y.H. Wong, J. Jiang, M. Venkatramesh, K.J. Gruys, F. Moshiri, et al. (2005). Metabolically engineered oilseed crops with enhanced seed tocopherol. *Metabolic Engineering* 7:384-400.

Keller, W.A., R.S.S. Datla, J.Z. Dong, F. Georges, A.A.K. Hussain, and G. Selvaraj (2004). Method and compositions for modifying levels of secondary metabolic compounds in plants. Patent application. US6703539.

Kinney, A.J., E.B. Cahoon, and W.D. Hitz (2002). Manipulating desaturase activities in transgenic crop plants. *Biochemical Society Transactions* 30:1099-1103.

Kinney, A.J., and G.M. Fader (2004). Suppression of specific classes of soybean seed protein genes. Patent application. US0139504.

Kohno-Murase, J., M. Murase, H. Ichikawa, and J. Imamura (1994). Effects of an antisense napin gene on seed storage compounds in transgenic *Brassica napus* seeds. *Plant Molecular Biology* 26:1115-1124.

Lee, T.T.T., M.M.C. Wang, R.C.W. Hou, L.J. Chen, R. Su, C. Wang, and J.T.C. Tzen (2003). Enhanced methionine and cysteine levels in transgenic rice seeds by the accumulation of sesame 2S albumin. *Bioscience Biotechnology and Biochemistry* 67:1699-1705.

Li, J. and R.L. Last (1996). The *Arabidopsis thaliana* try5 mutant has a feedback-resistant anthranilate synthase and elevated soluble tryptophan. *Plant Physiology* 110:51-59.

Liang, J., B. Rapp, L. Weaver, Q. Wang, L. Crow, J. Ream, B. Hill, W. Brown, T. Oulmassov, and K. Gruys (2003). Enhancing the nutritional value of soybean. *Plant Biology* abstract 38001, p. 31.

Martini, N., E. Neuhaus, and J. Schell (1999). Recombinant DNA molecules and method for increasing oil content in plants. Patent application WO9911805A1.

Nes, W.D. (1998). Transgenic plants with modified sterol biosynthetic pathways. Patent application. WO9845457A1.

Nordlee, J.A., S.L. Taylor, J.A. Townsend, L.A. Thomas, and R.K. Bush (1996). Identification of a Brazil-nut allergen in transgenic soybeans. *New England Journal of Medicine* 334:688-692.

Ogawa, T., M. Samoto, and K. Takahashi (2000). Soybean allergens and hypoallergenic soybean products. *Journal of Nutritional Science and Vitaminology* 46:271-279.

Pereira, S.L., Y.S. Huang, E.G. Bobik, A.J. Kinney, K.L. Stecca, J.C. Packer, and P. Mukerji (2004). A novel ω3-fatty acid desaturase involved in the biosynthesis of eicosapentaenoic acid. *Biochemical Journal* 378:665-671.

Porfirova, S., E. Bergmüller, S. Tropf, R. Lemke, and P. Dörmmann (2002). Isolation of an *Arabidopsis* mutant lacking vitamin E and identification of a cyclase essential for all tocopherol biosynthesis. *Proceedings of the National Academy of Sciences USA* 9:12495-12500.

Qi, Q., M. Hao, W. Wing-on Ng, S. Slater, S. Baszis, D.J. Weiss, and H.E. Valentin (2005). Application of the *Synechococcus nirA* promoter to establish an inducible expression system for engineering the *Synechocystis* tocopherol pathway. *Applied and Environmental Microbiology* 70:5678-5684.

Rapp, W.D., J. Peng, G. Nadig, and T. Venkatesh (2003). B-Conglycinins with increased essential amino acid content, transgenic plants producing them, and their use in feeds. Patent application. WO2003025153.

Renz, A., J. Bauer, M. StittNigel, R.M. Zrenner, O. Geigenberger, and H. Vigeolas (2003). Methods for increasing oil content in plants. Patent application. WO03095655A3.

Saito, K. (2000). Regulation of sulfate transport and synthesis of sulfur-containing amino acids. *Current Opinion in Plant Biology* 3:188-195.

Sanders, S.K., J.B. Morgan, D.M. Wulf, J.D. Tatum, S.N. Williams, and G.C. Smith (1997). Vitamin E supplementation of cattle and shelf-life of beef for the Japanese market. *Journal of Animal Science* 75:2634-2640.

Sato, S., A. Xing, X. Ye, B. Schweiger, A. Kinney, G. Graef, and T. Clemente (2004). Production of γ-linolenic acid and stearidonic acid in seeds of marker-free transgenic soybean. *Crop Science* 44:646-652.

Sattler, S.E., Z. Cheng, and D. Della Penna (2004). From *Arabidopsis* to agriculture: Engineering improved vitamin E content in soybean. *Trends in Plant Science* 9:365-367.

Savidge, B., J.D. Weiss, Y.H. Wong, M.W. Lassner, T.A. Mitsky, C.K. Shewmaker, D. Post-Beittenmiller, and H.E. Valentin (2002). Isolation and characterization of homogenization phytyltransferase genes from *Synechocystis* sp. 6803 and *Arabidopsis. Plant Physiology* 129:321-332.

Shintani, D. and D. Della Penna (1998). Elevating the vitamin E content of plants through metabolic engineering. *Science* 282:2098-2100.

Siehl, D. (1999). The biosynthesis of tryptophan, tyrosine, and phenylalanine from chorismate. In *Plant amino acids: Biochemistry and biotechnology,*, B.K. Singh (ed.). New York: Dekker, pp. 171-204.

Singh, S.P., X.-R. Zhou, Q. Liu, S. Stemne, and A.G. Green. (2005). Metabolic engineering of new fatty acids in plants. *Current Opinion in Plant Biology* 8:197-203.

Sun, S.S.M. (1999). Methionine enhancement in plants. In *Plant amino acids: Biochemistry and biotechnology*, B.K. Singh (ed.). New York: Dekker, pp. 509-522.

Tozawa, Y., H. Hasegawa, T. Terakawa, and K. Wakasa (2001). Characterization of rice anthranilate synthase α-subunit genes *OASA1* and *OASA2*. Tryptophan accumulation in transgenic rice expressing a feedback-insensitive mutant of *OASA1*. *Plant Physiology* 126:1493-1506.

Valentin, H.E., K. Lincoln, F. Moshiri, P.K. Jensen, Q. Qi, T.V. Venkatesh, B. Karunanandaa, S.R. Baszis, S.R. Norris, B. Savidge, et al. (2005). The *Arabidopsis vte5-1* mutant reveals a critical role for phytol kinase in seed tocopherol biosynthesis. *Plant Cell* 18:212-224.

Valentin, H.E., and Q. Qi (2005). Biotechnological production and application of vitamin E: Current state and prospects. *Applied Microbiology and Biotechnology* 68:436-444.

Van Eenennaam, A.L., K. Lincoln, T.P. Durrett, H.E. Valentin, C.K. Shewmaker, G.M. Thorne, J. Jiang, S.R. Baszis, C.K. Levering, E.D. Aasen, et al. (2003). Engineering vitamin E content: From *Arabidopsis* mutant to soy oil. *Plant Cell* 15:3007-3019.

Wandelt, C. (1992). Improved feed crops enriched in sulfur amino acids and methods for improvement. Patent application. WO92/14822.

Weaver, L.M., T.N. Oulmassov, G. Vaduva, J. Liang, M.J. Varagona and T. Venkatesh (2003). Transgenic high tryptophan plants. Patent application. 03/092363.

Zeh, M., A.P. Casazza, O. Kreft, U. Roessener, K. Bieberich, L. Willmitzer, R. Höfgen, and H. Hesse (2001). Antisense inhibition of threonine synthase leads to high methionine content in transgenic potato plants. *Plant Physiology* 127:792-802.

Chapter 6

# *Agrobacterium*-Mediated Genetic Transformation of Soybean

Randy D. Dinkins
Glenn B. Collins

## *INTRODUCTION*

Soybean [*Glycine max* (L.) Merrill] ranks as the world's primary source of vegetable protein and provides nearly 60 percent of the world's oil. The era of soybean genomics is reaching a point where it requires efficient gene transfer systems to assign and verify the biological function of cloned genes and to capitalize on the ability to deploy cloned genes in transgenic plants. Although soybean transformation today is several orders of magnitude more efficient than it was just a few years ago, it still lags far behind that of most other crops previously classified as recalcitrant. Soybean tissue culture and transformation continue to be laborious, and the frequency of recovery of transgenic plants is still very low. Ever since a simple method was introduced by Horsch et al. (1985) and subsequently examined extensively by others (Woodragen and Dons, 1992; Ishida et al., 1996; Hiei et al., 1997), *Agrobacterium tumefaciens*-mediated transformation has been the primary means for transformation of dicotyledonous species. Soybean was originally not considered to be a susceptible host for infection by *Agrobacterium* (De Cleene and De Ley, 1976), although it has been shown subsequently to exhibit significant *Agrobacterium* strain by genotype interaction to *Agrobacterium* infection (Byrne et al., 1987; Delzer and Somers, 1990). Thus, even at a time when previously recalcitrant species, such as rice and maize, are being routinely transformed via *Agrobacterium* (Ishida et al., 1996; Rashid et al., 1996; Hiei et al., 1997), soybean transformation with *Agrobacterium* remains relatively inefficient. Several factors have contributed to the slow use of *Agrobacterium*-mediated transformation in soybean, including the availability of an alternate particle bombardment method; however, improvements have resulted in a resurgence of *Agrobacterium* use for soybean transformation.

## AGROBACTERIUM-*SOYBEAN INTERACTION*

One of the first attempts to optimize *Agrobacterium*-mediated transformation in soybean was to identify *Agrobacterium* strains and soybean cultivar combinations that would increase transformation efficiency (Owens and Cress, 1985; Hinchee et al., 1988; Delzer and Somers, 1990; Bailey et al., 1994; Mauro et al., 1995; Meurer et al., 1998). The cultivar Peking was identified as one of the most susceptible genotypes to *Agrobacterium* infection based on its galling response, and it was the first cultivar used to successfully transform soybean (Hinchee et al., 1988). Attempts were made with limited success to identify the genetic basis for *Agrobacterium* susceptibility (Bailey et al., 1994; Mauro et al., 1995). It is now accepted that most, if not all, soybean cultivars are susceptible to *Agrobacterium*-mediated transformation, but that the transformation efficiency varies significantly among cultivars.

Wild-type *Agrobacterium* strains that efficiently produced galls on soybean, such as the A281 and Chry5, have now been disarmed and tested for soybean transformation. The A281 strain is a C58 background containing the pBo542-Ti plasmid, and Chry5 is a wild-type strain that was observed to be highly infective on soybean (Hood et al., 1984; Kovacs and Pueppke, 1993). The disarmed versions of the A281 strain are EHA101 and EHA105 (Hood et al., 1993) and the disarmed versions of Chry5 are KYRT1 and KPSF2 (Torisky et al., 1997; Palanichelvam et al., 2000). These disarmed *Agrobacterium* strains have been used successfully to produce transgenic soybean plants (Meurer et al., 1998; Zhang et al., 1999; Donaldson and Simmonds, 2000; Yan et al., 2000; Palanichelvam et al., 2000). Recent reevaluation of the KYRT1 strain has shown that a portion of the T-DNA is still present on the Ti plasmid; thus transformations using the KYRT1 strain can result in co-transfer of part of the wild-type T-DNA segment. The presence of the wild-type T-DNA segment can occasionally produce tumorigenic growth following transformation (Palanichelvam et al., 2000). The KPSF2 strain contains no additional T-DNA sequences and is thus completely disarmed. A comparison of the KYRT1 and KPSF2 strains on somatic embryos has suggested that the presence of the remaining T-DNA sequences in the KYRT1 strain enhanced the transformation process and was thus preferred to the completely disarmed KPSF2 strain (Ko et al., 2003, 2004). It should be interesting to further delimit the nature of this increased transformation in the Chry5 strain by more extensive comparisons of the KPSF2 and KYRT1 strains.

Another approach has been to improve the coculture conditions for *Agrobacterium* infection. This is done by the addition of compounds, such as acetosyringone, to induce the *Agrobacterium vir* genes, and by the use of *Agrobacterium* strains that constitutively express the *vir* genes (Stachel et al., 1985; Delzer and Somers, 1990; Hansen et al., 1994), and the selection of inoculum concentration or coculture temperatures that increase transformation rates (Fulner and Nester, 1996; Meurer et al., 1998).

## *CULTURE PROTOCOLS*

Regeneration of fertile transgenic soybean plants by *Agrobacterium*-mediated protocols has been reported using both the cotyledonary node system (Hinchee et al., 1988; Townsend and Thomas, 1994; Di et al., 1996; Zhang et al., 1999; Clemente et al., 2000; Donaldson and Simmonds, 2000; Olhoft et al., 2001) and the somatic embryogenic protocol (Yan et al., 2000; Ko et al., 2003). The cotyledonary node protocol is the easiest and thus amenable to a larger number of researchers or to those initiating a soybean transformation program. The somatic embryogenic protocol, while requiring more specialized technical expertise and a longer time in culture, has the advantage of ensuring that putative transgenic plants are not chimeric and offers the potential to regenerate a multiple number of plants derived from the same transformation event.

## *MERISTEMATIC CELL PROLIFERATION AND PLANT REGENERATION*

Explants placed on a high-cytokinin medium will produce a large number of shoots which after initiation and elongation, can be excised and rooted on an indole-3-butyric acid (IBA) and auxin-containing medium (Cheng et al., 1980; Lazzeri et al., 1985; Barwale et al., 1986; Wright et al., 1986, 1987). Plants have been obtained that were initiated in the meristem region of the apical dome (McCabe et al., 1988; Christou et al., 1988; Aragao et al., 2000), the small meristematic region of the cotyledonary node (Barwale et al., 1986; Wright et al., 1986) and epicotyl (Wright et al., 1987; Dan and Reichert, 1998). Apical or cotyledonary node meristematic tissue explants can be induced to regenerate whole plants very efficiently in soybean.

Cheng et al. (1980) first reported on organogenesis from cotyledonary explants derived from soybean seedlings. In this early work, soybean seeds were directly subjected to elevated levels of benzyl aminopurine (BAP) during germination. These high levels (> 2 μM) apparently were sufficient to overcome apical dominance in the developing seedling and led to multiple buds emanating from the axillary meristem (commonly referred to as the cotyledonary node). Subculturing excised bud-induced nodal region onto medium supplemented with BAP resulted in further development of shoots and eventual recovery of whole, in vitro-derived soybean plants (Cheng et al., 1980). The in vitro culture technique using the cotyledonary node explant was modified by Wright et al. (1986), who observed maximum in vitro response on a reduced-salt medium supplemented with 5 μM BAP from germination through successive subcultures. Other tissues and organs have also been used and reported to undergo organogenesis (Wright et al., 1987; McCabe et al., 1988; Dan and Reichert, 1998). However, the embryonic axis has been found to be one of the most responsive and easily obtained and maintained in culture.

Hinchee et al. (1988) were the first to report using the soybean cotyledonary node in transformation. Cotyledonary-node explants were prepared from soybean seeds pregerminated for 4-10 days on water agar and cultured on Gamborg's B5 medium (Gamborg et al., 1968) supplemented with 5 μM BAP. A nopaline strain of *A. tumefaciens* was used in the transformations, coupled with the *npt*II gene as the selectable marker. The cotyledonary-node explants were placed on a medium containing kanamycin (200-300 $mg \cdot l^{-1}$) for selection of transformed shoots. Following a shoot induction period on 5 μM BAP, the induced explants were subcultured to medium supplemented with reduced cytokinin levels for shoot elongation. Transformation frequencies ranged from 0.3 to 2.2 percent on a transformant per explant basis (Hinchee et al., 1988).

Di et al. (1996) implemented a number of modifications to the cotyledonary-node transformation system of Hinchee et al. (1988) that included the addition of 100 μM of the *Agrobacterium* virulence-inducing agent acetosyringone during inoculation, cocultivation, reduction of seed germination period to three days, and slicing the nodal region with a scalpel blade prior to inoculation. Under natural conditions, *Agrobacterium* is an opportunistic pathogen and some form of wounding is necessary for infection to occur. The *Agrobacterium* cotyledonary node transformation protocol explant preparation involves separating the cotyledons from the germinated seedlings by making a cut through the hypocotyl approximately 5 mm below the cotyledon. A vertical slice initiated between the cotyledons and continued through the center of the hypocotyl segment then separates the cotyledons. Using a razor or scalpel blade, the embryonic axis is removed and 8-12 slices parallel with the axis are made in order to wound the meristematic cells of the cotyledonary node. The skill of the researcher making the slices often determines the success or failure of the cotyledonary node transformation procedure (Trick and Finer, 1997; Meurer et al., 1998; Zhang et al., 1999).

The present basic *Agrobacterium*-mediated cotyledonary node protocol has changed little since the publications of Hinchee et al. (1988) and Di et al. (1996), except for the addition of thiol compounds to the medium (see below). Following the cocultivation period, explants were cultured in liquid shoot induction medium composed of Gamborg B5 salts (Gamborg et al., 1968), 5 μM BAP, supplemented with a selection agent and an antibiotic regime consisting of carbenicillin, cefotaxime, timentin, or vancomycin to eliminate *A. tumefaciens* cells. The liquid culture period is conducted for 3-5 days followed by a selection period on solid medium for up to 8 weeks. Cotyledonary nodes with developing shoots are subsequently transferred to elongation medium composed of Gamborg's basal medium (Gamborg et al., 1968) supplemented with 1.7 μM gibberellic acid ($GA_3$) and 10 μM IBA without any additional antibiotics to allow for root growth. An alternative phytohormone protocol for root induction and elongation was provided by Parrott and Clemente (2004) consisting of 2.8 μM zeatin riboside, 1.44 μM $GA_3$ and 570 nM indole-3-acetic acid (IAA). The authors stressed the exclusion of antibiotics in the root induction medium, suggesting that this inhibits root formation. Following root elongation, plantlets are transferred into soil.

Recent modifications to the soybean cotyledonary-node transformation system resulting in improvements have been focused on imbibition time prior to infection (Paz et al., 2006), the cocultivation and elongation steps, use of different antibiotic and herbicide selection agents and schemes (Zhang et al., 1999; Clemente et al., 2000; Olhoft et al., 2003), and the use of thiol compounds to decrease the action of polyphenol oxidase, which causes a browning reaction during culture (Olhoft et al., 2001; Olhoft and Somers, 2001). The presumed basis for the use of thiol compounds is that soybean induces a pathogen infection response during wounding or infection, which results in browning and necrosis of infected tissues. Thiol compounds probably act as inhibitors of the defense response. Using thiol compounds during cocultivation has shown increases in transient expression in soybean explants regardless of cultivar source and has resulted in increased overall transformation frequencies (Olhoft et al., 2001, 2003; Paz et al., 2004).

Kanamycin has not been a particularly effective selection agent for soybean transformation. The most common selectable marker presently in use in soybean transformation is the *hpt* gene that provides resistance to hygromycin. Though hygromycin is a harsh antibiotic, it has been used successfully in soybean transformation (Olhoft et al., 2003). The use of the herbicide marker genes *CP4* (resistance to glyphosate at 0.05 mM to 0.15 mM) or *bar* (resistance to glufosinate at 3-5 $mg{\cdot}l^{-1}$ or bialaphos at 4-6 $mg{\cdot}l^{-1}$) provides an alternative to antibiotic selection of transgenic soybeans (Zhang et al., 1999; Clemente et al., 2000; Zeng et al., 2004; Paz et al., 2004). The *CP4* gene has been shown to be very effective for selection in soybean. However, it is proprietary and not available for use by the general community.

## *SOMATIC EMBRYOGENESIS PROTOCOLS*

Somatic embryos induced on 2,4-D have a decided advantage in that they are able to proliferate continuously whenever the levels of 2,4-D in the culture medium are at or above 10 $mg{\cdot}l^{-1}$ (or 45 μM, although 2,4-D concentrations in the literature are normally in $mg{\cdot}l^{-1}$; Ranch et al., 1985), in a process known as repetitive or proliferative embryogenesis. The somatic embryogenic protocols involve inducing an explant to produce somatic embryos in tissue culture. Somatic embryos have both a shoot and root meristem and can germinate to form whole plants. Somatic embryogenic procedures involve several changes in media during the stages of (a) somatic embryo induction, (b) proliferative somatic embryo growth, (c) maturation, (d) germination, and (e) conversion to whole plants.

There have been numerous reports of soybean somatic embryo production in the literature (Christianson et al., 1983; Lippman and Lippman, 1984; Lazzeri et al., 1985; Ranch et al., 1985; Finer, 1988; Hartweck et al., 1988; Hepher et al., 1988; Parrott et al., 1988); however, somatic embryos produced using earlier protocols were not successful in producing transgenic soybean plants with stable inheritance of the transgene by the progeny. Altering the 1:1 ratio of ammonium to nitrate in the Murashige and Skoog (1962) basal salts to 1:4 in liquid medium, Finer (1988) described a somatic embryogenic system that used immature cotyledons as explants on high levels of 2,4-D (40 $mg{\cdot}l^{-1}$), which is the basis for most embryogenic systems in use today (Finer and McMullen, 1991). The first responsive cultivar identified was Fayette (Finer, 1988; Finer and Nagasawa, 1988), and subsequently the cultivar Jack was found to be highly regenerable (Stewart et al., 1996) and is now the standard for the somatic embryonic protocol. A screen of soybean genotypes also identified the cultivar Dwight as moderately efficient for somatic embryonic regeneration (Meurer et al., 2001). These three cultivars all share a common ancestor (PI88788) and attempts to elucidate a genetic basis have been unsuccessful, pointing to the fact that somatic embryogenesis is a complex trait (Bailey et al., 1993a; Tian et al., 1994; Simmonds and Donaldson, 2000). The combination of single (or few) cells and surface origin of proliferating embryos provides an attractive target for transgene introduction and selection.

Once somatic embryos have been induced on solid medium containing a high level of auxin, such as 2,4-D, continued proliferation and selection of transgenic somatic embryos following transformation is done on either solid, liquid, or an alternating sequence of solid and liquid media. Each method has its own advantages and disadvantages. Liquid medium protocols are initiated by placing clumps of the somatic embryos in liquid medium (Samoylov et al., 1998) and require weekly transfers. While this system often results in selection and identification of transgenic lines in 7-10 weeks, its prolonged use will induce undesirable mutations. Singh et al. (1998) demonstrated that the liquid tissue culture process was capable of inducing permanent cytogenetic and genetic aberrations in as little as 8 months with some cultivars. This heritable change was directly responsible for reduced fertility of tissue-cultured lines. More recent research has concentrated on how to maintain cultures in liquid proliferation and selection long enough to identify transformed lines without exceeding the culture time likely to cause abnormalities.

One of the major difficulties encountered in *Agrobacterium*-mediated transformation of immature cotyledons or somatic embryos is the treatment required to contain *Agrobacterium* overgrowth. A modification of the liquid suspension culture protocol that has resulted in the recovery of *Agrobacterium*-mediated transformed soybeans, as well as reducing the potential for *Agrobacterium* overgrowth along with circumventing the necessity for continual manipulation of the cultures, is the solid medium proliferation culture system (Yan et al., 2000; Ko et al., 2003). This procedure, while similar to the liquid medium protocol, requires only monthly transfers, compared to weekly for the liquid cultures, due to slower growth rates. Thus, a major advantage of the solid medium protocol is the simplicity of the tissue culture manipulations compared to the liquid suspension cultures. The primary disadvantage appears to be that the length of time required for selection of transgenic embryo lines (6-8 months) on solid medium can result in a high frequency of infertile or abnormal transgenic plants, as discussed above. One strategy to circumvent this problem has been to shorten the selection time. However, this can lead to an increase in the incidence of false positives or escapes. Thus, there appears to be a fine line (approximately six months on selection) between obtaining true transgenic embryogenic lines that will result in whole transformed fertile plants and transformed plants that have reduced fertility and are abnormal. This method has been used successfully to regenerate *Agrobacterium*-mediated transgenic soybean lines. Ko et al. (2003) pointed out that placement of the immature cotyledons with the adaxial side on the culture medium is critical for optimal success.

Somatic embryos must eventually undergo histodifferentiation and physiological maturation, followed by proper activation of the root and apical meristem during germination to permit proper conversion (Slawinska and Obendorf, 1991; Bailey et al., 1993a, 1993b; Samoylov et al., 1998). Media preparations for somatic embryo maturation, germination, and conversion usually do not contain exogenous plant growth regulators but do require the addition of charcoal to the medium, which presumably adsorbs any 2,4-D that may otherwise interfere with establishment of the proper polarity of the developing embryo (Cooke et al., 1993; Schmidt et al., 1994). Improvements in osmoticum, as well as adding exogenous amino acids, have been shown to greatly enhance and shorten the time for the recovery of soybean plants via somatic embryogenic methods in liquid cultures (Walker and Parrott, 2001). However, similar optimizations have not been done for the solid histodifferentiation and maturation medium, namely because these have worked sufficiently well. Taken together, the recovery of fertile transgenic soybean plants using the somatic embryogenic procedures is now common, although it still requires 6-9 months, and the frequency of recovery of transgenic plants continues to be low.

## *MERISTEMATIC CELL PROLIFERATION VERSUS SOMATIC EMBRYOGENESIS PROTOCOLS*

The regeneration of fertile transgenic soybean plants has been accomplished using both meristematic cell proliferation techniques and somatic embryogenesis protocols. Although the initial soybean transformation protocols utilizing the meristematic procedures (Hinchee et al., 1988) and somatic embryos (Parrott et al., 1989) were published during the same time period, the somatic embryogenic protocols have been primarily used in conjunction with particle bombardment, while meristematic explants have continued to be used in *Agrobacterium*-mediated transformation procedures. One of the primary advantages for the use of somatic embryo cycling in particle bombardment is the single-cell origin of somatic embryos on the epidermal surface of the immature cotyledons (Liu et al., 1992), providing a good target for bombardment (and potentially for *Agrobacterium* infection) and subsequent regeneration of nonchimeric plants. However, this potential has not been realized in *Agrobacterium*-mediated transformation of somatic embryos. In fact, it is presently not known if *Agrobacterium* can infect the rapidly proliferating somatic embryos in this system as none of the transformed plants obtained to date have utilized proliferating somatic embryos for infection, but rather the initial immature cotyledon explant.

Presently, the preference for a given system is dependent on each individual investigator. Meristematic cell proliferation protocols are much simpler and the plants initially regenerated are recovered much quicker than with the somatic embryogenic protocols. Cotyledonary node meristematic cells have been shown to be competent for transformation, and this transformation protocol has received considerable attention for that reason. Many researchers have evaluated a large number of soybean genotypes using the meristematic cell protocols, with the majority of genotypes providing at least some capability to regenerate multiple shoots (Byrne et al., 1987; Delzer and Somers, 1990; Meurer et al., 2001; Tomlin et al., 2002). Proliferative shoot clumps can be initiated directly from germinating seed with the caveat that only a select group of cells are capable of regeneration, and targeting of transformation must therefore be precise. Additionally, the targeted meristematic cells are often in different stages, and transformation often results in the recovery of chimeras (Meurer et al., 2001). Somatic embryogenic procedures, in contrast, rarely yield chimeric lines and can often yield multiple clones of the same transformation event. Thus, it is necessary to maintain the integrity of the lines to ensure that independent transgenic lines are produced (Yan et al., 2000). In addition, once a cell line is transformed, continued culture of some embryos arising from a transformation is possible, while other clones can be undergoing conversion or growth in the greenhouse. Thus, once a putative transformed line is identified, additional somatic embryos can be obtained to ensure that the transformation event is recovered; however, each of the individual plants obtained will have originated from the same transformation event and thus constitute a single transformation event.

## *CONCLUSION*

*Agrobacterium*-mediated transformation of soybean has lagged behind most of the other agronomically important crops. However, recent improvements in the cotyledonary-node and embryogenic protocols have brought soybean transformation to the point that it is now possible for scientists interested in performing soybean transformation to use the procedures with a good chance of success. While the use of particle bombardment has offered an additional method for soybean transformation, it has simultaneously impeded the progress of *Agrobacterium*-mediated transformation due to the division of effort. *Agrobacterium*-mediated transformation protocols have now been established for soybean that make it routine, although still at a lower frequency than would be desired.

## REFERENCES

Aragao, F.J.L., L. Sorkin, G.R. Vianna, and E.L. Rech (2000). Selection of transgenic meristematic cells utilizing a herbicidal molecule results in the recovery of fertile transgenic soybean [*Glycine max* (L.) Merril] plants at a high frequency. *Theoretical and Applied Genetics* 101:1-6.

Bailey, M.A., H.R. Boerma, and W.A. Parrott (1993a). Genotype effects on proliferative embryogenesis and plant regeneration of soybean. *In Vitro Cellular and Developmental Biology–Plant 29P*:102-108.

Bailey, M.A., H.R. Boerma, and W.A. Parrott (1993b). Genotype-specific optimization of plant regeneration from somatic embryos of soybean. *Plant Science* 93:117-120.

Bailey, M.A., H.R. Boerma, and W.A. Parrott (1994). Inheritance of tumor formation in response to *Agrobacterium tumefaciens* in soybean. *Crop Science* 34: 514-519.

Barwale, U.B., H.K. Kerns, and J.M. Widholm (1986). Plant regeneration from callus cultures of several soybean genotypes via embryogenesis and organogenesis. *Planta* 167:473-481.

Byrne, M.C., R.E. McDonnell, M.S. Wright, and M.G. Carnes (1987). Strain and cultivar specificity in the *Agrobacterium*-soybean interaction. *Plant Cell Tissue and Organ Culture* 8:3-15.

Cheng, T.Y., H. Saka, and T.H. Voqui-Dinh (1980). Plant regeneration from soybean cotyledonary node segments in culture. *Plant Science Letters* 19:91-99.

Christianson, M.L., D.A. Warnick, and P.S. Carlson (1983). A morphologically competent soybean suspension culture. *Science* 222:632-634.

Christou, P., D.E. McCabe, and W.F. Swain (1988). Stable transformation of soybean callus by DNA-coated gold particles. *Plant Physiology* 87:671-674.

Clemente, T.E., B.J. LaVallee, A.R. Howe, D. Conner-Ward, R.J. Rozman, P.E. Hunter, D.L. Broyles, D.S. Kasten, and M.A. Hinchee (2000). Progeny analysis of glyphosate selected transgenic soybeans derived from *Agrobacterium*-mediated transformation. *Crop Science* 40:797-803.

Cooke, T.J., R.H. Racusen, and J.D. Cohen (1993). The role of auxin in plant embryogenesis. *Plant Cell* 5:1494-1495.

Dan, Y. and N.A. Reichert (1998). Organogenic regeneration of the soybean from hypocotyl explants. *In Vitro Cellular and Developmental Biology–Plant* 34: 14-21.

De Cleene, M. and J. De Ley (1976). The host range of crown gall. *Botanical Gazette* 42:389-466.

Delzer, D.M. and D.A. Somers (1990). *Agrobacterium tumefaciens* susceptibility and plant regeneration of 10 soybean genotypes in maturity groups 00 to II. *Crop Science* 30:320-322.

Di, R., V. Purcell, G.B. Collins, and S.A. Ghabrial (1996). Production of transgenic soybean lines expressing the bean pod mottle virus coat protein precursor gene. *Plant Cell Reports* 15:746-750.

Donaldson, P.A. and D.H. Simmonds (2000). Susceptibility to *Agrobacterium tumefaciens* and cotyledonary node transformation in short-season soybean. *Plant Cell Reports* 19:478-484.

Finer, J.J. (1988). Apical proliferation of embryogenic tissue of soybean [*Glycine max* (L.) Merrill]. *Plant Cell Tissue and Organ Culture* 15:125-136.

Finer, J.J. and M.D. McMullen (1991). Transformation of soybean via particle bombardment of embryogenic suspension culture tissue. *In Vitro Cellular and Developmental Biology–Plant* 27:182.

Finer, J.J. and A. Nagasawa (1988). Development of an embryogenic suspension culture of soybean (*Glycine max* Merrill). *Plant Cell Tissue and Organ Culture* 15:125-136.

Fulner, J.K. and E.W. Nester (1996). Temperature affects the T-DNA transfer machinery of *Agrobacterium tumefaciens*. *Journal of Bacteriology* 178:1498-1504.

Gamborg, O.L., R.A. Miller, and K. Ojima (1968). Nutrient requirements of suspension cultures of soybean root cells. *Experimental Cell Research* 50:150-158.

Hansen, G., A. Das, and M.-D. Chilton (1994). Constitutive expression of the virulence genes improves the efficiency of plant transformation by *Agrobacterium*. *Proceedings of the National Academy of Sciences USA* 91:7603-7607.

Hartweck, L.M., P.A. Lazzeri, D. Cui, and G.B. Collins (1988). Auxin-orientation effects on somatic embryogenesis from immature soybean cotyledons. *In Vitro Cellular and Developmental Biology–Plant* 24:821-828.

Hepher, A., M.E. Boulter, N. Harris, and R.S. Nelson (1988). Development of a superficial meristem during somatic embryogenesis from immature cotyledons of soybean (*Glycine max* L.). *Annals of Botany* 62:513-519.

Hiei, Y., T. Komari, and T. Kubo (1997). Transformation of rice mediated by *Agrobacterium tumefaciens*. *Plant Molecular Biology* 35:205-218.

Hinchee, M.A., D.V. Conner-Ward, C.A. Newell, R.E. McDonnell, S.J. Sato, C.S. Gasser, D.A. Fischhoff, D.B. Re, R.T. Fraley, and R.B. Horsch (1988). Production of transgenic soybean plants using *Agrobacterium*-mediated gene transfer. *Nature Bio/Technology* 6:922.

Hood, E.E., S.B. Gelvin, L.S. Melchers, and A. Hoekema (1993). New *Agrobacterium* helper plasmids for gene transfer to plants. *Transgenic Research* 2:208-218.

Hood, E.E., G. Jen, L. Kayer, R.T. Fraley, and M.-D. Chilton (1984). Restriction endonuclease map of pTiBo542, a potential Ti plasmid vector for genetic engineering of plants. *Nature Bio/Technology* 2:702-709.

Horsch, R.B., J.E. Fry, N.L. Hoffman, D. Eicholtz, S.G. Rogers, and R.T. Fraley (1985). A simple and general method for transferring genes into plants. *Science* 227:1229-1231.

Ishida, Y., H. Saito, S. Ohta, Y. Hiei, T. Komari, and T. Kumashiro (1996). High efficiency transformation of maize (*Zea mays* L.) mediated by *Agrobacterium tumefaciens*. *Nature Bio/Technology* 14:745-750.

Ko, T.S., S. Lee, S.K. Farrand, and S.S. Korban (2004). A partially disarmed vir helper plasmid, pKYRT1, in conjunction with 2,4-dichlorophenoxyactic acid promotes emergence of regenerable transgenic somatic embryos from immature cotyledons of soybean. *Planta* 218:536-541.

Ko, T.S., S. Lee, S. Krasnyanski, and S.S. Korban (2003). Two critical factors are required for efficient transformation of multiple soybean cultivars: *Agrobacterium* strain and orientation of immature cotyledonary explant. *Theoretical and Applied Genetics* 107:439-447.

Kovacs, L. and S.G. Pueppke (1993). The chromosomal background of *Agrobacterium tumefaciens* Chry5 conditions high virulence on soybean. *Molecular Plant-Microbe Interactions* 6:601-608.

Lazzeri, P.A., D. Hildebrand, and G.B. Collins (1985). A procedure for plant regeneration from immature cotyledon tissue of soybean. *Plant Molecular Biology* 3:160-167.

Lippman, B. and G. Lippman (1984). Induction of somatic embryos in cotyledonary node tissue of soybean, *Glycine max* (L.) Merr. *Plant Cell Reports* 3:215-218.

Liu, W.N., M.J. Moore, and G.B. Collins (1992). Somatic embryogenesis in soybean via somatic embryo cycling. *In Vitro Cellular and Developmental Biology–Plant* 28P:153-160.

Mauro, A.O., T.W. Pfeiffer, and G.B. Collins (1995). Inheritance of soybean susceptibility to *Agrobacterium tumefaciens* and its relationship to transformation. *Crop Science* 35:1152-1156.

McCabe, D.E., W.F. Swain, B.J. Martinell, and P. Christou (1988). Stable transformation of soybean (*Glycine max*) by particle acceleration. *Nature Bio/Technology* 6:923-926.

Meurer, C.A., R.D. Dinkins, and G.B. Collins (1998). Factors affecting soybean cotyledonary node transformation. *Plant Cell Reports* 18:180-186.

Meurer, C.A., R.D. Dinkins, C.T. Redmond, K.P. McAllister, D.T. Tucker, D.R. Walker, W.A. Parrott, H.N. Trick, J.S. Essig, H.M. Frantz, et al. (2001). Embryogenic response of multiple soybean [*Glycine max* (L.) Merr.] cultivars across three locations. *In Vitro Cellular and Developmental Biology–Plant* 37:62-67.

Murashige, T. and E. Skoog (1962). A revised medium for rapid growth and bioassays with tobacco tissue cultures. *Physiologia Plantarum* 15:473-497.

Olhoft, P.M., L.E. Flagel, C.M. Donovan, and D.A. Somers (2003). Efficient soybean transformation using hygromycin B selection in the cotyledonary-node method. *Planta* 216:723-735.

Olhoft, P.M., K. Lin, J. Galbraith, N.C. Nielsen, and D.A. Somers (2001). The role of thiol compounds in increasing *Agrobacterium*-mediated transformation of soybean cotyledonary-node cells. *Plant Cell Reports* 20:731-737.

Olhoft, P.M. and D.A. Somers (2001). L-cysteine increases *Agrobacterium*-mediated T-DNA delivery into soybean cotyledonary-node cells. *Plant Cell Reports* 20:706-711.

Owens, L.D. and D.E. Cress (1985). Genotypic variability of soybean response to *Agrobacterium tumefaciens* strains harboring Ti or Ri plasmids. *Plant Physiology* 77:87-94.

Palanichelvam, K., P. Oger, S.J. Clough, C. Cha, A.F. Bent, and S.K. Farrand (2000). A second T-region of the soybean-super virulent chrysopine-type Ti plasmid pTiChry5, and construction of a fully disarmed vir helper plasmid. *Molecular Plant-Microbe Interactions* 13:1081-1091.

Parrott, W.A. and T.E. Clemente (2004). Transgenic soybean. In *Soybeans: Improvement, production, and uses, third ed., Agronomy Monograph 16,* H.R. Boerma and J.E. Specht (eds.). Madison WI: ASA, CSSA, and SSSA, pp. 265-302.

Parrott, W.A., L.M. Hoffman, D.F. Hildebrand, and G.B. Collins (1989). Recovery of primary transformants of soybean. *Plant Cell Reports* 7:615-617.

Parrott, W.A., E.G. Williams, D.F. Hildebrand, and G.B. Collins (1988). Effect of genotype on somatic embryogenesis from immature cotyledons of soybean. *Plant Cell Tissue and Organ Culture* 16:15-21.

Paz, M.M., J.C. Martinez, A.B. Kalvig, T.M. Fonger, and K. Wang (2006). Improved cotyledonary node method using an alterative explant derived from mature seed for efficient *Agrobacterium*-mediated soybean transformation. *Plant Cell Reports* 25:206-213.

Paz, M.M., H. Shou, Z. Guo, Z. Zhang, A.K. Banerjee, and K. Wang (2004). Assessment of conditions affecting *Agrobacterium*-mediated soybean transformation using the cotyledonary node explant. *Euphytica* 136:167-179.

Ranch, J.P., L. Oglesby, and A.C. Zielinski (1985). Plant regeneration from embryo-derived tissue cultures of soybean. *In Vitro Cellular and Developmental Biology–Plant* 21:635-658.

Rashid, H., S. Yokoi, K. Toriyama, and K. Hinata (1996). Transgenic plant production mediated by *Agrobacterium* in indica rice. *Plant Cell Reports* 15:727-730.

Samoylov, V.M., D.R. Tucker, F. Thibaud-Nissen, and W.A. Parrott (1998). A liquid medium-based protocol for rapid regeneration from embryogenic soybean cultures. *Plant Cell Reports* 18:49-54.

Schmidt, E.D.L., A.J. De Jong, and S.C. de Vries (1994). Signal molecules involved in plant embryogenesis. *Plant Molecular Biology* 26:1305-1313.

Simmonds, D.H. and P.A. Donaldson (2000). Genotype screening for proliferative embryogenesis and biolistic transformation of short-season soybean genotypes. *Plant Cell Reports* 19:485-490.

Singh, R.J., T.M. Klein, C.J. Mauvais, S. Knowlton, T. Hymowitz, and C.M. Kostow (1998). Cytological characterization of transgenic soybean. *Theoretical and Applied Genetics* 96:319-324.

Slawinska, J. and R.L. Obendorf (1991). Soybean somatic embryo maturation: Composition, respiration and water relations. *Seed Science Research* 1:624-629.

Stachel, S., E. Messens, M. Van Montagu, and P. Zambryski (1985). Identification of the signal molecules produced by wounded plant cells that activate T-DNA transfer in *Agrobacterium tumefaciens. Nature* 318:624-629.

Stewart, C.N., M.J. Adang, J.N. All, H.R. Boerma, G. Cardineau, D. Tucker, and W.A. Parrott (1996). Genetic transformation, recovery, and characterization of fertile soybean transgenic for a synthetic *Bacillus thuringiensis crylAc* gene. *Plant Physiology* 112:121-129.

Tian, L.N., D.C.W. Brown, H. Voldeng, and J. Webb (1994). In vitro response and pedigree analysis for somatic embryogenesis of long-day photoperiod adapted soybean. *Plant Cell Tissue and Organ Culture* 36:269-273.

Tomlin, E.S., S.R. Branch, D. Chamberlain, H. Gabe, M.S. Wright, and C.N. Stewart (2002). Screening of soybean, *Glycine max* (L.) Merrill, lines for somatic embryo induction and maturation capability from immature cotyledons. *In Vitro Cellular and Developmental Biology–Plant* 38:543-548.

Torisky, R.S., L. Kovacs, S. Avdiushko, J.D. Newman, A.G. Hunt, and G.B. Collins (1997). Development of a binary vector system for plant transformation based on the super virulent *Agrobacterium tumefaciens* strain Chry5. *Plant Cell Reports* 17:102-108.

Townsend, J.A. and L.A. Thomas (1994). Factors which influence the *Agrobacterium*-mediated transformation of soybean. *Journal of Cellular Biochemistry* 18A:78.

Trick, H.N. and J.J. Finer (1997). SAAT: Sonication-assisted *Agrobacterium*-mediated transformation. *Transgenic Research* 6:329-336.

Walker, D.R. and W.A. Parrott (2001). Effect of polyethylene glycol and sugar alcohols on soybean somatic embryo germination and conversion. *Plant Cell Tissue and Organ Culture* 64:55-62.

Woodragen, M.F. and H.J.M. Dons (1992). *Agrobacterium tumefaciens* mediated transformation of recalcitrant crops. *Plant Molecular Biology Reporter* 10:12-36.

Wright, M.S., S.M. Koehler, M.A. Hinchee, and M.G. Carnes (1986). Plant regeneration by organogenesis in *Glycine max*. *Plant Cell Reports* 5:150-154.

Wright, M.S., M.W. Williams, P.E. Pierson, and M.G. Carnes (1987). Initiation and propagation of *Glycine max* (L.) Merr. plants from tissue cultured epicotyls. *Plant Cell Tissue and Organ Culture* 8:83-90.

Yan, B., M.S.S. Reddy, G.B. Collins, and R.D. Dinkins (2000). *Agrobacterium* tumefaciens-mediated transformation of soybean [*Glycine max* (L.) Merrill.] using immature zygotic cotyledon explants. *Plant Cell Reports* 19:1090-1097.

Zeng, P., D.A. Vadnais, Z. Zhang, and J.C. Polacco (2004). Refined glufosinate selection in *Agrobacterium*-mediated transformation of soybean [*Glycine max* (L.) Merrill]. *Plant Cell Reports* 22:478-482.

Zhang, Z.Y., A.Q. Xing, P. Staswick, and T.E. Clemente (1999). The use of glufosinate as a selective agent in *Agrobacterium*-mediated transformation of soybean. *Plant Cell Tissue and Organ Culture* 56:37-46.

Chapter 7

# Genetic Transformation of Soybean Using Particle Bombardment and SAAT Approaches

John J. Finer
Kathryn M. Larkin

## *SOYBEAN TRANSFORMATION*

There are currently two main methods for successful and consistent transformation of soybean. Particle bombardment of proliferative embryogenic cultures (Finer and McMullen, 1991) and *Agrobacterium*-mediated transformation of cotyledonary nodes (Hinchee et al., 1988) have been studied and assimilated by most of the laboratories involved in the transformation of this important legume. Particle bombardment relies on physical introduction of DNA while *Agrobacterium* utilizes a biological vector for DNA delivery. These two systems present distinct and unique challenges, because both the target tissue and method for transgene insertion are different. Although these two transformation methods are very different, they do share similar transformation efficiencies. Additional methods exist and continue to be developed, but these two approaches remain the most often used. In this chapter, we present the theory, concepts, and methodologies developed for one of the two main soybean transformation methods, particle bombardment of embryogenic cultures, and our efforts to develop new techniques for *Agrobacterium*-mediated soybean transformation using embryogenic cultures and sonication-assisted *Agrobacterium*-mediated transformation (SAAT).

---

Salaries and research support were provided by state and federal funds appropriated to The Ohio State University/Ohio Agricultural Research and Development Center. Mention of trademark or proprietary products does not constitute a guarantee or warranty of the product by OSU/OARDC, and also does not imply approval to the exclusion of other products that may also be suitable.

## *EMBRYOGENIC CULTURES OF SOYBEAN AS A TARGET FOR GENE TRANSFER*

Although reports of embryogenesis in soybean extend back to the 1970s (Beversdorf and Bingham, 1977; Phillips and Collins, 1981), the utility of those reports was limited as embryo development was minimal and proliferative cultures were not obtained. Proliferative cultures are useful for transformation as selection pressure can be applied to targeted cultures to sort out transgenic materials from nontransformed tissues, while maintaining the tissue in a regenerative mode. In soybean, plants cannot yet be regenerated from true callus or mature leaf tissue, and young, meristematic, responsive tissue is absolutely required. Although shoots can be readily regenerated from leaves (Skoog and Miller, 1957; Murashige, 1974) or embryos generated from callus (Reinert, 1959; Ammirato, 1977) of other plants, this has not been clearly demonstrated for soybean.

The key reports to the establishment of proliferative embryogenic soybean cultures were those of Christianson et al. (1983), Ranch et al. (1985), and Finer and Nagasawa (1988). Christianson et al. (1983) were credited with developing an embryogenic suspension culture of soybean, which grew rapidly and was very responsive to manipulation. Although this work was based on recovery of only "one piece with numerous small embryoids," much was gleaned from this report. The importance of both 2,4-D and nitrogen balance as well as an excellent description of the morphology of a proliferative soybean culture were highlighted in that first work. The breakthrough work of Ranch et al. (1985) was significant as the benefits of the use of high 2,4-D levels for induction were solidly presented and many different cultivars were successfully evaluated.

Finer and Nagasawa (1988) used a combination of these procedures and others to establish proliferative embryogenic suspension cultures of soybean, which would eventually be used as a target for soybean transformation using both *Agrobacterium* (Trick and Finer, 1998) and the particle gun (Finer and McMullen, 1991). The main unique features of these key soybean embryogenesis reports are the use of very high 2,4-D levels for induction and proliferation of embryogenic cultures, the nodulated nature of the suspension cultures, and the histological characterization of the surface origin of embryos in the cultures.

Induction of embryogenesis in soybean requires the use of an auxin, usually 2,4-D, at very high levels. For other embryogenesis systems such as the model, wild carrot (*Daucus carota;* Reinert, 1959), or cereals such as wheat (Ozias-Akins and Vasil, 1982), rice (Wernicke et al., 1981), and corn (Armstrong and Green, 1985), 2,4-D was added to media at levels ranging from 0.1 to 1 mg·l$^{-1}$. For soybean, induction of embryogenesis is best with 20-40 mg·l$^{-1}$ 2,4-D, which is hundreds of times higher than the level used for these other plants. Embryogenesis can be induced in soybean using lower levels of 2,4-D (Lippmann and Lippmann, 1984; Lazzeri et al., 1985), but cultures were not proliferative and embryo development was quite abnormal. Since auxin induces embryogenesis and inhibits embryo development, these low-intermediate 2,4-D levels were sufficient to induce low numbers of embryos, but those embryos developed abnormally in the presence of 2,4-D. The hesitancy in the early years to evaluate high 2,4-D levels for induction of embryogenesis in soybean reflects an unfamiliarity with embryogenesis, an inability to interpret data (when embryo numbers increase with increasing auxin levels), and resistance to considering factors outside of the "normal" range.

The nodulated or clumpy nature of embryogenic soybean suspension cultures was also different from the embryogenesis models for the time. Finely divided suspension cultures, consisting of rapidly proliferating clumps of tissue 100-500 μm in diameter were the most common. Embryogenic soybean suspension cultures (Christianson et al., 1983; Finer and Nagasawa, 1988) consisted of 2-5 mm clumps of tissue that continually broke apart from the physical shaking of the culture. Efforts were made to modify the suspension culture growth parameters to mimic a finely divided suspension culture, but these were totally unsuccessful, and a clumpy suspension culture was finally accepted as normal for soybean.

The origin of soybean embryos in these cultures was unexpected, based on the literature (Rao and Narayanaswami, 1972), where a subepidermal origin was proposed. The term proembryonic cell masses (Haccius, 1978), which was used to describe specialized structures buried deep within the tissues, pervaded the literature at the time. The concept of a subepidermal embryo origin was not exclusively accepted (Dos Santos et al., 1983), but the idea still slowed progress. For soybean, histological analysis of proliferative cultures (Finer, 1988; Finer and Nagasawa, 1988) clearly demonstrated a surface origin of embryos, making this target suitable for transformation. This was later borne out by successful use of this tissue for transformation using both the particle gun (Finer and McMullen, 1991) and *Agrobacterium* (Trick and Finer, 1998). It is still not clear if embryo origin is epidermal or subepidermal for induction of embryogenesis from immature cotyledon explants of soybean (Hartwick et al., 1988; Santarém et al., 1997).

## DESCRIPTION OF SOYBEAN EMBRYOGENIC CULTURES

Although embryogenic soybean cultures have been extensively described in the literature (Christianson et al., 1983; Ranch et al., 1985; Finer and Nagasawa, 1988; Samoylov et al., 1998), initiation and maintenance of high-quality cultures remains problematic. The three main keys to maintaining a high-quality culture are use of an appropriate cultivar, high input maintenance with biweekly subculture, and visual selection. Other factors that are important but not critical are temperature (23-27°C), medium salt composition (10A40N, FN, or FN-lite; Finer and Nagasawa, 1988; Samoylov et al., 1998) and amino acid addition (5 mM asparagine—autoclaving is okay). Factors that are not important include photoperiod and light quality.

For those key factors influencing establishment of embryogenic cultures, cultivar selection is probably the most important. The most commonly used cultivar is Jack, which should initially be used by individuals with little experience with soybean embryogenesis. Once the proper culture morphology is identified using cv. Jack as the starting material, other cultivars can be evaluated. Embryogenic cultures of any cultivar can be established and the efficiency of initiation of embryogenesis is not always related to the ease of maintaining embryogenic cultures (Park and Finer, unpublished data). Once embryogenic cultures are initiated on the first induction medium containing 40 $mg{\cdot}l^{-1}$ 2,4-D, they are maintained on a medium containing 10-20 $mg{\cdot}l^{-1}$ 2,4-D. Biweekly subculture of established cultures is also important, with a rapid decline in culture quality if cultures are ignored. Embryogenic soybean cultures are apparently not as forgiving as other embryogenic cultures. Culture quality is most easily gauged by simple observation of the tissue. The careful observer needs to evaluate and weigh numerous qualitative characteristics for proper evaluation and selection of embryogenic materials. Those who cannot or will not attempt to carefully describe the different quality characteristics of soybean embryogenic tissues call the selection procedure "art." Observations of embryogenic soybean cultures need to be conducted under a good-quality dissecting microscope. Without magnification, most of the quality characteristics described below cannot be visualized.

The most important morphological characteristic in gauging the quality of embryogenic cultures of soybean is color. Those who have worked extensively with soybean tissue cultures describe these tissues as "healthy green," which is light green. The color is critical in the recovery of transgenic clones, where the green color of surviving, selected embryogenic tissue is easily seen. Very young (developmentally—not related to the age) proliferative embryos show little pigmentation or density. They are often clear and translucent and the embryos can be clearly seen proliferating as small globular embryos from the surface of the older, larger (but still globular in shape) embryos. Individual embryos in proliferative cultures do not advance past the globular stage. Old globular embryos can be large (up to 1 mm), light green, and opaque. Efficient proliferation of embryogenic cultures results in the production of numerous globular embryos to form a mass of embryos, resembling a small composite berry-shaped structure. With cultivars or conditions that are less favorable for embryo proliferation, fewer embryos will form and chains or branches of embryos growing from the apical surface of older embryos can be observed. With subculture, clumps of embryogenic tissue will break along weak seams, where the old internal embryos are joined (imagine concentric layering of embryos as they proliferate). Embryogenic tissues should be placed on fresh medium, while attempting to retain tissue orientation (proliferative embryo surface away from the medium). Tissues that contain nonembryogenic callus or necrotic areas should be avoided during subculture.

## *PARTICLE BOMBARDMENT*

Soybean transformation progressed rapidly after the introduction of the particle bombardment method (Sanford et al., 1987; Sanford, 1988) for the direct introduction of DNA. This success represented the maturation of both DNA introduction methodologies and tissue culture techniques. For particle bombardment, small tungsten or gold particles are coated with DNA and accelerated toward a suitable target tissue. The particles penetrate through the cell wall and end up either in or adjacent to the nucleus (Yamashita et al., 1991). The DNA then comes off the particles and integrates into the genome in a small percentage of target cells. Although most laboratories today use a commercially available particle bombardment device, the early successes in soybean transformation used homemade equipment (McCabe et al., 1988; Finer et al., 1992).

### *Bombardment of the Shoot Apex*

An electrical discharge particle bombardment device was used for the first successful production of transgenic soybean plants via particle bombardment (McCabe et al., 1988). In that first report, proliferative shoot meristematic tissues were targeted. Since the meristematic tissues were buried within the target tissue, a bombardment device (Christou et al., 1988) was specifically designed to allow deep particle penetration. With the conventional particle bombardment devices, particle penetration is mostly restricted to the first layer of cells, with occasional penetration to the second layer. For electrical discharge particle acceleration, an electrical charge is passed through a droplet of water, generating an extremely high explosive force, which is used to accelerate particles. The amount of voltage used can be varied to precisely control the extent of particle acceleration and depth of particle penetration. Although the particles were efficiently delivered to the appropriate meristematic target cells, most of the recovered plants were chimeric due to the multiple cell origin of the shoots. The progeny therefore needed to be carefully evaluated for the presence of the transgene (Christou, 1990). Methods were developed where the pattern of the chimera in the original plants (Christou and McCabe, 1992) could be reliably used to predict the events that would end up in germ line cells and subsequently inherited in the progeny. Although use of this approach for production of commercial transgenics required very high labor inputs with high throughput generation and screening of transgenics, this approach successfully yielded the event that led to the introduction of Roundup Ready soybean (Padgette et al., 1995), which has changed the face of the planet. An electrical discharge particle acceleration device was never made commercially available and this approach has not seen widespread use since its inception in the late 1980s.

### *Bombardment of Embryogenic Cultures*

Proliferative embryogenic cultures have become the target tissue of choice for soybean transformation via particle bombardment. This tissue is the most suitable as the embryogenic tissue multiplies very rapidly and the surface origin of the embryos makes them accessible to particle bombardment as well as other DNA delivery methods. The efficiency of transformation can be very high, with recovery of as many as 30 transgenic clones per bombardment. These 30 clones represent 30 independent transformation events from a single bombardment of 500 mg of embryogenic tissue. From each of these clones, many plants can be recovered through additional proliferation, development, and subsequent germination of transgenic embryos. Clone recovery averages three transgenic clones per bombardment, but the overall efficiency depends largely on the status of the target tissues and the selection scheme used. Problems reported in the early years of particle bombardment-mediated transformation of embryogenic cultures include sterility and genetic variation in transgenics (Finer and McMullen, 1991), low clone recovery, and high copy number integration of transgenes (Hadi et al., 1996). These problems have been largely controlled through rapid establishment of cultures, rapid generation of transgenics, and modifications to the transforming DNAs (Hazel et al., 1998). But after many years of advances, soybean transformation via bombardment of embryogenic cultures remains "inefficient but consistent."

As stated earlier, proliferative embryogenic soybean cultures are the main target tissue used for particle bombardment-mediated transformation of soybean. Both suspension cultures and D20 cultures, maintained on a semisolid medium, can be used. Since both targets are embryogenic in nature, they are obviously quite similar. But there are differences in the targets as well. Both target tissues need to be subcultured just prior to bombardment (3-7 days) to make certain that the cells in the target tissue are rapidly proliferating. Both targets benefit from physical drying prior to bombardment, to plasmolyze the cells and increase survival (Finer and McMullen, 1991; Vain et al., 1993). Clones from both targets are treated similarly for plant recovery, with embryo maturation, desiccation, and germination.

Generation of clones using either suspension cultures or D20 tissue is also different. Since suspension cultures grow very rapidly, they are typically more difficult to maintain. Visual selection of liquid cultures involves pouring the contents of suspension culture flasks into plates prior to observing the contents under a dissecting microscope. Individual clumps of tissue are picked up and subcultured using forceps. This increases the chance for culture contamination and tremendously increases labor input. Suspension cultures are the most rapidly growing embryogenic soybean culture system and must be subcultured frequently. The main benefit of using suspension cultures as the target is rapid clone recovery (4-6 weeks) due to more rapid growth of cultures. Clone recovery following bombardment of D20 tissues is slower (6-12 weeks), but the method seems to be more user friendly. The slower growth of the cultures allows more flexibility to the operator and rapid subculture is not as critical. Patience is needed as clone recovery can extend over 3 months. To individuals who are starting soybean transformation using bombardment of embryogenic cultures, the D20 target tissue may be more suitable.

## *A RECOMMENDED METHOD FOR SOYBEAN TRANSFORMATION VIA PARTICLE BOMBARDMENT OF EMBRYOGENIC TISSUES*

For the most current method, please see the Finer Laboratory Web site (http://www.oardc.ohio-state.edu/plantranslab/d20.htm). Although transformation of embryogenic suspension culture tissue and D20 tissue are pre-

TABLE 7.1. Media used for establishment and transformation of embryogenic soybean tissue.

| D40 | D20 | FN | M6AC |
|---|---|---|---|
| MS salts | MS salts | MS salts (-nitrates), | MS salts |
| B5 vitamins | B5 vitamins | 10 mM $NH_4NO_3$, | B5 vitamins |
| 3% sucrose | 3% sucrose | 30 mM $KNO_3$, | 6% maltose |
| 40 mg·l$^{-1}$ 2,4-D | 5·mM asparagine | B5 vitamins, | pH 5.7 |
| pH 7.0 | 20 mg·l$^{-1}$ 2,4-D | 3% sucrose, | 0.2% Gelrite |
| 0.2% Gelrite | pH 5.7 | 5 mM asparagine | 0.5% activated charcoal |
| | 0.2% Gelrite | 5 mg·l$^{-1}$ 2,4-D, | |
| | | pH 5.7 | |

sented separately above, the current method of choice utilizes a D20 target with a hybrid liquid-solid selection. This method is briefly outlined below.

### *Initiation of Proliferative Tissue from Immature Cotyledons*

- Select pods containing zygotic embryos, 2-3 weeks after anthesis. Immature embryos are 3-4 mm in length.
- Sterilize pods and remove the immature embryos. Discard the embryo axis. Plate immature cotyledons on D40 medium (Table 7.1) for 3-4 weeks at 25°C.

### *Proliferation of D20 Tissue*

- Transfer to D20 medium (Table 7.1). Select bright green, globular, proliferative embryos.
- Subculture every 2-3 weeks under conditions described for induction.

### *Bombardment and Selection*

- Subculture tissue (20-25 clumps/plate) 3-5 days prior to bombardment. Place tissue in the center of dish (containing medium) and dry by uncovering tissue for 15 minutes in the hood.
- Bombard using the Particle Inflow Gun (Finer et al., 1992) or another bombardment apparatus.
- For hybrid liquid-solid selection, place tissue in liquid culture (FN with 30 mg·l$^{-1}$ hygromycin, Table 7.1). Replace medium every week for 4-6 weeks. Remove hygromycin-resistant tissue and place on solid

D20 medium containing 15 mg·l$^{-1}$ hygromycin. Liquid selection reduces the selection period from 7-17 weeks to 4-6 weeks.
- Maintain on D20 Hyg 15. Keep track of growing sections. When there is enough tissue, extract DNA for PCR analysis.

### *Development and Maturation*

- Place clumps on M6AC (Table 7.1). From 6 to 9 clumps should be placed in a tall dish (100 × 25 mm) at 23°C. After 3-4 weeks, embryos are ready for desiccation.

### *Desiccation and Germination*

- Place 16-25 embryos in a dry 100 × 15 mm Petri dish, seal with Parafilm, and maintain at 25°C for 2-5 days.
- Germination should occur within one week after transfer to OMS (growth regulator-free MS medium).
- Place plants in soil once there is good root and shoot formation. Rinse OMS from the roots with water, to avoid fungal growth on residual medium on the roots. Keep under 24 hour lights, high humidity.
- To promote flowering and pod formation, plants can be placed directly under 16 hour days. To keep plants vegetative, maintain plants under a long day photoperiod (18-24 hour daylength) until they are 2 feet tall and then transfer to 16 hour or shorter days.

## AGROBACTERIUM-*MEDIATED TRANSFORMATION*

*Agrobacterium*-mediated transformation is a DNA introduction method utilizing the biological vector *Agrobacterium tumefaciens. A. tumefaciens* is a naturally occurring soilborne plant pathogen that has the unique ability to transfer a portion of its DNA, known as T-DNA (transferred DNA), to the genome of the plant. *Agrobacterium* was the first vector successfully used for introducing genes into the genome of plant cells (Horsch et al., 1985) and has continued to be the most commonly used method for DNA introduction in plants (Riva et al., 1998). Interestingly, the first report of *Agrobacterium*-mediated gene transfer in soybean (Hinchee et al., 1988) accompanied the first report of particle bombardment-mediated transformation of soybean (McCabe et al., 1988).

Despite this early report of transformation via *Agrobacterium,* soybean continues to be a somewhat difficult target for *Agrobacterium*-mediated transformation. Success with gene introduction via *Agrobacterium* in soybean has been reported using a variety of different tissues including seeds (Chee et al., 1989), somatic embryos (Parrott et al., 1989; Trick and Finer, 1998), cotyledonary nodes (Hinchee et al., 1988), and the embryonic shoot tip (Liu et al., 2004). However, the effort required for recovery of soybean transgenics using *Agrobacterium* is comparable to the effort required for particle bombardment-mediated transformation and the whole process remains inefficient.

For the process of DNA transfer and integration to occur with *Agrobacterium,* the DNA must be introduced into cells that are both susceptible to *Agrobacterium* and responsive to plant regeneration (Finer et al., 1996). Tissue type, age, genotype, and susceptibility to *Agrobacterium* all play a role in the effectiveness of bacterial infection. The T-DNA transfer process is controlled by *virulence* or *vir* genes. Induction of *vir* genes is dependent upon various factors that include the presence of phenolic compounds, media pH, and temperature (Godwin et al., 1991; Statchel et al., 1985; Townsend, 1993). Wounded plant tissues produce phenolic compounds such as acetosyringone, which induces *vir* genes (Statchel et al., 1985). However, some soybean genotypes may not produce enough acetosyringone to induce the *vir* genes (Godwin et al., 1991); therefore, to enhance DNA transfer, tissue culture media are often supplemented with this compound. Other options for enhancing transformation include using mutant bacteria that constitutively express their *vir* genes (Hansen et al., 1994).

The main method for *Agrobacterium*-mediated transformation of soybean (Hinchee et al., 1988) uses cotyledonary nodes as the target tissue. For this "cot-node" system, cotyledonary nodes were inoculated with *Agrobacterium* and induced to form shoots on a medium supplemented with cytokinin. The shoot initials, which are targeted for this method, are multicellular and buried deep within the tissue. Precision wounding of the cotyledonary nodes is therefore required to allow the *Agrobacterium* to reach the targeted cells. With the cot-node system, shoot meristems can be difficult to reach and transformation can result in chimeras. Improvements in the cot-node transformation system have been reported with the use of alternate selection agents (Zhang et al., 1999; Olhoft et al., 2003) and the addition of reducing agents to minimize pathogen-induced stress responses (Olhoft et al., 2001; Olhoft and Somers, 2001).

## *SONICATION-ASSISTED* AGROBACTERIUM-*MEDIATED TRANSFORMATION*

One of the main limiting factors for successful use of *Agrobacterium* for plant transformation is the ability to deliver the bacterium to the most appropriate target tissue. To enhance bacterial delivery, a method was developed that uses sonication for microwounding of the target tissue to provide multiple entry points for the bacterium (Trick and Finer, 1997). SAAT has been used for transforming various tissue types and plant species that are otherwise recalcitrant to transformation (Trick and Finer, 1997, 1999; Tang, 2003; Zaragozaa et al., 2004). For SAAT, plant tissue is sonicated in the presence of *Agrobacterium* for short periods of time. Ultrasound waves cause microwounds to form on the surface and deep within the plant tissue. Wounding due to sonication creates an entry point for the bacteria and may stimulate the production of signaling molecules (acetosyringone) involved in the T-DNA transfer process.

SAAT has been used successfully on both proliferative embryogenic suspension culture tissue (Trick and Finer, 1998) and proliferative D20 tissues (Larkin and Finer, unpublished data) of soybean. Concerns relating to target tissue use that were outlined previously for particle bombardment-mediated transformation also apply to use of SAAT. D20 tissues respond more slowly to manipulation and are a little more forgiving of neglect. Embryogenic suspension cultures are more challenging to develop but respond more quickly. It may also be easier to rid liquid cultures of bacteria following cocultivation of the bacteria with plant materials. Since tissues and cells are bathed in the liquid medium, use of antibiotics in the medium to eliminate the bacteria is more efficient.

## *SAAT OPTIMIZATION STRATEGIES*

Optimization of soybean transformation using SAAT requires evaluation of two different biological systems (*Agrobacterium* and proliferative embryogenic cultures) along with the physical effects of sonication, which need to be defined for each sonicator and target tissue. Under the proper conditions, sonication results in the production of microwounds on and within the target tissue. Ideally, wounds are small and deep, resulting in the production of nicks, small channels, and tunnels for bacterial entry. Insufficient sonication will yield minimal wounding and no enhancement in *Agrobacterium* introduction while aggressive sonication will result in extensive tissue disruption, which is undesirable. The goal of SAAT is to obtain a high level of bacterial infection, which increases the chances of stable transformation events without jeopardizing tissue health.

The microwounding observed in sonicated tissue is caused by cavitation, which varies depending on sonicator strength, duration of treatment, and resistance of target tissue to physical wounding. During cavitation, small bubbles form that implode, resulting in a localized shock wave. The shock wave can be sufficiently strong to cause mechanical damage (Suslick, 1988). Microwounds created by sonication can cover the surface of the treated tissue and range from 1 μm to 1 mm in size (Trick and Finer, 1998). The wounds on the surface of the plant material are clearly large enough for *Agrobacterium* to invade the wounded cells or tissues. Sonication of the plant tissue with *Agrobacterium,* as opposed to sonication of tissue before the application of *Agrobacterium,*, does not alter transformation efficiency. For ease of handling, sonication should be performed with *Agrobacterium.*

For optimization of SAAT, the following variables should be considered: type of sonicator, container used to hold target tissue, duration of sonication, length of coculture period, optical density of the bacterial suspension, antioxidant addition to the coculture medium, and temperature.

### *Sonicator Type*

Many different types of sonicators exist and efforts to evaluate them all would be extremely difficult. Bath sonicators are inexpensive, very common, and provide sufficient energy to cause microwounding. Probe sonicators are also effective but they tend to be high energy and very disruptive to the target tissue. Bath sonicators, with transducers (ultrasound speakers) mounted on the bottom and side of the bath, are preferred. Large industrial bath sonicators are not necessary. Care must be taken to maintain constant water levels in the bath and to place tissues in the same place within the bath for all experiments. The variation in ultrasound output across a bath can be visualized by placing a small sheet of thin aluminum foil in the bath directly and exposing the sheet to the output. Patterns of pitting of the sheet will indicate areas of high versus low sonic wave output.

### *Container Type*

Evaluation of various containers used to hold sonicated tissues is immensely important. Since the target tissues are never placed directly in the water bath, the containers need to be able to transmit the ultrasound through the container wall in a consistent manner. The results in Tables 7.2 to 7.4 represent results from SAAT of D20 soybean tissues and evaluation of tran-

TABLE 7.2. Effect of container type on DNA introduction used during SAAT in embryogenic soybean tissue.

| Container type | Number of GFP foci |
|---|---|
| 15 ml Falcon tube | 15.8 ± 3.8[ab] |
| 30 ml glass beaker | 4.8 ± 1.7[bc] |
| Microcentrifuge tube | 1.9 ± 1.0[c] |
| 13 mm × 100 mm borosilicate glass tube | 18.8 ± 3.8[a] |

*Notes:* Each value ± standard error represents the mean of three replicate samples with each replicate consisting of 9-12 pieces of D20 tissue. Means followed by the same letter within columns are not significantly different at the 0.05 probability level.

sient expression of an introduced *green fluorescent protein* gene (Ponappa et al., 1999). Tissues were initially placed in 1 ml of *Agrobacterium* suspension for 1 to 60 seconds. Following SAAT treatment, the tissue was blotted on sterile filter paper to remove excess *Agrobacterium* and plated on D20 medium containing 100 μM acetosyringone. Two days after coculture, tissue was placed on D20 media containing Timentin (400 $mg \cdot l^{-1}$) to suppress *Agrobacterium* growth. Gene expression was quantified after an additional 2 days, or a total of 4 days post SAAT. Table 7.2 presents results from evaluation of the following four different containers: 13 × 100 mm borosilicate glass tube, 30 ml glass beaker, Falcon 15 ml polypropylene round-bottom tube, and a 1.5 ml microcentrifuge tube.

Although ultrasound is reported to travel through harder material with higher efficiency, the shape and size of the container appeared to have greater influence on successful gene introduction than the composition. The best results were obtained with similarly sized polypropylene and glass containers, while the larger and thicker glass beaker and the smaller and thinner polypropylene microcentrifuge tube did not work very effectively.

### *Duration of Sonication*

Sonication duration has a strong effect on tissue damage and the efficiency of *Agrobacterium* introduction. Surprisingly, sonication (at the duration and levels used for SAAT) did not negatively affect *Agrobacterium* survival (Finer and Finer, 2000). The duration of the sonication treatment, which gives the optimum gene introduction, varies from 5 seconds (Larkin and Finer, unpublished data) to 10-60 seconds (Trick and Finer, 1998; Santarém et al., 1998). With softer tissues, tissue damage is more severe, and tissue survival can decline with these extended treatment durations. The extent of the damage can be easily and rapidly gauged by visual observation of pigmentation or cloudiness in the medium used to suspend the tissue for sonication. If the medium is pigmented or cloudy, tissue disruption is most likely severe. As SAAT causes microwounding and tissue disruption, it is important to optimize the duration of sonication to give some tissue wounding while not severely damaging the tissue.

### *Coculture Period Following SAAT*

For SAAT of soybean suspension culture tissue (Trick and Finer, 1998), a 2-day coculture using SAAT was optimum for transformation of embryogenic suspension cultures of soybean and for minimizing bacterial overgrowth. With SAAT of D20 tissue (Larkin and Finer, unpublished data), bacterial overgrowth on some pieces of tissue was extreme with a 2-day coculture and may have been detrimental to tissue survival. A 1-day coculture minimizes bacterial overgrowth, but DNA introduction declines. Apparently, a 1-day coculture period is insufficient to allow for T-DNA transfer into the plant nucleus. The best way to manage bacterial overgrowth from SAAT of D20 tissue may be to develop a repeated liquid wash of infected tissues with antibiotic-containing liquid medium (Finer and Larkin, unpublished data). A liquid wash of tissue has been used in *Agrobacterium*-mediated protocols for leaf explants of apple (De Bondt et al., 1994), somatic embryos of papaya (Fitch et al., 1993), and cotyledons of sunflower (Sankara Rao and Rohini, 1999).

### *Antioxidant Treatments*

Incorporation of antioxidants into the coculture medium has been very successfully used for enhancing *Agrobacterium*-mediated transformation of soybean using the cotyledonary node method (Olhoft et al., 2001) as well as other plants (Enríquez-Obregón et al., 1997, 1999). For SAAT of soybean embryogenic tissues, the following antioxidants were evaluated: 45 $\mu M$ $AgNO_3$; 2 $mg{\cdot}l^{-1}$ DTT; 100, 250, or 400 $mg{\cdot}l^{-1}$ L-cysteine; 100 $mg{\cdot}l^{-1}$ ascorbic acid; and 50 $mg{\cdot}l^{-1}$ citric acid (Table 7.3). Use of silver nitrate and DTT led to the lowest levels of GFP transient expression while use of citric acid gave higher delivery and expression of introduced DNA using SAAT.

Additional evaluation of 250 $mg{\cdot}l^{-1}$ L-cysteine and 50 $mg{\cdot}l^{-1}$ citric acid gave very high DNA introduction and GFP transient expression (Larkin and Finer, unpublished data). Because treatment with antioxidants during coculture seemed to have a beneficial effect on tissue health, this antioxidant

TABLE 7.3. Effect of various antioxidants during culture and on DNA introduction and expression in embryogenic soybean tissue following SAAT.

| Treatment | Number of GFP foci |
|---|---|
| Silver nitrate (45 μM) | 2.4 ± 1.3[b] |
| DTT (2 mg·l$^{-1}$) | 0.0 ± 0.0[b] |
| Cysteine (100 mg·l$^{-1}$) | 6.7 ± 2.0[b] |
| Ascorbic acid (100 mg·l$^{-1}$) | 7.1 ± 3.4[b] |
| Citric acid (50 mg·l$^{-1}$) | 35.4 ± 7.4[a] |
| No antioxidant | 12.8 ± 3.1[b] |

*Notes:* Each value ± standard error represents the mean of three replicate samples with each replicate consisting of 10 pieces of D20 tissue. Means followed by the same letter are not significantly different at the 0.05 probability level.

may lessen necrosis of the tissue subjected to SAAT and therefore promote cell division and proliferation. It is probable that a cell containing a stable integration event will more likely proliferate and develop into a cluster of transformed cells if the cells are actively dividing. Cysteine concentrations between 250 and 400 mg·l$^{-1}$ are recommended during coculture to improve tissue health. In addition, a mixture of citric acid and cysteine during coculture may result in even higher levels of transformation.

## *Optical Density of Bacterial Suspension*

The optical density of the initial bacterial suspension may play an important role in SAAT of embryogenic soybean tissues. Four optical densities were examined for their effect on successful DNA introduction, resulting in transient GFP expression (Table 7.4). An $OD_{600nm}$ at 0.3 gave the highest level of transient GFP expression and there was a significant difference between $OD_{600nm}$ at 0.3 and all other treatments.

## *Future of Particle Bombardment and SAAT of Soybean Embryogenic Tissues*

Embryogenic soybean tissue maintained in liquid culture and on a semisolid medium (D20) are clearly suitable for particle bombardment-mediated transformation and SAAT. Each of these different target tissues combined with the different methods of transformation has strengths and weaknesses.

TABLE 7.4. Effect of $OD_{600nm}$ on DNA introduction in embryogenic soybean tissue following SAAT.

| **Optical density of bacteria at 600 nm** | **Number of GFP foci** |
|---|---|
| 0.30 | 26.8 ± 6.1[a] |
| 0.10 | 9.5 ± 3.1[b] |
| 0.03 | 2.4 ± 1.0[b] |
| 0.01 | 9.9 ± 3.5[b] |

*Notes:* Each value ± standard error represents the mean of three replicate samples with each replicate consisting of 10 pieces of D20 tissue. Means followed by the same letter are not significantly different at the 0.05 probability level.

Embryogenic suspension cultures are more difficult to initiate and maintain but proliferate quickly and are quite responsive to manipulation. D20 tissue grows more slowly but is easier to initiate and maintain. Since particle bombardment is a physical method for DNA introduction, potential biological incompatibilities from use of *Agrobacterium* with some soybean lines are avoided. The complex DNA integration patterns resulting from particle bombardment-mediated transformation (Hadi et al., 1996; Kohli et al., 1999) are no longer problematic and can be minimized with the use of gene cassettes or fragments and lower concentrations of DNA. Delivery of particles to target cells buried deep within the tissue is still difficult, but this is not a concern for proliferative embryogenic tissue of soybean. The use of *Agrobacterium* for transformation is generally less expensive as it does not require specialized equipment such as a particle acceleration device. Removal of the bacteria post cocultivation can be frustrating, but tissue necrosis resulting from bacterial infection has been controlled to some extent by the use of reducing agents (Olhoft et al., 2001; Larkin and Finer, unpublished data).

It is intriguing that particle bombardment- and *Agrobacterium*-mediated transformation of soybean was first reported in 1988 and, although good progress has been made, transformation of this crop remains inefficient (but consistent). Transformation efficiencies for most other crops have improved tremendously, so that many laboratories can introduce DNAs of interest into many crops. Improvements in soybean transformation have been reported, but these have not been major breakthroughs. The potential of SAAT for soybean transformation is enormous but the acceptance and application of this technology has been slow. It is hoped that, with sustained efforts, soybean transformation efficiencies will continue to improve so that many different laboratories can be involved with DNA introduction into this crop.

## REFERENCES

Ammirato, P.V. (1977). Hormonal control of somatic embryo development from cultured cells of caraway; interactions of abscisic acid, zeatin, and gibberellic acid. *Plant Physiology* 59:579-586.

Armstrong, C.L. and C.E. Green (1985). Establishment and maintenance of friable, embryogenic maize callus and the involvement of L-proline. *Planta* 164:207-214.

Beversdorf, W.D. and E.T. Bingham (1977). Degrees of differentiation obtained in tissue cultures of *Glycine* species. *Crop Science* 17:307-311.

Chee, P.P., K.A. Fober, and J.L. Slightom (1989). Transformation of soybean (*Glycine max*) by infecting germinating seeds with *Agrobacterium tumefaciens. Plant Physiology* 91:1212-1218.

Christianson, M.L., D.A. Warnick, and P.S. Carlson (1983). A morphogenetically competent soybean suspension culture. *Science* 222:632-634.

Christou, P. (1990). Morphological description of transgenic soybean chimeras created by the delivery, integration and expression of foreign DNA using electric discharge particle acceleration. *Annals of Botany* 66:379-386.

Christou, P. and D.E. McCabe (1992). Prediction of germ-line transformation events in chimeric R0 transgenic soybean plantlets using tissue-specific expression patterns. *Plant Journal* 2:283-290.

Christou, P., D.E. McCabe, and W.F. Swain (1988). Stable transformation of soybean callus by DNA coated particles. *Plant Physiology* 87:671-674.

De Bondt, A., K. Eggermont, P. Druart, M. De Vil, I. Goderis, J. Vanderleyden, and W.F. Broekaert (1994). *Agrobacterium*-mediated transformation of apple (*Malus* x *domestica* Borkh.): An assessment of factors affecting gene transfer efficiency during early transformation steps. *Plant Cell Reports* 13:587-593.

Dos Santos, A.V.P., E.G. Cutter, and M.R. Davey (1983). Origin and development of somatic embryos in *Medicago sativa* L. (alfalfa). *Protoplasma* 117:107-115.

Enríquez-Obregón, G.A., D.L. Prieto-Samsonov, G.A. de la Riva, M. Perez, G. Selman-Housein, and R.I. Vazquez-Padron (1999). *Agrobacterium*-mediated Japonica rice transformation: A procedure assisted by an antinecrotic treatment. *Plant Cell Tissue and Organ Culture* 59:159-168.

Enríquez-Obregón, G.A., R.I. Vázquez-Padrón, D.L. Prieto-Samsnov, M. Pérez, and G. Selman-Housein (1997). Genetic transformation of sugarcane by *Agrobacterium tumefaciens* using antioxidant compounds. *Biotecnologia Aplicada* 14:169-174.

Finer, J.J. (1988). Apical proliferation of embryogenic tissue of soybean [*Glycine max* (L.) Merrill]. *Plant Cell Reports* 7:238-241.

Finer, J.J., T.S. Cheng, and D.P.S. Verma (1996). Soybean transformation: Technologies and progress. In *Soybean biotechnology,* D.P.S Verma and R. Shoemaker (eds.). Wallingford: CAB International, pp. 249-262.

Finer, J.J. and M.D. McMullen (1991). Transformation of soybean via particle bombardment of embryogenic suspension culture tissue. *In Vitro Cellular and Developmental Biology* 27:115-182.

Finer, J.J. and A. Nagasawa (1988). Development of an embryogenic suspension culture of soybean (*Glycine max* Merrill). *Plant Cell Tissue and Organ Culture* 15:125-136.

Finer, J.J., P. Vain, M.W. Jones, and M.D. McMullen (1992). Development of the particle inflow gun for DNA delivery to plant cells. *Plant Cell Reports* 11:232-238.

Finer, K.R. and J.J. Finer (2000). Use of *Agrobacterium* expressing green fluorescent protein to evaluate colonization of sonication-assisted *Agrobacterium*-mediated transformation-treated soybean cotyledons. *Letters in Applied Microbiology* 30:406-410.

Fitch, M.M.M., R.M. Manshardt, D. Gonsalves, and J.L. Slightom (1993). Transgenic papaya plants from *Agrobacterium*-mediated transformation of somatic embryos. *Plant Cell Reports* 12:245-249.

Godwin, I., G. Todd, B. Ford-Lloyd, and H.J. Newbury (1991). The effects of acetosyringone and pH on *Agrobacterium*-mediated transformation vary according to plant species. *Plant Cell Reports* 9:671-675.

Haccius, B. (1978). Question of unicellular origin of non-zygotic embryos in callus culture. *Phytomorphology* 28:74-81.

Hadi, M.Z., M.D. McMullen, and J.J. Finer (1996). Transformation of 12 different plasmids into soybean via particle bombardment. *Plant Cell Reports* 15:500-505.

Hansen, G., A. Das, and M.D. Chilton (1994). Constitutive expression of the virulence genes improves the efficiency of plant transformation by *Agrobacterium. Proceedings of the National Academy of Sciences USA* 91:7603-7607.

Hartwick, I.M., P.A. Lazzeri, D. Cui, G.B. Collins, and E.G. Williams (1988). Auxin-orientation effects on somatic embryogenesis from immature soybean cotyledons. In *Vitro Cellular and Developmental Biology–Plant* 24:821-828.

Hazel, C.B., T.M. Klein, M. Anis, H.D. Wilde, and W.A. Parrott (1998). Growth characteristics and transformability of soybean embryogenic cultures. *Plant Cell Reports* 17:765-772.

Hinchee, M.A.W., D.V. Connor-Ward, C.A. Newell, R.E. McDonnell, S.J. Sato, C.S. Gasser, D.A. Fischhoff, D.B. Re, R.T. Fraley, and R.B. Horsch (1988). Production of transgenic soybean plants using *Agrobacterium*-mediated gene transfer. *Nature Bio/Technology* 6:915-922.

Horsch, R.B., J.E. Fry, N.L. Hoffmann, D. Eichholtz, S.G. Rogers, and R.T. Fraley (1985). A simple and general method for transferring genes into plants. *Science* 227:1229-1231.

Kohli, A., S. Griffiths, N. Palacios, R.M. Twyman, P. Vain, D.A. Laurie, and P. Christou (1999). Molecular characterization of transforming plasmid rearrangements in transgenic rice reveals a recombination hotspot in the CaMV 35S promoter and confirms the predominance of microhomology mediated recombination. *Plant Journal* 17:591-601.

Lazzeri, P.A., D.F. Hildebrand, and G.B. Collins (1985). A procedure for plant regeneration from immature cotyledon tissue of soybean. *Plant Molecular Biology Reporter* 3:160-167.

Lippmann, B. and G. Lippmann (1984). Induction of somatic embryos in cotyledonary tissue of soybean, *Glycine max* (L.) Merr. *Plant Cell Reports* 3:215-218.

Liu, H.-K., C. Yang, and Z.-H. Wei (2004). Efficient *Agrobacterium tumefaciens*-mediated transformation of soybeans using an embryonic tip regeneration system. *Planta* 219:1042-1049.

McCabe, D.E., W.F. Swain, B.J. Martinell, and P. Christou (1988). Stable transformation of soybean (*Glycine max*) by particle acceleration. *Nature Bio/Technology* 6:923-926.

Murashige, T. (1974). Plant propagation through tissue cultures. *Annual Reviews in Plant Physiology* 25:135-165.

Olhoft, P.M., E.L. Flagel, C.M. Donovan, and D.A. Somers (2003). Efficient soybean transformation using hygromycin B selection in the cotyledonary-node method. *Planta* 216:723-735.

Olhoft, P.M., K. Lin, J. Galbraith, N.C. Nielsen, and D.A. Somers (2001). The role of thiol compounds in increasing *Agrobacterium*-mediated transformation of soybean cotyledonary-node cells. *Plant Cell Reports* 20:731-737.

Olhoft, P.M. and D.A. Somers (2001). L-cysteine increases *Agrobacterium*-mediated T-DNA delivery into soybean cotyledonary-node cells. *Plant Cell Reports* 20:706-711.

Ozias-Akins, P. and I.K. Vasil (1982). Plant regeneration from cultured immature embryos and inflorescences of *Triticum aestivum* L. (wheat): Evidence for somatic embryogenesis. *Protoplasma* 110:95-105.

Padgette, S.R., K.H. Kolacz, X. Delannay, D.B. Re, B.J. LaVallee, C.N. Tinius, W.K. Rhodes, Y.I. Otero, G.F. Barry, D.A. Eichholz, et al. (1995). Development, identification, and characterization of a glyphosate-tolerant soybean line. *Crop Science* 35:1451-1461.

Parrott, W.A., L.M. Hoffman, D.F. Hildebrand, E.G. Williams, and G.B. Collins (1989). Recovery of primary transformants of soybean. *Plant Cell Reports* 7:615-617.

Phillips, G.C. and G.B. Collins (1981). Induction and development of somatic embryos from cells suspension cultures of soybean. *Plant Cell Tissue and Organ Culture* 1:123-129.

Ponappa, T., A.E. Brzozowski, and J.J. Finer (1999). Transient expression and stable transformation of soybean using the jellyfish green fluorescent protein. *Plant Cell Reports* 19:6-12.

Ranch, J.P., L. Oglesby, and A.C. Zielinski (1985). Plant regeneration from embryo-derived tissue cultures of soybean. *In Vitro Cellular and Developmental Biology–Plant* 21:653-658.

Rao, P.S. and S. Narayanaswami (1972). Morphogenetic investigations in callus cultures of *Tylophora indica. Physiologia Plantarum* 27:271-276.

Reinert, J. (1959). Uber die kontrolle der Morphogenese und die induktion von Adventiveembryonen an Gewebenkulturen aus Karotten (Control of morphogenesis and induction of somatic embryogenesis in tissue cultures of carrot). *Planta* 53:318-333.

Riva, G.A., J. González-Cabrera, R. Vázquez-Padrón, and C. Ayra-Pardo (1998). *Agrobacterium tumefaciens*: A natural tool for plant transformation. *Nature Bio/Technology* 1(3).

Samoylov, V.M., D.M. Tucker, and W.A. Parrott (1998). Soybean [*Glycine max* (L.) Merrill] embryogenic cultures: the role of sucrose and total nitrogen content on proliferation. *In Vitro Cellular and Developmental Biology–Plant* 34:8-13.

Sanford, J.C. (1988). The biolistic process. *Trends in Biotechnology* 6:299-302.

Sanford, J.C., T.M. Klein, E.D. Wolf, and N. Allen (1987). Delivery of substances into cells and tissue using a particle bombardment process. *Particulate Science and Technology* 5:27-37.

Sankara Rao, K. and V.K. Rohini (1999). *Agrobacterium*-mediated transformation of sunflower (*Helianthus annuus* L.): A simple protocol. *Annals of Botany* 83:347-354.

Santarém, E.R., B. Pelissier, and J.J. Finer (1997). Effect of explant orientation, pH, solidifying agent and wounding on initiation of soybean somatic embryos. *In Vitro Cellular and Developmental Biology–Plant* 33:13-19.

Santarém, E.R., H.N. Trick, J.S. Essig, and J.J. Finer (1998). Sonication-assisted *Agrobacterium*-mediated transformation of soybean immature cotyledons: Optimization of transient expression. *Plant Cell Reports* 17:752-759.

Skoog, F. and C.O. Miller (1957). Chemical regulation of growth and organ formation in plant tissues cultivated in vitro. *Biological Action of Plant Growth Substances. 11th Symposium of Society for Experimental Biology* 11:118-131.

Statchel, S.E., E. Messens, M.V. Montagu, and P. Zambryski (1985). Identification of the signal molecules produced by wounded plant cells that activate T-DNA transfer in *Agrobacterium tumefaciens. Nature* 318:624-629.

Suslick, K. (1988). Homogenous sonochemistry. In *Ultrasound: Chemical, physical, and biological effects*, K. Suslick (ed.). Weinheim: VCH Publishers, pp. 123-161.

Tang, W. (2003). Additional virulence genes and sonication enhance *Agrobacterium tumefaciens*-mediated loblolly pine transformation. *Plant Cell Reports* 21:555-562.

Townsend, J.A. (1993). An improved method of *Agrobacterium*-mediated transformation of cultured soybean cells. US Patent WO 94/02620.

Trick, H.N. and J.J. Finer (1997). SAAT: Sonication-assisted *Agrobacterium*-mediated transformation. *Transgenic Research* 6:329-336.

Trick, H.N. and J.J. Finer (1998). Sonication-assisted *Agrobacterium*-mediated transformation of soybean [*Glycine max* (L.) Merrill] embryogenic suspension culture tissue. *Plant Cell Reports* 17:482-488.

Trick, H.N. and J.J. Finer (1999). Induction of somatic embryogenesis and genetic transformation of Ohio buckeye (*Aesculus glabra* Willd) *In Vitro Cellular and Developmental Biology–Plant* 35:57-60.

Vain, P., M.D. McMullen, and J.J. Finer (1993). Osmotic treatment enhances particle bombardment-mediated transient and stable transformation of maize. *Plant Cell Reports* 12:84-88.

Wernicke, W., R. Brettel, T. Wakizuka, and I. Potrykus (1981). Adventitious embryoid and root formation from rice leaves. *Zeitschrift für Pflanzenphysiologie* 103:361-365.

Yamashita, T., A. Iida, and H. Morikawa (1991). Evidence that more than 90% of β-glucuronidase-expressing cells after particle bombardment directly receive the foreign gene in their nucleus. *Plant Physiology* 97:829-831.

Zaragozaa, C., J. Munoz-Bertomeu, and I. Arrillaga (2004). Regeneration of herbicide-tolerant black locust transgenic plants by SAAT. *Plant Cell Reports* 22:832-838.

Zhang, Z., A. Xing, P. Staswick, and T.E. Clemente (1999). The use of glufosinate as a selective agent in *Agrobacterium*-mediated transformation of soybean. *Plant Cell Tissue and Organ Culture* 56:37-46.

Chapter 8

# Genetic Transformation as a Tool for Improvement of Pigeon Pea, *Cajanus Cajan* (L.) Millsp.

G. Lakshmi Sita
P. Venkatachalam

## *INTRODUCTION*

Pigeon pea, *Cajanus cajan* (L.) Millsp. is an important high-protein grain legume of the semiarid tropics and subtropics, and caters to the protein requirements of the majority of the population in the Indian subcontinent. Its protein content is about 22.3 percent and the seeds contain two globulins, cajanin and concojanin, accounting for 58 percent and 8 percent of the total nitrogen respectively, differing from each other in their sulfur and tryptophan content. The globulins, which form the chief proteins of the seed, appear to be characteristic of the genus. They are rich in tyrosine and moderately rich in cysteine, arginine, and lysine. The improvement of pigeon pea with regard to pests, lowering allergenic protein levels in seeds, and improving the quality of protein is therefore desirable. The use of transformation techniques for producing new breeding materials that would not be available in the germplasm among cross-compatible species holds great potential for attaining this goal. Genetic enhancement of this crop through biotechnological methods for quality improvement with desirable genes has not yet been achieved.

## *MAJOR CONSTRAINTS FOR THE CROP*

The most important yield constraint in pigeon pea cultivation is from the Lepidopteran pest *Helicoverpa armigera*. Another significant insect pest is the pod fly (*Melangromyza obtuse*). Fusarium wilt (caused by *Fusarium udum*), phytophthora stem blight (caused by *Phytophthora drechsleri* f.p. *cajani*) and sterility mosaic disease were identified as other major constraints on pigeon pea yield. Bacterial diseases like leaf spot and stem canker caused by *Xanthomonas cajani* and pigeon pea mosaic virus also cause damage to the crop. Development of extra early varieties and resistance to drought and water logging were identified as important breeding targets. In addition, improvement of seed quality is also much desired. Genetic engineering techniques combined with conventional breeding programs enhance both yield and nutritive value. However, genetic engineering of pigeon pea is still in its infancy, owing mainly to the difficulties with regeneration in tissue culture. For successful genetic engineering of any crop species, including pigeon pea, reproducible tissue culture protocols coupled with appropriate genes and the technology for their transfer to the recipient genome are two important prerequisites. Although pigeon pea is a recalcitrant crop, plant regeneration has been reported, but only a few protocols could be successfully utilized in genetic transformation (Geetha et al., 1999). Subsequently, other reports appeared (Lawrence and Koundal, 2001; Satyavathi et al., 2003; Dayal et al., 2003; Thu et al., 2003; Mohan and Krishnamurthy, 2003; Prasad et al., 2004) demonstrating that the crop is not as recalcitrant as presumed earlier and useful genes could be successfully introduced and expressed. The crop seems more amenable to genetic transformation and regeneration with the available methodology. We discuss the problems hindering pigeon pea improvement in general and available information on biotechnological methodologies developed so far toward reaching the goals of successful genetic transformation.

## *SCOPE FOR BIOTECHNOLOGICAL IMPROVEMENT*

A detailed survey has been done to identify crop improvement objectives in 12 important field crops, including pigeon pea, that are grown extensively in India (Grover and Pental, 2003). Need for transgenics was identified as one of the six objectives. Among these, need for (a) nutritional enhancement, (b) resistance to biotic and abiotic stresses, (c) herbicide resistance, and (d) value addition were identified as most urgent. Assessment revealed the priorities in terms of problems. The first observation was that the pigeon pea has not undergone any significant improvement in the past three decades. A number of biotic and abiotic stresses are severely affecting full realization of the yield potential. There is a need to increase productivity and enhance the nutritional value of this pulse crop. Although a large number of lines have been identified for resistance to fusarium wilt, sterility mosaic, and phytophthora stem blight, resistance to insect pests has been only partial and germplasm with absolute resistance is not available. While the importance of developing transgenics for pigeon pea improvement is widely acknowledged by breeders, efforts have not been made to coordinate with biotechnologists in finding required solutions. On the other hand, biotechnologists have made substantial efforts in this direction and considerable technology has been developed (Geetha et al., 1999; Lawrence and Koundal, 2001; Satyavathi et al., 2003; Dayal et al., 2003; Thu et al., 2003; Mohan and Krishnamurthy, 2003; Prasad et al., 2004) in transgenics.

## *DIFFERENT APPROACHES TO CROP IMPROVEMENT*

The classical approach for breeding cultivars is to select suitable phenotypes or mutants, which are then crossed, selfed, cloned, or combined with populations. A major drawback of this approach is that for most agronomically important traits, the phenotypic variance is the basis for selection, which is not only composed of the genotypic variance but also comprises an environmental component as well as interactions between genotype and environment. This tends to obscure selection progress and is one of the reasons why breeding for quantitative or polygenic traits is so tedious and time consuming. In addition, a combination of positively acting polygenes of one specific genetic background is difficult, if not impossible, because of recombinational dispersion of genes in each sexual generation. ICRISAT has been engaged in pigeon pea improvement research for the last 25 years. During this period, efforts were made to metamorphose the crop from a traditional medium-to-long duration, bushy, pest- and disease-ridden, and low-yielding subsistence crop to a novel early maturing, photoperiod-insensitive, and relatively high-yielding crop with the ability to escape and resist pest and disease attack. Efforts are underway to identify a cytoplasmic male sterility system to overcome problems of seed production common to hybrids based on a genetic male sterility system. Efforts are also made to improve resistance to key pests such as maruca, pod borer, and pod fly in addition to enhancing drought and water logging resistance (http://www.icrisat.org). In vitro culture has been undertaken using various explants like immature embryos, epicotyls, hypocotyls, nodal and internodal segments, meristems, leaves, and roots. Though fertile plants have been produced, limitations including poor regenerability and long-term callus phase, and somaclonal variations like phenotypic abnormalities, altered ploidy, and reduced plant fertility exist as in other legumes. Also, strong in vitro genetic differences exist in the morphogenetic capacity of different genotypes and explants.

## *IN VITRO APPROACHES*

Several constraints that limit crop production or quality have been addressed by conventional breeding and enhanced management, but there are situations where the existing germplasm lacks the required traits (Popelka et al., 2004). The in vitro approaches like organogenesis, somatic embryogenesis, protoplast regeneration, and genetic transformation (Sahoo et al., 2003) that exploit the regeneration capacity of plant cells and tissues provided the opportunity for genetic manipulation of elite cultivars by organogenesis or somatic embryogenesis. The in vitro developments of pigeon pea are discussed in the following sections.

## *PLANT REGENERATION VIA ORGANOGENESIS*

Improvement of this crop through biotechnological methods was initiated more than a quarter century ago. Shama Rao and Narayanaswami (1975) reported regeneration of pigeon pea from callus cultures of hypocotyls obtained from gamma-irradiated seeds, but failed to get regeneration from unirradiated controls. Subsequently, Kumar et al. (1983) reported on the optimal culture conditions for plant regeneration from callus of leaves and cotyledons of pigeon pea with an emphasis on creating genetic diversity. They also reported that multiple shoots were induced from excised cotyledonary segments when cultured on a medium with benzyl amino purine (BAP). Multiple shoot buds were also induced from cotyledonary node explants of pigeon pea (Mehta and Mohan Ram, 1980; Mohan and Krishnamurthy, 1998; Prakash et al., 1994). However, low frequencies of whole plant regeneration were obtained in all these cases. The above-mentioned literature suggests that the addition of a cytokinin to the regeneration medium could be the keystone for shoot bud differentiation in pigeon pea. These workers have not studied the factors that influence plant regeneration in detail. None of the protocols reported in the literature are suitable for the genetic transformation of pigeon pea because of the low frequency of regeneration (Sreenivasu et al., 1998). To fill in the gaps, in vitro regeneration and multiple shoot production of *Cajanus cajan* from different seedling explants and the subsequent transfer of the regenerated plantlets into soil were studied by Geetha et al. (1998).

### *Direct Multiple Shoot Bud Induction*

Regeneration of shoot buds from various explants of pigeon pea, such as leaves (Eapen and George, 1993; Eapen et al., 1998; Geetha et al., 1998;

Balarama Swamy Yadav and Padmaja, 2003), distal cotyledonary segments (Mohan and Krishnamurthy, 1998), cotyledonary node (Prakash et al., 1994; Geetha et al., 1998), shoot tips (Geetha et al., 1999), and multiple shoot induction from apical and axillary meristems (Franklin et al., 1998) have been reported. Only 36 percent of callus cultures obtained from primary leaves regenerated shoot buds (Eapen and George, 1993). Eapen et al. (1998) reported high-frequency shoot regeneration from primary leaf segments, but the shoot buds appeared only after 45-50 days of induction. Although shoot bud formation was observed in 83 percent of cultured cotyledon segments, only 56 percent of these shoot buds developed further, resulting in only 18 percent fully developed shoots (Mohan and Krishnamurthy, 1998). Cotyledonary nodes have been used to produce multiple shoot buds that not only may have been contaminated with axillary shoots but also failed to elongate further (Prakash et al., 1994). Franklin et al. (2000) reported regeneration from mature embryonal axis also. The dose of cytokinin is known to be critical in shoot organogenesis.

Geetha et al. (1998) compared the response of different seedling explants on various concentrations of BAP and kinetin. Direct shoot bud differentiation was observed after two weeks of culture initiation. Multiple shoots were initiated from all the explants after four weeks of culture. Both the type of explant and the choice of cytokinin and its concentration influenced the frequency of shoot formation. Explants on Murashige and Skoog (MS; 1962) basal medium without BAP or kinetin did not respond. The highest frequency of shoot bud regeneration was observed from cotyledonary node explants followed by epicotyl, hypocotyl, cotyledon, and leaflet explants. The maximum frequency of shoot formation from cotyledonary node explants was 93.2 percent on regeneration medium with BAP and 75.4 percent with kinetin. The lowest regeneration frequency from leaf explants induced by BAP was 65.4 percent and those induced by kinetin were 54.3 percent.

Of the two cytokinins tested, BAP was found to be more effective than kinetin in inducing shoot development and multiple shoot induction in all the explants. The regeneration frequency increased with an increase in the concentration of cytokinin, and 2.0 $mg \cdot l^{-1}$ was found to be an optimal concentration for maximum frequency of shoot bud formation. With increased cytokinin concentration (beyond 2.0 $mg \cdot l^{-1}$), the frequency of shoot bud regeneration decreased drastically. BAP was therefore concluded to be an effective cytokinin for pigeon pea shoot organogenesis. Similar observations were made for cotyledons of pigeon pea on B5 medium with BAP (Mehta and Mohan Ram, 1980). Gosal and Bajaj (1979) obtained multiple shoots from in vitro cultured zygotic embryos on B5 medium containing BAP (2.25 $mg \cdot l^{-1}$). Cheema and Bawa (1991) demonstrated that about two or three shoots resulted from cotyledon and hypocotyl explants cultured on MS medium having BAP and naphthalene acetic acid (NAA). George and Eapen (1994) reported only 26.6 percent direct shoot bud development from distal ends of cotyledon explants on MS medium fortified with BAP (1.0 $mg \cdot l^{-1}$) and indole acetic acid (IAA, 0.1 $mg \cdot l^{-1}$).

Eapen and George (1993) reported low-frequency regeneration from leaf callus cultures of pigeon pea. However, Kumar et al. (1983) reported 20 percent shoot bud regeneration from callus developed from cotyledon explants and 14 percent from leaf callus tissue on Blayde's medium with 2.25 $mg \cdot l^{-1}$ BAP. Subsequently, Kumar et al. (1984) produced multiple shoots from epicotyl segments, excised cotyledons, and shoot tips of pigeon pea on Blayde's medium with BAP. Sarangi and Gleba (1991) cultured the basal part of the embryo in which a part of the cotyledons adjacent to the embryo produced multiple shoots via callusing on MS medium containing 0.5 $mg \cdot l^{-1}$ BAP. They observed distinct variation within the regenerated plants. Naidu et al. (1995) observed multiple shoot differentiation from mature embryo axes and mature cotyledons of seven genotypes when cultured on MS medium supplemented with BAP (1, 2, 3, and 5 $mg \cdot l^{-1}$) alone or in combination with kinetin (0.1, 0.5, and 1 $mg \cdot l^{-1}$) or IAA (0.1 and 1 $mg \cdot l^{-1}$). They produced shoot buds on epicotyl explants derived from 10-12-day-old-seedlings by culturing on MS medium containing BAP (1.0 $mg \cdot l^{-1}$) and IAA (0.1 $mg \cdot l^{-1}$).

BAP was the most effective growth regulator in all these reports, indicating cytokinin specificity for multiple shoot induction in these tissues. Cytokinins in general are required to induce shoot buds from cultured tissues. Thidiazuron (TDZ) has been used effectively for shoot bud formation in leaf explants (leaf discs) of pigeon pea (Eapen et al., 1998). Sudarsana Rao et al. (2001) studied the role of amino acids in development of microshoots in pigeon pea. Direct shoot bud differentiation from leaf explants was reported by Misra (2002). Balarama Swamy Yadav and Padmaja (2003) described a protocol for plantlet regeneration via organogenesis from seedling leaf segments of pigeon pea on MS medium containing 5.0 $mg \cdot l^{-1}$ benzyl adenine. Chandra et al. (2003) reported that cotyledons from seedlings were cultured on media containing various cytokinins including TDZ, where organized shoot-like structures were noticed. However, subsequent regeneration of shoots from cotyledon explants was very poor. Although in most previous reports on pigeon pea BAP alone was shown to be potent for shoot bud differentiation (Eapen and George, 1993; George and Eapen, 1994; Mohan and Krishnamurthy, 1998; Geetha et al., 1998; Prakash et al., 1994), in our study a combination of BA and kinetin produced optimum and reproducible shoot bud differentiation. However, longer exposure to this cytokinin combination was detrimental to further elongation of shoot buds, and they had to be exposed to gibberellic acid ($GA_3$) for their further development.

### *Multiple Shoot Proliferation and Elongation*

In our study, BAP alone was found suitable for both multiple shoot bud induction and proliferation. However, multiple shoots obtained on various concentrations of BAP failed to elongate on the same medium, resulting in rosette shoot clumps. Hence it was necessary to develop suitable media for proliferation and elongation of shoot buds, which were not growing otherwise. Clumps of multiple shoots were transferred to test tubes containing various growth regulator combinations for elongation. Multiple shoot proliferation occurred from all five explants on all five concentrations of BAP within 4 weeks. It was found to be high in cotyledonary node followed by epicotyl, hypocotyl, cotyledon, and leaf explants. Explants required a higher concentration of BAP (2.0 $mg \cdot l^{-1}$) at the initial stage of shoot bud regeneration, but further growth and proliferation of the shoot buds was observed only after subculture to fresh medium with low levels of BAP. Addition of NAA (0.01-1.0 $mg \cdot l^{-1}$) enhanced the number of multiple shoots and shoot elongation. The optimum level of NAA was 0.1 $mg \cdot l^{-1}$ for highest number of multiple shoot production. Similar observations were also made for pigeon pea (Prakash et al., 1994). The problem of shoot elongation was overcome by transferring shoot cultures to a shoot-proliferating medium containing NAA and BAP. Incorporation of $GA_3$ (1-5 $mg \cdot l^{-1}$) along with BAP (1.0 $mg \cdot l^{-1}$) and NAA (0.1 $mg \cdot l^{-1}$) markedly increased the number of multiple shoots in all the explants. Moreover, shoots showed internode elongation and the highest number of multiple shoots was scored with BAP (1.0 $mg \cdot l^{-1}$) + NAA (0.1 $mg \cdot l^{-1}$) + $GA_3$ (3.0 $mg \cdot l^{-1}$) supplementations. Prakash et al. (1994) reported the proliferation of multiple shoot initials from the explant only in the presence of added IAA, suggesting that the cytokinin: auxin ratio is important for this response. This leads to the conclusion that media with lower concentrations of growth regulators favored elongation of shoot buds. The other two varieties used in our study also responded in the same manner with little variation.

Dayal et al. (2003) developed a protocol for efficient plant regeneration from leaf explants of pigeon pea for the production of transgenic plants. Leaf explants from 4-to 5-day-old in vitro raised seedlings were most efficient in producing multiple adventitious shoots in 90 percent of the explants on shoot induction medium (MS) with 1.125 $mg \cdot l^{-1}$ benzyl adenine and 1.0 $mg \cdot l^{-1}$ kinetin). Shoot buds originated from the petiolar cut end of the explants and elongated rapidly on a medium containing 0.2 $mg \cdot l^{-1}$ $GA_3$. In studies on the role of the lamina tissue in shoot bud regeneration from the petiolar cut end, it was found that leaf explants containing intact lamina were essential for the regeneration response, with shoot bud induction declining with reduced lamina tissue. A very low frequency of shoot regeneration occurred if the entire lamina was removed from the petiolar explants. The effect of the age of the explants on shoot regeneration was determined. Regeneration potential was also affected by explant origin, culture maintenance conditions, and age of the explants. Dolendro Singh et al. (2003) studied the effect of TDZ on seedlings of pigeon pea. Seedlings raised from decoated seeds on MS basal medium supplemented with low concentrations of TDZ (0.01-0.22 $mg \cdot l^{-1}$) induced multiple shoots. However, seedlings raised from seeds with intact seed coats failed to produce multiple shoots under the same conditions. Multiple shoots were also induced from seeds exposed for short duration (24 hours-14 days) on MS basal medium containing TDZ (2.2 $mg \cdot l^{-1}$).

Jain and Chaturvedi (2004) demonstrated that prolific differentiation of shoot buds was induced in the interveinal leaf lamina explants when cultured on filter paper bridges in a nutrient liquid medium supplemented with 0.25 $mg \cdot l^{-1}$ BAP and kinetin along with 0.05 $mg \cdot l^{-1}$ IAA and 40 $mg \cdot l^{-1}$ adenine sulfate producing an average of 50 shoots within 45 days. Development of normal shoots was noticed by subculture of explants with differentiated shoot buds on solidified nutrient medium containing 0.01 $mg \cdot l^{-1}$ each of BAP and IAA along with chloro choline chloride (10 $mg \cdot l^{-1}$), where more than 100 shoot buds were formed in the cultures. Srinivasan et al. (2004) described a simple and efficient de novo regeneration system from petioles explanted from young seedlings of pigeon pea. Multiple shoots were obtained when explants were cultured on MS medium containing NAA and BAP. They found that the response of the petiole decreased with an increase in the age of the leaf—that is, the older the leaf, the less the response. In general, younger seedlings (<5 days) provided explants that were highly regenerative, while the regeneration potential declined with increased age of explants. Hence, whole leaf explants from 4- to 5-day-old seedlings were used in the optimized protocol. One of the important features of regeneration in leaf explants of pigeon pea (Dayal et al., 2003) is a polarized regeneration response, where some of the tissues of an explant have a greater regeneration potential, and regeneration was restricted to petiolar tissues, thus suggesting the involvement of hormones or metabolites in the expression of cellular totipotency of the petiolar tissue. Regeneration reported by these authors is efficient for the regeneration of multiple shoots in pigeon pea, where 90 percent of cultures underwent differentiation of adventitious shoot buds from the petiolar cut ends of leaf explants that are devoid of any preexisting meristems. This shoot regeneration method appears to be applicable across a wide range of pigeon pea genotypes belonging to different maturity groups.

### *Rooting and Establishment of Plantlets*

In order to induce rooting from regenerated shoots, MS medium augmented with various concentrations of IAA, NAA, and IBA individually were tried in various experiments. All the three auxins tested induced rooting within 3 weeks of treatment. The frequency of rooting varied with different auxin concentrations. Among the various explants used, cotyledonary-node derived shoots responded well for root initiation, and maximum percentage was also observed in all three auxins tested. The frequency of rooting per shoot was significantly different among the treatments. The percentage of rooting increased with increase in the concentration of auxin. Formation of roots was observed in 69 percent of shoots on MS medium supplemented with IAA (0.2 $mg \cdot l^{-1}$) and 77 percent rooting of shoots with 0.2 $mg \cdot l^{-1}$ NAA. In pigeon pea, Kumar et al. (1983) observed rooting with IAA (1.0 $mg \cdot l^{-1}$) or NAA (0.01 $mg \cdot l^{-1}$) and George and Eapen (1994) obtained 90 percent of rooting with 0.2 $mg \cdot l^{-1}$ NAA. In our results, indole butyric acid (IBA) was found to be the best auxin for rooting, and 92 percent of shoots rooted with 0.2 $mg \cdot l^{-1}$ IBA treatment. Prakash et al. (1994) reported similar results in pigeon pea. Naidu et al. (1995) reported that the regenerated shoots were rooted individually on filter paper supports in half-strength (liquid) MS medium with 0.1 $mg \cdot l^{-1}$ IBA within 2 weeks and the percentage of rooting was 80-85 percent. Similarly, Mohan and Krishnamurthy (1998) also noticed that only 80-85 percent elongated shoots were rooted on half-strength MS medium with IBA (0.1 $mg \cdot l^{-1}$) and the survival percentage was 75-80 percent. Regenerated shoots were rooted on MS medium supplemented with 2.2 $mg \cdot l^{-1}$ IAA (Dayal et al., 2003). These authors claimed that there was 100 percent success in field establishment. Dolendro Singh et al. (2003) observed that shoots were rooted on MS medium containing 0.5 $mg \cdot l^{-1}$ IBA. Well-developed plants were established in soil and the survival rate was 95 percent. Of the two auxins used, IAA induced maximum rooting frequency in pigeon pea (Balarama Swamy Yadav and Padmaja, 2003). Jain and Chaturvedi (2004) reported that there was 100 percent rooting when shoots were cultured on medium containing 0.25 $mg \cdot l^{-1}$ of any of the three auxins used, IAA, IBA, and NAA. In vitro raised plants grew normally and produced viable seeds. Rooting obviously varies in different cultivars. Srinivasan et al. (2004) demonstrated that elongated shoots were rooted on MS medium supplemented with IBA and plants were established in soil.

Upon transfer of the rooted shoots to a hormone-free MS liquid medium, the roots showed elongation and numerous thin roots also developed. Plantlets with fully expanded leaves and well-developed roots after hardening were eventually established in soil. The percentage of the plantlets surviving transfer to soil was 90-95 percent. The regenerated plants did not show any detectable variation in morphology or growth characteristics when compared with the respective donor plants. Other researchers also reported similar survival percentages (George and Eapen, 1994; Prakash et al., 1994; Dolendro Singh et al., 2003) except Dayal et al. (2003). The protocol presented is substantially improved over the previous reports. In their investigation, aseptic seedlings were grown on hormone-free media, unlike the previous reports where seedlings were grown on MS medium supplemented with growth hormones like BAP. Regeneration was reported from callus cultures in all the previous reports except Prakash et al. (1994). However, regeneration frequency was much lower compared to that of Dayal et al (2003). Prakash et al. (1994) reported an in vitro propagation method using cotyledonary node explants, but the rate of multiplication was substantially low. In conclusion, efficient and reliable plant regeneration systems via direct shoot organogenesis from cotyledonary nodes, embryonal axis, and immature leaf explants, which are developed, can now be exploited for genetic transformation experiments. All reports agree on the fact that the use of in vitro-raised seedlings provides better regenerability than explants derived from mature tissue.

## *PLANT REGENERATION VIA SOMATIC EMBRYOGENESIS AND PROTOPLAST CULTURE*

Somatic embryogenesis has great potential in the near future for mass multiplication and genetic improvement of several leguminous species. Although significant progress has been made in the regeneration of plants via somatic embryogenesis in various legumes (Venkatachalam et al., 2003), it is not comparable to the induction of shoots in reproducibility. It has not yet been used for routine multiplication of transgenic plants. Similarly, protoplast culture, although reported, has not been successful in regenerating plantlets (Kulkarni and Krishnamurthy, 1980).

## *GENETIC TRANSFORMATION*

Having established efficient regeneration protocols, we have standardized the technology for efficient gene transfer. The gene transfer technology is initially standardized with different explants using marker genes like *npt*II and *uid*A and subsequently with desirable genes like rinderpest hemagglutinin (RPH) and hemagglutinin neuraminidase gene *(HN)* from peste des petits ruminants. We describe here the optimal conditions for genetic transformation in pigeon pea developed by us and other groups. Advantages and disadvantages of various gene transfer methods are well documented. Two methods are widely used for gene transfer to many crops successively. *Agrobacterium*-mediated gene transfer is most widely used in view of its easy access and affordability compared to microprojectile bombardment, which is more sophisticated and expensive.

### **Agrobacterium-*Mediated Gene Transfer***

A reproducible system for production of transgenic plants via *Agrobacterium* mediated transformation of pigeon pea was developed initially with marker genes. Shoot apices and cotyledonary node explants were transformed by cocultivation with *Agrobacterium tumefaciens* strain LBA4404. The strain harbors a binary vector carrying the reporter gene β-glucuronidase (*uid*A) and the marker gene neomycin phosphotransferase (*npt*II). Cocultivated explants were cultured on shoot regeneration medium with 2 $mg \cdot l^{-1}$ BAP and kanamycin (50 $\mu g \cdot ml^{-1}$ for selection, MS1). Approximately 45-62 percent of the explants produced putatively transformed shoots on the selection medium. Multiple shoots were selected repeatedly on a medium containing BAP (0.5 $mg \cdot l^{-1}$) and kanamycin (25 $\mu g \cdot ml^{-1}$). Elongated shoots were subsequently rooted on a medium supplemented with 25 $\mu g \cdot ml^{-1}$ kanamycin sulfate (MS3). The transgenic plants were later established in pots. Although transformation was achieved both with cotyledonary node and shoot apices, cotyledonary nodes responded better with 62 percent of the explants producing GUS-positive shoots after selection on MS2 medium. The presence of *uid*A and *npt*II genes in the transgenic plants was verified by PCR analysis. Integration of T-DNA into the genome of transgenic plants was further confirmed by Southern blot analysis. The method developed was subsequently used with two genes of high value for developing transgenics as a source of edible vaccines.

### *Development of Transgenics with* H *Gene via* Agrobacterium Tumefaciens*-Mediated Transfer*

Pigeon pea was used successfully as a model system for possible edible vaccine production against rinderpest disease, demonstrating that it is possible to express the required protein by genetic transformation. Briefly, our approach is described below. Hemagglutinin gene *(H)* of rinderpest virus was subcloned into binary vector pBI121. The *gus* gene of pBI121 was replaced by the *H* gene such that the gene is under the control of the CaMV 35S promoter. The binary vector pBI121 was digested with *Sma*I and *Sac*I to release the *gus* gene fragment to replace with full length *H* gene. The full-length *H* gene was obtained by digesting RBH3.4 plasmid DNA, which was derived from a cDNA library (kind gift from Dr. T. Barrett, Institute for Animal Health, Pirbright, UK), with *Eco*RI and *Bam*HI. Recombinants were checked for presence of the *H* gene in the correct orientation by *Xba*I digestion, which releases a 1.9-kb fragment. The recombinant binary vector was designated as pBIH. The binary vector was mobilized into *Agrobacterium tumefaciens* EHA105. Transformants were selected and maintained on Luria Bertani medium (Sambrook et al., 1989) containing 50 $\mu g \cdot ml^{-1}$ kanamycin and 25 $\mu g \cdot ml^{-1}$ rifampicin solidified with 1.5 percent agar. Embryonic axes and cotyledonary nodes were prepared as reported (Satyavathi et al., 2003) and aseptically cultured on MS medium supplemented with 2.0 $mg \cdot l^{-1}$ BAP (MS1). Subculture was done at an interval of 3 weeks on the same initial medium, and for subsequent subculturing MS medium with 0.5 $mg \cdot l^{-1}$ BAP (MS2) was used. For rooting, shoots were transferred to MS medium supplemented with 0.3 $mg \cdot l^{-1}$ IBA (MS3).

A single bacterial colony was inoculated into 25 ml of liquid LB medium with 50 $\mu g \cdot ml^{-1}$ kanamycin and 25 $\mu g \cdot ml^{-1}$ rifampicin, and incubated overnight (18-24 hours) at 28°C on a shaker at 120 rpm and used in the late-log phase ($A_{600}$ at 0.6) for transformation. Explants were precultured for 2 days on MS1 medium. Precultured explants were immersed in the *Agrobacterium* suspension in LB medium with gentle agitation (40 rpm) for 20 minutes at room temperature, drained on filter paper, and placed on the same regeneration medium (MS1) for cocultivation. The number of explants used for each transformation attempt ranged from 150 to 200. All explants were cocultivated for a period of 2 days in darkness at 25 ± 2°C. After cocultivation, explants were washed and cultured on selection medium (MS1) with antibiotics (50 $\mu g \cdot ml^{-1}$ kanamycin and 400 $\mu g \cdot ml^{-1}$ cefotaxime). After 3 weeks, shoots were subcultured on the same medium for a second passage and thereafter subcultured to MS2 medium under the same conditions for selection and further shoot elongation. Putative transformants (longer than

3 cm) were transferred to a rooting medium (MS3) in culture tubes containing 50 μg·ml$^{-1}$ kanamycin. Rooted putative transgenic plants were transferred to plastic cups containing 1:1 soil:sand and maintained in the greenhouse.

Among different explants of pigeon pea used for shoot organogenesis, embryonic axes and cotyledonary node (see section In Vitro Approaches) explants were found to be best for shoot induction (Geetha et al., 1998, 1999). Shoot regeneration occurred via adventitious proliferation from the axillary portion of the cotyledonary node and from the embryonic axes. Lawrence and Koundal (2001) reported recovery of pigeon pea transgenic for a protease inhibitor gene through regeneration of transgenic callus derived from embryonal axes. Mohan and Krishnamurthy (2003) described *Agrobacterium*-mediated genetic transformation of pigeon pea by using decapitated mature embryo axis explants using marker genes. Thu et al. (2003) obtained transgenic plants by *Agrobacterium tumefaciens*-mediated gene transfer. The presence of the transgenes in the pigeon pea genome was confirmed by GUS assays, polymerase chain reaction (PCR), and Southern hybridization. Rooted transgenic plants were successfully transferred to soil in the greenhouse. GUS and PCR assays of $T_1$ progenies confirmed that the transgenes were stably transmitted to the next generation.

In contrast, our method reports regeneration of transgenic shoots directly from cotyledonary explants as well as embryonal axes and thus confirms our earlier work (Geetha et al., 1999) on pigeon pea. The transformation efficiency obtained in our study was high compared to that of other reported legumes. In our study, explants like cotyledonary nodes and embryonic axes were more efficient for transformation compared to hypocotyls, epicotyls, and cotyledons. During the initial 2-day preculture period on a nonselective shoot regeneration medium, all explants retained green color. After cocultivation, almost 90 percent of pigeon pea explants retained a healthy green color and developed shoots on MS1 medium containing 50 μg·ml$^{-1}$ kanamycin. The surviving shoots were subcultured to fresh MS1 medium containing kanamycin. Vigorous shoot growth was observed in 3-4 weeks on this medium, producing an average number of three shoots per explant. Out of 675 cotyledonary node explants, 452 responded with higher transformation efficiency (67 percent) compared to embryonic axes. In the case of embryonic axes, 719 explants responded out of 1,400 explants (51 percent). Transformation efficiency is calculated as the number of explants survived on the selection medium developing shoots. Viable shoots on MS2 medium with 0.5 mg·l$^{-1}$ BAP and 50 μg·ml$^{-1}$ kanamycin elongated to 2-3 cm in length. Roots were initiated in 8-10 days, and sufficient roots were produced in 25 days. Different developmental stages (3-9 weeks) of the pigeon pea transformants regenerated from the cotyledonary node explants are clearly shown by the authors (Prasad et al., 2004). During rooting, only those shoots that were transgenic rooted and the escapes mostly did not root.

According to Draper et al. (1988), roots are generally much more sensitive to antibiotics, and thus the ability to root on selection medium containing high levels of selective agent is a strong indication of transformation. All transformants appeared morphologically normal in comparison with untransformed plants. About a 100 transformants were selected and transferred to soil. The survival rate was 90-95 percent. These plants grew to maturity, flowered, and set seed. The majority of putative transformed shoots showed the presence of the antibiotic resistance gene *npt*II as seen with the amplification of the predicted 0.7-kb internal fragment. Southern hybridization of the transgenics has shown 13 independent lines to be positive. In transformants, high-molecular-weight fragments ranging in size from 4.3 to 9.5 kb were present. No hybridization signal could be detected in the untransformed plant (Satyavathi et al., 2003).

Plants transformed with pBIH showed RT-PCR amplification of *npt*II. To rule out the possibility of amplification of contaminant DNA in samples, direct PCR amplification without reverse transcription was performed on RNA preparations. No amplified DNA fragments were detectable under those conditions, confirming RNA specificity of the reaction. The immunoblot analysis of protein from transgenic leaf samples performed with 10 independent transformants showed *H* expression, and the expected band of 72-78 kDa was detected in all lines. There was no protein corresponding to this band in the untransformed control plant. The expression level of the H protein ranged from 0.12 to 0.49 percent of the total soluble leaf protein. Compared to the levels of expression of the other proteins reported for vaccine production, the RPV H protein was expressed to a higher level in pigeon pea. Thus, this was the first report of expression of a viral protective antigen in plants to a high level from 35S promoter. About 20 primary transformants were grown to maturity. All plants flowered and set seeds. Some seeds were shriveled and did not germinate. Progeny from various independent transformations were tested for the presence of the *npt*II gene by PCR analysis. Out of 17 progeny tested for PCR, 12 were found to be positive. The western blot analysis in the $T_1$ progeny showed a protein in the range of 72-78 kDa, which was absent in the untransformed control. The procedure reported is very simple, efficient, and reproducible, and is applicable across diverse genotypes of pigeon pea.

These protocols were also used for successful transformation of pigeon pea (Prasad et al., 2004) with another gene of interest, namely, Peste des Petits Ruminant Virus (PPRV), which causes PPR disease in sheep and goats, with a high mortality rate. The two surface glycoproteins of PPRV hemagglutinin neuraminidase (HN) and fusion protein (F) confer protective immunity. We have successfully generated transgenic pigeon pea plants that expressed HN protein having biological activity. A 2.0 kb fragment containing the coding region of the *HN* gene from an Indian isolate was cloned into the binary vector pBI121 and mobilized into *Agrobacterium tumefaciens* strain GV3101. Cotyledonary nodes from germinated seeds of pigeon pea were used for transformation. The presence of transgenes *npt*II and *HN* in the plants was confirmed by PCR. The expression of HN protein in the transgenic lines was further confirmed by western blot analysis using polyclonal monospecific antibody to HN and, more important, plant-derived HN protein was shown to be biologically active, as demonstrated by neuraminidase activity. Transgenic plants were fertile and PCR of $T_1$ plants confirmed the inheritance of the transgene. We were successful in using direct transformation, that is, treating explants like embryos with *Agrobacterium* suspension for a 6-8 hour period and planting on the media after washing away the agro traces with sterile distilled water. These were highly regenerative and the plants expressed the introduced the *H* gene, as shown by the western blots (Prasad et al., 2003).

Surekha et al. (2005) developed transgenic plants in pigeon pea successfully via *Agrobacterium*-mediated transformation. In this study, a synthetic *cryI E-C* gene under a constitutive 35S promoter was used for transformation and the expression of the synthetic *cryI E-C* in transgenic pigeon pea plants conferred protection against Spodoptera larvae. Most recently, Sharma et al. (2006) have developed an efficient method to produce transgenic plants of pigeon pea by incorporating the *Cry1Ab* gene of *Bacillus thuringiensis* via *Agrobacterium tumefaciens*-mediated genetic transformation. PCR analysis showed the presence of single-copy inserts of transgene with 65 percent of the progenies from independent transformants. Further, it is reported that the level of Cry1Ab protein production varied with different tissues in the tested $T_2$ generation plants.

### *Biolistic Method of Transformation*

The biolistic method has been successfully used for studies on the genetic transformation of pigeon pea using *npt*II and *gus* as marker genes, where 50 percent of the selected plants showed gene integration and expression (Dayal et al., 2003). The shoot-forming petiolar region of the leaf explant was used to test the efficiency of gene transfer by using a biolistic particle device. The leaf explants were found to be efficient targets for gene transfer by microprojectile bombardment since they resulted in the production of a large number of putative transformants of pigeon pea. Although 90 percent of the bombarded explants exhibited transient expression of the *uid*A gene, 50 percent of the selected plants that were transferred to the glasshouse showed positive gene integration. The leaf explants were found to be efficient targets for gene transfer by microprojectile bombardment since they resulted in the production of a large number of putative transformants of pigeon pea. Although 90 percent of the bombarded explants exhibited transient expression of the *uid*A gene, 50 percent of the selected plants that were transferred to the glasshouse showed positive gene integration.

### *Development of Transgenics via Microprojectile DNA Delivery Method*

Plasmid pRT99-GUS containing the *uid*A and *npt*II genes, both under the control of the 35S promoter of cauliflower mosaic virus, was used to optimize genetic transformation of leaf explants. After bombardment, the explants were incubated on the same plate overnight and transferred to fresh plates containing shoot-inducing medium at a plating density of 10-12 explants per plate. The regeneration protocol used was as described by Dayal et al. (2003). Molecular analysis of the genomic DNA from the putative transformants was carried out for the presence of the introduced genes by PCR amplification of *uid*A and *npt*II, and Southern hybridization for the *npt*II gene, according to Sharma and Anjaiah (2000). For Southern blot hybridization, the genomic DNA was digested with *Xho*I, which has a unique site within the pRT99GUS plasmid DNA. To study the inheritance of the introduced genes in the $T_1$ generation, five seeds from eight selected primary transformants were germinated and PCR analysis to detect the *uid*A gene was carried out. According to the authors, 50 percent of the selected plants showed gene integration and expression. Thu et al. (2003) also developed transgenic plants using the biolistic method that were established in soil. Dayal's report showed significant improvement, with at least 50 percent of the selected transgenic plants showing positive gene integration and function as indicated in RT-PCR studies. Moreover, 50 percent of the transgenic plants showed single gene integration events. All the putative transgenic plants showed a Mendelian inheritance of the introduced genes in the $T_1$ generation, thus confirming the success of this regeneration and transformation protocol.

In conclusion, the most significant contributions in our investigations are twofold. Pigeon pea, which is known to be a highly recalcitrant crop for both transformation and regeneration, is now amenable to both. In spite of its economic importance as a highly consumed pulse and source of protein, it has not been exploited so far for quality improvement such as nutritional quality. In our investigation, we have now successfully demonstrated that desirable genes like hemagglutinin gene (*H*) of rinderpest virus and hemagglutinin-neuraminidase (*HN*) gene of PPRV can be expressed in pigeon pea for the development of an edible vaccine, thus showing clearly that technology is available for introducing any gene of interest. Recovery of pigeon pea transgenics with protease inhibitor gene through regeneration of transgenic callus derived from embryonal axes by Lawrence and Koundal (2001) also shows promise. Surekha et al. (2005) and Sharma et al. (2006) have developed transgenic plants of pigeon pea by incorporating the *cryI E-C* and *Cry1Ab* genes of *Bacillus thuringiensis* via *Agrobacterium tumefaciens*-mediated genetic transformation. Direct regeneration, however, will be a better method of choice for obtaining transgenics than regeneration from callus.

## *FUTURE PROSPECTS*

Developing transgenics in pigeon pea for quality improvement is a major long-term goal as it is one of the important grain legumes consumed extensively. It is advantageous to modify existing proteins of seeds to improve the composition of essential amino acids. It is not only essential for human consumption but also for animal consumption as many legumes are used as fodder crops. Hence, improvement of seed quality is one of the main targets of biotechnology. Biotechnology could make its most valuable contribution through the application of new techniques to traditional plant breeding. Plant proteins are an important component of the human diet and in some countries they represent the major food protein available to the population. Legume seed proteins consist of a storage fraction (globulins), which is the most abundant fraction, and a nonstorage fraction (albumin). Both these fractions are developmentally regulated and thus they accumulate inside the seed at different stages. They are important in determining the nutritional quality of seed proteins, having a well-balanced amino acid composition and providing a significant amount of the sulfur-containing amino acids present in the seed (Schroeder, 1984). Legume seed contains high levels of protein, which unfortunately has an unbalanced amino acid composition. Although rich in lysine and other essential amino acids, sulfur-containing amino acids such as cysteine and methionine are present in very low amounts, thus affecting the biological value of seeds. Therefore, the introduction of DNA sequences encoding methionine rich seed protein via gene transfer technologies is an attractive addition to classical methods for the improvement of protein quality of grain legumes. Some legumes, such as lentils and chickpea, have limited gene pools controlling protein content and composition, thus hindering the improvement of nutritional quality using classical approaches. The recent isolation of several seed protein genes and promoter regions and their utilization in gene manipulation and transformation experiments promise the possibility of new strategies to improve legumes. Combining biochemical molecular and biotechnological approaches developed during the last 5 years, pigeon pea can be easily manipulated by introducing appropriate genes to improve its nutritional quality.

## *CONCLUSION*

Genetic improvement of food crops affected by conventional plant breeding has brought about a phenomenal improvement in crop yields. Food fortification likely has played an important role in the current nutritional health and well-being of populations in industrialized countries (Misra et al., 2004). Since the early part of the 20th century, fortification has been used for specific health conditions. The nutrition and health communities have never considered using agriculture as a primary tool in their programs directed at alleviating poor nutrition and ill health globally. A potential application of genetic engineering of storage proteins is the improvement of the nutritional balance of essential amino acids, thereby correcting the deficiencies in the total protein compliment. The development of genetically enriched golden rice with vitamin A and iron is one of the most important success stories of all time. In addition to quality improvement, pigeon pea could be an ideal model system for the development of plant- derived vaccines for oral administration. This could revolutionize the world of vaccination and help bring to the poor the promise of existing and new vaccines for controlling infectious diseases. The experimental approach for transgenic development reported by us opens up new possibilities of introducing a variety of novel genes to enrich pigeon pea, which is an important food in many developing countries.

## REFERENCES

Balarama Swamy Yadav, P. and V. Padmaja (2003). Shoot organogenesis and plant regeneration from leaf segments of pigeon pea. *Plant Cell Tissue and Organ Culture* 73:197-200.

Chandra, A., V. Gupta, P. Burma, and D. Pental (2003). Patterns of morphogenesis from cotyledon explants of pigeon pea. *In Vitro Cellular and Developmental Biology-Plant* 39:514-519.

Cheema, H.K. and J. Bawa (1991). Clonal multiplication via multiple shoots in some legumes (*Vigna unguiculata* and *Cajanus cajan*). *Acta Horticulture* 289: 93-94.

Dayal, S., M. Lavanya, P. Devi, and K.K. Sharma (2003). An efficient protocol for shoot regeneration and genetic transformation of pigeon pea [*Cajanus cajan* (L.) Millsp.] using leaf explants. *Plant Cell Reports* 21:1072-1079.

Dolendro Singh, N., L. Sahoo, N.B. Sarin, and P.K. Jaiwal (2003). The effect of TDZ on organogenesis and somatic embryogenesis in pigeon pea [*Cajanus cajan* (L.) Millsp]. *Plant Science* 164:341-347.

Draper, J., R. Scott, and J. Hamil (1988). Transformation of dicotyledonous plant cells using the Ti plasmid of *Agrobacterium tumefaciens* and the Ri plasmid of *Agrobacterium rhizogenes*. In *Plant genetic transformation and gene expression: A laboratory manual,* J. Draper, R. Scott, P. Armitage, and R. Walden (eds.). Oxford, UK: Blackwell Scientific, pp. 69-160.

Eapen, S., and L. George (1993). Plant regeneration from leaf discs of peanut and pigeon pea: influence of benzyl adenine, indole acetic acid and indole acetic acid-amino conjugates. *Plant Cell Tissue and Organ Culture* 35:223-227.

Eapen, S., S. Tivarekar, and L. George (1998). Thidiazuron-induced shoot regeneration in pigeon pea (*Cajanus cajan* L.). *Plant Cell Tissue and Organ Culture* 53:217-220.

Franklin, G., R. Jeyachandran, R. Melchias, and S. Ignacimuthu (1998). Multiple shoot induction and regeneration of pigeon pea [*Cajanus cajan* (L.) Millsp] Vamban 1 from apical and axillary meristems. *Current Science* 74:936-937.

Franklin, G., R. Jeyachandran, R.G. Melchias, and S. Ignacimuthu (2000). Factors affecting regeneration of pigeon pea [*Cajanus cajan* (L.) Millsp] from embryonal axes. *Plant Growth Regulation* 30:31-36.

Geetha, N., P. Venkatachalam, and G. Lakshmi Sita (1999). *Agrobacterium* mediated genetic transformation of pigeon pea (*Cajanus cajan* L.) and development of transgenic plants via direct organogenesis. *Plant Biotechnology* 16:211-216.

Geetha, N., P. Venkatachalam, V. Prakash, and G. Lakshmi Sita (1998). High frequency induction of multiple shoots and plant regeneration from seeding explants of pigeon pea (*Cajanus cajan* L). *Current Science* 10:1036-1041.

George, L. and S. Eapen (1994). Organogenesis and embryogenesis from diverse explants in pigeon pea (*Cajanus cajan* L.). *Plant Cell Reports* 13:417-420.

Gosal, S.S. and Y.P.S. Bajaj (1979). Establishment of callus-tissue cultures and the induction of organogenesis in some grain legumes. *Crop Improvement* 6:154-160.

Grover, A. and D. Pental (2003). Breeding objectives and requirements for producing major field crops of India. *Current Science* 84:310-320.

Jain, M. and H.C. Chaturvedi (2004). In vitro cloning of *Cajanus cajan* var. Bahar through prolific shoot bud differentiation in leaf segments and production of fertile plants. *Indian Journal of Biotechnology* 3:258-262.

Kulkarni, D.D. and K.V. Krishnamurthy (1980). Isolation and culture of protoplasts of pigeon pea, *Cajanus cajan* (L.) Millsp. *Indian Journal of Experimental Biology* 27:939-942.

Kumar, A.S., T.P. Reddy, and G.M. Reddy (1983). Plantlet regeneration from different callus cultures of pigeon pea (*Cajanus cajan* L.). *Plant Science Letters* 32:271-278.

Kumar, A.S., T.P. Reddy, and G.M. Reddy (1984). Multiple shoots from cultured explants of pigeon pea and *Atylosia* species. *SABRAO Journal* 16:101-105.

Lawrence, P.K. and K.R. Koundal (2001). *Agrobacterium tumefaciens*–mediated transformation of pigeon pea [*Cajanus cajan* (L) Millsp.] and molecular analysis of regenerated plants. *Current Science* 80:1428-1432.

Mehta, U. and H.Y. Mohan Ram (1980). Regeneration of plantlets from the cotyledons of *Cajanus cajan. Indian Journal of Experimental Biology* 18:800-802.

Misra, B.K., R.K. Sharma, and S. Nagarajan (2004). Plant breeding: A component of public health strategy. *Current Science* 86:1210-1215.

Misra, P. (2002). Direct differentiation of shoot buds from leaf explants of *Cajanus cajan* L. *Biologia Plantarum* 45:347-351.

Mohan, M.L. and K.V. Krishnamurthy (1998). Plant regeneration in pigeon pea [*Cajanus cajan* (L.) Millsp.] by organogenesis. *Plant Cell Reports* 17:705-710.

Mohan, M.L. and K.V. Krishnamurthy (2003). In vitro morphogenesis in grain legumes: An overview. In *Improvement strategies of leguminosae biotechnology*, P.K. Jaiwal and R.P. Singh (eds.). Kluwer Academic Publishers, Dordrecht, Netherlands, pp. 23-63.

Murashige, T. and F. Skoog (1962). A revised medium for rapid growth and bioassays with tobacco tissue cultures. *Physiologia Plantarum* 15:473-497.

Naidu, R.B., D.D. Kulakarni, and K.V. Krishnamurthy (1995). Genotype-dependent morphogenetic potentiality of various explants of a food legume, the pigeon pea [*Cajanus cajan* (L.) Millsp.]. *In Vitro cellular and Developmental Biology–Plant* 31:26-30.

Popelka, C.J., N. Terryn, and T.J.V. Higgins (2004). Gene technology for grain legumes: Can it contribute to the food challenge in developing countries? *Plant Science* 167:195-206.

Prakash, N.S., D. Pental, and N.B. Sarin (1994). Regeneration of pigeon pea (*Cajanus cajan*) from cotyledonary node via multiple shoot formation. *Plant Cell Reports* 13:623-627.

Prasad, V., V.V. Satyavathi, Sanjaya, K.M. Valli, A. Khandelwal, M.S. Shaila, and G. Lakshmi Sita (2004). Expression of biologically active hemagglutinin-neuraminidase protein of *Peste des petits ruminants* virus in transgenic pigeon pea [*Cajanus cajan* (L.) Millsp.]. *Plant Science* 166:199-205.

Sahoo, L., N. Dolendro Singh, T. Sugla, R.P.Singh, and P.K. Jaiwal (2003). Genetic transformation in legumes. In *Plant genetic engineering: Applications and limitations. Volume 5, Improvement of Vegetables*, R.P. Singh and P.K. Jaiwal (eds.). USA: SCI TECH Publishers, pp. 267-322.

Sarangi, B. and Y.Y. Gleba (1991). Direct multiple regeneration in *Cajanus cajan* (L.) Millsp. *Acta Horticulture* 289:149-150.

Sambrook, J., E.F. Fritsch, and T. Maniatis (1989). *Molecular cloning: A laboratory manual.* Cold Spring Harbor, NY: Cold Spring Harbor Laboratory Press.

Satyavathi, V.V., V. Prasad, A. Khandelwal, M.S. Shaila, and G. Lakshmi Sita (2003). Expression of hemagglutinin of Rinder pest virus in transgenic pigeon pea [*Cajanus cajan* (L) Millsp.] plants. *Plant Cell Reports* 21:651-658.
Schroeder, H.E. (1984). Major albumins of *Pisum* cotyledons. *Journal of the Science of Food and Agriculture* 35:191-198.
Shama Rao, H. and S. Narayanaswami (1975). Effect of gamma irradiation on cell proliferation and regeneration in explanted tissues of pigeon pea [*Cajanus cajan* (L.) Millsp.]. *Radiation Botany* 15:301-305.
Sharma, K.K. and V. Anjaiah (2000). An efficient method for the production of transgenic plants of peanut (*Arachis hypogaea* L.) through *Agrobacterium tumefaciens*-mediated genetic transformation. *Plant Science* 159:7-19.
Sharma, K.K., M. Lavanya, and V. Anjaiah (2006). *Agrobacterium*-mediated production of transgenic pigeon pea [*Cajanus cajan* (L.) Millsp.] expressing the synthetic BT *Cry1Ab* gene. *In Vitro Cellular and Developmental Biology–Plant* 42:165-173.
Sreenivasu, K., S.K. Malik, P. Ananda Kumar, and R.P. Sharma (1998). Plant regeneration via somatic embryogenesis in Pigeon pea [*Cajanus cajan* (L.) Millsp.]. *Plant Cell Reports* 17:294-297.
Srinivasan, T., V.K. Verma, and P.B. Kirti (2004). Efficient shoot regeneration in pigeon pea, *Cajanus cajan* (L.) Millsp. using seedling petioles. *Current Science* 86:30-32.
Sudarsana Rao, G.V., R. Chandra, and R. Polisetty (2001). Role of amino acids in evolution of ethylene and methane, and development of micro shoots in *Cajanus cajan*. *Biologia Plantarum* 44:13-18.
Surekha, C., M.R. Beena, A. Arundhati, P.K. Singh, R. Tuli, A. Dutta-Gupta, and P.B. Kirti (2005). *Agrobacterium*-mediated genetic transformation of pigeon pea [*Cajanus cajan* (L.) Millsp.] using embryonal segments and development of transgenic plants for resistance against *Spodoptera*. *Plant Science* 169:1074-1080.
Thu, T.T., T.T.X. Mai, E. Dewaele, S. Farsi, Y. Tadesse, G. Angenon, and M. Jacobs (2003). In vitro regeneration and transformation of pigeon pea [*Cajanus cajan* (L.) Millsp.]. *Molecular Breeding* 11:159-168.
Venkatachalam, P., N. Geetha, P. Priya, N. Jayabalan, and G. Lakshmi Sita (2003). Somatic embryogenesis. In *Improvement strategies of leguminosae biotechnology*, P.K. Jaiwal and R.P. Singh (eds.). Dordrecht, Netherlands: Kluwer Academic Publishers, pp. 87-132.

Chapter 9

# Genetic Transformation of Chickpea (*Cicer arietinum* L.) Using Cotyledonary Node Explants

Indraneel Sanyal
D. V. Amla

## *INTRODUCTION*

Grain legumes are important crop plants for their protein-rich seeds as a major source of dietary protein for human and livestock consumption. Chickpea is the second most important pulse crop in the world, particularly in India and the African subcontinent, although it is grown in more than 40 countries. Despite large acreage of these crops, the total productivity remains low and consistent for the last few decades worldwide owing to physiological and biological constraints. Traditional breeding practices have been exhausted for further genetic improvement including agronomic characters and incompatibility barriers in the existing available germplasm (Ranalli and Cubero, 1997). Genetic transformation of grain legumes may provide solutions to certain constraints in widening their gene pool and genetic improvement for crop productivity (Popelka et al., 2004). However, grain legumes including chickpea are recalcitrant to tissue culture propagation and incompatible to DNA-mediated genetic transformation (Sharma et al., 2002; Somers et al., 2003). Stable transgenic plants of different large-seeded grain legumes have been produced using various plant tissues and DNA delivery methods including microprojectile bombardment of shoot meristems (McCabe et al., 1988), embryogenic suspension culture (Finer and McMullen, 1991) and *Agrobacterium tumefaciens*-mediated T-DNA delivery into immature cotyledons (Parrott et al., 1989; Yan et al., 2000), embryogenic suspension cultures (Trick and Finer, 1998) and cotyledonary nodes in several grain legumes (Hinchee et al., 1988; Meurer et al., 1998;

Muthukumar et al., 1996; Olhoft et al., 2003; Sanyal et al., 2003, 2005). Each system in the absence of efficient somatic embryogenesis in grain legumes has resulted in relatively inefficient production of transgenic plants owing to inefficient T-DNA delivery, selection of transformed cells, and plant regeneration. Among different explants available in grain legumes, the cotyledonary node is the preferred choice for direct organogenesis and genetic transformation in a number of grain legumes (Somers et al., 2003). Cotyledonary nodes of large-seeded grain legumes possess sectors of pre-existing axillary meristematic cells and also potentially regenerative cells in the L2 layer that are competent to differentiate into new meristematic cells or direct organogenesis subject to the application of a suitable combination of growth regulators and culture conditions. Considering the significance and versatile nature of cotyledonary node explants, their mode of regeneration, and interaction of *Agrobacterium* with target cells for T-DNA transfer along with screening and selection of putative transformants in major grain legumes. We have developed and optimized a reproducible procedure for preparation of cotyledonary nodes in chickpea and parameters for *A. tumefaciens* cocultivation that influence efficient genetic transformation in this recalcitrant grain legume with an idea to create optimal conditions for *Agrobacterium* interaction with excised cells, which are competent for de novo regeneration of shoots via an organogenic pathway.

The procedure for recovery of high-frequency stable transgenic chickpea plants including that of large-seeded grain legumes using cotyledonary node explants can be broadly divided into four steps. These steps require modifications and standardization in utilizing the direct organogenesis pathway for delivery of T-DNA, recovery of transformants, and development of stable transgenic plants.

1. Preparation of cotyledonary node explants for efficient in vitro direct organogenesis.
2. Conditions for *Agrobacterium* cocultivation of cotyledonary nodes.
3. Screening and selection of putative transformed shoots.
4. Establishment of complete stable transgenic plants and their molecular characterization.

## MATERIALS AND PROCEDURES

### Preparation of Cotyledonary Node Explants of Chickpea

Mature, healthy, intact breeder seeds of chickpea of different genotypes (C 235, BG 256, Pusa 362 and Pusa 372) after surface sterilization were used for preparation of cotyledonary node (CN) explants (Sanyal et al., 2003). The seeds are soaked overnight in sterile distilled water having 2.0 $mg \cdot l^{-1}$ 6-benzyl aminopurine (BAP) and germinated under axenic conditions on basal medium containing salts of Murashige and Skoog (MS, 1962) medium, B5 vitamins of Gamborg's medium (Gamborg et al., 1968), 2.0 $mg \cdot l^{-1}$ BAP, 0.05 $mg \cdot l^{-1}$ indolebutyric acid (IBA), gelled with 0.8 percent (w/v) agar, and incubated under normal culture conditions as described earlier (Sanyal et al., 2003). Except where stated otherwise, all media contained 3 percent (w/v) sucrose, solidified with 0.8 percent agar or 0.6 percent phytagel, adjusted to pH 5.8, and sterilized by autoclaving.

The straightforward conventional procedure of preparing CN is by excision of plumule, radicle portion of newly germinated seeds and the attached cotyledons with a surgical blade after 48-72 hours of incubation. We optimized a modified procedure for preparation of cotyledonary nodes of chickpea where the surface-sterilized seeds were soaked overnight in distilled water with 2.0 $mg \cdot l^{-1}$ BAP and seeded for germination on MS basal medium supplemented with BAP (2.0 $mg \cdot l^{-1}$) and IBA (0.05 $mg \cdot l^{-1}$). The cultures were monitored regularly for emergence of axillary shoots from the preexisting meristem. The incubated intact seeds after periodic intervals of 7, 14, 21, 28, and 35 days of culture were recovered for preparation of CNs. The main root, adventitious multiple shoots, and developed shoot buds adjoining the axillary and epicotyl region were excised as close as possible with a surgical blade. This was followed by precise transverse excision of 1.5-2.0 mm thick tissues of the surface layer, removing all the developed shoots and shoot buds along with preexisting axillary meristem of the cotyledonary region and exposing the regenerative cells of the L2 layer (Figure 9.1).

## Agrobacterium *Culture and Cocultivation Conditions*

*Agrobacterium tumefaciens* cultures with different binary vectors like pBI121, pRD400, pBIN200, and pRD401 harboring kanamycin resistance (*npt*II) as selection marker, bacterial *uid*A gene for β-glucuronidase (GUS) in pBI121, and *Bacillus thuringiensis* (Bt) insecticidal crystal protein gene(s) Bt-*cry1Ab* in pBIN200, *cry1Ac* in pRD400, both *cry1Ab,* and *cry1Ac* genes in pRD401plasmids respectively, were used in chickpea genetic transformation. The genes were driven by CaMV35S constitutive promoter with double enhancer and *nos* terminator in all the constructs. The T-DNA regions of various binary vectors used for chickpea transformation with different gene cassettes are shown in Figure 9.2. Bacterial cultures were grown overnight in an incubator shaker at 28°C in 25 ml yeast extract

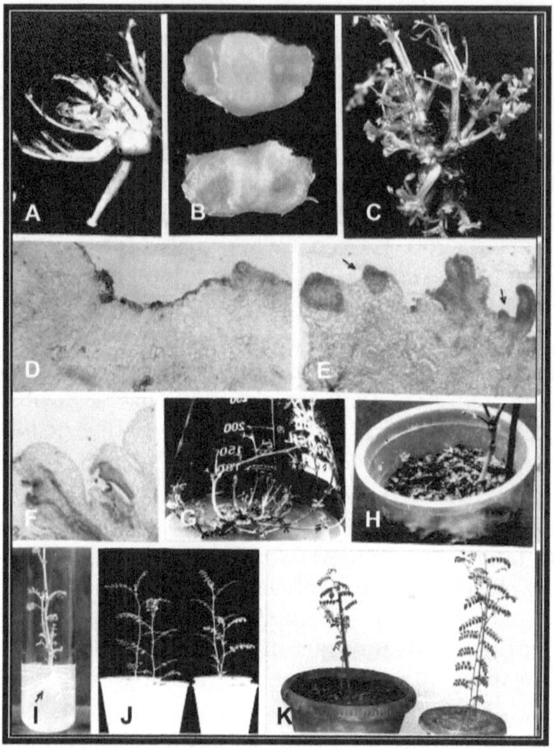


FIGURE 9.1. Preparation of cotyledonary node of chickpea and in vitro regeneration of plantlets: A, seed germinated for 21 days on MS basal medium plus 2.0 mg·l$^{-1}$ BAP before excision; B, top view of prepared cotyledonary node after excision of germinated shoot buds; C, shoot regeneration from the cotyledonary node; D-F, histological sections showing explant immediately after excision and development of shoots on the excised surface; G, cotyledonary node with numerous regenerated shoots; H, micrografting of scion on rootstock; I, root development; J, hardening of plantlets; and K, well established mature plants in glasshouse.

broth (YEB) medium, containing the desired combination of antibiotics like rifampicin (50 mg·l$^{-1}$), kanamycin (50 mg·l$^{-1}$), and streptomycin (50 mg·l$^{-1}$). The exponential cultures were centrifuged and resuspended either in YEB or cocultivation medium consisting of MS basal medium with 1.0 mg·l$^{-1}$ BAP, 200 mg·l$^{-1}$ L-cysteine and 100 μM acetosyringone. The excised CNs were preincubated (preconditioned) for 24 to 72 hours on MS

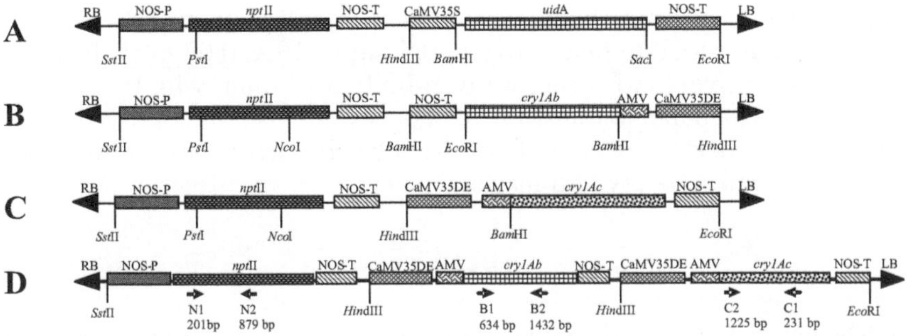


FIGURE 9.2. T-DNA region of various binary vectors harboring reporter gene *uid*A, plant selection marker *npt*II, and Bt-insecticidal crystal protein genes *cry1Ab* and *cry1Ac* for plant transformation. A- pBI121; B- pBIN200; C- pRD400, and D- pRD401. RB and RL- right and left border sequences; *npt*II- gene coding for neomycin phosphotransferase II; CaMV35S- Cauliflower mosaic virus 35S promoter; CaMV35DE- CaMV35S promoter with double enhancer; AMV- alfalfa mosaic virus 5'UTR sequence; *cry1Ab*- 1.845 kb modified sequence of *cry1Ab* gene; *cry1Ac*- 1.845 kb modified sequence of *cry1Ac* gene; NOS-P- promoter of nopaline synthase; NOS-T- terminator sequence of nopaline synthase, and solid arrows indicate the annealing sites for set of PCR primers of respective genes.

medium supplemented with BAP (1.0 $mg \cdot l^{-1}$), 4 μM L-glutamine, 4 μM L-arginine, 200 $mg \cdot l^{-1}$ L-cysteine, and solidified with 1 percent (w/v) agar.

About 12-15 excised preconditioned CNs were incubated in glass tubes with 10 ml *A. tumefaciens* suspension of optical density (OD) at 600 nm between 0.7 and 0.8. The tubes were first sonicated at 42 kHz for 30 to 60 seconds in a water bath sonicator (Branson 2510, United States) followed by 10 to 20 minutes of vacuum treatment. Cocultivated explants were blotted dry on sterile filter paper and plated adaxial side down on cocultivation medium containing 200 $mg \cdot l^{-1}$ L-cysteine, 100 $mg \cdot l^{-1}$ dithiothreitol (DTT) and 100 μM acetosyringone in solidified MS medium for up to 3 days.

The cocultivated explants were then washed with 50 ml of liquid MS medium containing 1.0 $mg \cdot l^{-1}$ BAP and 500 $mg \cdot l^{-1}$ cefotaxime or carbenicillin. The washed explants were plated on selective medium, same as the washing medium except containing 100 $mg \cdot l^{-1}$ kanamycin and 0.02 $g \cdot l^{-1}$ silver nitrate for proliferation of shoots. After 15 days of culture, explants were transferred to fresh selective medium with kanamycin (75-100 $mg \cdot l^{-1}$) and allowed to grow for 30 days with subculturing after every 15 days on fresh medium. The kanamycin concentration was increased to 100 to 125 $mg \cdot l^{-1}$ for two subsequent cycles for individual regenerated shoots. Resulting putative transgenic shoots were rooted in half-strength MS medium with one-fourth total nitrate nitrogen, 0.5 $mg \cdot l^{-1}$ IBA, 0.02 $g \cdot l^{-1}$ silver nitrate, and 1 percent (w/v) sucrose in solidified medium with 0.6 percent (w/v) agar. The putative transformed shoots were also micrografted on 8-day-old rootstock for rapid and consistently high-frequency survival, development, and recovery of transgenic plants.

## *RESULTS*

The cotyledonary node explants of chickpea were prepared at periodic intervals of 7 days, according to the modified procedure, and used for in vitro regeneration and *Agrobacterium*-mediated transformation. The excised cotyledonary nodes were cultured on shoot induction medium consisting of MS basal medium fortified with 1 to 2 $mg \cdot l^{-1}$ BAP or thidiazuron (TDZ), 4 μM L-glutamine, 4 μM L-arginine, 200 $mg \cdot l^{-1}$ L-cysteine, and 0.02 $g \cdot l^{-1}$ silver nitrate when *Agrobacterium* cocultivation was performed. The results obtained are summarized in Table 9.1, which clearly showed the development of maximum number of adventitious shoots (12.66 ± 0.64) per responding CN prepared from 21 days germinated seedling of chickpea, following incubation on MS medium fortified with 1.0 $mg \cdot l^{-1}$ BAP, 0.05 $mg \cdot l^{-1}$ IBA, L-glutamine, and L-arginine (Figure 9.1A-C). The treatment and combination also revealed statistically best response as compared to CNs prepared from the seeds germinated for 7, 14, 28, and 35 days in all four genotypes of chickpea: C 235, BG 256, Pusa 362, and Pusa 372 (Table 9.1).

The histological sections of CNs at periodic intervals revealed direct adventitious origin of shoots from the non-meristematic subepidermal cells of

TABLE 9.1. Effect of age of cotyledonary nodes of chickpea variety Pusa 362 on shoot development.

| BAP | TDZ | IBA | Pusa 362 | | | |
|---|---|---|---|---|---|---|
| ($mg \cdot l^{-1}$) | ($mg \cdot l^{-1}$) | ($mg \cdot l^{-1}$) | 7 d | 14 d | 21 d | 28 d |
| 1.0 | - | 0.05 | 3.54 ± 0.44 | 5.08 ± 0.30 | 10.55 ± 0.44 | 4.64 ± 0.12 |
| 2.0 | - | 0.05 | 4.54 ± 0.52 | 8.45 ± 0.76 | 12.66 ± 0.64 | 8.54 ± 0.32 |
| 5.0 | - | 0.05 | 4.12 ± 0.10 | 6.82 ± 0.34 | 10.04 ± 0.85 | 7.80 ± 0.54 |
| - | 0.2 | 0.05 | 3.20 ± 0.21 | 5.32 ± 0.22 | 10.20 ± 0.67 | 7.54 ± 0.24 |
| - | 0.5 | 0.05 | 4.30 ± 0.76 | 6.44 ± 0.66 | 11.58 ± 0.52 | 6.32 ± 0.40 |
| - | 1.0 | 0.05 | 3.38 ± 0.08 | 5.38 ± 0.46 | 8.84 ± 0.24 | 5.20 ± 0.80 |
| Control (without PGR) | | | 1.24 ± 0.52 | 1.58 ± 0.30 | 1.88 ± 0.46 | 2.24 ± 0.50 |

*Note:* PGR, Plant growth regulator.

the L2 layer of the CN explant (Figure 9.1D-F). There were no visible signs of an intervening callus phase during shoot development and development of shoot buds initiated after 2 weeks of culture (Figure 9.1G). The individual shoots were rooted with 80 to 86 percent efficiency on unmodified MS medium having reduced nitrogen (0.084 $g \cdot l^{-1}$ $NH_4NO_3$ and 0.096 $g \cdot l^{-1}$ $KNO_3$), 0.5 $mg \cdot l^{-1}$ IBA, 1 percent (w/v) sucrose, and 0.6 percent (w/v) agar (Figure 9.1I). The conversion frequency of regenerants into complete plantlets (Figure 9.1J and K) through the process of in vitro acclimatization and hardening was 38 to 42 percent while micrografting of young shoots directly on 7-day-old stock of the same genotype resulted in 82 to 85 percent survival and complete development of plants in the glasshouse (Figure 9.1H).

Transient as well as stable expression of the *uid*A (GUS) marker gene was used to monitor and optimize *Agrobacterium*-mediated DNA delivery and recovery of transformants using cotyledonary node explants in all four genotypes of chickpea. The enzymatic activity of GUS was monitored by the positive histochemical assay as evidenced by the blue foci on the surface and developed in the thin sections of CNs and emerging shoots during antibiotic screening. About 85 to 92 percent of the CNs, which survived in the selection medium containing kanamycin, expressed GUS. A strong GUS-positive signal was detected in the leaflets and shoots of putative transformed plants established after three cycles of antibiotic screening (Figure 9.3A-D).

Among the four different strains (GV 2260, GV 3850, LBA 4404, and EHA 105) of *Agrobacterium* tested, we found strain LBA 4404 showed the

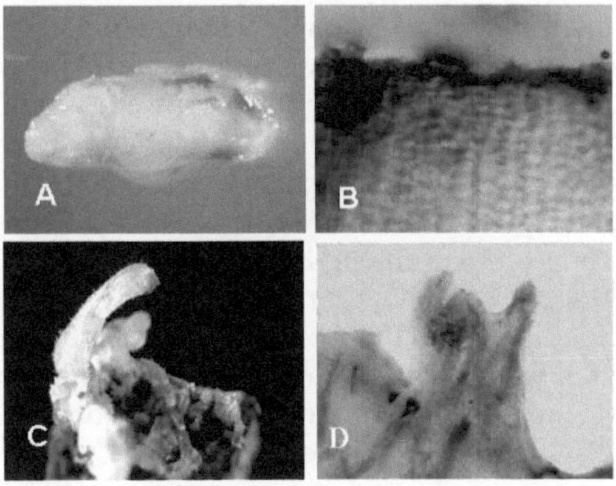


FIGURE 9.3. Histochemical GUS assay of cotyledonary node explants cocultivated with *A. tumefaciens* harboring pBI121; A, showing expression of GUS foci; B, expression of *uid*A gene in excised cells of CNs; C-D, GUS foci in developing new shoots from CNs. (See also color gallery.)

maximum number of GUS foci per explant, which was statistically significant compared to GV 2260, GV 3850 and EHA 105, that resulted into larger GUS smears and severe necrosis of explants. Therefore, strain LBA 4404 was preferred over others. The addition of acetosyringone (100 μM) as compared to syringaldehyde, syringic acid, and salicylic acid into the inoculum and cocultivation media increased the efficiency of T-DNA delivery both in terms of the number of explants displaying expression of *uid*A gene and the average number of GUS foci per explant. An increase in the level of acetosyringone beyond 200 μM significantly affected the regeneration frequency and size of GUS foci, whereas complete withdrawal of acetosyringone reflected a significant decrease in the frequency of explants expressing GUS and the mean number of GUS spots per explant in all the genotypes of chickpea. The concentration of *Agrobacterium* cell density in the cocultivation suspension with excised explants revealed that the optical density of bacterial culture between 0.8 and 1.0 (5-9 × $10^9$ cells/ml) produced maximum GUS spots, while increasing or decreasing the density caused a substantial decrease in the number and size of GUS spots. The preconditioning (preculture) of the excised explants on agar plates played an important role in increasing the transformation frequency. Explants preconditioned for 24 hours, before cocultivation for 48 hours produced a higher number of transformation events than explants preconditioned more than 24 hours. The duration of cocultivation with *Agrobacterium* was another important factor for increasing the transformation efficiency. Explants cocultivated for 48 hours yielded optimum efficiencies compared to those cocultivated for 24 and 72 hours. Taking into account the preconditioning time versus cocultivation, we found that explants cocultivated for 48 hours after 24 hours of preconditioning showed significantly better results than explants preconditioned for more than 24 hours at $p < 0.05$. The duration of sonication of the cotyledonary nodes with bacterial cells followed by vacuum treatment also showed a dramatic effect on the extent of transient expression of GUS in cotyledonary nodes. Sonication treatment ranging from 30 to 60 seconds to preconditioned CNs resulted in microwounding, efficient interaction with *Agrobacterium* cells, and a maximum level of transient expression with minimum inhibition of in vitro shoot regeneration. Sonication of CNs for more than 60 seconds showed a significant decrease in frequency of regeneration and GUS foci per responding explant. Based on optimized parameters, cocultivation of 24 hour preconditioned chickpea cotyledonary nodes prepared by the modified procedure was performed with *A. tumefaciens* strain LBA 4404 harboring different binary vectors like pRD400, pBIN200, and pRD401 harboring modified *Bacillus thuringiensis* crystal protein gene(s) *cry1Ab, 1Ac,* and both the genes together. Following three successive cycles of screening on kanamycin-supplemented selection medium, several putative $T_0$ primary transformed shoots were recovered with transformation frequencies ranging between 1.5 and 2.7 percent for individual *Bt-cry* genes and about 1.85 percent for coexpression of both the *cry* genes (Table 9.2). The procedure and parameters optimized with GUS reporter gene were found optimum for transfer and expression of *Bt-cry* genes in chickpea.

## *DISCUSSION*

The axillary meristems at the junction of the cotyledons and embryo axis contain cells that are competent for regeneration and therefore an amenable target for genetic transformation. Cotyledonary nodes from mature seeds in most of the grain legumes have been the most responsive explants for the induction of multiple shoots via organogenesis (Parrott et al., 1992; Polisetty et al., 1997; Chandra and Pental, 2003). The conventional procedure for preparation of CNs is by excision of plumule, radicle, and two cotyledons with a surgical blade of the newly germinated mature seeds of chickpea and other grain legumes as well. Large-seeded grain legumes are extremely recalcitrant to genetic transformation owing to inadequate availability of competent target cells in explants or embryogenic cultures for delivery of foreign DNA and regeneration of the transformed cells into plantlets (Somers et al., 2003; Amla and Sanyal, 2004). Most of the grain legumes are deprived of induction of high-frequency efficient somatic embryogenesis, which is easily amenable to DNA-mediated transformation (Parrott et al.,

TABLE 9.2. Genetic transformation of chickpea using cotyledonary node explants with *A. tumefaciens* strain LBA 4404 harboring different binary vectors with *cry1Ab* and *1Ac* genes.

| | | | Molecular analysis | | | |
|---|---|---|---|---|---|---|
| | | | PCR | Southern | | |
| **Binary vectors** | **Chickpea genotypes** | **Shoots** | ***cry1Ab/1Ac* + *nptII*** | **Bt** | ***nptII*** | **Cotransformation frequency (%)** |
| pRD400 | Pusa 362 | 22 | 21 | 22 | 21 | 1.86 |
| pBIN200 | Pusa 362 | 33 | 32 | 11 | 11 | 2.77 |
| | BG 256 | 11 | 10 | 11 | 10 | 1.49 |
| pRD401 | Pusa 362 | 33 | 29 | 29 | 30 | 1.85 |

*Note:* Bt, *Bacillus thuringiensis;* PCR, polymerase chain reaction.

1995). Therefore, a direct organogenic system derived from cotyledonary node in most of the grain legumes is the frequently used system for *Agrobacterium*-mediated T-DNA delivery into regenerable axillary meristematic cells (Hinchee et al., 1988; Meurer et al., 1998; Olhoft et al., 2003). However, production of transgenic plants remains low, most likely due to inefficient T-DNA delivery into meristematic cells of CNs, proliferation of transgenic cells, and difficulty in plant regeneration. Incubation of CNs on medium supplemented with cytokinins like BAP and TDZ results in shoot proliferation and micropropagation from the preexisting meristems instead of de novo formation of meristem. However, the number of shoots recovered from these explants is low, ranging between four and seven under the optimum growth conditions and combinations of phytohormones.

Considering several components of the cotyledonary node-based transformation system, the modified procedure for preparation of CN as well as *Agrobacterium* cocultivation conditions seem to provide favorable conditions for delivery of T-DNA into metabolically active cells as a consequence of excision and sonication. The preexisting mature meristematic cells were provoked to proliferate for 21 days and excised to expose cells of the L2 layer that may possess few young embedded meristem and layers of cells that may generate new meristematic activity for organogenesis on cytokinin-supplemented medium. Wounding of CNs results in the release of phenolics to induce *vir* function in *A. tumefaciens* (Bolton et al., 1986), stimulation of host DNA replication for facilitating T-DNA integration (Kudirka et al., 1986; Villemont et al., 1997), and tissue disruption for de novo shoot production from the wounded surface (Wright et al., 1986) that provides access to *Agrobacterium* for interaction with the target cells. Addition of thiol compounds L-cysteine along with L-glutamine and L-arginine appears to improve T-DNA delivery by inhibiting the activity of polyphenol oxidases induced as a result of plant pathogen infection and wound-induced response due to excision (Olhoft et al., 2001). Repeated selection on increasing threshold levels of antibiotic results in screening for escapes and non-chimeric transformed cells due to the improvised selection pressure. Using different *Bt-cry* gene constructs for cocultivation of preconditioned CNs of chickpea, a fairly high frequency of putative transgenic shoots was recovered as compared to earlier reports in grain legumes (Babaoglu et al., 2000).

In conclusion, the complexities of *Agrobacterium*-mediated T-DNA delivery in recalcitrant grain legumes via organogenic pathways of regeneration can be possibly overcome with improved conditions of bacterial interaction with the competent regenerative target cells in cotyledonary node prepared and processed by the modified procedure as described herein.

## REFERENCES

Amla, D.V. and I. Sanyal (2004). Complexities of in vitro regeneration and genetic transformation in large seeded grain legumes for developing stable transgenic plants. In *Golden Jubilee Volume*, P. Pushpangadan, S. Kumar, and V.K. Kochhar (eds.). National Botanical Research Institute, Lucknow, India, pp. 215-220.

Babaoglu, M., M.R. Davey, and J.B. Power (2000). Genetic engineering of grain legumes: Key transformation events. *AgBiotechNet* 2:1-8.

Bolton, G.W., E.W. Nester, and M.P. Gordon (1986). Plant phenolic compounds induce expression of *Agrobacterium tumefaciens* loci needed for virulence. *Science* 232:983-985.

Chandra, A. and D. Pental (2003). Regeneration and genetic transformation of grain legumes: An overview. *Current Science* 84:381-387.

Finer, J.J. and M.D. McMullen (1991). Transformation of cotton (*Gossypium hirsutum*) via particle bombardment of embryogenic suspension culture tissue. *In Vitro Cellular and Developmental Biology–Plant* 27:175-181.

Gamborg, O.L., R.A. Miller, and K. Ojima (1968). Nutrient requirements of suspension cultures of soybean root cells. *Experimental Cell Research* 50:150-158.

Hinchee, M.A.W., D.V. Connor-Ward, C.A. Newell, R.E. McDonnell, S.J. Sato, C.S. Gasser, D.A. Fischhoff, D.B. Re, R.T. Fraley, and R.B. Horsch (1988). Production of transgenic soybean plants using *Agrobacterium* mediated DNA transfer. *Nature Bio/Technology* 6:912-922.

Kudirka, D.T., S.M. Colburn, M.A. Hinchee, and M.S. Wright (1986). Interactions of *Agrobacterium tumefaciens* with soybean [*Glycine max* (L.) Merr.] leaf explants in tissue culture. *Canadian Journal of Genetics and Cytology* 28:808-817.

McCabe, D.E., W.F. Swain, B.J. Martinell, and P. Christou (1988). Stable transformation of soybean (*Glycine max*) by particle acceleration. *Nature Bio/Technology* 6:923-926.

Meurer, C.A., R.D. Dinkins, and G.B. Collins (1998). Factors affecting soybean cotyledonary node transformation. *Plant Cell Reports* 18:180-186.

Murashige, T. and F. Skoog (1962). A revised medium for rapid growth and bioassays with tobacco tissue cultures. *Physiologia Plantarum* 15:473-497.

Muthukumar, B., M. Mariamma, K. Veluthambi, and A. Gnanam (1996). Genetic transformation of cotyledon explants of cowpea [*Vigna unguiculata* (L.) Walp.] using *Agrobacterium tumefaciens*. *Plant Cell Reports* 15:980-985.

Olhoft, P.M., L.E. Flagel, C.M. Donovan, and D.A. Somers (2003). Efficient soybean transformation using hygromycin-B selection in the cotyledonary node method. *Planta* 216:723-735.

Olhoft, P.M., K. Lin, J. Galbraith, N.C. Nielsen, and D.A. Somers (2001). The role of thiol compounds in increasing *Agrobacterium*-mediated transformation of soybean cotyledonary-node cells. *Plant Cell Reports* 20:731-737.

Parrott, W.A., M.A. Bailey, R.E. Durham, and H.V. Mathews (1992). Tissue culture and regeneration of legumes. In *Biotechnology and Crop Improvement in Asia,*

J.P. Moss (ed.). Patancheru, India, International Crop Research Institute for Semi Arid Tropics (ICRISAT), pp. 115-148.

Parrott, W.A., R.E. Durham, and M.A. Bailey (1995). Somatic embryogenesis in legumes. In *Biotechnology in agriculture and forestry, somatic embryogenesis and synthetic seed II,* Y.P.S. Bajaj (ed.). Berlin: Springer-Verlag, Vol. 31, pp. 199-227.

Parrott, W.A., L.M. Hoffman, P. Hildebrandt, E.G. Williams, and G.B. Collins (1989). Recovery of primary transformants of soybean. *Plant Cell Reports* 7:615-617.

Polisetty, R., V. Paul, J.J. Deveshwar, S. Khetarpal, K. Suresh, and R. Chandra (1997). Multiple shoot induction by benzyl adenine and complete plant regeneration from seed explants of chickpea (*Cicer arietinum* L.). *Plant Cell Reports* 16: 565-571.

Popelka, J.C., N. Terryn, and T.J.V. Higgins (2004). Gene technology for grain legumes: Can it contribute to the food challenge in developing countries? *Plant Science* 167:195-206.

Ranalli, P. and J.I. Cubero (1997). Bases for genetic improvement of grain legumes. *Field Crops Research* 53:69-82.

Sanyal, I., A.K. Singh, and D.V. Amla (2003). *Agrobacterium tumefaciens*-mediated transformation of chickpea (*Cicer arietinum* L.) mature embryo axis and cotyledonary nodes. *Indian Journal of Biotechnology* 2:524-532.

Sanyal, I., A.K. Singh, M. Kaushik, and D.V. Amla (2005). *Agrobacterium*-mediated transformation of chickpea (*Cicer arietinum* L.) with *Bacillus thuringiensis cry1Ac* gene for resistance against pod borer insect *Helicoverpa armigera. Plant Science* 168:1135-1146.

Sharma, H.C., J.H. Crouch, K.K. Sharma, N. Seetharama, and C.T. Hash (2002). Applications of biotechnology for crop improvement: Prospects and constraints. *Plant Science* 163:381-395.

Somers, D.A., D.A. Samac, and P.M. Olhoft (2003). Recent advances in legume transformation. *Plant Physiology* 131:892-899.

Trick, H.N. and J.J. Finer (1998). Sonication-assisted *Agrobacterium* mediated transformation of soybean [*Glycine max* (L.) Merr.] embryogenic suspension culture tissue. *Plant Cell Reports* 17:482-488.

Villemont, E., F. Dubois, R.S. Sangwan, G. Vasseur, Y. Bourgeois, and B.S. Sangwan-Norreel (1997). Role of host cell cycle in the *Agrobacterium*-mediated genetic transformation of *Petunia*: Evidence of S-phase control mechanism for T-DNA transfer. *Planta* 201:160-172.

Wright, M.S., S.M. Koehler, M.A. Hinchee, and M.G. Carnes (1986). Plant regeneration by organogenesis in *Glycine max. Plant Cell Reports* 5:150-154.

Yan, B., M.S.S. Reddy, G.B. Collins, and R.D. Dinkins (2000). *Agrobacterium tumefaciens*-mediated transformation of soybean [*Glycine max* (L.) Merrill.] using immature zygotic cotyledon explants. *Plant Cell Reports* 19:1090-1097.

Chapter 10

# Genetic Transformation of *Phaseolus* and *Vigna* Species

Nancy Terryn
Marc van Montagu

## *INTRODUCTION*

This review presents the state of the art in *Phaseolus* and cowpea (*Vigna unguiculata*) genetic transformation. Since both *Phaseolus* and cowpea are members of the same "tropical tribe" within the family Fabaceae, they are discussed here together. Also, approaches successful for genetic transformation of *Phaseolus,* as described further, might offer possibilities to transform cowpea. *Phaseolus vulgaris,* or the common bean, is the most important grain legume for direct food use, as it produces seeds that are consumed worldwide as dietary protein source by millions of people, mainly in South America, India, and China. Its annual production is around 24 million metric tons (Popelka et al., 2004). Due to the high variability within the *Phaseolus* family, genetic improvement by classical breeding has been quite successful (Singh, 1999). Main targets are, as for most crops, to breed varieties with high yield and tolerance or resistance to diseases, pests, and drought. Other *Phaseolus*-related breeding objectives include better taste, low flatulence factors, pod shattering, and plant architecture as well as pod distribution that allow easy mechanical harvesting.

Cowpea, or black-eyed pea, on the other hand, is the major food legume for Africa (mainly Niger and Nigeria), also grown in some countries in Latin America, like northeastern Brazil. Cowpea is quite a drought-tolerant crop and its annual production is close to 4 million metric tons per year (Popelka et al., 2004).

---

The authors wish to thank Marcelle Holsters, Sylvia Burssens, Fernand Lambein and Dulce de Oliveira for reading the manuscript and giving valuable comments.

Regeneration and transformation of large-seeded grain legumes are still troublesome. Regeneration ability seems to be dependent on genotype, explant, and tissue culture conditions. Yet the establishment and optimization of genetic transformation protocols for legumes is a highly important research topic. Transformation is a very valuable tool to applications in both agriculture and industry as discussed later, and as a research tool for large-scale genomic programs. Transgenic plants are needed in functional genomics to create a library of insertion mutants to verify or elucidate the function of newly identified genes through, for example, silencing or overexpression, or to follow their expression through promoter-marker gene fusions.

## *STATE OF THE ART OF GENETIC TRANSFORMATION IN COWPEA AND* PHASEOLUS

In general, a prerequisite for a stable transformation protocol is a highly efficient regeneration system. Regeneration of different cowpea explants through direct organogenesis has been described using mainly hypocotyls, cotyledons, or thin cell layer (Pelligrineschi, 1997; Brar et al., 1999; Choi et al., 2003; Le et al., 2002). Although somatic embryogenesis has been obtained from different explants such as cell suspensions, callus, cotyledons, and embryonic axes (Anand et al., 2001; Amitha and Reddy, 1996 and references therein), it has been difficult to obtain regenerated plants at a high frequency. As for the method of DNA transfer, both *Agrobacterium* infection and particle gun methodology of cowpea has been demonstrated (Garcia et al., 1986; Muthukumar et al., 1996; Ikea et al., 2003) but no transformants have been obtained beyond the primary ones. One group with vast expertise on grain legume transformation has published a protocol for stable genetic transformation for cowpea, based on *Agrobacterium* transformation of the cotyledonary nodes (Popelka et al., 2006).

Until today, a stable and reproducible protocol was not available for *Agrobacterium*-mediated *Phaseolus vulgaris* transformation in spite of the fact that there has been much more research on the transformation of *Phaseolus* as compared to cowpea. An initiative like Phaseomics, a genome program, will need genetic transformation as a tool (Broughton et al., 2003, and http://www.phaseolus.net). The focus in this program is to physically map the genome (Blair et al., 2003) with the ultimate goal of sequencing it.

Several regeneration protocols, as well as genotypes susceptible to *Agrobacterium* are available (Svetleva et al., 2003). *Phaseolus* explants have been shown to be susceptible to *Agrobacterium,* although this is genotype dependent (Lewis and Bliss, 1994; Brasileiro et al., 1996). Most attempts to produce stable transgenic *Phaseolus* plants resulted in nonregenerable transgenic callus or chimeric tissues (Franklin et al., 1993; Dillen et al., 1995). The research group of the Department of Molecular Genetics, Ghent University, Belgium, has a long track record in *Phaseolus* transformation, albeit not the common bean, *Phaseolus vulgaris,* but the tepary bean, *Phaseolus acutifolius*. Transformation of the latter is stable and reproducible (Dillen et al., 1997; De Clercq et al., 2002, 2008; Zambre et al., 2005). The protocol is based on the generation of callus tissue from either floral meristems or the cotyledonary node region. This callus, after several passages on a medium containing thidiazuron and indoleacetic acid, is cocultivated with *A. tumefaciens,* followed by selection on genticin-containing medium. After several rounds of selection, Gus-positive callus is transferred onto shoot induction medium. Fully developed shoots are allowed to set roots, and plantlets are then transferred to soil. A detailed description of the protocol is published elsewhere (Zambre et al., 2006). In this way, several transgenic *P. acutifolius* lines have been generated, aimed at either insect tolerance (Zambre et al., 2005) or increased methionine content (De Clercq et al., 2008). In addition, *P. acutifolius* has been evaluated as a potential crop for molecular farming (De Jaeger et al., 2002).

Since *P. acutifolius* can be crossed with *P. vulgaris,* albeit through embryo rescue (Mejía-Jiméez et al., 1994), this can be a possible route to genetically improve *P. vulgaris,* as long as a reproducible transformation system is not available for the latter species. In addition *P. acutifolius* can serve very well as a model to study gene function or transgene expression in *Phaseolus,* as we have shown in our work.

Through particle bombardment, several transgenic lines have been obtained with traits of interest to agriculture (Aragão et al., 1998, 2002). Transgenic bean plants with tolerance to bean golden mosaic virus are currently in field trials for evaluation and will eventually be commercialized (Embrapa, Brazil). However, this protocol has not been applied by other groups. Also, the efficiency inherent to a method based on particle bombardment of the meristem is quite low.

## Agrobacterium rhizogenes

It might be useful to mention here that in some cases genetic transformation through *A. rhizogenes* to create transgenic hairy roots might be a useful alternative to stable transformation. Many agricultural problems relate to processes in the root system. Transgenic hairy roots can thus be useful to study the genes involved in these processes or to evaluate candidate genes through transgenic approaches. Although this idea dates back to the early days of nitrogen fixation studies, there is a renewed interest as more genomic data are available and more labs want a quick tool for genetic analysis. In *Medicago* a protocol has been described that uses so-called composite plants to study nodulation-related events (Boisson-Dernier et al., 2001). Composite plants are created by inoculating young seedlings, of which the roots have been removed, with *A. rhizogenes*. Hairy roots are formed at the site of the cut and replace the natural root system. If *A. rhizogenes* also carries the T-DNA with the gene of interest and a selectable marker, transgenic hairy roots expressing these gene can be selected for. These composite plants, with a transgenic root system and wild-type upper parts, can consequently be used as study material. This protocol has recently been adapted and even optimized to be used in important legumes such as *Phaseolus* and soybean (Colpaert et al., 2007; Estrada et al., 2006; Collier et al., 2005).

## *SOME PROBLEMS OF BEAN AND COWPEA THAT COULD BE SOLVED BY GENETIC ENGINEERING*

In general, the productivity of legumes is very low in comparison to cereals, making the crops unattractive for the farmer. To change this trend, it is important to increase legume productivity. Unfortunately, legume crop improvements as well as improved agricultural practices were not able to solve some specific problems. Plant breeding approaches are more successful when they can rely on wide genetic variability. If, however, no genes conferring a trait of specific interest (e.g., pest resistance) are found within the crop as a primary gene pool, and its wild relatives, as a secondary gene pool, the implementation of novel biotechnological solutions should be considered. A few examples are given for problems of bean and cowpea cultivation for which biotechnological solutions are sought.

### *Insect Resistance*

For cowpea, the major challenge is to increase yield through insect resistance, as a wide range of insect pests attack cowpea, even during the postharvest period. These include the cowpea leaf beetle (*Ootheca mutabilis*), the pod borer (*Maruca vitriata*), and the pod-sucking insect (*Clavigralla tomentosicollis*) as well as aphids and thrips. Yield losses can be up to 90 percent depending on region, year, and cultivar. *Maruca* attack is a particularly important threat: it is estimated that in Niger, one of the major cowpea-growing countries, yield loss can be as high as 40 to 60 percent (Gressel et al., 2004). In addition, foods prepared with infected grains have an unpleasant flavor and have low value on the market. A low level of resistance to the *Maruca* pod borer has been found in the wild relative *Vigna vexillata*, but classical transfer of the resistance and backcrossing has not been successful. The use of cowpea engineered to express the *Bacillus thuringiensis* Cry proteins is thus considered a valuable solution for this problem.

*Phaseolus* also suffers from bean pod borers, bruchids, and weevils that can be devastating, also at seed storage level (Singh, 1999). The heterologous expression of *Phaseolus Arcelin* genes, isolated from wild *P. vulgaris* varieties, has been evaluated with regard to insect tolerance. *P. acutifolius* lines have been obtained with high levels of *Arcelin* expression, but these show only a marginally less susceptibility to the bean bruchid *Zabrotes subfasiatis* (Goossens et al., 2000; Zambre et al., 2005).

### *Virus Resistance*

Two worldwide viruses, the bean golden mosaic virus and the bean yellow mosaic virus, are very troublesome for bean cultivation. Virus resistance has been shown to be a trait that can easily be introduced. The strategy is based on the expression of virus-related sequences that trigger the virus-induced silencing system of the plant (Tenllado et al., 2004). This has been used for beans by the Brazilian research center of Embrapa to obtain bean golden mosaic virus-tolerant transgenic plants (Aragão et al., 1998).

### *Parasitic Plants*

*Striga* and *Orobanche* species are parasitic plants that can attack legumes, such as faba bean and cowpea (Gressel et al., 2004). Conventional technologies of control of these weeds are quite labor intensive and ineffective. Potential solutions could be offered by the development of herbicide-resistant crops, as has been shown for corn (Kanampiu et al., 2003). Another biotechnological option would be to secrete a parasite weed toxic compound from the legume's roots (Duke, 2003).

### *Nutritional Aspects of* **Phaseolus** *and Cowpea*

Besides agricultural problems, nutritional aspects of beans and cowpea could also be enhanced through genetic engineering. Legumes have a suboptimal amino acid composition with respect to their consumption by humans and animals, as they are low in the essential sulfur-containing amino acids. Breeding efforts to enhance these levels in legume seeds have been unsuccessful so far. The heterologous expression of methionine-rich proteins has been shown to be possible (Molvig et al., 1997; Dinkins et al., 2001; De Clercq et al., 2008). However, care should be taken so that these proteins will not cause allergic reactions, as has been shown for the methionine-rich 2S albumin (Nordlee et al., 1996). Also, expressing a mutant gene coding for a key enzyme of the methionine biosynthetic pathway or combining both strategies has been done (Demidov et al., 2003). Higher methionine-containing legumes are also important for feed, since it would no longer require synthetic methionine supplementation (Ravindran et al., 2002). Besides the suboptimal amino acid composition, legumes also contain antinutritional factors like phytic acid, tannins, and trypsin inhibitors. Another undesirable trait is the high content of certain highly soluble oligosaccharides such as raffinose, which cause flatulence. Humans and other monogastric animals cannot digest these compounds, and reducing their content through biotechnological means would be desirable.

## *NEW AVENUES FOR GENETIC TRANSFORMATION*

Apart from the methods described above, non-tissue-culture-based legume transformation protocols such as flower dip or seedling infiltration techniques are a very attractive research area. However, for bean and cowpea, the protocol will need to be very efficient to make it easily applicable. Extensive efforts with soybean have not yet met with any success (Bent et al., 2002).

The improvement of culture-based transformation protocols usually comprises testing various culture conditions, media compositions, hormones, and selection systems. Although this is often a robust way to enhance transformation efficiency, it is quite time consuming. However, the data accumulating on tissue culture-based plant development, and more precisely the recently gained knowledge on organogenesis and embryogenesis, will soon allow the development of alternative approaches to improve legume transformation. One example is the use of the *RepA* gene of a mastre virus, known to interact with cell cycle regulators, that has been shown to increase transformation frequency in maize (Gordon-Kamm et al., 2002). Other interesting examples are genes promoting vegetative to embryogenic transitions, such as *Wushel* (Zuo et al., 2002) and *Baby Boom* (Boutilier et al., 2002). These genes might be evaluated for their usefulness in transformation.

The new tools of system biology, that combine genomics, proteomics and metabolomics would allow a better understanding of the *Agrobacterium* transformation process. The identification of genes with a positive effect on transformation-related steps could be very useful in the development of *Agrobacterium* transformation systems for recalcitrant species and varieties. It would be an enormous step forward if *Agrobacterium* could be engineered in such a way that it becomes a perfect gene delivery system for all plant systems.

## REFERENCES

Amitha, K., and T.P. Reddy (1996). Induction of somatic embryogenesis and regeneration in cowpea (*Vigna sinensis* L.). *Advances in Plant Science* 9:23-28.

Anand, R.P., A. Ganapathi, G. Vengadesan, N. Selvaraj, V.N. Anbazhagan, and S. Kulothungan (2001). Plant regeneration from immature cotyledon-derived callus of *Vigna unguiculata* (L.) Walp. (cowpea). *Current Science* 80:671-674.

Aragão, F., S. Ribeiro, L. Barros, A. Brasileiro, D. Maxwell, E. Rech, and J. Faria (1998). Transgenic beans (*Phaseolus vulgaris* L.) engineered to express viral antisense RNAs show delayed and attenuated symptoms to bean golden mosaic geminivirus. *Molecular Breeding* 4:491-499.

Aragão, F.J., G.R. Vianna, M.M.C. Albino, and E.L. Rech (2002). Transgenic dry bean tolerant to the herbicide glufosinate ammonium. *Crop Science* 42:1298-1302.

Bent, A., A. McGill, K. Schroeder, C. Desfeux, and S. Clough (2002). Soybean molecular genetics: Germ line transformation attempt, new efforts at virus-induced gene silencing. *Abstracts ICLGG meeting,* June 2-6, Minneapolis-St. Paul, p. 29.

Blair, M.W., F. Pedraza, H.F. Buendia, E. Gaitan-Solis, S.E. Beebe, P. Gepts, and J. Tohme (2003). Development of a genome-wide anchored microsatellite map for common bean (*Phaseolus vulgaris* L.). *Theoretical and Applied Genetics* 107: 1362-1374.

Boisson-Dernier, A., M. Chabaud, F. Garcia, G. Becard, C. Rosenberg, and D.G. Barker (2001). *Agrobacterium rhizogenes*-transformed roots of *Medicago truncatula* for the study of nitrogen-fixing and endomycorrhizal symbiotic associations. *Molecular Plant-Microbe Interactions* 14:695-700.

Boutilier, K., R. Offringa, V.K. Sharma, H. Kieft, T. Ouellet, L.M. Zhang, J. Hattori, C.M. Liu, A.A. van Lammeren, B.L.A. Miki et al. (2002). Ectopic expression of *BABY BOOM* triggers a conversion from vegetative to embryonic growth. *Plant Cell* 14:1737-1749.

Brar, M.S., J.M. Al-Khayri, T.E. Morelock, and E.J. Anderson (1999). Genotypic response of cowpea *Vigna unguiculata* (L.) to in vitro regeneration from cotyledon explants. *In Vitro Cellular and Developmental Biology–Plant*. 35:8-12.

Brasileiro, A.C, F.J. Aragão, S. Rossi, D.M. Dusi, L.M.G. Barros, and E.L. Rech (1996). Susceptibility of common and tepary beans to *Agrobacterium* strains and improvement of *Agrobacterium* mediated transformation using microprojectile bombardment. *Journal of the American Society of Horticultural Science* 121: 810-815.

Broughton, W.J., Hernandez, G., Blair, M., Beebe, S., Gepts, P., Vanderleyden. J. (2003) Beans (*Phaseolus spp.*)-model food legumes. Plant and Soil 252:55-128.

Choi, P.S., D.Y. Cho and W.Y. Soh (2003). Plant regeneration from immature embryo cultures of *Vigna unguiculata. Biologia Plantarum* 47:305-308.

Collier, R., B. Fuchs, N. Walter, W. K. Lutke, and C.G. Taylor (2005). Ex vitro composite plants: An inexpensive, rapid method for root biology. *Plant Journal* 43:449-457.

Colpaert, N., Tilleman, S., Gheyssen, G., Terryn, N. (2007). The use of *Agrobacterium rhizogenes* as a tool to study trangene expression in root systems of *Phaseolus* and cowpea. Manuscript in preparation.

De Clercq, J., T. Hamelryck, M. Zambre, A. Goossens, M. Van Montagu, N. Terryn, and G. Angenon (2008). Increasing the methionine content of *Phaseolus* beans through modification of the *Phaseolus vulgaris* seed protein Arcelin 5a. *African Journal of Biotechnology,* in press.

De Clercq, J., M. Zambre, M. Van Montagu, W. Dillen, and G. Angenon (2002). An optimized *Agrobacterium*-mediated transformation procedure for *Phaseolus acutifolius* A Gray (tepary bean). *Plant Cell Reports* 21:333-340.

De Jaeger, G., S. Scheffer, A. Jacobs, M. Zambre, O. Zobell, A. Goossens, A. Depicker, and G. Angenon (2002). Boosting heterologous protein production in transgenic dicotyledonous seeds using *Phaseolus vulgaris* regulatory sequences. *Nature Bio/Technology* 20:1265-1268.

Demidov, D., C. Horstmann, M. Meixner, T. Pickardt, I. Saalbach, G. Galili, and K. Muntz (2003). Additive effects of the feed-back insensitive bacterial aspartate kinase and the Brazil nut 2S albumin on the methionine content of transgenic narbon bean (*Vicia narbonensis* L.). *Molecular Breeding* 11:187-201.

Dillen, W., J. De Clercq, A. Goossens, M. Van Montagu, and G. Angenon (1997). *Agrobacterium*-mediated transformation of *Phaseolus acutifolius* A Gray. *Theoretical Applied Genetics* 94:151-158.

Dillen, W., G. Engler, M. Van Montagu, and G. Angenon (1995). Electroporation-mediated DNA delivery to seedling tissues of *Phaseolus vulgaris* L. (common bean). *Plant Cell Reports* 15:119-124.

Dinkins, R.D., M.S.S. Reddy, C.A. Meurer, B. Yan, H. Trick, F. Thibaud-Nissen, J.J. Finer, W.A. Parrott, and G.B. Collins (2001). Increased sulfur amino acids in soybean plants overexpressing the maize 15 kD zein protein. *In Vitro Cellular and Developmental Biology–Plant* 37:742-747.

Duke, S.O. (2003). Weeding with transgenes. *Trends in Biotechnology* 21:192-194.

Estrada, G., E. Murillo, J.L. Reyes, C. Diaz, J.L. Olivares, G. Guillén, O. Santana, X. Alvarado, L. Cardenas, C. Quinto, et al. (2006). Transgenic *Phaseolus vulgaris* hairy roots induced by *Agrobacterium rhizogenes* can be efficiently nodulated by rhizobia. Abstract presented at the 3rd ICLGG Meeting, Brisbane, Australia, 9-13 April 2006, p. 58.

Franklin, C.I., T.N. Trieu, B.G. Cassidy, R.A. Dixon, and R.S. Nelson (1993). Genetic transformation of green bean callus via *Agrobacterium* mediated DNA transfer. *Plant Cell Reports* 12:74-79.

Garcia, J.A., J. Hillie, and R. Goldbach (1986). Transformation of cowpea *Vigna unguiculata* cells with a full length DNA copy of cowpea mosaic virus m-RNA. *Plant Science* 44:89-98.

Goossens, A., C. Quintero, W. Dillen, R. De Rycke, J.F. Valor, J. De Clercq, M. Van Montagu, C. Cardona, and G. Angenon (2000). Analysis of bruchid resistance in the wild common bean accession G02771: no evidence for insecticidal activity of arcelin 5. *Journal of Experimental Botany* 51:1229-1236.

Gordon-Kamm, W., B.P. Dilkes, K. Lowe, G. Hoerster, X.F. Sun, M. Ross, L. Church, C. Bunde, J. Farrell, P. Hill, et al. (2002). Stimulation of the cell cycle and maize transformation by disruption of the plant retinoblastoma pathway. *Proceedings of the National Academy of Sciences USA* 99:11975-11980.

Gressel, J., A. Hanafi, G. Head, W. Marasas, B. Obilana, J. Ochanda, T. Souissi, and G. Tzotzos (2004). Major heretofore intractable biotic constraints to African food security that may be amenable to novel biotechnological solutions. *Crop Protection* 23:661-689.

Ikea, J., I. Ingelbrecht, A. Uwaifo, and G. Thottappilly (2003). Stable gene transformation in cowpea (*Vigna unguiculata* L. Walp.) using particle gun method. *African Journal of Biotechnology* 2:211-218.

Kanampiu, F.K., V. Kabambe, C. Massawe, L. Jasi, D. Friesen, J.K. Ransom, and J. Gressel (2003). Multi-site, multi-season field tests demonstrate that herbicide seed-coating herbicide-resistance maize controls *Striga* spp. and increases yields in several African countries. *Crop Protection* 22:697-706.

Le, B.V., M.H. Cruz de Carvalho, Y. Zuily-Fodil, A.T.P. Thi, and K.T.T. Van (2002). Direct whole plant regeneration of cowpea [*Vigna unguiculata* (L.) Walp.] from cotyledonary node thin cell layer explants. *Journal of Plant Physiology* 159:1255-1258.

Lewis, M.E. and F.A. Bliss (1994). Tumor formation and beta-glucuronidase expression in *Phaseolus vulgaris* inoculated with *Agrobacterium tumefaciens*. *Journal of the American Society of Horticultural Science* 119:361-366.

Mejía-Jiménez, A., C. Munoz, H.J. Jacobsen, W.M. Roca, and S.P. Singh (1994). Interspecific hybridization between common and tepary beans increased hybrid embryo growth, fertility, and efficiency of hybridization through recurrent and congruity backcrossing. *Theoretical and Applied Genetics* 88:324-331.

Molvig, L., L.M. Tabe, B.O. Eggum, A.E. Moore, S. Craig, D. Spencer, and T.J.V. Higgins (1997). Enhanced methionine levels and increased nutritive value of seeds of transgenic lupins (*Lupinus angustifolius* L.) expressing a sunflower seed albumin gene. *Proceedings of the National Academy of Sciences USA* 94:8393-8398.

Muthukumar, B., M. Mariamma, K. Veluthambi, and A. Gnanam (1996). Genetic transformation of cotyledon explants of cowpea [*Vigna unguiculata* (L.) Walp.] using *Agrobacterium tumefaciens*. *Plant Cell Reports* 15:980-985.

Nordlee, J.A., S.L. Taylor, J.A. Townsend, L.A. Thomas, and R.K. Bush (1996). Identification of a Brazil nut allergen in transgenic soybean. *New England Journal of Medicine* 334:688-692.

Pellegrineschi, A. (1997). In vitro regeneration via organogenesis of cowpea [*Vigna unguiculata* (L.) Walp.]. *Plant Cell Reports* 17:89-95.

Popelka, J.C., S. Gollasch, and T.J.V. Higgins (2006). Genetic transformation of cowpea (*Vigna unguiculata* L.) and stable transgene transmission to sexual progeny. *Plant Cell Reports* 25:304-309.

Popelka, J.C., N. Terryn, and T.J.V. Higgins (2004). Gene technology for grain legumes: Can it contribute to the food challenge in developing countries? *Plant Science* 167:195-206.
Ravindran, V., L.M. Tabe, L. Molvig, T.J.V. Higgins, and W.L. Bryden (2002). Nutritional evaluation of transgenic high-methionine lupins (*Lupinus angustifolius* L) with broiler chickens. *Journal of the Science of Food and Agriculture* 82:280-285.
Singh, S.P. (1999). Integrated genetic improvement. In *Common bean improvement in the 21st century,* S.P. Singh (ed.). Dordrecht: Kluwer Academic Publishers, pp. 133-165.
Svetleva, D., M. Velcheva, and G. Bhowmik (2003). Biotechnology as a useful tool in common bean (*Phaseolus vulgaris* L.) improvement. *Euphytica* 131:189-200.
Tenllado, F., C. Llave, and J.R. Diaz-Ruiz (2004). RNA interference as a new biotechnological tool for the control of virus diseases in plants. *Virus Research* 102:85-96.
Zambre, M., A. Goossens, C. Cardona, M. Van Montagu, N. Terryn, and G. Angenon (2005). An efficient and reproducible *Agrobacterium*-mediated transformation system for cultivated *Phaseolus acutifolius* (tepary bean). *Theoretical and Applied Genetics* 110:914-924.
Zambre, M., M. Van Montagu, G. Angenon, and N. Terryn (2006). Tepary bean (*Phaseolus acutifolius*). In *Agrobacterium protocols, methods in molecular biology vol. 343,* K. Wang (ed.). Totowa, NJ: Humana Press, pp. 407-414.
Zuo, J.R., Q.W. Niu, G. Frugis, and N.H. Chua (2002). The *WUSCHEL* gene promotes vegetative-to-embryonic transition in *Arabidopsis. Plant Journal* 3:349-359.

Chapter 11

# Genetic Engineering in Azuki Bean

Masao Ishimoto

## *INTRODUCTION*

Azuki bean [*Vigna angularis* (Willd.) Ohwi & Ohashi] is a traditional crop in East Asia (Lumpkin and McClary, 1994). Wild azuki bean, *V. angularis* var. *nipponensis* (Ohwi) Ohwi and Ohashi, the presumed progenitor of the cultivated species, is distributed from northeastern China, through the Korean peninsula and Japan to the Himalayan region of India, Nepal, and Bhutan (Tomooka et al., 2002). The domestication of azuki bean is estimated to have occurred more than once in these regions (Zong et al., 2003). Azuki bean is not an economically important legume worldwide, but it is widely cultivated in eastern Asia, including China and Japan. In Japan, azuki bean is the second most important grain legume after soybean [*Glycine max* (L.) Merrill] and it is cultivated all over the country except for Okinawa prefecture (Tomooka et al., 2002). Azuki bean is a member of the genus *Vigna,* which includes many economically important legume crops such as cowpea [*Vigna unguiculata* (L.) Walpers] and mung bean [*Vigna radiata* (L.) Wilczek], and is taxonomically close to soybean and common bean (*Phaseolus vulgaris* L.; Choi et al., 2004). This crop shows an attractive competence for transformation and regeneration by the *Agrobacterium*-mediated transformation system with kanamycin or hygromycin selection (Yamada et al., 2001; Hanafy et al., 2006).

## *GENETIC AND MOLECULAR FEATURES OF AZUKI BEAN*

Azuki bean possesses several advantageous properties as a model crop of grain legumes; a smaller genome size (539 Mbp) than soybean (1,120 Mbp), peanut (*Arachis hypogaea* L., 2,280 Mbp) and pea (*Pisum sativum*

L., 4,880 Mbp; Parida et al., 1990; Arumuganathan and Earle, 1991); a lower number of chromosomes ($n = 11$) compared to soybean ($n = 20$) and peanut ($n = 20$); a short generation period and an easy artificial cross-pollination. In addition, there has been a continuous addition of literature on molecular markers and genetic linkage maps, which is essential in developing map-based cloning and marker-assisted breeding strategies in azuki bean. Two different linkage maps have been developed based on restriction fragment length polymorphism (RFLP) and random amplified polymorphic DNA (RAPD) (Kaga et al., 1996, 2000). More recently, many simple sequence repeat (SSR) markers were developed (Wang et al., 2004) and applied to the construction of a linkage map with the amplified fragment length polymorphism (AFLP), RFLP, and SSR markers in addition to agronomic traits (Wang et al., 2004; Somta et al., 2005; Han et al., 2005). These features demonstrate that azuki bean is amenable to molecular genetic analysis and advanced molecular breeding programs.

## *PLANT REGENERATION FROM EXPLANTS AND PROTOPLASTS*

Plant regeneration of azuki bean was first reported from epicotyl explants of the seedlings (Ozaki, 1985). Epicotyl explants placed on the MS medium (Murashige and Skoog, 1962) containing 1 to 2 mg·l$^{-1}$ BAP (benzyl aminopurine) and 0.05-0.1 mg·l$^{-1}$ NAA ($\alpha$-naphthalene acetic acid) induced calli on the cutting surface. Adventitious shoot formation was observed from the compact calli on the MS medium supplemented with 0.2-1.0 mg·l$^{-1}$ BAP. The organogenesis from epicotyl explants occurred efficiently in a number of azuki bean varieties (Adachi et al., 1990; Sato et al., 1990; Yamada et al., 2001). Sato (1995) confirmed shoot regeneration from epicotyls of 15 Japanese varieties in varying frequencies, and observed a high frequency of shoot formation (92 percent) in the cultivar Beni-dainagon. The root formation from regenerated shoots was stimulated on the MS medium without plant growth regulators.

Plant regeneration was also achieved from immature first (primary) leaves of 3-9-day-old seedlings with lower frequency compared with the regeneration from epicotyls (Ozaki, 1986). Callusing was observed on the ridge of the explants on a medium with 2.0 mg·l$^{-1}$ BAP and 0.5 mg·l$^{-1}$ NAA, and shoots were regenerated from the compact calli on the MS medium supplemented with 0.2 mg·l$^{-1}$ BAP.

Protoplasts were isolated from mesophyll cells of azuki bean seedling (Ge et al., 1989) or epicotyl-derived calli (Sato et al., 1993). The isolated protoplasts regenerated cell walls divided continuously in liquid or on solidified media, resulting in the formation of calli. Plantlets were obtained from the protoplast-derived calli on media containing the combination of plant growth regulators, BAP as cytokinin, and NAA or 2,4-D (2,4-dichlorophenoxyacetic acid) as auxin. The protoplast-derived calli could be maintained for 10 months, but the regeneration frequency is as low as 2.5 to 3 percent (Sato et al., 1993). The appropriate azuki bean genotypes and culture conditions should be selected to improve the efficiency of plant regeneration from protoplasts.

## AGROBACTERIUM-*MEDIATED TRANSFORMATION USING EPICOTYL EXPLANTS*

Epicotyl explant, one of the best transforming materials, is easily obtained and efficiently regenerated through callus formation. These epicotyl explants have been used successfully in genetic transformation (Sato and Matsukawa, 1990; Sato, 1995; Ishimoto et al., 1996; Yamada et al., 2001; Chen et al., 2005). Sato (1995) extensively studied the conditions required for successful *Agrobacterium*-mediated transformation by using epicotyl explants. Callus induction from epicotyls on the regeneration medium was hampered by the addition of kanamycin above 50 $mg{\cdot}l^{-1}$. The addition of 500 $mg{\cdot}l^{-1}$ of carbenicillin effectively eliminated the bacteria from inoculated explants with little influence on adventitious shoot formation. Application of acetosyringone during inoculation and cocultivation had no effect on transformation efficiency. Finally, he produced several independent transgenic plants of cv. Erimo-shozu, which is one of the leading Japanese varieties, using neomycin phosphotransferace II (NPT II) gene as a selective marker. Yamada et al. (2001) suggested a number of modifications to the epicotyl explant transformation system, including the combination of a high concentration of BAP (10 $mg{\cdot}l^{-1}$) and acetosyringone to increase transformation efficiency, and the inhibition of the growth of highly infective *Agrobacterium* strains such as AGL1 (Lazo et al., 1991) and EHA105 (Hood et al., 1993). The transformation efficiency was approximately 2 percent. Regeneration of fertile transgenic azuki bean plants was reported using hygromycin phosphotransferace (HPT) gene as a selective marker for hygromycin with comparable transformation efficiency (3.6 percent) to the selection by kanamycin (Hanafy et al., 2006).

## A TRANSFORMATION PROTOCOL FOR AZUKI BEAN

### Preparation of Explants

Dry seeds of azuki bean were sterilized in 70 percent ethanol for 30 seconds followed by 1 percent sodium hypochlorite for 15 minutes and washed twice with sterile distilled water. The seeds are highly sensitive to chlorine gas and lose germination ability easily. Sodium hypochlorite solution must be used for seed sterilization. The sterile seeds were plated on basal MS medium (pH 5.8; Murashige and Skoog, 1962) containing 30 $g \cdot l^{-1}$ of sucrose and 8 $g \cdot l^{-1}$ of agar. Seeds were germinated at 25°C in the dark. Seven to ten days after plating, elongated epicotyls of etiolated seedlings were cut into pieces about 10 mm long with a scalpel blade and 25 explants were plated longitudinally on cocultivation medium in 90 × 20 mm plastic plates.

### *Inoculation of* Agrobacterium

Cocultivation was conducted on MS medium containing 30 $g \cdot l^{-1}$ of sucrose, 8 $g \cdot l^{-1}$ of agar, 10 $mg \cdot l^{-1}$ of BAP, and 100 μM of acetosyringone. A single colony of the *A. tumefaciens* strain EHA105 or AGL1 harboring an appropriate binary plasmid was transferred to 5 ml of liquid Luria-Bertani (LB) media supplemented with 10 $\mu g \cdot ml^{-1}$ of rifampicin and an adequate concentration of the selective agent such as 100 $\mu g \cdot ml^{-1}$ spectinomycin for the binary plasmid and grown at 28°C overnight. Bacterial cells were collected by centrifugation at 3,000 rpm for 10 minutes and resuspended to a final $OD_{600}$ of 0.1 to 0.2 in MS liquid medium containing 15 $g \cdot l^{-1}$ of glucose; 2 ml of the suspension were placed on the wounded site of each explant (Figure 11.1A). After two days of cocultivation at 25°C in the dark, explants were washed twice with MS liquid medium. Subsequently, explants were blotted dry using sterile filter paper and plated on solid MS medium containing 1 $mg \cdot l^{-1}$ of BA, 100 $mg \cdot l^{-1}$ of kanamycin (Wako Pure Chemical Industries, Ltd., Osaka, Japan) or 15 $mg \cdot l^{-1}$ of hygromycin B (Roche Diagnostics, Mannheim, Germany), and 500 $mg \cdot l^{-1}$ of lilacillin (Takeda Pharmaceutical Co., Ltd., Osaka, Japan), and incubated at 25°C under cool white fluorescent light (16/8 light regime, 50-60 $\mu mol \cdot m^{-2} \cdot s^{-1}$).

### Recovery of Transgenic Plants

The explants were transferred to fresh medium at two-week intervals. During 2 to 4 weeks' selection, most of the explants gradually turned dark brown with the antibiotic selection, and almost all explants formed white

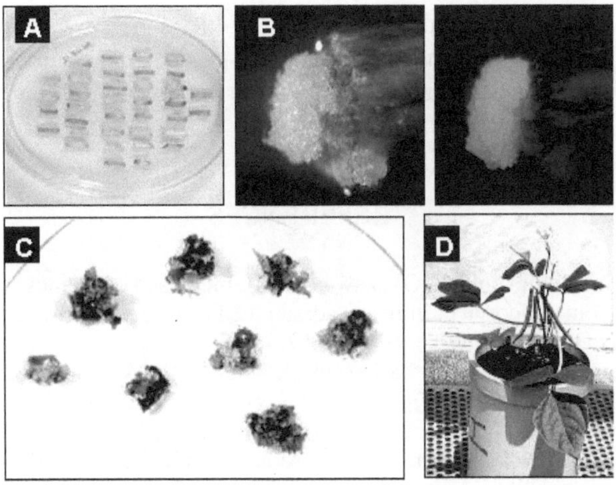


FIGURE 11.1. Production of fertile transgenic azuki bean plants. (A) Cocultivation of epicotyl explants; (B) induction of transformed organogenic calli (left panel), expression of *gfp* by green fluorescence (right panel); (C) shoot induction from organogenic calli; (D) a fertile transgenic azuki bean plant grown in soil and in a glasshouse. (See also color gallery.)

calli from both edges of the explants (Figure11.1B). The adventitious shoots were regenerated from the induced calli on the upper side of epicotyl explants in 4 to 5 weeks (Figure 11.1C). When adventitious shoots developed more than two leaves, they were excised and transferred to MS medium without plant growth regulators but containing 500 $mg{\cdot}l^{-1}$ of lilacillin and the selective agent. Rooted shoots were excised and repeatedly selected on the same medium. After substantial root growth, plants were directly transplanted to soil in the glasshouse, which was controlled at 25 to 28°C under the natural light condition. Azuki bean plants produced fertile flowers and pods that contained viable seeds in 2 to 3 months (Figure 11.1D).

## *CONCLUSION*

*A. tumefaciens*-mediated transformation of azuki bean is a highly reproducible and efficient genetic engineering system. The high competence of the azuki bean transformation system could facilitate its adoption as a model system in understanding the function of genes in grain legumes in addition to the molecular breeding of azuki bean. Any desired gene of which the function confirms in vivo in transgenic azuki bean plants encourages us to transfer the same to soybean or other economically important grain legumes.

## REFERENCES


Adachi, T., Y. Kikuta, and Y. Okazawa (1990). Increased efficiency in plant regeneration from epicotyl culture of adzuki bean variety "Erimo-shozu" (*Vigna angularis* Ohwi & Ohashi). *Hokuno* 57:63-65.

Arumuganathan, K. and E.D. Earle (1991). Nuclear DNA content of some important plant species. *Plant Molecular Biology Reporter* 9:208-219.

Chen, R., S. Tsuda, K. Matsui, M. Fukuchi-Mizutani, M. Ochiai, S. Shimizu, E. Sakuradani, T. Aoki, R. Imaizumi, S. Ayabe, and Y. Tanaka (2005). Production of $\gamma$-linolenic acid in *Lotus japonicus* and *Vigna angularis* by expression of the $\Delta$-6-fatty-acid desaturase gene isolated from *Mortierella alpina. Plant Science* 169:599-605.

Choi, H.-K., J.H. Mun, D.J. Kim, H. Zhu, J.-M. Baek, J. Mudge, B. Roe, N. Ellis, J. Doyle, G.B. Kiss, N.D. Young, and D.R. Cook (2004). Estimating genome conservation between crop and model legume species. *Proceedings of the National Academy of Sciences USA* 101:15289-15294.

Ge, K., Y. Wang, X. Yuan, P. Huang, J. Yang, Z. Nie, D. Testa, and N. Lee (1989). Plantlet regeneration from protoplasts isolated from mesophyll cells of adzuki bean (*Phaseolus angularis)* weight. *Plant Science* 63:209-216.

Han, O.K., A. Kaga, T. Isemura, X.W. Wang, N. Tomooka, and D.A.Vaughan (2005). A genetic linkage map for azuki bean [*Vigna angularis* (Willd.) Ohwi & Ohashi]. *Theoretical and Applied Genetics* 111:1278-1287.

Hanafy, M.S., M.S. Rahman, M.M. Khalafalla, H.A. El-Shemy, Y. Nakamoto, M. Ishimoto, and K. Wakasa (2006). Accumulation of free tryptophan in azuki bean (*Vigna angularis*) induced by expression of a gene (*OASA1D*) for a modified $\alpha$-subunit of rice anthranilate synthase. *Plant Science* 171:670-676.

Hood, E.E., S.B. Gelvin, L.S. Melchers, and A. Hoekema (1993). New *Agrobacterium* helper plasmids for gene transfer to plants. *Transgenic Research* 2:208-218.

Ishimoto, M., T. Sato, M.J. Chrispeels, and K. Kitamura (1996). Bruchid resistance of transgenic azuki bean expressing seed $\alpha$-amylase inhibitor of common bean. *Entomologia Experimentalis et. Aplicata* 79:309-315.

Kaga, A., T. Ishii, K. Tsukimoto, E. Tokoro, and O. Kamijima (2000). Comparative molecular mapping in *Ceratotropis* species using an interspecific cross between azuki bean (*Vigna angularis*) and rice bean (*V. umbellata*). *Theoretical and Applied Genetics* 100:207-213.

Kaga, A., M. Ohnishi, T. Ishii, and O. Kamijima (1996). A genetic linkage map of azuki bean constructed with molecular and morphological markers using an interspecific population (*Vigna angularis* x *V. nakashimae*). *Theoretical and Applied Genetics* 93:658-663.

Lazo, G.R., P.A. Stein, and R.A. Ludwig (1991). A DNA transformation-competent *Arabidopsis* genomic library in *Agrobacterium. Nature Bio/Technology* 9:963-967.

Lumpkin, T.A. and D.C. McClary (1994). Azuki bean: Botany, production and uses. Wallingford, UK: CAB International.

Murashige, T. and F. Skoog (1962). A revised medium for rapid growth and bioassays with tobacco tissue culture. *Physiologia Plantarum* 15:473-497.

Ozaki, K. (1985). Plantlet formation from epicotyl culture of azuki bean. *Plant Tissue Culture Letters* 2:59-62.

Ozaki, K. (1986). Plantlet formation from the calli of primary leaf of azuki bean (*Vigna angularis*). *Japanese Journal of Breeding* 36:416-419.

Parida, A., S.N. Raina, and R.K.J. Narayan (1990). Quantitative DNA variation between and within chromosome complements of *Vigna* species (Fabaceae). *Genetica* 82:125-133.

Sato, T. (1995). Basic study of biotechnology in adzuki bean (*Vigna angularis* Ohwi & Ohashi). *The Report of Hokkaido Prefectural Agricultural Experiment Stations* 87:1-74.

Sato, T., D. Asaka, T. Harada, and I. Matsukawa (1990). Plant regeneration from adzuki bean epicotyl and callus derived from epicotyl (in Japanese with English abstract). *The Bulletin of Hokkaido Prefectural Agricultural Experiment Stations* 61:51-60.

Sato, T., D. Asaka, T. Harada, and I. Matsukawa (1993). Plant regeneration from protoplasts of adzuki bean (*Vigna angularis* Ohwi & Ohashi). *Japanese Journal of Breeding* 43:183-191.

Sato, T. and I. Matsukawa (1990). *Agrobacterium*-mediated transformation of azuki bean (*Vigna angularis* Ohwi & Ohashi) (in Japanese). *2nd Plant Tissue Culture Colloquium Abstracts*: 124-125.

Somta, P., A. Kaga, N. Tomooka, K. Kashiwaba, T. Isemura, B. Chaitieng, P. Srinivas, and D.A. Vaughan (2005). Development of an interspecific *Vigna* linkage map between *Vigna umbellata* (Thunb.) Ohwi & Ohashi and *V. nakashimae* (Ohwi) Ohwi & Ohashi and its use in analysis of bruchid resistance and comparative genomics. *Plant Breeding* 125:77-84.

Tomooka, N., D.A. Vaughan, H. Moss, and N. Maxted (2002). *The Asian Vigna,* Vigna *subgenus* Ceratotropis *genetic resources.* Dordrecht: Kluwer Academic Publishers.

Wang, X.W., A. Kaga, N. Tomooka, and D.A. Vaughan (2004). The development of SSR markers by a new method in plants and their application to gene flow studies in azuki bean [*Vigna angularis* (Willd.) Ohwi & Ohashi]. *Theoretical and Applied Genetics* 109:352-360.

Yamada, T., M. Teraishi, K. Hattori, and M. Ishimoto (2001). Transformation of azuki bean by *Agrobacterium tumefaciens*. *Plant Cell Tissue and Organ Culture* 64:47-54.
Zong, X.X., A. Kaga, N. Tomooka, X.W. Wang, O.K. Han, and D.A. Vaughan (2003). The genetic diversity of the *Vigna angularis* complex in Asia. *Genome* 46:647-658.

Chapter 12

# Somaclonal Variation and Genetic Transformation in *Lathyrus sativus*

S. L. Mehta
I. M. Santha

## *INTRODUCTION*

*Lathyrus sativus* or grass pea, popularly known as Khesari in India, is an annual pulse crop belonging to the family Fabaceae. It is grown as a winter pulse crop for animal feed and human consumption in the Middle East, France, Spain, India, Bangladesh, China, Pakistan, Nepal, and African countries. However, it is cultivated in Australia, Europe, and North America as a fodder crop only. It is considered nutritionally important due to the high protein content of 18 to 34 percent in seeds and 17 percent in mature leaves, and its high lysine content.

*Lathyrus* is a hardy crop and has the ability to grow under various adverse agricultural conditions, including both drought and waterlogging, in cool and at high altitudes, and to adapt to saline, alkaline, clay, or otherwise poor soils. Being a pulse crop, it also enriches the soil by adding around 67 $kg \cdot ha^{-1}$ of nitrogen to the soil in a single season. The potential of this legume crop has not been fully exploited due to the presence of a toxic compound, β-N-oxalyl L-amino alanine (BOAA), also known as β-N-oxalyl-L-α, β-diamino propionic acid (ODAP), which has been suspected to cause a paralytic disease known as lathyrism on prolonged consumption. In a recently reviewed article, Vaz Patto et al. (2006) discussed the recent trends in *Lathyrus* improvement.

Various efforts have been made in the past by plant breeders to develop *Lathyrus* cultivars with no or low levels of ODAP. In our institute, a *Lathyrus* cultivar P-24, was developed with low ODAP content of 0.3 percent, compared to 0.6 to 1.2 percent in commercially grown cultivars. But the major bottleneck was the instability of the low-ODAP P-24 and other strains at different locations. Subsequent studies in our laboratory have shown that this instability is due to cross-pollination by honeybees, which was ignored by the plant breeders earlier. ODAP is a small molecule of an amino acid derivative with the following molecular structure:

```
      COOH
      |
      CO
      |
      NH
      |
      CH2
      |
H2N- CH-COOH
```

synthesized by terminal two enzymes

$$\text{Oxalic acid} + \text{CoA} + \text{ATP} \xrightleftharpoons[\text{Mg2+}]{\text{Oxalyl-CoA synthetase}} \text{Oxalyl-CoA+AMP+PPi}$$

$$\text{Oxalyl- CoA} + \text{DAP} \xrightarrow[\text{ODAP synthase}]{} \text{ODAP+CoA}$$

Developments in molecular biology and biotechnology offer the opportunity of silencing genes by antisense RNA technology or cloning of genes for degradation of the toxin. Since *Lathyrus* is a hardy crop, developing a variety with zero or extremely low ODAP could be a boon to farmers in arid regions of the world. Therefore, we tried three approaches, namely 1) somaclonal variation; 2) antisense cloning of any one enzyme involved in the biosynthesis of ODAP; and 3) isolation, characterization, and cloning of an ODAP-degrading gene from a soil microbe and transformation of *L. sativus.*

## *EXPLOITATION OF SOMACLONAL VARIATION*

### *Regeneration of* **L. sativus**

Availability of good regeneration protocols is a prerequisite for transformation of any crop plant. In general, grain legumes are very recalcitrant to shoot and root regeneration, and successful regeneration protocols are not yet available for many of the legumes. In the past, many researchers have tried to regenerate *L. sativus* but with little success. Mukhopadhyaya and Bhojwani (1978) and Gharyal and Maheswari (1983) got shoot formation in calli from shoot apices and leaf explants, but rooting was not obtained. Sinha et al. (1983) reported the development of callus and shoot regeneration from stem explants but could not show evidence of rooting and development of regenerated plants bearing viable seeds. Hence, efforts in our lab were initially directed at developing successful in vitro regeneration protocols for *L. sativus*. Regeneration of *L. sativus* cultivar P-24 using leaf, root explants, and internode (Roy et al., 1991, 1992, 1993) was attempted. Barik et al. (2004) have shown that cotyledonary node segments of *Lathyrus* cultured on Murashige and Skoog (MS) medium containing 2.0 $mg \cdot l^{-1}$ benzyl adenine (BA) proliferated shoots on average 11.3 per cultured explant of the cultivar IC120487. Up to 81.8 percent of the shoots developed roots following transfer to half-strength MS medium containing 0.5 $mg \cdot l^{-1}$ (2.85 $\mu M$) indole-3-acetic acid (IAA). Plantlets were successfully acclimatized and established in soil.

### *Plant Regeneration from Leaf Discs of* L. sativus

Leaf explants, each 0.7-0.8 cm long, 0.4-0.5 cm wide, and 3-4 mg in weight of aseptically grown 6-7-day old seedlings of *L. sativus* cultivar P-24 seedlings were inoculated for callusing on sterile Murashige and Skoog (MS) or B5 medium containing 30 $g \cdot l^{-1}$ sucrose and solidified with 8 $g \cdot l^{-1}$ agar and supplemented with different combinations of growth regulators (Roy et al., 1991) with pH adjusted to 5.8. The callus was subcultured in MS and B5 media with different combinations of growth regulators for regeneration. The cultures were incubated under 70 $\mu E \cdot m^2 \cdot S^{-1}$ incandescent light in a photoperiod cycle of 16 hours light/8 hours darkness at 25 ± 1°C. Rooting was obtained by inoculating healthy shoots in half strength MS medium containing 2 percent sucrose and 0.01 $m \cdot l^{-1}$ indole butyric acid (IBA) and covering the lower portion of the tube with carbon paper to keep them in the dark. With this, 91 percent of shoots produced good rooting. Good rooting was obtained in 28 to 30 days, after which plants were transferred to 1/10 MS liquid for 10 days, later to water, and hardened for an additional 7 days. Hardened plantlets were then transferred to soil in paper cups supplemented with 5 ml 1/10 MS medium each day for 15 days. Healthy plants were transferred to field conditions to complete their life cycle (72-80 days). Similar experiments were also carried out using root and internode as explants by carefully controlling the phytohormone levels in the basal media (Roy et al., 1992, 1993).

Using root explants, callus and shoot regeneration was obtained only in MS medium supplemented with 10.7 µM naphthalene acetic acid (NAA) and increased concentration of kinetin (0.9 µM for 14 days to 1.4 µM for 18 days) during callusing. Shoots obtained were rooted in ½ MS supplemented with 0.5 µM IBA. Root induction was 48 percent and reproducible (Roy et al., 1992). In case of internode explants, callus was raised on B5 medium supplemented with 10.7 µM NAA + 2.2 µM BA and shoot regeneration on modified MS medium containing 10.7 µM NAA + 2.2 µM BA. Rooting was attempted in ½ MS, ¾ MS, and MS in combination with 0.05, 0.5, and 0.75 µM IBA. Successful root induction was obtained in ½ MS supplemented with 0.5 µM IBA four weeks after transfer. Root induction was 81 percent and reproducible (Roy et al., 1993). The above studies showed that the requirement for growth regulator for getting successful regeneration of *L. sativus* is very specific and narrow, which varies for different explants.

Thus, development of tissue culture methods to get viable plants regenerated through callus culture led us to exploit the property of somaclonal variation for getting lines with reduced levels of ODAP. In our study, about 300 in vitro regenerated plants, obtained by controlling the media and concentration of growth regulators very critically, were taken to the field, out of which 102 plants survived and produced viable seeds. Progeny plants were analyzed for chemical as well as phenotypic variation. ODAP content varied from as low as 0.03 percent to as high as 0.89 percent. The seeds from low-ODAP-containing (less than 0.1 percent) plants were grown separately in subsequent generations. Phenotypic variation observed as well as ODAP content have remained unchanged over generations. The low-ODAP lines (0.03-0.08 percent) showed segregation when analyzed for ODAP in the $R_2$ generation. The majority of plants had ODAP content less than 0.01 percent but in some cases it was more than 0.1 percent and as high as 0.6 percent. The segregation for ODAP content was possibly because of cross-pollination caused by honeybees.

## *BIOCHEMICAL AND MOLECULAR CHARACTERIZATION*

Altogether, 18 somaclones with low ODAP content (<0.1 percent) have been fully characterized at the biochemical and molecular level using isozyme analysis, RAPD, RFLP, mitochondrial specific genes, and photosynthetic efficiency (Ali et al., 2000; Chakrabarti et al., 1999; Mandal et al., 1996; Mehta et al., 1994; Tiwari et al., 1995).

### *Isolation of ODAP-Degrading Genes from Soil Microbe*

Plant genetic engineering techniques can be used as tools to develop toxin-free *L. sativus* by incorporating a gene, which can degrade the toxin in the plant. With this in mind, we tried to isolate soil microbes that utilize ODAP as a sole source of carbon and nitrogen. We were successful in isolating soil microorganisms from the drain near the IARI campus that showed the property of utilizing ODAP as a sole source of carbon and nitrogen. Three strains of soil microorganisms designated as BYAI, BYKI, and BYT, depending upon their resistance to the antibiotics ampicillin, kanamycin, and tetracycline, respectively, were isolated. BYA1 has been identified as *Enterobacter cloacae* (Yadav et al., 1992) and BYT1 and BYK1 as *Pseudomonas stutzeri* (Praveen et al., 1994; Sachdev et al., 1995). Each of these strains has a plasmid and the property of degradation of ODAP in all these strains is borne on the plasmid present in them. The plasmid from pBYA1 has a size over 50 kb. In order to isolate the gene responsible for degradation of ODAP, a partial *Sau*3AI library of the plasmid was prepared in the vector pUC18. The recombinant clone with the property of degrading ODAP was selected by growing the recombinant clones on minimal agar plates containing ODAP as the sole source of carbon and nitrogen. One of the recombinant clones that could utilize ODAP as sole source of carbon and nitrogen was selected and analyzed further. It had an insert size of 1.8 kb, which was sequenced (Sukanya et al., 1993). Sequence analysis of the clone showed the largest open reading frame of 630 nucleotides in one of the frames. It had sequences similar to other *E. coli* promoters at the 5' end, representing the –43,–35, and –10 sequences and a start codon GUG in place of the normal AUG. The gene coded for a polypeptide of 199 amino acids. This fragment has been cloned in an expression vector pMalC$_2$ and was expressed to encode a fusion protein with maltose binding protein. This protein, on cleavage with the end peptidase, Xa-factor, followed by purification and SDS-PAGE analysis, was found to be approximately 20.9 KD and corresponded to that revealed from sequence analysis (Nair et al., 1994). Further studies using this purified protein will reveal the metabolic pathway of ODAP utilization. So far, there have been no reports on the development of transgenic *Lathyrus sativus*.

## *TRANSFORMATION OF* LATHYRUS SATIVUS

### *Use of Wild-Type* Agrobacterium *Strain A208*

Transformation of *Lathyrus sativus* was attempted using wild-type strain A208 of *Agrobacterium tumefaciens*. Various explants such as leaf, internode, and root were cocultured with the wild-type strain A208. Callus was formed within a fortnight on Gamborg's (B5) basal medium, which continued to proliferate and could be maintained without exogenous supply of growth regulators. The callus could not be regenerated into shoots. Crown galls were raised by infecting the lower internodes of 10-day-old seedlings with *A. tumefaciens* A208 in another experiment. These were formed within 10 days and were transferred to Gamborg's basal medium lacking growth regulators. After several subcultures, shoots were regenerated from this callus and occasionally the shoots rooted in the same media. The shoots after rooting in the rooting media were transferred to the field after hardening and produced viable seeds.

### *Use of* Agrobacterium *Strain C58 Carrying the Binary Vector pBI121*

An *Agrobacterium* strain carrying the pBI121 plasmid vector, which has an *npt*II gene as well as a gene coding for the β-glucuronidase (*uid*A) driven by CaMV35S promoter, was used to transform *L. sativus*. Various explants such as leaf (7-8-day-old) internode and root segments (of 5-day-old seedling) were cocultured with the *Agrobacterium* strain by suspending the explants in the bacterial suspension for 1 to 2 seconds, 2 minutes, 5 minutes, and 10 minutes. Then these were blotted dry on sterile filter paper and cultured on callusing media. After 48 to 72 hours of cocultivation, explants were transferred to the same medium containing kanamycin (50 $\mu g \cdot ml^{-1}$) and carbenicillin (500 $\mu g \cdot ml^{-1}$) for selection and arresting the bacterial growth. After every fortnight, they were transferred to fresh media of the same composition for the first 8 weeks and then were subcultured every 3 to 4 weeks. Root explants turned pale brown and did not form any callus. Internodes callused on the selection media but the frequency was very low. The leaf explants responded and the callus was initiated after 5 to 6 weeks of cocultivation and fresh leaves gave greater frequency of putatively transformed callus (50 percent) than precultured leaf segments (16.5 percent). Callus grew green on kanamycin selection media, indicating that transformation has occurred. However, further attempts to regenerate shoots from these calli were not successful. Attempts were also made to transform *Lathyrus sativus* with a construct carrying the ODAP degrading gene. In this case also, transformed calli were obtained on kanamycin selection media, but further regeneration was not obtained.

Barik et al. (2005) developed transgenic *Lathyrus* plants expressing the *npt*II and *gus*-intron genes using *Agrobacterium tumefaciens* based vectors. They have used epicotyl explants for developing the transformation protocol and showed that the strain LBA4404 was more efficient in transforming

*Lathyrus* compared to EHA105. Highest transformation rate was achieved by hand pricking the epicotyl explants and treating them with the bacterial suspension at an optical density ($OD_{600}$) of 0.6 for ten minutes followed by cocultivation for four days on the culture medium at pH 5.6. Transformed shoots rooted on 0.5 $mg \cdot l^{-1}$ IAA. Kanamycin resistance segregated in a ratio of 3:1 in the $T_1$ generation indicating the sexual transmission of transgenes.

## *SOMATIC EMBRYOGENESIS*

Somatic embryogenesis and organogenesis were successful with *L. sativus* (Barna and Mehta, 1995). Somatic embryogenesis was induced by culturing immature leaf segments and internodal segments of cultivar P-24 on MS medium supplemented with NAA alone or in combination with kinetin for 8 to 10 days followed by 4 weeks on growth-regulator-free medium. Direct embryo formation was observed on the explants without intervening callus phase, only on medium with 2,4-dichlorophenoxy acetic acid. The embryos formed were bottle shaped to spherical. Supplementation of MS media with glutamine, asparagine, or abscisic acid resulted in shape differences of the embryo, having bottle shaped in the former and spherical in the latter two cases. About 98 percent of individual embryos on transfer to hormone-free medium germinated to form roots with slight elongation of hypocotyls, but only 2 percent germinated to form shoots.

### *Isolation of Oxalyl CoA Synthetase*

Application of antisense RNA/ribozyme technology to silence or block the biosynthetic pathway of ODAP synthesis is another strategy to develop an ODAP-free cultivar of *L. sativus*. Two terminal steps involving two different enzymes, oxalyl CoA synthetase and ODAP synthase, are proposed for the biosynthesis of ODAP. Therefore, blocking the expression of any one of these enzymes will indirectly reduce the synthesis of ODAP. This can be achieved by applying antisense or ribozyme technology. The foremost need for applying any of these technologies is to have the genes encoding them isolated and characterized. The method followed was to isolate and purify the more stable enzyme oxalyl CoA synthetase, raise antibodies against it, and use the antibody developed to screen a cDNA library of *L. sativus* prepared in an expression vector. The enzyme oxalyl CoA synthetase has been isolated, purified, and antibody raised (Sehgal et al., 1992), and a cDNA library of *L. sativus* has been prepared. Our goal in the project was to develop low-ODAP strains of *L. sativus* by any one of the possible methods, which was achieved by exploiting somaclonal variations where no ethical issues were involved. We did not further follow the other methods of developing transgenic plants. A sufficiently large quantity of dal obtained from *Lathyrus* variety Ratan (Bio-L-212) was given to Central Toxicology Research Institute, CSIR, at Lucknow, India. Two years of feeding trials with monkeys did not show any adverse effect and from that study it was concluded that feeding even 400 g *Lathyrus sativus* per day is safe. The low-ODAP somaclones developed in the present study have been shown to be stable with respect to all the characteristics they are selected for.

## *FUTURE PROSPECTS*

By carefully selecting the growth media and controlling the growth hormone concentrations, we have been able to develop a successful regeneration protocol for *L. sativus* using leaf, internode, and root explants. A large number of somaclonal variants with a lot of phenotypic variations have been developed. Some of the somaclones developed had very low toxin content (<0.07 percent) combined with high-yielding characteristics. One of the low-toxin somaclones, Bio L212, after extensive trials at various locations, has been released for cultivation by the Government of India, but the ban imposed on the sale of this grain legume should be lifted in order to fully exploit the potential of this golden crop.

## REFERENCES

Ali, K., K.C. Bansal, A. Sachdev, N. Tiwari, I.M. Santha, and S.L. Mehta (2000). Physiological and biochemical analysis of *Lathyrus sativus,*, somaclones derived from root and internode explants. *Journal of Plant Biochemistry and Biotechnology* 9:27-30.

D. Barik, D.P. S. K. Naik, U. Mohapatra, and P. K. Chand (2004). High-frequency plant regeneration by in vitro shoot proliferation in cotyledonary node explants of grasspea (*Lathyrus sativus* L). *In Vitro Cellular and Developmental Biology–Plant* 40:467-470.

Barik, D.P., U. Mohapatra, and P. K. Chand (2005). Transgenic grass pea (*Lathyrus sativus* L.): Factors affecting *Agrobacterium* mediated regeneration and transformation. *Plant Cell Reports 24*:523-531.

Barna, K.S. and S.L. Mehta (1995). Genetic transformation and somatic embryogenesis in *Lathyrus sativus. Journal of Plant Biochemistry and Biotechnology* 4:67-71.

Chakrabarti, A., I.M. Santha, and S.L. Mehta (1999). Molecular characterization of low ODAP somaclones of *Lathyrus sativus. Journal of Plant Biochemistry and Biotechnology* 8:25-29.

Gharyal, P.K. and S.C. Maheswari (1983). Genetic and physiological influences on differentiation in tissue cultures of legume *Lathyrus sativus. Theoretical and Applied Genetics* 66:23-126.

Mandal, P.K., I.M. Santha, and S.L. Mehta (1996). RAPD analysis of *Lathyrus sativus* somaclones. *Journal of Plant Biochemistry and Biotechnology* 5:83-86.

Mehta, S.L., K. Ali, and K.S. Barna (1994). Somaclonal variation in food legume—*Lathyrus sativus. Journal of Plant Biochemistry and Biotechnology* 3:73-77.

Mukhopadhyaya, A. and S.S. Bhojwani (1978). Shoot-bud differentiation in tissue cultures of leguminous plants. *Zeitschrift fuer Pflanzen Physiologie* 88:263-268.

Nair, A.J., G.S. Khatri, I.M. Santha, and S.L. Mehta (1994). Cloning of ODAP degrading gene and its expression as fusion protein in *Escherichia coli. Journal of Plant Biochemistry and Biotechnology* 3:103-106.

Praveen, S., R.P. Johari, and S.L. Mehta (1994). Cloning and expression of OXDAPRO degrading gene from soil microbe. *Journal of Plant Biochemistry and Biotechnology* 3:25-29.

Roy, P.K., K. Ali, A. Gupta, G.K. Barat, and S.L. Mehta (1993). β-N-oxalyl-α,β diamino propionic acid in somaclones derived from internode explants of *Lathyrus sativus. Journal of Plant Biochemistry and Biotechnology* 2:9-13.

Roy, P.K., G.K. Barat, and S.L. Mehta (1992). In vitro plant regeneration from callus derived from root explants of *Lathyrus sativus. Plant Cell Tissue and Organ Culture* 29:135-138.

Roy, P.K., B. Singh, S.L. Mehta, G.K. Barat, N. Gupta, P.B. Kirti, and V.L. Chopra (1991). Plant regeneration from leaf disc of *Lathyrus sativus. Indian Journal of Experimental Biology* 29:327-330.

Sachdev, A., M. Sharma, R.P. Johari, and S.L. Mehta (1995). Characterization and cloning of ODAP degrading gene from a soil microbe. *Journal of Plant Biochemistry and Biotechnology* 4:33-36.

Sehgal, D., I.M. Santha, and S.L. Mehta (1992). Purification of oxalyl CoA synthetase enzyme from *Lathyrus sativus* and raising of antibodies. *Journal of Plant Biochemistry and Biotechnology* 1:97-100.

Sinha, R.R., K. Das, and S.K. Sen (1983). Plant regeneration from stem derived callus of the seed legume, *Lathyrus sativus* L. *Plant Cell Tissue and Organ Culture* 2:67-76.

Sukanya, R., I.M. Santha, and S.L. Mehta (1993). Cloning and characterization of a gene responsible for the *Lathyrus sativus* neurotoxin degradation. *Journal of Plant Biochemistry and Biotechnology* 2:77-82.

Tiwari, N., K.C. Bansal, K. Ali, A. Sachdev, I.M. Santha, and S.L. Mehta (1995). Physiological and biochemical analysis of somaclones derived from leaf explants of *Lathyrus sativus. Journal of Plant Biochemistry and Biotechnology* 4:85-89.

Vaz Patto, M.C., B. Skiba, E.C.K. Pang, S.J. Ochatt, F. Lambein, and D. Rubiales (2006). *Lathyrus* improvement for resistance against biotic and abiotic stresses: from classical breeding to marker assisted selection. *Euphytica* 147: 133–147.

Yadav, V.K., I.M. Santha, M.P. Timko, and S.L. Mehta (1992). Metabolism of the *Lathyrus sativus* L. neurotoxin β-N-oxalyl-α,β diamino propionic acid by pure culture of a soil born microbe. *Journal of Plant Biochemistry and Biotechnology* 1:87-92.

## Chapter 13

# *Agrobacterium*-Mediated Transformation of Pea (*Pisum sativum* L.)

Anna Nadolska-Orczyk

## *INTRODUCTION*

Pea (*Pisum sativum* L.) has several favorable biological features that determine its agricultural importance. Its climate and soil requirements are modest enough to enable it to adapt to cultivation in a wide range of environments. Pea is predominantly grown in Europe, China, India, and North America. Like the other legume species, it has the ability to fix atmospheric nitrogen and enrich soil with nitrogen, and it has seeds with high protein content and of good quality. Green peas contain 5 to 8 percent protein, 0.5 percent fat, and 10 to 15 percent carbohydrate, and the mature seeds contain 20 to 25 percent protein, 1 to 3 percent fat, and 60 percent carbohydrate. Peas are limited in their contents of the sulfur-containing amino acids methionine and cysteine, but are rich in lysine and other essential amino acids. As a crop, they are used as a vegetable for human consumption, fodder and as a green manure, while the dry seeds are used for feeds. Besides the pea's high nutritional value, its other main attributes are high net yield in a relatively short growth period (short-term crop) and its good intake characteristics. In terms of worldwide production levels, it ranked fourth among the grain legumes after soybean, peanut, and the beans (FAO, 1994). The chromosome number of the whole genus *Pisum* is low ($2n = 14$); however, the genome is very large, estimated for 1C as $4.4 \times 10^9$ bp (Dolezel et al., 1998). Only about 10 to 25 percent of the DNA comprises large, gene-rich regions, called the gene space, separated by repetitive sequences that are gene-empty (Barakat et al., 2000). This genome organization might influ-

This work was supported by a statutory grant from the Plant Breeding and Acclimatization Institute.

ence the distribution and expression of transgenes introduced by direct (biolistic) and indirect (*Agrobacterium*-mediated) methods (Kohli et al., 2003).

Work on the genetic transformation of pea is focused on improving variety and on the fundamental study of gene function. The goals of variety improvement studies are to achieve better seed quality (protein composition and quality, lowering antinutritional factors) and a higher level of disease, especially virus resistance.

Since the discovery of the phenomenon of RNA interference (Hamilton and Baulcombe, 1999), genetic transformation has been used not only to increase variability by adding new genes but also to silence genes by suppressing gene transcription (TGS) or translation (PTGS; reviewed by Paszkowski and Whitham, 2001). This approach makes it possible to knock out genes coding some unfavorable products. A newly developed tool for studying gene function in pea is virus-induced gene silencing (VIGS). Constantin et al. (2004) efficiently applied PEBV vectors, which used pea early browning virus to study genes involved in a wide range of biological processes in this species.

## *TOWARD* AGROBACTERIUM-*MEDIATED TRANSFORMATION*

### *In Vitro Regeneration and Plant Genotype*

The requirements for the successful application of the *Agrobacterium*-mediated transformation method are a well-working, short-term regeneration procedure and a responsive plant genotype. Until the late 1980s, pea plant regeneration was achieved by shoot-tip (meristem), embryo, callus, and suspension culture (reviewed by Griga and Novak, 1990). However, these systems were not easily applicable to direct or *Agrobacterium*-mediated transformation. In the case of shoot-tip and embryo culture, plants were developed from axillary buds, which was a type of in vitro vegetative propagation. The genetic transformation of such tissue would give rise to chimerical transgenic plants. In vitro regeneration through callus or suspension culture is a long-term procedure generating a higher frequency of somaclonal variant plants and difficulties with rooting and seed set. Methods developed from hypocotyl explants (Nielsen et al., 1991) and nodal thin-cell layer segments (Nauerby et al., 1991) seemed to be more suitable, although never employed for genetic transformation. Efficient plant regeneration was obtained using immature cotyledons (Nadolska-Orczyk, 1991; Nadolska-Orczyk et al., 1994; Ozcan et al., 1992, 1993). In two different procedures based on the same explants, various results were obtained. Direct regeneration of many buds developing to shoots was observed in the basal, cut part of cotyledons (Nadolska-Orczyk et al., 1994). The quantified ability of six pea cultivars to form shoots after 7 weeks of culture ranged from 3.5 to 9.9 and the percentage of responding explants was higher than 73. This procedure has also been routinely used with other breeding lines or cultivars in our lab (Nadolska-Orczyk, unpublished data). Some cultivars were also able to regenerate plants via somatic embryogenesis; however, the method was inefficient and cultivar dependent. Additionally, somatic embryos were of multicellular origin, as shown by Griga (2002). Ozcan et al. (1992, 1993) also described two types of regeneration via organogenesis and somatic embryogenesis. Shoot regeneration occurred at the proximal end of cotyledons following an initial callus growth (Ozcan et al., 1992). The best reported frequency for two tested cultivars was 11.7 and 15.6 shoots per explant after 6 weeks of culture for at least 75 percent of responding explants. In the method of somatic embryogenesis, Ozcan et al. (1993) obtained from 1 to 9.3 somatic embryos per explant depending on the cultivar and medium composition. There was no information on how many of these embryoids were viable for development to plants.

Sanago et al. (1996) indicated the potential use of thidiazuron as a growth regulator promoting plant regeneration of pea. They obtained an average of up to 20 shoots, which were formed from the hypocotyl explants of two out of four tested cultivars. The same growth regulator was used to obtain multiple shoot regeneration from nodal tissue of pea (Tzitzikas et al., 2004). Subculture of these shoots resulted in the formation of small buds, developing in the next steps of culture to fertile plants.

The described methods might be used for *Agrobacterium*-mediated transformation. Some of them were already proved to be suitable for the procedure: the regeneration method of Schroeder et al. (1993) used by Shade et al. (1994) to obtain transgenic pea seeds resistant to bruchid beetles; or our regeneration procedure (Nadolska-Orczyk et al., 1994), used to study factors influencing *Agrobacterium*-mediated transformation (Nadolska-Orczyk and Orczyk, 2000).

### *Preliminary Research on* **Agrobacterium** *Strain and Ti Plasmid*

Hobbs et al. (1989) tested combinations of three wild-type strains, A281, C58, and Ach5, and different pea genotypes for potential use in transformation. A281 was the most virulent of these as determined by size and number of tumors. This strain was reported to be hypervirulent on solanaceous plants (Hood et al., 1986). The differences between genotypes in terms of their response to inoculation were considerable. Subsequently, nontumorigenic, disarmed *Agrobacterium tumefaciens* strains were used. They were: succinamopine strain EHA101, a derivative of A281; nopaline A208 with a C58 chromosomal background; octopines GV3111, GV3101; and LBA4404, derivative of C58 (Lulsdorf et al., 1991). The succinamopine strain EHA101 (pBI1042) produced the highest percentage of transformed callus lines independently of the type of selection (kanamycin or hygromycin). However, for the less virulent strain LBA4404 (pBI1042), the percentage of transformed calli was significantly higher for hygromycin selection (63 percent) than for kanamycin selection (17 percent).

## *METHODS OF* AGROBACTERIUM-*MEDIATED PEA TRANSFORMATION*

The methods of *Agrobacterium*-mediated pea transformation in which fertile, transgenic plants were obtained are listed in Table 13.1. Puonti-Kaerlas et al. were the first to generate transgenic plants in pea in 1990. They reported on the selection of transgenic pea plants. An analysis of transmission of the transferred DNA in the progeny of these plants was presented two years later (Puonti-Kaerlas et al., 1992). This method was not promising because the obtained plants were tetraploid. The authors suggested two possible reasons for this: the long culture period and high cytokinin and auxin concentrations in the medium.

### *Cultivars, Types of Explant, and In Vitro Culture*

Many researchers have already documented that, for a large number of plant species, there is a genotype dependence of sensitivity to inoculation with *Agrobacterium*. This is also true in the case of *Agrobacterium*-mediated pea transformation. Different genotypes (cultivars or breeding lines) that were potentially of breeding importance were used for research. Puonti-Kaerlas et al. (1990) tested five cultivars. Only two, Stivo and Puget, regenerated transgenic plants. In other studies, the efficiency of transformation of various cultivars or breeding lines differed, but all the tested genotypes gave transgenic plants (Grant et al., 1995, 1998, 2003; Nadolska-Orczyk and Orczyk, 2000).

Highly regenerable explants were used for successful pea transformation and regeneration. Two of these consisted of meristematic, undifferentiated cells. There were sections from the embryonic axes of immature seeds (Schroeder et al., 1993) and the lateral cotyledonary meristems in germinating seeds (Davies et al., 1993; Bean et al., 1997). The main advantage of the latter type of explant was the ease of availability of the plant material. The lateral buds are relatively undifferentiated and contain fewer cells than

TABLE 13.1. *Agrobacterium*-mediated pea transformation procedures, in which fertile transgenic plants were recovered.

| Cultivar | Explant | Strain (vector) | Selection ($mg \cdot l^{-1}$) | Research group |
|---|---|---|---|---|
| Stivo<br>Puget | Seedling epicotyls and shoot cultures | GV3101 (pGV2260::pGV1503)<br>GV3101 (pGV3850::pCAP212)<br>GV3101 (pGV3850::pLD1) | Hygromycin B (15)<br>Kanamycin (75) | Puonti-Kaerlas et al., 1990, 1992 |
| Green feast<br>Rondo | Embryonic axis of immature seeds | AGL1 (pSLJ1561) | Phosphinothricin (15) | Schroeder et al., 1993 |
| Puget | Lateral cotyledonary meristems from germinating seeds | C58/3 (SLJ1911)<br>EHA105 (SLJ1561) | Kanamycin (60-100)<br>Phosphinothricin (2.5-5) | Davies et al., 1993<br>Bean et al., 1997 |
| Bolero<br>Trounce<br>Bohatyr<br>Huka | | AGL1 (pLN27, based on pGA643) | Phosphinothricin (10) | Grant et al., 1995 |
| six cultivars or breeding lines | Immature cotyledons | AGL1 (four different, based on pGA643) | Kanamycin (75) | Grant et al., 1998 |
| Bolero<br>Lincoln<br>97-B19 line | | AGL1: (pTGUS)<br>(pMG)<br>KYRT1: (pTGUS)<br>(pMG) | Kanamycin (75) | Grant et al., 2003 |
| Heiga<br>Laser | Immature cotyledons | EHA105: (pGPTV-KAN)<br>(pGPTV-BAR)<br>(pGPTV-HPT)<br>(pGPTV-DHFR)<br>LBA4404 (p35SGUSINT)<br>C58C1 (p35SGUSINT) | Kanamycin (50)<br>Phosphinothricin (2-4)<br>Hygromycin (100)<br>Methotrexate (1) | Nadolska-Orczyk and Orczyk, 2000 |

the meristem of the primary shoot. The first putative transformants were recovered 12 weeks post inoculation, and they were generally chimeric. Nonchimeric transgenic plants were obtained 16 weeks postinoculation and no morphological aberrations were seen in $T_0$. Two groups, Grant et al. (1995, 1998, 2003) and Nadolska-Orczyk and Orczyk (2000), successfully used immature cotyledons for *Agrobacterium*-mediated transformation, based on different regeneration methods. In the procedure developed by Grant et al. (1995), explants formed calli at the cotyledon attachment scar, and shoots developed from these calli. Embryo-like structures were also formed, suggesting a mixture of organogenesis and embryogenesis. In our procedure, multiple de novo buds developed directly in the cut, basal part of the explant (Figure 13.1), which was the same cotyledon attachment scar as mentioned above. Both methods were very efficient and cultivar independent. The transgenic plants were morphologically normal and fertile. It is possible that chimeric plants, although not reported on, were obtained in these methods.

### *Strains of* Agrobacterium

As already shown in preliminary research, the virulence of a given *Agrobacterium* strain against a particular plant genotype is a key factor in the plant's successful transformation. The hypervirulence of EHA105, which is a disarmed strain of wild-type A281 (Hood et al., 1986), and of

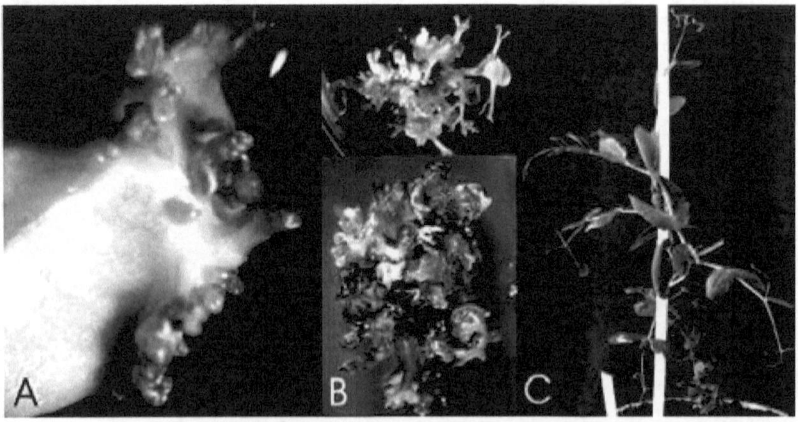


FIGURE 13.1. In vitro pea regeneration from immature cotyledons. (A) Initiation of buds on the cut surface of the cotyledon after 5 weeks of culture. (B) Multiple shoot formation after 8 weeks of culture in different cultivars. (C) In vitro plant growing in soil after 12 weeks of culture. (See also color gallery.)

AGL1, which facilitates DNA transfer to many dicotyledonous plants (Lazo et al., 1991) was confirmed for pea (Table 13.1). Transgenic pea plants were also obtained using C58C1 and LBA4404 (Nadolska-Orczyk and Orczyk, 2000) and KYRT1 (Grant et al., 2003). In our study, the ratios of transformation efficiencies of the strains tested (octopine LBA4404, nopaline C58C1, and hypervirulent succinamopine EHA105) were 1:2.2:8.2. Grant et al. (2003) compared the efficiency of two *A. tumefaciens* strains, AGL1 and KYRT1, which had been shown to be highly tumorigenic on soybean. KYRT1 was on average three times more efficient than AGL1.

### *Selection Cassettes and Agents*

The results of the successful use of selection agents for recovering pea transformants are contradictory. The first transgenic plants obtained were hygromycin resistant, and that trait was inherited stably as a single dominant trait (Puonti-Kaerlas et al., 1992). These plants were tetraploids, suggesting that chromosome doubling had taken place before the integration of *hpt* into pea nuclear DNA. Many of the selected $T_0$ plants were aberrant at the phenotypic level and did not produce any seeds. There were no non-transformed escapes in the method. We compared hygromycin selection with kanamycin, phosphinothricin, and methotrexate (Nadolska-Orczyk and Orczyk, 2000). The efficiency of transformation was high with phosphinothricin and kanamycin selection, but failed with the other two. Davies et al. (1993) reported that there were a number of shoots that escaped kanamycin selection, but that increasing the kanamycin concentration reduced the effectiveness of the procedure. Compared with phosphinothricin, kanamycin selection was unsatisfactory due to its lengthy nature and the frequent production of phenotypically abnormal plants (Bean et al., 1997). Schroeder et al. (1993) and Grant et al. (1995) found kanamycin selection to be ineffective. However, when Grant et al. (1998) retested the use of kanamycin in their next study, they found it to have a relatively high level of selection efficiency for a range of transformed pea cultivars. This positive data for kanamycin selection was confirmed by us (Nadolska-Orczyk and Orczyk, 2000) and later by Grant et al. (2003). A negative feature was our finding that kanamycin selection led to an approximately 5 percent production of phenotypically abnormal plants, which was earlier observed by Davies (unpublished results; Bean et al., 1997). By contrast, all the phosphinothricin-resistant plants were phenotypically unchanged (Figure 13.2, Nadolska-Orczyk et al., 2000). There were no contradictory results in the case of phosphinothricin as selection agent, as demonstrated by several researchers (see Table 13.1). The efficiency of selection was good. Another

FIGURE 13.2. Pea shoot and plant selection after *Agrobacterium*-mediated transformation. (A) Shoot formation on the medium containing 2 mg·l$^{-1}$ phosphinothricin (top) and plant growth on the medium containing 4 mg·l$^{-1}$ phosphinothricin, beside visible nongrowing, browning shoots. (B) Flowering and seed-setting transgenic plant. (C) Segregation of resistant, transgenic plant (left) and susceptible plant (right) in $T_1$ generation. (See also color gallery.)

very important advantage of using the *bar* gene is that the resistance determined by the gene is environmentally safe.

The promoter driving the gene can strongly influence the level of expression. Only two kinds of promoters (P-Nos and P-35S) were used in selection cassettes. Additionally, in the research of Davies et al. (1993), the double enhancer of the 35S promoter fused to the *aph3*'II gene coding for resistance to kanamycin was tested. A comparison of the selection cassettes containing these promoters was not possible due to the differences in the experimental conditions used by the researchers. In their research, nos/*hpt* or nos/*bar* cassettes, and as a last resource 35S/*bar* were proved to be useful.

## TRANSFORMATION EFFICIENCY, EXPRESSION, AND INHERITANCE

Transgenes introduced by *Agrobacterium*-mediated transformation of pea were reported as stably transmitted to the next generation in a Mendelian inheritance pattern. A detailed analysis of copy numbers via Southern blot was done by Davies et al. (1993), and it revealed a single copy at one locus in one plant, multiple tandem copies at a single locus in another and multiple copies at several loci in two others. Those authors also noticed that the *Gus* gene was introduced in one plant in multiple rearranged copies. Despite this, the plant could express GUS activity. Using Southern blot analysis, one to three transgene loci were revealed by Grant et al. (1995) and Nadolska-Orczyk and Orczyk (2000). Grant et al. (1998) reported a relatively high copy number in one or multiple loci and additional smaller fragments than expected, for which partial gene transfer was suggested. The T-DNA integration patterns and the copy number for KYRT1-derived plants and AGL1-derived plants were similar. The transgenes were inserted into one to several loci (Grant et al., 2003). The expression of *hpt* (Puonti-Kaerlas et al., 1992), *bar* and/or *npt*II (Schroeder et al., 1993; Bean et al., 1997; Grant et al., 1995; Nadolska-Orczyk and Orczyk, 2000), and *gus* (Davies et al., 1993) was proved in both the primary transgenic pea plants and in the next generation (the progeny plants). A positive correlation between a low copy number and its expression was observed by Davies et al. (1993). A much higher level of GUS expression was noticed in the case of single copy or multiple tandem copies at a single locus than in that of multiple loci.

The transformation efficiency ranged from 0.16 to 16.7 percent of independent transformants, and was dependent on all the factors described in the text above. According to the procedures listed (excluding the first one), transgenic, seed-bearing primary regenerants were obtained 6 to 8 months after explant inoculation by bacteria. Taking into consideration the system of regeneration of pea through shoot organogenesis, the first putative transformants might be chimeric.

## *GENETICALLY IMPROVED PEAS*

Shade et al. (1994) reported that two developments enabled them to obtain transgenic pea plants resistant to bruchid beetles. The first was the evidence that the growth of larvae of two seed-eating beetles of the family Bruchidae was inhibited when the diet of the larvae contained a low level of α-amylase inhibitor (α-AI). The second was the development of the procedure of pea transformation by Schroeder et al. (1993), which made it possible to introduce the α-amylase inhibitor gene, isolated from the common bean, into the garden pea. Expression of the gene was driven by the promoter of the bean phytohemagglutinin-L gene (*dlec2*). This construct gave a very good level of expression in pea seeds, which contained up to 1.0 to 1.2 percent of α-AI-Pv and were resistant to cowpea weevils (*Callosobruchus maculates*) and azuki bean weevils (*C. chinensis*).

Charity et al. (1999) used proteinase inhibitors to increase resistance to insect pests in transgenic pea plants. A cDNA clone encoding a multi-domain proteinase inhibitor precursor from *Nicotiana alata* (Na-PI) under the control of the promoter of a ribulose-1,5-bisphosphate carboxylase small subunit gene was transferred into pea and tobacco. The gene segregated as a dominant Mendelian trait. Larvae of *Helicoverpa armigera* that ingested transgenic plant tissue exhibited a higher mortality rate or a delayed growth rate relative to control larvae.

Nifantova et al. (2005) produced transgenic pea plants by expressing a mutant *ahas/als* (acetolactate synthase) gene and analyzed the transgenics for the introduced transgene. Prescott et al. (2005) showed that transgenic expression of a plant protein (alpha-amylase inhibitor (AI)-1 from the common bean, *Phaseolus vulgaris*) in transgenic pea led to the synthesis of a structurally modified form of this inhibitor. Employing the models of inflammation in mice, they showed that the consumption of the modified alphaAI concurrently with other heterogeneous proteins promoted immunological cross priming, which then elicited specific immunoreactivity of these proteins. Rolletscheck et al. (2005) expressed an amino acid permease gene (*VfAAP1*) from *Vicia faba* in seeds of *Vicia narbonensis* and *Pisum sativum* under legumin B4 promoter. They observed that *VfAAP1* expression increased the seed sink strength for nitrogen, improved plant nitrogen status, and lead to higher seed protein. These studies concluded that seed protein synthesis is nitrogen limited in the test crops and that seed uptake activity for nitrogen is rate limiting for storage protein synthesis. Richter et al. (2006) expressed genes for polygalacturonase inhibiting protein (PGIP) from raspberry, and stilbene synthase under its own elicitor inducible promoter in pea plants and confirmed the integration of the foreign genes. De Sousa-majer et al. (2007) have developed transgenic pea plants with resistance to pea weevil larvae by the expression of bean alpha-amylase inhibitor.

## *RESULTS*

1. The bottleneck in *Agrobacterium*-mediated pea transformation is the regeneration step.
2. Among the explants tested, immature cotyledons were the best in regard to transformation efficiency, genotype independence, and obtaining phenotype-normal transgenic plants. However, the only time-saving type of explant is described in the procedure of Davies et al. (1993).

3. The hypervirulent *Agrobacterium* strains KYRT1, AGL1, and EHA105 were the most successful in pea transformation.
4. Constructions of nos/*bar*, nos/*npt*II, or 35S/*npt*II were effective for the selection of transgenic pea plants.
5. The recommended selecting agent is phosphinothricin, as it is efficient and environmentally safe, and selects phenotypically normal, fertile plants.

## *PROTOCOLS*

### *In Vitro Plant Regeneration: Immature Cotyledons*

1. Excise 9-11-day-old pods and sterilize in 70 percent ethanol for 30 seconds, followed by a 15-minute treatment with 7.5 percent (v/v) sodium hypochlorite containing one drop of Tween 20. Rinse four times with sterile water.
2. Isolate 3-7 mm long immature seeds, and remove the embryo axis. Place four to five cotyledons on their abaxial side on the 8 cm Petri dish containing the primary medium.
3. The composition of the media is:

- Primary medium: basal MS (MS minerals and vitamins; Murashige and Skoog, 1962) + 5 mg·l$^{-1}$ IBA + 0.5 mg·l$^{-1}$ BAP + 30 g·l$^{-1}$ sucrose solidified with 2 g·l$^{-1}$ gelrite
- Shoot regeneration medium: basal MS + 0.25 mg·l$^{-1}$ NAA + 1 mg·l$^{-1}$ BAP + 20 g·l$^{-1}$ sucrose and 2 g·l$^{-1}$ gelrite
- Shoot rooting medium: ½ MS minerals + 1 mg·l$^{-1}$ IAA (filtered) + 15 g·l$^{-1}$ sucrose and 2 g·l$^{-1}$ gelrite

4. After 4-5 weeks of culture on the primary medium, cut off morphogenic tissue with buds from the rest of the explant and transfer the tissue to the shoot regeneration medium. Every week thereafter, separate developed shoots from the explant and move to the shoot-rooting medium.
5. The culture conditions are a temperature of 22°C, 16-hour photoperiod under white fluorescent light.
6. Rooted plants are transferred to soil and kept under a regime of 18/15°C day/night, 16-hour photoperiod under 350 μmol·s$^{-1}$·m$^{-1}$ light.

## *Agrobacterium-Mediated Transformation: Immature cotyledons*

1. Take an *A. tumefaciens* strain from a glycerol stock and streak on a solid medium (LB, AB, MG/L) supplemented with antibiotics for plasmid/strain selection, to form single bacterial colonies over 2 days at 28°C.
2. Select a single colony to inoculate the appropriate liquid medium and culture for 2 days at 28°C. The optical density of bacteria should be between 0.6 and 1.2 ($OD_{600}$).
3. Preculture explants on solidified primary medium for 2 days.
4. Place the culture of *A. tumefaciens* on ice for 20 minutes, and then centrifuge for 10 minutes at 6,000 rpm and at room temperature. Discard the supernatant and resuspend pellet in the same volume of liquid primary medium.
5. Transfer 20 explants per 25 ml to the liquid primary medium in 100 ml Erlenmeyer flasks. Add 100 μM acetosyringone (optional, depending on pea genotype). Mix the culture with 1 ml resuspended *A. tumefaciens* inoculum. Shake the coculture of bacteria and explants at 120 rpm at 22°C for 2 days.
6. Place the explants on solidified primary medium supplemented with 150 $mg{\cdot}l^{-1}$ timentin and an appropriate selective agent. The suggested concentration of kanamycin is 50 $mg{\cdot}l^{-1}$ and of phosphinothricin is 2 $mg{\cdot}l^{-1}$.
7. After 4-5 weeks of culture, separate growing, morphogenic, or callus tissue from browning tissue and continue subculture on the regeneration medium containing the selecting agent. Continue the selection of developing buds and shoots with doubled concentration (4 $mg{\cdot}l^{-1}$) in the case of phosphinothricin.

## *Differences Compared to the Method of Grant et al. (1995)*

1. Immature pods were harvested at the eating pea stage (green seeds at maximum size, before starting to dry). The embryonic axis and distal half of the immature cotyledon (to the embryonic axis) were discarded.
2. The cotyledon segments were immersed for one hour in an overnight culture of *A. tumefaciens*.
3. Explants were plated on B5 (Gamborg et al., 1968) medium containing 1.3 $mg{\cdot}l^{-1}$ BAP, 30 $g{\cdot}l^{-1}$ sucrose, 8 $g{\cdot}l^{-1}$ agar (Difco), and 200 μM acetosyringone, pH 5.5. After 6 days of cocultivation, the explants were washed with sterile water and rinsed in 400 $mg{\cdot}l^{-1}$ timentin. Cot-

yledons were transferred every 2 weeks to fresh B5 medium containing the selection agent (10 $mg \cdot l^{-1}$ phosphinothricin).

4. After three to four transfers, the cotyledon tissue was cut away from the growing shoots and callus. Shoots longer than 10 mm were excised and transferred to the root initiation medium: B5 with 1 $mg \cdot l^{-1}$ IBA + sucrose and agar as above + 150 $mg \cdot l^{-1}$ timentin, pH 5.8.
5. After 7 days, shoots were transferred to B5 medium without plant growth regulators, containing selection agent (10 $mg \cdot l^{-1}$ phosphinothricin) and 150 $mg \cdot l^{-1}$ timentin.

## **Agrobacterium-*Mediated Transformation of Cotyledonary Lateral Buds (According to Davies et al., 1993; Modified by Bean et al., 1997)***

1. Sterilized seeds were germinated for 24 hours at 20°C in 25 ml of sterile water and then placed on sterile grit moistened with sterile water for a further 48 hours. The testa, primary shoot, and root were excised and the lateral cotyledonary meristems inoculated using surgical scalpel blades coated in *Agrobacterium*.
2. The peas were cocultivated for 4 days, at 20°C with a 16-hour photoperiod, on filter paper moistened with water.
3. The peas were transferred onto B5 medium supplemented with 4.5 $mg \cdot l^{-1}$ BAP, 0.1 $mg \cdot l^{-1}$ IBA, 500 $mg \cdot l^{-1}$ augmentin, 0.7 percent (wt/vol) agarose, 0.5 $g \cdot l^{-1}$ 2-(N-morpholino) ethane sulfonic acid, pH 5.7, and 2.5 $mg \cdot l^{-1}$ of phosphinothricin.
4. After 3 weeks, the explants were placed on the same B5 medium with the concentration of phosphinothricin increased to 5 $mg \cdot l^{-1}$.
5. After 12 weeks, the explants were transferred to B5 containing 1 $mg \cdot l^{-1}$ BAP, 1 $mg \cdot l^{-1}$ IBA, and 0.5 $mg \cdot l^{-1}$ gibberellic acid.
6. Shoots were excised and placed in the same medium as used for the explants containing the selection agent.
7. The authors continued the culture of shoots unaffected by the selection agent on the selection medium for 2-3 weeks. Putative transformants were rooted using ½ MS medium supplemented with 1.5 percent sucrose, 0.8 percent agarose, and 2 $mg \cdot l^{-1}$ IAA or grafted on rootstock seedlings.

### *Comments on the Procedure*

At step 6, I would suggest the use of shoot-rooting medium containing selection agent (according to the first protocol). Shoots growing or growing and rooting on this medium are putative transformants.

## REFERENCES

Barakat, A., P. Gallois, M. Raynal, D. Mestre-Ortega, C. Sallaud, E. Guiderdoni, M. Delseny, and G. Bernardi (2000). The distribution of T-DNA in the genomes of transgenic *Arabidopsis* and rice. *FEBS Letters* 471:161-164.

Bean, S.J., P.S. Gooding, and P.M. Mullineaux (1997). A simple system for pea transformation. *Plant Cell Reports* 16:513-519.

Charity, J.A., M.A. Anderson, D.J. Bittisnich, M. Whitecros, and T.J.V. Higgins (1999). Transgenic tobacco and peas expressing a proteinase inhibitor from *Nicotiana alata* have increased insect resistance. *Molecular Breeding* 5:357-365.

Constantin, G.D., B.N. Krath, S.A. MacFarlane, M. Nicolaisen, I.E. Johansen, and O.S. Lund (2004). Virus-induced gene silencing as a tool for functional genomics in a legume species. *Plant Journal* 40:622-631.

Davies, D.R., J. Hamilton, and P. Mullineaux (1993). Transformation of peas. *Plant Cell Reports* 12:180-183.

de Sousa-Majer, M.J., D.C. Hardie, N.C. Turner, and T.J. Higgins (2007). Bean alpha amylase inhibitor in transgenic peas inhibit the development of pea weevil larvae. *Journal of Economic Entomology* 100:1416-1422.

Dolezel, J., J. Greilhuber, S. Lucretti, A. Meister, M.A. Lysak, L. Nardi, and R. Obermayer (1998). Plant genome size estimation by flow cytometry: Inter-laboratory comparison. *Annals of Botany* 82:17-26.

Food and Agriculture Organization of the United Nations (FAO) 1994. Production Year Book. Rome, Italy: FAO.

Gamborg, O.L., R.A. Miller, and K. Ojima (1968). Nutrient requirements of suspension cultures of soybean root cells. *Experimental Cell Research* 50:150-158.

Grant, J.E., P.A. Cooper, B.J. Gilpin, S.J. Hoglund, J.K. Reader, M.D. Pither-Joyce, and G.M. Timmerman-Vaughan (1998). Kanamycin is effective for selecting transformed peas. *Plant Science* 139:159-164.

Grant, J.E., P.A. Cooper, A.E. McAra, and T.J. Frew (1995). Transformation of peas (*Pisum sativum* L.) using immature cotyledons. *Plant Cell Reports* 15:254-258.

Grant, J.E., L.M.J. Thomson, M.D. Pither-Joyce, T.M. Dale, and P.A. Cooper (2003). Influence of *Agrobacterium tumefaciens* strain on the production of transgenic peas (*Pisum sativum* L.). *Plant Cell Reports* 21:1207-1210.

Griga, M. (2002). Morphology and anatomy of *Pisum sativum* somatic embryos. *Biologia Plantarum* 45:173-182.

Griga, M. and F.J. Novak (1990). Pea (*Pisum sativum* L.). In *Biotechnology in agriculture and forestry, legumes and oilseed crops I, Vol. 10,* Y.P.S. Bajaj (ed.). Berlin-Heidelberg: Springer-Verlag, pp. 65-99.

Hamilton, A.J. and D.C. Baulcombe (1999). A species of small antisense RNA in posttranscriptional gene silencing in plants. *Science* 286:950-952.

Hobbs, S.L.A., J.A. Jackson, and J.D. Mahon (1989). Specificity of strain and genotype in the susceptibility of pea to *Agrobacterium tumefaciens. Plant Cell Reports* 8:274-277.

Hood, E.E., G.L. Helmer, R.T. Fraley, and M.-D. Chilton (1986). The hypervirulence of *Agrobacterium tumefaciens* A281 is encoded in a region of pTiBo542 outside of T-DNA. *Journal of Bacteriology* 168:1291-1301.

Kohli, A., R.M. Twyman, R. Abranches, E. Wegel, E. Stoger, and P. Christou (2003). Transgene integration, organization and interaction in plants. *Plant Molecular Biology* 52:247-258.

Lazo, G.R., P.A. Stein, and R.A. Ludwig (1991). A DNA transformation-competent *Arabidopsis* genomic library in *Agrobacterium. Nature Bio/Technology* 9: 963-967.

Lulsdorf, M.M., H. Rempel, J.A. Jackson, D.S. Baliski, and S.L.A. Hoobs (1991). Optimizing the production of transformed pea (*Pisum sativum* L.) callus using disarmed *Agrobacterium tumefaciens* strains. *Plant Cell Reports* 9:479-483.

Murashige, T. and F. Skoog (1962). A revised medium for rapid growth and bioassays with tobacco tissue cultures. *Physiologia Plantarum* 15:473-497.

Nadolska-Orczyk, A. (1991). Some aspects of plant regeneration in pea (*Pisum sativum* L.). In *Proceedings of EUCARPIA symposium on genetic manipulation in plant breeding,* Reus/Salou, Spain, p. 21.

Nadolska-Orczyk, A., L. Miłkowska, and W. Orczyk (1994). Two ways of plant regeneration from immature cotyledons of pea. *Acta Societatis Botanicorum Poloniae* 63:153-157.

Nadolska-Orczyk, A. and W. Orczyk (2000). Study of the factors influencing *Agrobacterium* mediated transformation of pea (*Pisum sativum* L.). *Molecular Breeding* 6:185-194.

Nauerby, B., M. Madsen, J. Christiansen, and R. Wyndaele (1991). A rapid and efficient regeneration system for pea (*Pisum sativum*), suitable for transformation. *Plant Cell Reports* 9:676-679.

Nielsen, S.V.S., G.B. Poulsen, and M.E. Larsen (1991). Regeneration of shoots from pea (*Pisum sativum*) hypocotyl explants. *Physiologia Plantarum* 82:99-102.

Nifantova, S.N., I.u.V. Simonenko, I.K. Komarnitskii, and N.V. Kuchuk (2005). Production of transgenic pea (*Pisum sativum* L.) plants resistant to herbicide pusuit. *Tsitilogiia Genetika* 39:16-21.

Ozcan, S., M. Barghchi, S. Firek, and J. Draper (1992). High frequency adventitious shoot regeneration from immature cotyledons of pea (*Pisum sativum* L.). *Plant Cell Reports* 11:44-47.

Ozcan, S., M. Barghchi, S. Firek, and J. Draper (1993). Efficient adventitious shoot regeneration and somatic embryogenesis in pea. *Plant Cell Tissue and Organ Culture* 34:271-277.

Paszkowski, J. and S. Whitham (2001). Gene silencing and DNA methylation processes. *Current Opinion in Plant Biology* 4:123-129.

Prescott, V.E., P.M. Campbell, A. Moore, J. Mattes, M.E. Rothenberg et al. (2005). Transgenic expression of bean alpha amylase inhibitor in pea results in altered structure and immunogenicity. *Journal of Agricultural and Food Chemistry* 16: 9023-9030.

Puonti-Kaerlas, J., T. Eriksson, and P. Engstrom (1990). Production of transgenic pea (*Pisum sativum* L.) plants by *Agrobacterium tumefaciens*-mediated gene transfer. *Theoretical and Applied Genetics* 80:246-252.

Puonti-Kaerlas, J., T. Eriksson, and P. Engstrom (1992). Inheritance of a bacterial hygromycin phosphotransferase gene in the progeny of primary transgenic pea plants. *Theoretical and Applied Genetics* 84:443-450.

Richter, A., H.J. Jacobsen, A. de Kathen, G. de Lorenzo, K. Briviba, R. Hain, and H. Kiesecker (2006). Transgenic peas (*Pisum sativum*) expressing polygalcturonase inhibiting protein from raspberry and stilbene synthase from grapevine (*Vitis vinifera*). *Plant Cell Reports* 25:1166-1173.

Rolletscheck, H., F. Hosein, M. Miranda, U. Heim, K.P. Goetz et al. (2005). Ectopic expression of an amino acid transporter (VfAAP1) in seeds of *Vicia narbonensis* and pea increases storage proteins. *Plant Physiology* 137:1236-1249.

Sanago, M.H.M., V.I. Shattuck, and J. Strommer (1996). Rapid plant regeneration of pea using thidiazuron. *Plant Cell Tissue and Organ Culture* 45:165-168.

Schroeder, H.E., A.H. Schotz, T. Wardley-Richardson, D. Spencer, and T.J.V. Higgins (1993). Transformation and regeneration of two cultivars of pea (*Pisum sativum* L.). *Plant Physiology* 101:751-757.

Shade, R.E., H.E. Schroeder, J.J. Pueyo, L.M. Tabe, L.L. Murdock, T.J.V. Higgins, and M.L. Chrispeels (1994). Transgenic pea seeds expressing the α-amylase inhibitor of the common bean are resistant to bruchid beetles. *Nature Bio/Technology* 12:793-796.

Tzitzikas, E.N., M. Bergervoet, K. Raemakers, J.-P. Vincken, A. van Lammeren, and R.G.F. Visser (2004). Regeneration of pea (*Pisum sativum* L.) by a cyclic organogenic system. *Plant Cell Reports* 23:453-460.

## Chapter 14

# Genetic Transformation of Pea by Microprojectile Bombardment

Z. Molnar

## *INTRODUCTION*

Pea (*Pisum sativum* L.) is an important crop, owing to its symbiotic nitrogen-fixing capability and high nutritive value. Its seeds contain protein up to 40 percent of their dry weight (Monti and Grillo, 1983). Pea storage proteins tend to be rich in arginine, asparagine, glutamine, and lysine, while essential amino acids like methionine and cysteine are underrepresented. The major objectives in pea breeding are resistance against viruses, fungi, and insects as well as environmental stresses. The genetic basis of the present pea varieties is fairly narrow. In vitro cultures of different pea tissues have been routinely used for several years, but the apparent recalcitrance of pea in de novo regeneration of whole plants has prevented the application of genetic engineering in its genetic improvement breeding (Puonti-Kaerlas, 1993). Several promising regeneration systems have been elaborated (Grant et al., 1995; Ochatt et al., 2000; Tzitzikas et al., 2004), indicating the potential for genetic modification. An organogenic/meristematic tissue is formed from nodal tissue, which can be maintained in a cyclic fashion, giving it callus-like properties (Tzitzikas et al., 2004) making it an ideal system for transgenic plant production.

## *GENE TRANSFER TO PLANTS USING BIOLISTICS*

Several direct gene transfer methods have been developed to transform plant species. The most promising of them is gene shooting or biolistics (= biological ballistics). The process of gene delivery by the bombardment of DNA into intact plant cells has been described in detail (Klein et al., 1987;

Sanford, 1990), and several types of shooting devices have been reported (Kikkert, 1993; McCabe and Christou, 1993; Vain et al., 1993; Jenes et al., 1992). The process involves high-velocity acceleration of microprojectiles carrying foreign DNA, their penetration through the cell wall and membrane, and delivery into plant cells. Jenes et al. (1992) developed a gas-driven gun (GENEBOOSTER) for the bombardment of living tissues with DNA. The device was built with an automatic electronic control system, a newly designed stopping plate, macroprojectiles, convenient and continuous settings of the required shooting parameters, and fast reloading potentials to fulfill the requirements of high transformation frequency. Transient gene expression can be observed in differentiated plant tissues. The expression of foreign genes is examined by visual and quantitative methods, and the study of genes responsible for plant ontogeny can be possible.

## *FACTORS AFFECTING TRANSFORMATION BY MICROPROJECTILE BOMBARDMENT*

Microprojectile bombardment has a number of desirable attributes. The methodologies employed are simple, efficient, and essentially identical, regardless of the nature of the target cells and DNA used. It is also possible to wound plant tissues, allowing more efficient transformation via *Agrobacterium* (agrolistics; Southgate et al., 1995). Considerable research has concentrated on the utilization of microprojectile bombardment resulting in the development and refinement of several other versions of the apparatus. Most of these follow the same principle developed by Sanford et al. (1987). Minor alterations in the standard biolistic protocol have yielded tremendous improvements, including proper preculture of the explant material, the use of baffling screens and small microprojectiles, and subjecting the tissue to an osmotic pretreatment by either partial drying in a laminar flow hood or culturing in a medium containing an osmotic agent, such as mannitol (Hansen and Wright, 1999).

Optimizing conditions for delivering DNA is necessary to minimize tissue damage and bombardment variability. Among parameters that have been commonly optimized for developing efficient transformation technique are size, quantity, and velocity of microcarrier; cell sizes; concentration of calcium chloride and spermidine (for coating DNA onto microcarrier); DNA-microcarrier volume per bombardment; vacuum pressure; distance between macrocarrier to stopping plate and stopping plate to target tissue; and effect of multiple bombardment (Klein et al., 1988). These parameters were found to be species and very often tissue-type dependent.

Transient gene expression requires a gene product, which can be easily and rapidly monitored. The GUS system, based on the *gus* (*uid*A) gene coding for the enzyme β-glucuronidase, has been used extensively to optimize microprojectile bombardment parameters. Introduced foreign genes usually require a promoter in order to be expressed within recipient cells. The CaMV 35S promoter has been used extensively with various degrees of success in many plant species, sometimes as a double or triple promoter to increase gene expression. This promoter works well in the majority of dicots, such as pea. Warkentin et al. (1992) found that transient GUS enzyme activity was influenced by the promoter-leader sequence driving the *gus* gene, as measured by fluorometric and histochemical assays. Promoters can be tissue or developmentally specific. For example, the CaMV 35S promoter can be tissue specific in transgenic soybean (Yang and Christou, 1990). The maize alcohol dehydrogenase (*Adh1*) promoter (which is often used in monocot transformation) has been reported to be developmentally regulated and organ specific (Perl et al., 1992). Constructs coding for antibiotic or herbicide resistance can be used to select transformed cells from which plants can subsequently be regenerated. The genes generally used are *npt*II (coding neomycin phosphotransferase) and *bar* (coding for the enzyme phosphinothricin acetyl transferase). The use of the *npt*II gene enabled the first recovery of transgenic callus following microprojectile bombardment of soybean (Christou et al., 1988).

## *GENETIC TRANSFORMATION IN PEA*

Although some protocols of pea transformation by agroinfection have been elaborated, the regeneration of transgenic peas still is not routine. Two main factors are necessary to improve the pea indirect transformation efficiency: an efficient regeneration system and an appropriate *Agrobacterium tumefaciens* strain (Pniewski and Kapusta, 2005). Transgenic peas have been regenerated mainly via organogenesis de novo (Puonti-Kaerlas et al., 1990) and also directly from immature embryos (Grant et al., 1995; Bean et al., 1997; Nadolska-Orczyk and Orczyk, 2000). Transgenic pea plants were obtained after transformation conducted by hypervirulent *Agrobacterium* strains like EHA105 (Bean et al., 1997; Nadolska-Orczyk and Orczyk, 2000) or AGL1 (Grant et al., 1995), but also by other strains. Unfortunately, the rate of pea transformation remains relatively low. For this reason, an efficient transformation protocol is still being searched for. The identification of suitable conditions of regeneration and transformation is especially necessary for cultivars of particular or practical importance or for adaptation to specific environmental conditions (Polowick et al., 2000).

Because of the above-mentioned problems of agroinfection in pea, the real advantage of biolistics to produce transgenic pea plants lies in its application. Fertile transgenic soybean plants were produced after meristem transformation by use of biolistics nearly 2 decades ago (Christou et al., 1989), but only some studies on particle bombardment on different pea explants have been reported so far (Jordan et al., 1992; Warkentin et al., 1992; Molnar et al., 1999). However, because of its wide applicability, microprojectile-mediated DNA delivery can overcome some of the limitations associated with genome-specific tissue culture and thus has the potential to transform any variety, species, or genus.

## *PROTOCOL FOR PEA TRANSFORMATION BY MICROPROJECTILE BOMBARDMENT*

### *Plant Material*

The pea (*Pisum sativum* L.) genotype *Akt,* derived from our breeders at the Faculty of Agricultural and Food Sciences in Mosonmagyaróvár (University of West Hungary, Hungary), was selected according to its proper in vitro regeneration capacity (Molnar, 1991).

### *Explants for Bombardment*

Mature seeds were surface sterilized by immersing in 70 percent ethanol for 1 minute, and in 2 percent NaOCl for 15 minutes. The sterilization process was followed by rinsing three times with sterile distilled water. Mature embryos were dissected after 24 hours soaking in sterile distilled water.

### *Culture Conditions*

The plumules and radicles of mature embryos were cultured separately on the surface B5 (Gamborg et al., 1968) medium solidified with 0.8 percent agar, supplemented with 2 percent sucrose and 2 $mg \cdot l^{-1}$ NAA + 1 $mg \cdot l^{-1}$ BAP. The pH value of culture medium was set to 5.8 prior to autoclaving in 100 kPa at 121°C for 15 minutes. Gene transfer took place after 3 days of incubation.

### *Plasmid DNA*

The plasmids pRT103*gus* (β-glucuronidase-*gus*-gene), pFF19H (hygromycin resistance-*hph*-gene) obtained from the Agricultural Biotechnology Center (ABC) at Godollo (Hungary) were introduced into the target tissues. All plasmids were created carrying the CaMV 35S promoter.

### *Gene Transfer*

The GENEBOOSTER particle accelerating device, produced by ELAK Ltd. Co. (Budapest, Hungary), was used for transformation of pea. It consists of a barrel, a vacuum chamber for bombardment, and an electronic unit. The vacuum chamber is made of 3 mm steel covered by a plastic coating. The electronic control unit includes the vacuum and pressure sensors, electric valves, the controlling microprocessors, and a stainless steel secondary container. A vacuum pump (Sartorius, type SM 16692) is connected to the device to provide vacuum in the chamber. The macroprojectiles (macrocarriers) are made of plastic bonomit (provided by ELAK Ltd. Co., Budapest, Hungary) and the stopping plate is made of steel. Between the stopping plate and the target tissue there is usually a stainless steel screen of variable mesh size (tissue smashing screen, Sigma) mounted. DNA precipitation onto tungsten microprojectiles (microcarriers) was done according to the instructions of the device. The tungsten microcarriers (0.7-2.2 μm diameter on average, SILVANA-GTE, Towanda, PA) were stored in absolute ethanol (50 $mg \cdot ml^{-1}$). Suspensions were vortexed vigorously for one to two minutes to disperse aggregated particle lumps. The suspension was spun for 1 minute at 10,000 rpm and the supernatant was discarded. The pellet was resuspended in 1 ml of sterile distilled water, vortexed, spun for 1 minute and the supernatant was discarded. The final pellet was resuspended in 1 ml of sterile distilled water and 50 μl aliquoted (enough for six to eight bombardments) each in microtubes while vortexing of the suspension continued. A 5 μl sample of DNA solution (1 $\mu g \cdot \mu l^{-1}$), 50 μl of $CaCl_2$ (2.5 M) and 20 μl of spermidine (0.1 M, free base form) were added one by one to the 50 μl particle suspension. The mixture was vortexed for 3 minutes, spun for 15 seconds at 10,000 rpm, and the supernatant discarded. The pellet was washed with 250 μl of absolute ethanol. The final pellet was resuspended in 60 μl of sterile distilled water. A 6-7 μl sample of the aliquot was loaded onto the center of the macroprojectile and bombarded. The investigated physical parameters of GENEBOOSTER included the appropriate pressure of nitrogen gas for bombardment of the actual tissue, the shooting distance (i.e., distance from stopping plate to target tissue), and the size of tungsten particles used as microprojectiles. Two controls were also incorporated: tissues without bombardment and bombardment of microcarrier without DNA. Five replicates were used for obtaining reliable data, as suggested by Sanford et al. (1993). The explants were incubated for 2 days at 24°C in darkness prior to transient GUS assay.

### *Histochemical Assay for GUS*

The bombarded tissue was histochemically assayed for transient *Gus* gene expression following a modified protocol by Jefferson (1987) using X-Gluc (5-bromo-4-chloro-3-indolyl-β-D-glucuronic acid) as a substrate for the β-glucuronidase enzyme and the indigo blue color was observed in the cells and tissues. At 2 days after bombardment the tissues were stained overnight with GUS buffer at 37°C and blue spots were scored visually using a stereoscopic microscope.

### *Selection for Transgenics*

The plumules and radicles bombarded with plasmid DNA carrying the *hph* selectable marker gene were cultured on the same B5 medium that was used in the three-days preculture with the selective agent, 50 $mg{\cdot}l^{-1}$ hygromycin B (Sigma).

### *Plant Regeneration*

The explants surviving the selection procedure were transferred to B5 medium supplemented with 10 μM thidiazuron to obtain shoot regeneration. Rooting of shoots was carried out on B5 medium with 5 μM NAA and 5 μM IBA.

### *Analysis of Data*

Analysis of variance (ANOVA) and Duncan's multiple-range test using STATISTICA (StatSoft Inc.) software were used to analyze the data.

## *RESULTS*

Bombardment of the microprojectiles without DNA and control tissues did not show any transient GUS gene expression. Optimization was done using transient GUS gene expression as an indicator for efficiency of the parameters studied (Figure 14.1). Each blue spot arising from the histochemical localization of GUS activity, whether in a single cell or a group of cells, was considered as one expression unit (Parveez et al., 1997).

### *The Effect of Nitrogen Gas Pressure*

The various levels of nitrogen pressure applied had significant effects on shooting. The device needed at least 13 bars to perform a weak shoot, 16 bars to achieve penetration of about 10 percent of the particles into the target cells, and 18 to 24 bars to achieve penetration of 30 to 80 percent of the particles in the cells of plumules and radicles (Table 14.1). It was observed that reducing the pressure resulted in significant reduction of the expression, which may be due to the reduction of the force of microprojectiles in penetrating the cells. Higher pressure (28-30 bars) caused major damage on the surface of the target cells and tissues.

Increasing the pressure not only caused the tissues to be dislodged but also reduced the GUS expression significantly.

### *The Size of Microprojectiles*

The size of tungsten particles has an effect on their penetration into the living cells. Efficient gene transfer was obtained using microcarriers with 1.5 μm diameter, since the highest number of blue cells was visible after this treatment, when the cells were stained with X-Gluc for the histochemical assay of GUS (Table 14.2). These results can be explained by the fact that the smaller particles (0.7 μm diameter) do not have enough energy when ex-

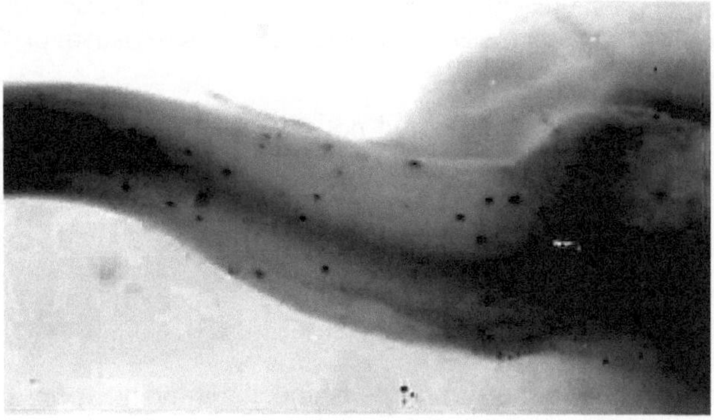

FIGURE 14.1. Histochemical GUS staining of radicules after 2 days of microprojectile bombardment. (See also color gallery.)

TABLE 14.1. Effect of nitrogen gas pressure on the efficiency of penetration into the mature embryo tissues of pea.

| Nitrogen pressure (bars) | Penetration | Notes |
|---|---|---|
| 13 | Nondetectable | The device shoots with a delay and with no power |
| 16 | Detectable | About 10% of the particles penetrated into the cells |
| 18 | Detectable | About 30% of the particles penetrated into the cells |
| 20 | Detectable | About 60-80% of the particles penetrated into the cells |
| 24 | Detectable | About 30-50% of the particles penetrated into the cells |
| 28 | Detectable | |
| 30 | Detectable | Visible injuries on the tissues at the hitting site of the particles |

TABLE 14.2. Efficiency of gene delivery according to the size of the microprojectiles.

| Diameter of tungsten particles (μm) | Number of blues pots/cm$^2$ | Notes |
|---|---|---|
| 0.7 | 1-54 | Considerable amount of particles can be observed on the surface of the tissues |
| 1.5 | 320-940 | |
| 1.8 | 70-180 | |
| 2.2 | 15-42 | Severe damage on the surface cell layer can be observed |

*Note:* Nitrogen pressure: 20 bars. The data represent the minimum and maximum of 10 shots with each size of tungsten particles. Plasmid: pRT103gus.

posed, so that many of them could not penetrate the cell wall. The bigger sizes (1.8 and 2.2 μm) have a relatively big surface compared to the cell size and cause major damage when penetrating the cells. These cells did not show any transient gene expression because they died very soon.

*The Optimal Shooting Distance*

While investigating the shooting distance, all nine positions of the target holder of the vacuum chamber in GENEBOOSTER were considered and used. The results of the shootings are summarized in Table 14.3. The suitable shooting distances in the case of the pea cells were the fifth (175 mm)

TABLE 14.3. The effect of shooting distance on the size of shooting area and efficiency of gene delivery.

| Level of target holder | Shooting distance (mm) | Shot area ($cm^2$) | Notes |
|---|---|---|---|
| 1. (bottom) | 265 | 13.85 | Shooting distance is too long. The bombarding particles do not have sufficient energy to penetrate at a high rate. |
| 2. | 240 | 11.05 | |
| 3. | 215 | 10.17 | |
| 4. | 195 | 8.80 | Shooting distance and the size of hit area are suitable. |
| 5. | 175 | 7.06 | |
| 6. | 155 | 5.82 | Shooting distance is too short; the size of area hit is small. Too much damage on the surface of cell layers. |
| 7. | 135 | 4.52 | |
| 8. | 115 | 3.66 | |
| 9. (top) | 95 | 2.54 | |

*Note:* Target tissue: embryo tissues of pea. Microprojectiles: tungsten particles (1.5 μm diameter). Plasmid: pRT103gus.

and sixth (155 mm) positions of the sample holder. No difference in transient GUS expression was observed when the distance was increased or decreased. However, bombarding at a shorter distance caused massive tissue dislocation. When the holder was put closer or at a more distant position, the shot caused either major damage to the tissues, or the efficiency of penetration needed for gene delivery dropped dramatically. Similar results were reported in *Phaseolus vulgaris* L. (Eissa Ahmed, 2002): increasing the distance reduced the efficiency as well as tissue damage. On the other hand, bombarding at much shorter distance increased tissue damage even though transient expression increased. Optimization of the distance before the stopping plate to target tissue is necessary to allow even spreading of the DNA microcarriers onto the target tissue without causing damage to the tissues due to the blast of compressed air and acoustic shock generated by the device (Russell et al., 1992).

### *Cost of GENEBOOSTER Application*

Calculating the cost of bombardment per shot showed an average of \$1US/shot (not including the price of the device), which is quite low compared with the running cost of other available particle devices, which varies between \$3-4 US for each shot (Jenes et al., 1992).

***Transformation Results***

The optimized conditions were based not only on the number of transient GUS expression units but also on physical impact on the tissues and shot-to-shot variability. Bombardment conditions resulting in massive tissue dislocation were not used even when they resulted in higher transient GUS expression because of the physical damage to the cells.

***These experiments have led to the following optimizations:***

- The optimal nitrogen gas pressure for GENEBOOSTER is about 20-22 bars (Table 14.1).
- The best tungsten particle (microprojectile) size to penetrate the embryo tissues of pea is 1.5 μm (Table 14.2, Figure 14.1).
- The optimal shooting distance to penetrate the cell wall and membrane of the target in vitro pea tissues is 155 mm (Table 14.3).
- Plumules and radicles cotransformed with plasmids pRT103*gus* and pFF19H and subcultured to selective medium produced calli and shoots, the sign of *hph* gene activity.
- Shoot regeneration occurred in some cases on calli derived from plumules on regeneration medium.

The regenerated pea seedlings seem to carry the introduced foreign genes during the selection and DNA hybridization as a molecular proof of stable transformation. A total of 36 putative transgenic pea plants were regenerated out of 3,350 plumule and radicle explants on selective medium. These numbers represent 1.07 percent transformation efficiency on the regenerated plantlet level referring to the *hph* gene. Out of the 36 regenerated plants, 20 could be transferred to soil and grown to flowering in greenhouse conditions. This gives a final transformation efficiency of 0.6 percent. Molecular analysis of plants obtained by biolistic techniques shows transformation generally revealed a complex pattern of transgene integration. In addition, delivery of long-fragment DNA is challenging because breaks can occur in the delivered DNA. Although the fate of introduced DNA is not clear, ligation of the transgenic DNA fragments before integration is proposed to account for the observation of arrays of transgenic DNA integrated at the same site into the plant genome (Hansen and Wright, 1999).

## REFERENCES


Bean, S.J., P.S. Gooding, P.M. Mullineaux, and D.R. Davies (1997). A simple system for pea transformation. *Plant Cell Reports* 16:513-519.

Christou, P., D.E. McCabe, and W.F. Swain (1988). Stable transformation of soybean callus by DNA-coated gold particles. *Plant Physiology* 87:671-674.

Christou, P., W.F. Swain, N.S. Yang, and D.E. McCabe (1989). Inheritance and expression of foreign genes in transgenic soybean plants. *Proceedings of the National Academy of Sciences USA* 86:7500-7504.

Eissa Ahmed, E. (2002). Tissue culture, in vitro bud induction and transformation of bean (*Phaseolus vulgaris* L.). PhD thesis, Faculty of Horticultural Sciences, Saint Stephanus University, Godollo, Hungary.

Gamborg, O.L., R.A. Miller, and K. Ojima (1968). Nutrient requirements of suspension cultures of soybean root cells. *Experimental Cell Research* 50:151-198.

Grant, J.E., P.A. Cooper, A.E. McAra, and T.J. Frew (1995). Transformation of peas (*Pisum sativum* L.) using immature cotyledons. *Plant Cell Reports* 15:254-258.

Hansen, G. and M.S. Wright (1999). Recent advances in the transformation of plants. *Trends in Plant Science* 4:226-231.

Jefferson, R.A. (1987). Assaying chimeric genes in plants: The GUS gene fusion system. *Plant Molecular Biology Reporter* 5:387-405.

Jenes, B., H. Moore, J. Cao, W. Zhang, and R. Wu (1992). Techniques for gene transfer. In *Transgenic Plants, Vol. 1,*, S.D. Kung and R. Wu (eds.). San Diego, CA: Academic Press, pp. 125-146.

Jordan, M.C., H. Rempel, and S.L.A. Hobbs (1992). Genetic transformation of *Pisum sativum* L. via *Agrobacterium tumefaciens* or particle bombardment. Proceedings of the 1st European Conference on Grain Legumes, AEP Editors, Angers, pp. 115-116.

Kikkert, J.R. (1993). The Biolistic PDS-1000/He device. *Plant Cell, Tissue and Organ Culture* 33:221-226.

Klein, T.M., E.C. Harper, Z. Svab, J.C. Sanford, M.E. Fromm, and P. Maliga (1988). Stable genetic transformation of intact *Nicotiana* cells by the particle bombardment process. *Proceedings of the National Academy of Sciences USA* 85:8502-8505.

Klein, T.M., E.D. Wolf, R. Wu, and J.C. Sanford (1987). High-velocity microprojectiles for delivering nucleic acids into living cells. *Nature* 327:70-73.

McCabe, D. and P. Christou (1993). Direct DNA transfer using electric discharge particle acceleration (ACCELL technology). *Plant Cell Tissue and Organ Culture* 33:227-236.

Molnar, Z. (1991). Borso (*Pisum sativum* L.) szovettenyesztesi vizsgalatok. I. Genotipusok in vitro tesztelese (Pea (*Pisum sativum* L.) tissue culture studies. I. In vitro genotype tests; in Hungarian). *Acta Ovariensis* 33:11-19.

Molnar, Z., B. Jenes, and V. Ordog (1999). Genetic transformation of pea (*Pisum sativum* L.) via particle bombardment. *Biologia (Bratislava)* 54(Suppl. 7):50.

Monti, L.M. and S. Grillo (1983). Legume seed improvement for protein content and quality. *Plant Foods Human Nutrition* 32:253-266.

Nadolska-Orczyk, A. and W. Orczyk (2000). Study of the factors influencing *Agrobacterium*-mediated transformation of pea (*Pisum sativum* L.). *Molecular Breeding* 6:185-194.

Ochatt, S.J., C. Pontecaille, and M. Rancillac (2000). The growth regulators used for bud regeneration and shoot rooting affect the competence for flowering and seed set in regenerated plants of protein peas. *In Vitro Cell and Developmental Biology–Plant* 36:188-193.

Parveez, G.K.A., M.K.U. Chowdhury, N.M. Saleh (1997). Physical parameters affecting transient GUS gene expression in oil palm (*Elaeis guineensis* Jacq.) using the biolistic device. *Industrial Crops and Products* 6:41-50.

Perl, A., H. Kless, A. Blumenthal, G. Galili, and E. Galun (1992). Improvement of plant regeneration and GUS expression in scutellar wheat calli by optimization of culture conditions and DNA-microprojectile delivery procedures. *Molecular and General Genetics* 235:279-284.

Pniewski, T. and J. Kapusta (2005). Efficiency of transformation of Polish cultivars of pea (*Pisum sativum* L.) with various regeneration capacities by using hypervirulent *Agrobacterium tumefaciens* strains. *Journal of Applied Genetics* 46: 139-147.

Polowick, P.L., J. Quandt, and J.D. Mahon (2000). The ability of pea transformation technology to transfer genes into peas adapted to western Canadian growing conditions. *Plant Science* 153:161-170.

Puonti-Kaerlas, J. (1993). Genetic engineering in pea crop improvement. *Acta Agriculturae Scandinavia Section B, Soil and Plant Science* 43:65-73.

Puonti-Kaerlas, J., T. Eriksson, and P. Engström (1990). Production of transgenic pea (*Pisum sativum* L.) plants by *Agrobacterium tumefaciens*–mediated gene transfer. *Theoretical and Applied Genetics* 80:246-252.

Russell, J.A., M.K. Roy, and J.C. Sanford (1992). Physical trauma and tungsten toxicity reduce the efficiency of biolistic transformation. *Plant Physiology* 98:1050-1056.

Sanford, J.C. (1990). Biolistic plant transformation. *Physiologia Plantarum* 79: 206-209.

Sanford, J.C., T.M. Klein, E.D. Wolf, and N. Allen (1987). Delivery of substances into cells and tissues using a particle bombardment process. *Journal of Particle Science and Technology* 5:27-37.

Sanford, J.C., F.D. Smith, and J.A. Russel (1993). Optimizing the biolistic process for different biological applications. *Methods in Enzymology* 217:483-509.

Southgate, E.M., M.R. Davey, J.B. Power, and R. Marchant (1995). Factors affecting the genetic engineering of plants by microprojectile bombardment. *Biotechnology Advances* 13:631-651.

Tzitzikas, E.N., M. Bergervoet, K. Raemakers, J.P. Vincken, A. Van Lammeren, and R.G.F. Visser (2004). Regeneration of pea (*Pisum sativum* L.) by a cyclic organogenic system. *Plant Cell Reports* 23:453-460.

Vain, P., N. Keen, J. Murillom, C. Rathusm, C. Nemes, and J.J. Finer (1993). Development of the particle inflow gun. *Plant Cell Tissue and Organ Culture* 33: 237-246.

Warkentin, T.D., M.C. Jordan, and S.L.A. Hobbs (1992). Effect of promoter-leader sequences on transient reporter gene expression in particle bombarded pea (*Pisum sativum* L.) tissues. *Plant Science* 87:171-177.

Yang, N.S. and P. Christou (1990). Cell type specific expression of CaMV 35S-Gus gene in transgenic soybean plants. *Developmental Genetics* 11:289-293.

Chapter 15

# *In Planta* Transformation of *Pisum sativum* L. Using Microinjection and Electroporation

Paul F. Lurquin

## *INTRODUCTION*

Electroporation is widely used to introduce several classes of molecules, including DNA, into prokaryotic and eukaryotic living cells. The technique was first used to transform murine cells (Neumann et al., 1982) and subsequently plant protoplasts (Fromm et al., 1985). Although successful, these first experiments were attempted before significant understanding of the physical and biological phenomena underlying cellular electropore formation had been gained. This situation has changed much in the intervening years, but many questions remain unanswered or are only partially answered.

It is now rather clear that DNA transfer mediated by electric discharges is critically dependent on first achieving membrane breakdown at high enough electric field strength (about 200 V/cm or higher for eukaryotic cells, depending on their size) and second, dissipating enough electrical energy into the electroporation cell (15 J/ml or more; Lurquin, 1997; Chen et al., 1998; Lurquin et al., 2001). Proper values of electric field strength are required to achieve dipole alignment in the cell membrane and electrocompression, while adequate energy release as heat at the level of the membrane (but not that of the electroporation cell as a whole) might be necessary to disrupt lipid ordering or force the donor DNA to enter the recipient cells.

One of the remaining questions pertains to the mechanism of DNA uptake by cells subjected to an electric discharge. It has been suggested that DNA enters living cells through an endocytocis-like mechanism during electroporation (Chernomordik et al., 1990; Tsong, 1991). In this scenario, membrane disruption is such that parts of it encapsulate cell-bound donor DNA, the complex then being engulfed in an endocytotic fashion. Results obtained with large liposomes mimicking eukaryotic cells do not support this interpretation (Lurquin and Athanasiou, 2000), although it remains to be seen precisely how living cells behave in that regard. So far, a few results obtained with *Avena sativa* and *Nicotiana tabacum* protoplasts do not support an invagination/endocytosis model for DNA uptake either (Mehrle et al., 1985; Nagata, 1989). Therefore, it may well be that DNA is internalized by diffusing through electropores, although more work is needed to resolve this question.

Another poorly understood aspect of electroporation concerns the effects of the cell wall on the electropermeability of plant and other cells. Results obtained with *Chlamydomonas reinhardtii* demonstrate that the presence of a cell wall does not constitute an insurmountable barrier to DNA uptake (Brown et al., 1991). Indeed, comparisons between wild-type *Chlamydomonas* and the wall-less *cw-15* mutant show reduced electrotransformation efficiency of the wild type, but not absence thereof. Thus, the presence of a cell wall does not abolish DNA uptake. Electroporation of wild-type *Chlamydomonas* was also observed through uptake of fluorescent ATP derivatives that, because of their electric charge, do not spontaneously penetrate these cells (Athanasiou and Lurquin, unpublished data). Comparisons between intact higher plant cells and plant cells devoid of a cell wall (protoplasts) in electroporation experiments have not been conducted because intact plant cells cannot be grown in single-cell cultures.

In view of the observations made with *Chlamydomonas* (and other single-celled organisms equipped with a cell wall), however, one should not expect that electrotransformation of intact plant cells is impossible. Indeed, the first positive report to that effect was published in 1987 (Lindsey and Jones, 1987) and was met with much skepticism. However, the Ghent, Belgium laboratory subsequently showed that rice and maize tissues could be electrotransformed and regenerated (in the case of maize) into transgenic plants (Dekeyser et al., 1990; D'Halluin et al., 1992). Since that time, many reports have shown that organized plant tissues, including embryos, can be electrotransformed (Penza et al., 1992; Akella and Lurquin, 1993; Kloti et al., 1993; Songstad et al., 1993; Xu and Li, 1994; Arencibia et al., 1995; Dillen et al., 1995; Gustafson et al., 1995; Jardinaud et al., 1995; Luong et al., 1995; Pescitelli and Sukhapinda, 1995; Rao, 1995; Hansch et al., 1996; Xiayi et al., 1996; Mitchell et al., 1998; Arencibia et al., 1998; Wu and Feng, 1999; Sorokin et al., 2000).

Given that a whole variety of undifferentiated and differentiated plant tissues in many species are amenable to electrotransformation, it is legitimate to wonder whether tissues *in planta* are also electrotransformable. Indeed, transforming isolated tissues, embryos, microspores, and so on still means that a regeneration step in vitro is necessary to produce transgenic plants. *In planta* transformation, that is, transformation of meristematic cells *in situ,* has the potential to produce offspring transgenic plants without the need for a tissue culture step. In this report, we show that such an approach, involving electroporation, is possible with *Pisum sativum*. Our technique, explained below, provides an alternative to *Agrobacterium tumefaciens*-mediated gene transfer, always a rather difficult proposition with legumes. This chapter is mostly based on work previously published by our laboratory (Chowrira et al., 1995, 1996, 1998). It should be noted that a very broad intact cell/whole plant/plant tissue electroporation patent was granted to Dev and Hayakawa in 1999.

## *METHODOLOGY, ANALYTICAL PROCEDURES, AND RESULTS*

### *Preparing the Donor DNA*

Plasmid DNA at a concentration of about 200 μg per ml in Murashige and Skoog (MS) salts is mixed gently with Lipofectina reagent (150 μl of reagent for each 1 ml of DNA solution). The mixture is then incubated at room temperature for 15 minutes prior to use in electroporation experiments. Lipofectin acts as a DNA protectant against nuclease attack.

### *Processing the Plants*

In general we use 3-week-old plants (var. Sparkle) grown at 25°C day/20°C night with a 10-hour night/14-hour day photoperiod. On the day preceding electroporation, plants are decapitated close to the node of a fully expanded leaf. Only the top nodal bud is retained and all other axillary buds are removed. The top nodal bud is then allowed to grow overnight. On the next day, 2 μL of the donor DNA mixture is injected into the bud using a Hamilton microsyringe (Chowrira et al., 1995).

### *Conditions for Electroporation*

We use the Progenetor electroporation system manufactured by Hoefer Scientific (San Francisco, California). This system delivers square pulses. The Progenetor comes equipped with a 6 mm gap circular electrode. We use the electrode in the upside down position with a plexiglass ring surrounding it (Chowrira et al., 1996). This ring is used as the electroporation cell. Prior to electroporation, 2 ml of plasmid DNA/Lipofectin mixture is added to the cell. This mixture can be used to electroporate several plants. It is replaced when the solution turns slightly green as a result of chlorophyll being electrophoresed out of the pulsed nodal buds. Two square pulses of 99 ms duration are delivered at 200 V (electric field strength is thus 333 V/cm). Control plants are injected with DNA but not electroporated, or injected with vector DNA or MS salts and electroporated. We have not tried our technique with parallel, flat electrodes. It should be remembered that a circular electrode generates an inhomogeneous electric field (Lurquin, 1997). After electroporation, plants are returned to the greenhouse, used for analysis, or selfed for further study. We have conducted studies up to the $R_4$ generation.

### *Donor Plasmids*

We use plasmid DNA in supercoiled configuration. Tested donor plasmids either contained a plant-expressible *uid*A reporter gene or a plant-expressible pea enation mosaic virus (PEMV) coat protein gene.

### *Molecular Analysis of Electroporated Plants*

In general, we recover a few chimeric $R_0$ individuals from a few dozen electroporated plants, which allow us to identify transgenic individuals without the need for selection. While using the *uid*A gene as a reporter, we identified $R_0$ individuals staining positive for GUS activity in a variety of tissues including petals, pollen grains, and stem (Chowrira et al., 1995). Some of these individuals showed GUS staining in all tissues, when selfed. Figure 15.1 shows that $R_1$ offspring obtained from two independently generated chimeric individuals contain the GUS-intron construct integrated at different genomic sites (Chowrira et al., 1996). Figure 15.2 demonstrates the same phenomenon with three $R_1$ plants transgenic for the PEMV coat protein gene (Chowrira et al., 1998). Northern blots performed with RNA isolated from $R_1$ plants transgenic for *uid*A showed the presence of both the unprocessed and processed transcripts (Chowrira and Lurquin, unpublished data).

Polymerase chain reaction (PCR) analysis for transgenic pea plants was performed with the PEMV coat protein gene only. Experiments confirmed that transgenic pea plants, up to the $R_4$ generation, contained a PCR-detectable coat protein gene as well as the PEMV coat protein (Chowrira et al., 1998).

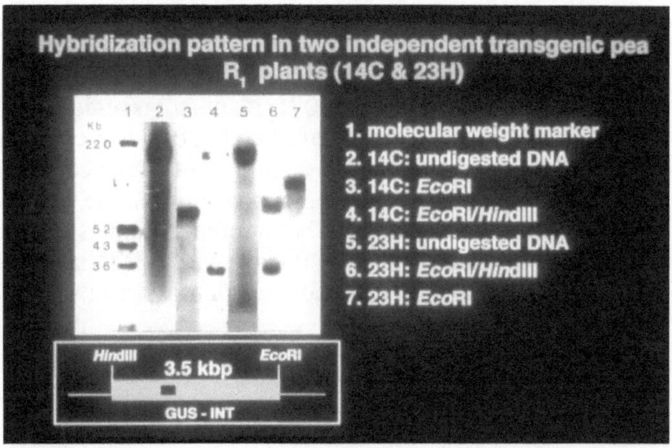


FIGURE 15.1. Southern blots of DNA extracted from two transgenic pea plants containing the GUS-intron construct. The probe was an *Ssp*I-*Sst*I fragment containing a portion of the intron and all the downstream *uid*A gene coding sequence. Double digestion with *Hin*dIII and *Eco*RI releases the 3.5 kb *uid*A cassette.

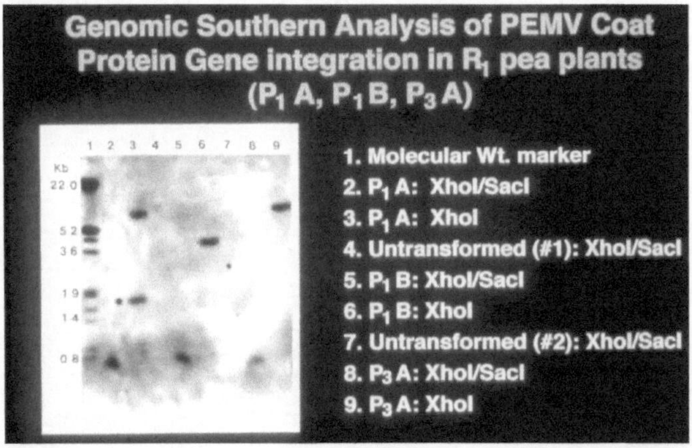


FIGURE 15.2. Southern blots of DNA extracted from three transgenic pea plants containing the PEMV coat protein construct. The probe was an *Xho*I-*Sac*I fragment containing the PEMV coat protein coding sequence. Double digestion with these two enzymes releases the 595 bp fragment corresponding to the coat protein gene. Lane 3 shows that this particular transgenic contains two copies of the transgene.

## *RESPONSE OF PEMV TRANSGENICS TO VIRAL CHALLENGE*

Our results showed that nearly 1,000 $R_2$ plants transgenic for the PEMV coat protein gene displayed a variety of phenotypes when challenged with PEMV. Symptoms ranged from full susceptibility to delayed symptoms, to recovery, to full resistance. Figure 15.3 shows an example of a transgenic individual displaying retarded growth but not the extreme stunting characteristic of the disease. ELISA tests conducted with PEMV antibody correlated well with visual analysis of phenotypes (Chowrira et al., 1998).

## *CONCLUSION*

The technique described in this chapter presents an alternative to *Agrobacterium*- and biolistic-mediated transformation of *Pisum sativum*. We assume that electroporation of DNA into terminal buds results in the transformation of meristematic cells, some of which will eventually differentiate into gametes. We have indeed detected pollen grains staining positive for GUS (Chowrira et al., 1996) in $R_0$ plants. We also have preliminary evidence that other grain legumes can be transformed in the same way (Chowrira et al., 1996). For example, lentils electroporated with the coat protein gene from the bean yellow mosaic virus test positive for expression as shown

FIGURE 15.3. Pea plants challenged with PEMV (25 days postinoculation). Left, uninoculated plant; middle, inoculated transgenic; right, inoculated control. (See also color gallery.)

by western blotting (Chowrira, Gupta, and Lurquin, unpublished data). More work is needed to determine the stability of transgenes in transformed pea plants over many generations.

## REFERENCES

Akella, V. and P.F. Lurquin (1993). Expression in cowpea seedlings of chimeric transgenes after electroporation into seed-derived embryos. *Plant Cell Reports* 12:110-117.

Arencibia, A., E. Gentinetta, S. Castaglione, A. Kohli, P. Vain, M. Leech, P. Christou, and F. Sala (1998). Molecular analysis of the genome of transgenic rice (*Oryza sativa* L.) plants produced via particle bombardment or intact cell electroporation. *Molecular Breeding* 4:99-109.

Arencibia, A., P.R. Molina, G. de la Riva, and G. Selman-Housein (1995). Production of transgenic sugarcane (*Saccharum officinarum* L.) plants by intact cell electroporation. *Plant Cell Reports* 14:305-309.

Brown, L.E., S.L. Sprecher, and L.R. Keller (1991). Introduction of exogenous DNA into *Chlamydomonas reinhardtii* by electroporation. *Molecular and Cell Biology* 11:2328-2332.

Chen, G.Y., A.J. Conner, J. Wang, A.G. Fautrier, and R.J. Field (1998). Energy dissipation as a key factor for electroporation of protoplasts. *Molecular Biotechnology* 10:209-216.

Chernomordik, I.V., A.V. Sokolov, and V.G. Budker (1990). Electrostimulated uptake of DNA by liposomes. *Biochim Biophys Acta* 1024:179-183.

Chowrira, G.M., V. Akella, E.P. Fuerst, and P.F. Lurquin (1996). Transgenic grain legumes obtained by *in planta* electroporation-mediated gene transfer. *Molecular Biotechnology* 5:85-96.

Chowrira, G.M., V. Akella, and P.F. Lurquin (1995). Electroporation-mediated gene transfer into intact nodal meristems *in planta. Molecular Biotechnology* 3:17-23.

Chowrira, G.M., T.D. Cavileer, S.K. Gupta, P.F. Lurquin, and P.H. Berger (1998). Coat protein-mediated resistance to pea enation mosaic virus in transgenic *Pisum sativum* L. *Transgenic Research* 7:265-271.

Dekeyser, R.A., B. Claes, R.M.U. De Rycke, M.E. Habets, M. Van Montagu, and A.B. Caplan (1990). Transient gene expression in intact and organized rice tissues. *Plant Cell* 2:591-602.

Dev, S.B. and Y. Hayakawa (1999). Electroporation-mediated molecular transfer in intact plants. U.S. Patent 5,859,327 (Assignee: Genetronics, Inc., San Diego, California).

D'Halluin, K., E. Bonne, M. Bossut, M. De Beuckeleer, and J. Leemans (1992). Transgenic maize plants by tissue electroporation. *Plant Cell* 4:1495-1505.

Dillen, W., G. Engler, and M. Van Montagu (1995). Electroporation-mediated DNA delivery to seedling tissues of *Phaseolus vulgaris* L. (common bean). *Plant Cell Reports* 115:119-124.

Fromm, M., L.P. Taylor, and V. Walbot (1985). Expression of genes transferred into monocot and dicot plant cells by electroporation. *Proceedings of National Academy of Sciences USA* 82:5824-5828.

Gustafson, V.D., P.S. Baenziger, and A. Mitra (1995). Electroporation of wheat anther culture-derived embryoids. *Cereal Research Communications* 23:207-231.

Hansch, R., T. Koprek, and H. Heydemann (1996). Electroporation-mediated transient gene expression in isolated scutella of *Hordeum vulgare. Physiologia Plantarum* 98:20-27.

Jardinaud, M.F., A. Souvre, and M. Beckert (1995). Optimization of DNA transfer and transient β-glucuronidase expression in electroporated maize (*Zea mays* L.) microspores. *Plant Cell Reports* 15:55-58.

Kloti, A., V.A. Iglesias, J. Wunn, P.K. Burkhardt, S.K. Datta, and I. Potrykus (1993). Gene transfer by electroporation into intact scutellum cells of wheat embryos. *Plant Cell Reports* 12:671-675.

Lindsey, K. and M.G.K. Jones (1987). Transient gene expression in electroporated protoplasts and intact cells of sugar beet. *Plant Molecular Biology* 10:43-52.

Luong, H.T., P.R. Shewry, and P.A. Lazzeri (1995). Transient gene expression in cassava somatic embryos by tissue electroporation. *Plant Science* 107:105-115.

Lurquin, P.F. (1997). Gene transfer by electroporation. *Molecular Biotechnology* 7:5-35.

Lurquin, P.F. and K. Athanasiou (2000). Electric field-mediated DNA encapsulation into large liposomes. *Biochemical Biophysical Research Communications* 267:838-841.

Lurquin, P.F., G. Chen, and A. Conner (2001). Electroporation and cell energy factor. In *Transgenic plants and crops,* G.G. Khatchatourians, A. McHughen, R. Scorza, W.-K. Nip, and Y.H. Hui (eds.). New York: Marcel Dekker, pp. 145-149.

Mehrle, W., U. Zimmermann, and R. Hampp (1985). Evidence for asymmetrical uptake of fluorescent dyes through electro-permeabilized membranes of *Avena* mesophyll protoplasts. *FEBS Letters* 185:89-94.

Mitchell, T.D., A.S. Bhagsari, P. Ozias-Akins, and S.K. Dhir (1998). Electroporation-mediated transient gene expression in intact cells of sweet potato. *In Vitro Cellular and Developmental Biology–Plant* 34:319-324.

Nagata, T. (1989). Cell biological aspects of gene delivery into plant protoplasts by electroporation. *International Review of Cytology* 116:229-255.

Neumann, E., M. Schaefer-Ridder, Y. Wang, and P.H. Hofschneider (1982). Gene transfer into mouse lyoma cells by electroporation in high electric fields. *EMBO Journal* 1:841-845.

Penza, R., V. Akella, and P.F. Lurquin (1992). Transient expression and histological localization of a *Gus* chimeric gene after direct transfer to mature cowpea embryos. *BioTechniques* 13:576-578.

Pescitelli, S.M. and K. Sukhapinda (1995). Stable transformation via electroporation into maize type II callus and regeneration of fertile transgenic plants. *Plant Cell Reports* 14:712-716.

Rao, K.V. (1995). Transient gene expression in electroporated immature embryos of rice (*Oryza sativa* L.). *Journal of Plant Physiology* 147:71-74.

Songstad, D.D., F.G. Halaka, D.L. De Boer, C.L. Armstrong, M.A.W. Hinchee, C.G. Ford-Santino, S.M. Brown, M.E. Fromm, and R.B. Horsch (1993). Transient expression of GUS and anthocyanin constructs in intact immature maize embryos following electroporation. *Plant Cell Tissue and Organ Culture* 33: 195-201.

Sorokin, A.P., X. Ke, D. Chen, and M.C. Elliott (2000). Production of fertile transgenic wheat plants via tissue electroporation. *Plant Science* 156:227-233.

Tsong, T.Y. (1991). Electroporation of cell membranes. *Biophysical Journal* 60: 297-306.

Wu, F.S. and T.Y. Feng (1999). Delivery of plasmid DNA into intact plant cells by electroporation of plasmolyzed cells. *Plant Cell Reports* 18:381-386.

Xiayi, K., Z. Xiuen, and S. Heping (1996). Electroporation of immature maize zygotic embryos and regeneration of transgenic plants. *Transgenic Research* 5:219-221.

Xu, X. and B. Li (1994). Fertile transgenic indica rice plants obtained by electroporation of the seed embryo cells. *Plant Cell Reports* 13:237-242.

Chapter 16

# *Agrobacterium*-Mediated Genetic Transformation of Peanut

P. Bhatnagar-Mathur
V. Anjaiah
P. B. Kirti
Kiran K. Sharma

## *INTRODUCTION*

Legumes are important sources of dietary proteins and fats in developing countries of the semiarid tropics where peanut (*Arachis hypogaea* L.) is one of the important food legume crops. It is a rich source of protein (23 percent) and edible oil (43 percent; Norden, 1980) and is considered one of the principal economic crops of the world. The genus *Arachis* belongs to the subfamily Papilionaceae of the family Fabaceae. It is native to South America and comprises diploid ($2n = 20$), tetraploid ($2n = 40$) and octaploid ($2n = 80$) species.

Peanut is a seed-propagating, self-pollinating crop originating from Brazil. The geographical classification of peanut is delineated into six regions: the Americas, Africa, Asia, Near East, Europe, and Oceania (Gregory et al., 1980). The total area under peanut cultivation is over 24.8 million hectares and the world's production is over 32.8 million tons per year, with an average yield of 1.32 tons per hectare (Rao and Nigam, 2001). India is the major producer of peanut, with a total production of 8.9 million tons per year. Peanuts are utilized in several ways; the edible oil is important for human consumption and the meal is used for livestock feed. It is also used directly for food in industrial countries including the United States, Canada, and the European Union.

Since the mid 1970s edible peanuts have increased in both domestic consumption and export trade. In contrast, production in Africa has declined by 17 percent over the last two decades. The major reasons for such low production are various biotic and abiotic stresses. Resource-poor farmers who obtain low yields of 500-800 $kg{\cdot}ha^{-1}$ due to various biotic and abiotic constraints grow about 93.8 percent of the world's production of peanut. *Aspergillus flavus,* which produces aflatoxins, for which no adapted resistant genotype is available, adversely affects peanut quantity and quality. Foliar diseases such as early and late leaf spot caused by *C. arachidicola* and *Phaeoisariopsis personata* (= *C. personata*) respectively are the most damaging diseases (Subramanyam et al., 1985). Among the insect pests, Spodoptera, legume pod borer, aphids, and thrips cause the greatest losses to the peanut crop (Wightman and Ranga Rao, 1993).

Conventional plant breeding techniques and methodologies have not been proven successful in imparting resistance against various biotic and abiotic stresses due to species barriers in the natural system. Peanut improvement has been limited due to the lack of integration of resistance to many diseases and pests from wild species of peanut (Stalker and Moss, 1987) because of problems with sterility barriers and genomic incompatibilities associated with traditional breeding. Genetic engineering approaches have been shown to be comparatively fast, leading to better isolation and cloning of genes controlling desired traits and their introduction into crop plants for combating biotic and abiotic stresses. Several gene transfer approaches have been employed to improve stress tolerance in different crop plants (Holmberg and Bulow, 1998). The application of recombinant DNA technologies for crop improvement in the semiarid tropics has shown great potential (Sharma and Ortiz, 2000). The advent of biotechnological tools including marker-assisted selection and gene transfer across the species barrier has opened up novel opportunities for enhancing seed quality, disease and pest resistance, viral resistance, abiotic stress tolerance, and nutritional improvement that are not accessible normally by conventional breeding, that is, they are limited by sexual incompatibility (Sharma and Ortiz, 2000). However, for successful genetic transformation, a reliable and effective regeneration system adaptable to transformation methods is needed. The transformation protocols for peanut are now well established and development of transgenic peanut expressing desirable foreign genes is going to be a reality soon (Sharma and Anjaiah, 2000). This chapter briefly reviews progress with tissue culture and genetic transformation of peanut and its possible applications for improvement of this important legume crop.

## *TISSUE CULTURE METHODS*

Advances in plant tissue culture techniques have been exploited for in vitro regeneration of peanut plants. In vitro regeneration in peanut occurs through organogenesis or embryogenesis. Regeneration by organogenesis occurs by the development of shoots directly on the surface of cultured explants (McKently et al., 1991; Hazra et al., 1989) or by an intervening callus phase, that is, the development of shoots from the callus tissue (Bajaj et al., 1981; Bajaj and Gosal, 1983, 1988). In earlier reports, the organogenic systems that regenerated shoots from immature leaflets, seed explants, de-embryonated cotyledons, hypocotyls, epicotyls, and anther-derived callus (Li et al., 1994; McKently et al., 1990; Mroginski and Fernandez, 1980; Mroginski et al., 1981; Narasimhulu and Reddy 1983; Pittman et al., 1983; Willcox et al., 1991) had a low frequency and plants were not realized in good frequencies. There are numerous reports of tissue culture and regeneration of peanut from diverse explants with various combinations of phytohormones in culture media (Table 16.1). However, not much success with genetic transformation of *Arachis* species was achieved until recently (Sharma and Anjaiah, 2000) due to the lack of efficient protocols to obtain whole plants through in vitro regeneration of adventitious shoot buds from the transformed tissues. This has prompted some workers to adopt non-tissue-culture-based approaches that do not depend on the regeneration of adventitious shoot buds for generating transgenic plants of peanut (Rohini and Rao, 2000). Direct regeneration systems have advantages due to the rapidity of morphogenesis and no requirement for frequent subculture; besides, de novo production of shoot primordia is extremely rapid and initially synchronous with the period of cellular differentiation. Such a regeneration system favors easy accessibility for *Agrobacterium*-mediated genetic transformation. Sharma and Anjaiah (2000) successfully obtained high-frequency direct shoot regeneration from mature cotyledon explants in various peanut genotypes. Shoot organogenesis and plants were also successfully obtained using immature leaflets (McKently et al., 1991; Sharma et al., unpublished data). Regeneration via somatic embryogenesis has also been reported (Cucco and Jaume, 2000; Gill and Saxena, 1992; Zhuang et al., 1999), which has been used in transformation studies in peanut (Ozias-Akins and Branch, 1990; Sellars et al., 1990; Chengalrayan et al., 1994, 1997). However, conversion of somatic embryos into plants remains inefficient and limits the application of somatic embryogenesis in many systems, including genetic transformation (Wetzstein and Baker, 1993).

TABLE 16.1. Responses of various explants and hormones on in vitro shoot regeneration in peanut.

| Explant | Genotype/ cultivar | Medium | Growth regulators | Morphogenic response | Reference |
|---|---|---|---|---|---|
| Apical meristem | | MS + B5 vitamins | NAA (10 μM) + BA (0.1 μM) | Single shoots with many roots | Kartha et al., 1981 |
| | | | NAA (10 μM) + BA (1 μM) | Shoots without any further development | Kartha et al., 1981 |
| Mesocotyl | | MS | IAA (11 μM) + kinetin (2.3 μM) | Shoots with roots | Bajaj, 1982 |
| Epicotyl | | MS | Casein hydrolysate | Multiple shoots, roots | Bajaj, 1982 |
| | New Mexico Valencia | MS | BA (10 $mg{\cdot}l^{-1}$) + NAA (1 $mg{\cdot}l^{-1}$) | Organogenesis | Cheng et al., 1992 |
| Hypocotyl | ICG 4367, US 48, TMV 2, TG 19B | MS | IAA (2 $mg{\cdot}l^{-1}$) + kinetin (2 $mg{\cdot}l^{-1}$) | Shoots | Narasimhulu and Reddy, 1983 |
| De-embryonated cotyledons | ICG 4367, US 48, TMV 2, TG 19B | MS | 2,4-D (2 $mg{\cdot}l^{-1}$) + kinetin (2 $mg{\cdot}l^{-1}$) | Multiple shoots | Narasimhulu and Reddy, 1983 |
| | TG- 17 | Moist cotton wool | BA (1 $mg{\cdot}l^{-1}$) | Multiple shoots | Bhatia et al., 1985 |
| Cotyledonary nodes | | MS | NAA (1 $mg{\cdot}l^{-1}$) + BA (3 $mg{\cdot}l^{-1}$) | Multiple shoots | Banerjee et al., 1988 |

| | | | | | |
|---|---|---|---|---|---|
| Mature cotyledons | JL-24, J-11, ICGS-11, ICGS 44, Robut-33-1 | MS + B5 organics | BA (20 μM) + 2,4-D (10 μM) | Adventitious shoot buds | Sharma and Anjaiah, 2000 |
| Immature leaflets | | MS + Gamborg vitamins | NAA (1 $mg \cdot l^{-1}$) + BA (1 $mg \cdot l^{-1}$) | 50% shoots | Pittman et al., 1983 |
| | JL-24 | MS | NAA (4 $mg \cdot l^{-1}$) + BA (5 $mg \cdot l^{-1}$) | | Chengalrayan et al., 1994 |
| | NC-7 | MS | NAA (2 $mg \cdot l^{-1}$) + BA (4 $mg \cdot l^{-1}$) | Shoots | Utomo et al., 1996 |
| Embryo axis | | MS | None | Shoots regenerated into plantlets | Atreya et al., 1984 |
| Immature embryos | New Mexico Valencia | - | TDZ (10 $mg \cdot l^{-1}$) | - | Kanyand et al., 1994 |
| | Several varieties | B5 | Picloram (0.5-1 $mg \cdot l^{-1}$) | Shoots with roots | Ozias-Akins et al., 1993 |
| Embryos | MK 374, M 13, TMV 2, Robut-33-1 | MS or White's medium | None | Whole plants | Sastri et al., 1980 |
| | Span cross, Dixie Spanish | MS | IAA (0.5 $mg \cdot l^{-1}$) + $GA_3$ (0.05 $mg \cdot l^{-1}$) + Zeatin (0.5 $mg \cdot l^{-1}$) | Shoots | Ozias-Akins and Branch, 1990 |
| Ovules | | MS | Kinetin + $GA_3$ | Shoots and roots | Martin, 1970 |
| Ovaries | MK 374, M 13, TMV 2, Robut-33-1 | MS | BA (0.5 $mg \cdot l^{-1}$) + NAA (2 $mg \cdot l^{-1}$) | | Sastri et al., 1980 |

TABLE 16.1 *(Continued)*

| Explant | Genotype/ cultivar | Medium | Growth regulators | Morphogenic response | Reference |
|---|---|---|---|---|---|
| Epicotyl | ICG 4367, US 48, TMV 2, TG 19B | MS | None | 9-28% shoots | Narasimhulu and Reddy, 1983 |
| De-embryonated cotyledons | MK 374, M 13, TMV 2, Robut-33-1 | MS | Zeatin (4 $mg \cdot l^{-1}$) or kinetin (4 $mg \cdot l^{-1}$) | Multiple shoots | Sastri et al., 1980 |
| Leaflets | | MS | NAA (1 $mg \cdot l^{-1}$) + BA (1 $mg \cdot l^{-1}$) | Organogenic callus | Mroginski et al., 1981 |
| | TMV 2 | MS | BA (2 $mg \cdot l^{-1}$) + NAA (0.5 $mg \cdot l^{-1}$) | Shoot primordia | Venkatachalam et al., 1999 |
| Plumule | Okrun | MS | BA (30 μM) + NAA (5 μM) + brassin (1 μM) | Multiple shoots | Ponsamuel et al., 1998 |

## *GENETIC TRANSFORMATION*

Genetic transformation for incorporation of novel genes into the peanut gene pool has opened up new opportunities for crop improvement in this important legume. The transformation and regeneration protocols for peanut are now well established. Transformation efficiencies frequently are directly related to the tissue culture response and therefore highly regenerative cultures are often transformation competent. The developments in genetic transformation in peanut have emboldened researchers to pursue the development of transgenic peanut plants capable of producing high-quality peanuts resistant to various diseases, insect pests, and abiotic stresses (Sharma and Anjaiah, 2000; Rohini and Rao, 2001). Peanut tissue is susceptible to infection by wild-type strains of *A. tumefaciens* (Lacorte et al., 1991). Several methods for DNA transfer are used for the genetic transformation of peanut (Table 16.2). Novel genes can be introduced into actively growing peanut cells biologically through *Agrobacterium*-mediated gene transfer or through direct and physical DNA delivery methods such as electroporation or microprojectile bombardment. However, *Agrobacterium*-mediated gene transfer is the most widely applied system in peanut. Increasingly, there is a trend toward the use of *A. tumefaciens* for DNA delivery in crop improvement programs compared with microprojectile bombardment. This is driven by the development of highly virulent strains and binary vectors that are useful for genetic transformation and their ease of use and researcher familiarity. There is also the consensus that because *A. tumefaciens* generally delivers only the T-DNA, transgene loci resulting from *A. tumefaciens* infection are less complex than those produced via direct DNA delivery methods. A unique advantage of *Agrobacterium* T-DNA transfer is the accurate processing of the T-DNA between the right and left borders and its precise transfer and integration into the plant genome. Historically, both microprojectile bombardment and *A. tumefaciens* have been used for DNA delivery into either organogenic or embryogenic cultures of peanut (Table 16.2).

### *Key Elements of Efficient Transformation*

Despite significant advances over the past decade, development of efficient transformation methods can take many years of painstaking research. Peanut transformation, like all other transformation systems, relies on common key elements. The major components for the development of transgenic plants are:

1. the development of reliable tissue culture regeneration systems
2. preparation of gene constructs and transformation with suitable vectors
3. efficient techniques of transformation for the introduction of genes into the crop plants
4. recovery and multiplication of transgenic plants
5. molecular and genetic characterization of transgenic plants for stable and efficient gene expression
6. transfer of genes to elite cultivars by conventional breeding methods if required
7. evaluation of transgenic plants for their effectiveness in alleviating the biotic and abiotic stresses in the field condition
8. biosafety assessments including health, food, and environmental safety
9. deployment of genetically modified plants.

Transformation of plants involves the stable introduction of DNA sequences, usually into the nuclear genome of cells capable of giving rise to a whole transformed plant. Transformation without regeneration and regeneration without transformation are of limited value. The very basis of regeneration in tissue cultures is the recognition that somatic plant cells are totipotent (i.e., capable of giving rise to a whole plant) and can be stimulated to regenerate into whole plants in vitro, via organogenesis or somatic embryogenesis, provided they are given the correct hormonal and nutritional conditions (Skoog and Miller, 1957). Adventitious shoots or somatic embryos are thought to arise from single cells and thus provide identifiable totipotent cells that are both competent and accessible for gene transfer and will give rise directly to nonchimeric transformed plants. Transformation techniques reliant on plant regeneration from in vitro cultured tissues have been described for many species (Lindsey and Jones, 1989; Dale et al., 1993; Birch, 1997). There are numerous reports of tissue culture and transformation of peanut from various explants (Kartha et al., 1981; Sastri and Moss, 1982; Kanyand et al., 1994). Regeneration via somatic embryogenesis has also been reported as one of the promising methods for transformation studies in peanut (Ozias-Akins et al., 1993; Sellars et al., 1990; Baker and Wetzstein, 1995; Chengalrayan et al., 1994, 1997).

A suitable system for selection of transgenic tissues and plants is one of the most important aspects of any transformation system. The utility of any particular gene construct as a transformation marker varies depending on the plant species and explant involved.

TABLE 16.2. Genetic transformation in peanut.

| Explant | Mode of gene transfer | Strain/plasmid | Gene of interest | Transformation frequency | Reference |
|---|---|---|---|---|---|
| Cotyledon | At | | | 3.30% | Rohini and Rao, 2000 |
| Leaf | At | pBI121 | *gus, npt*II | 0.2-0.3% | Cheng et al., 1997 |
| Embryonic axis | At | EHA 101/ pMON9793 | *uidA, npt*II | 9% | McKently et al., 1995 |
| | Pb | | | 0.9-1% | Brar and Cohen, 1994 |
| | Pb | pAC2MR/ pACH2MR | *MerApe*9, *hph*/*MerApe*9, mercuric ion reductase | | Yang et al., 2003 |
| Embryonic axis, cotyledon, leaf, petiole explants | At | pTiBo542/ pTiT37 | *uid*A, *npt*II | | Lacorte et al., 1991 |
| Mature cotyledons | At | PBI121/ pROKII:IPCVcp | *IPCV (coat protein)* | 55% | Sharma and Anjaiah, 2000 |
| | At | | *H protein gene* | | Khandelwal et al., 2003 |
| Cotyledons | At | LBA4404/pBI121 | *uid*A, *npt*II | 47% | Venkatachalam et al., 2000 |
| | Pb | pCAMBIA-1301 | *uid*A, *hph* | 1.6% | Yang et al., 2001 |
| | Pb | pMOG617/pxVGH | *uid*A, *hph* | 168 hygomycin resistant lines recovered | Wang et al., 1998 |

TABLE 16.2 *(Continued)*

| | | | | | |
|---|---|---|---|---|---|
| Embryogenic callus | Pb | | | 1% | Ozias-Akins et al., 1993 |
| | Pb | pDO432/pHyg$^r$/ pGIN | *Luc, hph hph* | 54 independent transgenic lines | Livingstone and Birch, 1995 |
| Immature cotyledon | Pb | | *cry1Ac* | | Singsit et al., 1997 |
| Somatic embryos | Pb | pCB13-N$^+$ pCB13-N$^{++}$ | *hph* gene nucleocapsid protein gene of TSWV | 52 hygromycin resistant cell lines | Yang et al., 1998 |
| Embryonic leaflets | El | | | | Padua et al., 2000 |
| Leaf, epicotyl | At | EHA 101 | *uid*A | 12-36% (leaves) 15-42% (epicotyl) | Egnin et al., 1998 |
| Leaf discs | At | pBI121 | *gus, npt*II | 6.7% putative shoots; 20 confirmed sterile transgenic plants | Eapen and George, 1994 |
| Epicotyl | Pb | pKYLX80-N11 pTRA140 | *uid*A, *hph* | | Magbanua et al., 2000 |

*Note:* At: *Agrobacterium tumefaciens*; Pb: particle bombardment; El: electroporation.

### *Selection System*

Most vectors used for the genetic transformation of plants carry marker genes that allow the recognition of transformed cells by either selection or screening. The most popular selectable marker genes used in plant transformation vectors include constructs providing resistance to antibiotics such as kanamycin and hygromycin, and genes that allow growth in the presence of herbicides such as phosphinothricin, glyphosate, bialaphos, and several other chemicals (Wilmink and Dons, 1993). For successful selection, the target plant cells must be susceptible to relatively low concentrations of the antibiotic or herbicide in a nonleaky manner. Choice of antibiotic and selective concentration varies across different explants and genotypes in peanut. Clemente et al. (1992) have shown kanamycin to be an effective selection agent to select stably transformed callus tissue obtained from immature leaflets of peanut. To date, kanamycin resistance is the most widely used selectable marker. Judicious choice of selection levels may be an important criterion for the recovery of transformed cells, because too high a level would be deleterious even to the transformed cells at initial stages of screening. Preculture of inoculated explants for 2 weeks in the absence of selection was important for enhanced efficiency of transformation, although absence of selection at initial stages may also result in very low recovery of transformants (Moloney et al., 1989).

Screenable marker reporter genes have also been developed from bacterial genes coding for easily assayed enzymes, such as chloramphenicol acetyl transferase (*CAT;* Herrera-Estrella et al., 1983), β-glucuronidase (*GUS;* Jefferson, 1987), luciferase (*LUX;* Olsson et al., 1988), green fluorescent protein (*GFP;* Reichel et al., 1996), nopaline synthase, and octopine synthase (Herrera-Estrella et al., 1988). However, in peanut transformation β-glucuronidase (Jefferson, 1987) is the most widely used screenable marker. The optimization of selection and identification systems is crucial for improving transformation efficiency. The development of a selection system based on hygromycin B greatly increased transgenic soybean production and reduced both the number of nontransformed escapes and time in culture (Olhoft et al., 2003).

## Agrobacterium-*Mediated Genetic Transformation*

The naturally evolved unique system of *Agrobacterium* transfers the foreign DNA sequences precisely into plant cells using Ti plasmids. An *Agrobacterium*-based DNA transfer system offers many unique advantages in plant transformation: (1) higher frequency of stable transformation with many single-copy insertions, (2) a precise transfer and integration of DNA sequences with defined ends, (3) a linked transfer of genes of interest along with a transformation marker, (4) a reasonably low incidence of transgene silencing, and (5) the ability to transfer long stretches of T-DNA.

Preliminary evidence in peanut transformation suggests gene transfer into the calli on coculturing seedling-derived hypocotyl explants with *Agrobacterium* (Dong et al., 1990; Lacorte et al., 1991; Mansur et al., 1993). *Agrobacterium*-mediated transformation using leaf explants of peanut resulted in a transformation frequency of 2 percent (Eapen and George, 1994). The immature embryonic axis has also been employed as an explant for *Agrobacterium*-mediated transformation in peanut. McKently (1995) developed a procedure whereby embryonic axes from mature seeds of peanut cocultivated with *A. tumefaciens* were stably transformed.. Cheng et al. (1997) obtained fertile transgenic plants with 0.3 percent frequency using leaf segments. However, a high transformation frequency was obtained with cotyledonary node explants precultivated on medium for 3 days followed by 4 days of cocultivation with *A. tumefaciens* strain LBA4404 carrying marker *gus* and *npt*II genes (Venkatachalam et al., 1998). Shoot regeneration occurred within 4 weeks. Besides, Yang and co-workers (1998) introduced the nucleocapsid gene of tomato spotted wilt virus along with the *uid*A and *npt*II marker genes in a sense orientation, into peanut variety New Mexico Valencia, using *Agrobacterium*-mediated transformation.

Precultured peanut cotyledons cocultivated for 2 days with *Agrobacterium* strain LBA 4404, harboring pBI121 containing *uid*A and *npt*II genes, followed by transfer on an embryo induction medium containing NAA, BA, kanamycin, and cefotaxime resulted in transformed embryos, which efficiently gave rise to shoots (47 percent) on MS medium containing BA and kanamycin (Venkatachalam et al., 2000). A non-tissue-culture-based transformation method involving direct cocultivation of cotyledon-attached embryo axis with *Agrobacterium* treated with wounded tobacco leaf extract resulted in a stable 3 percent transformation frequency (Rohini and Rao, 2000). An efficient system with high transformation frequency, above 55 percent, based on cotyledon explants forming adventitious shoot buds (>90 percent) has been developed by Sharma and Anjaiah (2000). A number of independently transformed peanut plants with coat protein gene of IPCV were produced by this method. Besides, *Agrobacterium*-mediated transgenic peanut plants expressing the hemagglutinin (H) protein of rinderpest virus have also been developed as an expression system for the delivery of recombinant subunit vaccine through fodder as a means of mass immunization of domestic ruminants as well as wildlife (Khandelwal et al., 2003). More recently, Swathi Anuradha et al. (2006) produced promoter tagged transgenic plants of peanut using the cotyledonary nodes as explants and a promoterless fusion gene *npt*II:*gus*.

## *DIRECT GENE TRANSFER*

Direct DNA transfer methods can circumvent the genotype dependence of *Agrobacterium* infection. Direct gene transfer has been accomplished by several methods such as microprojectile bombardment, electroporation of protoplasts and intact tissues, microinjection of protoplasts or meristems, and polyethylene glycol–mediated transformation of protoplasts. Of these methods, microprojectile bombardment is the most widely deployed method for genotype-independent genetic transformation. Microprojectile bombardment or particle gun bombardment has a number of characteristics that make it an attractive alternative for DNA delivery in peanut and has been demonstrated as a practical means of introducing a number of agronomically important genes.

Particle bombardment, developed by Sanford and his co-workers (Sanford, 1990; Sanford et al., 1987; Klein et al., 1988), has been successfully used for direct introduction of genes into a number of plant species including peanut. Choice of the explant for bombardment can be made on the basis of criteria such as regeneration potential, favorable metabolic conditions for the expression of a particular genetic construction, or cellular organization that facilitates unambiguous selection of the transformants (Schnall and Weissinger, 1995). Transient expression (Li et al., 1995) and stable transformation have been observed in callus lines from immature peanut leaflet tissue bombarded with microcarrier particles carrying plasmid DNA (Clemente et al., 1992). Of 875 leaflets of the cultivar UPL PN 4 bombarded, 202 kanamycin-resistant calli were recovered but only one untransformed shoot was produced. Similar observations were reported by Schnall and Weissinger (1995) where regenerated plants from slow-growing brown callus as well as green clusters formed by bombarding leaflets did not show any stable transformation. However, bombardment of 1-2-year-old embryogenic callus derived from immature embryos followed by stepwise selection for resistance to hygromycin B in solid and liquid media produced transgenic shoots at a frequency of 1 percent (Ozias-Akins et al., 1993), while the shoot meristems of mature embryonic axis produced transgenic plants at a relatively low transformation frequency of 0.9-1.0 percent (Brar and Cohen, 1994). Transgenic peanut plants expressing the *cry1Ac* gene for resistance to the cornstalk borer (*Elasmopalpus lignosellus*) have been reported (Singsit et al., 1997) by using the somatic embryos from immature cotyledons of peanut bombarded with vectors containing the codon-modified *cry1Ac* gene along with the *hpt* gene for antibiotic resistance with an efficiency of 0.85 to 2.3 transgenic lines per bombardment. ELISA of Cry1Ac protein from the putatively transformed plants showed the expression of Cry1Ac protein up to 0.18 percent of the total soluble protein. Insect bioassays conducted at a temperature of 27°C, light/dark cycle of 16:8 hours and 70 percent relative humidity also indicated various levels of resistance to *E. lignosellus*. The transient gene expression as assayed by GUS assay has been found to be affected by both particle size and amount of DNA used for coating and to be positively correlated with gene copy number (Lacorte et al., 1997). Livingstone and Birch (1995) efficiently transformed both Spanish and Virginia types of peanut by particle bombardment into embryogenic callus derived from mature seeds, followed by single-step selection for hygromycin B resistance resulting in 3 to 6 independent transformants per bombardment of 10 $cm^2$ embryogenic calluses with copy number ranging from 1 to 20 with a mean of 4 copies. Recent reports show further increased transformation efficiencies, ranging from $2.6 \pm 3.5$ to $19.8 \pm 18.5$ hygromycin B-resistant lines per bombardment (5 $cm^2$) with fertility rates of 32 percent (Wang et al., 1998).

Among the different genes that have been introduced by particle gun bombardment is the 2S albumin gene from Brazil nut (Lacorte et al., 1997). A high-frequency transformation and regeneration of somatic embryos via microprojectile bombardment has been achieved with constructs containing the *hpt* gene and the nucleocapsid protein (*N*) gene of the lettuce isolate of tomato spotted wilt tospovirus (Yang et al., 1998). The primary transformant containing a single copy of the transgene expressing the N protein, indicating a gene-silencing mechanism operating in the primary transgenic lines with multiple gene integration, has been observed. More recently, peanut lines exhibiting high levels of resistance to peanut stripe virus (PStV) were obtained following cobombardment of embryogenic callus derived from mature seeds of the commercial cultivars Gajah and NC7 with the hygromycin resistance gene and one of two forms of the PStV coat protein (*CP*) gene, an untranslatable full-length sequence (*CP2*) or a translatable gene encoding a *CP* with an N-terminal truncation (*CP4;* Higgins et al., 2004). More recently, Bhatnagar-Mathur et al. (2007) developed transgenic peanut plants by expressing the *AtDREB1A* gene under the stress inducible *rd29A* gene promoter and demonstrated that one of the transgenic events showed 40 percent higher transpiration efficiency than the control plants under water limiting conditions.

Preculture and osmotic treatments have important effects on transformation. Rinsing leaf and epicotyl explants of var. New Mexico in half-strength MS medium prior to infection has been reported as more conducive to

*Agrobacterium* transformation than the runner-type cultivars. The transient transformation efficiency significantly increased from 12 percent to 36 percent for leaf explants and 15 percent to 42 percent for epicotyls (Egnin et al., 1998). The preculture process influences the competence for transformation of bombarded epidermic cells and subepidermic cells on the adaxial surface of peanut cotyledons. Cotyledons precultured for 3 days on half-strength MS medium followed by 3 hours treatment in osmosis medium before particle bombardment with a plasmid containing a chimeric *hph* gene conferring resistance to hygromycin and a chimeric intron-*gus* gene resulted in a high transformation frequency (Yang et al., 2001). The biolistic-based systems for gene delivery into embryogenic calluses and embryo axes are labor intensive and require the bombardment of a large number of explants to obtain a few transformed cell lines (1 percent) which produce transgenic plants at low frequencies that are often chimeric or result from a few transformation events.

The advantages of particle bombardment system are: (1) DNA may be transferred without using specialized vectors; (2) the introduction of multiple DNA fragments or plasmids can be accomplished by cobombardment, thus eliminating the necessity of constructing a single large plasmid containing multiple transforming sequences; and (3) organelle transformation is achieved only by particle bombardment. Though the biolistic gene delivery system has been successfully used to create transgenics, certain drawbacks of the technique have been observed, for example, high copy number and rearrangements of transgenes, thus causing gene silencing or genomic rearrangements.

Different methods based on biological or direct DNA transfer have been developed for the production of transgenic peanut over the last few years. Padua et al. (2000) employed an electroporation method for direct gene transfer into intact embryonic leaflets of peanut in a modified electroporation buffer (EPRm) supplemented with 75 μM NaCl. A positive effect on the number of shoots and regeneration efficiency was observed using electric strengths of 500-625 v/cm. Research is being carried out globally with single or multiple gene introductions to produce pest-resistant, healthier, and high-quality peanuts.

## *GENETICALLY ENGINEERED PLANTS FOR PEANUT IMPROVEMENT*

Genes for transformation can be broadly divided into those that will be used to overcome agronomic limitations (high yield potential, resistance to biotic and abiotic stresses) and ones that could be used to enhance value-added traits (Schnall and Weissinger, 1995). Although major emphasis is currently being placed on improving the primary constraints, the manipulation of value-added traits, such as flavor and nutrition will be of much concern for peanut improvement using transgenic technology. Transgenic technology could conceivably be used in peanut for the introduction of disease and pest resistance as well as value-added traits such as improved vitamin, protein, and oil quality, enhancing the crop product value, quality, and safety. The genus *Arachis,* which itself is a repository for most of the valuable pest- and disease-resistance genes, could be used to transform cultivated peanut varieties. Current efforts include incorporating immunity or very high resistance to several viral and fungal diseases through transformation of peanut cultivars that have very high demand for which no adapted resistant peanut genotypes are available. Improved crop protection through the transfer and expression of disease resistance genes will decrease or eliminate the usage of pesticides, which are costly to the grower and may be harmful to the environment.

### *Fungal and Bacterial Disease Resistance*

Peanuts are susceptible to aflatoxin contamination; peanuts contaminated with aflatoxins cannot be used for human consumption. In addition to their direct impact on the yield and quality of agricultural crops, they are often indirectly related to the introduction of plant pathogens, which produce aflatoxins, a group of potent carcinogens. Development and integration of plant resistance, biological control, and genetic approaches for application in localized and area-wide pest management programs is a must for the eradication of this fungus. Peanut produces stilbene phytoalexins in response to fungal infection. Stilbenes inhibit fungal growth and spore germination of *Aspergillus* species and aflatoxin contamination does not occur as long as kernels have the capacity to produce stilbenes. Stilbene synthase has been identified as the key enzyme for the biosynthesis of stilbene. The gene encoding of this enzyme has already been characterized and even successfully expressed in tobacco. Organ-specific expression of multiple copies of a gene for stilbene synthesis is likely to enhance production of stilbenes in peanut kernels and hence make them less prone to colonization by *Aspergillus flavus* and coincident aflatoxin contamination. Hydrolytic enzymes such as chitinases and glucanases, which degrade the fungal cell wall, also pose as attractive candidates for development of disease-resistant peanut plants (Eapen, 2003). A novel approach of introducing microbial toxins (phytotoxins) such as tabtotoxin acetyl transferase and glucose oxidase into the plant has emerged as an efficient way to develop resistance in a wide range of host species (Eapen, 2003). This approach can be conveniently used to impart resistance against bacterial wilt of peanut caused by *Burkholderia solanacearum,* formerly known as *Pseudomonas solanacearum.*

### *Virus Resistance*

Viruses pose a great threat to peanut production throughout the world. Viruses such as the peanut clump virus, peanut bud necrosis virus (IPCV), groundnut rosette assistor virus (GRAV), peanut mottle virus (PMV), peanut stripe virus (PStV), tobacco streak virus (TSV), and tomato spotted wilt virus (TSVV) cause considerable damage to the crop. Genetic transformation has been used to develop peanut varieties with total resistance and not just tolerance to these viral diseases. The insertion of genetic material from the virus confers resistance to infection by preventing virus replication and spread. The development of genetically transformed peanut cultivars with resistance to viruses and other biotic constraints should have tremendous impact on crop productivity, especially in the resource-poor agricultural systems of the semiarid tropics. Franklin et al. (1993) reported transformed callus expressing the PStV coat protein gene through *Agrobacterium*-mediated genetic transformation. The introduction of coat protein gene of IPCV by using *Agrobacterium*-mediated transformation has led to the production of virus-resistant peanut plants (Sharma and Anjaiah, 2000). Besides, peanut transgenics for resistance to GRAV, TSV, and PBNV are being produced and evaluated in ICRISAT (K.K. Sharma, unpublished results).

### *Biofortification*

Peanut is deficient in the essential amino acid methionine, besides lysine, threonine, and isoleucine, which lowers its dietary and nutritional value. The nutritional quality of peanut can be improved by either raising the level of sulphur-containing amino acids of storage proteins or by changing the proportion of methionine-rich proteins already present in the peanut seed. High methionine levels cannot be produced by conventional breeding methods because of their failure to detect genotypes containing desirable levels of methionine. Hence, genetic transformation is an alternative approach for developing methionine-rich-peanuts. Attempts have been made to produce transgenic peanut plants with improved protein quality by transferring genes like the Brazil nut 2S albumin gene (Lacorte et al., 1997). The regenerated transgenic plants are being tested for the incorporation of the methionine-rich protein genes. The success in peanut transformation technology enabled researchers to address more complex and important aspects of biofortification in peanut for enhanced levels of beta-carotene (pro-vitamin A) by using bacterial *crtB* and maize *psy1* genes (ICRISAT, unpublished results).

### *Improvement in Quality of Oil*

The long chain saturated fatty acids (LSFAs) amyl arachidic (20:0), behenic (22:0), and lignoceric (24:0), present predominantly in the sn-3 position, have been reported to contribute to arteriosclerosis. If further elongation of stearic acid can be prevented, peanut oil would be free from these hazardous fatty acids. The elongation of the chain behind C18 is catalyzed by membrane-bound enzyme steoryl-CoA-: α-ketoeicosanoyl-CoA synthetase. Engineering a gene coding for antisense RNA in peanut may help reduce activity of this enzyme and hence of LSFA. For enhancing shelf-life of peanut products, a higher oleic/linoleic (O/L) ratio is considered desirable. Increasing the protein of oleic acid in peanut oil can attain this. The introduction of the first double bond in the plant fatty acids occurs by the action of enzyme stearoyl-ACP desaturase. Expression of additional copies of the gene for this enzyme may enhance the content of oleic acid and hence the O/L ratio.

## *CONCLUSION*

Plant regeneration from somatic cells is essential for successful in vitro genetic manipulation techniques, since transformation efficiencies are directly related to tissue culture response. Genetic transformation offers a complementary means to conventional crop breeding, especially for characteristics that are rare or may not be available in the genetic resources of peanut. We consider genetic transformation to be a tool that may allow the breaking of gene transfer barriers for high productivity and nutritional quality of the crop. With the rapid progress in genetic mapping and the isolation of new genes from various organisms, there will be new opportunities to modify plants using a range of genetic strategies. It is important that internationally accepted biosafety standards and local regulatory capacities be strengthened within developing countries. Development and deployment of transgenic plants in an effective manner will be an important prerequisite for sustainable use of biotechnology for crop improvement. The gains in crop productivity through scientific advancement will help to achieve sustainable food security, poverty reduction, and environmental protection. Research on transgenic crops provides new tools to improve agriculture in areas of the world where low rainfall and biotic stress are the major constraints on crop productivity.

## REFERENCES

Atreya, C.D., J.P. Rao, and N.C. Subrahmanyam (1984). In vitro regeneration of peanut (*Arachis hypogaea* L.) plantlets from embryo axes and cotyledon segments. *Plant Science Letters* 34:379-383.

Bajaj, Y.P.S. (1982). Regeneration of plants from pollen embryos of *Arachis*, *Brassica* and *Triticum* species cryo-preserved for one year. *Current Science* 52:484-486.

Bajaj, Y.P.S. and S.S. Gosal (1983). Somatic hybridization and embryo culture studies on *Arachis hypogaea* and *Arachis villosa. American Journal of Botany* 70:83-84.

Bajaj, Y.P.S. and S.S. Gosal (1988). Isolation and fusion of protoplasts of *Arachis hypogaea* and *Arachis villosa. International Arachis Newsletter* 3:13-14.

Bajaj, Y.P.S., P. Kumar, K.S. Labana, and M.M. Singh (1981). Regeneration of plants from seedling explants and callus cultures of *Arachis hypogaea* L. *Indian Journal of Experimental Biology* 19:1026-1029.

Baker, C.M. and H.Y. Wetzstein (1995). Repetitive somatic embryogenesis in peanut cotyledon cultures by continual exposure to 2,4-D. *Plant Cell Tissue and Organ Culture* 40:249-54.

Banerjee, S., S. Bandyopadhyay, and P.D. Ghosh (1988). Cotyledonary node culture and multiple shoot formation in peanut: Evidences for somatic embryogenesis. *Current Science* 57:252-257.

Bhatia, C.R., G.S.S. Murty, and V.H. Mathews (1985). Regeneration from de-embryonated peanut (*Arachis hypogaea*) plants from elite cultivars using ACCELL technology. *Plant Journal* 5:745-753.

Bhatnagar-Mathur, P., M.J. Devi, D.S. Reddy, M. Lavanya, V. Vandez, K. Yamaguchi-Shinozaki, and K.K. Sharma (2007). Stress inducible expression of *AtDREB1A* in transgenic peanut (*Arachis hypogaea* L.) increases transpiration efficiency under water limiting conditions. *Plant Cell Reports* (EPub-ahead of publication).

Birch, R.G. (1997). Plant transformation: Problems and strategies for practical application. *Annual Review of Plant Physiology and Molecular Biology* 48:297-326.

Brar, G.S. and B.A. Cohen (1994). Recovery of transgenic peanut plants from elite cultivars utilizing ACCELL technology. *Plant Journal* 5:745-753.

Cheng, M., D.C.H. His, and G.C. Phillips (1992). In vitro regeneration of Valencia-type peanut (*Arachis hypogaea* L.) from cultured petiolules, epicotyl sections and other seedling explants. *Peanut Science* 19:82-87.

Cheng, M., R.L. Jarret, Z. Li, and J.W. Demski (1997). Expression and inheritance of foreign genes in transgenic peanut plants generated by *Agrobacterium* mediated transformation. *Plant Cell Reports* 16:541-544.

Chengalrayan, K., V.B. Mhaske, and S. Hazra (1997). High-frequency conversion of abnormal peanut somatic embryos. *Plant Cell Reports* 16:783-786.

Chengalrayan, K., S.S. Sathaye, and S. Hazra, S. (1994). Somatic embryogenesis from mature embryo-derived leaflets of peanut (*Arachis hypogaea* L.). *Plant Cell Reports* 13:578-581.

Clemente, T.E., D. Robertson, T.G. Isleib, M.K. Beute, and A.K. Weissinger (1992). Evaluation of peanut (*Arachis hypogaea* L.) leaflets from mature zygotic embryos as recipient tissue for biolistic gene transfer. *Transgenic Research* 1:275-284.

Cucco, M.F. and A.D.R. Jayme (2000). Protocol for regeneration in vitro of *Arachis hypogaea* L. *Electronic Journal of Biotechnology* 3:154-160.

Dale, P.J., J.A. Irwin, and J.A. Scheffler (1993). The experimental and commercial release of transgenic crop plants. *Plant Breeding* 111:1-22.

Dong, J.D., Y.P. Bi, L.S. Xia, S.M. Sun, Z.H. Song, B.T. Guo, X.C. Jiang, and Q.Q. Shao (1990). Teratoma induction and nopaline synthase gene transfer in peanut. *Acta Genetica Sinica* 17:13-16.

Eapen, S. (2003). Regeneration and genetic transformation in peanut: current status and future prospects. In *Applied genetics of leguminosae biotechnology,* P.K. Jaiswal and R.P. Singh (eds.). Dordrecht: Kluwer Academic Publishers, pp. 165-186.

Eapen, S. and L. George (1994). *Agrobacterium tumefaciens* mediated gene transfer in peanut (*Arachis hypogaea* L.). *Plant Cell Reports* 13:582-586.

Egnin, M., A. Mora, and C.S. Prakash (1998). Factors enhancing *Agrobacterium tumefaciens*-mediated gene transfer in peanut (*Arachis hypogaea* L.). *In Vitro Cellular and Developmental Biology–Plant* 34:310-318.

Franklin, C.I., K.M. Shorrosh, A.N. Trieu, B.G. Cassidy, and R.S. Nelson (1993). Stable transformation of peanut callus via *Agrobacterium*-mediated DNA transfer. *Transgenic Research* 2:321-324.

Gill, R. and P.K. Saxena (1992). Direct somatic embryogenesis and regeneration of plants from seedling explants of peanut (*Arachis hypogaea* L.): Promotive role of TDZ. *Canadian Journal of Botany* 70:1186-192.

Gregory, N.C., A. Krapovickas, and P.M. Gregory (1980). Structural variation, evolution and classification in *Arachis*. In *Advances in Legume Science*, R.J. Summerfield and A.H. Bunting (eds.). Kew, UK: Royal Botanical Garden, pp. 469-481.

Hazra, S., S.S. Sathaye, and A.P. Mascarenhas (1989). Direct somatic embryogenesis in peanut (*Arachis hypogaea* L.). *Nature Bio/Technology* 7:949-951.

Herrera-Estrella, L., A. Depicker, M. Van Montagu, and J. Schell (1983). Expression of chimeric genes transferred into plant cells using a Ti-plasmid-derived vector. *Nature* 303:209-213.

Herrera-Estrella, L., T.H. Teeri, and J. Simpson (1988). Use of reporter genes to study gene expression in plant cells. In *Plant molecular biology manual: B1,* S.B. Gelvin, R.A. Schilperoort, and D.P.S. Verma (eds.). Dordrecht: Kluwer Academic Publishers, pp. 1-22.

Higgins, C.M., R.M. Hall, N. Mitter, A. Cruickshank, and R.G. Dietzgen (2004). Peanut stripe potyvirus resistance in peanut (*Arachis hypogaea* L.) plants carrying viral coat protein gene sequences. *Transgenic Research* 138:1-9.

Holmberg, N. and L. Bulow (1998). Improving stress tolerance in plants by gene transfer. *Trends in Plant Science* 3:61-66.

Jefferson, R.A. (1987). Assaying chimeric genes in plants: The *Gus* gene fusion system. *Plant Molecular Biology Reporter* 5:387-405.

Kanyand, M., A.P. Desai, and C.S. Prakash (1994). Thidiazuron promotes high frequency regeneration of peanut (*Arachis hypogaea*) plants in vitro. *Plant Cell Reports* 14:1-5.

Kartha, K.K., K. Pahl, N.L. Leung, and L.A. Mroginski (1981). Plant regeneration from meristems of grain legumes, soybean, cowpea, peanut, chickpea and bean. *Canadian Journal of Botany* 59:1671-1679.

Khandelwal, A., K.J.M. Vally, N. Geeta, P. Venkatachalam, M.S. Shaila, and G. Lakshmi Sita (2003). Engineering hemagglutinin (H) protein of rinder pest virus into peanut (*Arachis hypogaea* L.) as a possible source of vaccine. *Plant Science* 165:77-84.

Klein, T.M., M.E. Fromm, and T. Gradziel (1988). Factors influencing gene delivery into *Zea mays* cells by high velocity microprojectiles. *Nature/Biotechnology* 6:923-926.

Lacorte, C., F.J.L. Aragu, E.R. Almeida, E. Mansur, and E.L. Rech (1997). Transient expression of GUS and 2S albumin gene from Brazil nut in peanut (*Arachis hypogaea* L.) seed explants using particle bombardment. *Plant Cell Reports* 16:619-623.

Lacorte, C., E. Mansur, B. Timmerman, and A.R. Cordeiro (1991). Gene transfer into peanut (*Arachis hypogaea* L.) by *Agrobacterium tumefaciens. Plant Cell Reports* 10:354-357.

Li, Z., R.L. Jarret, M. Cheng, and J.W. Demski (1995). Improved electroporation buffer enhances transient gene expression in *Arachis hypogaea* protoplasts. *Genome* 38:58-863.

Li, Z., R.L. Jarret, R.N. Pittman, and J.W. Demski (1994). Shoot organogenesis from cultured seed explants of peanut (*Arachis hypogaea* L.) using thidiazuron. *In Vitro Cell and Developmental Biology–Plant* 30:187-191.

Lindsey, K. and M.G.K. Jones (1989). Plant biotechnology in agriculture. Milton, UK: Open University Press.

Livingstone, D.M. and R.J. Birch (1995). Plant regeneration and microprojectile-mediated gene transfer in embryonic leaflets of peanut (*Arachis hypogaea* L.). *Australian Journal of Plant Physiology* 22:585-591.

Magbanua, Z.V., H.D. Wilde, J.K. Roberts, K. Chowdhury, J. Abad, J.W. Moyer, H.Y. Wetzstein, and W.A. Parrott (2000). Field resistance to tomato spotted wilt virus in transgenic peanut (*Arachis hypogaea* L.) expressing an antisense nucleocapsid gene sequence. *Molecular Breeding* 6:227-236.

Mansur, E., C. Lacorte, and V.G. De Freitas (1993). Regulation of transformation efficiency of peanut (*Arachis hypogaea* L.) explants by *Agrobacterium tumefaciens. Plant Science* 99:89-91.

Martin, J.P. (1970). A contribution to the study of certain hereditary characters of agronomic importance in the groundnut. *Oleagineux* 22:673-676.

McKently, A.H. (1995). Effect of genotype on somatic embryogenesis from axes of mature peanut embryos. *Plant Cell Tissue and Organ Culture* 42:251-254.

McKently, A.H., G.A. Moore, H. Doostdar, and R.P. Niedz (1995). *Agrobacterium*-mediated transformation of peanut (*Arachis hypogaea* L.) embryo axes and the development of transgenic plants. *Plant Cell Reports* 14:699-703.

McKently, A.H., G.A. Moore, and F.P. Gardner (1990). In vitro plant regeneration of peanut from seed explants. *Crop Science* 30:192-196.

McKently, A.H., G.A. Moore, and F.P. Gardner (1991). Regeneration of peanut and perennial peanut from cultured leaf tissue. *Crop Science* 31:833-837.

Moloney, M.M., J.M. Walker, and K.K. Sharma (1989). An efficient method for *Agrobacterium*-mediated transformation in *Brassica napus* cotyledon explants. *Plant Cell Reports* 8:238-242.

Mroginski, L.A. and A. Fernandez (1980). Obtencion de plantulas por cultivo in vitro de anteras de escpecies silvestres *Arachis* (Leguminosae). *Oleagineux* 35:89-92.

Mroginski, L.A., K.K. Kartha, and J.P. Shyluk (1981). Regeneration of peanut plantlets by in vitro culture of immature leaves. *Canadian Journal of Botany* 59:826-830.

Narasimhulu, S.B. and G.M. Reddy (1983). Plantlet regeneration from different callus cultures of *Arachis hypogaea* L. *Plant Science Letters* 31:157-163.

Norden, A.J. (1980). Use of the pedigree breeding method to develop multiline peanut varieties. *Proceedings of Peanut Breeding Symposium.* American Peanut Research and Education Society 1980. Research Report No. 80, North Carolina State University, Raleigh.

Olhoft, P.M., L.E. Flagel, C.M. Donovan, and D.A. Somers (2003). Efficient soybean transformation using hygromycin B selection in the cotyledonary-node method. *Planta* 216:723-735.

Olsson, O., C. Koncz, and A.A. Szalay (1988). The use of the *luxA* gene of the bacterial luciferase operon as a reporter gene. *Molecular and General Genetics* 215:1-9.

Ozias-Akins, P. and W.D. Branch (1990). Incompatibility during late embryogeny in some crosses of *Arachis hypogaea-A. stenosperma* and the utility of the tissue culture methods for hybrid rescue. *Proceedings of American Peanut Research and Education Society* 22:51.

Ozias-Akins, P., J.A. Schnall, W.F. Anderson, C. Singsit, T.E. Clemente, M.J. Adang, and A.L.Weissinger (1993). Regeneration of transgenic peanut plants from stably transformed embryogenic callus. *Plant Science* 93:185-194.

Padua, V.L.M., M.C. Pestana, M. Margis-Pinheiro, and D.E. De Oliveira (2000). Electroporation of intact embryonic leaflet of peanut: Gene transfer and stimulation of regeneration capacity. *In Vitro Cell and Developmental Biology–Plant* 36:374-378.

Pittman, R.N., D.G. Banks, J.S. Kirby, E.D. Mitchell, and P.E. Richardson (1983). In vitro culture of immature peanut leaves: Morphogenesis and plantlet regeneration. *Peanut Science* 10:21-25.

Ponsamuel, J., D.V. Huhman, B.G. Cassidy, and D. Post-Beittenmiller (1998). In vitro regeneration via caulogenesis and brassin-induced shoot conversion of dormant buds from plumular explants of peanut (*Arachis hypogaea* L. cv 'Okrun'). *Plant Cell Reports* 11:373-378.

Rao, R.C.N. and S.N. Nigam (2001). Genetic options for drought management in groundnut. In *Management of agricultural drought—agronomic and genetic options,* N.P. Saxena (ed.). New Delhi: Oxford & IBH Publishing, pp. 123-142.

Reichel, C., J. Mathur, and P. Eckes, K. Langenkemper, C. Koncz, J. Schell, B. Reiss, and C. Maas (1996). Enhanced green fluorescence by the expression of an *Aequorea victoria* green fluorescent protein mutant in mono- and dicotyledonous plant cells. *Proceedings of National Academy of Sciences USA* 93:5888-5893.

Rohini, V.K. and K. S. Rao (2000). Transformation of peanut (*Arachis hypogaea* L.): A non-tissue culture based approach for generating transgenic plants. *Plant Science* 150:41-49.

Rohini, V.K. and K. S. Rao (2001). Transformation of peanut (*Arachis hypogaea* L.) with tobacco chitinase gene: Variable response of transformants to leaf spot disease. *Plant Science* 160:889-898.

Sanford, J.C. (1990). Biolistic plant transformation. *Physiologia Plantarum* 79: 206-209.

Sanford, J.C., T.M. Klein, E.D. Wolf, and N. Allen (1987). Delivery of substances to the cells and tissues using a particle gun bombardment process. *Particle Gun Technology* 5:27-37.

Sastri, D.C. and J.P. Moss (1982). Effects of growth regulators on incompatible crosses in the genomes *Arachis hypogaea* L. *Journal of Experimental Botany* 33:1293-1301.

Sastri, D.C., M. Nalini, and J.P. Moss (1980). *Arachis* ovary and ovule culture in vitro. In *Proceedings of symposium on plant tissue culture, genetic manipulation and somatic hybridization of plant cells,* P.S. Rao, M.R. Heble, and M.S. Chadah (eds.). Trombay, India: Bhabha Atomic Research Centre, pp. 366-374.

Schnall, J.A. and A.K. Weissinger (1995). Genetic transformation in *Arachis hypogaea* L. Vol 34, In *Biotechnology in agriculture and forestry,* Y.P.S. Bajaj (ed.). Heidelberg: Springer-Verlag, pp. 135-144.

Sellars, R.M., G.M. Southward, and G.C. Philips (1990). Adventitious somatic embryogenesis and cultured immature zygotic embryos of peanut and soybean. *Crop Science* 30:408-414.

Sharma, K.K. and V. Anjaiah (2000). An efficient method for the production of transgenic plants of peanut (*Arachis hypogaea* L.) through *Agrobacterium tumefaciens*-mediated genetic transformation. *Plant Science* 159:7-19.

Sharma, K.K. and R. Ortiz (2000). Program for the application of genetic transformation for crop improvement in the semi-arid tropics. *In Vitro Cell and Developmental Biology–Plant* 36:83-92.

Singsit, C., M.J. Adang, R.E. Lynch, A. Wang, G. Cardineau, and P. Ozias-Akins (1997). Expressions of a *Bacillus thuringiensis cry IAc* gene in transgenic peanut plants and its efficiency against lesser cornstalk borer. *Transgenic Research* 6:169-176.

Skoog, F. and C.O. Miller (1957). Chemical regulation of growth and organ formation in plant tissues cultured in vitro. *Symposium of the Society of Experimental Biology* 11:118-131.

Stalker, H.T. and J.P. Moss (1987). Speciation, genetics, and utilization of *Arachis* species. *Advances in Agronomy* 41:1-40.

Subramanyam, P., J.P. Moss, D. McDonald, P.V. Subba Rao, and V.R. Rao (1985). Resistance to leaf spot caused by *Cercosporidium personatum* in wild *Arachis* species. *Plant Disease* 69:951-954.

Swathi Anuradha, T., S.K. Jami, R.S. Datla, and P.B. Kirti (2006). Genetic transformation of peanut (*Arachis hypogaea* L.) using cotyledonary node as explant and a promoterless *gus::npt*II fusion vector. *Journal of Biosciences* 31:1-12.

Utomo, S.D., A.K. Weissinger, and T.G. Isleib (1996). High efficiency peanut regeneration using a non-imbibed immature leaflet culture method. *Peanut Science* 23:71-75.

Venkatachalam, P., N. Geeta, and N. Jayabalan (1998). Induction of somatic embryos and plantlet development in cell suspension cultures of *Arachis hypogaea* L. *Breeding Science* 48:231-236.

Venkatachalam, P., N. Geetha, A. Khandelwal, M.S. Shaila, and G. Lakshmi Sita (1999). Induction of direct somatic embryogenesis and plant regeneration from mature cotyledon explants of *Arachis hypogaea* L. *Current Science* 77:269-273.

Venkatachalam, P., N. Geeta, A. Khandelwal, M.S. Shaila, and G. Lakshmi Sita (2000). *Agrobacterium*-mediated genetic transformation and regeneration of transgenic plants from cotyledon explants of groundnut (*Arachis hypogaea* L.) via somatic embryogenesis. *Current Science* 78:1130-1136.

Wang, A., H. Fan, C. Singsit, and P. Ozias-Akins (1998). Transformation of peanut with a soybean VspB promoter-*uid*A chimeric gene. 1. Optimization of a transformation system and analysis of GUS expression in primary transgenic tissues and plants. *Physiologia Plantarum* 102:38-48.

Wetzstein, H.Y. and C.M. Baker (1993). The relationship between somatic embryo morphology and conversion in peanut (*Arachis hypogaea* L.). *Plant Science* 92:81-89.

Wightman, J.A. and G.V. Ranga Rao (1993). Groundnut insects. In *The groundnut crop,* J. Smart (ed.). London: Chapman Hall, pp. 394-469.

Willcox, M.C., S.M. Reed, J.A. Burns, and J.C. Wynne (1991). Effect of microspore stage and media on anther culture of peanut (*Arachis hypogaea* L.). *Plant Cell Tissue and Organ Culture* 24:25-28.

Wilmink, A. and J.J.M. Dons (1993). Selective agents and marker genes for use in transformation of monocotyledonous plants. *Plant Molecular Biology Reporter* 11:165-185.

Yang, H.Y, J. Nairn, and P.Ozias-Akins (2003). Transformation of peanut using a modified bacterial mercuric ion reductase gene driven by an actin promoter from *Arabidopsis thaliana. Journal of Plant Physiology* 160:945-952.

Yang, H.Y., C. Singsit, A. Wang, D. Gonsalves, and P. Ozias-Akins (1998). Transgenic plants containing nucleocapsid protein gene of tomato spotted wilt *tospovirus* show divergent levels of gene expression. *Plant Cell Reports* 17:693-699.

Yang, D.X., Z.M. Wei, and H.L. An (2001). Transgenic peanut plants obtained by particle bombardment via somatic embryogenesis regeneration system. *Cell Research* 11:156-160.

Zhang, W.J., S.B. Zhang, S.H. Liu, and L.L. Cai (1999). Somatic embryogenesis and plant regeneration from axes of peanut embryos. *Journal of Tropical and Subtropical Botany* 7:153-158.

Chapter 17

# Cotyledonary Node and Embryo Axes as Explants in Legume Transformation with Special Reference to Peanut

T. Swathi Anuradha
S. K. Jami
M. R. Beena
P. B. Kirti

## *INTRODUCTION*

Grain legumes are one of the most important groups of crop plants next to cereals. They have been transformed by various methods, including *Agrobacterium*-mediated gene transfer, microinjection, electroporation, and particle bombardment. There are a few reviews updating the genetic transformation of grain legumes (Atkins and Smith 1997, Christou 1997; Babaoglu et al., 2000; Chandra and Pental, 2003; Somers et al. 2003).

The inability of *Agrobacterium* to penetrate and adhere to the walls of cells capable of undergoing transformation and shoot regeneration is probably one of the main factors limiting the transformation in many plants, including grain legumes (Babaoglu et al., 2000). However, the current trend is an increased use of *A. tumefaciens* for gene delivery in crop improvement programs. This is particularly due to the development of advanced binary vectors and hypervirulent strains. By increasing *Agrobacterium*-mediated T-DNA delivery, there has been considerable progress in legume biotechnology (Olhoft and Somers, 2003). The most successful transformation systems in grain legumes leading to the recovery of transgenic plants include cotyledonary nodes and embryonic axes (Babaoglu et al., 2000). Cotyledonary nodes and embryonic axes are emerging as new tools in legume biotechnology. These explants have terminal or axillary meristems that serve as sources of totipotent cells and are considered the best targets for transformation.

## *GENETIC TRANSFORMATION OF GRAIN LEGUMES USING COTYLEDONARY NODE AS AN EXPLANT*

A good regeneration protocol is a prerequisite for achieving success in genetic transformation studies. Various constraints in the transformation of groundnut include poor regeneration, complicated and long-term regeneration via a callus phase, and influence of the genotype on the regeneration system. A regeneration method that avoids these problems is required (Beena et al., 2005).

Cotyledonary nodes have been viewed as most successful for the induction of multiple shoots via organogenesis among grain legumes (Chandra and Pental, 2003). The range of genotypes that have been transformed via the *Agrobacterium* based cotyledonary node method is steadily growing (Olhoft and Somers, 2001). Regeneration of shoots from the cotyledonary node after *Agrobacterium* cocultivation is emerging as a rapid and comparatively efficient method of transformation in a good number of grain legume species (Olhoft and Somers, 2003). Cotyledonary nodes have been found to regenerate shoots in soybean with preconditioning the seedlings by germinating the seeds on a medium containing benzyl amino purine (BAP; Cheng et al., 1980; Wright et al., 1986) and is also used as an assay system to assess the regeneration potential of different genotypes of soybean (Barwale et al., 1986).

*Agrobacterium*-mediated T-DNA delivery of genes into regenerable cells in the axillary meristems of the cotyledonary node was reported for the first time in soybean (Hinchee et al., 1988). Genotype-independent *Agrobacterium*-mediated transformation of cotyledonary nodes and regeneration of shoots have also been patented (Townsend and Thomas, 1993). Later, transformation using cotyledonary node as an explant with low frequencies has been published (Di et al., 1996). The efficiency in these transformation experiments is low due to inefficient selection of transformed cells giving rise to shoots. By using suitable *Agrobacterium* strains, a higher frequency (2 percent) of transformation has been achieved (Meurer et al., 1998). A 3 percent transformation frequency is attained by using EHA 105 and EHA 101 strains of *Agrobacterium* (Zhang et al., 1999).

Improvements for the soybean cot-node system have been reported. Cocultivation medium supplemented with a mixture of thiol compounds like L-cysteine, dithiothreitol, and sodium thiosulfate has enhanced transformation frequency (Olhoft and Somers, 2001). An efficient system based on hygromycin B selection for obtaining transgenic shoots with a very low frequency of escapes has also been reported for soybean (Olhoft and Somers, 2003). These newer methods could successfully increase the transformation efficiency of soybean cot-node system to more than 16 percent of treated explants.

In pigeon pea (*Cajanus cajan*), multiple shoots were produced at a high frequency (43 ± 8.9) from cotyledonary nodes (Prakash et al., 1994). This system is not only highly reproducible but also variety independent. Moreover, the time taken for the complete regeneration of plants under in vitro conditions is only 50 days. This method is similar to that reported for soybean (Cheng et al., 1980). When cotyledonary nodes and shoot apices were treated with *Agrobacterium,* the former responded better with 62 percent of the explants producing GUS-positive shoots after selection (Geetha et al., 1999). When both cotyledonary nodes and embryonic axes were used as explants for genetic transformation, it was reported that cotyledonary nodes responded better in terms of regeneration and transformation as compared to embryonic axes, resulting in a transformation frequency of 67 percent (Satyavathi et al., 2003).

In chickpea, much of the work was focused on embryo axes compared to cotyledonary node. However, an average of 12.6 shoots per node were regenerated on the medium supplemented with BAP (Subhadra et al., 1998). Transformation of chickpea using cotyledonary node explants has been achieved by Sarmah et al. (2004).

Recently, Indurker et al. (2007) have used epicotyl explants from 7-day-old seedlings of the chickpea varieties, Chaffa and PG12 (Mahatma Phule Krishi Vishwavidyalay, Rahuri, India), ICCC37 and ICCC32 (ICRISAT, Patencheru, India) were used in bombardment experiments using a vector carrying *npt*II and *cryIAc* genes for developing resistance in chickpea against the pod borer, *Heliothes armigera* and the generalist herbivore, *Spodoptera litura*. They have demonstrated that a pressure of 900 psi and 1.0 μm gold particles as microcarriers were efficient in bombarding the epicotyl explants. The authors have also suggested that they could achieve the highest ever transformation frequency in this protocol for chickpea (18.0 percent) and the transgenic plants showed enhanced levels of resistance against the target insects compared to control plants.

Cotyledonary nodes from mature seeds have been most responsive for the induction of multiple shoots via organogenesis in pea (Jackson and Hobbs, 1990). Transformation using cotyledonary node as an explant was reported earlier in pea (Davies et al., 1993). A simple pea transformation protocol based on cotyledonary node that overcame many of the difficulties recognized with previously discussed protocols has been reported (Bean et al., 1997); but the overall transformation efficiency was as low as 1.1 percent. Later on, an improved pea transformation protocol with 3.5 percent transformation frequency using lateral cotyledonary meristems was developed (Hellens et al., 2000).

Like other large-seeded legumes, *Vigna* species were recalcitrant to in vitro regeneration until the establishment of the cotyledonary node system. In mung bean [*Vigna radiata* (L.) Wiczek], regeneration has been achieved successfully using cotyledonary node explants (Gulati and Jaiwal, 1994). This protocol has proved to be simple, genotype independent, and applicable to several cultivars of *V. radiata*. Under the influence of suitable cytokinins such as BAP and kinetin, regeneration of shoots and whole plants from cotyledonary nodes of *V. radiata* was achieved. When hypocotyls, primary leaves, and cotyledonary nodes with both the proximal halves of the cotyledons attached to the embryonic axes were excised from 2-4-day-old in vitro dark grown seedlings and infected with *Agrobacterium*, transgenic plants could be produced successfully from cotyledonary node explants only (Jaiwal et al., 2001). The authors discussed the advantages of the *Vigna* cot-node system over other regeneration systems. The susceptibility of cotyledonary nodes of asparagus bean, *Vigna sesquipedalis*, was shown earlier. Successful transformation of *V. sesquipedalis* using cotyledonary node explants has led to the production of transgenic shoots that were GUS positive. Chi-square analysis has revealed that the progeny of the plants segregated in a Mendelian fashion confirming stable integration of the transgenes (Ignacimuthu, 2000).

Cotyledonary nodes were employed to regenerate plants via organogenesis in *V. mungo* (Ignacimuthu et al., 1997; Ignacimuthu and Franklin, 1999; Avenido and Hattori, 1999). *V. mungo* has been shown to be susceptible to *Agrobacterium tumefaciens* and transgenic callus lines have been generated (Karthikeyan et al., 1996), but there was no good report on the recovery of transgenic *V. mungo* plants until 2002 (Sahoo et al., 2002). For the first time, a transformation protocol for the successful production of transgenic *V. mungo* has been established via cotyledonary nodes without cotyledons (Saini and Jaiwal, 2003). The transformation frequency achieved was 1 percent and analysis of $T_0$ plants showed the expression and integration of the *gus* gene into the genome. Transmission and segregation of the transgene from the parent to the progeny were confirmed in the study. One of the clear advantages of cotyledon-based explants in genetic manipulation studies is their availability throughout the year. The success in the production of stably transformed plants from cotyledonary nodes may be due to the high regeneration potential of cells, which can resist the selection pressure. The transformation procedure involves direct shoot organogenesis and it takes less time to produce transgenic plants from them.

This simple system should be applicable to any crop for transferring desirable agronomic traits. A key feature of this cot-node system is the regenerability of the preexisting meristematic cells wounded during *Agrobacterium* cocultivation. As these cells are readily accessible to the *Agrobacterium,* transformed shoots can be obtained from them at high frequencies. Thus, this cot-node system is highly reproducible and variety independent and can be adapted to recalcitrant legumes.

We have generated putative promoter-tagged lines in *Arachis hypogaea* cv. JL-24 using cotyledonary node as an explant and a promoterless *gus::npt*II bifunctional fusion gene (Datla et al., 1991) mediated by *Agrobacterium*-based vectors. Parameters enhancing genetic transformation such as seedling age, *Agrobacterium* genetic background, and cocultivation period were studied by using the binary vector p35SGUS-INT. Approximately 31 percent and 3.54 percent transformation frequency are achieved with p35SGUSINT and promoterless construct, respectively. The progeny of 10 $T_0$ plants obtained with promoterless construct followed a Mendelian inheritance pattern of segregation in the $T_1$ generation (Swathi Anuradha et al., 2006). The total number of responding explants as well as the number of shoots per explant is still low, thus limiting the regeneration and transformation efficiency of cotyledonary nodes to a certain extent. Though the lateral meristems of cotyledonary nodes contain highly regenerable cells, the presence of a limited number of meristematic cells in the axil of the cotyledonary node whose capacity for regeneration is short-lived will act as a limitation to their transformation efficiency. Irrespective of these minor limitations, cotyledonary nodes could still be considered as potential explants for gene delivery, particularly in legumes. The protocol is rapid, genotype independent, and cost effective (Table 17.1).

## *EMBRYO AXES*

To some extent, the limitations imposed by the culture of cotyledon-based explants can be overcome by using shoot apex and embryonic axis explants with fully exposed and broad preexisting meristems that have a high regeneration potential to rapidly form adventitious shoots with minimal tissue culture in a genotype-independent fashion. The efficiency of shoot apical transformation system is low, apparently due to inefficient selection of transgenic cells and low rates of transgenic shoot regeneration and plant establishment. The use of embryonic axes has many advantages over other systems: a simple protocol with high regeneration frequency and

TABLE 17.1. Genetic transformation in legumes using cotyledonary node explants.

| Plant | Strain | Plasmid | Transformation frequency (%) | Selection marker | Agent | Reference |
|---|---|---|---|---|---|---|
| Soybean | EHA 105 | pPTN101, pPTN105, pPTN140 | 3 | *bar* | PPT | Zhang et al. (1999) |
| Soybean | LBA 4404 | pTOk233<br>pGPTV-HPT | 16.4 | *uid*A<br>*hpt* | GUS<br>HYG | Olhoft et al. (2003) |
| Lentil | Particle bombard-ment | Binary with *ALS* gene | – | *als* | Chloro-sulfuron | Gulati et al. (2002) |
| Mung bean | LBA 4404<br>C58C1<br>EHA 105 | pTOK233, pIG121Hm, pBin19: GUS-INT | 0.9 | *npt*II | Kan | Jaiwal et al. (2001) |
| Asparagus bean | EHA 101 | pBI-GUS-INT | 2 | *bar*<br>*npt*II | PPT<br>Kan | Ignacimuthu (2000) |
| Pigeon pea | LBA 4404 | pBI121 | 62 | *npt*II<br>*Uid*A | Kan<br>GUS | Geetha et al. (1999) |
| Pea | C58/3 | pSLJ1911 | 1.44 | *npt*II<br>*uid*A | Kan<br>GUS | Davies et al. 1993 |
| Pea | EHA 105 | pSLJ1561 | 1.1 | *bar* | PPT | Bean et al. (1997) |
| Black gram | EHA 105 | pCAMBIA 2301 | 1 | *npt*II<br>*uid*A | Kan<br>GUS | Saini and Jaiwal (2003) |
| Peanut cv JL-24 | GV2260, LBA4404 | p35S-GUS-INT, promoterless *gus::npt*II construct in pRD 400 | 35 and 3.54 | *npt*II<br>*uid*A | Kan<br>GUS | Swathi Anuradha et al. (2006) |

formation of adventitious shoots, reduced explant necrosis after cocultivation, efficient T-DNA transfer, and simpler plantlet establishment after transfer to soil (Liu et al., 2004).

There is a current trend toward increasing the use of embryo axes over cotyledonary node among grain legumes for genetic transformation. In the years 2003-2005, the majority of the transformation reports on grain legumes showed the utility of this explant in gene technology. In soybean, the first report of embryonic axis regeneration and transformation showed that the embryonic axes yielded highest frequency of regeneration (88.6 percent), followed by the hypocotyl segments (56.4 percent), followed by the cot-node system (40.3 percent; Hai Kun Liu et al., 2004). But the overall transformation efficiency was 15.8 percent, which is not an improvement over the 16.4 percent efficiency reported by using cot-node explants (Olhoft et al., 2003). Even though cot-node transformation efficiency is high in some cases, researchers prefer embryo axes to cotyledonary nodes. However, cot-node transformation has not lost its significance in grain legume transformation to date. Thus, one can only say that the use of the embryonic tip is a useful and effective new method for genetic transformation (Aragao et al., 2000).

McKently et al. (1995) used embryo axes for generating transgenic plants in peanut. They have used the embryo axes for cocultivation using the *Agrobacterium* strain EHA101 with a binary vector carrying *npt*II and *gus* genes. Transgenic plants were confirmed by GUS and polymerase chain reaction analysis followed by Southern hybridization. Using shoot meristems from mature embryonic axes, Brar et al. (1994) have generated transgenic plants in the elite varieties, Florunner and Florigiant. They have used bombardment method of gene delivery to the shoot meristems and utilized the *bar* gene for the selection of transformants. They have demonstrated the transgenic nature by Southern analysis and transmission of the foreign genes to the progeny. Similarly, Livingstone et al. (2005) used somatic embryos in particle bombardment mediated generation of transgenic peanut plants and generated transgenic peanut expressing oxalate oxidase gene for enhanced resistance against *Sclerotinia minor*. Athmaram et al. (2006) used zygotic embryos for developing somatic embryos, which were used in particle bombardment for generating transgenic peanut expressing the Bluetongue outer coat protein gene *VP2*.

In pigeon pea (*Cajanus cajan*), a high frequency of plantlet formation from decapitated embryo axes was achieved by culturing explants on Murashige and Skoog (MS) medium supplemented with BAP and indoleacetic acid (IAA: Rathore and Chand, 1999). The authors showed that the removal of 2 mm of shoot apex can increase the regeneration capability of decapitated embryos and removal of sections smaller or larger than 2.0 mm would only result in reduction of the shoot regeneration percentage. Around 55 to 60 percent of the inoculated embryos succeeded in regenerating de novo adventitious shoots from the cut surface and six or seven shoots were obtained from each explant.

Genetic transformation of pigeon pea embryo axis explants (cv. Pusa 855) using a cowpea protease inhibitor gene in plasmid pBI121 has been reported, and the authors reported a transformation frequency between 30 and 59 percent (Lawrence and Koundal, 2001). Both embryo axis and cot-node explants were used for the transformation in cv. Hyderabad (Satyavathi et al., 2003, Table 17.2). Transgenic plants were developed from embryonal segments by using the strain GV2260 harboring a binary vector pPK202 carrying *npt*II and a synthetic *cry IE-C* under CaMV35s promoter. Bioassays and Western analysis done on $T_1$ and $T_2$ generation plants showed stable integration and transmission of transgenes (Surekha et al., 2005). But the percentage of confirmed transgenic plants appears to be low compared to previous reports. The difference in transformation efficiencies in different species has been attributed to the genotypic differences in crop species, type of explant cultured, cocultivation procedure followed, and the genetic background of the *Agrobacterium* strains used in the transformation experiments (Surekha et al., 2005).

In chickpea (*Cicer arietinum*), embryo axes deprived of the apical meristem were shown to regenerate adventitious shoots on MS medium supplemented with kinetin. Genetic transformation with this system did successfully produce transformants at 4 percent transformation efficiency (Fontana et al., 1993). Kar et al. (1996) reported a transformation frequency of 1 to 5 percent with embryonic axes. But both of these reports have not provided any substantial evidence for stable integration into the progeny of the primary transformants. Kar et al. (1996, 1997) developed transgenic chickpea plants in the varieties ICCV-1 and ICCV-6 overexpressing a chimeric truncated *cryIAc* using embryo axes for resistance against larvae of the insect *Heliothis armigera*. An efficient and reliable nonantibiotic selection using the phosphinothricin-acetyl transferase (*pat*) and aspartate kinase (*AK*) genes for the production of transgenic chickpea plants has been developed (Tewari-Singh et al., 2004). Cotyledons are split open and shoot and root apices are removed and then the decapitated embryo explants with one of the cotyledons are used for transformation.

The development of an efficient transformation protocol for genotypes using *A. tumefaciens* and embryonic axes with 5.1 percent transformation efficiency has helped in stable integration and expression of marker genes through the $T_1$ and $T_2$ generations (Senthil et al., 2004). The successful re-

TABLE 17.2. Genetic transformation in legumes using embryonic axis explants.

| Plant material | Strain | Plasmid | Marker | Agent | Transformation frequency (%) | Reference |
|---|---|---|---|---|---|---|
| Soybean | EHA 105 | pCAMBIA 2301 | *npt*II/*Uid*A | Kan GUS | 15.8 | Liu et al. (2004) |
| Soybean | EHA 105 | — | *AHAS* | | — | Aragao et al. (2000) |
| Lupin | AgLO | — | *bar* | PPT | 0.4-2.8 | Pigeare et al. (1997) |
| Bean | MB | pB5/35S-bar | *bar* | PPT | 0.6, 0.5 | Aragao et al. (2002) |
| Pea | EHA 105 | — | *bar, npt*II *ALS* | PPT, Kan, chlorsulfuron | 0.1-2.4 | Polowick et al. (2000) |
| Pea | AgLO, AgLI, EHA 105 | pP35SGIB | *uid*A | GUS | 0.7-4.1 | Tomasz et al. (2005) |
| Chickpea | LBA 4404 | pBI121 | *uid*A | GUS | 1-5 | Fontana et al. (1993) |
| Chickpea | GV2260 | p35GUS-INT | *uid*A/*bar* | GUS PPT | 0.4 | Krishnamurthy et al. (2000) |
| Chickpea | AgII | pGIN1 | *uid*A | GUS | 5.1 | Senthil et al. (2004) |
| Chickpea | LBG66 | pCAMBIAI301 | *gus:npt*II | GUS, Kan | 3.1 | Polowick et al. (2004) |

TABLE 17.2 *(continued)*

| Plant material | Strain | Plasmid | Marker | Agent | Transformation frequency (%) | Reference |
|---|---|---|---|---|---|---|
| Pigeon pea | GV2260 | pBI121 | *npt*II | Kan | 30-59 | Lawrence and Koundal (2001) |
| Pigeon pea | EHA 105 | pBI121 | *npt*II | Kan | 67-CN<br>51-EA | Satyavathi et al., (2003) |
| Pigeon pea | GV2260 | pPK202 | *npt*II | Kan | 15 | Surekha et al. (2005) |
| Black gram | EHA 105 | pCAMBIA2301 | *npt*II/*uid*A | Kan<br>GUS | 6.5 | Saini and Jaiwal (2005) |
| Peanut (in planta) | LBA 4404 | pKIWI105 | *npt*II | Kan | 3.3 | Rohini and Rao (2000) |

covery of transgenic plants was made possible by the use of a phase of regeneration without selection, followed by selection pressure combined with rapid proliferation of shoots on medium containing thidiazuron during the initial stages of selection.

Immature embryos have been reported as an appropriate source of explants in pea. Different types of embryo-derived explants have been used, such as slices of embryo axes (Schroeder at al. 1993; Polowick et al., 2000), and basal parts of the embryo (Grant et al., 1995). The rate of pea transformation is relatively low. Polowick et al. (2000) examined the transformation susceptibility of eight Canadian pea genotypes. They described how the plant genotype affected the transformation rate. Grant et al. (2003) studied the influence of *Agrobacterium tumefaciens* strains AGLI and KYRT1 and found that KYRTI is three times more efficient than AGLI for producing transgenic plants. An *Agrobacterium*-based method of pea transformation has been developed for several edible and fodder cultivars of pea (Pniewski and Kapusta, 2005). The regeneration capacity has differed distinctly in different cultivars. The highest average number of adventitious buds formed by a single explant (5.1-5.8) is a characteristic for Wiato, Grapis, Pionier, Konserwowy, and IHAR. Transformation efficiency has varied from 0.7 to 4.1 percent depending on the type of cultivar and *Agrobacterium* strain. Researchers have observed that the AGL1 strain provides higher transformation efficiency (1.4-4.1 percent) than EHA 105 (0.6-0.9 percent). Approximately 75 percent of progeny plants proved to be polymerase chain reaction positive for *uid*A and *bar* genes, proving Mendelian inheritance of transgenes.

In black gram, *V. mungo,* transgenic plants are produced using cotyledonary nodes with 1 percent transformation efficiency, as discussed earlier. Transformation frequency increased significantly from an average of 1 percent to 6.5 percent by using shoot apices excised from embryonic axes precultured on 10 μM for 3 days and wounded prior to infection with *Agrobacterium* strain EHA 105 carrying the binary vector pCAMBIA2301 (Saini and Jaiwal, 2005). In this study, preconditioning and wounding of small shoot apices resulted in a significant increase in the transformation frequency. This is because it circumvents the attachment step or releases phenolic inducers of *Agrobacterium vir* functions. The majority of the $T_0$ plants transmitted transgenes in a 3:1 ratio to their progeny. Cotyledonary node and embryonic axis systems yield efficiencies of transformation about an order of magnitude better than most other systems, demonstrate repeatability, and describe the expression of transgenes through $T_0$, $T_1$, and $T_2$ generations in fertile and morphologically normal plants.

## *GENETIC TRANSFORMATION OF PEANUT USING COTYLEDONARY NODE EXPLANTS*

### *Plant Tissue Culture*

- Surface sterilizing agents: 70 percent alcohol, 0.1 percent mercuric chloride and double distilled water.
- SIM (shoot induction medium) for cocultivation: Murashige and Skoog basal medium, 30 $g \cdot l^{-1}$ sucrose, 4.5 $g \cdot l^{-1}$ agargel, 4 $mg \cdot l^{-1}$ BAP (Sigma), 0.1 $mg \cdot l^{-1}$ naphthalene acetic acid (Sigma), pH 5.75-5.8.
- SIM I: SIM containing 250 $mg \cdot l^{-1}$ cefotaxime.
- SIM II: SIM containing 250 $mg \cdot l^{-1}$ cefotaxime and 150 $mg \cdot l^{-1}$ kanamycin (Sigma), pH 5.75-5.8.
- SEM (shoot elongation medium): MS basal medium, 30 g sucrose, 2 $mg \cdot l^{-1}$ BAP, 0.1 $mg \cdot l^{-1}$ naphthalene acetic acid (Sigma), 150 $mg \cdot l^{-1}$ kanamycin, pH 5.8.
- RIM (root induction medium): MS basal medium, 4 g agargel (Sigma), 30 g sucrose supplemented with 0.8 $mg \cdot l^{-1}$ naphthalene acetic acid, 250 $mg \cdot l^{-1}$ cefotaxime, and 50 $mg \cdot l^{-1}$ kanamycin, pH 5.8.

### *Preparation of* Agrobacterium *Culture*

- Inoculate freshly grown *Agrobacterium* in 50 ml of Luria and Bertani (LB) containing antibiotics for bacterial selection. Incubate at 28°C and 250 rpm for 24 hours. *Agrobacterium* strains GV2260, EHA 105, and LBA 4404 are suggested for genetic transformation of peanut; AGL1 and EHA 105 are supervirulent and cause necrosis of the explants if not used carefully.
- Inoculate 1 ml of this culture into 100 ml of LB containing antibiotics for bacterial selection. Incubate at 28°C with shaking for 6-7 hours.
- Assess bacterial growth and when the $OD_{600}$ value is 0.8-1.0, the *Agrobacterium* is ready to use for transformation (OD should not exceed 1.4).
- Centrifuge *Agrobacterium* cultures in a tabletop centrifuge for 5 minutes at 5,000 rpm and discard the supernatant.
- Resuspend the bacterial pellet in 50 ml of sterile ½ MS solution (without sucrose) and store the culture at 4°C for 2 hours prior to *Agro* infection.

***Seed Sterilization***

- Shell mature *Arachis hypogaea* cv. JL-24 pods and use 100 seeds for each experiment.
- In a laminar flow hood, transfer seeds to a sterile 500 ml conical flask and add approximately 50 ml of 70 percent alcohol and swirl gently for 1 minute. Discard the ethanol.
- Add 100 ml of 0.1 percent and shake for 7 minutes. Shaking should be done continuously without any time interval. Soaking the explant in mercuric chloride solution hinders the regeneration ability of the explants.
- Discard the mercuric chloride and rinse the seeds five to seven times with 80 ml sterile double distilled water 1 minute each.
- Add approximately 200 ml of sterile double distilled water (SDDW) and cover the vessel with aluminum foil. Leave the seeds in culture room for 3 hours for imbibition. Germinate cot-node explants in tubes having filter paper wicks. Embryo axis explants can be prepared immediately with the soaked seedlings. For cot-node explants, germinate seeds in glass tubes with liquid MS medium and filter paper wicks (see Hinchee et al., 1988; Townsend and Thomas, 1993; Di et al., 1996) for the method of explant preparation). Embryo axis explants may be prepared following Liu et al. (2004) for soybean.
- Treat explants with *Agrobacterium* and transfer explants onto SIM plates with sterile forceps, without letting them dry out. Insert them into the medium by hypocotyl end. Do not treat the explants for more than 5 minutes for EHA and AGL strains. The total time for vacuum infiltration and transferring of explants to SIM should not exceed 30 minutes for embryo axes.
- Cocultivate on SIM for 72 hours in light. A cocultivation period of 48 hours is enough when the explant is infected with EHA strain. If bacterial overgrowth is observed, then the explants should be washed with sterile double distilled water containing 250 $mg \cdot l^{-1}$ cefotaxime.
- Transfer the explants from the SIM onto Petri plates (110 mm) containing SIM I using sterile forceps and maintained in the culture conditions.
- After 5 days carefully transfer the explants from SIM I to SIM II. Cut 1mm toward the radicle end of embryo explants.
- Subculture both types of explants every 2 weeks for a period of 6 weeks by transferring them individually onto fresh Petri dishes containing SIM II. All these steps should be carried out carefully under aseptic conditions only.

### *Shoot Elongation*

- Cut the multiple shoot bunches into three or four bits and then transfer each part to SEM. When the shoots reach a height of 3-4 cms and have two or three internodes they can be cultured on RIM. Cutting should be done carefully; otherwise, developing shoots will be damaged.
- Carry out two subcultures on this medium in the same manner.

### *Root Induction*

- Excise the developing transgenic shoots at the internode region and transfer to sterile tubes containing 10-15 ml RIM.
- Roots will develop within 15-20 days. If callus forms at the root shoot junction, transgenic shoots can be excised and transferred to RIM.

### *Transfer to Soil*

- Gently remove plantlets with well-formed root systems from culture tubes and wash the roots under running tap water for 5 minutes carefully.
- Transfer each plantlet to a small pot containing soil and vermiculite in a 1:1 ratio and cover the pot to retain 85 percent humidity.
- These plants can be maintained under culture room conditions for a couple of weeks.
- Transfer them to bigger pots under glasshouse conditions and allow them to flower and set seed.
- After complete pod formation is over, watering should be ceased for a period of 10-15 days and then collect the seeds.
- Dry the seeds and store them in paper covers.

## NOTE

Chu et al. (2007) generated transgenic peanut expressing human *Bcl-xL* gene and showed that one of the transgenic lines showed enhanced tolerance to the herbicide, Paraquat. Dodo et al. (2007) developed transgenic peanut for reduced peanut allergy using the RNAi approach by downregulating the *Ara h2* gene. This has resulted in reduced Ara h2 and reduced allergenicity in peanut kernels. Similarly, Bhatnagar-Mathur et al. (2007) developed transgenic peanut plants by expressing the *AtDREB1A* gene under the stress inducible *rd29A* gene promoter and demonstrated that one of the transgenic events showed 40 percent higher transpiration efficiency than the control plants under water limiting conditions. Beena et al. (2008) have developed transgenic peanut expressing two genes, the synthetic *cryI E-C* and a rice chitinase cDNA under independent CaMV35S promoters using cotyledonary node system of transformation. The transgenic plants showed resistance to the insect Spodoptera and enhanced resistance to the late leaf spot fungus *Phaeoisariopsis personata,* which are both serious problems in peanut cultivation.

## REFERENCES


Aragao, F.J.L., L. Sarokin, G.R. Vianna, and E.L. Rech (2000). Selection of transgenic meristematic cells utilizing a herbicidal molecule results in recovery of fertile transgenic soybean [*Glycine max* (L.) Merrill] plants at a high frequency. *Theoretical and Applied Genetics* 101:1-6.

Aragao, F.J.L., G.R. Vianna, M.M.C. Albino, and E.L. Rech (2002). Transgenic dry bean tolerant to the herbicide glufosinate ammonium. *Crop Science* 42:1298-1302.

Atkins, C.A. and P.M.C. Smith (1997). Genetic transformation and regeneration of legumes. In *Biological fixation of nitrogen for ecology and sustainable agriculture.* A. Legocki, H. Bothe, and A. Puhler (eds.). Berlin: Springer Verlag, pp. 283-304.

Athmaram, T.N., G. Bali, and K.M. Devaiah (2006). Integration and expression of Bluetongue *VP2* gene in somatic embryos of peanut through particle bombardment. *Vaccine* 24:2994-3000.

Avenido, R.A. and K. Hattori (1999). Differences in shoot regeneration response from cotyledonary node explants in Asiatic *Vigna* species support genomic grouping within subgenus *Ceratotropis* (Piper) Verdc. *Plant Cell Tissue and Organ Culture* 58:99-110.

Babaoglu, M., M.R. Davey, and J.B. Power (2000). Genetic engineering of grain legumes: Key transformation events. *AgBiotech Net* (June): ABN 050.

Barwale, U.B., M.M. Meyer, and J.M. Widholm (1986). Screening of *Glycine max* and *Glycine soja* genotypes for multiple shoot formation at the cotyledonary node. *Theoretical and Applied Genetics* 72:423-428.

Bean, S.J., P.S. Gooding, P.M. Mullineaux, and D.R. Davies (1997). A simple system for pea transformation. *Plant Cell Reports* 16:513-519.

Beena, M.R., S.K. Jami, T. Srinivasan, T. Swathi Anuradha, G. Padmaja, and P.B. Kirti (2005). Efficient direct shoot regeneration from cotyledonary node explants of peanut (*Arachis hypogaea* L. cv JL-24). *Indian Journal of Plant Physiology* 10:205-210.

Beena, M.R., R. Tuli, A. Datta-Gupta, and P.B. Kirti (2008). Transgenic peanut (*Arachis hypogaea* L.) plants expressing *cryI E-C* and rice chitinase cDNA (*Chi 11*) exhibit resistance against insect pest, *Spodoptera litura* and fungal pathogen *Phaeoisariopsis personata.* (Revised version, submitted).

Bhatnagar-Mathur, P., M.J. Devi, D.S. Reddy, M. Lavanya, V. Vandez, K. Yamaguchi-Shinozaki, and K.K. Sharma (2007). Stress inducible expression of *AtDREB1A* in transgenic peanut (*Arachis hypogaea* L.) increases transpiration efficiency under water limiting conditions. *Plant Cell Reports* (EPub-ahead of publication).

Brar, G.S., B.A. Cohen, C.L. Vick, and G.W. Johnson (1994). Recovery of transgenic peanut (*Arachis hypogaea* L.) plants from elite cultivars utilizing ACCELL technology. *Plant Journal* 5:745-753.

Chandra, A. and D. Pental (2003). Regeneration and genetic transformation of grain legumes: An overview. *Current Science* 84:381-387.

Cheng, T.-Y., H. Saka, and T.H. Voqui-Dinh (1980). Plant regeneration from soybean cotyledonary node segments in culture. *Plant Science Letters* 19:91-99.

Christou, P. (1997). Biotechnology applied to grain legumes. *Field Crops Research* 53:83-97.

Chu, Y., X.Y. Deng, P. Faustinelli, and P. Ozias-Akins (2007). *Bcl-xL* transformed peanut (*Arachis hypogaea* L.) exhibits paraquat tolerance. *Plant Cell Reports* (EPub-ahead of publication).

Datla, R.S.S., J. Hammerlindl, L. Pelcher, W.L. Crosby, and G. Selvaraj (1991). A bifunctional fusion between β-glucuronidase and neomycin phosphotransferase: A broad spectrum marker enzyme for plants. *Gene* 101:2139-2246.

Davies, D.R., J. Hamilton, and P. Mullineaux (1993). Transformation of peas. *Plant Cell Reports* 12:180-183.

Di, R., V. Purcell, G.B. Collins, and S.A. Ghabrial (1996). Production of transgenic soybean lines expressing the bean pod mottle virus coat protein precursor gene. *Plant Cell Reports* 15:746-750.

Dodo, H.M., K.N. Konan, F.C. Chen, M. Egnin, and O.M. Viquez (2007). Alleviating peanut allergy using genetic engineering: The silencing of the immunodominant allergen *Ara h2* leads to its significant reduction and a decrease in allergenicity. *Plant Biotechnology Journal* (EPub-ahead of publication).

Fontana, G.S., L. Santini, S. Caretto, G. Frugis, and D. Mariotti (1993). Genetic transformation in the grain legume *Cicer arietinum* (chick pea). *Plant Cell Reports* 12:194-199.

Geetha, N., P. Venkatachalam, and G. Lakshmi Sita (1999). *Agrobacterium* mediated genetic transformation of pigeon pea (*Cajanus cajan* L.) and development of transgenic plants via direct organogenesis. *Plant Biotechnology* 16:213-218.

Grant, J.E., P.A. Cooper, A.E. McAra, and T.J. Frew (1995). Transformation of peas (*Pisum sativum* L.) using immature cotyledons. *Plant Cell Reports* 15:254-258.

Grant, J.E., L.M.J. Thomson, M.D. Pither Joyce, T.M. Dale, and P.A. Cooper (2003). Influence of *Agrobacterium tumefaciens* strain on the production of transgenic peas (*Pisum sativum* L.) *Plant Cell Reports* 21:1207-1210.

Gulati, A. and P.K. Jaiwal (1994). Plant regeneration from cotyledonary node explants of mungbean [*Vigna radiata* (L.) Wilczek]. *Plant Cell Reports* 13:523-527.

Gulati, A., P. Schryer, and A. McHughen (2002). Production of fertile transgenic lentil (*Lens culinaris* Medik) plants using particle bombardment. *In Vitro Cellular and Developmental Biology–Plant* 38:316-324.

Hellens, R.P., E.A. Edwards, N.R. Leyland, S. Bean, and P.M. Mullineaux (2000). pGreen: A versatile and flexible binary vector for *Agrobacterium* mediated plant transformation. *Plant Molecular Biology* 42:819-832.

Hinchee, M.A.W., D.V. Connor-Ward, C.A. Newell, R.E. McDonnel, S.J. Sato, C.S. Gasser, D.A. Fischoff, B. Reo, R.T. Fraley, and R.B. Horsch (1988). Production of transgenic soybean plants using *Agrobacterium*-mediated DNA transfer. *Nature Bio/Technology* 6:915-922.

Ignacimuthu, S. (2000). *Agrobacterium*-mediated transformation of *Vigna sesquipedalis* Koern (asparagus bean). *Indian Journal of Experimental Biology* 38: 493-498.

Ignacimuthu, S. and G. Franklin (1999). Regeneration of plantlets from cotyledon and embryonic axis explants of *Vigna mungo* (L.) Hepper. *Plant Cell Tissue and Organ Culture* 55:75-80.

Ignacimuthu, S., G. Franklin, and G. Melchias (1997). Multiple shoot formation and in vitro fruiting from cotyledonary nodes of *Vigna mungo* (L.) Hepper. *Current Science* 73:733-735.

Indurker, S., H.S. Misra, and Susan Eapen (2007). Genetic transformation of chickpea (*Cicer arietinum* L.) with insecticidal crystal protein gene using particle gun bombardment. *Plant Cell Reports* 26:755-763.

Jackson, J.A. and S.L.A. Hobbs (1990). Rapid multiple shoot production from cotyledonary node explants of pea (*Pisum sativum* L.). *In Vitro Cellular and Developmental Biology–Plant* 26:835-838.

Jaiwal, P.K., R. Kumari, S. Ignacimuthu, I. Potrykus, and C. Sauter (2001). *Agrobacterium tumefaciens* mediated genetic transformation of mungbean [*Vigna radiata* (L.) Wilczek]—a recalcitrant grain legume. *Plant Science* 161:239-247.

Kar, S., D. Basu, S. Das, N.A. Ramakrishnan, P. Mukherjee, P. Nayak, and S.K. Sen (1997). Expression of *cryIA* (c) gene of *Bacillus thuringiensis* in transgenic chickpea plants inhibit development of pod-borer (*Heliothis armigera*) larvae. *Transgenic Research* 6:177-185.

Kar, S., T.M. Johnson, P. Nayak, and S.K. Sen (1996). Efficient transgenic plant regeneration through *Agrobacterium*-mediated transformation of chickpea (*Cicer arietinum* L.). *Plant Cell Reports* 16:32-37.

Karthikeyan, A.S., K.S. Sarma, and K. Veluthambi (1996). *Agrobacterium tumefaciens* mediated transformation of *Vigna mungo* (L.) Hepper. *Plant Cell Reports* 15:328-331.

Krishnamurthy, K.V., K. Suhasini, A.P. Sagare, M. Meixner, A. De Kathen, T. Pickardt, and O. Scheider (2000). *Agrobacterium* mediated transformation of chickpea (*Cicer arietinum* L.) embryo axes. *Plant Cell Reports* 19:235-240.

Lawrence, P.K. and K.R. Koundal (2001). *Agrobacterium tumefaciens*-mediated transformation of pigeon pea [*Cajanus cajan* (L) Mill sp] and molecular analysis of regenerated plants. *Current Science* 80:1428-1432.

Liu, H.K., Y. Cho, and Z.M. Wei (2004). Efficient *Agrobacterium tumefaciens*-mediated transformation of soybeans using an embryonic tip regeneration system. *Planta* 219:1042-1049.

Livingstone, D.M., J.L. Hampton, P.M. Phipps, and E.A. Graban (2005). Enhancing resistance to *Sclerotinia minor* in peanut by expressing a barley oxalate oxidase gene. *Plant Physiology* 137:1354-1362.

McKently, A.H., G.A. Moore, H. Doostdar, and R.P. Niedz (1995). *Agrobacterium* mediated transformation of peanut (*Arachis hypogaea* L.) embryo axes and the development of transgenic plants. *Plant Cell Reports* 14:699-703.

Meurer, C.A., R.D. Dinkins, and G.B. Collins (1998). Factors affecting soybean cotyledonary node transformation. *Plant Cell Reports* 18:180-186.

Olhoft, P.M. and D.A. Somers (2001). L-Cysteine increases *Agrobacterium* mediated T-DNA delivery into soybean cotyledonary node cells. *Plant Cell Reports* 20:706-711.

Pigeare, A., D. Abernethy, P.M. Smith, K. Simpson, N. Fletcher, and C.-Y. Lu (1997). Transformation of a grain legume (*Lupinus angustifolius* L.) via *Agrobacterium tumefaciens* gene transfer to shoot apices. *Molecular Breeding* 3:341-349.

Pniewski, T. and J. Kapusta (2005). Efficiency of transformation of Polish cultivars of pea (*Pisum sativum* L.) with various regeneration capacities by using hypervirulent *Agrobacterium tumefaciens* strains. *Journal of Applied Genetics* 46: 139-147.

Polowick, P.L., D.S. Baliski, and D. Mahon (2004). *Agrobacterium tumefaciens*-mediated transformation of chickpea (*Cicer arietinum* L.): Gene integration, expression and inheritance. *Plant Cell Reports* 23:485-491.

Polowick, P.L., J. Quandt, and J.D. Mahon (2000). The ability of pea transformation technology to transfer genes into peas adapted to western Canadian growing conditions. *Plant Science* 153:161-170.

Prakash, N.S., D. Pental, and N.B. Sarin (1994). Regeneration of pigeon pea (*Cajanus cajan*) from cotyledonary node via multiple shoot formation. *Plant Cell Reports* 13:623-627.

Rathore, R.S. and L. Chand (1999). Plantlet regeneration from decapitated embryonic axes of pigeon pea [*Cajanus cajan* (L.) Mill sp]. *Indian Journal of Experimental Biology* 37:496-498.

Rohini, V.K. and K. S. Rao (2000). Transformation of peanut (*Arachis hypogaea*): A non tissue culture approach for generating transgenic plants. *Plant Science* 15:41-49.

Sahoo, L., T. Sugla, and P.K. Jaiwal (2002). In vitro regeneration and genetic transformation of *Vigna* species. In *Biotechnology for the improvement of legumes,* P.K. Jaiwal and R.P. Singh (eds.). Dordrecht: Kluwer Academic Publishers, pp. 1-48.

Saini, R.S. and P.K. Jaiwal (2003). Stable genetic transformation of *Vigna mungo* (L.) Hepper via *Agrobacterium tumefaciens. Plant Cell Reports* 21:851-859.

Saini, R. and P.K. Jaiwal (2005). Transformation of a recalcitrant grain legume, *Vigna mungo* (L.) Hepper using *Agrobacterium tumefaciens* mediated gene transfer to shoot apical meristem cultures. *Plant Cell Reports* 24:164-171.

Sarmah, B.K., A. Moore, W. Tate, L. Molvig, R.L. Morton, D.P. Rees, P. Chiaiese, M.J. Chrispeels, L.M. Tabe, and T.J.V. Higgins (2004). Transgenic chickpea seeds expressing high level of a bean alpha-amylase inhibitor. *Molecular Breeding* 14:73-82.

Satyavathi, V.V., V. Prasad, A. Khandelwal, M.S. Shaila, and G. Lakshmi Sita (2003). Expression of hemagglutinin protein of rinder pest virus in transgenic pigeon pea plants [*Cajanus cajan* (L.) Millsp.] plants. *Plant Cell Reports* 21:651-658.

Schroeder, H.E., A.H. Schotz, T. Wardley-Richardson, D. Spencer, and T.J.V. Higgins (1993). Transformation and regeneration of two cultivars of pea (*Pisum sativum* L.). *Plant Physiology* 101:751-757.

Senthil, G., B. Williamson, R.D. Dinkins, and G. Ramsay (2004). An efficient transformation system for chick pea (*Cicer arietinum* L.). *Plant Cell Reports* 23:297-303.

Somers, D.A., D.A. Samac, and P.M. Olhoft (2003). Recent advances in legume transformation. *Plant Physiology* 131:892-899.

Subhadra, R., K. Vashisht, J.B. Chowdhury, M. Singh, and P.K. Sareen (1998). Multiple shoots from cotyledonary node of non modulating genotype (ICC435M) of chick pea, *Cicer arietinum. Indian Journal of Experimental Biology* 36:1276-1279.

Surekha, C., M.R. Beena, A. Arundhati, P.K. Singh, R. Tuli, A. Datta-Gupta, and P.B. Kirti (2005). *Agrobacterium* mediated genetic transformation of pigeon pea [*Cajanus cajan* (L.) Millsp.] using embryonal segments and development of transgenic plants for resistance against Spodoptera. *Plant Science* 169:1077-1084.

Swathi Anuradha, T., S.K. Jami, R.S. Datla, and P.B. Kirti (2006). Genetic transformation of peanut (*Arachis hypogaea* L.) using cotyledonary node as explant and a promoterless *gus::npt*II fusion vector. *Journal of Biosciences* 31:1-12.

Tewari-Singh, N., J. Sen, H. Kiesecker, V.S. Reddy, H.J. Jacobsen, and S. Guha-Mukherjee (2004). Use of a herbicide or lysine plus threonine for non-antibiotic selection of transgenic chickpea. *Plant Cell Reports* 22:576-583.

Townsend, J.A. and L.A. Thomas (1993). An improved method of *Agrobacterium* mediated transformation of cultured soybean cells. U.S. patent W094/02620.

Wright, M.S., H. Koehler, M.A. Hinchee, and H.G. Carnes (1986). Plant regeneration by organogenesis in *Glycine max. Plant Cell Reports* 5:150-154.

Zhang, Z.-Y., A.-Q. Xiang, Q. Staswick, et al. (1999). The use of glufosinate as a selective agent in *Agrobacterium*-mediated transformation of soybean. *Plant Cell Tissue and Organ Culture* 56:37-46.

Chapter 18

# A Novel *In Planta* Approach to Gene Transfer for Legumes

K. Sankara Rao
V. K. Rohini

## *INTRODUCTION*

The study of legume biology is rapidly undergoing a revolutionary transformation due to the application of genomic methods. The pace of discovery in legume biology has accelerated, fueled by the development of two model systems from within the forage legume group, namely, *Lotus japonicus* and *Medicago truncatula* for molecular genetic analysis and by the application of genomic approaches (VandenBosch and Stacey, 2003a). The impetus for the development of legume models has come primarily from researchers interested in the *Rhizobium*-legume symbiosis (see VandenBosch and Stacey, 2003b). Both *L. japonicus* and *M. truncatula* are now chosen as models for investigation by a large number of laboratories involved in genomics. An international multi-institutional collaboration effort to develop a complete inventory and functional analysis of the *Medicago* genome is underway (see VandenBosch and Stacey, 2003b). Advances have also been made in transcriptome, proteome, and metabolome analysis for legume research using *L. japonicus* (VandenBosch and Stacey, 2003b). The results are expected to expedite the discovery of agronomically important genes, in both model and crop legumes and to enhance understanding of gene and genome evolution within the Leguminosae (VandenBosch and Stacey, 2003b).

A key component of most functional genomics approaches is a high-frequency transformation system useful for developing various gene identification strategies. A valuable somatic embryogenesis system useful for transformation was developed for *Medicago truncatula* using leaf callus into culture (Rose and Nolan, 1995). High-throughput transformation systems are being rapidly developed for *L. japonicus* (Oger et al., 1996) and *M. truncatula* (Trieu and Harrison, 1996; Trieu et al., 2000).

## *TARGETS FOR GENETIC MODIFICATION OF LEGUMES*

The ability to transfer genes into plant species is central to both basic and applied molecular biology. *Agrobacterium in planta* transformation procedures have been a success with nonleguminous *Arabidopsis*. Similar transformation success reported for the legume *Medicago truncatula* (Trieu et al., 2000) shows that *in planta* methods can be adapted to legumes as well. Development of robust transformation systems for several useful legumes is desirable for their use in various gene identification strategies, for crop improvement, and to pursue a number of important biological questions unique to legumes. Objectives and priorities for transgenic improvement of legumes vary from crop to crop. Deficiencies and needs for which conventional breeding cannot provide solutions should be accorded priority in transgenic approaches. There is a need to increase productivity and enhance nutritional value of pulses and other large-seed legume crops. Cultivars resistant to biotic and abiotic stresses and which have better protein quality and quantity are badly needed. The great potential of these approaches is still largely unrealized but should yield seed protein with enhanced quality in the future. Other targets for legume seed protein improvement include improved digestibility, removal of antinutritional factors and activities that generate undesirable flavors, and removal of potential allergens. Soybean and groundnut improvement requires reduction of the saturated forms of lipid, mainly palmitate (Kinney, 1998).

Insect pests and diseases caused by viruses and fungal pathogens are major problems of legumes worldwide. Development of resistance to drought and water logging were also identified as important breeding targets for leguminous crops. Further, there is an opportunity to use molecular methods for producing male sterile and restorer lines for hybrid seed production in crops like pigeon pea (Grover and Pental, 2003).

## *TRANSFORMATION OF LEGUMES*

Legumes are regarded as difficult-to-transform plants, the limitation mostly being their poor regeneration ability in vitro. Gene transfer into these species can therefore be challenging. Even among the few legumes that are apparently amenable to transformation, the process is often slow and the number of plants generated from each explant following DNA transfer is often low. For legumes such as pigeon pea, peanut, and chickpea that have been regarded as "more difficult to transform," regeneration is highly genotype specific, although many legumes like alfalfa (*Medicago sativa*), barrel medic (*M. truncatula*), and red clover (*Trifolium pratense*), for example, which were previously classified as recalcitrant, have now been shown to be amenable to transformation (see Chandra and Pental, 2003; Somers et al., 2003). Progress in transformation of large-seeded legumes has been extensively reviewed (Christou, 1997; Nagl et al., 1997; Trick et al., 1997). DNA delivery systems developed for several legumes since or in addition to Atkins and Smith (1997) and Babaoglu et al. (2000) have been summarized in Somers et al. (2003).

Although in vitro plant regeneration methods such as organogenesis and embryogenesis have been developed for several economically important legumes, these methods have not provided regeneration robust enough for use in transformation protocols, which is a serious limitation to the exploitation of gene transfer to its full potential. However, considerable success has been achieved in the transformation of large-seed legumes like soybean [*Glycine max* (L.) Merril], common bean [*Phaseolus vulgaris* L.], chickpea [*Cicer arietinum* L.], pigeon pea [*Cajanus cajan* (L.) Millsp.], mungbean [*Vigna radiata* (L.) Wilczek], cowpea [*Vigna unguiculata* (L.) Walp] and black gram [*Vigna mungo* (L.) Hepper], and forage legumes like barrel medic (*Medicago truncatula*), *Lotus japonicus,* alfalfa (*Medicago sativa*) and white clover (*Trifolium repens*) (Olhoft and Somers, 2001; Trieu and Harrison, 1996; Trinh et al., 1998; Larkin et al., 1996; Voisey et al., 1994; Delgado-Sanchez et al., 2006; Sanyal et al., 2005; Surekha et al., 2005; Jaiwal et al., 2001; Ramakrishnan et al., 2005; Saini and Jaiwal, 2005) that has enabled research to advance from expression of marker genes to evaluation of genes for crop improvement. Soybean has emerged as the most amenable among seed legumes for transformation and is one of the four major global transgenic crops (James, 2003).

Commercialization of the first transgenic forage crop, Round up Ready alfalfa, is also underway (Somers et al., 2003). Because inducing somatic embryogenesis or organogenesis in many legume species is difficult and genotype dependent, regeneration of shoots from cotyledonary node or from other meristematic regions after *Agrobacterium* infection is emerging as a rapid and relatively efficient method of transformation for a number of legume species including soybean (*Glycine max;* Olhoft and Somers, 2001), *Lotus japonicus* (Oger et al., 1996), barrel medic (*Medicago truncatula;*

Trieu and Harrison, 1996; Trinh et al., 1998), white clover (*Trifolium repens;* Larkin et al., 1996), pigeon pea (*Cajanus cajan;* Surekha et al., 2005), chickpea (*Cicer arietinum;* Sanyal et al., 2005), black gram (*Vigna mungo;* Saini and Jaiwal 2005) and common bean (*Phaseolus vulgaris;* Delgado-Sanchez et al., 2006). Several other interesting approaches and methodologies have come out of intense efforts on gene transfer into legumes in the last 30 years. Currently, there is considerable interest in developing transformation methods that do not depend on tissue culture regeneration or those that substantially eliminate the intervening tissue culture step. Transformation without involvement of tissue culture has clear advantages. It can provide a high-throughput method that requires minimal labor, expense and expertise. Rates of unintended mutagenesis are also reduced. More important among the tissue culture-independent transformations, *in planta* methods that are incredibly simple have been developed for the nonleguminous species, *Arabidopsis* (Clough and Bent, 1998). Early success in *Arabidopsis in planta* transformation came from the work of Ken Feldman and David Marks. They applied *Agrobacterium* to *Arabidopsis* seeds, raised plants in the absence of any selection, then collected progeny seeds and germinated them on antibiotic-containing media to identify transformants (Feldmann and Marks, 1987; Feldmann, 1992). Although the procedure was difficult to reproduce consistently, successive rounds produced transformants at a high enough rate that thousands of transformed lines were produced in a few years. These insertional mutagenesis lines helped expeditious gene cloning by the *Arabidopsis* community (Azipiroz-Leehan and Feldmann, 1997). Other laboratories later succeeded in generating transformed *Arabidopsis* lines by "clip 'n' squirt" methods (Chang et al., 1994; Katavic et al., 1994). Reproductive inflorescences were clipped off, *Agrobacterium* was applied to the center of the plant rosettes, new inflorescences that formed a few days later were again removed, *Agrobacterium* was reapplied, and plants were then allowed to develop and set seed. Transformants were obtained more reliably by this method than with the seed treatment method, but the methods were only marginally more productive than traditional tissue culture approaches to *Arabidopsis* transformation (Valvekens et al., 1988). A third stage of the revolution in *Arabidopsis* transformation came when Georges Pelletier, Nicole Bechtold, and Jeff Ellis reported success in transformation by following vacuum infiltration (Bechtold et al., 1993; Bechtold and Pelletier, 1998). *Arabidopsis* plants at the early stages of flowering were uprooted and placed in a jar in a solution of *Agrobacterium.* A vacuum was applied and then released, causing air trapped within the plant to bubble off and be replaced with the *Agrobacterium* solution. Plants were rooted back into soil, grown to seed, and in the next generation stably transformed lines were selected using the antibiotic or the herbicide appropriate for the selectable marker gene. Transformation rates often exceeded 1 percent of the seeds tested. Variations of this method have been widely adopted by *Arabidopsis* researchers.

One significant variation resulting from these efforts is the floral dip method, which eliminates vacuum infiltration in the protocol, and the transformants are obtained by merely treating the protruding inflorescences by dipping in *Agrobacterium* solution (Clough and Bent, 1998). By this method, plants did not need to be uprooted for *Agrobacterium* treatment and replanted. Many different *Arabidopsis* ecotypes were transformable and many different *Agrobacterium* strains could be used, although notable differences in efficiency were observed (Bent, 2000).

Application of similar transformation methods seems likely with at least some legumes. Two *in planta* transformation systems were described for the forage legume, *Medicago truncatula* (Trieu et al., 2000); one method is based on infiltration of flowers with *Agrobacterium tumefaciens,* similar to the *Arabidopsis* flower infiltration protocol, and the other on infiltration of seedlings. Both methods were reported to result in high transformation frequencies (Somers et al., 2003). Thus, the majority of reports of nontissue-culture approaches are focused on *Agrobacterium tumefaciens*-mediated transformation. This is probably due to the simplicity of the transformation system and precise integration of transgenes. Also, such developments and improvements as the use of hypervirulent or modified *Agrobacterium* strains and adding thiol compounds to the cocultivation medium have enhanced transformation frequencies (Veluthambi et al., 2003; Hood et al., 1993; Trick and Finer, 1997; Ke et al., 2001; Olhoft and Somers, 2001). More recently, Weeks et al. (2007) have developed an *in planta* method of transformation of alfalfa by simply vortexing the seedlings, which are cut at the apical node region, with a suspension of *Agrobacterium* cells containing sterile sand, and the treated seedlings were grown to maturity. The progenies of 7 percent of treated plants segregated for the transgene, caffeic acid o-methyltransferase (*Comt*) indicating the efficiency of the method. The T-DNA region of the binary vector did not carry any marker gene. Hence, this has resulted in the development of marker free alfalfa transgenic plants.

Multiple plant species have now been successfully transformed using *Agrobacterium in planta* approaches. Pursuit of similar strategies would facilitate development of transformation systems for difficult-to-transform legumes as well. Further variations of *in planta* transformation are also being introduced to design the most time-and labor-efficient methods for regeneration of transgenic plants.

## *EMBRYO TRANSFORMATION: AN* IN PLANTA *METHOD FOR LARGE SEED LEGUMES*

We have aimed at establishing a system for large seed legumes to overcome the constraints of in vitro plant regeneration following *Agrobacterium* infection and investigated the use of the entire embryo axis of the germinating seed as the target tissue for transformation and for its subsequent growth directly into a transformed plant. *Agrobacterium* infection is directed toward the plumule, cotyledonary node, and the surrounding regions of the peanut embryo, where one cotyledon is broken off to provide a wound site. The embryo axis is later allowed to grow into a mature plant, which is monitored for the presence of transgenes. The initial assessment of transformation by this method was based on *uid*A gene expression. The *uid*A reporter gene in pKIWI105 that expresses only upon transfer to plant cells is used for the optimization of the transformation protocol and GUS expression is taken as a direct measure of transformation. Molecular analysis of transformants carried out later is used to correlate the above observations. The use of wounded tobacco leaf extract at the time of infection is introduced to increase transformation efficiency. The other crops selected to test this strategy were pigeon pea (Sankara Rao et al. 2008, in press) and field bean (Sankara Rao and Rohini, unpublished results) and recalcitrant nonlegumes, sunflower and safflower (Sankara Rao and Rohini, 1999; Rohini and Sankara Rao, 2000a). The method is basically developed on the same lines as that in soybean (Chee et al., 1989). The detailed methodology of embryo transformation of peanut that follows is improvised over the one developed earlier (Rohini and Sankara Rao, 2000b). This improvised protocol is routinely being used in our experiments to transform other large seed legumes as well.

Peanut (*Arachis hypogaea* L.) seeds (embryos) of cultivar TMV-2 with one of the cotyledons removed are soaked overnight and later surface sterilized with 0.1 percent mercuric chloride for 5 to 7 minutes, followed by thorough rinses with sterile water. They are then germinated on wet filter paper for 48 hours. The cotyledonary node and the plumule region of the embryo axes are pricked randomly with a sterile sewing needle and immersed in the suspension of *Agrobacterium* in Winans AB medium (Winans et al., 1988) to which tobacco leaf extract is added. The infection is carried out by gentle agitation at 28 to 30°C for 16 hours. The time of infection appeared to affect transformation efficiency. The infection period is deliberately prolonged because the quantity of phenolics exuding from the wounded embryo axes appeared to be much less compared to the amount exuding from seedlings or from mature plant not specific for *vir* gene transcription. With prolonged infection, it is likely that the bacteria remain persistent in the apoplast of the embryonal axis, in which case they might continue to interact with the plant cells and enhance the chances of transformation.

*Agrobacterium* strain LBA 4404 harboring the binary vector pKIWI105 is used for transformation. The plasmid carries genes for β-glucuronidase (*uid*A) and neomycin phosphotransferase (*npt*II) driven by CaMV 35S and nopaline synthase promoters respectively. The *uid*A gene in pKIWI105 lacks a functional bacterial ribosome-binding site that prevents its expression in the bacteria. *Agrobacterium* strain LBA 4404/pKIWI105 is grown overnight at 29 to 30°C in Luria and Bertani medium (pH-7.0) containing 50 $\mu g \cdot ml^{-1}$kanamycin. The bacterial cells are later resuspended in Winans AB medium (Winans et al., 1988; pH 5.2) and grown for 18 hours. Wounded tobacco leaf extract (tobacco leaves, 2 g in 2 ml sterile water) is then added to this suspension. Following infection, the germlings are blot dried, washed thoroughly with 500 $\mu g \cdot ml^{-1}$ of cefotaxime for about 18 hours, and placed on autoclaved soilrite (Vermiculite equivalent) for germination to progress under aseptic conditions in capped bottles. Five or six days after the germlings have developed two or three foliage leaves, they are transferred to soilrite in pots and allowed to grow under growth room conditions for at least 10 days before they are transferred to the greenhouse. The pots are initially covered with polythene bags to maintain humidity. The growth chamber is maintained at 26 to 28°C under a 14 hour photoperiod with fluorescent light of intensity of 35 $\mu mol \cdot m^{-2} \cdot s^{-1}$. Several embryos are infected simultaneously and the experiments are repeated to generate a large number of transformants. After acclimatization to greenhouse conditions, all the plants have shown healthy growth and set seed normally. The seeds of the $T_0$ generation plants are harvested and germinated for analysis of integration and inheritance of the transgenes in the tissues of progeny ($T_1$ generation) plants. GUS enzyme activity is assessed following the method of Jefferson's histochemical assay (Jefferson, 1987; Figure 18.1). The cotransformation of *npt*II (neomycin phosphotransferase II) transgene is also verified by the NPT II and PAGE assay performed according to Reiss et al. (1984). The transformed nature of the $T_1$ generation plants is further evaluated by polymerase chain reaction analysis of *uid*A gene in the transformants and by Southern analysis.

The feasibility of the transformation strategy adopted in the study is initially evaluated by monitoring the number of peanut embryo axes germinating into normal seedlings following wounding by excision of one cotyledon and by pricking with a needle, infection with *Agrobacterium,* and decontamination treatment. A 16 hour period of infection was chosen on the basis of the observations from the preliminary experiments on the optimization of

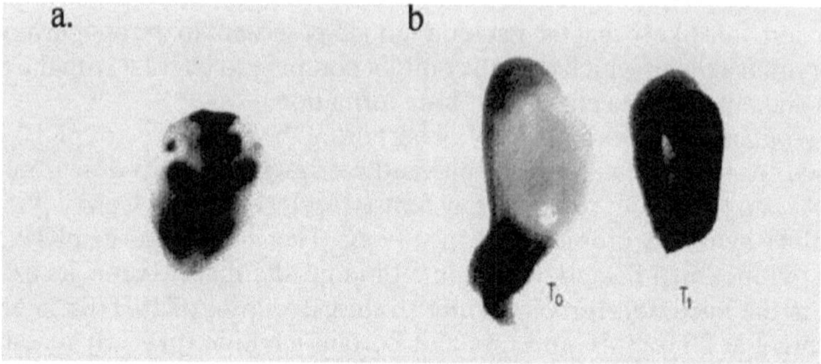


FIGURE 18.1. GUS histochemical analysis: a, GUS expression in the leaf of primary transformant; b, GUS expression in the embryo of $T_0$ and $T_1$ plants. (See also color gallery.)

infection time. As the infection treatment given is for an extended period, and any further exposure of explants to bacteria appears deleterious, a discrete cocultivation step that generally follows infection is not included in the protocol. Embryos of peanut cv. TMV-2 when infected in the absence of acetosyringone or tobacco leaf extract did not show GUS expression. In the experiments performed initially to determine the tolerance of peanut to kanamycin, it was observed that uninfected embryos (control) did not germinate beyond 150 $\mu g \cdot ml^{-1}$ of kanamycin. Further, there is a reduction in the germination rate when wounded uninfected as well as infected embryos are germinated in the presence of kanamycin. Therefore, selection on kanamycin is eliminated in the protocol. A large percentage (up to 90 percent) of embryos subjected to this procedure survived wounding and 16 hour infection, and germinated into normal healthy plants. The transformed tissue sectors have persisted into gametes at reasonable frequencies. The procedure was not difficult to reproduce and it was possible to generate and screen a large number of putative transformants. Successful rounds of experiments produced transformants at a high enough rate that in a matter of few months a good number of transformed lines could be identified.

The outline of the embryo transformation protocol is shown in Figure 18.2. One major change in this transformation protocol when applied to different crops would be the infection and cocultivation period. The exudation of phenolics upon wounding varies from plant to plant and tissue to tissue. It was seen that the phenolics in the peanut embryo axes are much less when compared to those in sunflower and safflower. When the same conditions as those applied to peanut were used on sunflower and safflower, the latter could not be revived. Therefore, in the case of sunflower and safflower, 10

Seeds with one of the cotyledons removed were soaked

↓

Surface decontaminated

↓

Germinated on wet filter paper for two days

↓

Embryo axes of germinating seeds pricked randomly with sterile needle and immersed in the suspension of *Agrobacterium* in Winans' AB medium with wounded tobacco leaf extract

↓

Gentle agitation at 28°C

↓

Decontamination with cefotaxime at 500 $mg \cdot l^{-1}$

↓

Germination in soilrite in pots grown for ten days under growth room conditions

↓

Seedlings transferred to soil in greenhouse

FIGURE 18.2. *In planta* transformation protocol for peanut.

minute infection and 24 hour cocultivation periods were employed. Selection on kanamycin is eliminated in the procedure because it is possible that the transformed sectors would be lost when the untransformed embryonal tissues begin to die when subjected to kanamycin selection.

The *in planta* method described has been effectively used in introducing traits of agronomical importance into peanut (Rohini and Sankara Rao, 2001). Introduction and expression of a heterologous tobacco chitinase transgene in peanut by this method has led to a high level of accumulation of the protein in some of the transgenic plants. Small-scale field tests have indicated the ability of these plants to resist the fungal pathogen *Cercospora arachidicola*.

## *CONCLUSION*

Genetic manipulation of legumes entails a major investment that can generate considerable economic benefits over the long term, besides its scientific advantages. Successful production of transgenic plants relies on combining a suitable transformation protocol with a robust plant regeneration step. Routine transformation protocols are limited in most legumes. The low success rate has been attributed to poor regeneration ability of legumes in tissue cultures, especially via callus, and also lack of compatible gene delivery methods. Much effort has been devoted in recent years to developing and optimizing efficient in vitro regeneration systems for legumes to facilitate gene transfer. Parallel efforts are also continuing to evolve transformation procedures, which seek to bypass the tissue culture phase for legume transformation. Novel legume-specific, non-tissue-culture systems are being developed in a number of legume species (Somers et al., 2003). The embryo transformation strategy described here that has the potential to generate transgenics in large numbers with considerable ease is yet another step forward in our progress in legume transformation. Embryo transformation strategy presents certain advantages:

- Does not require in vitro plant regeneration
- Allows rapid screening of large numbers of transformants
- Genotype and variety independent
- Precludes culture-induced variations

The procedure can be applied to all those genotypes and cultivars of large seeded legumes that are susceptible to *Agrobacterium tumefaciens* infection, although the efficiency might vary between them. The degree of susceptibility of these crops to *Agrobacterium* infection could be the reason for this difference.

Further, embryo transformation is relatively simple, highly reproducible, and less labor intensive. Incorporation of such methods of gene transfer will accelerate crop improvement considerably.

## REFERENCES

Atkins, C.A. and P.M.C. Smith (1997). Genetic transformation and regeneration of legumes. In *Biological fixation of nitrogen for ecology and sustainable agriculture*, A. Legoki, H. Bothe, and A. Puhler (eds.). Berlin: Springer-Verlag, pp. 283-304.

Azipiroz-Leehan, R. and K.A. Feldmann (1997). T-DNA insertion mutagenesis in *Arabidopsis*: Going back and forth. *Trends in Genetics* 13:152-156.

Babaoglu, M., M.R. Davey, and J.B. Power (2000). Genetic engineering of grain legumes: Key transformation events. *Agbiotech Net* 2000, vol. 2.

Bechtold, N., J. Ellis, and G. Pelletier (1993). *In planta Agrobacterium*-mediated gene transfer by infiltration of adult *Arabidopsis thaliana* plants. *Comptes Rendus de l'Académie des Sciences-Paris* 316:1194-1199.

Bechtold, N. and G. Pelletier (1998). *In planta Agrobacterium*-mediated transformation of adult *Arabidopsis thaliana* plants by vacuum infiltration. *Methods in Molecular Biology* 82:259-266.

Bent, A.F. (2000). *Arabidopsis in planta* transformation: Uses, mechanisms and prospects for transformation of other species. *Plant Physiology* 124:1540-1547.

Chandra, A. and D. Pental (2003). Regeneration and genetic transformation of grain legumes: An overview. *Current Science* 84:381-387.

Chang, S.S., S.K. Park, B.C. Kim, B.J. Kang, D.U. Kim, and H.G. Nam (1994). Stable genetic transformation of *Arabidopsis thaliana* by *Agrobacterium* inoculation *in planta. Plant Journal* 5:551-558.

Chee, P.P., A.K. Fober, and J.L. Slightom (1989). Transformation of soybean (*Glycine max* L.) by infecting germinating seeds with *Agrobacterium tumefaciens. Plant Physiology* 91:1212-1218.

Christou, P. (1997). Biotechnology applied to grain legumes. *Field Crop Research* 53:83-97.

Clough, S.J. and A.F. Bent (1998). Floral dip: A simplified method for *Agrobacterium*-mediated transformation of *Arabidopsis thaliana. Plant Journal* 16: 735-743.

Delgado-Sanchez, P., M. Sancedo-Ruiz, S.H. Guzman-Maldonado, E. Villordo-Pineda, M. Gonzalez Chavira, S. Fraire-Velazquaz, J.A. Acosta-Gallegos, and A. Mora-Aviles (2006). An organogenic plant regeneration system for common bean (*Phaseolus vulgaris* L.). *Plant Science* 170:822-827.

Feldmann, K.A. (1992). T-DNA insertion mutagenesis in *Arabidopsis* seed infection transformation. In *Methods in Arabidopsis research,* C. Koncz, N.-H. Chua, and J. Schell (eds.). Singapore: World Scientific, pp. 274-289.

Feldmann, K.A. and M.D. Marks (1987). *Agrobacterium*-mediated transformation of germinating seeds of *Arabidopsis thaliana:* A non-tissue culture approach. *Molecular and General Genetics* 208:1-9.

Grover, A. and D. Pental (2003). Breeding objectives and requirements for producing transgenics for major field crops of India. *Current Science* 84:310-320.

Hood, E.E., S.B. Gelvin, L.S. Melchers, and A. Hoekema (1993). New *Agrobacterium* helper plasmids for gene transfer to plants. *Transgenic Research* 2:208-218.

Jaiwal, P.K., R. Kumari, S. Ignacimuthu, I. Potrykus, and C. Sautter (2001). *Agrobacterium tumefaciens*-mediated genetic transformation of mungbean [*Vigna radiata* (L.) Wilczek]—a recalcitrant grain legume. *Plant Science* 161:239-247.

James, C. (2003). Global review of commercialized transgenic crops. *Current Science* 84:303-310.

Jefferson, R. (1987). Assaying chimeric genes in plants, the GUS gene fusion system. *Plant Molecular Biology Reporter* 5:387-405.

Katavic, V., G.W. Haughn, D. Reed, M. Martin, and L. Kunst (1994). *In planta* transformation of *Arabidopsis thaliana. Molecular and General Genetics* 245: 363-370.

Ke, J., R. Khan, T. Johnson, D.A., Somers, and A. Das (2001). High-efficiency gene transfer to recalcitrant plants by *Agrobacterium tumefaciens. Plant Cell Reports* 20:150-156.

Kinney, A.J. (1998). Plants as industrial chemical factories: New oils from genetically engineered soybeans. *Fett/Lipid* 100:173-176.

Larkin, P.J., J.M. Gibson, U. Mathesius, J.J. Weinman, E. Gartner, E. Hall, G.J. Tanner, B.G. Rolfe, and M.A. Djordjevic (1996). Transgenic white clover: Studies with the auxin-responsive promoter, GH3, in root gravitropism and lateral root development. *Transgenic Research* 5:325-335.

Nagl, W., S. Ignacimuthu, and J. Becker (1997). Genetic engineering and regeneration of *Phaseolus and Vigna:* State of the art and new attempts. *Journal of Plant Physiology* 150:625-644.

Oger, P., A. Petit, and Y. Dessaux (1996). A simple technique for direct transformation and regeneration of the diploid legume species *Lotus japonicus. Plant Science* 116:159-168.

Olhoft, P.M. and D.A. Somers (2001). L-Cysteine increases *Agrobacterium*-mediated T-DNA delivery into soybean cotyledonary node cells. *Plant Cell Reports* 20:706-711.

Ramakrishnan, K., R. Gnanam, P. Sivakumar, and A. Manickam (2005). In vitro somatic embryogenesis from cell suspension cultures of cowpea, *Vigna unguiculata* (L.) Walp. *Plant Cell Reports* 24:449-461.

Reiss, B., R. Sprengel, H. Will, and H. Schaller (1984). A new sensitive method for qualitative and quantitative assay of neomycin phosphotransferase in crude cell extracts. *Gene* 30:211-218.

Rohini, V.K. and K. Sankara Rao (2000a). Embryo transformation, a practical approach for realizing transgenic plants of safflower (*Carthamus tinctorius* L.). *Annals of Botany* 86:1043-1049.

Rohini, V.K. and K. Sankara Rao (2000b). Transformation of peanut (*Arachis hypogaea* L.): A non-tissue culture based approach for generating transgenic plants. *Plant Science* 150:41-49.

Rohini, V.K. and K. Sankara Rao (2001). Transformation of peanut (*Arachis hypogaea* L.) with tobacco chitinase gene: Variable response of transformants to leaf spot disease. *Plant Science* 160:883-892.
Rose, R.J. and K.E. Nolan (1995). Regeneration of *Medicago truncatula* from protoplasts isolated from kanamycin-sensitive and kanamycin resistant plants. *Plant Cell Reports* 14:349-353.
Saini, R. and P.K. Jaiwal (2005). Transformation of a recalcitrant grain legume, *Vigna mungo* (L.) Hepper using *Agrobacterium tumefaciens*-mediated gene transfer to shoot apical meristem. *Plant Cell Reports* 24:164-171.
Sankara Rao, K. and V.K. Rohini (1999). *Agrobacterium*-mediated transformation of sunflower (*Helianthus annuus* L.): A simple protocol. *Annals of Botany* 83:347-354.
Sankara Rao, K., S. Rohini, D.P. Sharma, E. Keshamma and M. Udaya Kumar (2008). *In planta* transformation of pigeon pea: A method to overcome the recalcitrancy of the crop to regeneration in vitro. *Physiology and Molecular Biology of Plants* (In press).
Sanyal, I.K., Aditya Singh, M. Kaushik, and D.V. Amla (2005). *Agrobacterium tumefaciens*-mediated transformation of chick pea (*Cicer arietinum* L) with Bt (*cry*1Ac) for resistance against pod borer insect *Helicoverpa armigera. Plant Science* 168:1135-1146.
Somers, D.A., A. Deborah, P. Samac, and P.M. Olhoft (2003). Recent advances in legume transformation. *Plant Physiology* 131:892-899.
Surekha, C., M.R. Beena, A. Arundhati, P.K. Singh, R. Tuli, A. Datta-Gupta, and P.B. Kirti (2005). *Agrobacterium*-mediated genetic transformation of pigeon pea [*Cajanus cajan* (L.) Millsp.] using embryonic segments and development of transgenic plant for resistance against *Spodoptera. Plant Science* 169:1074-1080.
Trick, H.N., R.D. Dinkins, E.R. Santarem, R. Di, V. Samoylov, C. Meurer, D. Walker, W.A. Parrott, J.J. Finer, and G.B. Collins (1997). Recent advances in soybean transformation. *Plant Tissue Culture and Biotechnology* 3:9-26.
Trick, H.N. and J.J. Finer (1997). SAAT: Sonication-assisted *Agrobacterium*-mediated transformation. *Transgenic Research* 6:329-337.
Trieu, A.T., S.H. Burleigh, I.V. Kardailsky, I.E. Maldonado-Mendoza, W.K. Versaw, L.A. Blaylock, H. Shin, T.-J. Chiou, H. Katagi, and G.R. Dewbre (2000). Transformation of *Medicago truncatula* via infiltration of seedlings or flowering plants with *Agrobacterium. Plant Journal* 22:531-541.
Trieu, A.T. and M.J. Harrison (1996). Rapid transformation of *Medicago truncatula:* Regeneration via shoot organogenesis. *Plant Cell Reports* 16:6-11.
Trinh, T.H., P. Ratet, E. Kondorosi, P. Durand, K. Kamate, P. Baner, and A. Kondorosi (1998). Rapid and efficient transformation of diploid *Medicago truncatula* and *Medicago sativa* ssp. *falcata* lines improved in somatic embryogenesis. *Plant Cell Reports* 17:345-355.
Valvekens, D., M.V. Montague, and M.V. Lijsbettens (1988). *Agrobacterium tumefaciens*-mediated transformation of *Arabidopsis thaliana* root explants by

using kanamycin selection. *Proceedings of the National Academy of Sciences USA* 85:5536-5540.

VandenBosch, K.A. and G. Stacey (2003a). Advances in legume biology. Editorial. *Plant Physiology* 131:839.

VandenBosch, K.A. and G. Stacey (2003b). Summaries of legume genomics projects from around the globe. Community resources for crops and models. *Plant Physiology* 131:840-865.

Veluthambi, K., A.K. Gupta, and A. Sharma (2003). The current status of plant transformation technologies. *Current Science* 84:368-380.

Voisey, C.R., D.W.R. White, B. Dudas, R.D. Appleby, P.M. Ealing, and A.G. Scott (1994). *Agrobacterium*-mediated transformation of white clover using direct shoot organogenesis. *Plant Cell Reports* 14:309-314.

Weeks, J.T., J. Ye, and C.M. Rommen (2007). Development of an *in planta* method for transformation of alfalfa (*Medicago sativa*). Transgenic Research (EPub-ahead of publication).

Winans, S.C., R.A. Kerstetter, and E.W. Nester (1988). Transcriptional regulation of the *vir* A and *vir* G genes of *Agrobacterium tumefaciens. Journal of Bacteriology* 170:4047-4054.

## Chapter 19

# *Agrobacterium*-Mediated Genetic Transformation of Faba Bean

M. Hanafy
H.-J. Jacobsen
P. Böttinger
T. Pickardt

## *INTRODUCTION*

The development of gene transfer techniques for faba bean is of commercial interest as they will facilitate the production of cultivars with improved characteristics such as resistance to biotic and abiotic stress, and enhancing the nutritional value. Chocolate spot (*Botrytis fabae*) and ascochyta blight (*Ascochyta fabae*) are the most widespread and devastating fungal diseases in all production areas. Viral diseases and the parasitic weed *Orobanche crenata* are the most important factors contributing to losses in faba bean production in the West Asian and North African region (Hanounik et al., 1993; Robertson and Saxena, 1993). Abiotic stresses like limited cold tolerance and the susceptibility to drought stress are major constraints on yield. Today, breeding programs of *Vicia faba* could be supplemented by recombinant DNA technology, which requires the development of reproducible protocols for its genetic transformation. The combination of conventional breeding approaches, plant tissue culture techniques, and recombinant DNA technology opens tremendous avenues for improvement of faba bean in a short time.

## *GENETIC TRANSFORMATION*

### *State of the Art*

Like most other grain legumes, faba bean exhibits a rather low competence for in vitro culture, mainly due to difficulties in regeneration from callus tissues and the high content of phenolic compounds, which causes cell death (Selva et al., 1989; Bieri et al., 1984). Unfortunately, efficient in vitro techniques are limited for faba bean compared with other major economically important crops. Several investigators have worked extensively on plant regeneration and transformation (Schiemann and Eisenreich 1989; Quandt et al., 1993; Jelenic et al., 2000). No transgenic faba bean plants were reported in any of these studies. Faba bean is shown to be susceptible to transformation so far only by our group, which reported the production of fertile transgenic plants using *Agrobacterium*-mediated transformation. The first transgenic plants were recovered using a system based on de novo regeneration using thidiazuron (TDZ; Böttinger et al., 2001). The second protocol was based on direct shoot organogenesis from meristematic cells of mature or immature embryo axes (Hanafy et al., 2005).

### *Protocol I: De Novo Regeneration Protocol*

This *Agrobacterium*-based transformation protocol makes use of the de novo regeneration of shoot initials from callus that was adopted from a plant regeneration protocol for protoplast-derived calli (Tegeder et al., 1995). We recovered transgenic plants by inoculation of stem segments with *Agrobacterium* strain EHA 101 or EHA 105, harboring different binary vectors, followed by callus induction on MS (Murashige and Skoog, 1962) medium containing 0.5 $mg \cdot l^{-1}$ each of TDZ, 2,4-dichlorophenoxyacetic (2,4-D) and naphthalene acetic acid (NAA) and 100 $mg \cdot l^{-1}$ kanamycin or 2 $mg \cdot l^{-1}$ DL-phosphinothricin (PPT) as a selective agent. Subsequently, transgenic shoots were regenerated via organogenesis using a high concentration of TDZ (7.5 $mg \cdot l^{-1}$) and 0.75 $mg \cdot l^{-1}$ NAA. Finally, plants were recovered by micrografting. This process requires 16 to 24 months to get seed producing primary transformants, is laborious and time consuming because the quantity of regenerable tissue is very limited, and has a relatively low efficiency regeneration that is prone to somaclonal variation.

### *Protocol II: Embryo Axes Transformation System*

In order to improve the transformation efficiency of *Vicia faba* and overcome the difficulties previously reported by Böttinger et al. (2001), we developed an *Agrobacterium*-mediated transformation system based upon direct shoot organogenesis after transformation of meristematic cells derived from zygotic embryo axes to regenerate stable fertile transgenic plants. The use of embryonic axes, which were cultivated on media containing high concentrations of cytokinins in combination with low auxin concentrations, resulted in direct shooting without an intermediate callus phase. By this method the possibility of somaclonal variation is reduced to the minimal level. This transformation system was an adaptation of the protocol previously reported by Schroeder et al. (1993) in *Pisum sativum* and allowed the recovery of seed-bearing primary transformants within approximately 9 to 10 months using two faba bean cultivars Mythos and Albatross.

## *METHODOLOGY*

### *Chemicals and Culture Conditions*

The MS basal medium (Murashige and Skoog, 1962) and B5 Gamborg medium (Gamborg et al., 1968) provided by Duchefa (M0222, G0209, respectively) are generally used in our experiments, supplemented with various growth regulators and 3 percent sucrose. Media are solidified by 0.3 percent Gelrite (Roth/Germany) and adjusted to pH 5.7 prior to autoclaving for 15 minutes at 120°C. Plant growth regulators NAA, 2,4-D, and 6-benzyl amino purine (BAP) are obtained from Duchefa, and TDZ from Riedel-de-Haen. Antibiotics are obtained from Duchefa (kanamycin) and Smith Kline-Beecham (Betabactyl, a combination of ticarcillin and potassium clavunate, identical to timentin) and Combactam (Sulbactam-Na Pfizer). PPT was offered by Hoechst/Germany (now BAYER Crop Science). As culture vessels, Petri plates (9 cm) and 250 ml glass containers (diameter 8 cm, covered with glass lids) are generally used. The following abbreviations apply for the different plant media:

- TNZ: MS + 3 percent sucrose, 0.5 mg·l$^{-1}$ 2,4-D, 0.5 mg·l$^{-1}$ NAA, 0.5 mg·l$^{-1}$ TDZ.
- TNZ1: MS + 3 percent sucrose, 0.5 mg·l$^{-1}$ 2,4-D, 0.5 mg·l$^{-1}$ NAA, 0.5 mg·l$^{-1}$ TDZ, 500 mg·l$^{-1}$ Betabactyl.
- TNZ2K: MS + 3 percent sucrose, 0.5 mg·l$^{-1}$ 2, 4-D, 0.5 mg·l$^{-1}$ NAA, 0.5 mg·l$^{-1}$ TDZ, 300 mg·l$^{-1}$ Betabactyl, 100 mg·l$^{-1}$ kanamycin.

- TNZ2P: MS + 3 percent sucrose, 0.5 $mg \cdot l^{-1}$ 2,4-D, 0.5 $mg \cdot l^{-1}$ NAA, 0.5 $mg \cdot l^{-1}$ TDZ, 300 $mg \cdot l^{-1}$ Betabactyl, 2 $mg \cdot l^{-1}$ PPT.
- MTN: MS + 3 percent sucrose, 7.5 $mg \cdot l^{-1}$ TDZ, 0.75 $mg \cdot l^{-1}$ NAA.
- MB1: MS + 3 percent sucrose, 1 $mg \cdot l^{-1}$ BAP, 1 $mg \cdot l^{-1}$ gibberellic acid ($GA_3$), 100 $ml \cdot l^{-1}$ coconut milk, 100 $mg \cdot l^{-1}$ ticarcillin, and 50 $mg \cdot l^{-1}$ Combactam.
- CCM: Gamborg B5 medium + 30 $g \cdot l^{-1}$ sucrose, 0.5 $mg \cdot l^{-1}$ kinetin, and 1 $mg \cdot l^{-1}$ 2,4-D.
- SRM: MS basal salt medium supplemented with B5 vitamins (Gamborg et al., 1968), 2 $mg \cdot l^{-1}$ NAA, 2 $mg \cdot l^{-1}$ BAP, 150 $mg \cdot l^{-1}$ ticarcilline, and 100 $mg \cdot l^{-1}$ Combactam.
- SIM1: MS basal salt medium supplemented with B5 vitamins, 4.5 $mg \cdot l^{-1}$ BAP, and 2 $mg \cdot l^{-1}$ PPT.
- SIM: MS basal salt medium supplemented with B5 vitamins, 4.5 $mg \cdot l^{-1}$ BAP, 0.1 $mg \cdot l^{-1}$ NAA, 100 $mg \cdot l^{-1}$ ticarcillin, 50 $mg \cdot l^{-1}$ Combactam, and 2 $mg \cdot l^{-1}$ PPT.

Antibiotics and phosphinothricin are added to the medium after autoclaving. Cultures are kept at 21°C under cool white fluorescent light (80 $\mu mol \cdot m^{-2} \cdot s^{-1}$).

### *Plant Material*

Transformation experiments were conducted using three cultivars of faba bean: Mythos, Albatross (obtained from Norddeutsche Pflanzenzucht/W. Lemke, Hohenlieth, Germany) and Giza 2 (obtained from Agricultural Research Center, Egypt). Mature seeds are surface sterilized by immersion for 1 minute in 70 percent ethanol and for 8 minutes in sodium hypochlorite solution (4 percent active chlorine), followed by washing four or five times with sterilized tap water, and soaked overnight in sterile tap water with shaking at 90 to 95 rpm (protocol I and II). Immature pods of the faba bean cultivars at 30 to 35 days after pollination were surface sterilized with 70 percent ethanol for 3 minutes and subsequently washed with sterilized distilled water four to five times (protocol II). The sterilized seeds were germinated in the dark at 20°C on ½ concentrated MS basal medium (lacking sucrose). After approximately 10 days, internodal segments of the arising main shoot are used for cocultivation with *A. tumefaciens*. The remaining seedlings are kept in the dark, and secondary shoots arising from the cotyledonary bud during the following weeks are used as an explant source for further transformation experiments as well. Thus each seed finally provided 30 to 40 internodal segments.

## **Agrobacterium *Strain and Plasmid***

Disarmed hypervirulent, succinamopine-type *Agrobacterium tumefaciens* strain EHA 105 and EHA 101 (Hood et al., 1986, 1993) harboring various binary vectors have been used. As a plant selectable marker, T-DNA in the plasmid vector used so far contains the kanamycin resistance gene *npt*II (driven by the nopaline synthase promoter or the bi-directional TR1/2 promoter; see Böttinger et al., 2001) or the *bar* gene of *Streptomyces hygroscopicus* encoding phosphinothricin acetyltransferase (PAT) under the control of nopaline synthase promoter (Hanafy et al., 2005), which confers resistance to the herbicide PPT and its derivatives, the active component of BASTA by acetylation (Thompson et al., 1987).

## **Agrobacterium *Culture Medium***

- *Agrobacteria* are grown either in yeast extract broth (YEB) medium (Van Larebeke et al., 1977) or Luria-Bertani (LB) medium (Luria et al., 1960) containing appropriate antibiotics.
- Bacterial cultures for weekly experiments are initiated from stock plates that are stored up to four weeks at 4°C before being refreshed from long-term glycerol stocks (–70°C).
- The day before the experiment, a single bacterial colony harboring the binary vector is grown in 25 ml LB or YEB liquid medium containing appropriate antibiotics (14-16 hours) at 28°C on a shaker with 180 rpm to $OD_{650}$ of 0.8. The bacterial colonies should be from a fresh plate.
- Bacteria are harvested by centrifugation and resuspended in an equal volume of liquid TNZ medium (de novo regeneration protocol), or the bacterial culture is diluted at the rate of 1:5 with liquid CCM (embryo axis transformation system).

## *DE NOVO REGENERATION PROTOCOL*

### *Inoculation Procedure*

Faba bean cultivars Mythos and Giza 2 are used in the transformation experiments using this protocol. Epicotyls are cut into segments of 0.2 to 0.4 cm length in the bacterial suspension and incubated for 30 minutes, then transferred into 250 ml glass containers with solidified TNZ medium and cocultivated for 48 hours at 20°C in the dark.

### *Selection and Plant Regeneration*

Following cocultivation, explants are thoroughly washed in sterilized distilled water and cultured in a glass container on TNZ1 in the dark at 20°C. After 3 to 4 days for recovery without selection pressure, the explants are subsequently transferred to plastic Petri dishes containing TNZ2K or TNZ2P medium (depending on the selectable marker). The explants are subcultured every 2 weeks on fresh medium for a period of 3 to 4 months. Resistant calli reaching a diameter of approximately 5 mm are transferred to MTN medium in 250 ml glass containers. Calli are subcultured every 3 to 4 weeks on this medium for a period of more than 12 months. The appearance of shoot primordia varies between 4 and 12 months. Shoot primordia are transferred to elongation medium (MB1) and subcultured every 3 to 4 weeks.

## *EMBRYO AXIS TRANSFORMATION SYSTEM*

### *Explant Preparation and Infection*

- Place about 10 ml of *Agrobacterium* inoculum into a Petri plate (6 cm).
- Embryo axes of both mature and immature seeds are wounded by the removal of the root tips and slicing of the embryo axes to three or four segments longitudinally with a sharp razor blade wetted by the *Agrobacterium* suspension.
- Prepare 30-40 explants per *Agrobacterium* inoculum plate. The explants should be inoculated for 15-20 minutes (immature embryos) and 30 minutes (mature embryos) with occasional agitation. The explants should be covered by the inoculum.

### *Cocultivation and Recovery Period*

- After infection, transfer the explants (30-40 per plate) to solid cocultivation medium (CCM). Make sure that the explants are collected in the center of the plate.
- Wrap the plates with Parafilm and place them at 25°C in the dark for 3-4 days.
- After the cocultivation period, briefly wash the explants in sterilized distilled water.
- Place the explants on recovery medium (SRM) for 2 weeks.

- Wrap the plates with vent tape and incubate at 25°C, 18:6 photoperiod.

### *Selection and Elongation*

- After 2 weeks on SRM, transfer explants to shoot induction medium (SIM).
- Remove any dead tissues and make a horizontal fresh cut at the base of the explants.
- Transfer the explants to fresh SIM and allow tissues to be incubated at 25°C, 18:6 photoperiod, for 3-6 months.
- Subculture the tissue to fresh SIM every 2-3 weeks. At each transfer, make a fresh cut at the base of the explant.

### *Regeneration of Transgenic Plants and Micrografting*

Healthy resistant shoots more than 1 cm in length from both transformation systems are excised and grafted onto etiolated seedlings of untransformed faba bean about 7 to 10 days old (see Pickardt et al., 1995 for details).

### *Plant Acclimatization*

- After 2-3 weeks, when the shoot develops new leaves, take the plant gently from the medium and wash roots with tap water and transplant the plantlet into a small pot (6 cm diameter) containing autoclaved soil and cover with a transparent plastic bag.
- Grow the plantlets at 24°C, 18 hour photoperiod for 2 weeks. Remove the plastic bag gradually during the second week, once the plantlet grows and develop new leaves.
- Transplant the plantlet to a larger pot (20 cm diameter) and cultivate under greenhouse conditions.

### *Molecular Analysis of Herbicide-Resistant Transgenic Plants*

Expression of the selectable marker *bar* gene was confirmed by painting the leaflets of the transgenic plants with BASTA (a commercial formulation of PPT containing 200 $mg \cdot l^{-1}$ ammonium glufosinate, Hoechst Ltd.) diluted at a concentration of 300 to 400 $mg \cdot l^{-1}$ ammonium glufosinate. The opposite leaflet of each pair was marked and left untreated as a control. $T_1$, $T_2$, and $T_3$ were tested by the same method, or young plants (around 3-4 weeks after germination) were sprayed with the same BASTA solution. Resistance of leaflets or plants was scored after 7 to 10 days.

Genomic DNA was isolated from young leaf tissues using DNeasy Plant Mini Kit (Qiagen). DNA samples (25 μg) were restricted with *Hin*dIII, separated in 0.8 percent (w/v) agarose gel, and blotted onto positively charged nylon membrane (Boehringer/Germany) according to manufacturer's instructions. A 750 bp *Eco*RI/*Bam*HI fragment containing an *SFA8* sequence was excised from the transformation vector pGLsfa (pGPTV-*bar* derivative, containing *bar,* sulphur rich sunflower albumin, *SFA8* under *V. faba* legumin B promoter) and used as a probe. The latter was labeled using DIG High Prime Labeling Kit (Boehringer/Germany).

## *RESULTS*

Our methods for regeneration of transgenic plants have been based on two alternative protocols: protocol I based on de novo regeneration using TDZ (Böttinger et al., 2001), and protocol II based on direct-shoot organogenesis from meristematic cells of mature or immature embryo axes (Hanafy et al., 2005). In both transformation systems, the control experiments showed that PPT (2 $mg \cdot l^{-1}$) totally suppressed callus development from wild-type faba bean epicotyl segments cultured on TNZ or embryo axes cultured on SIM1. All the explants died. With reference to the de novo regeneration protocol, in a series of transformation experiments, resistant callus started proliferating after about 1 to 2 months on the surface of a number of explants on TNZ2P medium. Within 3 to 4 months after culturing the explants on TNZ2P medium, 4.30 to 31.59 percent of the explants produced resistant calli (Figure 19.1). PPT-resistant calli with a diameter of about 5 to 10 mm were transferred to the MTN medium to increase the callus viability to regenerate (Figure 19.1B). Shoot regeneration occurred after 6 to12 months on MTN medium (Figure 19.1C). Shoot primordia were transferred to shoot elongation medium MB1.

Because of the very low rooting percentage of the regenerates, shoots reaching a suitable size (within 2 to 4 months) were micrografted onto nontransgenic rootstocks of *V. faba* to recover a whole plant. Afterward, the plants were transferred to soil for acclimatization. Regenerated plants were transferred to the greenhouse for further plant development and production of $T_1$ seeds. Some regenerated clones showed morphological abnormalities, such as dwarfing and formation of abnormal flowers and subsequently ab-

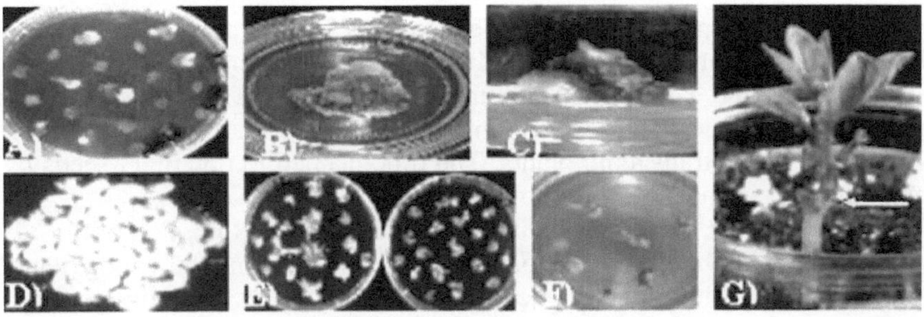


FIGURE 19.1. (A) Initiation of resistant callus from stem explants under selection pressure of 2 mg·l$^{-1}$ PPT on TNZ2P medium. (B) Callus proliferation on MTN medium. (C) Shoot regeneration on MTN medium. (D) Explant segments derived from embryonic axes. (E) Multiple shoot regeneration under selection pressure (left) all the control explants (WT) were dead (right). (F) Further selection between PPT-resistant and susceptible regenerated shoots on medium containing 2 mg·l$^{-1}$ PPT. (G) Recovering the transgenic plant by micrografting. (See also color gallery).

normal (or no) pods. Some plants showed narrow leaves and very weak stems with weak apical dominance. Only four clones of two cultivars (Mytos and Giza 2) produced normal flowers and pods with seeds. The time needed to obtain $T_1$ seeds by this process is about 16 to 24 months.

On the other hand, the feasibility of the transformation strategy based on embryo axis transformation was initially evaluated by monitoring the number of regenerated shoots from the embryo axis explants cultured on medium with high BAP concentration (4.5 mg·l$^{-1}$) where routinely four to five shoots regenerated from each explant. In a series of transformation experiments, the explants (immature or mature embryonic axes) were inoculated with *Agrobacterium* strain EHA105/pGlsfa (harboring *SFA8* and *bar* genes) alone or cotransformed with EHA101/pAN109, which carried a mutated *lysC* gene from *E. coli* and *npt*II (Böttinger et al., 2001; Figure 19.1D). After 3 to 4 weeks of culturing on selective medium, all control explants had died. On the other hand, the transformed explants started regenerating (via organogenesis) and about three to four shoots appeared from each explant on SIM (Figure 19.1E). The shoots selected for 4 to 6 months (Figure 19.1F) were grafted in vitro and finally transferred to the greenhouse to set seeds (Figure 19.1G). A total of seven stable independent transformants (containing *SFA8* linked to *bar* from cv. Mythos and Albatross) have been recovered.

Transformation frequencies ranged from 0.15 to 2.0 percent. In cotransformation experiments, transformed plants ($T_1$) were screened for the presence of both the T-DNAs by polymerase chain reaction analysis. It was found that only the T-DNA encoding the selectable marker PPT has been integrated in the transgenic plants. The time needed to obtain $T_1$ seeds by this protocol is about 9 to 10 months.

The transgenic plants were analyzed by testing the expression of the *bar* gene in the greenhouse by leaf-painting assay or spraying the young plants with the same dilution of the herbicide. Within 2 days necrotic spots appeared on the untransformed leaves (Figure 19.2). Ten days after BASTA application, the treated transgenic plants and leaflets showed complete tolerance, in contrast to nontransformed plants, which showed leaflets that were completely necrotic. Southern blot analyses were performed to confirm the integration of the transgenes. DNA digestion was performed with *Hin*dIII and the fragments were subjected to Southern hybridization analysis using a DIG-labeled 750 bp *SFA8* fragment. Line ME1/2/1 contained a single copy of integrated T-DNA, whereas line G2/1/2 (recovered from de novo regeneration protocol) harbored two copies of T-DNA; no hybridization signal was observed in the wild-type (WT) (Figure 19.2B).

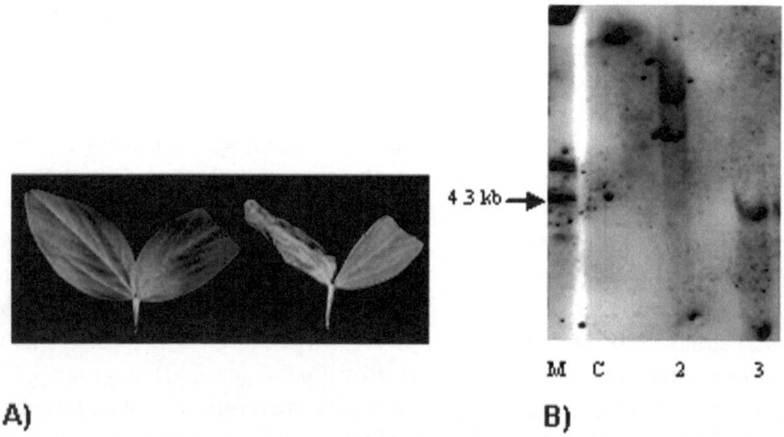


FIGURE 19.2. (A) Herbicide leaf painting test showing the resistance of transgenic leaf to BASTA application (left). (B) Southern blot analysis of DNA isolated from transformed plants of faba bean. *M,* DNA marker size; *C,* Control plant; *Lane 2* and *3, Hin*dIII digested total genomic DNA from $T_1$ progenies of clones G2/1/2 cv. Giza 2 (Lane 2) and ME1/2/1 cv. Mythos (Lane 3). (See also color gallery.)

## *DE NOVO REGENERATION VERSUS EMBRYO AXIS TRANSFORMATION PROTOCOLS*

The production of transgenic plants requires an efficient regeneration system. Two successful systems in faba bean transformation have been developed. However, the regeneration of transgenic faba bean is still labor intensive for both transformation systems and faba bean is considered a recalcitrant species. We have recovered different transgenic clones from three cultivars, Mythos, Albatross, and Giza 2. Each transformation system has its advantages and drawbacks and the preference for a system is dependent on the individual scientists. The major advantage in using de novo regeneration protocol is regeneration of nonchimeric plants.

This process, however, is more time consuming and the main constraints are poor regeneration ability following a callus phase, reduced fertility, and the high percentage of phenotypic abnormalities in regenerated plants. This can be ascribed to the fact that explants and callus cells of *Vicia faba* tend to produce high amounts of phenolic compounds, resulting in subsequent toxicity to the cultured tissue (Bieri et al., 1984; Selva et al., 1989), which is possibly due to the long cultivation time in vitro (around 7-16 months; McClintock, 1984). The recovery of seed-producing putative transformed plants under these circumstances took about 16 to 24 months, which is a considerable period. Transformation strategies that minimize the in vitro culture period and avoid the callus phase would therefore be advantageous in this case. Moreover, transformation of pre-excited meristematic cells on the embryo axes is much simpler. The major success in legume transformation was achieved by methods based on transformation of the preexisting meristems on the embryo axes, cotyledonary nodes, shoot tips, or nodal explants. This approach has been successfully applied in combination with *Agrobacterium* in different grain legumes such as pea (Schroeder et al., 1993; Bean et al., 1997), chickpea (Krishnamurthy et al., 2000), *Lupinus angustifolius* (Pigeaire et al., 1997), and soybean (Yan et al., 2000; Olhoft et al., 2003). In combination with direct gene transfer systems, transformation success was reported in soybean (McCabe et al., 1988; Sato et al., 1993), peanut (Brar et al., 1994), and *Phaseolus vulgaris* (Aragão et al., 1996).

An *Agrobacterium*-mediated genetic transformation system using pre-excited meristematic cells on the embryo axes is an effective method for the production of transgenic faba bean (cv. Mythos and Albatross), which provides a useful strategy for efficient insertion of economically useful genes into faba bean cultivars. We have demonstrated that the embryo axes of faba bean are competent for production and multiplication of shoots and inherited transformation. The recent achievements also emphasize the benefits of improvements in faba bean biotechnology. Commercially defined faba bean production is limited not only by diseases, pests, and abiotic stress but also by an imbalance in nutritional quality of the seed protein, all of which cannot be solved by traditional breeding programs. However, the exploitation of potentially useful transgenes is likely to overcome these difficulties, resulting in rapid commercial improvement of faba bean cultivars.

## REFERENCES

Aragãno, F.J.L., L.M.G. Barros, A.C.M. Brasileiro, S.G. Ribeiro, F.D. Smith, J.C. Sanford, J.C. Faria, and E.L. Rech (1996). Inheritance of foreign genes in transgenic bean (*Phaseolus vulgaris* L.) co-transformed via particle bombardment. *Theoretical Applied Genetics* 93:142-150.

Bean, S.J., P.S. Gooding, P.M. Mullineaux, and D.R. Davies (1997). A simple system for pea transformation. *Plant Cell Reports* 16:513-519.

Bieri, V., J. Schmid, and E.R. Keller (1984). Shoot tip culture in Vicia faba L. In Efficiency in plant breeding: Proceedings of the 10th congress of the European association for research on plant breeding, W. Lange, A.C. Zeven, and N.F. Hogenboom (eds.). Wageningen, Netherlands: EUCARPIA, p. 295.

Böttinger, P., A. Steinmetz, O. Schieder, and T. Pickardt (2001). *Agrobacterium* mediated transformation of *Vicia faba. Molecular Breeding* 8:243-254.

Brar, G.S., B.A. Cohen, C.L. Vick, and G.W. Johnson (1994). Recovery of transgenic peanut (*Arachis hypogaea* L.) plants from elite cultivars utilizing ACCELL® technology. *Plant Journal* 5:745-753.

Gamborg, O.L., R. Miller, and K. Ojima (1968). Nutrient requirements of suspension cultures of soybean root cells. *Experimental Cell Research* 50:148-151.

Hanafy, M., T. Pickardt, H. Kiesecker, and H.J. Jacobsen (2005). *Agrobacterium*-mediated transformation of faba bean (*Vicia faba* L.) using embryo axes. *Euphytica* 142:227-236.

Hanounik, S.B., G.J. Jellis, and M.M. Hussein (1993). Screening for disease resistance in faba bean. In *Breeding for stress tolerance in cool-season food legumes,* K.B. Singh, and M.C. Saxena (eds.). Chichester: John Wiley, pp. 97-106.

Hood, E.E., S.B. Gelvin, L.S. Melchers, and A. Hoekema (1993). New *Agrobacterium* helper plasmids for gene transfer to plants. *Transgenic Research* 2:208-218.

Hood, E.E., G.L. Helmer, R.T. Fraley, and M.D. Chilton (1986). The hypervirulence of *Agrobacterium tumefaciens* A281 is encoded in a region of pTiBo542 outside of the T-DNA. *Journal of Bacteriology* 168:1291-1301.

Jelenic, S., P.T. Mitrikeski, D. Papes, and S. Jelaska (2000). *Agrobacterium*-mediated transformation of broad bean *Vicia faba* L. *Food Technology and Biotechnology* 38:167-172.

Krishnamurthy, K.V., K. Suhasini, A.P. Sagare, M. Meixner, A. de Kathen, T. Pickardt, and O. Schieder (2000). *Agrobacterium* mediated transformation of chickpea (*Cicer arietinum* L.) embryo axes. *Plant Cell Reports* 19:235-240.

Luria, S.E., J.N. Adam, and R.C. Teng (1960). Transduction of lactose utilizing ability among strains of *Escherichia coli* and *Shigella dysenteriae* and the properties of the transducing phage particles. *Virology* 12:348-390.

McCabe, D., W. Swain, B. Martinell, and P. Christou (1988). Stable transformation of soybean (*Glycine max*) by particle acceleration. *Nature Bio/Technology* 6:923-926.

McClintock, B. (1984). The significance of responses of the genome to challenge. *Science* 226:792-801.

Murashige, T. and F. Skoog (1962). A revised medium for rapid growth and bioassays with tobacco tissue cultures. *Physiologia Plantarum* 15:473-497.

Olhoft, P.M., L.E. Flagel, M. Christopher, and D.A. Somers (2003). Efficient soybean transformation using hygromycin B selection in the cotyledonary-node method. *Planta* 216:723-735.

Pickardt, T., I. Saalbach, D.Waddell, M. Meixner, K. Müntz and O. Schieder (1995). Seed specific expression of the 2S albumin gene from Brazil nut (*Bertholletia excelsa*) in transgenic *Vicia narbonensis. Molecular Breeding* 1:295-301.

Pigeaire, A., D. Abernethy, P.M. Smith, K. Simpson, N. Fletcher, C.-Y. Lu , C.A. Atkins, and E. Cornish (1997). Transformation of a grain legume (*Lupinus angustifolius* L.) via *Agrobacterium tumefaciens*-mediated gene transfer to shoot apices. *Molecular Breeding* 3:341-349.

Quandt, H.J., A. Pühler, and I. Broer (1993). Transgenic root nodules of *Vicia hirsuta*: A fast and efficient system for the study of gene expression in indeterminate-type nodules. *Molecular Plant-Microbe Interactions* 6:699-706.

Robertson, L.D. and M.C. Saxena (1993). Problems and prospects of stress resistance breeding in faba bean. In *Breeding for stress tolerance in cool-season food legumes,* K.B. Singh and M.C. Saxena (eds.). Chichester: John Wiley, pp. 37-50.

Sato, S., C. Newell, K. Kolacz, L. Tredo, J. Finer, and M. Hinchee (1993). Stable transformation via particle bombardment in two different soybean regeneration systems. *Plant Cell Reports* 12:408-413.

Schiemann, J. and G. Eisenreich (1989). Transformation of field bean *Vicia faba* L. cells expression of a chimeric gene in cultured hairy roots and root-derived callus. *Biochemie Physiologie Pflanzen* 185:135-140.

Schroeder, H.E., A.H. Schotz, T. Wardley-Richardson, D. Spencer, and T.J.V. Higgins (1993). Transformation and regeneration of two cultivars of pea (*Pisum sativum* L). *Plant Physiology* 101:751-757.

Selva, E., M. Stouffs, and M. Briquet (1989). In vitro propagation of *Vicia faba* L. by micro-cutting and multiple shoot induction. *Plant Cell Tissue and Organ Culture* 18: 167-179.

Tegeder, M., D. Gebhardt, O. Schieder, and T. Pickardt (1995). Thidiazuron-induced plant regeneration from protoplasts of *Vicia faba* cv. Mythos. *Plant Cell Reports* 15:164-169.

Thompson, C.J., N. Rao Movva, R. Tizard, R. Crameri, E.J. Davies, M. Lauwereys, and J. Bottermann (1987). Characterization of the herbicide-resistance gene *bar* from *Streptomyces hygroscopicus*. *EMBO Journal* 6:2519-2523.

Van Larebeke, N., C. Genetello, J. Hernalsteens, A. De Picker, I. Zaenen, E. Messens, M. Van Montagu, and J. Schell (1977). Transfer of Ti-plasmids between *Agrobacterium* strains by mobilization with conjugative plasmid RP4. *Molecular and General Genetics* 152:119-124.

Yan, B., M.S. Srinivasa Reddy, G.B. Collins, and R.D. Dinkins (2000). *Agrobacterium tumefaciens*-mediated transformation of soybean (*Glycine max* L. Merrill) using immature zygotic cotyledon explants. *Plant Cell Reports* 19: 1090-1097.

FIGURE 3.1. An alfalfa callus producing green embryos on B5H with 25 $mg \cdot l^{-1}$ kanamycin and 500 $mg \cdot l^{-1}$ cefotaxime.

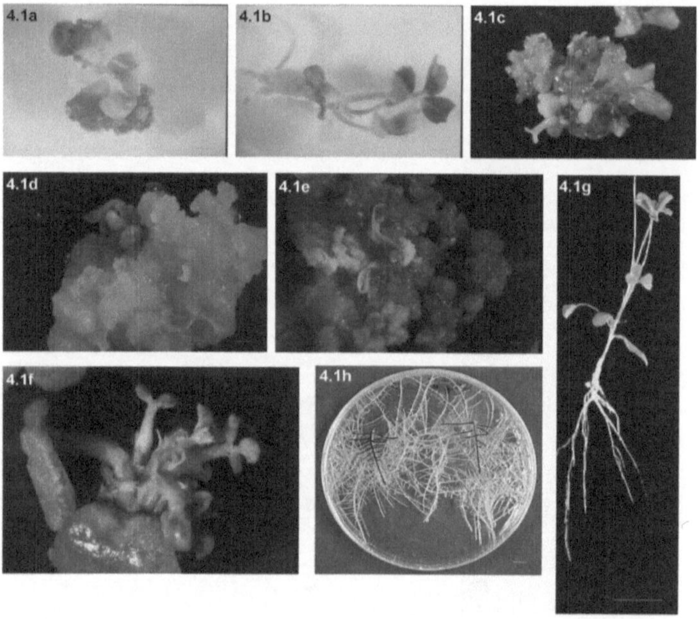


FIGURE 4.1. Transformation and regeneration of *M. truncatula.* a: Green kanamycin-resistant calli formed on a 2HA foliole following transformation. b: Regenerated plantlet. c: Secondary embryogenesis on embryo-development medium. d-e: Comparison of embryogenic calli from different Jemalong lines. Note that embryogenic calli are brown. The A17 line with less than five embryos per callus. e: Callus derived from the highly embryogenic genotype 2HA is covered in globular embryos. f: Shoot formation through organogenesis on a half-cotyledon of Jemalong A17. g: "Composite plant" generated following transformation with *A. rhizogenes*, with Ri T-DNA transformed roots developing from the site of radicle sectioning. Note that transformed roots present a normal phenotype. h: In vitro propagation of excised Ri T-DNA transformed roots. *Source:* Reprinted from Chabaud et al., 1996, with permission. (Bar represents 1 mm in a, b, c, d, e, f, and 1 cm in g, h.)

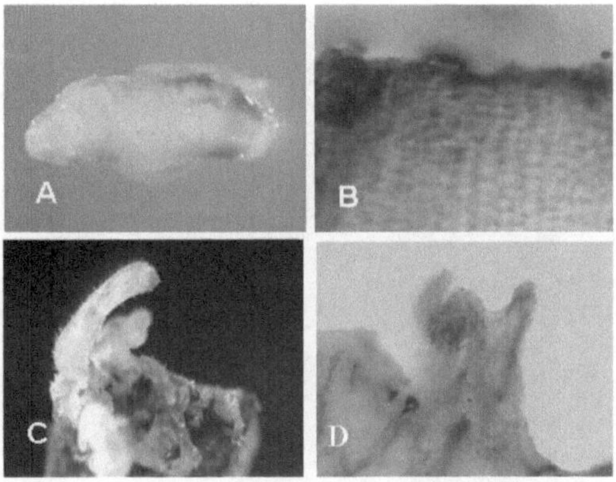


FIGURE 9.3. Histochemical GUS assay of cotyledonary node explants co-cultivated with *A. tumefaciens* harboring pBI121; A, showing expression of GUS foci; B, expression of *uid*A gene in excised cells of CNs; C-D, GUS foci in developing new shoots from CNs.

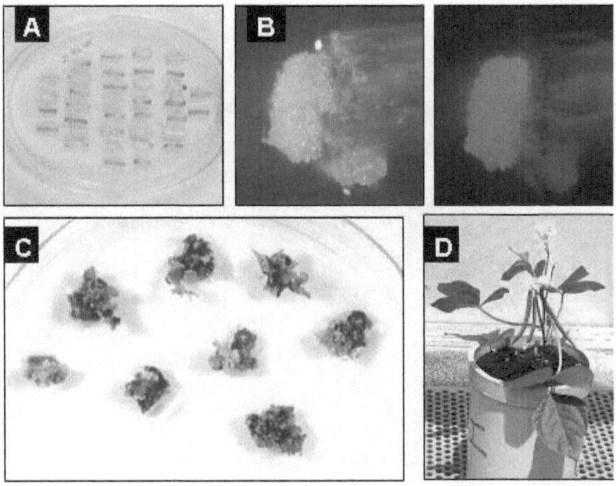


FIGURE 11.1.Production of fertile transgenic azuki bean plants. (A) Cocultivation of epicotyl explants; (B) induction of transformed organogenic calli (left panel), expression of *gfp* by green fluorescence (right panel); (C) shoot induction from organogenic calli; (D) a fertile transgenic azuki bean plant grown in soil and in a glasshouse.

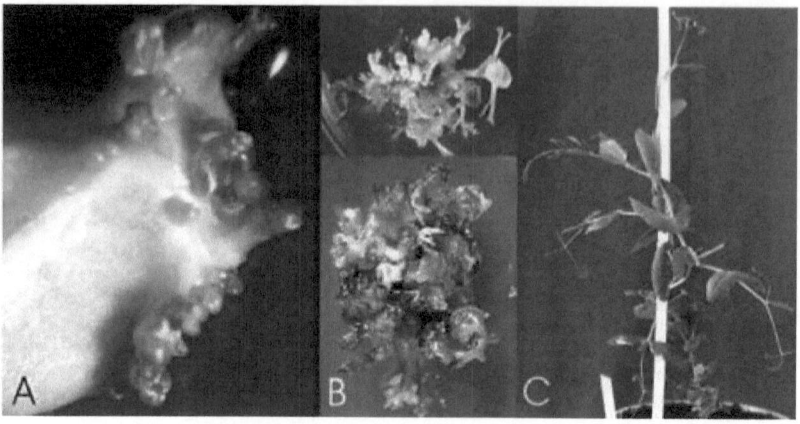


FIGURE 13.1. In vitro pea regeneration from immature cotyledons. (A) Initiation of buds on the cut surface of the cotyledon after 5 weeks of culture. (B) Multiple shoot formation after 8 weeks of culture in different cultivars. (C) In vitro plant growing in soil after 12 weeks of culture.

FIGURE 13.2. Pea shoot and plant selection after *Agrobacterium*-mediated transformation. (A) Shoot formation on the medium containing 2 $mg \cdot l^{-1}$ phosphinothricin (top) and plant growth on the medium containing 4 $mg \cdot l^{-1}$ phosphinothricin, beside visible nongrowing, browning shoots. (B) Flowering and seed-setting transgenic plant. (C) Segregation of resistant, transgenic plant (left) and susceptible plant (right) in $T_1$ generation.

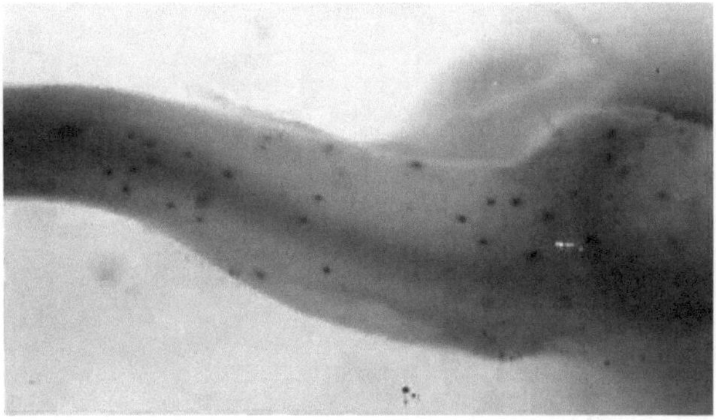

FIGURE 14.1. Histochemical GUS staining of radicules after 2 days of microprojectile bombardment.

FIGURE 15.3. Pea plants challenged with PEMV (25 days postinoculation). Left, uninoculated plant; middle, inoculated transgenic; right, inoculated control.

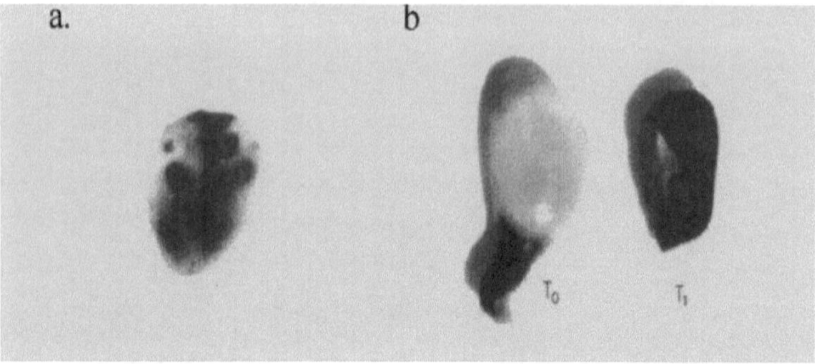


FIGURE 18.1. GUS histochemical analysis: a, GUS expression in the leaf of primary transformant; b, GUS expression in the embryo of $T_0$ and $T_1$ plants.

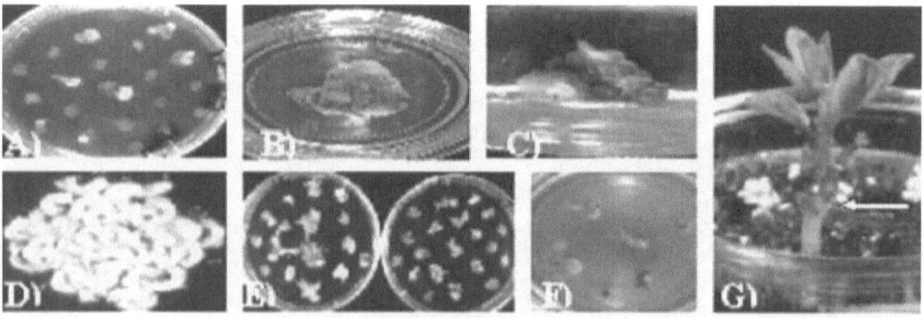


FIGURE 19.1. (A) Initiation of resistant callus from stem explants under selection pressure of 2 mg·l$^{-1}$ PPT on TNZ2P medium. (B) Callus proliferation on MTN medium. (C) Shoot regeneration on MTN medium. (D) Explant segments derived from embryonic axes. (E) Multiple shoot regeneration under selection pressure (left) all the control explants (WT) were dead (right). (F) Further selection between PPT-resistant and susceptible regenerated shoots on medium containing 2 mg·l$^{-1}$ PPT. (G) Recovering the transgenic plant by micrografting.

A) B)

FIGURE 19.2. (A) Herbicide leaf painting test showing the resistance of transgenic leaf to BASTA application (left). (B) Southern blot analysis of DNA isolated from transformed plants of faba bean. *M,* DNA marker size; *C,* Control plant; *Lane 2* and *3, Hin*dIII digested total genomic DNA from $T_1$ progenies of clones G2/1/2 cv. Giza 2 (Lane 2) and ME1/2/1 cv. Mythos (Lane 3).

FIGURE 20.1. Direct regeneration from hairy roots in *L. corniculatus.* Hairy roots profusely growing in liquid culture regenerate through shoot organogenesis. The insert shows details of a single regenerating hairy root. Courtesy of Drs. P. Morris and M. P. Robbins of IGER, Aberystwyth.

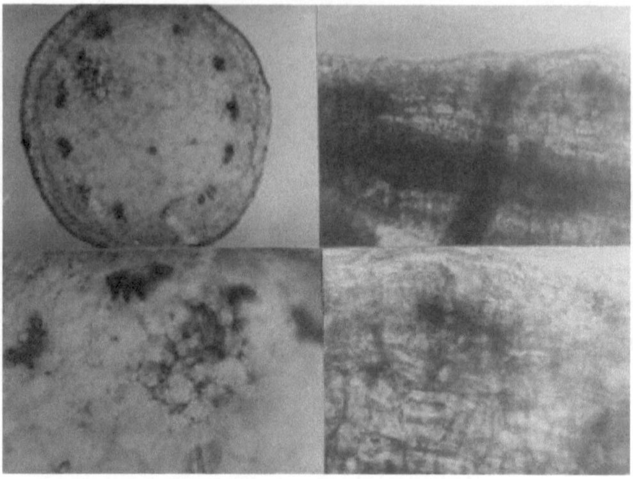

FIGURE 21.2. Histochemical GUS assay of the transformants of *L. leucocephala* cv. K-29 for analysis of *gus* gene in stem sections (left panel) and leaves (right panel) indicating expression of the gene (blue sections). The lower panel is a magnified view of the upper panel.

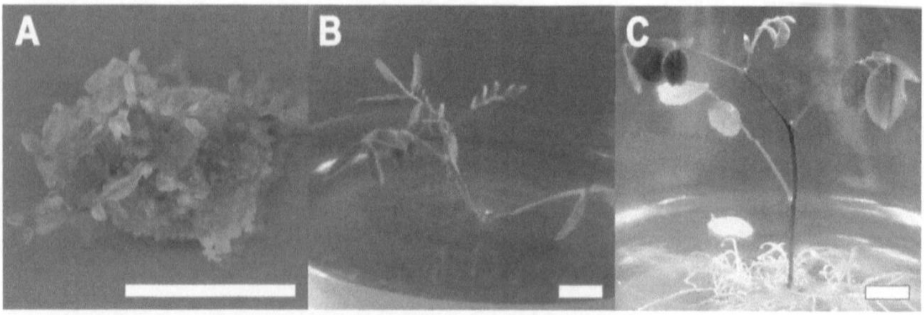


FIGURE 22.1. Regeneration of transgenic *R. pseudoacacia*. (A) Shoot organogenesis from transgenic callus. (B) A transgenic shoot. (C) A transgenic plantlet. Bars: 1 cm, approximately 24 percent (Table 22.1).

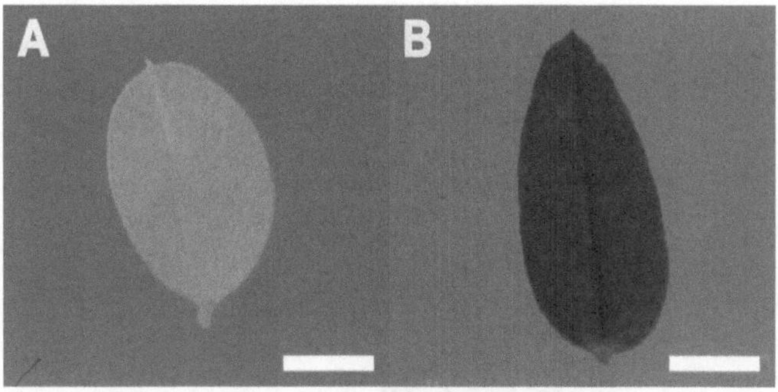


FIGURE 22.2. Histochemical assay of GUS activity in transgenic *R. pseudoacacia*. A control leaf (A) and a leaf of randomly selected transgenic plants (B) were subjected to histochemical staining with 5-bromo-4-chloro-3-indolyl glucuronide (X-Gluc). Bars: 1 cm.

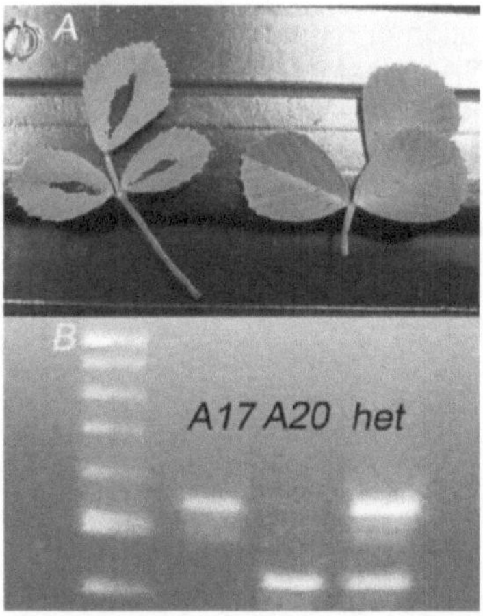


FIGURE 23.1. Morphological and molecular markers in *Medicago truncatula*. (A) Leaves of the two ecotypes most commonly used for mapping. Left, A17 ecotype with chevron markings on leaves; right, A20 ecotype with speckles on the leaves. (B) Agarose gel electrophoresis of a cleaved amplified polymorphic sequence marker for use in mapping in an A17 × A20 segregating population. The PCR product contains a restriction site in the A20 ecotype that is absent in A17, allowing one to score the codominant marker in each plant. (A small fragment resulting from the digestion in A20 and the heterozygous plant is not shown on the gel picture.)

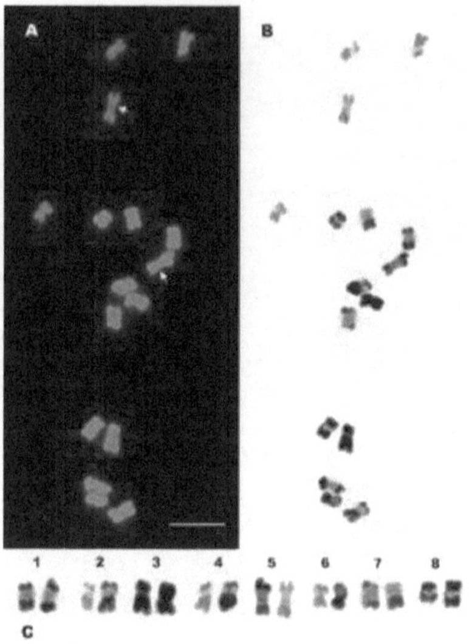

FIGURE 25.1. Identification of metaphase chromosomes of Jemalong A17 by using FISH with 5S rDNA (red signal) and *MtR1* (green signal). (A) Metaphase chromosomes are stained with DAPI. (B) A merged image of chromosomes and the FISH signals of 5S rDNA and *MtR1*. (C) Karyotype of metaphase chromosomes. Chromosomes are digitally sorted out of metaphase complement, ordered, and numbered according to their corresponding linkage groups. Arrows indicate NORs. Bar = 5 μm.

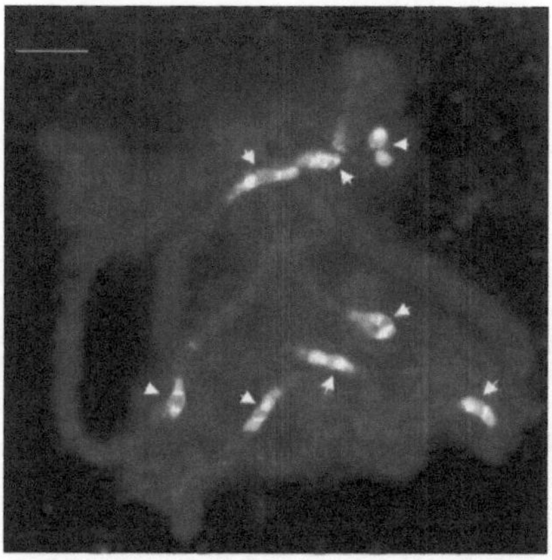

FIGURE 25.2. Localization of *MtR2* and *MtR3* satellites on *M. truncatula* Jemalong A17 pachytene chromosomes by FISH. *MtR2* (red signal) is located in pericentromeric heterochromatin of all chromosomes. *MtR3* signals (green) coincide with primary constrictions (indicated by arrowheads) of all chromosomes. Chromosomes are stained with DAPI. Bar = 5 μm.

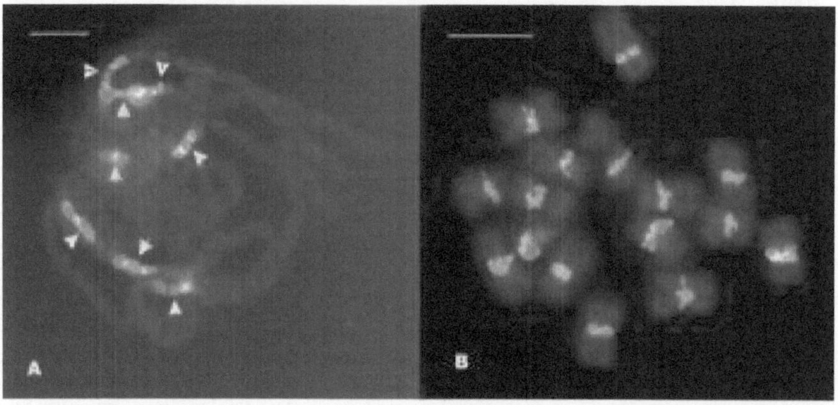


FIGURE 25.3. Localization of *MtR1*, *MtR2,* and *MtR3* satellites on chromosomes of *M. truncatula* DZA315.16 by FISH. (A) Pachytene complement. *MtR3* signals (red) are located at centromeres (indicated by arrowheads) and *MtR2* signals (green) appear at pericentromeric heterochromatin of all chromosomes. (B) Metaphase chromosomes. *MtR1* signals (green) are located in pericentromeric heterochromatin of four chromosomes and *MtR2* (red signals) are at pericentromeric heterochromatin of all chromosomes. Chromosomes are stained with DAPI. Bar = 5 μm.

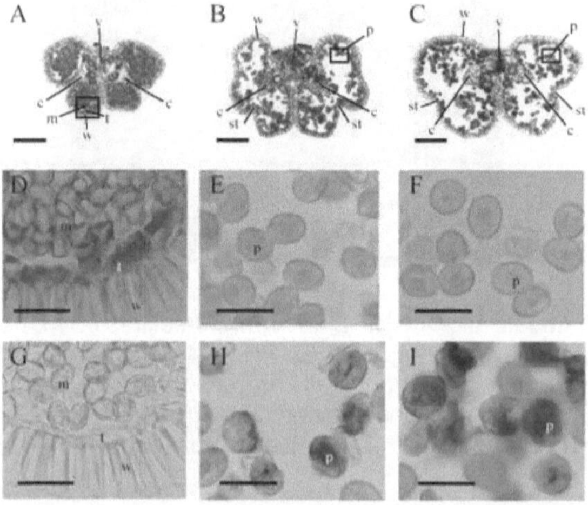


FIGURE 31.1. Cross-section of the anther at different developmental stages and *in situ* localization of anther-specific genes during anther development of *L. japonicus* is shown in (A) to (C). Each cross-section of anther was stained with toluidine blue. (A) Anther at stage 1 contained uninucleate microspores and tapetal cells. (B) Anther at stages 1-2 contained binucleate pollen grains and tapetal cells. Tapetal cells started degenerating in this stage. (C) Anther at stage 2 contained mature pollen grains. Tapetal cells disappeared during this stage just before dehiscence. c, connective; m, microspore; p, pollen grain; st, stomium; v, vascular bundle; w, anther wall. Dig-labeled antisense RNA probe was hybridized to a cross-section of the anther tissues at different developmental stages. A cross-section of the anther locule at stages 1, 1-2, and 2 are shown in (D) to (I). The gene-encoding lipid transfer protein (LjImfb-R39) was specifically expressed in immature anther tapetum (D), and not in mature pollen grains (E, F). In the case of gene-encoding late embryogenesis abundant protein (LjMfb-U92), a hybridization signal was specifically detected in pollen grains (H, I), but not in anther tapetum (G). Bar = 100 µm.

# Chapter 20

# Genetic Transformation of *Lotus* Species

Francesco Damiani
Francesco Paolocci
Sergio Arcioni

## *INTRODUCTION*

Over the last 30 years, the study of biology has acquired a very fast pace. Goals that were unthinkable 50 years ago have been reached and now serve as a basis for novel accomplishments and suggestions. The whole genome sequencing of many species, including humans, and the possibility of reproducing organs and individuals from a few or single cells are advancements that touch our sensibility and upset our conscience but surely offer perspectives for improving the quality of life and human health. With respect to the plant kingdom, the progress is even more appealing because, aside from the innovative perspectives, the ethical concern about their exploitation is limited compared to animals.

Genetic transformation is the most controversial application of novel biology to plants, but also the one with the highest potential for reshaping the agricultural scenario. Many species can be modified in their genetic structure: the insertion in their genomes of novel heterologous sequences makes it possible to introduce new traits or inhibit unwanted biosynthetic pathways. The final application of genetic transformation is the production of transgenic crops, but the power of this methodology is such that it has been a tool to better understand molecular and physiological mechanisms underlying the basic aspects of plant growth and production. The ability of a species to integrate foreign genes into its genome was at first correlated with the capability to regenerate plants from undifferentiated and differentiated tissues, and for this reason *Lotus* spp. have been fully exploited. In fact, several species belonging to this genus provide, to a moderate agricultural use, several useful features so that *Lotus* has been regarded as a model system for several studies such as those relating to plant-*Rhizobium* symbiosis, biosynthesis of secondary metabolites like condensed tannins, and modification of protein composition in forage. Since the early 1990s, a species of the genus, *L. japonicus,* has been proposed as a model for genomic studies (Handberg and Stougaard, 1992).

## *THE GENUS* LOTUS

The genus contains 44 species that show a high polymorphism for several aspects: the chromosome base number ranging from 5 to 7, the ploidy level (diploids and tetraploids), the mode of pollination from selfing to autoincompatibility, the attitude to agamic reproduction, and finally the ability to accumulate secondary metabolites; some species show a wide polymorphism for the level and the tissue of condensed tannins accumulation.

The agronomic relevance of the genus is limited; however, while some species have no productive advantage, others are quite interesting as producers of high-quality nonbloating forage. In comparison with its principal alternative crops, alfalfa and clover, *Lotus* has the advantage of being pastured, but for hay production and persistence, it is largely inferior. However, its adaptation to some stress conditions such as flooding, acidic and low-fertility soils, and high salinity helps some *Lotus* spp. replace clover and alfalfa as forage producers and for reclamation and revegetation of land areas.

Three species are utilized for agricultural application: *L. corniculatus, L. glaber* (syn. = *tenuis*), and *L. uliginosus* (= *pedunculatus*). *L. corniculatus* is the most important one and several varieties have been bred in North America, Australia, and Europe, from where it originated (Grant, 2004). The two other species have their main utility in restricted areas where particular edaphic conditions occur, such as *L. glaber* in the Pampa Salada Deprimida in the province of Buenos Aires (Locatelli et al., 1997) or *L. uliginosus* in fragipan soils (Kaiser and Heath, 1990). *L. alpinus* may have some role in the restoration of high-altitude environments above the timberline (Urbanska, 1994).

## *IN VITRO MANIPULATION OF* LOTUS *SPECIES*

Although not very many species of the genus have agricultural relevance, this genus is one of the most utilized for developing novel technologies. The responsiveness to in vitro manipulation, the small genome size, and its relevance as donor of useful traits to important forage crops made it the target for many studies developing biotechnological methods. *L. corniculatus,* being the most cultivated species, is also the species receiving maximum attention. When novel technologies began to be developed in plants, *L. corniculatus* was one of the first species to work with, and this has resulted in positive response to callus production and plant regeneration. As a matter of fact, callus-derived plants were obtained from several varieties: Viking, Empire, B354 (Nizeki and Grant, 1971), Leo (Swanson and Tomes, 1980), Franco (Mariotti et al., 1984), and Rodeo (Petit et al., 1987). Such regenerants allowed studies on somaclonal variation (Damiani et al., 1985, 1990; Pezzotti et al., 1985), application of in vitro selection methods (Swanson and Tomes, 1980; Orshinsky and Tomes, 1984), and development of somatic hybridization procedures to overcome interspecific sexual barriers such as the ploidy level (Aziz et al., 1990) or to transfer useful agronomic traits from wild *Lotus* species to cultivated ones (Wright et al., 1987).

Attempts were also made to utilize somatic hybridization and *Lotus* spp. as a donor partner in intergeneric combinations aimed at transferring the genes for the accumulation of condensed tannins in leaves and roots of more important forage species. However, both symmetric and asymmetric hybridizations did not succeed in these attempts. To carry out such experiments, minor species of the genus such as *L. pedunculatus* (Pupilli et al., 1990), *L. conimbricensis* (Wright et al., 1987), and *L. angustissimus* (Nenz et al., 1996) received the attention of plant biotechnologists. Since the early 1980s, *Lotus* spp. were also investigated as model systems in molecular biology experiments and mainly, but not only, focused on the process of nitrogen fixation (Stougaard et al., 1986; Forde et al., 1989). Handberg and Stougaard (1992) proposed a minor species of the genus, *Lotus japonicus,* as a model system for classical and molecular genetics studies. In fact, this species satisfies several needs of scientists: small genome size, ploidy level ($2n = 2x$), reproduction through selfing, ease in handling flowers, and a short cycle from seeding to seed set. Kawaguchi et al. (2005) proposed *L. burttii* as a crossing partner of *L. japonicus* cv. Gifu for increasing the number of codominant markers in molecular biology studies.

## *GENETIC TRANSFORMATION OF* LOTUS *SPECIES*

The ability of *Agrobacterium* to infect dicotyledonous plants was largely exploited in transforming *Lotus* species. Similarly, the ability of *A. rhizo-*

*genes* to produce transgenic differentiated tissues within a few days from transformation was utilized in analyzing the suitability of anthocyanin regulators as novel reporter genes in several *Lotus* spp. (Damiani et al., 1998). It was also utilized in producing kanamycin-resistant calli to be utilized as a donor of selectable protoplasts to be involved in somatic hybridization experiments (Aziz et al., 1990). The utilization of hairy root transformation is highly recommended, when an efficient selectable reporter is not available, as in the case of alfalfa, where the most common selection systems such as hygromycin, kanamycin, and phosphinothricin are not totally effective. However, this is not the case in *Lotus* spp., which are highly sensitive to kanamycin (Handberg and Stougaard, 1992; Nenz et al., 1996), hygromycin (Handberg and Stougaard, 1992; Damiani et al., 1993), and phosphinothricin (Lohar et al., 2001). However, the hairy root transformation system is still utilized because of the possibility of identifying independent transformation events and the simplicity and rapidity of obtaining transgenic plants from transgenic roots via direct regeneration (Robbins et al., 2003). Such a method, described in detail in a subsequent section of this chapter, works with *Lotus corniculatus,* but not with the other *Lotus* species where, also in the case of hairy root transformation, plant regeneration occurs after callus dedifferentiation and hormone treatment (Damiani et al., 1993). Nevertheless, transgenic plants were obtained through hairy root transformation in *Lotus glaber* (Damiani et al., 1993) and *L. angustissimus* (Nenz et al., 1996).

Although transformed plants obtained from hairy roots do not show any alteration in the nitrogenase levels and the ability to fix nitrogen (Morris et al., 1999), they were utilized in studying the genetics and physiology of nitrogen symbiotic fixation, hairy root transformed plants show an altered phenotype: wrinkled leaves, shorter internodes, low fertility, shallow root apparatus, and reduced persistence (Damiani et al., 1993). Such traits may be useful for some agricultural purposes, but modifications induced by *rol* genes may hide the effect of the transgene or of the tagging elements. Therefore, for some applications, disarmed *A. tumefaciens* may be more effective. The common strains of *A. tumefaciens* C58 and LBA 4404 proved effective in producing transgenic calli, and plant regeneration occurred easily from *L. corniculatus* cultivars (Webb, 1986; Webb et al., 1996), quite independently from their origin, but not in the other species of major agronomic interest, *L. glaber*. Actually, only *L. corniculatus* and *japonicus* have been transformed with *A. tumefaciens*. Difficulties occur in transforming *L. glaber* and, in fact, apart from the report of Damiani et al. (1993) no other transformed plants of such species have been reported. Probably in this species, unlike *L. corniculatus,* plant regeneration ability is limited to fewer genotypes of Italian origin.

The increasing relevance of genomic studies calls for an increased exploitation of model systems to, in turn, fully exploit the new molecular and genetic boundaries. *L. japonicus* is, together with *M. truncatula,* the model for legumes and therefore for the study of symbiotic nitrogen fixation. For this reason, experiments of transformation of such species are growing exponentially since the pioneer work of Handberg and Stougaard (1992). A model species for genomics must satisfy the following requirements: high efficiency of transformation, the absence of plant modifications induced by the transforming vector, the absence of somaclonal variation, and the production of single integration events. Growth conditions for plantlets, induction of hairy roots, and nodulation of composite plants were optimized for large-scale screening in Petri dishes (Stiller et al., 1997). *A. tumefaciens* hypocotyl transformation protocols were proposed by Handberg and Stougaard (1992) and Stiller et al. (1997) in order to avoid lateral effects due to hairy root transformation. Lombari et al. (2003) proposed a different method based on the transformation of roots in order to reduce the time of culture and consequently the undesired somaclonal variation, and increase the single integration events. Further, the improvement of transformation efficiency was obtained by culturing different explants in the presence of the growth regulator thidiazuron, which is able to induce somatic embryogenesis on primary regenerants (Barbulova et al., 2005). Somatic embryogenesis is a valuable method of regeneration and the single-cell origin of embryos prevents the production of regenerants of a chimeric nature, and makes transformants more stable and less affected by the alteration produced during the in vitro callus phase. However, to obtain full profit from the method proposed by Babulova and co-workers, it is a prerequisite for such plants regenerated through somatic embryogenesis to reproduce through seeds in order to establish a novel genotype suited for regeneration through somatic embryogenesis. However, previous experiments in alfalfa indicated that such a strategy is not very practical since epigenetic effects play a major role in the expression of such a trait (Crea et al., 1995).

Kanamycin and hygromycin have been utilized as selective agents of transgenic calli and tissues, but Lohar et al. (2001) claimed that the use of the *bar* gene, conferring resistance to phosphinothricin, may reduce somaclonal variation and plant infertility. Streptothricin was also proposed as a novel selectable marker by Jelenska et al. (2000).

## *APPLICATION OF GENETIC TRANSFORMATION IN* LOTUS *SPECIES*

*Lotus* transformation is mainly performed for theoretical studies aimed at investigating genetic and physiological themes from the molecular biology point of view. Genomics is the principal reason for *L. japonicus* transformation. T-DNA and transposing tagging have been successfully attempted. Schauser et al. (1998) identified two putatively symbiotic loci from two mutant lines identified in progeny from 1,112 primary transformants, obtained after *Agrobacterium tumefaciens* T-DNA-mediated transformation. Thykjaer et al. (1995) showed the ability of the maize transposon Ac to excise and reinsert in *L. japonicus,* demonstrating the feasibility of insertion mutagenesis with such transposable elements. Transformation with the promoterless *gus* (*uid*A) gene was performed to take advantage of enrichment for gene tagging events associated with the quickly assayable activation of a *uid*A promoterless construct (Martirani et al., 1999; Buzas, 2005). Activation tagging is also profitably applied in *L. japonicus* (Imaizumi et al., 2005). Poor results were obtained, on the contrary, when a gene targeting approach was attempted using large flanking regions for homologous recombination, and a positive-negative selection method (Thykjaer et al., 1997).

Most of the genomic approaches applied in *Lotus* were aimed at isolating the genes involved in symbiotic nitrogen fixation. With transposon tagging, mutants arrested at the stage of bacterial recognition were obtained and the inactivated genes coded for regulatory genes controlling the formation of infection threads and the initiation of nodule primordia (Schauser et al., 1998). Expressed sequence tag (EST) analysis also provides a powerful method of investigation of such complex functions and allows identifying cDNA fragments specifically expressed during nodule development (Kouchi et al., 2004). Genetic transformation aimed at silencing gene expression is a tool for investigating the real involvement of individual sequences in the process in depth and suitable vectors have been produced and assayed (Kumagai and Kouchi, 2003).

*Lotus* has also been utilized as model system in studying the biosynthetic pathway of condensed tannins (Robbins et al., 1992). Since early work on transformation of *L. corniculatus* (Carron et al., 1994), it was attempted to modify the tannin biosynthesis to produce mutants for isolating unknown genes of the pathway. Structural genes in sense and antisense orientations were utilized to transform different genotypes of *L. corniculatus* (Colliver et al., 1997; Bavage et al., 1997) and the effect of transformation tested in hairy roots and plants (Robbins et al., 1998). Amazing results were sometimes observed, such as the increase of tannin synthesis and the expression of endogenous genes in some lines transformed with *DFR,* a key structural gene of the pathway, in an antisense orientation (Robbins et al., 1998). This phenomenon was attributed to a compensative mechanism that triggers different expression levels among members of such a gene family (Paolocci et al., 2005). An alternative strategy of transformation consisted of utilizing regulatory genes to modify the pathway and produce solid mutants to be used for isolating genes with subtractive or comparative cDNA strategies (Damiani et al., 2000). The transformation of *L. corniculatus* with *Sn,* a positive regulator of the anthocyanin biosynthesis in maize, reduced the tannin levels in leaves of some transgenic plants because of a probable silencing effect between the endogenous and exogenous regulators (Damiani et al., 1999). This effect was dependent on both the genotype and the copy number of the transgene. In fact, transformants of the same genotype (S50) may be enhanced or suppressed depending on the number of the transgene copies integrated in the host genome, while *Sn* was unable to modulate tannin levels in other genotypes of the same *L. corniculatus* variety (Paolocci et al., 1999; Robbins et al., 2003).

Legumes are the highest plant protein producers; however, their composition is poor in sulfur-rich amino acids, which have a relevant impact on the efficiency of dairy, meat, and wool production. Experiments of genetic transformation with genes coding for sulfur-rich proteins have been performed in several legumes. In *L. corniculatus,* an attempt was made to induce the synthesis of zeins, the maize endosperm storage proteins that are rich in the essential sulfur-rich amino acids (Bellucci et al., 1997, 2000, 2002). In such experiments, the expression of different zein coding genes triggered the accumulation in leaves of the corresponding proteins to 0.3 percent of total extractable proteins. This amount is, however, not enough to significantly improve the nutritive value of such forages. Hence, other strategies, such as chloroplast transformation, need to be exploited to achieve agronomically important results.

More success was obtained when the modification of fatty acid composition was attempted. In fact, by transferring the fungal gene Δ6-fatty-acid desaturase through *Agrobacterium* based vectors, lines accumulating up to 12.82 and 13.75 percent of the total fatty acids in the leaves were obtained (Ren et al., 2005). Bellucci et al. (2004) attempted to provide a new ammonium assimilation pathway by inserting a bacterial gene coding for asparagine synthetase in *L. corniculatus* with the goal of increasing plant vigor and production. However, transgenic plants showed a faster growth cycle resulting in premature flowering, which negatively affected biomass production and photosynthetic activity. *L. japonicus* was also used as a model for experiments on oral allergen-specific immunotherapy, facilitating it to produce allergenic proteins from the house dust mite *Dermatophagoides farinae* (Tomoaki et al., 2005).

The above-cited works represent only a fraction of the massive scientific literature appearing, when *Lotus* and *genetic transformation* are utilized as key words for bibliographic searches. Therefore most scientific work utilizing genetic transformation in *Lotus* for studying traits of wide plant biology relevance is not cited here. We mention only two practical applications of *Lotus* transformation: (1) the study of pollen flow between transgenic and nontransgenic species in the genus *Lotus* to determine if pollen contamination occurs and how it is affected by distance (De Marchis et al., 2003); and (2) the reclamation of the Salado River Basin, an area located in the province of Buenos Aires (Argentina) featuring halomorphic soils, where sodium is the predominant cation, and *L. glaber* is frequently used by farmers to increase forage yield and quality of pasture (Remis et al., 1995). To improve plant adaptation to such environmental conditions, attempts to modify polyamine synthesis have been performed (Chiesa et al., 2004).

## *PROTOCOLS*

In this chapter, protocols for transformation of the three main species are reported.

### L. corniculatus-*Hairy Root Transformation*

*A. rhizogenes* wild-type strain is utilized for hypocotyl inoculation using a hypodermic needle. Damiani et al. (1993) also inoculated stems, petioles, and leaves. Hairy roots are produced within 2 to 4 weeks, detached, and placed on B5 (Gamborg et al., 1968) medium with 2 percent sucrose and cefotaxime (150 $\mu g \cdot ml^{-1}$) or carbenicillin (1.0 $mg \cdot ml^{-1}$) and the eventual selective agent, in the dark at 20 to 25°C for 2 to 3 weeks. Plates are then transferred to light (50 $\mu g \cdot ml^{2} \cdot s$) and from each root a shoot is produced (Figure 20.1). Plantlets are then moved to MS (Murashige and Skoog, 1962) ½ strength supplemented with the same antibiotics and after 1 to 2 months to soil in a mist chamber until completely acclimatized to greenhouse conditions.

*A. tumefaciens* strains C58 and LBA 4404 work perfectly for *L. corniculatus* transformation. Axenic stems or leaves are incubated for 5 to 30 minutes in bacterial suspension grown overnight, blotted on sterile filter paper, and placed for 3 days in nonselective MSN1 medium (Damiani et al., 1993).

FIGURE 20.1. Direct regeneration from hairy roots in *L. corniculatus*. Hairy roots profusely growing in liquid culture regenerate through shoot organogenesis. The insert shows details of a single regenerating hairy root. Courtesy of Drs. P. Morris and M. P. Robbins of IGER, Aberystwyth. (See also color gallery.)

After 4 days of cocultivation, explants are washed five times with sterile water and transferred to the same medium supplemented with selective agents for bacterium (cefotaxime, timentin, and carbenicillin) and for transgenic cells. Subculture is performed every 3 weeks and after two subcultures, calli start organogenesis. Shoots are then removed from callus and placed on MS growth regulator free for rooting. Plantlets are handled as previously described for hairy root transformation (Figure 20.1).

## *Genetic Transformation of* L. glaber

In *L. glaber*,, hairy root transformation has only been performed by Damiani et al. (1993). Infection is performed with *A. rhizogenes* as described for *L. corniculatus* and hairy roots are then placed on selective MSN1 and subcultured for 2 to 3 weeks. Regeneration, rooting, and plant transfer to soil were performed as described above.

## *Genetic Transformation of* L. japonicus

*A. tumefaciens* strains LBA 4404, AGL0, AGL1, and C58 derivative (Oger et al., 1996) are all efficient to transform the variety Gifu. The explants utilized for infection are hypocotyls, derived from 7-day-old seedlings, cut longitudinally or transversally, or roots (Lombari et al., 2003) explanted from 3-week-old plantlets. Roots after 5 days of pretreatment in CIM (B5 supplemented with 3 $mg \cdot l^{-1}$ indole acetic acid [IAA], 0.15 $mg \cdot l^{-1}$ 2,4-dichlorophenoxy acetic acid, 0.6 $mg \cdot l^{-1}$ benzyl adenine, 0.3 $mg \cdot l^{-1}$ isopentenyladenine [IPA], 1 percent agar) are dipped in the *Agrobacterium* culture and wounded by cutting into 0.5 cm pieces, blotted in sterile filter paper, and placed in CIM. After 2 days, explants are washed in B5 medium and transferred to CIM selective medium. Three weeks later explants are moved to selective SIM (shoot-inducing medium): B5 salts supplemented with 0.2 $mg \cdot l^{-1}$ BA and 10 mM $(NH_4)_2SO_4$. Emerging shoots are moved to shoot elongation medium consisting of B5 salts and 1 percent sucrose. For root induction, shoots are placed in half-strength B5 salts plus 1 percent sucrose and 0.1 $mg \cdot l^{-1}$ naphthalene acetic acid (NAA). Root elongation occurred on the same medium, but NAA free.

With the described protocols, Lombari and co-workers (2003) claim to have increased the efficiency of transformation with respect to the protocol based on hypocotyl transformation, even reducing the time necessary for plant regeneration and, more important, reducing the time of callus induction and consequently the risk of unwanted somaclonal variation.

*A. rhizogenes* worked perfectly in *L. japonicus,* producing hairy roots from the infection point independently from the tissue wounded. Several bacterial strains were successfully tested. Derived hairy roots were either cultured in vitro for further studies or utilized for regenerating transgenic plants with a protocol similar to that previously described for *A. tumefaciens* transformation.

Selective agents utilized in the described experiments are mainly kanamycin and hygromycin with a concentration of 50 $mg \cdot l^{-1}$ for the former and 15-20 $mg \cdot l^{-1}$ for the latter.

## CONCLUSION

*Lotus* is a model system for the more advanced applications of genetic manipulation and genomic studies. For these reasons, a lot of emphasis has been placed on varied investigations and the number of publications coming out of these efforts is increasing continuously. Several ongoing projects are based on the exploitation of the potentialities of the genus (VandenBosch and Stacey, 2003). Nevertheless, we should not neglect the importance of this genus for practical purposes such as forage production and land reclamation in some areas of our planet.

## REFERENCES


Aziz, M.A., P.K. Chand, J.B. Power, and M.R. Davey (1990). Somatic hybrids between the forage legumes *L. corniculatus* L. and *L. tenuis* Waldst et Kit. *Journal of Experimental Botany* 41:471-479.

Barbulova, A., E. D'Apuzzo, A. Rogato, and M. Chiurazzi (2005). Improved procedures for *in vitro* regeneration and for phenotypic analysis in the model legume *Lotus japonicus. Functional Plant Biology* 32:529-536.

Bavage, A.D., I.G. Davies, M.P. Robbins, and P. Morris (1997). Expression of an *Antirrhinum* dihydroflavonol reductase gene results in changes in condensed tannin structure and accumulation in root cultures of *Lotus corniculatus* (bird's foot trefoil). *Plant Molecular Biology* 35:443-458.

Bellucci, M., A. Alpini, and S. Arcioni (2002). Zein accumulation in forage species (*Lotus corniculatus* and *Medicago sativa*) and coexpression of the δ-zein:KDEL and β-zein:KDEL polypeptides in tobacco leaf. *Plant Cell Reports* 20:848-856.

Bellucci, M., A. Alpini, F. Paolocci, L. Cong, and S. Arcioni (2000). Accumulation of δ-zein and β-zein: KDEL to high levels in tobacco leaves and differential synthesis of BiP in transformants. *Theoretical and Applied Genetics* 101:796-804.

Bellucci, M., L. Ederli, F. De Marchis, S. Pasqualini, and S. Arcioni (2004). Transformation of *Lotus corniculatus* plants with *E. coli* asparagine synthetase A: Effect on nitrogen assimilation and plant development. *Plant Cell Tissue and Organ Culture* 78:139-150.

Bellucci, M., B. Lazzari, A. Viotti, and S. Arcioni (1997). Differential expression of a γ-zein gene in *Medicago sativa, Lotus corniculatus* and *Nicotiana tabacum. Plant Science* 127:161-169.

Buzas, D.M., D. Lohar, S. Sato, Y. Nakamura, S. Tabata, C.E. Vickers, J. Stiller, and P.M. Gresshoff (2005). Promoter trapping in *Lotus japonicus* reveals novel root and nodule GUS expression domains. *Plant and Cell Physiology* 46:1202-1212.

Carron, T.R., M.P. Robbins, and P. Morris (1994). Genetic modifications of condensed tannin biosynthesis in *Lotus corniculatus.* 1. Heterologous antisense dihydroflavonol reductase down-regulates tannin accumulation in "hairy roots" cultures. *Theoretical and Applied Genetics* 87:1006-1015.

Chiesa, M.A., O.A. Ruiz, and D.H. Sánchez (2004). *Lotus* hairy roots expressing inducible arginine decarboxylase activity. *Biotechnology Letters* 26:729-733.

Colliver, S.P., P. Morris, and M.P. Robbins (1997). Differential modification of flavonoid and isoflavonoid biosynthesis with an antisense chalcone synthase construct in transgenic *Lotus corniculatus. Plant Molecular Biology* 35:509-522.

Crea, F., M. Bellucci, F. Damiani, and S. Arcioni (1995). Genetic control of somatic embryogenesis in alfalfa (*M. sativa* L. cv. Adriana). *Euphytica* 81:151-155.

Damiani, F., D. Mariotti, M. Pezzotti, and S. Arcioni (1985). Variation among plants regenerated from tissue culture of *Lotus corniculatus* L. *Zeitschrift für Pflanzenzüchtung* 94:332-339.

Damiani, F., E. Nenz, F. Paolocci, and S. Arcioni (1993). Introduction of hygromycin resistance in *Lotus* spp. through *Agrobacterium rhizogenes* transformation. *Transgenic Research* 2:330-335.

Damiani, F., F. Paolocci, P.D. Cluster, S. Arcioni, G.J. Tanner, R.G. Joseph, Y.G. Li, J. de Majnik, and P.J. Larkin (1999). The maize transcription factor Sn alters proanthocyanidin production in transgenic *Lotus corniculatus* plants. *Australian Journal of Plant Physiology* 26:159-169.

Damiani, F., F. Paolocci, G. Consonni, G. Crea, C. Tonelli, and S. Arcioni (1998). A maize anthocyanin transactivator induces pigmentation in several transgenic dicotyledonous species. *Plant Cell Reports* 17:339-344.

Damiani, F., F. Paolocci, V. Turchetti, and S. Arcioni (2000). Toward the identification of the genes for the synthesis of condensed tannins in forage legumes In *Phytosphere '99 highlights on European plant biotechnology research and technology transfer,* G.E. de Vries and K. Metzlaff (eds.). Amsterdam: Elsevier, pp. 257-263.

Damiani, F., M. Pezzotti, and S. Arcioni (1990). Somaclonal variation in *Lotus corniculatus* L. in relation to plant breeding purposes. *Euphytica* 46:35-41.

De Marchis, F., M. Bellucci, and S. Arcioni (2003). Measuring gene flow from two birdsfoot trefoil (*Lotus corniculatus*) field trials using transgenes as tracer markers. *Molecular Ecology* 12:1681-1685.

Forde, B.G., H.M. Day, J.F. Turtin, S. Wen-jun, J.V. Cullimore, and J.E. Oliver (1989). Two glutamine synthase genes in *Phaseolus vulgaris* L. display with contrasting developmental and spatial patterns of expression in transgenic *Lotus corniculatus* plants. *Plant Cell* 1:391-401.

Gamborg, O.L., R.A. Miller, and K. Ojima (1968). Nutrient requirements of suspension cultures of soybean root cells. *Experimental Cell Research* 50:151-158.

Grant, W.F. (2004). List of *Lotus corniculatus* (birdsfoot trefoil), *L. uliginosus/L. pedunculatus* (big trefoil), *L. glaber* (narrow leaf trefoil) and *L. subbiflorus* cultivars. Part 1. Cultivars with known or tentative country origin. *Lotus Newsletter* 34:12-26.

Handberg, K. and J. Stougaard (1992). *Lotus japonicus,* an autogamous, diploid legume species for classical and molecular genetics. *Plant Journal* 2:487-496.

Imaizumi, R., S. Sato, N. Kameya, I. Nakamura, Y. Nakamura, S. Tabata, S. Ayabe, and T. Aoki (2005). Activation tagging approach in a model legume, *Lotus japonicus. Journal of Plant Research* 118:391-399.

Jelenska, J., E. Tietze, J. Tempé, and J. Brevet (2000). Streptothricin resistance as a novel selectable marker for transgenic plant cells. *Plant Cell Reports* 19:298-303.

Kaiser, C.J. and M.E. Heath (1990). Big trefoil: A new legume for pastures on fragipan soils. In *Advances in new crops,* J. Janick and J.E. Simon (eds.). Portland, OR: Timber Press, pp. 191-194.

Kawaguchi, M., A. Pedrosa-Harand, K. Yano, M. Hayashi, Y. Murooka, K. Saito, T. Nagata, K. Namai, H. Nishida, D. Shibata, et al. (2005). *Lotus burttii* takes a position of the third corner in the *Lotus* molecular genetics triangle. *DNA Research* 12:69-77.

Kouchi, H., K. Shimomura, S. Hata, A. Hirota, G.J. Wu, H. Kumagai, S. Tajima, N. Suganuma, A. Suzuki, T. Aoki, et al. (2004). Large-scale analysis of gene expression profiles during early stages of root nodule formation in a model legume, *Lotus japonicus*. *DNA Research* 11:263-274.

Kumagai, H. and H. Kouchi (2003). Gene silencing by expression of hairpin RNA in *Lotus japonicus* roots and root nodules. *Molecular Plant-Microbe Interactions* 16:663-688.

Locatelli, M.L., J.L. Remis, and A.A. Iglesias (1997). Evaluation of forage quality attributes for *Lotus* spp. grown in the Salado River Basin. *Lotus Newsletter* 28:17-21.

Lohar, D., K. Schuller, D.M. Buzas, P.M. Gresshoff, and J. Stiller (2001). Transformation of *Lotus japonicus* using the herbicide resistance *bar* gene as selectable marker. *Journal of Experimental Botany* 52:1697-1702.

Lombari, P., E. Ercolano, H. El Alaoui, and M. Chiurazzi (2003). A new transformation-regeneration procedure in the model legume *Lotus japonicus:* Root explants as a source of large numbers of cells susceptible to *Agrobacterium*-mediated transformation. *Plant Cell Reports* 21:771-777.

Mariotti, D., M. Pezzotti, E. Falistocco, and S. Arcioni (1984). Plant regeneration from leaf-derived callus of *Lotus corniculatus* cv. Franco. *Genetics and Agriculture* 38:219-223.

Martirani, L., J. Stiller, R. Mirabella, F. Alfano, A. Lamberti, S.E. Radutoiu, M. Iaccarino, P.M. Gresshoff, and M. Chiurazzi (1999). T-DNA tagging of nodulation- and root-related genes in *Lotus japonicus*. Expression patterns and potential for promoter-trapping and insertional mutagenesis. *Molecular Plant-Microbe Interactions* 12:275-284.

Morris, P., K.J. Webb, M.P. Robbins, L. Skot, and B. Jørgensen (1999). Application of biotechnology to *Lotus* breeding. In *Trefoil: the Science and technology of Lotus,* P. Beuselinck, J. Nelson, and C. Hoveland (eds.). ASA-CSSA. Special Publication 28, pp. 199-228.

Murashige, T. and F. Skoog (1962). A revised medium for rapid growth and bioassays with tobacco tissue cultures. *Physiologia Plantarum* 15:473-497.

Nenz, E., F. Pupilli, F. Paolocci, F. Damiani, C.A. Cenci, and S. Arcioni (1996). Plant regeneration and genetic transformation of *Lotus angustissimus* L. *Plant Cell Tissue and Organ Culture* 45:145-152.

Nizeki, M. and W.F. Grant (1971). Callus, plantlet formation, and polyploidy from cultured anthers of *Lotus* and *Nicotiana*. *Canadian Journal of Botany* 49:2041-2051.

Oger, P., A. Petit, and Y. Dessaux (1996). A simple technique for direct transformation and regeneration of the diploid legume species *Lotus japonicus*. *Plant Science* 116:159-168.

Orshinsky, B.R. and D.T. Tomes (1984). Comparisons of plants derived from cuttings, node cultures, and ethyl methane sulfonate treated node cultures of birdsfoot trefoil (*Lotus corniculatus*). *Canadian Journal of Botany* 62:1501-1504.

Paolocci, F., T. Bovone, N. Tosti, S. Arcioni, and F. Damiani (2005). Light and an exogenous transcription factor qualitatively and quantitatively affect the biosynthetic pathway of condensed tannins in *Lotus corniculatus* leaves. *Journal of Experimental Botany* 56:1093-1103.

Paolocci, F., R. Capucci, S. Arcioni, and F. Damiani (1999). Birdsfoot trefoil: A model for studying the synthesis of condensed tannins. In *Plant polyphenols 2. Chemistry, biology, pharmacology,* G.G. Groos, R.W. Hemingway, and T. Yoshida (eds.). New York: Kluwer Academic, pp. 343-356.

Petit, A., J. Stougaard, A. Kuhle, K.A. Marcker, and J. Tempé (1987). Transformation and regeneration of the legume *Lotus corniculatus:* A system for molecular studies of symbiotic nitrogen fixation. *Molecular and General Genetics* 207: 245-250.

Pezzotti, M., S. Arcioni, F. Damiani, and D. Mariotti (1985). Time related behaviour of phenotypic variation in *Lotus corniculatus* regenerants under field conditions. *Euphytica* 34:619-623.

Pupilli, F., S. Arcioni, F. Damiani, and M. Pezzotti (1990). Plant regeneration from callus cultures of *Lotus pedunculatus* Cav. *Plant Cell Tissue and Organ Culture* 23:193-199.

Remis, J.L., O.A. Ruiz, R.A. Ugalde, and A.A. Iglesias (1995). Evaluation of *Lotus* spp. growth in the Salad River Basin. *Lotus Newsletter* 25:17-20.

Ren, C., S. Tsuda, K. Matsui, M. Fukuchi-Mizutani, M. Ochiai, S. Shimizu, E. Sakuradani, T. Aoki, R. Imaizumi, S. Ayabe, and Y. Tanaka (2005). Production of $\gamma$-linolenic acid in *Lotus japonicus* and *Vigna angularis* by expression of the $\Delta 6$-fatty-acid desaturase gene isolated from *Mortierella alpina. Plant Science* 169:599-605.

Robbins, M.P., A.D. Bavage, C. Strudwicke, and P. Morris (1998). Genetic manipulation of condensed tannins in higher plants. II. Analysis of birdsfoot trefoil plants harbouring antisense dihydroflavonol reductase constructs. *Plant Physiology* 116:1133-1144.

Robbins M.P., T.R. Carron, and P. Morris (1992). Transgenic *Lotus corniculatus*: A model system for the modification and genetic manipulations of condensed tannin biosynthesis. In *Plant polyphenols: Synthesis, properties, significance,* R.A. Hemingway and P.E. Laks (eds.). New York: Plenum Press, pp. 111-133.

Robbins, M.P., F. Paolocci, J.W. Hughes, V. Turchetti, G. Allison, S. Arcioni, P. Morris, and F. Damiani (2003). *Sn,* a maize bHLH gene, transactivates anthocyanin and condensed tannin pathways in *Lotus corniculatus. Journal of Experimental Botany* 54:239-248.

Schauser, L., K. Handberg, N. Sandal, J. Stiller, T. Thykjaer, E. Pajuelo, A. Nielsen, and J. Stougaard (1998). Symbiotic plant mutants identified after T-DNA transformation of *Lotus japonicus. Molecular and General Genetics* 259:414-423.

Stiller, J., L. Martirani, S. Tuppale, R. Chian, M. Chiurazzi, and P.M. Gresshoff (1997). High frequency transformation and regeneration of transgenic plants in the model legume *Lotus japonicus. Journal of Experimental Botany* 48:1357-1365.

Stougaard, J., K.A. Marcker, L. Otten, and J. Schell (1986). Nodule-specific expression of chimeric soybean leghaemoglobin gene in transgenic *Lotus corniculatus. Nature* 321:669-674.

Swanson, E.B. and D.T. Tomes (1980). Plant regeneration from cell cultures of *Lotus corniculatus* and the selection and characterization of 2,4-D tolerant cell lines. *Canadian Journal of Botany* 58:1205-1209.

Thykjaer, T., J. Finnemann, L. Schauser, L. Christensen, C. Poulsen, and J. Stougaard (1997). Gene targeting approaches using positive-negative selection and large flanking regions. *Plant Molecular Biology* 35:523-530.

Thykjaer, T., J. Stiller, K. Handberg, J. Jones, and J. Stougaard (1995). The maize transposable element *Ac* is active in the legume *Lotus japonicus. Plant Molecular Biology* 27:981-993.

Tomoaki, K., G. Yuij, O. Kazuhiss, H. Makoto, and Y. Murooka (2005). Expression of a major house dust mite allergen gene from *Dermatophagoides farinae* in *Lotus japonicus* accession Miyakojima MG-20. *Journal of Bioscience and Bioengineering* 99:165-168.

Urbanska, K.M. (1994). Use of *Lotus alpinus* in alpine ecosystem restoration. *Proceedings of the First International Lotus Symposium.* University of Missouri Ext. Publication MX411. Columbia: University of Missouri, pp. 172-176.

VandenBosch, K.A. and G. Stacey (2003). Summaries of legume genomics projects from around the globe. Community resources for crops and models. *Plant Physiology* 131:840-865.

Webb, K.J. (1986).Transformation in the forage legumes. In Genetic manipulation in plant breeding: Proceedings, international symposium organized by EUCARPIA, September 8-13, 1985, Horn et al. (eds.), Berlin, pp. 831-833.

Webb, K.J., M.J. Gibbs, S. Mizen, L. Skøt, and J.A. Gatehouse (1996). Genetic transformation of *Lotus corniculatus* with *Agrobacterium tumefaciens* and the analysis of the inheritance of transgenes in the T1 generation. *Transgenic Research* 5:303-312.

Wright, R.L., D.A. Somers, and R.L. McGraw (1987). Somatic hybridization between birdsfoot trefoil (*Lotus corniculatus* L.) and *L. conimbricensis* Willd. *Theoretical and Applied Genetics* 75:151-156.

Chapter 21

# Genetic Transformation of Tree Legume, *Leucaena leucocephala*

Smita Rastogi
U. N. Dwivedi

## *INTRODUCTION*

Leguminous trees are commercially valuable forestry species having an incredibly diverse array of uses. They offer a promising future for the production of timber, fodder, fuel wood, industrial energy, and raw material for pulp and paper manufacture, wasteland reclamation and erosion control, nitrogen fixation, and soil fertility enrichment.

## *IN VITRO REGENERATION OF* LEUCAENA

Quite a lot of advances have been made in tissue culture methodologies related to *L. leucocephala* (Table 21.1). In vitro regeneration studies for *L. leucocephala* were initiated in the early 1980s. Earlier studies on in vitro tissue culture of *L. leucocephala* were conducted using explants such as hypocotyls, cotyledons, protoplasts, and callus derived from the seedling explants (Glovack and Greatbach, 1980; Peasley and Collins, 1980; Nagmani and Venkateswaran, 1983; Venkateswaran and Gandhi, 1982; Ravishankar et al., 1983; Kulkarni et al., 1984; Nataraja and Sudhadevi, 1984).

---

Financial support from the Department of Biotechnology, New Delhi, and Senior Research Fellowship to SR from the Council of Scientific and Industrial Research, New Delhi, and grants from the Department of Science and Technology, New Delhi, under the FIST program are gratefully acknowledged. The authors also thank Dr. Srinivasan, National Research Center on Biotechnology, IARI, New Delhi, for the strain of *A. tumefaciens* C58 and the vector p35SGUSINT.

TABLE 21.1. Plant regeneration in *L. leucocephala* using different explants.

| Explant source | Explants | Method | Result | References |
|---|---|---|---|---|
| Seedling | Hypocotyl | Organogenesis | Plantlet | Glovack and Greatbach, 1980<br>Peasley and Collins, 1980<br>Nagamani and Venkateswaran, 1983<br>Ravishankar et al., 1983<br>Nataraja and Sudhadevi, 1984 |
| Seedling | Cotyledonary node, nodal segment | Organogenesis | Plantlet, field planting | Dhawan and Bhojwani, 1984, 1985 |
| Mature tree | Nodal segment | Shoot proliferation | Plantlet, field planting | Dhawan and Bhojwani, 1984; Kulkarni et al., 1984; Datta and Datta, 1985 |
| Mature tree | Lateral bud | Shoot proliferation | Plantlet field planting | Goyal et al., 1985 |

Goyal et al. (1985) were the first to report shoot multiplication, rooting, and hardening using lateral bud explants of mature trees of *L. leucocephala* K-67. Datta and Datta (1985) reported in vitro clonal multiplication of *L. leucocephala* using nodal explants of mature trees. However, these plantlets did not survive for more than 30 to 40 days of culturing. The inclusion of a higher dose of $KNO_3$ (double the concentration to that of Murashige and Skoog [MS]) in addition to indole acetic acid (IAA) or naphthalene acetic acid (NAA) in the MS medium partially circumvented this problem and led to the production of more vigorous rooting, better growth, and a higher percentage of long-term survival of regenerants. Similarly, in vitro regeneration of *L. leucocephala* cv. K-8 using both seedling-raised explants containing cotyledonary node segments and explants raised from adult plants containing nodal segments resulted in only 40 percent survival of the plants due to transplantation shock (Dhawan and Bhojwani, 1984, 1985). Morphophysiological studies confirmed that the cultured plants had poor control of water loss as their leaves bore very little epicuticular wax and lacked starch grains. During hardening, the appearance of starch grains preceded normalization of mesophyll cells (Dhawan, 1993). Furthermore, a pretransplant hardening in medium containing only inorganic nutrients (Dhawan and Bhojwani, 1987a), in vitro nodulation by *Rhizobium* (Dhawan and

Bhojwani, 1987b), and transfer of nodulated in vitro-raised *Leucaena* to garden soil by laying inoculum of either *Glomus fasiculatum* or *G. macrocarpum*, vesicular arbuscular mycorrhizal fungi, around their roots (Puthur et al., 1998; Naqvi and Mukerji, 1998) have been described to increase resistance to transplantation shock, thereby enhancing transplantation efficiency of the micropropagated plants. Somaclonal variants of *L. leucocephala* with high $CO_2$-assimilating potential have been produced in vitro (Pardha Saradhi and Alia, 1995) and plantlets were successfully transferred to soil in soilrite inoculated with *Rhizobium*. The variants differed in terms of leaf size, green pigmentation, growth rate, extent of branching, net photosynthesis capacity, intracellular $CO_2$ concentration, stomatal conductance, and rate of transpiration.

## *TRANSFORMATION PROTOCOL FOR* LEUCAENA LEUCOCEPHALA

### *Materials and Methods*

#### *Plant Material*

Juvenile shoots of the mature tree of *Leucaena leucocephala* cv. K-8, K-29, and K-850 procured from the departmental garden and cut into single-node explants (1 inch), containing axillary buds, without leaves, were used as explants. These explants were thoroughly washed with mild detergent (0.5 percent v/v labolene), running tap water, and finally with distilled water. The explants were surface sterilized by a quick dip in absolute alcohol (5 seconds) followed by treatment with 1 percent v/v sodium hypochlorite for 15 minutes. Subsequently, they were washed extensively with sterile water and treated with 0.5 percent w/v mercuric chloride. After 5 minutes of treatment, they were rinsed three times with sterile water. Surface disinfected explants were decapitated to remove proximal meristematic ends and incised gently with a scalpel at various regions.

For cotyledonary node explants, seeds of *L. leucocephala* cv. K-8, K-29, and K-850 were soaked in boiling water for 5 minutes and then treated in the same way as described for nodal explants from mature tree. Seeds were then allowed to germinate aseptically for 2 days in sterile Petri plates with moistened filter paper at 29 ± 0.2°C in the dark. Cotyledonary nodes without cotyledons and radicles were incised with a scalpel and used as explants.

*Bacterial Strain*

An *Agrobacterium tumefaciens* binary vector, p35SGUSINT (derived from pBIN19), containing the *gus* reporter gene interrupted with a plant intron (Vancanneyt et al., 1990) under the control of CaMV35S promoter (P-35S) and a polyadenylation signal (PA) and kanamycin resistance selection marker gene, neomycin phosphotransferase (*npt*II) under the control of the promoter of *nos* gene (P-NOS) and the polyadenylation signal (nPA) was used for plant transformation. The binary vector was mobilized into *A. tumefaciens* C58 MP90 strain by the freeze-thaw method (Sambrook et al., 1989).

*Plant Regeneration Medium*

Shoot regeneration medium comprised full-strength MS medium (Murashige and Skoog, 1962) supplemented with 3 percent w/v sucrose, 100 $mg \cdot l^{-1}$ myoinositol, 100 $mg \cdot l^{-1}$ glutamine, 6-benzyl aminopurine (BAP; 20.9 μM; 5 $mg \cdot l^{-1}$), and NAA (5.37 μM; 1 $mg \cdot l^{-1}$), and solidified with 0.8 percent w/v agar. Putative transformants were selected on shoot regeneration medium supplemented with 25 to 50 $mg \cdot l^{-1}$ kanamycin, depending on the cultivar. Kanamycin was filter sterilized through a 0.2 μm filter before its addition to sterile medium. Root induction was achieved on half-strength MS medium containing 2 percent w/v sucrose, 100 $mg \cdot l^{-1}$ myoinositol, indole 3-butyric acid (IBA; 14.76 μM; 3 $mg \cdot l^{-1}$), kinetin (kin; 0.232 μM; 0.05 $mg \cdot l^{-1}$) and solidified with 0.8 percent w/v agar. The pH of the media was adjusted to 5.7 prior to autoclaving at 15 psi for 20 minutes.

*Plant transformation*

*A. tumefaciens* MP90 C58 (p35SGUSINT), harboring *npt*II and *Gus* genes, was grown overnight at 28°C in 40 ml liquid Luria Bertani (LB) medium containing 100 $\mu g \cdot ml^{-1}$ kanamycin. Bacterial cells were pelleted at 4,000 rpm for 5 minutes and suspended in 150 ml of liquid MS medium. Surface sterilized and wounded nodal or cotyledonary node explants were cocultivated with the culture of *Agrobacterium* for 5 minutes in a sterile Petri plate. Explants were blotted dry on sterile filter papers and implanted horizontally on shoot regeneration medium containing kanamycin (25 $mg \cdot l^{-1}$) for selection of putative transformants and cefotaxime (250 $mg \cdot l^{-1}$) for suppression of excessive growth of *Agrobacterium*. Both kanamycin and cefotaxime were filter sterilized through a 0.2 μm filter and added aseptically to the autoclaved medium. After two days of culture in MS medium containing kanamycin and cefotaxime (both filter sterilized through a 0.2 μm filter and added aseptically to the autoclaved MS medium), explants were transferred to fresh medium. During this period, the bulk of the phenolics leached out into the medium. After 7 days, the explants were placed on selection medium containing 25 to 50 $mg{\cdot}l^{-1}$ kanamycin (depending on the cultivars) only. The explants, without cocultivation, were allowed to regenerate directly on shoot regeneration medium without kanamycin to generate control plants. Cultures were kept at 29 ± 0.2°C with a photoperiod of 16 hours (at a light intensity of about 3,000 lux) and 55 percent relative humidity. Green surviving shoots were subjected to two or three passages of selection by repeated excision of branches from the basal tissues and their direct exposure to the selection medium. Rooting in these putative transformants was induced by transferring onto root induction medium without kanamycin under similar conditions of light, humidity, and temperature as were employed for shoot proliferation. Rooted plantlets were subjected to acclimatization under soil conditions, inside the culture room, in a pot covered with a polythene sheet having pinholes for gaseous exchange. After 2 weeks of hardening, pots were transferred to a covered greenhouse, watered twice daily to field capacity, and received natural lighting. A standardized protocol is detailed in Figure 21.1.

### *Histochemical and Fluorogenic GUS Assay*

Histochemical and fluorogenic assays of GUS expression were performed on tissues as described by Jefferson (1987) with slight modifications. For histochemical GUS assay, a reaction mixture containing 2 mM 5-bromo-4-chloro-3-indolyl glucuronide (X-gluc) in 20 μl N,N-dimethyl formamide buffered with 50 mM sodium phosphate buffer (pH 7.0), 0.1 percent Triton X-100, 10 mM $Na_2EDTA$, and 0.5 mM potassium ferrocyanide/ferricyanide was infiltrated under vacuum into the leaf and stem sections of kanamycin-resistant plants. Control samples were incubated at 37°C for 48 hours and β-glucuronidase (GUS) activity was detected microscopically after decoloration of tissue with absolute ethanol (Figure 21.2).

Fluorogenic assay of Gus activity was performed in leaf extracts of putative transformants using 4-methyl umbelliferyl β-D-glucuronide (MUG) as substrate. Leaves were homogenized in a Gus extraction buffer comprising 50 mM sodium phosphate buffer (pH 7.0), 10 mM β-mercaptoethanol, 10 mM $Na_2EDTA$, 0.1 percent sodium lauryl sarcosine and 0.1 percent Triton X-100. Next, 20 μl of leaf homogenate (0.1 mg $tissue{\cdot}\mu l^{-1}$ GUS extraction buffer) was added to 0.5 ml of assay buffer (1 mM MUG in GUS extraction buffer). Control samples were derived from nontransformed control plants.

*A. tumefaciens* MP90-C58 harboring *nptII* and *gus* genes

↓

Inoculate 40 ml LB medium containing 100 μg·ml$^{-1}$ kanamycin

↓

Grow overnight at 28°C

↓

Pellet bacterial cells by centrifugation at 4000 rpm for 5 min

↓

Suspend pellet in 150 ml liquid MS medium

↓

Cocultivate surface sterilized and incised nodal or cotyledonary node explants with *A. tumefaciens* culture for 5 min in sterile Petri plate

↓

Blot dry explants on sterile filter paper

↓

Transfer explants on shoot regeneration medium containing 25 mg·l$^{-1}$ kanamycin (for selection of putative explants) and 250 mg·l$^{-1}$ cefotaxime (for suppression of excessive growth of *Agrobacterium*)

FIGURE 21.1. A standardized protocol for the transformation of *Leucaena*.

The system was incubated at 37°C. Aliquots were withdrawn at different time intervals. The reaction was stopped with 0.2 M $Na_2CO_3$ in a final volume of 1 ml and the intensity of fluorescence was visualized under ultraviolet light.

*Southern Hybridization Analysis*

The integration of the *gus* gene into the plant genome of kanamycin-resistant and Gus-positive lines was analyzed by Southern blot hybridization

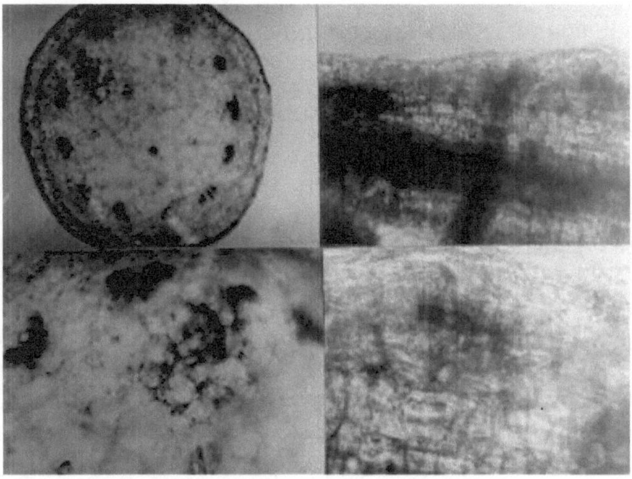

FIGURE 21.2. Histochemical GUS assay of the transformants of *L. leucocephala* cv. K-29 for analysis of *gus* gene in stem sections (left panel) and leaves (right panel) indicating expression of the gene (blue sections). The lower panel is a magnified view of the upper panel. (See also color gallery.)

as described by Sambrook et al. (1989) with slight modifications. Genomic DNA was isolated from control and transformed plant tissues by the method described by Davis et al. (1980). Next, 10 μg DNA from each sample (control and transformants) was digested overnight with *Hin*dIII at 37°C, fractionated on 0.8 percent w/v agarose gel, and transferred to nylon membrane (Nytran 0.45, Schleicher and Schuell, Germany). The membrane was baked at 80°C. The hybridization probe consisted of the *gus* gene-coding sequence. Standard molecular biology protocols were used in Southern hybridization.

## *RESULTS*

### *Plant Regeneration and Transformation*

A minimal inhibitory concentration of kanamycin for primary screening of transformants, which was lethal for nontransformants, was found to range between 25 and 50 $mg \cdot l^{-1}$ depending on the variety used (25 $mg \cdot l^{-1}$ for K-8 and 50 $mg \cdot l^{-1}$ for K-29 and K-850). In case of putative transformants, 4 to 6 cm shoots were produced within 3 to 4 weeks of cocultivation without an intermediary callus phase. However, a marked variation in the shoot initiation response depending upon plant variety as well as explant source was observed. Among all the cultivars tested, K-29 was the most responsive, as number of shoots per explant was the highest in this case. Within 3 to 4 weeks, about 5 to 6 shoots per nodal explant developed in K-29, while the number ranged from 3 to 4 shoots per explant in K-8 and K-850 (Figure 21.3). Moreover, for all the cultivars, nodal explants exhibited a high regeneration frequency, forming 3 to 6 shoots, while cotyledonary node explants exhibited a lower frequency forming only 2 to 3 shoots per explant.

During selection, nontransformed shoots displayed browning of tissues, whereas transformed shoots survived and appeared green and healthy. After three weeks of culturing in the presence of kanamycin, green shoots along with yellowish white shoots were excised from the basal explant tissue and subcultured on fresh selection medium. A few shoots became chlorotic, suggesting further elimination of untransformed tissues, while green shoots proliferated on selection medium and were subjected to further subculturing. Expression of the *npt*II gene was further examined by placing leaves and shoot pieces from putative transformants and nontransformed control plants directly onto the selection medium containing kanamycin at a concentration ranging from 25 to 50 $mg \cdot l^{-1}$ depending on the cultivar. Callusing was observed in case of transformants within 1 to 2 weeks, whereas nontransformants turned necrotic and dead with no signs of callusing. Based on the number of surviving explants after 30 to 35 days of selection on kanamycin-containing medium, a transformation frequency on the order of 13 to 20 percent was achieved. Kanamycin-resistant shoots were then transferred to root initiation medium. Root initiation was observed within

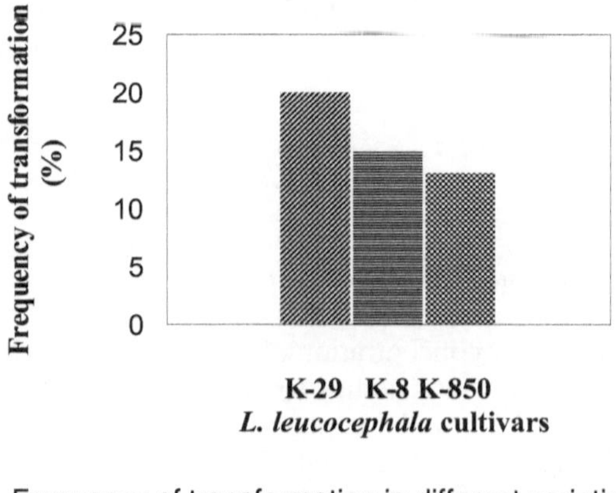


FIGURE 21.3. Frequency of transformation in different varieties of *Leucaena*.

15 to 21 days of culture. About 70 to 80 percent of shoots produced well-developed roots. After a short period of acclimatization, the rooted plantlets, morphologically indistinguishable from field-grown plants maintained under the same growth conditions, were successfully established in soil in the greenhouse ($T_0$ plants). The transplanted plants successfully adapted to natural environmental conditions with a survival rate of more than 75 percent. On average, it took just 1.5 to 2 months from explant inoculation to establishment of plants in soil. A few of the transformed plants displayed precocious flowering and pod development compared to normal control plants.

Direct shoot regeneration via organogenesis without an intervening callus phase from explants was achieved on MS medium supplemented with BAP (20.9 μM; 5 $mg \cdot l^{-1}$) and NAA (5.37 μM; 1 $mg \cdot l^{-1}$). The efficacy of BAP to induce multiple shoots from explants has also been demonstrated in other tree legumes (Gharyal and Maheshwari, 1990; Pradhan et al., 1998; Sonia et al., 2000). The number of regenerated shoots varied with explant type and genotype. Initially, premature leaf fall was frequently observed during in vitro regeneration of *L. leucocephala*. This was, however, prevented by inclusion of glutamine in the medium. This also complemented the greening of leaves. Moreover, the regeneration ability of explants was also enhanced. Dhawan and Bhojwani (1985) have also reported usage of adenine or phloroglucinol to circumvent this problem of leaf fall. Another problem encountered during in vitro culture was leaching of phenolics onto the medium from incised ends of the explants. This problem was surmounted by transferring the explants after 2 days of culture inoculation onto the fresh medium. The shoots obtained were transferred to half-strength MS medium for root induction. The quality and number of roots formed per shoot was quite significant in media containing IBA (14.76 μM; 3 $mg \cdot l^{-1}$) and kinetin (0.232 μM; 0.05 $mg \cdot l^{-1}$). Other auxins and growth regulators tested, such as IAA, NAA, and 2,4-D, were not found to be conducive to root induction. Pardha Saradhi and Alia (1995) also suggested the suitability of IBA as a plant growth regulator for improving rooting frequency in *L. leucocephala*. Rooted plantlets were successfully transferred to soil after 2 weeks of hardening under polythene sheets with pinholes to provide high atmospheric humidity. This helped in overcoming water stress, which is often a problem in tissue-culture raised plants, leading to transplantation shock and a low survival rate (Dhawan, 1993). Thus, a survival rate of more than 75 percent was achieved. The method of transplantation used in the present study was simple with a very high survival rate, while the regenerated plants reported in earlier studies either suffered from a low survival rate or employed tedious and time-consuming procedures such as in vitro nodulation by *Rhizobium* (Dhawan and Bhojwani, 1987b) and inoculation with vesicular arbuscular mycorrhizal fungi (Puthur et al., 1998).

Analysis of selection regimes revealed that direct placement of explants on medium containing kanamycin at a concentration above 50 mg·l$^{-1}$ was detrimental for explant survival, whereas in the presence of 25 to 50 mg·l$^{-1}$ kanamycin, shoot formation from explant and plant regeneration were optimal. The cultivars K-29 and K-850 could resist a higher dosage of kanamycin (50 mg·l$^{-1}$) compared to K-8 (25 mg·l$^{-1}$). The genotypic variation in the resistance to kanamycin in the cultivars of *L. leucocephala* was due to differences in innate tolerance to kanamycin. Varying concentrations of kanamycin ranging from as low as 9 mg·l$^{-1}$ (Mullins et al., 1997) to as high as 100 mg·l$^{-1}$ (Tsai et al., 1994; Cervera et al., 1998) have been employed for selection of transformants in other tree species. Similarly, higher concentrations of kanamycin (100 mg·l$^{-1}$ or above) were used for selection of transformants in grain legumes (Cheng et al., 1996; Hinchee et al., 1988; Krishnamurthy et al., 2000; Sharma and Anjaiah, 2000). In our case, none of the explants survived more than 50 mg·l$^{-1}$ of kanamycin. The transformation efficiency (number of putative transformants per number of explants treated) of *L. leucocephala* was on the order of 13 to 20 percent depending on the cultivars. A similarly high frequency of transformation (approximately 24 percent) from stem segments has also been achieved in another tree legume, *Robinia pseudoacacia* (Igasaki et al., 2000), while a low transformation frequency (<2 percent) has been reported for most of the grain legumes (Kar et al., 1996; Yan et al., 2000; Krishnamurthy et al., 2000; Jaiwal et al., 2001). In another tree species, *Eucalyptus camaldulensis,* only 3 to 6 percent explants have been reported to regenerate into transformants (Mullins et al., 1997).

Thus, a simple and reproducible protocol for genetic transformation and in vitro plant regeneration of *L. leucocephala* has been established (Rastogi and Dwivedi (2003). This protocol has also been used successfully in our lab for alteration of lignin content and composition of *L. leucocephala,* leading to its enhanced applicability as easily digestible and nutritious fodder as well as having better pulping efficiency. Using the above protocol, Rastogi and Dwivedi (2006) have generated transgenic *Leucaena* plants expressing the gene for an O-methyltransferase (*OMT*) in antisense orientation and demonstrated that the transgenic plants had up to a 60 percent reduction in the activity of the enzyme O-methyltransferase and a 28 percent reduction in lignin content. This protocol also envisions the possibility of manipulating *L. leucocephala* for improving other quality attributes, such as reduced mimosine content and resistance to psyllid insects.

## REFERENCES

Cervera, M., J. Juarez, A. Navarro, J.A. Pina, N. Duran-Vila, L. Navarro, and L. Pena (1998). Genetic transformation and regeneration of mature tissues of woody fruit plants bypassing the juvenile stage. *Transgenic Research* 7:51-59.

Cheng, M., R.L. Jarret, Z. Li, A. Xing, and W. Demski (1996). Production of fertile transgenic peanut (*Arachis hypogaea* L.) plants using *Agrobacterium tumefaciens. Plant Cell Reports* 15:653-657.

Datta, K. and S.K. Datta (1985). Auxin + $KNO_3$ induced regeneration of leguminous tree—*Leucaena leucocephala* through tissue culture. *Current Science* 54:248-250.

Davis, R.W., M. Thomas, J. Cameron, T.P. St. John, S. Scherer, and R. Padgett (1980). Rapid DNA isolations for enzymatic and hybridization analysis. *Methods in Enzymology* 65:404-411.

Dhawan, V. (1993). Micropropagation of nitrogen fixing trees. In *Micropropagation of woody plants,* M.R. Ahuja (ed.). Amsterdam: Kluwer Academic Publishers, pp. 303-315.

Dhawan, V. and S.S. Bhojwani (1984). Reduction in cost of tissue culture of *Leucaena leucocephala* (Lam.) deWit. by replacing AR grade sucrose by sugar cubes. *Current Science* 53:1159-1161.

Dhawan, V. and S.S. Bhojwani (1985). In vitro vegetative propagation of *Leucaena leucocephala* (Lam) deWit. *Plant Cell Reports* 4:315-318.

Dhawan, V. and S.S. Bhojwani (1987a). Hardening in vitro and morphological changes in the leaves during acclimatization of micropropagated plants of *Leucaena leucocephala* (Lam) de Wit. *Plant Science* 53:65-72.

Dhawan, V. and S.S. Bhojwani (1987b). In vitro nodulation of seedlings and micropropagated plants of the tropical tree legume *Leucaena leucocephala* by *Rhizobium. Proceedings of Indian National Science Academy* 53B:351-357.

Gharyal, P.K. and S.C. Maheshwari (1990). Differentiation in explants from mature leguminous trees. *Plant Cell Reports* 8:550-553.

Glovack, L. and W. Greatbach (1980). Successful tissue culture of *Leucaena. Leucaena Research Reports* 3:81.

Goyal, Y., R.L. Bingham, and P. Felker (1985). Propagation of the tropical tree, *Leucaena leucocephala* K67 by in vitro bud culture. *Plant Cell Tissue and Organ Culture* 4:3-10.

Hinchee, M.A.W., D.V. Connor-Ward, C.A. Newell, R.E. McDonnell, S.J. Sato, C.S. Gasser, D.A. Fischoff, D.B. Re, R.T. Fraley, and R.B. Horsch (1988). Production of transgenic soybean plants using *Agrobacterium*-mediated DNA transfer. *Biotechnology* 6:915-922.

Igasaki, T., T. Mohri, H. Ichikawa, and K. Shinohara (2000). *Agrobacterium tumefaciens* mediated transformation of *Robinia pseudoacacia. Plant Cell Reports* 19:448-453.

Jaiwal, P.K., R. Kumari, S. Ignacimuthu, I. Potrykus, and C. Sautter (2001). *Agrobacterium tumefaciens* mediated genetic transformation of mungbean (*Vigna radiata* L. Wilczek)—a recalcitrant grain legume. *Plant Science* 161:239-247.

Jefferson, R.A. (1987). Assaying chimeric genes in plants: The *GUS* gene fusion system. *Plant Molecular Biology Reporter* 5:387-405.

Kar, S., T.M. Johnson, P. Nayak, and S.K. Sen (1996). Efficient transgenic plant regeneration through *Agrobacterium*-mediated transformation of chickpea (*Cicer arietinum*). *Plant Cell Reports* 16:32-37.

Krishnamurthy, K.V., K. Suhasini, A.P. Sagare, M. Meixner, A. deKathen, T. Pickardt, and O. Scheider (2000). *Agrobacterium*-mediated transformation of chickpea (*Cicer arietinum* L.) embryo axes. *Plant Cell Reports* 19:235-240.

Kulkarni, D.K., P.K. Gupta, and A.F. Mascarenhas (1984). Tissue culture studies of *Leucaena leucocephala. Leucaena Research Reports* 5:37-39.

Mullins, K.V., D.J. Llewellyn, V.J. Hartney, S. Strauss, and E.S. Dennis (1997). Regeneration and transformation of *Eucalyptus camaldulensis. Plant Cell Reports* 16:787-791.

Murashige, T. and F. Skoog (1962). A revised medium for rapid growth and bioassays with tobacco tissue culture. *Physiologia Plantarum* 15:473-497.

Nagmani, R. and S. Venkateswaran (1983). In vitro culture of hypocotyl and cotyledon segments of *Leucaena. Leucaena Research Reports* 4:88-89.

Naqvi, N.S. and K.G. Mukerji (1998). Mycorrhization of micropropagated *Leucaena leucocephala* (Lam.) deWit. *Symbiosis* 24:103-114.

Nataraja, K. and A.M. Sudhadevi (1984). In vitro induction of plants from seedling explants of subabul (*Leucaena leucocephala* Lam.). *Indian Journal of Plant Physiology* 27:255-258.

Pardha Saradhi, P. and Alia. (1995). Production and selection of somaclonal variants of *Leucaena leucocephala* with high carbon dioxide assimilating potential. *Energy Conservation and Management* 36:759-762.

Peasley, E.L. and G.B. Collins (1980). Development of an in vitro culture system for *Leucaena* K-8. *Leucaena News Letter* 1:54.

Pradhan, C., S. Kar, S. Pattnaik, and P.K. Chand (1998). Propagation of *Dalbergia sissoo* Roxb. through in vitro shoot proliferation from cotyledonary nodes. *Plant Cell Reports* 18:122-126.

Puthur, J.T., K.V.S.K. Prasad, P. Sharmila, and P. Pardha Saradhi (1998). Vesicular arbuscular mycorrhizal fungi improves establishment of micropropagated *Leucaena leucocephala* plantlets. *Plant Cell, Tissue and Organ Culture* 53:41-47.

Rastogi, S. and U.N. Dwivedi (2003). *Agrobacterium tumefaciens*-mediated transformation of *Leucaena leucocephala*-a multipurpose tree legume *Physiology and Molecular Biology of Plants* 9:207-216.

Rastogi, S. and U.N. Dwivedi (2006). Down-regulation of lignin biosynthesis in transgenic *Leucaena leucocephala* harboring O-methyltransferase gene. *Biotechnology Progress* 22:609-616.

Ravishankar, G.A., A. Wali, and S. Grewal (1983). Plantlet formation through tissue cultures of *Leucaena leucocephala. Leucaena Research Reports* 4:37.

Sambrook, J., E.F. Fritsch, and T. Maniatis (1989). *Molecular cloning: A laboratory manual.* Cold Spring Harbor, New York: Cold Spring Harbor Laboratory Press.

Sharma, K.K. and V. Anjaiah (2000). An efficient method for the production of transgenic plants of peanut (*Arachis hypogaea* L.) through *Agrobacterium tumefaciens*-mediated transformation. *Plant Science* 159:7-19.

Sonia, L., A. Gulati, S. Dahiya, R.P. Singh, and P.K. Jaiwal (2000). In vitro multiplication of a multipurpose tree legume, *Tamarindus indica* from cotyledonary nodes. *Physiology and Molecular Biology of Plants* 6:21-25.

Tsai, C.J., G.K. Podila, and V.L. Chiang (1994). *Agrobacterium*-mediated transformation of quaking aspen (*Populus tremuloides*) and regeneration of transgenic aspen. *Plant Cell Reports* 14:94-97.

Vancanneyt, G., R. Schmidt, A. O'Connor-Sanchez, L. Willmitzer, and M. Rocha-Sosa (1990). Construction of an intron-containing marker gene: Splicing of the intron in transgenic plants and its use in monitoring early events in *Agrobacterium*-mediated plant transformation. *Molecular General Genetics* 220: 245-250.

Venkateswaran, S. and V. Gandhi (1982). Mass propagation and genetic improvement of forest trees for biomass production by tissue culture. *Biomass* 2:5-15.

Yan, B., M.S. Srinivasa Reddy, G.B. Collins, and R.D. Dinkins (2000). *Agrobacterium tumefaciens*-mediated transformation of soybean [*Glycine max* (L.) Merrill] using immature zygotic cotyledon explants. *Plant Cell Reports* 19: 1090-1097.

# Chapter 22

# Genetic Transformation of Tree Legume, *Robinia*

Tomohiro Igasaki
Kenji Shinohara

## *INTRODUCTION*

*Robinia pseudoacacia* is a leguminous tree species that has been planted worldwide. In Japan, it is used mainly for landscaping and production of honey. Much research related to in vitro regeneration systems of *R. pseudoacacia* has been reported (Han, Keathley, and Gordon, 1993). However, to our knowledge, only one example of *Agrobacterium rhizogenes*-mediated transformation of *R. pseudoacacia* has been documented (Han, Keathley, Davis et al., 1993). The reported transformation resulted in abnormal leaf morphology. We established a new, simple, and reliable procedure for the transformation of *R. pseudoacacia* by disarmed *A. tumefaciens* (Igasaki et al., 2000). The leaves of transgenic plants generated by our method exhibited no apparent morphological abnormalities.

## *PROCEDURE FOR TRANSFORMATION OF* R. PSEUDOACACIA

### *Plant Material*

Shoot cultures derived from one mature seed were maintained on a medium that contained Murashige and Skoog's basal salts (Murashige and Skoog, 1962), Gamborg's B5 vitamins (Gamborg et al., 1968) and 2 percent (w/v) sucrose (MSB5S medium) supplemented with 0.3 percent (w/v) gellan gum. Shoot cultures were incubated at 25°C under cool white fluorescent light (30 μmol·$m^{-2}$·$s^{-1}$, 16-hour photoperiod) and subcultured every 2 months.

### *Binary Vector and Bacterial Strain*

The binary vector pSMAH621, which contains genes for β–glucuronidase (GUS) and hygromycin phosphotransferase (HPT) in its T-DNA region, was used in this study (Igasaki et al., 2000). Three disarmed strains of *A. tumefaciens,* namely, LBA4404 (pAL4404; Hoekema et al., 1983), EHA101 (pEHA101; Hood et al., 1986) and GV3101 (pMP90; Koncz and Schell, 1986), that harbored the binary vector were used in transformation experiments. *A. tumefaciens* was grown overnight at 28°C in liquid Luria-Bertani medium (Sambrook et al., 1989) in the presence of 100 mg·$l^{-1}$ spectinomycin. For transformation of tissues of *R. pseudoacacia,* an overnight culture was diluted to 5 × $10^8$ cells/ml with liquid *R. pseudoacacia* regeneration medium (RPR medium; MSB5S medium containing 0.05 mg·$l^{-1}$ 2,4-dichlorophenoxyacetic acid, 2.5 mg·$l^{-1}$ *trans*-zeatin, and 2.5 mg·$l^{-1}$ 6-benzyladenine) supplemented with 20 μM acetosyringone (3',5'-dimethoxy-4'-hydroxyacetophenone; Aldrich, Milwaukee, WI).

### *Transformation and Regeneration of Plants*

Leaf and stem segments from shoot cultures of *R. pseudoacacia* were vacuum infiltrated for 20 minutes in the above-mentioned suspension of *A. tumefaciens* (Horsch et al., 1985). Tissues were then blotted with sterile filter paper and incubated for 2 days on RPR medium supplemented with 0.3 percent (w/v) gellan gum and 20 μM acetosyringone. Each segment was washed three times with liquid RPR medium and then once with liquid RPR medium that contained 500 mg·$l^{-1}$ carbenicillin (disodium salt; Sigma, St. Louis, MO) and 500 mg·$l^{-1}$ cefotaxime (sodium salt; Sigma). The tissues were blotted with sterile filter paper and placed on RPR medium containing 20 mg·$l^{-1}$ hygromycin B, 500 mg·$l^{-1}$ carbenicillin, 500 mg·$l^{-1}$ cefotaxime, and 0.3 percent (w/v) gellan gum. The shoots that formed on the calli formed on the tissue segments were excised from calli and transferred to MSB5S medium supplemented with 20 mg·$l^{-1}$ hygromycin B, 500 mg·$l^{-1}$ carbenicillin, 500 mg·$l^{-1}$ cefotaxime, and 0.3 percent (w/v) gellan gum.

## *IDENTIFICATION OF SUITABLE SELECTION MARKER,* A. TUMEFACIENS *STRAIN AND PLANT TISSUE FOR TRANSFORMATION OF* R. PSEUDOACACIA

### *Suitable Selection Marker*

We examined the effects of kanamycin, geneticin, and hygromycin B on the survival of leaf discs of *R. pseudoacacia* in light, which was determined by the extent of chlorosis of leaf discs. Hygromycin B at low concentrations killed leaf discs more effectively than kanamycin at similar concentrations, and geneticin had intermediate effects on survival (Igasaki et al., 2000). These results suggested that the *hpt* gene was the most suitable selection marker for transformation. When we used the binary vector containing the gene for neomycin phosphotransferase (*npt*II), many nontransgenic calli grew up in the presence of kanamycin. Therefore, we used the binary vector pSMAH621, which included *hpt* gene in its T-DNA region.

### *Suitable* A. tumefaciens *Strain*

The frequency of transformation often depends on the strain of *A. tumefaciens* that is used (Bechtold et al., 1993; Wenck et al., 1999). We tested three different strains for transformation of *R. pseudoacacia* (Table 22.1). Strain GV3101 (pMP90) gave the highest frequency of transformation and the frequency was much higher than that obtained with strains LBA4404 (pAL4404) and EHA101 (pEHA101).

TABLE 22.1. Frequency of transformation of *R. pseudoacacia* with different strains of *A. tumefaciens.*

| ***Agrobacterium* strain** | **Type of explant** | **Transformation frequency** |
|---|---|---|
| LBA4404 | Leaf | 0.0 |
| (pAL4404) | Stem | 0.9 |
| EHA101 | Leaf | 1.1 |
| (pEHA101) | Stem | 3.2 |
| GV3101 | Leaf | 14.6 |
| (pMP90) | Stem | 24.1 |

### *Suitable Plant Tissue*

We examined various tissues of *R. pseudoacacia* in terms of their susceptibility to *A. tumefaciens*-mediated transformation and identified stem explants as the best tissues for such transformation (Table 22.1). The percentage of leaf and stem segments that produced hygromycin-resistant calli after transformation was much higher than those of petiole and root segments. Hygromycin-resistant calli derived from leaf and stem segments were obtained on the selection medium within 2 weeks after transformation.

## *CHARACTERIZATION OF TRANSGENIC PLANTS*

### *Regeneration of Transgenic Plants*

Han, Keathley, and Gordon (1993) reported a system for the regeneration of plantlets from calli derived from cambial tissues of *R. pseudoacacia*. Using their system with slight modifications, we were able to regenerate transgenic *R. pseudoacacia* from hygromycin-resistant calli (Figure 22.1 A-C). The frequency of regeneration of transgenic plants from hygromycin-resistant calli was more than 90 percent. Shoots of transgenic *R. pseudoacacia* were induced from the hygromycin-resistant calli on selective RPR medium within 4 weeks. These shoots subsequently formed roots on selective MSB5S medium within 4 weeks. The frequency of regeneration of transgenic *R. pseudoacacia* plants from stem segments was approximately 24 percent (Table 22.1). The morphological features of the transgenic *R. pseudoacacia* plants that had the introduced T-DNA of pSMAH621 in the

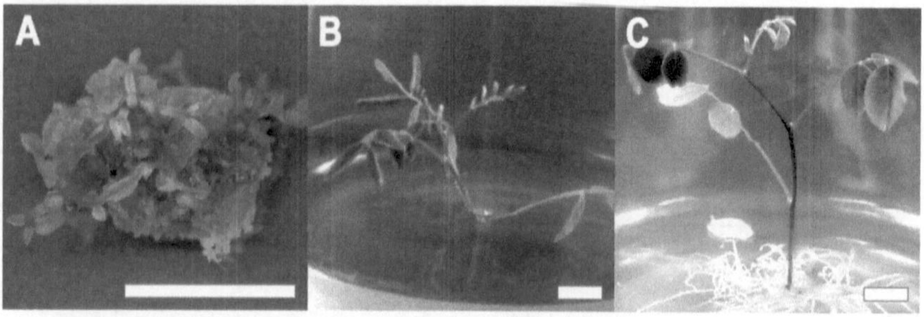


FIGURE 22.1. Regeneration of transgenic *R. pseudoacacia*. (A) Shoot organogenesis from transgenic callus. (B) A transgenic shoot. (C) A transgenic plantlet. Bars: 1 cm, approximately 24 percent (Table 22.1). (See also color gallery.)

genome did not differ from those of nontransgenic plants (Igasaki et al., 2000).

### *Expression of GUS Activity*

Histochemical and fluorimetric assays of GUS activity in transgenic plants were performed as described by Jefferson et al. (1987). The leaves of transgenic plants were strongly positive for GUS activity (Figure 22.2). The GUS activity in the transgenic plants was 38- to 110-fold higher than that in controls (Igasaki et al., 2000). These results suggested that an integrated GUS gene was expressed at high levels under the control of the 35S promoter of cauliflower mosaic virus.

### *Detection of Integrated HPT and GUS Genes in the Genome of Transgenic Plants*

Successful transformation of *R. pseudoacacia* was also confirmed by Southern blotting analysis (Igasaki et al., 2000). No variations in the number of copies of the *hpt* gene were observed among the transgenic plants. The appearance of single bands at different respective positions on Southern blots indicated the presence of approximately one copy of the inserted *hpt* and *gus* genes per diploid genome, as well as the random integration of the T-DNA at different sites in the genome of *R. pseudoacacia* (Igasaki et al., 2000). The variation in GUS activity among the transgenic plants was most likely the result of the random integration of the T-DNA into the ge-

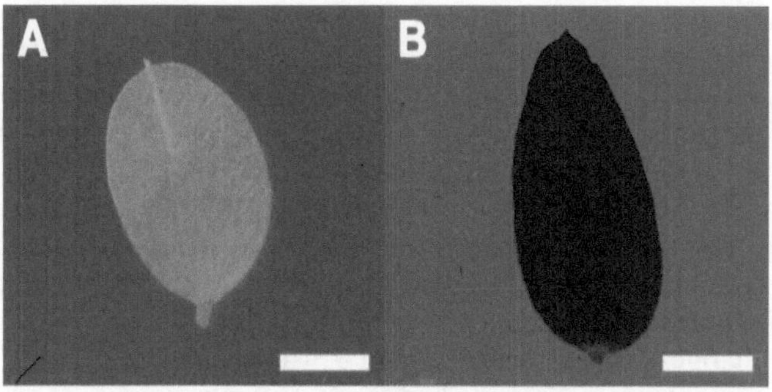


FIGURE 22.2. Histochemical assay of GUS activity in transgenic *R. pseudoacacia.* A control leaf (A) and a leaf of randomly selected transgenic plants (B) were subjected to histochemical staining with 5-bromo-4-chloro-3-indolyl glucuronide (X-Gluc). Bars: 1 cm. (See also color gallery.)

nome of *R. pseudoacacia,* and probably reflected the effects of adjacent genomic sequences on the expression of the *gus* gene.

## *GENERAL CONSIDERATIONS*

We established a simple and reliable procedure for the regeneration of transgenic *R. pseudoacacia*. In this system, hygromycin was the most suitable antibiotic for selection. Hygromycin allowed the selection of transgenic *R. pseudoacacia* cells exclusively and eliminated chimeric and non-transgenic cells. Our results were somewhat inconsistent with the previous report by Han, Keathley, Davis, et al. (1993), who used kanamycin for *A. rhizogenes*-mediated transformation of *R. pseudoacacia*. Transformation using *A. rhizogenes* resulted in leaves with abnormal morphology (Han, Keathley, Davis, et al., 1993). By contrast, transformation using *A. tumefaciens* did not result in any morphological changes in the transgenic plants (Figure 22.1C). The absence of such changes is very important for future genetic engineering of this woody plant because it should allow selective improvement of single traits without the loss of any of the desired traits of the parental line.

Efficient and reproducible transformation systems have been reported for a limited number of broad-leaved trees, in particular, poplars and aspens. These systems have allowed the selective improvement of individual traits, such as resistance to herbicides (De Block, 1990), pests (McCown et al., 1991), and mercury (Rugh et al., 1998). Alterations in the characteristics of wood and in the patterns of growth of aspen have been induced by the constitutive expression of genes for the biosynthesis of indoleacetic acid (Tuominen et al., 1995) and gibberellin (Eriksson et al., 2000). Repression of lignin biosynthesis has been induced in transgenic aspen by down-regulation of the gene for 4-coumarate: coenzyme A ligase, resulting in the promotion of cellulose accumulation and growth (Hu et al., 1999). Weigel and Nilsson (1995) generated an early-flowering variety of transgenic hybrid aspen by inducing expression of the LEAFY gene of *Arabidopsis*. This strategy should be very useful for reducing the generation time of woody plant by inducing precocious flowering. Our group has induced morphological changes in Lombardy poplar (Mohri et al., 1999) and *R. pseudoacacia* (Igasaki, unpublished results) by introduction of a homeobox gene from rice. The present system for the efficient transformation of *R. pseudoacacia* should allow the selective improvement of single traits by the introduction of economically relevant genes that regulate morphological traits and resistance to insects and disease.

## REFERENCES

Bechtold, N., J. Ellis, and G. Pelletier (1993). *In planta Agrobacterium* mediated gene transfer by infiltration of adult *Arabidopsis thaliana* plants. *Comptes Rendus de l'Académie des Sciences–Paris* 316:1194-1199.

De Block, M. (1990). Factors influencing the tissue culture and the *Agrobacterium tumefaciens*-mediated transformation of hybrid aspen and poplar clones. *Plant Physiology* 93:1110-1116.

Eriksson, M.E., M. Israelsson, O. Olsson, and T. Moritz (2000). Increased gibberellin biosynthesis in transgenic trees promotes growth, biomass production and xylem fiber length. *Nature Bio/Technology* 18:784-788.

Gamborg, O.L., R.A. Miller, and K. Ojima (1968). Nutrient requirement of suspension cultures of soybean root cells. *Experimental Cell Research* 50:151-158.

Han, K.H., D.E. Keathley, J.M. Davis, and M.P. Gordon (1993). Regeneration of transgenic woody legume (*Robinia pseudoacacia* L., black locust) and morphological alterations induced by *Agrobacterium rhizogenes* mediated transformation. *Plant Science* 88:149-157.

Han, K.H., D.E. Keathley, and M.P. Gordon (1993). Cambial tissue culture and subsequent shoot regeneration from mature black locust (*Robinia pseudoacacia* L.). *Plant Cell Reports* 12:185-188.

Hoekema, A., P.R. Hirsch, P.J.J. Hooykaas, and R.A. Shilperoort (1983). A binary plant vector strategy based on separation of vir- and T-regions of the *Agrobacterium tumefaciens* Ti-plasmid. *Nature* 303:179-180.

Hood, E.E., G.L. Helmer, R.T. Fraley, and M.D. Chilton (1986). The hypervirulence of *Agrobacterium tumefaciens* A281 is encoded in a region of pTiBo542 outside of T-DNA. *Journal of Bacteriology* 168:1291-1301.

Horsch, R.B., J.E. Fry, N.L. Hoffmann, D. Eichholtz, S.G. Rogers, and R.T. Fraley (1985). A simple and general method for transferring genes into plants. *Science* 227:1229-1231.

Hu, W.-J., S.A. Harding, J. Lung, J.L. Popko, J. Ralph, D.D. Stokke, C.J. Tsai, and V.J. Chiang (1999). Repression of lignin biosynthesis promotes cellulose accumulation and growth in transgenic trees. *Nature Bio/Technology* 17:808-812.

Igasaki, T., T. Mohri, H. Ichikawa, and K. Shinohara (2000). *Agrobacterium tumefaciens*-mediated transformation of *Robinia pseudoacacia. Plant Cell Reports* 19:448-453.

Jefferson, R.A., T.A. Kavanagh, and M.W. Bevan (1987). GUS fusions: β-glucuronidase as a sensitive and versatile gene fusion marker in higher plants. *EMBO Journal* 6:3901-3907.

Koncz, C. and J. Shell (1986). The promoter of TL-DNA gene 5 controls the tissue-specific expression of chimeric genes carried by a novel type of *Agrobacterium* binary vector. *Molecular and General Genetics* 204:383-396.

McCown, B.H., D.E. McCabe, D.R. Russell, D.J. Robinson, K.A. Barton, and K.F. Raffa (1991). Stable transformation of *Populus* and incorporation of pest resistance by electric discharge particle acceleration. *Plant Cell Reports* 9:590-594.

Mohri, T., T. Igasaki, N. Futamura, and K. Shinohara (1999). Morphological changes in transgenic poplar induced by expression of the rice homeobox gene *OSH1*. *Plant Cell Reports* 18:816-819.

Murashige, T. and F. Skoog (1962). A revised medium for rapid growth and bioassays with tobacco cultures. *Physiologia Plantarum* 15:473-497.

Rugh, C.L., J.F. Senecoff, R.B. Meagher, and S.A. Merkle (1998). Development of transgenic yellow poplar for mercury phytoremediation. *Nature Bio/Technology* 16:925-928.

Sambrook, J., E.F. Fritsch, and T. Maniatis (1989). *Molecular cloning: A laboratory manual.* 2nd ed. Cold Spring Harbor, NY: Cold Spring Harbor Laboratory Press.

Tuominen, H., H. Sitbon, C. Jacobsson, G. Sandberg, O. Olsson, and B. Sundburg (1995). Altered growth and wood characteristics in transgenic hybrid aspen expressing *Agrobacterium tumefaciens* T-DNA indoleacetic acid-biosynthetic genes. *Plant Physiology* 109:1179-1189.

Weigel, D. and O. Nilsson (1995). A developmental switch sufficient for flower initiation in diverse plants. *Nature* 377:495-500.

Wenck, A.R., M. Quinn, R.W. Whetten, G. Pullman, and R. Sederoff (1999). High-efficiency *Agrobacterium*-mediated transformation of Norway spruce (*Picea abies*) and loblolly pine (*Pinus taeda*). *Plant Molecular Biology* 39:407-416.

Chapter 23

# *Medicago truncatula* As a Model System in Plant Molecular Biology

Julia Frugoli*

## *MODEL SYSTEM*

Encarta *World English Dictionary* defines the word model as "a simplified version of something complex used, for example, to analyze and solve problems or make predictions." Indeed, the term model system is exceedingly popular in biology. A PubMed search with the keywords "model AND system" brings up almost 100,000 papers and if one tries to limit the results by adding "AND biology" to the search, the number is still in the thousands. This does not include the many additional works that use model systems without using those words. Despite the popularity of the term, a true model system has defined characteristics. In *Molecular Biology of the Gene,* Watson et al. (2004) suggested characteristics of model systems including (1) the availability of powerful tools of traditional and molecular genetics, and (2) the attraction of a critical mass of investigators. For the study of processes unique to legume biology and those better understood in legumes, *Medicago truncatula* deserves the label model system because it meets these criteria.

Inherent in the definition above are several features a model system must incorporate. A model system must be easier to study and manipulate than the systems it represents. A model system must also display the traits and characters one wishes to study. Finally, one must be able to generalize predictions made from a model system to other systems. Ease of manipulation and prediction in a system do no good if the model system is too far re-

*100 Jordan Hall, Department of Genetics and Biochemistry, Clemson University, Clemson, SC 29634, USA (e-mail: jfrugol@CLEMSON.EDU).

moved from the systems it is meant to model to have predictive value. Along these lines, there are a number of processes in plant molecular biology for which legumes in general and *M. truncatula* in particular are the best system for study. The initial development of *M. truncatula* as a model grew out of a need for molecular analysis of symbiotic interactions, both with rhizobia and mycorrhizae (Harrison, 1998). Both these processes are cases in which it is evident that while *Arabidopsis* is a wonderful model system for much of plant molecular biology, it simply cannot be used for studying symbiotic interactions because it does not participate in these symbiotic relationships. In addition, there are other developmental and interactive processes that, while they can be studied in *Arabidopsis,* are more amenable to study in *M. truncatula* because of its life cycle and morphology (size), the range of pathogens and pests to which it is susceptible, and the specific biochemical pathways in legume metabolism. For example, a range of studies including examining defense responses (Chapter 36, this volume), herbivory (Gomez et al., 2005), secondary metabolites (Broeckling et al., 2005), and metal accumulation (Lopez-Millan et al., 2004) were well suited to *M. truncatula*. A third set of processes is different enough in legumes that the *Arabidopsis* system is too far removed to have the necessary predictive value, although it can aid the study. Examples of processes in this category with recent research investigations in *M. truncatula* include legume seed and pod development (Wang and Grusak, 2005), legume floral development (Benlloch et al., 2003), and nodule nutrient exchange and physiology (Carvalho et al., 2003).

## *THE AVAILABILITY OF POWERFUL TOOLS OF TRADITIONAL AND MOLECULAR GENETICS*

### *Traditional Genetics*

Gregor Mendel worked with garden pea and therefore legumes can truly claim the title of the original genetic model system. Any legume supplanting pea as a genetic model system must be at least as amenable to genetic manipulation as pea and *M. truncatula* is. A self-fertile diploid relative of alfalfa with prolific seed production, *M. truncatula* lends itself to classical genetic analysis, but with a generation time of 8 to 10 weeks depending on genotype and vernalization, *M. truncatula* is more amenable than pea to laboratory-scale genetics. Several genetic tools are available, including ecotypes distinguishable by both morphological and molecular markers (Figure 23.1), a male sterile mutant useful for making controlled crosses

FIGURE 23.1. Morphological and molecular markers in *Medicago truncatula*. (A) Leaves of the two ecotypes most commonly used for mapping. Left, A17 ecotype with chevron markings on leaves; right, A20 ecotype with speckles on the leaves. (B) Agarose gel electrophoresis of a cleaved amplified polymorphic sequence marker for use in mapping in an A17 × A20 segregating population. The PCR product contains a restriction site in the A20 ecotype that is absent in A17, allowing one to score the codominant marker in each plant. (A small fragment resulting from the digestion in A20 and the heterozygous plant is not shown on the gel picture.) (See also color gallery.)

(Penmetsa and Cook, 2000), and a community-recognized registration process for mutant gene names (Frugoli and VandenBosch, 2001).

*M. truncatula* is nodulated by the well-characterized and completely sequenced symbiont *Rhizobium meliloti,* with the benefit of the availability of hundreds of *R. meliloti* mutants and several synthetic Nod factors (Dénarié and Debelle, 1996; Long, 1996). *M. truncatula* mutants have been generated by gamma irradiation, ethyl methyl sulfonate (EMS), and fast neutron bombardment, but researchers have also taken advantage of differences in traits between the numerous ecotypes documented at http://www.noble.org/ medicago/ecotypes.html to perform genetic analyses.

### *Reverse Genetics*

Reverse genetics approaches identify mutants in a gene of interest based on sequence differences, and such approaches based on classical mutagenesis are available in *M. truncatula*. One approach, TILLING (Targeting Induced Local Lesions in Genomes), has proven useful in *Arabidopsis* and other plants (reviewed in Henikoff et al., 2004). The system has been developed in the Cook Lab at University of California, Davis for a population of EMS mutagenized *M. truncatula* plants (VandenBosch and Stacey, 2003). Briefly, the sequence of a gene of interest is analyzed with a computer program that determines the consequence of all possible EMS mutations (primarily G to A transitions) on the amino acid sequence of the deduced protein. Regions are chosen for PCR polymerase chain reaction (PCR) amplification based on the concept that those regions most likely to result in altered protein function are highly conserved domains in proteins because they are more likely to be required for function. PCR primers are designed to amplify these regions of DNA from each plant and the PCR products are analyzed for single base pair changes on a sequencing gel using an enzyme that detects and cleaves single base mismatches in DNA. Using high throughput methodology and a well well-characterized and curated population of mutagenized plants, a plant containing a lesion in the gene of interest can be identified in days. A second approach, which takes advantage of a curated population of fast-neutron-generated mutants, is being developed at the Noble Foundation and the John Innes Centre. A high-throughput PCR-based screen allows one to identify plants with short deletions in the gene of interest. Both approaches have required a significant initial commitment of time and money by the laboratories developing them, but the ability to rapidly screen for multiple mutations in a single gene makes such resources invaluable.

Recently, researchers have identified a retrotransposon, *Tnt1,* that moves randomly in *M. truncatula* but only upon passage through tissue culture (d'Erfurth et al., 2003). This retrotransposon can be used to generate a large population of plants with tagged mutation sites, an important resource for both forward and reverse genetics. A population mutagenized by *Tnt1* can be used for reverse genetic screens by sequencing of tagged sites and forward genetic screens by observation of phenotypes (Brocard et al., Chapter 24, this volume).

In addition to the reverse genetic resources that rely on a population of mutagenized plants described above, the ability to transform *M. truncatula* allows the use of techniques such as RNA interference to silence genes of interest. The efficacy of this technique has been documented (Limpens et al., 2004) including use in whole plants and in transformed roots. A large-scale project to identify gene function by silencing in *M. truncatula* is underway (http://www.nsf.gov/awardsearch/showAward.do?AwardNumber=0421676).

### *Molecular Genetics Tools*

What sets *M. truncatula* apart as a model system are the wealth of molecular genetic tools, including transformation, molecular markers, physical, genetic, and cytogenetic maps, genome sequencing, and informatics resources available in this species. A partial list of genome-scale projects in *M. truncatula* is available in (VandenBosch and Stacey, 2003) and multiple articles in the April 2006 edition of *Current Opinion in Plant Biology.*

### *Transformation*

Critical to a molecular genetic system is the ability to transform plants with foreign DNA. *M. truncatula* is transformable by *Agrobacterium tumefaciens* via tissue culture and regeneration (Trieu and Harrison, 1996; Zhou et al., 2004; Chabaud et al., Chapter 4, this volume) to obtain transgenic plants in about 3 months. For more rapid analysis of gene action in roots, *Agrobacterium rhizogenes*-induced hairy roots can be generated easily in 3 to 4 weeks (Boisson-Dernier et al., 2001). Finally, an initial report of transformation via vacuum infiltration of *Agrobacterium tumefaciens* (Trieu et al., 2000) suggests rapid, large-scale transformations may be possible, although the experiment has proven difficult to repeat.

### *Markers and Maps*

Many polymorphic ecotypes have been gathered (Bonin et al., 1996) and hundreds of codominant PCR markers, SNPs, and AFLP markers are distributed throughout the genome, so that a detailed genetic map has emerged (Kulikova et al., 2001; Schnabel et al., 2003; Choi et al., 2004). Four BAC libraries generated using three different restriction enzymes have been constructed and are publicly available through the Clemson University Genomics Institute (https://www.genome.clemson.edu/cgi-bin/orders?page=service Browse&service=bacrc), and a BAC-based physical map of the genome has been generated (Choi et al., 2004). FISH analysis of chromosomes (Kulikova et al., 2001; and Kulikova Chapter 25, this volume) and a web database (www.medicago.org) allow the community to link molecular and genetic data. The synteny between *M. truncatula* and other legumes has been used to clone genes in less tractable legumes, such as the *SYM2* locus in pea (Gualtieri et al., 2002; Limpens et al., 2003) and the *NORK* locus in alfalfa (Endre et al., 2002). When the *M. truncatula* genome is completed, opportunities to use *M. truncatula* information to clone genes in other legumes will increase.

### *Genome Sequencing*

The National Science Foundation in the USA and the European Union have committed to sequencing the gene space in *M. truncatula* by 2008. The result of this international program will be the first complete legume genome, a critical resource for comparative work among legumes. The sequencing has been divided by linkage group, with the EU Medicago Consortium sequencing linkage groups 3 and 5, and the U.S. Consortium (TIGR, the University of Oklahoma, and the University of Minnesota) sequencing the rest of the linkage groups. The project was jump-started by a substantial initial investment by the Noble Foundation for trial sequencing of a set of anchor BACs by Bruce Roe at the University of Oklahoma. The project has proceeded rapidly, and as of October 2006, approximately 60 percent of the gene space had been completed, with 40 to 50 percent of the overall genome captured in the sequence assembly (Nevin Young, personal communication.). Sequence is helpful, but the bioinformatics tools described below, which are also a part of the sequencing effort, make the information accessible and useful for all researchers in *M. truncatula* and other legumes. Annotation of the genome is an international effort (Town, 2006).

### *Transcriptomics*

Large-scale expressed sequence tag (EST) sequencing is essential for functional genomics studies, permitting the direct identification of large gene collections and setting the stage for further analysis, such as those using DNA microarray technology. Several large EST projects have been completed (Covitz et al., 1998; Györgyey et al., 2000; Bell et al., 2001; Journet et al., 2002). The analysis of the over 100,000 ESTs isolated from many different libraries constructed from diverse stages and treatments that came out of these projects is facilitated by searchable databases such as MtDB2 (Lamblin et al., 2003) and the TIGR Gene Index (http://www.tigr.org). Both microarray and macroarray analyses of gene expression changes during symbiosis have already been published (Liu et al., 2003; El Yahyaoui et al., 2004; Lohar et al., 2006), and an Affymetrix chip with 16,000 bioinformatically optimized oligonucleotides is available. *M. truncatula* microarray chips might also be useful in the study of closely related species such as *M. sativa,* which shares high sequence identity with *M. truncatula.*

Another transcriptomic approach, Serial Analysis of Gene Expression (SAGE), is a method for comprehensive analysis of gene expression patterns using short sequence tags obtained from a unique position within each transcript (10-14 bp) to uniquely identify a transcript. The expression level of the corresponding transcript is determined by quantifying of the number of times a particular tag is observed. A project applying SAGE to *M. truncatula* is underway at the Center for *Medicago* Genomics Research at the Nobel Foundation (http://www.noble.org/medicago/GEP.html).

### *Proteomics*

In order for proteomic approaches to be useful in a system, a large sequence resource is necessary to match the sequences of peptides generated in tryptic digests to their proteins of origin. The growing sequence resource in all legumes allows identification of *M. truncatula* proteins by their mass spectra. At present, the only plants with enough sequence data to make efficient use of this technology are *M. truncatula,* rice, poplar and *Arabidopsis*. A comprehensive review of considerations important in proteomics technology and applications in *M. truncatula* and *Arabidopsis* was recently published (Agrawal et al., 2005a, 2005b). Proteomics approaches in *M. truncatula* include analyses of seed development (Gallardo et al., 2003), pathogen interactions (Colditz et al., 2004), symbiosome membranes (Catalano et al., 2004), root microsomes (Valot et al., 2004), and other organ/or tissue tissue-specific approaches (Mathesius et al., 2001; Watson et al., 2003; Imin et al., 2004).

### *Metabolomics*

Alfalfa produces a number of secondary metabolites of great interest because of their contributions to human health and animal forage quality. The principle behind metabolomics is that metabolic profiling on a genomic scale offers a view of the metabolic status of an organism, which can lend insight to study gene function or whole plant biology (Trethewey et al., 1999). Successful attempts to link proteomics, transcriptomics, and metabolomics for cell cultures in *M. truncatula* have emerged from these studies (Broeckling et al., 2005; Suzuki et al., 2005).

Metabolomics is a new and evolving science, and requires specialized equipment and multifaceted technical strategies. The Nobel Foundation employs a strategy that utilizes sequential or selective extraction followed by parallel analyses. The parallel analyses achieve a comprehensive view of the metabolome with high-performance liquid chromatography (HPLC), capillary electrophoresis (CE), gas chromatography (GC), mass spectrometry (MS), and various combinations of the above techniques such as GC/

MS, LC/MS, and CE/MS. In addition to studying biological responses to biotic and abiotic elicitors in *M. truncatula* cell cultures, these techniques are being applied to the study of natural variants in *M. truncatula, M. truncatula* development, lignin biosynthesis, and legume-insect interactions.

### *Bioinformatics Tools*

All of the "omics"-scale tools discussed above necessitate a strong bioinformatics infrastructure for the species. *M. truncatula* bioinformatics resources are multitude. A good place to begin is the Medicago Consortium Web site, http://www.medicago.org. Linked from this page are informatics tools such as Ensemble, which allow a real time view of the annotation of the genome, and tools allowing browsing of the genome for markers, genes, the location of BACs, the status of the sequencing project, or the sequence status of any individual BAC. Users can also view the contigs assembled for sequencing, and make comparisons to other legumes through the Legume Information System (Gonzales et al., 2005) and the Consensus Legume Database (legumes.org). Tools are also available through links from the medicago.org Web site for examining ESTs (TIGR, MtDB2, MENS), and in the future, examining microarray data. *In silico* approaches in *M. truncatula* have led to important insights, such as the identification of a large family of small legume-specific transcripts with conserved cysteine motifs, whose function continues to be investigated (Fedorova et al., 2002; Mergaert et al., 2003; Graham et al., 2004).

## *THE RESEARCH COMMUNITY*

The strength of a model lies not only in its ability to solve problems or make predictions, but also in the number of people who adopt the model. Systems such as *Drosophila* and yeast are strong models for many biological processes not only because of their ease of use, but because of the enormous amount of data accumulated over time about numerous aspects of the system. Each group that uses the model to answer a question adds to the total knowledge base, enriching the power of the model. This is true for *Medicago truncatula* as well.

An international steering committee, elected by *M. truncatula* researchers plans biannual *M. truncatula* meetings, identifies community needs, and works to address them. It was the international committee that first identified the need for a set of rules for the naming of mutants, leading to a community standard (Frugoli and VandenBosch, 2001). The committee also works with members of the community to encourage large-scale projects, such as the development of the Affymetrix chip and current efforts to establish an *M. truncatula* stock center and an online protocol handbook. The committee consists of a mix of researchers from all over the globe and several levels of seniority, reflecting the diversity of the *M. truncatula* research community.

Resources available to the scientific community at large are gathered at Medicago.org. Members of the steering committee are listed on the Web site, as well as protocols for *M. truncatula truncatula*-specific techniques, a collection of important links to other legume resources, *M. truncatula*-specific literature search tools, and a "phonebook" of *M. truncatula* researchers. A listserve has also been established and directions for joining are given at the site.

In the past few years, *M. truncatula*'s power as a model has increased as more researchers adopted it for study, allowing "omics" level biology to come to legumes. An examination of the number of research reports in PubMed using *M. truncatula* over the past ten 10 years shows an exponen-

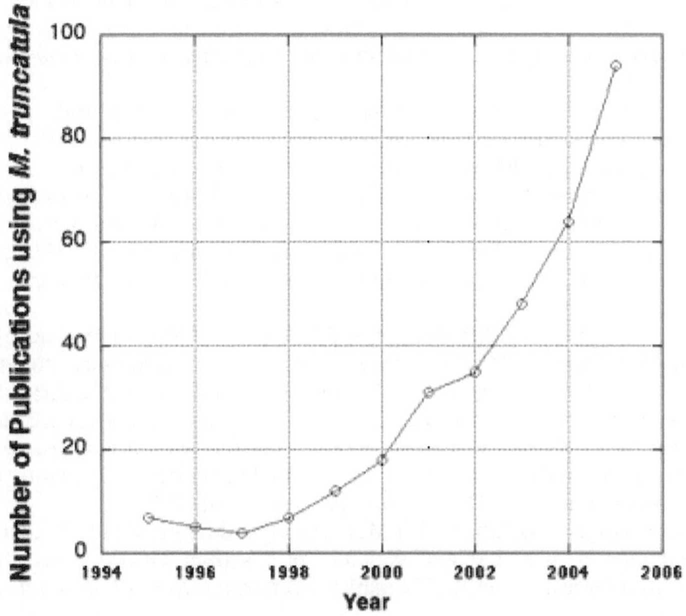


FIGURE 23.2. Number of publications cited in NCBI's PubMed that make reference to *Medicago truncatula* for the period 1994-2006. Data were obtained by sequential searching of PubMed by year, using the term *Medicago truncatula.*

tial increase in the number of papers, with no signs of leveling off, reflecting the popularity of the system (Figure 23.2). Many of the papers in the last few years were in high-impact journals such as *Science* and *Nature*. The result of the increasing number of people using the *M. truncatula* system directly or indirectly to address biological problems in legumes is an international research community that communicates, coordinates, and collaborates for the benefit of individual scientists and legume biology in general.

## *ACKNOWLEDGMENTS*

The author would like to thank Dr. Harry D. Kurtz Jr., Elise Schnabel, and the late Dr. Shu-hua Cheng for helpful comments on the manuscript.

## REFERENCES

Agrawal, G.K., M. Yonekura, Y. Iwahashi, H. Iwahashi, and R. Rakwal (2005a). System, trends and perspectives of proteomics in dicot plants, Part I: Technologies in proteome establishment. *Journal of Chromatographic Biochemistry and Analytical Technology in the Biomedical and Life Sciences* 5:109-123.

Agrawal, G.K., M. Yonekura, Y. Iwahashi, H. Iwahashi, and R. Rakwal (2005b). System, trends and perspectives of proteomics in dicot plants, Part II: Proteomes of the complex developmental stages. *Journal of Chromatographic Biochemistry and Analytical Technology in the Biomedical and Life Sciences* 5:125-136.

Bell, C.J., R.A. Dixon, A.D. Farmer, R. Flores, J. Inman, R.A. Gonzales, M.J. Harrison, N.L. Paiva, A.D. Scott, J.W. Weller, and G.D. May (2001). The *Medicago* genome initiative: A model legume database. *Nucleic Acids Research* 29:114-117.

Benlloch, R., C. Navarro, J.P. Beltran, and L.A. Canas (2003). Floral development of the model legume *Medicago truncatula:* Ontogeny studies as a tool to better characterize homeotic mutations. *Sexual Plant Reproduction* 15:231-241.

Boisson-Dernier, A., M. Chabaud, F. Garcia, G. Becard, C. Rosenberg, and D.G. Barker (2001). *Agrobacterium-rhizogenes*-transformed roots of *Medicago truncatula* for the study of nitrogen-fixing and endomycorrhizal symbiotic associations. *Molecular Plant-Microbe Interactions* 14:695-700.

Bonin, I., T. Huguet, M. Gheratdi, J.M. Prosperi, and I. Olivieri (1996). High level of polymorphism and spatial structure in a selfing plant species *Medicago truncatula* Gaertn. using RAPD markers. *American Journal of Botany* 83:843-855.

Broeckling, C.D., D.V. Huhman, M.A. Farag, J.T. Smith, G.D. May, P. Mendes, R.A. Dixon, and L.W. Sumner (2005). Metabolic profiling of *Medicago truncatula* cell cultures reveals the effects of biotic and abiotic elicitors on metabolism. *Journal of Experimental Botany* 56:323-336.

Carvalho, H.G., I.A. Lopes-Cardoso, L.M. Lima, P.M. Melo, and J.V. Cullimore (2003). Nodule-specific modulation of glutamine synthetase in transgenic *Medicago truncatula* leads to inverse alterations in asparagine synthetase expression. *Plant Physiology* 133:243-252.

Catalano, C.M., W.S. Lane, and D.J. Sherrier (2004). Biochemical characterization of symbiosome membrane proteins from *Medicago truncatula* root nodules. *Electrophoresis* 25:519-531.

Choi, H.K., D.J. Kim, T. Uhm, E. Limpens, H. Lim, J.H. Mun, P. Kalo, R.V. Penmetsa, A. Seres, O. Kulikova, et al. (2004). A sequence-based genetic map of *Medicago truncatula* and comparison of marker colinearity with *M. sativa*. *Genetics* 166:1463-1502.

Colditz, F., O. Nyamsuren, K. Niehaus, H. Eubel, H.P. Braun, and F. Krajinski (2004). Proteomic approach: Identification of *Medicago truncatula* proteins induced in roots after infection with the pathogenic oomycete *Aphanomyces euteiches*. *Plant Molecular Biology* 55:109-120.

Covitz, P.A., L.S. Smith, and S.R. Long (1998). Expressed sequence tags from a root-hair-enriched *Medicago truncatula* cDNA library. *Plant Physiology* 117: 1325-1332.

Dénarié, J. and F. Debelle (1996). *Rhizobium* lipo-chitooligosaccharide nodulation factors: Signaling molecules mediating recognition and morphogenesis. *Annual Review of Biochemistry* 65:503-535.

d'Erfurth, I., V. Cosson, A. Eschstruth, H. Lucas, A. Kondorosi, and P. Ratet (2003). Efficient transposition of the *Tnt1* tobacco retrotransposon in the model legume *Medicago truncatula*. *Plant Journal* 34:95-106.

El Yahyaoui, F., H. Kuster, B. Ben Amor, N. Hohnjec, A. Puhler, A. Becker, J. Gouzy, T. Vernie, C. Gough, A. Niebel, et al. (2004). Expression profiling in *Medicago truncatula* identifies more than 750 genes differentially expressed during nodulation, including many potential regulators of the symbiotic program. *Plant Physiology* 136:3159-3179.

Endre, G., A. Kereszt, Z. Kevei, S. Mihacea, P. Kalo, and G.B. Kiss (2002). A receptor kinase gene regulating symbiotic nodule development. *Nature* 417:962-966.

Fedorova, M., J. van de Mortel, P.A. Matsumoto, J. Cho, C.D. Town, K.A. VandenBosch, J.S. Gantt, and C.P. Vance (2002). Genome-wide identification of nodule-specific transcripts in the model legume *Medicago truncatula*. *Plant Physiology* 130:519-537.

Frugoli, J. and K. VandenBosch (2001). Standards and guidelines for genetic nomenclature for the model legume *Medicago truncatula*. *Molecular Plant-Microbe Interactions* 13:1364-1367.

Gallardo, K., C. Le Signor, J. Vandekerckhove, R.D. Thompson, and J. Burstin (2003). Proteomics of *Medicago truncatula* seed development establishes the time frame of diverse metabolic processes related to reserve accumulation. *Plant Physiology* 133:664-682.

Gomez, S., M. Cox, J. Bede, K. Inoue, H. Alborn, J. Tumlinson, and K. Korth (2005). Lepidopteran herbivory and oral factors induce transcripts encoding novel terpene synthases in *Medicago truncatula. Archives of Insect Biochemistry and Physiology* 58:114-127.

Gonzales, M.D., A. Farmer, K. Gajendran, D. Grant, R. Shoemaker, W.D. Beavis, and M.E. Waugh (2005). The Legume Information System (LIS): An integrated information resource for comparative legume biology. *Nucleic Acids Research* 33:660-665.

Graham, M.A., K.A. Silverstein, S.B. Cannon, and K.A. VandenBosch (2004). Computational identification and characterization of novel genes from legumes. *Plant Physiology* 135:1179-1197.

Gualtieri, G., O. Kulikova, E. Limpens, D.J. Kim, D.R. Cook, T. Bisseling, and R. Geurts (2002). Microsynteny between pea and *Medicago truncatula* in the *SYM2* region. *Plant Molecular Biology* 50:225-235.

Györgyey, J., D. Vaubert, J.L. Jiménez-Zurdo, C. Charon, T. Troussard, A. Kondorosi, and E. Kondorosi (2000). Analysis of *Medicago truncatula* nodule-expressed sequence tags. *Molecular Plant-Microbe Interactions* 13:62-71.

Harrison, M.J. (1998). Development of the arbuscular mycorrhizal symbiosis. *Current Opinion in Plant Biology* 1:360-365.

Henikoff, S., B.J. Till, and L. Comai (2004). TILLING: Traditional mutagenesis meets functional genomics. *Plant Physiology* 135:630-636.

Imin, N., F. De Jong, U. Mathesius, G. van Noorden, N.A. Saeed, X.D. Wang, R.J. Rose, and B.G. Rolfe (2004). Proteome reference maps of *Medicago truncatula* embryogenic cell cultures generated from single protoplasts. *Proteomics* 4:1883-1896.

Journet, E.P., D. van Tuinen, J. Gouzy, H. Crespeau, V. Carreau, M.J. Farmer, A. Niebel, T. Schiex, O. Jaillon, O. Chatagnier, et al. (2002). Exploring root symbiotic programs in the model legume *Medicago truncatula* using EST analysis. *Nucleic Acids Research* 30:5579-5592.

Kulikova, O., G. Gualtieri, R. Geurts, D.J. Kim, D. Cook, T. Huguet, J.H. de Jong, P.F. Fransz, and T. Bisseling (2001). Integration of the FISH pachytene and genetic maps of *Medicago truncatula. Plant Journal* 27: 49-58.

Lamblin, A.-F. J., J.A. Crow, J.E. Johnson, K.A.T. Silverstein, T.M. Kunau, A. Kilian, D. Benz, M. Stromvik, G. Endré, K.A. VandenBosch, et al. (2003). MtDB: A database for personalized data mining of the model legume *Medicago truncatula* transcriptome. *Nucleic Acids Research* 31:196-201.

Limpens, E., C. Franken, P. Smit, J. Willemse, T. Bisseling, and R. Geurts (2003). LysM receptor kinases regulating rhizobial Nod-factor induced infection. *Science* 302:630-633.

Limpens, E., J. Ramos, C. Franken, V. Raz, B. Compaan, H. Franssen, T. Bisseling, and R. Geurts (2004). RNA interference in *Agrobacterium rhizogenes*-transformed roots of *Arabidopsis* and *Medicago truncatula. Journal of Experimental Botany* 55:983-992.

Liu, J., L.A. Blaylock, D. Endre, J. Cho, C.D. Town, K. VandenBosch, and M.J. Harrison (2003). Transcript profiling coupled with spatial expression analyses reveals genes involved in distinct developmental stages of an arbuscular mycorrhizal symbiosis. *Plant Cell* 15:2106-2123.

Lohar, D.P., N. Sharopova, G. Endre, S. Penuela, D. Samac, C. Town, K.A. Silverstien, and K.VandenBosch (2006). Transcript analysis of early nodulation events in *Medicago truncatula. Plant Physiology* 140:221-234.

Long, S. (1996). *Rhizobium* symbiosis: Nod factors in perspective. *The Plant Cell* 8:1885-1889.

Lopez-Millan, A. F., D.R. Ellis, and M.A. Grusak (2004). Identification and characterization of several new members of the ZIP family of metal ion transporters in *Medicago truncatula. Plant Molecular Biology* 54:583-596.

Mathesius, U., G. Keijzers, S.H.A. Natera, J.J. Weinman, M.A. Djordjevic, and B.G. Rolfe (2001). Establishment of a root proteome reference map for the model legume *Medicago truncatula* using the expressed sequence tag database for peptide mass finger printing. *Proteomics* 1:1424-1440.

Mergaert, P., K. Nikovics, Z. Kelemen, N. Maunoury, D. Vaubert, A. Kondorosi, and E. Kondorosi (2003). A novel family in *Medicago truncatula* consisting of more than 300 nodule-specific genes coding for small, secreted polypeptides with conserved cysteine motifs. *Plant Physiology* 132:161-173.

Penmetsa, R.V. and D.R. Cook (2000). Isolation and characterization of diverse developmental mutants in *Medicago truncatula. Plant Physiology* 123:1387-1397.

Schnabel, E., O. Kulikova, T. Bisseling, R. V. Penmetsa, D. Cook, and J. Frugoli (2003). An integrated physical, genetic and cytogenetic map around the *sunn* locus of *M. truncatula. Genome* 46:665-672.

Suzuki, H., M.S. Reddy, M. Naoumkina, N. Aziz, G.D. May, D.V. Huhman, L.W. Sumner, J.W. Blount, P. Mendes, and R.A. Dixon (2005). Methyl jasmonate and yeast elicitor induce differential transcriptional and metabolic re-programming in cell suspension cultures of the model legume *Medicago truncatula. Planta* 220:696-707.

Town, C. (2006). Annotating the genome of *Medicago truncatula. Current Opinion in Plant Biology* 9:95-98.

Trethewey, R.N., A.J. Krotzky, and L. Willmitzer (1999). Metabolic profiling: A rosetta stone for genomics? *Current Opinion in Plant Biology* 2:83-85.

Trieu, A.T., S.H. Burleigh, I.V. Kardailsky, I.E. Maldonado-Mendoza, W.K. Versaw, L.A. Blaylock, H. Shin, T.J. Chiou, H. Katagi, G.R. Dewbre, et al. (2000). Transformation of *Medicago truncatula* via infiltration of seedlings or flowering plants with *Agrobacterium. Plant Journal* 22:531-541.

Trieu, A.T. and H.M. J. Harrison (1996). Rapid transformation of *Medicago truncatula*: Regeneration via shoot organogenesis. *Plant Cell Reports* 16:6-11.

Valot, B., S. Gianinazzi, and D.G. Eliane (2004). Sub-cellular proteomic analysis of a *Medicago truncatula* root microsomal fraction. *Phytochemistry* 65:1721-1732.

VandenBosch, K. and G. Stacey (2003). Summaries of legume genomics projects from around the globe. Community resources for crops and models. *Plant Physiology* 131:840-865.

Wang, H.L. and M.A. Grusak (2005). Structure and Development of *Medicago truncatula* pod wall and seed coat. *Annals of Botany* 95:737-747.

Watson, B.S., V.S. Asirvatham, L. Wang, and L.W. Sumner (2003). Mapping the proteome of barrel medic (*Medicago truncatula*). *Plant Physiology* 131:1104-1123.

Watson, J., T. Baker, S. Bell, A. Gann, M. Levine, and R. Losnick (2004). *Molecular Biology of the Gene.* San Francisco, CA: Pearson Benjamin Cummings & Cold Spring Harbor Laboratory Press.

Zhou, X., M.B. Chandrasekharan, and T.C. Hall (2004). High rooting frequency and functional analysis of GUS and GFO expression in transgenic *Medicago truncatula* A17. *New Phytologist* 162:813-822.

## Chapter 24

# Reverse Genetic Approaches in *Medicago truncatula*

Lysiane Brocard
Isabelle d'Erfurth
Adam Kondorosi
Pascal Ratet

### *REVERSE GENETICS IN PLANTS*

Reverse genetics is a powerful tool to elucidate gene functions and to contribute, for example, to the elucidation of the function of many genes revealed by sequencing the genome of a model organism. This gene function discovery is generally done by altering the expression or by disrupting the gene under study. Several methodologies are used to carry out these reverse genetic studies. In both yeast, *Saccharomyces cerevisiae,* and the mouse, homologous recombination can be used specifically to delete genes of interest and to analyze the effect of the mutation on growth and development. This approach has been successfully used to identify essential genes in *S. cerevisiae* (Ross-Macdonald et al., 1999; Lucau-Danila et al., 2000) and similar programs have been initiated in the case of the mouse (Zambrowicz et al., 1998). In the worm *Caenorhabditis elegans,* RNAi technology was applied to inactivate all single genes of this organism (Kamath et al., 2001). Thus, reverse genetics is complementary and may be as important as genome sequencing for the molecular identification of biological functions.

Although systematic disruption by homologous recombination is not possible in higher plants, reverse genetics can be efficiently applied to the model species using genetic and genomic tools. For example, the very efficient flower dip transformation method using the soil bacterium *Agrobacterium tumefaciens* in *Arabidopsis thaliana* has allowed the construction of large T-DNA mutant collections that can also be used in reverse genetic studies (Krysan et al., 1999; Sallaud et al., 2004). In other plants, insertion mutagenesis using the T-DNA or transposons or retrotransposons in rice (*Oryza sativa*), maize (*Zea mays*), and *Antirrhinum* (Carpenter and Coen, 1990) have been very helpful tools to identify developmental mutants and can also be used as reverse genetic tools to identify gene functions (Tissier et al., 1999; Hirochika, 2001; Szabados et al., 2002). Thus, reverse genetics in plants was historically based mainly on the use of insertion mutagenesis and more recently on RNAi technology (Baulcombe, 2004), chemical mutagenesis for TILLING (Targeting Induced Local Lesions in Genomes; Henikoff et al., 2004), or gamma ray and fast neutron mutagenesis for PCR-based mutation screening (Li et al., 2001).

Efficient molecular tools were only recently developed for legume plants despite their long-standing agronomic interest for human and animal nutrition. These tools help elucidate biological functions in this group of plants such as symbiosis with bacteria and fungi (mycorrhiza) as well as production of secondary metabolites that are absent in *Arabidopsis*.

## *DEVELOPMENT OF MOLECULAR TOOLS FOR MODEL LEGUMES*

Despite the pioneering work of Mendel with pea (*Pisum sativum*) to set up the basis of genetics, legume plants have lagged behind in molecular genetic studies. More recently, the development of genetic tools in legumes was triggered by studies of plant-microbe symbiotic interactions and resulted in the development of two plant systems, *Lotus japonicus* and *Medicago truncatula* (Cook et al., 1997; Stougaard, 2001; Tadege et al., 2005; Sato and Tabata, 2006), that allowed through molecular studies the identification of a few mutant legume genes (Endre et al., 2002; Krusell et al., 2002; Nishimura et al., 2002; Stracke et al., 2002).

*L. japonicus* is a diploid, autogamous species with a good seed set and a generation time of approximately 3 months. *L. japonicus* is susceptible to *Agrobacterium* transformation and can be regenerated through tissue culture (Handberg and Stougaard, 1992). Using T-DNA as an insertion tool, this capacity allowed better understanding of nodule organogenesis and identification of new genes (Schauser et al., 1998).

*M. truncatula* is an annual relative of the cultivated alfalfa (*Medicago sativa*) and pea (*P. sativum*) with a broad genetic variability represented by numerous ecotypes (Bonin et al., 1996) and with various symbiotic specificities with different *Sinorhizobium meliloti* strains (Tirichine et al., 2000). Its small, diploid genome (Blondon et al., 1994) and efficient transformation capacity (Cosson et al., 2006) make this plant a suitable model for studies on various aspects of legume molecular genetics and genomics (Oldroyd and Geurts, 2001; Stacey et al., 2006; www.isv.cnrs-gif.fr/embo01/). In addition, about 225,129 expression sequence tag (EST) sequences are now available (Frugoli and Harris, 2001; www.ncbi.nlm.nih.gov/dbEST/dbEST_ summary.html) and sequencing of the euchromatic part of its genome has been initiated (www.genome.ou.edu/*Medicago*.html).

As large scale T-DNA mutagenesis necessary for construction of large mutant collections in model legume plants is not yet feasible, alternative routes for mutagenesis need to be explored and developed. Insertion mutagenesis with heterologous transposable elements is one alternative way to identify further legume symbiotic genes as well as genes of agronomic importance. Tilling on ethyl methyl sulfonate (EMS) treated populations and PCR-based screening of fast neutron populations are other reverse genetic tools that are developed to exploit more efficiently the traditional large mutant collections.

Next we describe the state of art of these different technologies in *M. truncatula*.

## *REVERSE GENETICS TOOLS IN LEGUMES*

### *T-DNA Technology*

T-DNA technology uses the natural capacity of the pathogenic bacterium *Agrobacterium* to transfer part of its DNA (called the T-DNA) into the nuclear genome of plant cells (reviewed in Tinland, 1996). Two types of *Agrobacterium* strains are used for genetic transformation. In wild-type *A. tumefaciens* strains, the T-DNA genes encode oncogenes that will induce the formation of a tumor on the infected plant tissue. In the *A. rhizogenes* strains, the T-DNA genes encode oncogenes that will induce the production of adventitious roots called hairy root tissue. This is used to produce rapidly chimeric plants (Boisson-Dernier et al., 2001) with the untransformed aerial part and transgenic roots cotransformed with the *Ri* T-DNA and the construct of interest. These chimeric plants can be used to study biological problems restricted to the root system.

The T-DNA transformation process is used to introduce genes in plants but also to inactivate plant genes by insertion mutagenesis. The transfer of the T-DNA to the plant nucleus depends on the expression of the bacterial *vir* genes that delimit the extent of the DNA sequence transferred to the nucleus, by recognizing specific sequences called T-DNA right and left borders (RB and LB). In between these borders any DNA construct can be introduced and transferred into the plant genome. Thus, in order to construct transgenic plants, the oncogenes are deleted from the T-DNA and replaced by a gene of interest generally associated to a selectable gene. The integration of the T-DNA in the genome probably depends on the plant DNA repair machinery. Generally one copy of the T-DNA is inserted randomly in the plant genome, preferentially in transcribed regions or in their vicinity.

For *Arabidopsis,* efficient *in planta* transformation protocols were developed (Bechtold and Pelletier, 1998; Clough and Bent, 1998; Bent, 2000). These simplified *in planta* transformation protocols (flower dip) have allowed the large-scale production of transgenic plants necessary for T-DNA tagging strategies resulting in the production of large mutant collections (Ostergaard and Yanofsky, 2004). In legumes, such *in planta* protocols do not work and the T-DNA plant transformation protocols mostly rely on in vitro regeneration of the transformed tissues (see Chapter 4, this volume) and vary from one plant to another. In *L. japonicus,* T-DNA tagging was initiated (Schauser et al., 1998; Webb et al., 2000) allowing the isolation of developmentally regulated *gus*-gene fusions. However, this did not result in the isolation of disrupted mutants. In this work, mutant plants were characterized, but the mutations resulted from somaclonal variations.

Several laboratories have reported the in vitro transformation and regeneration of *M. truncatula* cv. Jemalong through somatic embryogenesis or direct organogenesis (Nolan et al., 1989; Chabaud et al., 1996, 2003; Trieu and Harrison, 1996; Iantcheva et al., 2001; Zhou et al., 2004). Our laboratory in Gif sur Yvette has developed a very efficient transformation/regeneration protocol based on somatic embryogenesis for another *M. truncatula* line, R108-1 (Cosson et al., 2006; www.isv.cnrs-gif.fr/embo01/index.html). With this protocol, production of transgenic plants can be done routinely in the laboratory using *A. tumefaciens* strain EHA105 (Hellens et al., 2000). Other laboratories used other disarmed *A. tumefaciens* strains with the exception of strain LBA4404 (Hellens et al., 2000), which is normally used for tobacco transformation. This latter strain is not efficient at all for *Medicago* transformation.

Using the protocol described in Kamaté et al. (2000), we have created a small collection of *M. truncatula* transgenic lines (1,000 plants; Scholte et al., 2002, unpublished data) carrying the pGKB5 T-DNA construct used to establish a T-DNA tagged library of *Arabidopsis*. This T-DNA allows the construction of gene fusions with the *gus* reporter gene. Using this T-DNA tagged population, several lines expressing *gus* fusion specifically in some plant organs like the nodules have been obtained. However, these generally did not correspond to fusions of the reporter gene with symbiotic genes (Scholte et al., 2002) and may represent artifacts inherent to this technology. Symbiotic and developmental mutants were isolated in this population (4 out of 271 lines) but genetic studies indicated that the mutations were not linked to the T-DNA insert and probably corresponded to somaclonal mutations (Brocard et al., 2006). The presence of such somaclonal variations generated by the in vitro culture conditions was also observed with other plants, including the model plant, *L. japonicus* (Schauser et al., 1998). Thus, the T-DNA-mediated gene fusion technology can provide developmental markers in *M. truncatula* but cannot be used efficiently for large-scale forward genetic studies. If a large-scale transformation protocol by vacuum infiltration could be set up in *M. truncatula,* reverse genetic studies by PCR screening for T-DNA disrupted genes or large-scale sequencing of T-DNA borders could then be applied.

### *Gene Silencing-Based Technologies*

RNAi (interfering RNA) technology is an alternative to insertion mutagenesis for reverse genetic studies in various organisms. This technology does not require the construction of mutant collections and in plants it only requires the introduction of a transgene via T-DNA. This transgene should carry a hairpin sequence with at least 22 to 25 nucleotides homologous to the gene of interest. Transcripts with homology to this hairpin sequence can be integrated to a protein complex called RISC where they are degraded in small 22 to 25 nucleotide RNA molecules. The efficiency of this technology seems to depend on the construct (vector and promoter) used for the inactivation (Wesley et al., 2001) and on the plant growth conditions like temperature (Szittya et al., 2003). In addition, it was shown in *C. elegans* and *A. thaliana* that the RNAi efficiency depends also on the characteristics of the target gene and on expression level and pattern (Kerschen et al., 2004). An advantage of this technology comes from the different phenotypes that can be obtained in one experiment, as the result of the partial to complete inactivation of the target gene by the hairpin construct. However, domain/function and complementation studies are not possible and care should be taken that the inactivating RNA should target only the gene under study. This is particularly true when studying multigene families.

RNAi-induced silencing was successfully applied in *Arabidopsis* (Kariola et al., 2005) and initiated in legume plants, where it was successfully used to demonstrate the role of the symbiotic *LYK* genes (Limpens et al., 2003) in the formation of the infection threads during the early stages of the interaction with rhizobia. In this work, the hairy root transformation method was used because it has the advantage of being rapid. However, this technique is restricted to the analysis of root- or nodule-expressed genes and the transgenic material cannot be further analyzed or stored because only the root system is transgenic (chimeras). It was also used to demonstrate the role of the *KOJAK* and *CRE1* genes in *Medicago* root development (Limpens et al., 2004; Gonzalez-Rizzo et al., 2006).

We have produced transgenic *M. truncatula* R108 plants carrying an RNAi construct based on the pHannibal vectors directed against the *MtPistillata* Mad box gene and demonstrated through alteration of the flower development that gene inactivation by interfering RNA is very efficient in this plant (R. Benlloch, personal communication). In addition, production of transgenic plants allows study of the effect of the gene inactivation over several generations.

## *TILLING*

The development of TILLING technology (McCallum et al., 2000) made possible the efficient reverse genetic screening of mutant collections produced originally by EMS for forward genetic studies. By using high-throughput technology, the screening of point mutations in any organism has now become possible.

TILLING technology is based on the capacity of the S1 nuclease *Cel*I from celery to detect single base pair mismatches in a heteroduplex DNA molecule. PCR amplification of a target gene in DNA pools from plants carrying wild-type and mutated loci results in the production of PCR fragments differing by one base pair. This PCR mix is denatured, and reannealed heteroduplexes are produced that can be detected by this *Cel*I enzyme that cleaves the 3' end of mismatches while homoduplexes remain intact. The different fragments can be separated on LI-COR gel analysis system (Middendorf et al., 1992; Henikoff et al., 2004). When mismatches are detected in a pool, the PCR fragment corresponding to each individual of the pool is amplified and sequenced to detect the mutant plant. Theoretically, by analyzing a large number of DNA pools of an EMS population, point mutations can be found in any given gene or region of the genome. Because the number of plants per pool is restricted, due to the technique, the number of DNA pools to analyze should be important and may require high throughput technology in order to be efficient. Generally, it results in the isolation of several alleles for each analyzed gene, revealing the power of this technology. However, it should be noted that most of these point mutations will be silent because they will not change the protein sequence (degenerated amino acid code) or induce amino acid changes with no effect on the protein activity. In addition, it should be noted here that more than 1,000 mutations per genome can be present in these EMS-treated plants and that care should be taken in attributing a phenotype to a given mutation. Multiple alleles of the analyzed gene resulting in a similar phenotype or genetic analysis of the mutation following backcrosses to wild-type plants will indicate whether the phenotype is the consequence of the mutation in the analyzed gene. As *M. truncatula* is autogamous, $T_1$ plants can be directly used for genomic DNA screening.

TILLING programs were initiated in *M. truncatula* by the laboratories of D. Cook (VandenBosh and Stacey, 2003), J. Clarck (John Innes genome lab), and R. Thompson (INRA, Dijon; www.eugrainlegumes.org/).

### *Fast Neutron Technology*

Reverse genetic approaches by screening fast neutron populations is another technological development that allows the use of mutant populations previously utilized for forward genetic screens. It is an attractive alternative to EMS-induced point mutations because it induces deletions in the genome of the treated plant, and these deletions might be useful for the analysis of clustered genes with redundant function. These large deletions can also be a disadvantage by inducing double mutants. Fast neutron bombardment results in an average of 10 deletions per genome in *A. thaliana* (*Arabidopsis* Genome Initiative, 2000) and the size of each deletion can be between 0.8 and 12 kb. This lower number of mutations per genome as compared to EMS-treated plants should facilitate the genetic analysis of the mutant plants.

PCR technology is well suited for the detection of these deletions in rather large DNA pools of $T_1$ plants and thus allows rapid analysis of large mutant populations. In *A. thaliana,* the screening is done on megapools of 2,600 mutant lines by nested PCR. For this analysis, two oligonucleotides couples are designed on the gene of interest in order to amplify a rather big fragment. If in the population deletions exist in the amplified region, the PCR reaction will generate a fragment of small size. Using PCR conditions that will favor the synthesis of this small fragment, the mutated plant will be easily detected. The laboratories of Sharon R. Long (http://cmgm.stanford.edu/biology/long/index.html) and Giles Oldroyd (www.eugrainlegumes.org/) and scientists at the Noble Foundation (www.noble.org) have set up this technology for *M. truncatula* to discover novel alleles of known genes.

### *Class II Transposable Elements Are Not Suitable for Reverse Genetics in* **M. truncatula**

Transposable elements represent an attractive alternative for constructing insertion mutant collections (Sundaresan, 1996) because the number of transgenic lines to be produced is not a limiting factor when using these mobile elements, or it is less limiting.

In plants, two classes of transposable elements have been used for insertion mutagenesis. Class I elements, also called retrotransposons, transpose through an RNA intermediate that is reverse-transcribed into a linear double stranded DNA before its integration into the host genome (copy and paste mechanism). Class II elements are also called DNA transposons, as they transpose through a DNA intermediate (cut and paste mechanism).

Among the DNA transposable elements, the maize *En/Spm* element was shown to transpose efficiently in several plants including the model plant *Arabidopsis,* where it was used to develop large-scale mutagenesis. However, our work (d'Erfurth, Cosson, Eschstruth, Rippa, et al., 2003) indicates that this element is poorly active in *M. truncatula* and therefore not suitable for setting up insertion mutagenesis in *M. truncatula*. Similar results were obtained using the *Arabidopsis Tag1* transposable element (d'Erfurth et al., 2006).

Interestingly, the symbiotic *nin* mutant (Schauser et al., 1999) was isolated in *L. japonicus* using the maize *Ac/Ds* element, indicating that this element is functional in this plant. However, this heterologous transposable element was not further used for large-scale mutagenesis in this plant, probably because of low transposition activity in the regenerated plants. Thus legume plants may have developed efficient systems to control the activity of Class I transposable elements in their genome.

### *Retrotransposons for Reverse Genetic Approaches*

More recently, several retrotransposons (Class I transposable elements) were used as mutagens in plants (see below). As members of this class of elements transpose via a copy/paste mechanism, they are able to invade different host genomes. In the plant kingdom, for example, at least 70 percent of the maize genome is composed of LTR retrotransposons (Bennetzen, 2000). Transposition of some of these elements is activated by biotic and abiotic stresses, and interestingly in tissue culture. Therefore, different retrotransposons were used for large-scale insertion mutagenesis in different model plants using this capacity. For example, *Tos17,* an endogenous retrotransposon of rice, and *Tnt1* and *Ttol,* two retrotransposons of tobacco, were used for gene tagging respectively in *A. thaliana* and in rice (Okamoto and Hirochika, 2000; Courtial et al., 2001; Yamazaki et al., 2001). These studies demonstrated that retrotransposons efficiently transpose into genes of the heterologous hosts and that the target site sequences exhibit moderate or no consensus. In addition, they do not transpose in the vicinity of their original location (like DNA transposons) but are rather dispersed in the genome as a result of their mode of transposition requiring a cytoplasmic phase. Thus, these elements are good candidates for gene tagging in leguminous plants.

*Tnt1* was isolated following its transposition into the *NiaD* gene of tobacco (Grandbastien et al., 1989). It is an autonomous 5.3 kb copia-like LTR element that creates a target site duplication of 5 bp upon insertion. *Tnt1* transposition was induced in cultures of tobacco protoplasts (Pouteau et al., 1991) and during in vitro transformation of *A. thaliana* (Courtial et al., 2001). Its integration seems to be random as no obvious site specificity was observed and genes were frequently hit. The insertions are stable and frequently unlinked due to the replication cycle of LTR retrotransposons (Courtial et al., 2001). All these characteristics make *Tnt1* a good candidate for insertion mutagenesis also in other plants.

We have demonstrated that *Tnt1* transposes actively during in vitro transformation of *M. truncatula* R108 and Jemalong lines (d'Erfurth, Cosson, Eschstruth, Lucas, et al., 2003), generating from 4 to 40 insertions per regenerated plant. These insertions are stable during the life cycle of *M. truncatula* and most of them are genetically independent and can be separated by recombination. In the small tagged population generated in this plant, developmental as well as symbiotic *Tnt1*-tagged mutants were already isolated (d'Erfurth, Cosson, Eschstruth, Lucas, et al., 2003; Benlloch et al., 2006; Marsh et al., 2007). In addition, in *Medicago, Tnt1* seems to transpose preferentially into *Medicago* genes and *Tnt1* multiplication by transposition can be reinduced by tissue culture, making this element an attractive candidate to initiate large-scale insertion mutagenesis programs in *M. truncatula* (www.noble.org/Medicago/; www.igv.cnr.it/bis/Porceddu/; www.eugrainlegumes.org/).

The mutated plants generated in these programs can be used for forward as well as reverse genetics, which can be conducted by PCR screening on DNA pools or by sequencing *Tnt1* insertion sites in the transgenic lines. By randomly sequencing I-PCR fragments, we have shown that it is possible to identify more than 30 *Tnt1* insertion sites from a single I-PCR reaction performed on one transgenic line. Thus, it should be possible in the future to isolate the majority of the insertion sites in one transgenic plant. A variation of this technique could be done using restriction sites cutting inside the transposable element and resulting in the amplification of only one border. This will be suitable when the sequence of the euchromatic regions of the *Medicago* genome is available. Furthermore, as the techniques used for the isolation of these borders can be automated, it should be possible in the future to get the position of most of the *Tnt1* inserts in the *Medicago* genome, find your favorite mutants on the Web site, and order them from stock centers. In the future, the development of such a sequence database corresponding to the majority of the *Tnt1* inserts in the population may represent a valuable tool for the scientific community.

## *ANALYSIS OF THE* TNT1-*INDUCED MUTANTS*

The efficiency of *Tnt1* transposition during the regeneration process results in *Medicago* lines carrying multiple *Tnt1* inserts (d'Erfurth, Cosson, Eschstruth, Lucas, et al., 2003) and, by consequence, possibly multiple mutations. The genetic link of one particular insertion to any observed phenotype should then be analyzed in detail. Two possibilities can be envisaged. First, *Tnt1* insertion sites are sequenced from one line and Your Favorite Gene (YFG) is disrupted. The original line should be backcrossed to wild-type plants and the progeny analyzed for cosegregation of a putative phenotype with the insertion. The disrupted region can be easily followed in the progeny by PCR, using one oligonucleotide corresponding to the *Tnt1* end and one oligonucleotide in the sequence of the *YFG* gene. Using one oligonucleotide on each side of the insertion site will in addition indicate for each plant if the YFG locus is wild-type, heterozygous, or homozygous for the insertion. Using this simple PCR analysis, several successive backcrosses can be performed and the disrupted locus detected in the plants independently of the phenotype. Genetic links between phenotype and homozygous mutant loci can then be confirmed in a segregating population. Second, one line shows an interesting phenotype (Altered Leaf Development, ALD, for example) but the *Tnt1* insertion sites have not been characterized yet in this line. One straightforward possibility is to isolate various insertion sites in this line, using the I-PCR technique described above. Once these loci are sequenced, the genetic link between the mutant loci and the ALD phenotype can be analyzed by PCR in a segregating population. Alternatively, transposon display technology (Melayah et al., 2001) can be applied on a segregating population to find the *Tnt1* border linked to the phenotype. This technique might be more time consuming but also more powerful for the characterization of the tagged locus.

## *CONCLUSION*

In this review, we suggest that the diploid *M. truncatula* is a suitable model plant to investigate the biology of the legume family through reverse genetic approaches. Indeed, most of the novel molecular biology tools can

TABLE 24.1. Comparison of the different reverse genetic tools developed in *M. truncatula*.

| | **Transposon (tn) mutagenesis** | **Retrotransposon mutagenesis** | **T-DNA mutagenesis** | **EMS mutagenesis** | **Fast neutron mutagenesis** | **RNAi** |
|---|---|---|---|---|---|---|
| Time required for the realization of the technique | Long | Long | Long | Short | Short | Short |
| Cost of the technique | High | High | High | Low | Low | Low |
| Space required to generate the mutants | Important | Important | Important | Important | Important | Small |
| Number of mutations generated per plant | 1-4 | 20-50 | 1-4 | >1,000 | >10 | 1 |
| Use of in vitro regeneration | Yes (heterologous tn)/no (endogenous tn) | Yes | Yes | No | No | Yes |
| Somaclonal mutations | Yes | Yes | Yes | No | No | Yes |
| Stable mutation | No/yes (defective tn) | Yes | Yes | Yes | Yes | No |
| Reverse genetics possible | Yes | Yes | Yes | Yes | Yes | - |
| Forward genetics possible | Yes | Yes | Yes | Yes | Yes | - |
| Generation of large deletions affecting several genes | No | No | No | No | Yes | - |
| Possibility to do local mutagenesis | Yes | No | No | No | No | - |
| Mutation specificity | High | High | High | High | High | Low |
| Easy isolation of the mutations identified after phenotypic screening | Yes | Yes | Yes | No | No | - |

be applied to *M. truncatula*. The advantages and disadvantages of these various approaches have been summarized in Table 24.1.

The original criteria for the choice of this plant were: (1) its close relationship to cultivated legumes, and (2) the successful transformation and regeneration by tissue culture via both *Agrobacterium tumefaciens* and *rhizogenes* (Barker et al., 1990). Using these regeneration capacities, insertion mutagenesis approaches could be initiated either by T-DNA or by retrotransposon technologies. The T-DNA approach seems still limited because of the lack of an efficient in vivo transformation system and the number of somaclonal mutations detected in the T-DNA transgenic lines.

Even if T-DNA technology cannot be efficiently used for insertion mutagenesis, it is still an excellent tool for reverse genetics in *M. truncatula* via RNAi technology. When knockout or EMS mutants do not exist for the gene studied, the silencing approach could be used successfully via either *A. tumefaciens* or *A. rhizogenes,* keeping in mind that RNAi efficiency depends on the characteristics of the target gene. The development of the TILLING method makes possible the screening by reverse genetics of EMS-mutated collections originally constructed for forward genetic studies. Similarly, fast neutron mutant collections are well suited for reverse genetic approaches when the sequence of the gene under study is known. Gene identification through forward genetic studies is less obvious with these collections.

Transposable elements are good alternatives to T-DNA for insertion mutagenesis. The efficiency of the tobacco retrotransposon *Tnt1* transposition during the regeneration process in *M. truncatula* and the demonstration of its efficient gene disruption capacities resulted in the initiation of several international programs for the production of large *Tnt1*-tagged collections.

In the near future, all these tools will help the rapid elucidation of biological functions specific to legume plants.

## REFERENCES

*Arabidopsis* Genome Initiative (2000). Analysis of the genome sequence of the flowering plant *Arabidopsis thaliana. Nature,* 408:796-815.

Barker, D.G., S. Bianchi, F. Blondon, Y. Dattée, G. Duc, S. Essad, P. Flament, P. Gallusci, G. Génier, G. Pierre, et al. (1990). *Medicago truncatula,* a model plant for studying the molecular genetics of the *Rhizobium*-legume symbiosis. *Plant Molecular Biology Reporter* 8:40-49.

Baulcombe, D. (2004). RNA silencing in plants. *Nature* 431:356-363.

Bechtold, N. and G. Pelletier (1998). *In planta Agrobacterium*-mediated transformation of adult *Arabidopsis thaliana* plants by vacuum infiltration. *Methods in Molecular Biology* 82:259-266.

Bennetzen, J.L. (2000). Transposable element contributions to plant gene and genome evolution. *Plant Molecular Biology* 42:251-269.

Benlloch R., d'Erfurth I., Ferrandiz C., Cosson V., Beltran J.P., Canas L.A., Kondorosi A., Madueno F., Ratet P. (2006). Isolation of mtpim proves *Tnt1* a useful reverse genetics tool in *Medicago truncatula* and uncovers new aspects of *AP1*-like functions in legumes. *Plant Physiology,* 142:972-983.

Bent, A.F. (2000). *Arabidopsis in planta* transformation. Uses, mechanisms, and prospects for transformation of others species. *Plant physiology* 124:1540-1547.

Blondon, F., D. Marie, S. Brown, and A. Kondorosi (1994). Genome size and base composition in *Medicago sativa* and *Medicago truncatula* species. *Genome* 37:264-270.

Boisson-Dernier, A., M. Chabaud, F. Garcia, G. Becard, C. Rosenberg, and D.G. Barker (2001). *Agrobacterium rhizogenes*-transformed roots of *Medicago truncatula* for the study of nitrogen-fixing and endomycorrhizal symbiotic associations. *Molecular Plant-Microbe Interactions* 14:695-700.

Bonin, I., T. Huguet, M. Gherardi, J.M. Prosperi, and I. Olivier (1996). High level of polymorphism and spatial structure in a selfing plant species, *Medicago truncatula* (Leguminosae) shown using RAPD markers. *American Journal of Botany* 83:843-855.

Brocard, L., M. Schultze, A. Kondorosi, and P. Ratet (2006). T-DNA mutagenesis in the model plant *Medicago truncatula*: Is it efficient enough for legume molecular genetics? *CAB Reviews* 23.

Cosson V., Durand P., d'Erfurth I., Kondorosi A., Ratet P. (2006). *Medicago truncatula* transformation using leaf explants. *Methods in Molecular Biology,* 343:115-28.

Carpenter, R. and E.S. Coen (1990). Floral homeotic mutations produced by transposon-mutagenesis in *Antirrhinum majus. Genes and Development* 4:1483-1493.

Chabaud, M., F. de Carvalho-Niebel, and D.G. Barker (2003). Efficient transformation of *Medicago truncatula* cv. Jemalong using the hypervirulent *Agrobacterium tumefaciens* strain AGL1. *Plant Cell Reports* 22:46-51.

Chabaud, M., C. Larsonneau, C. Marmouget, and T. Huguet (1996). Transformation of barrel medic (*Medicago truncatula* Gaertn.) by *Agrobacterium tumefaciens* and regeneration via somatic embryogenesis of transgenic plants with the *MtENOD12* nodulin promoter fused to the *Gus* reporter gene. *Plant Cell Reports* 15:305-310.

Clough, S.J. and A.F. Bent (1998). Floral dip: A simplified method for *Agrobacterium*-mediated transformation of *Arabidopsis thaliana. Plant Journal* 16: 735-743.

Cook, D., K. VandenBosch, F. deBruijn, and T. Huguet (1997). Model legume gets the nod. *Plant Cell* 8:275-281.

Courtial, B., F. Feuerbach, S. Eberhard, L. Rohmer, H. Chiapello, C. Camilleri, and H. Lucas (2001). *Tnt1* transposition events are induced by in vitro transformation of *Arabidopsis thaliana,* and transposed copies integrate into genes. *Molecular Genetics and Genomics* 265:32-42.

d'Erfurth, I., V. Cosson, A. Eschstruth, H. Lucas, A. Kondorosi, and P. Ratet (2003). Efficient transposition of the *Tnt1* tobacco retrotransposon in the model legume *Medicago truncatula. Plant Journal* 34:95-106.

d'Erfurth, I., V. Cosson, A. Eschstruth, S. Rippa, E. Messinese, P. Durand, H. Trinh, A. Kondorosi, and P. Ratet (2003). Rapid inactivation of the maize transposable element *En/Spm* in *Medicago truncatula. Molecular Genetics and Genomics* 269:732-745.

d'Erfurth I., Cosson V., Mondy S., Brocard L., Kondorosi A., Ratet P. (2006). The low level of activity of the *Arabidopsis thaliana Tag1* transposon correlates with the absence of two minor transcripts in *Medicago truncatula. Molecular Breeding,* 17:317-328.

Endre, G., A. Kereszt, Z. Kevei, S. Mihacea, P. Kalo, and G.B. Kiss (2002). A receptor kinase gene regulating symbiotic nodule development. *Nature* 417:962-966.

Frugoli, J. and J. Harris (2001). *Medicago truncatula* on the move! *Plant Cell* 13:458-463.

Gonzalez-Rizzo, S., M. Crespi, and F. Frugier (2006). The *Medicago truncatula CRE1* cytokinin receptor regulates lateral root development and early symbiotic interaction with *Sinorhizobium meliloti. Plant Cell* 18:2680-2693.

Grandbastien, M.A., A. Spielmann, and M. Caboche (1989). *Tnt1,* a mobile retroviral-like transposable element of tobacco isolated by plant cell genetics. *Nature* 337:376-380.

Handberg, K. and J. Stougaard (1992). *Lotus japonicus,* an autogamous diploid species for classical and molecular genetics. *Plant Journal* 2:487-496.

Hellens, R.P., E.A. Edwards, N.R. Leyland, S. Bean, and P.M. Mullineaux (2000). pGreen: A versatile and flexible binary Ti vector for *Agrobacterium*-mediated plant transformation. *Plant Molecular Biology* 42:819-832.

Henikoff, S., B.J. Till, and L. Comai (2004). TILLING. Traditional mutagenesis meets functional genomics. *Plant Physiology* 135:630-636.

Hirochika, H. (2001). Contribution of the *Tos17* retrotransposon to rice functional genomics. *Current Opinion in Plant Biology* 4:118-122.

Iantcheva, A., M. Vlahova, T.H. Trinh, S.C. Brown, A. Slater, M.C. Elliott, and A. Atanasov (2001). Assessment of polysomaty, embryo formation and regeneration in liquid media for various species of diploid annual *Medicago. Plant Science* 160:621-627.

Kamaté K., I.D. Rodriguez-Llorente, M. Scholte, P. Durand, P. Ratet, E. Kondorosi, A. Kondorosi, T.H. Trinh (2000). Transformation of floral organs with GFP in *Medicago truncatula. Plant Cell Reports,* 19:647-653.

Kamath, R.S., M. Martinez-Campos, P. Zipperlen, A.G. Fraser, and J. Ahringer (2000). Effectiveness of specific RNA-mediated interference through ingested double-stranded RNA in *Caenorhabditis elegans. Genome Biology 2*: Research 0002.1.

Kariola, T., G. Brader, J. Li, and E.T. Palva (2005). Chlorophyllase 1, a damage control enzyme, affects the balance between defence pathways in plants. *Plant Cell* 17:282-294.

Kerschen, A., C.A. Napoli, R.A. Jorgensen, and A.E. Muller (2004). Effectiveness of RNA interference in transgenic plants. *FEBS Letters* 566:223-228.
Krusell, L., L.H. Madsen, S. Sato, G. Aubert, A. Genua, K. Szczyglowski, G. Duc, T. Kaneko, S. Tabata, F. De Bruijn, et al. (2002). Shoot control of root development and nodulation is mediated by a receptor-like kinase. *Nature* 420:422-426.
Krysan, P.J., J.C. Young, and M.R. Sussman (1999). T-DNA as an insertional mutagen in *Arabidopsis. Plant Cell* 11:2283-2290.
Li, X., Y. Song, K. Century, S. Straight, P. Ronald, X. Dong, M. Lassner, and Y. Zhang (2001). A fast neutron deletion mutagenesis-based reverse genetics system for plants. *Plant Journal* 27:235-242.
Limpens, E., C. Franken, P. Smit, J. Willemse, T. Bisseling, and R. Geurts (2003). LysM domain receptor kinases regulating rhizobial Nod factor-induced infection. *Science* 302:630-633.
Limpens, E., J. Ramos, C. Franken, V. Raz, B. Compaan, H. Franssen, T. Bisseling, and R. Geurts (2004). RNA interference in *Agrobacterium rhizogenes*-transformed roots of *Arabidopsis* and *Medicago truncatula. Journal of Experimental Botany* 55:983-992.
Lucau-Danila, A., R. Wysocki, T. Roganti, and F. Foury (2000). Systematic disruption of 456 ORFs in the yeast *Saccharomyces cerevisiae. Yeast* 16:547-552.
Marsh J.F., A. Rakocevic, R.M. Mitra, L. Brocard, J. Sun, A. Eschstruth, S.R. Long, M. Schultze, P. Ratet, and G.E.D. Oldroyd (2007). *Medicago truncatula NIN* is essential for rhizobial-independent nodule organogenesis induced by autoactive calcium/calmodulin-dependent protein kinase. *Plant Physiology,* 144:324-335.
McCallum, C.M., L. Comai, E.A. Greene, and S. Henikoff (2000). Targeted screening for induced mutations. *Nature Bio/Technology* 18:455-457.
Melayah, D., E. Bonnivard, B. Chalhoub, C. Audeon, and M.A. Grandbastien (2001). The mobility of the tobacco *Tnt1* retrotransposon correlates with its transcriptional activation by fungal factors. *Plant Journal* 28:159-168.
Middendorf, L.R., J.C. Bruce, R.C. Bruce, R.D. Eckles, D.L. Grone, S.C. Roemer, G.D. Sloniker, D.L. Steffens, S.L. Sutter, J.A. Brumbaugh, et al. (1992). Continuous, on-line DNA sequencing using a versatile infrared laser scanner/electrophoresis apparatus. *Electrophoresis* 13:487-494.
Nishimura, R., M. Ohmori, H. Fujita, and M. Kawaguchi (2002). A *Lotus* basic leucine zipper protein with a RING-finger motif negatively regulates the developmental program of nodulation. *Proceedings of the National Academy of Sciences USA* 99:15206-15210.
Nolan, K.E., R.J. Rose, and J.R. Gorst (1989). Regeneration of *Medicago truncatula* from tissue culture: Increased somatic embryogenesis using explants from regenerated plants. *Plant Cell Reports* 8:278-281.
Okamoto, H. and H. Hirochika (2000). Efficient insertion mutagenesis of *Arabidopsis* by tissue culture-induced activation of the tobacco retrotransposon *Tto1*. *Plant Journal* 23:291-304.
Oldroyd, G.E. and R. Geurts (2001). *Medicago truncatula*, going where no plant has gone before. *Trends in Plant Science* 6:552-554.

Ostergaard, L. and M.F. Yanofsky (2004). Establishing gene function by mutagenesis in *Arabidopsis thaliana. Plant Journal* 39:682-696.

Pouteau, S., E. Huttner, M.A. Grandbastien, and M. Caboche (1991). Specific expression of the tobacco *Tnt1* retrotransposon in protoplasts. *EMBO Journal* 10:1911-1918.

Ross-Macdonald, P., P.S. Coelho, T. Roemer, S. Agarwal, A. Kumar, R. Jansen, K.H. Cheung, A. Sheehan, D. Symoniatis, L. Umansky, et al. (1999). Large-scale analysis of the yeast genome by transposon tagging and gene disruption. *Nature* 402:413-418.

Sallaud, C., C. Gay, P. Larmande, M. Bes, P. Piffanelli, B. Piegu, G. Droc, F. Regad, E. Bourgeois, D. Meynard, et al. (2004). High throughput T-DNA insertion mutagenesis in rice: A first step towards *in silico* reverse genetics. *Plant Journal* 39:450-464.

Sato S. and Tabata S. (2006). *Lotus japonicus* as a platform for legume research. *Current Opinion in Plant Biology,* 9:128-132.

Schauser, L., K. Handberg, N. Sandal, J. Stiller, T. Thykjaer, E. Pajuelo, A. Nielsen, and J. Stougaard (1998). Symbiotic mutants deficient in nodule establishment identified after T-DNA transformation of *Lotus japonicus. Molecular and General Genetics* 259:414-423.

Schauser, L., A. Roussis, J. Stiller, and J. Stougaard (1999). A plant regulator controlling development of symbiotic root nodules. *Nature* 402:191-195.

Scholte, M., I. d'Erfurth, S. Ripa, S. Mondy, V. Cosson, P. Durand, C. Breda, T. Trinh, I. Rodriguez-Llorente, E. Kondorosi, et al. (2002). T-DNA tagging in the model legume *Medicago truncatula* allows efficient gene discovery. *Molecular Breeding* 10:203-215.

Stacey, G., M. Libault, L. Brechenmacher, J. Wan, and G.D. May (2006). Genetics and functional genomics of legume nodulation. *Current Opinion in Plant Biology* 9:110-121.

Stougaard, J. (2001). Genetics and genomics of root symbiosis. *Current Opinion in Plant Biology* 4:328-335.

Stracke, S., C. Kistner, S. Yoshida, L. Mulder, S. Sato, T. Kaneko, S. Tabata, N. Sandal, J. Stougaard, K. Szczyglowski, and M. Parniske (2002). A plant receptor-like kinase required for both bacterial and fungal symbiosis. *Nature* 417:959-962.

Sundaresan, V. (1996). Horizontal spread of transposon mutagenesis: New uses for old elements. *Trends in Plant Science* 1:184-190.

Szabados, L., I. Kovacs, A. Oberschall, E. Abraham, I. Kerekes, L. Zsigmond, R. Nagy, M. Alvarado, I. Krasovskaja, M. Gal, et al. (2002). Distribution of 1000 sequenced T-DNA tags in the *Arabidopsis* genome. *Plant Journal* 32:233-242.

Szittya, G., D. Silhavy, A. Molnar, Z. Havelda, A. Lovas, L. Lakatos, Z. Banfalvi, and J. Burgyan (2003). Low temperature inhibits RNA silencing-mediated defence by the control of siRNA generation. *EMBO Journal* 22:633-640.

Tadege, M., P. Ratet, and K.S. Mysore (2005). Insertional mutagenesis: A Swiss Army knife for functional genomics of *Medicago truncatula. Trends in Plant Science* 10:229-235.

Tinland B. (1996). The integration of T-DNA into plant genome. *Trends in Plant Science,* 1:178-184.

Tirichine, L., F. de Billy, and T. Huguet (2000). *Mtsym6,* a gene conditioning *Sinorhizobium* strain-specific nitrogen fixation in *Medicago truncatula. Plant Physiology* 123:845-851.

Tissier, A.F., S. Marillonnet, V. Klimyuk, K. Patel, M.A. Torres, G. Murphy, and J.D.G. Jones (1999). Multiple independent defective *suppressor-mutator* transposon insertions in *Arabidopsis:* A tool for functional genomics. *Plant Cell* 11:1841-1852.

Trieu, A. and M.J. Harrison (1996). Rapid transformation of *Medicago truncatula:* Regeneration via shoot organogenesis. *Plant Cell Reports* 16:6-11.

VandenBosh, K.A. and G. Stacey (2003). Summaries of legume genomics projects from around the globe. Community resources for crops and models. *Plant Physiology* 131:840-865.

Webb, K.J., L. Skot, M.N. Nicholson, B. Jorgensen, and S. Mizen (2000). *Mesorhizobium loti* increases root-specific expression of a calcium-binding protein homologue identified by promoter tagging in *Lotus japonicus. Molecular Plant-Microbe Interactions* 13:606-616.

Wesley, S.V., C.A. Helliwell, N.A. Smith, M.B. Wang, D.T. Rouse, Q. Liu, P.S. Gooding, S.P. Singh, D. Abbott, P.A. Stoutjesdijk, et al. (2001). Construct design for efficient, effective and high-throughput gene silencing in plants. *Plant Journal* 27:581-590.

Yamazaki, M., H. Tsugawa, A. Miyao, M. Yano, J. Wu, S. Yamamoto, T. Matsumoto, T. Sasaki, and H. Hirochika (2001). The rice retrotransposon *Tos17* prefers low-copy-number sequences as integration targets. *Molecular Genetics and Genomics* 265:336-344.

Zambrowicz, B.P., G.A. Friedrich, E.C. Buxton, S.L. Lilleberg, C. Person, and A.T. Sands (1998). Disruption and sequence identification of 2,000 genes in mouse embryonic stem cells. *Nature* 392:608-611.

Zhou, X., M.B. Chandrasekharan, and T.C. Hall (2004). High rooting frequency and functional analysis of GUS and GFP expression in transgenic *Medicago truncatula* A17. *New Phytologist* 162:813-822.

Chapter 25

# Fluorescence *In Situ* Hybridization on *Medicago truncatula* Chromosomes

Olga Kulikova
Hans de Jong
Ton Bisseling

Fluorescence *in situ* hybridization (FISH) is a molecular cytogenetic technique that allows localization of DNA sequences by hybridizing their probes on chromosomal targets. The results can be visualized under the microscope using a probe directly labeled with a fluorescent dye or a probe containing a hapten molecule that is further detected biochemically or with antibodies conjugated with fluorescent dye molecules. Initially, FISH was confined to localization studies of repetitive DNA sequences on highly condensed mitotic metaphase chromosomes and interphase nuclei (reviews in Trask, 1991; Jiang et al., 1996). However, for most detailed chromosome mapping studies, highly condensed metaphase complements are insufficient to distinguish neighbor DNA targets less than 1 to 2 Mb apart. Higher spatial resolution values of up to 50 to 100 kb can be achieved with FISH on interphase nuclei, but these targets lack the information of the native chromosomal structure. A better alternative for hybridizations in diploid plant species is the less contracted pachytene chromosomes, which generally measure 10 to 40 times longer than their mitotic metaphase counterparts and display a differentiated pattern of heterochromatin blocks (de Jong et al., 1999). Modern cytogenetic maps are mostly based on FISH position of genetically defined single-copy sequences on the pachytene chromosomes and are highly informative to support the construction of physical maps for map-based cloning projects and to position genes in pericentromere heterochromatin regions where recombination estimates are unreliable due to very low frequency of crossover events (Roberts, 1965; Zhong et al., 1999).

## *DEVELOPMENT OF FISH TECHNOLOGY IN* M. TRUNCATULA

The nitrogen fixation model legume plant *Medicago truncatula* ($2n = 2x = 16$) is considered an excellent plant for cytogenetic studies (Gerbah et al., 1999; Kulikova et al., 2001). It has a relatively small genome of 560 to 580 Mb per 1C and contains eight chromosome pairs in diploid cells (Blondon et al., 2004; Agarwal and Gupta, 1983). FISH was applied to *M. truncatula* for karyotyping metaphase chromosomes of the accessions R108-1 and Jemalong (Gerbah et al., 1999). Identification of small metaphase chromosomes on the basis of relative arm length and centromere position was found unreliable. Hybridization with 45S rDNA and 5S rDNA as probes assisted in the identification of three chromosomes. It showed that the Jemalong genotype has a single NOR and three 5S rDNA loci, and R108-1 has one NOR locus but also two 5S rDNA loci. The 5S rDNA locus on the short arm of NOR chromosome in Jemalong is absent in R108-1.

*M. truncatula* Jemalong A17 has been used as the standard for constructing pachytene karyotypes. Jemalong A17 is of special interest because the genome of this single-seed descent line was selected for sequencing. Identification of chromosomes was based on length of chromosome arms, size of pericentromere and nucleolar organizing regions, other diagnostic heterochromatin blocks, location of tandem repeats 5S, 45S rDNA, and *Mt*R1 (Kulikova et al., 2001). *Medicago*'s pachytene chromosomes clearly display the typical heterochromatin pattern of *Arabidopsis* (Koornneef et al., 2003) with large heterochromatic blocks around all centromeres and the NOR. Linkage groups were assigned to the cytogenetic map by FISH with few BAC clones per chromosome. The chromosomes were numbered according to their corresponding linkage maps. By now, many more BAC clones have been mapped and each of them can also be useful in the identification of specific chromosome regions (Ane et al., 2002; Gualtieri et al., 2002; Schnabel et al., 2003; Choi et al., 2004).

The *Mt*R1 satellite repeat has a 166-bp motif organized in long tandem arrays with typical head-to-tail orientation. One typical FISH experiment on metaphase chromosomes with 5S rDNA, 45S rDNA, and *Mt*R1 as probes is shown in Figure 25.1B. The 5S rDNA loci are located on chromosomes 2, 5, and 6, and the 45S rDNA region on chromosome 5, whereas *Mt*R1 resides on six pericentromere heterochromatin blocks, namely on the long arm of chromosome 1, on the same arm of chromosome 2 where 5S rDNA region is also present, on the short arms of chromosome 4, on both arms of chromosome 7, and on the short arm of chromosome 8. The combi-

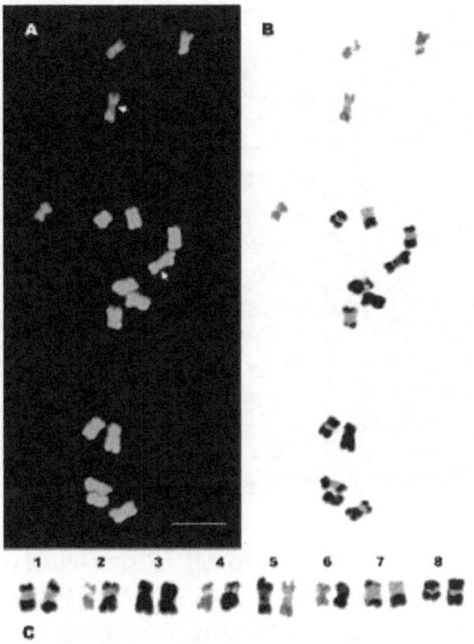


FIGURE 25.1. Identification of metaphase chromosomes of Jemalong A17 by using FISH with 5S rDNA (red signal) and *MtR1* (green signal). (A) Metaphase chromosomes are stained with DAPI. (B) A merged image of chromosomes and the FISH signals of 5S rDNA and *MtR1*. (C) Karyotype of metaphase chromosomes. Chromosomes are digitally sorted out of metaphase complement, ordered, and numbered according to their corresponding linkage groups. Arrows indicate NORs. Bar = 5 μm. (See also color gallery.)

nation of only 5S rDNA and *Mt*R1 is sufficient to distinguish all eight pairs of chromosomes in the Jemalong genome (Figure 25.1C).

Two more tandem repeats, *Mt*R2 and *Mt*R3 were mapped on *M. truncatula* chromosomes (Kulikova et al., 2004). *Mt*R2 was found in the pericentromere regions of all 16 chromosome arms, whereas *Mt*R3 was localized in functional centromeres or primary constrictions of all chromosomes (Figure 25.2).

*Mt*R1, *M*tR2, and *M*tR3 were also mapped on the chromosomes of R108-1 and DZA315.16 accessions. The genome size of R108-1 is 20 percent smaller than that of Jemalong A17 (Blondon et al., 1994). The DZA315.16 DNA content is identical to that of Jemalong (Thoquet et al., 2002). In contrast to *Mt*R3, which also occupies centromere positions in R108-1, *Mt*R1 and *Mt*R2 are entirely absent. However, all three repeats did occur in the DZA315.16 genome. *Mt*R3 also has a centromeric location, and *Mt*R1 and

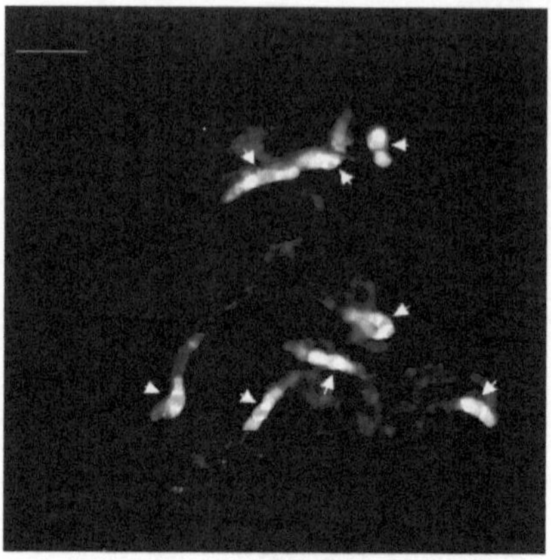

FIGURE 25.2. Localization of *MtR2* and *MtR3* satellites on *M. truncatula* Jemalong A17 pachytene chromosomes by FISH. *MtR2* (red signal) is located in pericentromeric heterochromatin of all chromosomes. *MtR3* signals (green) coincide with primary constrictions (indicated by arrowheads) of all chromosomes. Chromosomes are stained with DAPI. Bar = 5 μm. (See also color gallery.)

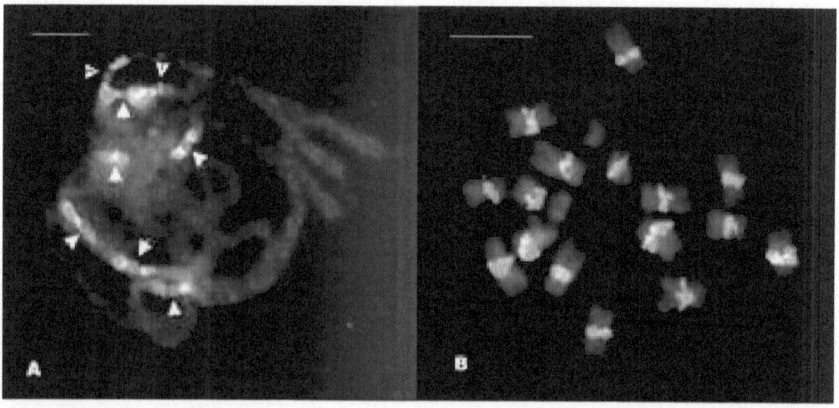


FIGURE 25.3. Localization of *MtR1*, *MtR2*, and *MtR3* satellites on chromosomes of *M. truncatula* DZA315.16 by FISH. (A) Pachytene complement. *MtR3* signals (red) are located at centromeres (indicated by arrowheads) and *MtR2* signals (green) appear at pericentromeric heterochromatin of all chromosomes. (B) Metaphase chromosomes. *MtR1* signals (green) are located in pericentromeric heterochromatin of four chromosomes and *MtR2* (red signals) are at pericentromeric heterochromatin of all chromosomes. Chromosomes are stained with DAPI. Bar = 5 μm. (See also color gallery.)

*Mt*R2 are located in pericentromeric heterochromatin but have a different distribution in comparison to Jemalong (Figures 25.3).

## *GENERAL TECHNICAL ASPECTS OF FISH ON* M. TRUNCATULA *CHROMOSOMES*

The FISH protocol presented in this chapter is adopted from the *Arabidopsis* technique (Fransz et al., 1998; Zhong et al., 1999). The following steps were discerned.

### *Plant Material Fixation and Spread Preparations*

It is imperative for reliable and reproductive FISH signals that much attention is paid to production of high-quality chromosome preparations. The chromosomes and nuclei should be well separated and free of cytoplasm and cell wall debris, as this may cause excessive nonspecific signals. Careful digestion of the cell walls by pectolytic enzymes is an essential step in the procedure. Pectolytic enzymes generally consist of various combinations of cellulases and pectinases. Cell wall digestion for pollen mother cells requires an additional callase (β-1,3 glucanase), which can be found in the snail gut enzyme cytohelicase. We work routinely with a general-purpose mixture of cellulase RS, pectolyase Y-23, and cytohelicase for making the preparations of both mitotic and meiotic chromosomes.

#### *Mitotic Chromosome Preparations*

- Germinate seeds of *M. truncatula* on wet filter paper in Petri dishes until roots are about 1 cm long. To arrest cells at metaphase, treat seedlings with a saturated aqueous solution of 1-bromonaphthalene for 4 hours at 20°C.
- Excise roots from seedlings and fix them in freshly prepared Carnoy's fixative (acetic acid-ethanol 1:3) for 3-4 hours. This mixture is prepared less than 30 minutes before use to minimize deleterious effects of formed esters in the fixative. It penetrates the tissue very rapidly and a fixation time of 3-4 hours will normally be sufficient. The amount of fixative is at least 10 times the volume of the plant material. Fixed material can be used immediately or stored at –20°C for months.

- Place fixed roots in a small watch glass and wash for 2 × 5 minutes in distilled water.
- Replace the water with citrate buffer (10 mM citrate buffer, pH 4.5) and wash 1 × 5 minutes.
- Incubate the material in enzyme mixture containing 0.15% (w/v) cellulase RS, 0.15% (w/v) pectolyase Y23, and 0.15% (w/v) cytohelicase in citrate buffer) for 75 minutes in a moist chamber at 37°C. We mostly use a small container with tightly closed lid and wet filter paper on the bottom.
- After two rinses with water, the soft and fragile roots are kept on ice.
- Dissect a root tip and put it on a clean slide in 2 μl water. Tap the root tip with a blunt dissection needle and examine under a binocular microscope to monitor until a fine suspension has been obtained.
- Adding 15 μl of 60% acetic acid will clear up the suspension and make the cells sticky. The droplet should be carefully stirred with the needle, without touching the glass surface.
- Place the slide on a hot block at 45°C, add another 15 μl of 60% acetic acid, and keep stirring for 30-60 seconds. In this step, acid-soluble proteins and various cytoplasmic components will be dissolved in the acetic acid and clear up the spread preparation.
- Now carefully put some drops of freshly prepared Carnoy's fixative in a circle around the cell suspension and wait until the fixative covers the entire glass slide surface. Discard excessive solution by tilting the slide and pipette more fixative onto the preparation.
- Place the slide vertically and air dry.
- Examine the preparation under the phase contrast microscope at 40× magnification and select only the best slides for *in situ* hybridization. Selection criteria for high quality are spreading of the chromosomes, free from cytoplasm and cell walls, debris, and dirt. It is good to check the first slides to monitor the results of the spreading procedure. If cytoplasm surrounding nuclei is still visible the next slide needs to be treated with more 60% acetic acid. Slides can be used immediately or stored before use in a tightly closed box in a dry and cool place (+4°C). For longer storage, wrap the box in aluminum foil and store at –80°C.

*Meiotic Pachytene Chromosome Preparations*

The production of high-quality pachytene spreads in comparison to metaphase chromosomes is technically more demanding and degree of satisfactory results may differ between related species and even between genotypes of the same species (review by de Jong et al., 1999). However, in most cases partially spread chromosomes can be sufficient for localizing DNA sequences of interest if the FISH includes a marker probe with known chromosome position.

- Collect whole inflorescences in the morning into tubes filled with Carnoy fixative and leave for 6-7 hours. The fixative should be refreshed a few times until the flower buds are yellow-white and the solution remains clear.
- Select young, unopened flower buds whose anthers are still green or partly transparent before they were fixed. Yellow anthers contain mature pollen. As anthers at pachytene (less than 2 mm) are very small and meiotic divisions not synchronized, we use entire flower buds instead of single anthers for the preparation of pachytene chromosome spreads.
- Place fixed flower buds in a small watch glass with water and wash for 2 × 5 minutes at room temperature with distilled water.
- Replace the water by citrate buffer and wash for 5 minutes.
- Incubate the material in enzyme mixture containing a two times higher enzyme concentration than used for mitotic material (0.3 percent each of cellulase RS, pectolyase Y23, and cytohelicase in citrate buffer) for 2¼ hours in a moist chamber at 37°C.
- Wash flower buds for 2 × 5 minutes with water.
- Select a flower bud and put it on a clean slide in 2 µl water. Remove parts of connective tissue with needles.
- Transfer the flower bud to a clean slide in a 2 µl water droplet.
- Tap the flower bud with the tip of a blunt needle and monitor under the dissecting microscope until a fine suspension has formed.
- Add about 15 µl of 60 percent acetic acid to clear the suspension and make the cells sticky. Place the slide on a hot block at 45°C, add another 15 µl of 60 percent acetic acid, and keep stirring for 30-60 seconds.
- Cells are spread on glass by carefully pipetting drops of the Carnoy's fixative around the droplet with cleared cell suspension and wait until the fixative covers the whole glass slide. Discard excessive solution by tilting the slide and pipette more fixative onto the preparation and air dry the preparation.
- Examine the preparation under the phase contrast microscope without a cover glass.

### *Labelling Probes*

The quality and properties of the probes is the second critical step in FISH experiments. In the direct method, the label that has been incorporated into a probe can be examined by fluorescence microscope immediately after the hybridization. The most common fluorochromes for direct labeling include fluorescein isothiocyanate (FITC), rhodamine (TRITC), Texas red (TR), 7-amino-4-methyl-coumarin-3-acetic acid (AMCA), Cy3 (indocarbocyanine), and Cy5 (indodicarbocyanine). An advantage of direct labeling is the easy visualization of the hybridized probe signal on the chromosomes, but their signals may not be bright enough to detect hybridizations on small chromosomal targets. The signal can be amplified further in the latter case by fluorescein-conjugated antibodies. In the indirect method, a nonfluorescent hapten (either biotin or digoxigenin) is incorporated into the DNA. After *in situ* hybridization, the label has to be detected with either avidin or streptavidin (in the case of biotin) or antibodies raised against digoxigenin. The (strept)avidin and antidigoxigenin molecules are linked with one of the fluorochromes mentioned above and can be visualized by fluorescent microscopy.

Plasmid DNA isolation including BAC DNA can be done by using a High Pure Plasmid Isolation Kit according to the manufacturer's protocol (Roche) or according to the standard alkaline lysate protocol (Sambrook et al., 1989) and labeled by using a BIOTIN or DIG Nick Translation Kit (Roche) according to the instructions of the manufacturer. Clones containing small inserts (less than 1 kb) can be PCR labeled with biotin-16-dUTP or digoxigenin-11-dUTP according to the instructions of the manufacturer (Roche) or using the High Primed Labeling Kit (Roche).

It is important to find out if the DNA to be used as a probe for *in situ* hybridization is sufficiently labeled, because this may limit detection. The labeling can be detected using a color reaction based on alkaline phosphatase conjugate and nitro blue tetrazolium on a Southern dot blot (standard Roche protocol).

### *Hybridization and Probe Detection*

#### *DNA Probe Mixture*

- Take the required amount of each DNA probe (200-250 ng) per slide. Add 50 μg fragmented salmon sperm DNA and 2-2.5 volume of ice-cold ethanol. Keep the solution on ice for 15 minutes.

- Spin down for 30 minutes at 4°C.
- Remove supernatant and wash pellet in 70 percent ethanol.
- Spin down for 15 minutes at room temperature.
- Remove supernatant and resuspend the pellet in 10 μl hybridization buffer (HB50), containing 50 percent deionized formamide, 2xSSC, pH 7.0.
- Dissolve DNA for 15 minutes in a 37°C water bath.
- Add 10 μl of 20 percent dextran sulfate in HB50 (preheated at 42°C).
- Denature DNA by placing it for 5 minutes in boiling water.
- Place DNA on ice for 3 minutes and spin down briefly.

*High-Complexity Probe Preparation (Containing Repeats Giving High Level of Background)*

Retrotransposons and transposable elements belong to the class of dispersed repeats that exist abundantly in heterochromatin and euchromatin. FISH mapping of BAC clones harboring genes of interest but also such repeats may produce high amounts of background signals due to hybridization of the dispersed repeats in the probe, and so can make the detection of the single-copy region of the BAC impossible. In this case, preannealing of the BAC DNA probe and fragmented genomic DNA is required as an extra step before hybridization.

- Take the required amount of probe DNA (for one slide). Add 100-fold excess of competitor DNA (genomic DNA of *M. truncatula,* fragmented to an average size 200-500 bp by autoclaving at 105°C for 5 minutes or by sonication), 50-fold excess of salmon sperm DNA, and 2-2.5 volume of ice-cold ethanol. Mix well. Keep the solution on ice for 15 minutes.
- Spin down for 30 minutes at 4°C.
- Remove supernatant and wash pellet in 70 percent ethanol.
- Spin down for 15 minutes at room temperature.
- Remove supernatant and resuspend the pellet in 10 μl HB50.
- Dissolve DNA for 15 minutes in a 37°C water bath.
- Add 10 μl of 20 percent dextran sulfate in HB50 (preheated to 42°C).
- Denature DNA by placing it for 10 minutes in water bath at 80°C.
- Place DNA for 3 minutes on ice and spin down.
- Let preannealing take place by keeping the sample for 2 hours at 37°C.

*Pretreatment of Slides*

- Dry slides at 60°C for 30 minutes or at 37°C overnight.
- Pipette 100 µl of 100 $\mu g \cdot ml^{-1}$ RNaseA solution (DNAase-free) in 2x SSC (0.3 M sodium chloride, 0.03 M sodium citrate, pH 7.0) and cover with a 24 × 50 mm cover slip. RNaseA incubation is performed to remove endogenous RNA and improve signal-to-noise ratio of the hybridization.
- Incubate at 37°C for 60 minutes and wash in 2x SSC (saline sodium citrate) at root temperature for 2 × 5 minutes.
- Wash in PBS (10 mM sodium phosphate pH7.0, 140 mM NaCl) for 5 minutes.
- Fix in 1 percent formaldehyde in PBS for 10 minutes.
- Wash in PBS for 2 × 5 minutes.
- Dehydrate slides through an ethanol series (70 percent, 90 percent, 100 percent), each step 2 minutes, and air dry the preparations.

*Hybridization*

- Denature chromosomes by placing slides in a preheated formamide solution (70 percent in 2x SSC) at 70°C for 2 minutes, then dehydrate them in an ice-cold ethanol series (70 percent, 90 percent, 100 percent, 2 minutes each), and air dry the preparations.
- Drop hybridization mix on a target area of the slide and cover the solution with 24 × 24 mm cover slip, avoiding formation of air bubbles.
- Place the slides in a moist chamber and incubate overnight at 37°C.

*Post-hybridization washings (next day)*

- Prepare three Coplin jars containing HB50 and place them in a water bath at 42°C.
- Gently remove cover slips and incubate slides in each jar for 5 minutes.
- Wash slides in 2x SSC 2 × 5 minutes.
- Wash the slides in TN (100 mM Tris-HCL pH 7.5, 150 mM NaCl), 0.05 percent (v/v) Tween-20 for 5 minutes.

*Detection of Two-Color Hybridization Signals*

- Add 100 µl TNB buffer (TN buffer, containing 1 percent blocking reagent (Roche) (see Table 25.1).

- Incubate at 37°C for 30 minutes to prevent nonspecific binding of antibodies to the glass surface and rinse briefly with TNT buffer (TN + 0.05 percent Tween-20).
- Add avidin~Texas red (Vector Laboratories, Burlingame, CA) or streptavidin~Cy3 (Jackson ImmunoResearch Laboratories) 100 μl TNB, cover with 25 × 50 mm cover slip, and incubate in a moist chamber at 37°C for 30-60 minutes.
- Wash in TNT for 3 × 5 minutes.
- Add goat-antiavidin~biotin (or goat-streptavidin~biotin, Vector Laboratories) and sheep-antidigoxigenin (Roche) antibodies in 100 μl TNB, and cover with 25 × 50 mm cover slip.
- Incubate in a moist chamber at 37°C for 30-60 minutes.
- Wash in TNT buffer for 3 × 5 minutes.
- Add avidin~Texas red and rabbit-antisheep~FITC (Jackson ImmunoResearch Laboratories) antibodies in 100 μl TNB, cover with 25 × 50 mm cover slip, and incubate in a moist chamber at 37°C for 30-60 minutes and wash in TNT for 3 × 5 minutes.
- Dehydrate slides in an ethanol series of 70 percent, 90 percent, and 100 percent, 2 minutes each, air dry, and mount with 1-2 $\mu g \cdot ml^{-1}$ DAPI (4',6-diamidino-2-phenylindole) in Vectashield antifade solution (Vector Laboratories).

### *Imaging and Analysis*

We studied most of the chromosome preparations under a Zeiss Axioskop microscope equipped with high numerical aperture planapochromatic optics and epifluorescence illumination for DAPI, FITC, and Cy3. We captured the images with a black and white Photometrics Sensys 1,305 × 1,024 pixel CCD camera using Genius Image Analysis software (Applied Imaging). All fluorescent images were captured as RAW files, processed, and

TABLE 25.1. Two-color FISH detection system.

| Label | First incubation | Second incubation | Third incubation |
|---|---|---|---|
| Digoxigenin (green) | | Sheep-anti-Dig (1:200) | Rabbit-anti-sheep~FITC (1:500) |
| Biotin (red) | Avidin~Texas red (1:1000) or Streptavidin~Cy3 (1:250) | Biotinylated-anti-avidin (1:25) Biotinylated-anti-streptavidin (1:50) | Avidin~Texas red (1:1000) Streptavidin~Cy3 (1:250) |

pseudocolored separately and merged into a multilayer RGB image. The DAPI images were displayed in gray or light blue to retain differentiation of the heterochromatin and euchromatin. In a few cases we applied extra sharpening on the DAPI using high-Gauss 5 × 5 or 7 × 7 filters. For three-color FISH we used transparent channels for the red and green fluorescence in order to display overlapping of the FITC and Cy3 fluorescence. Cropping, resolution, and contrast optimization can be carried out in Adobe Photoshop.

## REFERENCES


Agarwal, K. and P.K. Gupta (1983). Cytological studies in the genus *Medicago* Linn. *Cytologia* 48:781-793.

Ane, J.-M., J. Levi, P. Thoquet, O. Kulikova, F. de Billy, V. Penmetsa, D.-J. Kim, F. Debelle, C. Rosenberg, D.R. Cook, et al. (2002). Genetic and cytogenetic mapping of *DMI1, DMI2* and *DMI3* genes of *Medicago truncatula* involved in NOD factor transduction, nodulation, and mycorrhization. *Molecular Plant-Microbe Interactions* 15:1108-1118.

Blondon, F., D. Marie, S. Brown, and A. Kondorosi (1994). Genome size and base composition in *Medicago sativa* and *M. truncatula* species. *Genome* 37:264-275.

Choi, H.-K., D. Kim, T. Uhm, E. Limpens, H. Lim, J.-H. Mun, P. Kalo, R. Varma Penmetsa, A. Seres, et al. (2004). A sequence-based genetic map of *Medicago truncatula* and comparison of marker colinearity with *M. sativa. Genetics* 166:1463-1502.

de Jong, H., P. Fransz, and P. Zabel (1999). High resolution FISH in plants—techniques and applications. *Trends in Plant Science* 4:258-263.

Fransz, P., S. Armstrong, C. Alonso-Blanco, T.C. Fischer, R.A. Torrez-Ruiz, and G. Jones (1998). Cytogenetics for the model system *Arabidopsis thaliana. Plant Journal* 13:867-876.

Gerbah, M., Z. Kevei, S. Siljak-Yakovlev, E. Kondorosi, A. Kondorosi, and T.H. Trinh (1999). FISH chromosome mapping allowing karyotype analysis in *Medicago truncatula* lines Jemalong J5 and R-108-1. *Molecular Plant-Microbe Interactions* 12:947-950.

Gualtieri, G., O. Kulikova, E. Limpens, D.-J. Kim, D.R. Cook, T. Bisseling, and R. Geurts (2002). Microsynteny between pea and *Medicago truncatula* in the SYM2 region. *Plant Molecular Biology* 50:225-235.

Jiang, J., S.H. Hulbert, B.S. Gill, and D.C. Ward (1996). Interphase fluorescence *in situ* hybridization mapping: A physical mapping strategy for plant species with large complex genomes. *Molecular and General Genetics* 252:497-502.

Koornneef, M., P. Fransz, and H. De Jong (2003). Cytogenetic tools for *Arabidopsis thaliana. Chromosome Research* 11:183-194.

Kulikova, O., R. Geurts, M. Lamine, D.-J. Kim, D.R. Cook, J. Leunissen, H. De Jong, B.A. Roe, and T. Bisseling (2004). Satellite repeats in the functional centromere and pericentromeric heterochromatin of *Medicago truncatula. Chromosoma* 113:276-283.

Kulikova, O., G. Gualtieri, R. Geurts, D.-J. Kim, D. Cook, T. Huguet, H. de Jong, P. Fransz, and T. Bisseling (2001). Integration of the FISH-pachytene and genetic maps of *Medicago truncatula. Plant Journal* 27:49-58.

Roberts, P.A. (1965). Difference in the behavior of eu- and heterochromatin: Crossing over. *Nature* 205:725-726.

Sambrook, J., E.F. Fritsch, and T. Maniatis (1989). *Molecular cloning: A laboratory manual.* 2nd ed. Cold Spring Harbor: Cold Spring Harbor Laboratory Press.

Schnabel, E., O. Kulikova, R.V. Penmetsa, T. Bisseling, D.R. Cook, and J. Frugoli (2003). An integrated physical, genetic and cytogenetic map around the *sunn* locus of *Medicago truncatula. Genome* 46:665-672.

Thoquet, P., M. Gherardi, E.-P. Journet, A. Kereszt, J.-M. Ane, J.-M. Prosperi, and T. Huguet (2002). The molecular genetic linkage map of the model legume *Medicago truncatula:* An essential tool for comparative legume genomics and the isolation of agronomically important genes. *BMC Plant Biology* 2:1-13.

Trask, B.J. (1991). Fluorescence *in situ* hybridization: Applications in cytogenetics and gene mapping. *Trends in Genetics* 7:149-154.

Zhong, X.B., J. Bodeau, P.F. Fransz, V.M. Williamson, A. van Kammen, J.H. de Jong, and P. Zabel (1999). FISH to meiotic pachytene chromosomes of tomato locates the root-knot nematode resistance gene *Mi-1* and the acid phosphatase gene *Aps-1* near the junction of euchromatin and pericentromeric heterochromatin of chromosome arms 6S and 6L respectively. *Theoretical and Applied Genetics* 98:365-270.

Zhong, X.B., P.F. Fransz, J. Wennekes-van Eden, P. Zabel, A. van Kammen, and H. de Jong (1996). High-resolution mapping of pachytene chromosomes and extended DNA fibers by fluorescence *in situ* hybridization. *Plant Molecular Biology Reporter* 14:232-242.

Chapter 26

# Molecular and Genetic Analyses of Transgenic Plants

S. R. Bhat
R. Srinivasan

## *INTRODUCTION*

Development of plant tissue culture and plant cell transformation techniques has revolutionized plant biology. Although plant transformation is technically less demanding and is now within the reach of any modest laboratory, reproducible and efficient protocols in a number of legumes are yet to be established. In all such cases where transformation protocols are established, researchers from a variety of disciplines can employ this approach to find answers to specific questions. At the outset, the cause and effect relations in transformation experiments appear straightforward. However, in practice, all effects observed are not directly attributable to the transgene introduced through transformation. While a large body of literature is available on transformation of various plant species, discussion on genetic and molecular analyses of transgenic plants is limited. Since the methods of analysis are not specific to legumes, in this chapter we discuss the appropriate methods and aspects that need to be considered while analyzing the results of transformation experiments.

## *CAUSES OF VARIATION AMONG TRANSGENICS*

Variations are often observed among transgenic plants even derived from the same experiment. These variations may be due to (1) the method of transformation, (2) the effect of introduced DNA, or (3) the combined effect of both. The present methods of transformation can deliver DNA into an individual cell rather than transforming each and every cell of a fully differentiated plant. Thus, the single transformed cell will have to be multiplied and differentiated to obtain a complete transformed plant. This generally requires a period of in vitro culture lasting 6 weeks to more than 3 months (Puonti-Kaerlas et al., 1992; Schroeder et al., 1993, 1995; Pigeare et al., 1997). Tissue culture is known to cause heritable and nonheritable changes (somaclonal variation; Larkin and Scowcroft, 1981). Hence, some of the variations observed among transgenics may arise from tissue culture (Sala et al., 2000; Labra et al., 2001, 2004). For example, all pea transgenics recovered in a study were tetraploids (Puonti-Kaerlas et al., 1992). Similarly, Imai et al. (1993) reported a high frequency of tetraploids in transgenics derived from *Agrobacterium*-mediated transformation of potato tuber discs. Current methods of transformation lead to insertion of foreign DNA at random locations within the host chromosomes (Barakat et al., 2000). Since the expression of a transgene is not only controlled by its promoter but also influenced by flanking host genomic regions, individual transgenics may exhibit variation due to position effect (Dean et al., 1988). Further, the number of insertions and the integrity of the introduced DNA sequences are variable among independent transgenics, and thus contribute to variation (Perrin et al., 2000; Aragao, 1999; De Block, 1993). Chimerism may also contribute to variation (Christou, 1990; Schroeder et al., 1991). Therefore, analysis of variation among transgenics assumes importance to separate transgene effects from other associated effects.

## *METHODS OF ANALYSIS OF TRANSGENIC PLANTS*

Techniques used for analysis of transgenic plants depend on the purpose of the experiment. Usually researchers are interested in finding answers to some or all of the following while performing transformation experiments:

1. Whether transformation has occurred and, if so, the frequency of transformation.
2. Whether the foreign gene has been incorporated into the host genome. If so,
   a. Number of copies introduced
   b. Site(s) of integration
   c. Intactness of introduced DNA
3. Whether the introduced gene is expressed, if so, the specificity and levels of expression.
4. Is the product biologically functional?

5. Is the foreign gene inherited and expressed as per expectation?
6. Whether the changes observed are due to transgene or due to other causes.

Different techniques appropriate for the above are listed in Table 26.1. In the following pages we discuss the important aspects to be considered while employing the above techniques and interpretation of the results.

### *Occurrence of Transformation*

Since only a few of the several hundred cells subjected to transformation actually get transformed, and it takes a long time (up to 4 months) to realize transformed plants (Pigeaire et al, 1997; Puonti-Kaerlas et al., 1992), re-

TABLE 26.1. Techniques employed for the analysis of transgenic plants.

| Stage | Purpose | Technique | Remarks |
|---|---|---|---|
| Cell/tissue culture | Occurrence of transformation | Histochemical (GUS, GFP)<br>PCR | Indicative<br>Indicative |
| $T_0$ plants | Transformation | PCR, RT-PCR | Fairly reliable<br>Confirmatory |
| | Transformation determination of copy number & DNA rearrangements | Southern hybridization<br>FISH | Confirmatory test |
| | Site of transgene integration | DNA sequencing of clones obtained from<br>1. Genomic DNA library<br>2. Inverse PCR<br>3. Plasmid rescue<br>4. SON PCR | |
| | Transgene expression | RT-PCR<br>Northern hybridization | Highly reliable |
| | Transgene expression (protein level) | Western hybridization<br>ELISA | Reliable but sensitive |
| | Functional activity of transgene | Specific biochemical tests, bioassay | Reliable |
| | Localization of protein | *In situ* hybridization | |
| $T_1/BC_1$ generation | Transgene inheritance | Growth on selective medium, PCR, Southern hybridization | |
| | Transgene expression | Same as above<br>Field trials | Reliable assessment of transgene effects |

searchers often seek an early assessment of the occurrence and frequency of transformation. Almost all vectors have marker genes linked to the genes of interest (we refer to them as transgenes here to distinguish them from marker genes). The marker gene product helps either visual identification of transformed cells (scorable marker) or selective growth of transformed cells on a specific selection medium (selectable marker). With visual or scorable markers such as green fluorescent protein (GFP) or β-glucuronidase (GUS), one can examine the cells or tissues following 24 to 48 hours of transformation and estimate the number of cells or groups of cells (spots) showing marker gene expression. This test gives only a rough estimate of transformation, because marker gene expression can occur even without actual integration of the gene into the host genome. Further, leaky expression from the *Agrobacterium* can also give wrong signals. In some vectors, therefore, an intron is introduced into the marker gene so that its expression is seen only when the gene is expressed from the plant cell. Among the popular scorable markers, the Gus assay is a destructive method whereas with GFP one can visualize and follow growth and development of transformed cells.

With selectable markers, transformation becomes apparent only when colonies of cells or green shoots emerge from the treated cells or explants. However, even with selectable markers, nontransgenic escapes are almost invariably found (Todd and Tague, 2001). The polymerase chain reaction (PCR) technique is widely used to screen shoots recovered on a selection medium (McKersie et al. 1993, 1999). For this, DNA is isolated from the recovered shoots and used in PCR with primers that specifically bind to sequences in the transferred DNA. Amplification of a fragment of expected size in the PCR reaction indicates that the shoots are transformed. This, however, is not a confirmatory test because complete elimination of *Agrobacterium* from $T_0$ plants is almost impossible (Nauerby et al., 1997). Thus DNA contamination from *Agrobacterium* or DNA used for direct transformation can yield false positives. Oger and Dessaux (2003) suggested an additional PCR with primers specific to *vir* genes to check whether bacterial DNA contamination is contributing to false positives.

Nain et al. (2005) have devised a way to overcome the problem of DNA contamination from bacteria interfering with PCR assay. This technique takes advantage of the difference in DNA reassociation kinetics between plant and bacterial or plasmid DNA. Following denaturation, the DNA is allowed to reassociate. The less complex plasmid DNA reassociates faster than complex plant DNA. Therefore, the former will be a substrate for the restriction enzyme. Such restricted DNA will not serve as a template for PCR reactions. The transgene incorporated into the host genome will remain as single-stranded DNA and escape restriction. Thus PCR amplification will be observed only if the transgene is incorporated into plant DNA. The efficacy of this method, however, is yet to be widely tested.

Another way to get a reliable result of transformation is to perform reverse transcriptase-polymerase chain reaction (RT-PCR; D'Haullin et al., 1990; McKersie et al., 1993). Since marker genes are generally constitutively expressed, RNA isolated from test samples can be used for RT-PCR amplification of the marker gene. DNA contamination can be easily checked by performing a PCR with RNA as template, but without a reverse transcriptase reaction. If there is an intron interrupting the marker gene, the size of the amplicon will help in resolving whether the amplification is from RNA or contaminating DNA. RT-PCR is a simple and well-established technique but has not been widely employed in assessing transgenics. In our opinion, this approach deserves a wider application.

Enzyme assays for marker genes have also been employed to detect transformants. For example, Molvig et al. (1997) tested glucuronidase (encoded by the *uid*A gene) and *PAT* (encoded by the *bar* gene) activity in $T_0$ lupins to identify positive transformants. Similarly, Schroeder et al. (1993) employed PAT assay and RT-PCR for *npt*II in pea for screening of transgenics.

### *Integration and Copy Number Determination*

The Southern hybridization technique is considered ideal for confirming the integration of foreign DNA into the host genome. Since a considerable amount of DNA is required for preparing a Southern blot, this can be undertaken when a large amount of tissue is available for sampling. Usually when transgenic plants are multiplied in vitro or established in soil, sampling is done.

A single Southern blot can be probed at least two or three times with different probes and can thus be used to derive different information pertaining to foreign gene integration. Therefore, care should be taken while selecting the restriction enzymes and probe sequences for Southern hybridization. Before we consider this further, it may be appropriate to briefly discuss the fate of foreign DNA after it is introduced into a plant cell. As stated earlier, foreign DNA gets integrated at random locations in the host chromosome (Tinland, 1996). It has been found that on average each primary transgenic plant carries 1.5 to 2.5 copies of the transgene (Feldman, 1991; Alonso et al., 2003). These transgenes may be linked or unlinked. *Agrobacterium*-mediated transformation is considered more precise, where only the DNA flanked by the T-DNA border sequences is transferred to the host chromosome. However, in reality, besides T-DNA, vector backbone sequences are also found incorporated in more than 40 percent of cases (Kononov et al., 1997; De Buck et al., 1999; Porsch et al., 1998). Similarly, in particle bombardment, where a large piece of DNA, and sometimes the whole plasmid DNA, is used, there is no defined region that will be incorporated into the host genome. In addition, rearrangement of foreign DNA is more frequent in transgenics obtained via particle bombardment (Reddy et al., 2003; Kohli et al., 2003). Thus, a Southern blot may show a variety of hybridization patterns depending on the exact nature of foreign gene integration. The strategy and approach to Southern hybridization to elicit information pertaining to transgene integration using a T-DNA vector is illustrated in Figure 26.1.

In the first place, Southern hybridization is employed to prove the integration of foreign DNA into the host genome. For this, a restriction enzyme that cuts only once within the foreign DNA, *Sal*I, *Hin*dIII, *Kpn*I or *Xba*I in the above case, should be used for digesting the DNA isolated from the putative transgenic plants. A sequence from the T-DNA that does not carry an internal restriction site for the enzyme used for Southern blot preparation should be chosen to probe the blot. In the DNA from transformed plants, a restriction fragment containing the probe sequence will be released following two cuts, one within the T-DNA and the second in the flanking host DNA. Since foreign genes are integrated independently at random sites, each insertion is unique with respect to the flanking plant DNA. Hence, upon Southern hybridization with the probe, a polymorphic banding pattern will be observed. Furthermore, each band will correspond to one integration

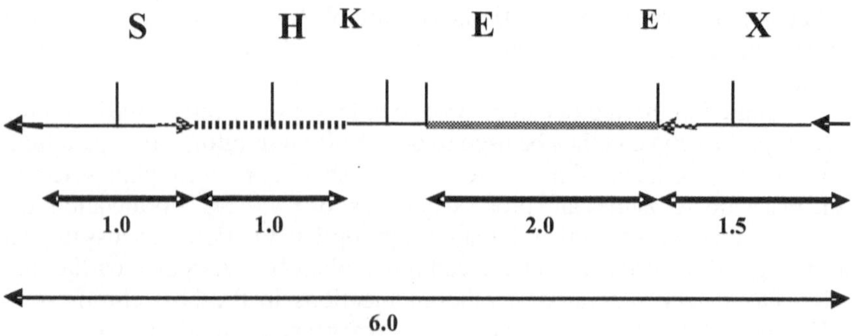


FIGURE 26.1. T-DNA fragment with various restriction enzyme sites (top). Sizes of various fragments (in kilo base pairs) are shown in middle and lower panels. S, *Sal*I; H, *Hin*dIII; K, *Kpn*I; E, *Eco*RI; X, *Xba*I.

site of foreign DNA in multicopy integration or rearrangements. Using the probe and enzyme information, the minimum expected size of the band could be calculated. For example, in the above case, if the *Eco*RI fragment of the transgene is used to probe the blot prepared with *Kpn*I, each complete integration of T-DNA will show a band longer than 3.5 kb. Now, if the same blot is probed with the marker gene sequence, a new restriction fragment length polymorphism (RFLP) pattern will be observed with each band longer than about 2.5 kb. However, the number of bands in each case will be the same as the previous one. Any discrepancy will indicate incomplete transfer of T-DNA or DNA rearrangement. Pickardt et al. (1995) used a similar strategy to analyze transgenic *Vicia narbonensis*. In their study, they restricted the DNA with an enzyme, which released an internal T-DNA fragment containing the transgene 2S-albumin. By probing the blot with the marker gene *npt*II, the junction fragment with plant DNA was revealed showing one to five integrations of T-DNA in different transgenics. When the same blot was probed with 2-S albumin, a single 1.2 kb fragment was revealed. However, the intensity of the signal was different among independent transgenics, reflecting variations in transgene copy number. If multiple copies of transgenes are inserted at a location, more complex RFLP patterns may be visualized. Some more informative strategies are explained in Bhat and Srinivasan (2003).

In some cases, the host genome may carry sequences homologous to the gene of interest (e.g., antisense constructs or constructs with genes derived from related species). If such a sequence is used as a probe, the band corresponding to the one observed in the control should not be counted as a proof of transformation. In general, there is always some unique sequence in the inserted DNA (such as marker gene, promoter, or presequence), which can be used as a probe in Southern analysis. In most cases, single-copy transgenics are sought for a detailed study of transgene effects. Fortunately, single-copy transgenics can be relatively easily identified through Southern hybridization.

Slot blot or dot blot hybridization can be used for determining the exact number of copies of the transgene in multiple-copy transgenics. The method essentially consists of loading a known amount of probe DNA into a known amount of DNA from the untransformed plant so as to make what are called single-and-multiple copy reconstitutions.

The intensity of the hybridization signal obtained from the transformed plants is compared with the signal obtained with this reconstituted sample to determine the copy number. The critical factor in this method is an accurate determination of the DNA concentration in the samples. An illustration of copy number determination in transgenic chickpea samples using slot

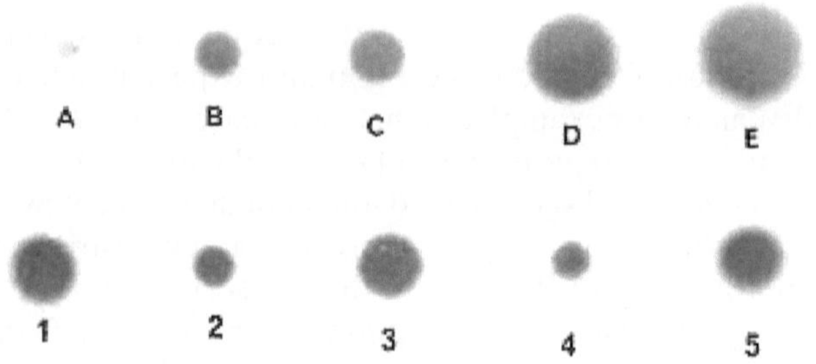


FIGURE 26.2. Dots A, B, C, D, and E are 1,2, 3, 5 and 10 copies of the reconstituted *gus* gene equivalent to that in 5 mg of chickpea genomic DNA; Dots 1, 2, 3, 4, and 5 are DNA (5 µg) from transgenic chickpea samples. Based on the comparison of areas of the spots of the transgenic sample with reconstituted dots as reference, it was determined that the transgenic samples 1-5 contain 3, 2, 4, 2, and 3 copies of the *gus* gene, respectively (data from Sahoo, 1996).

blot hybridization is given in Figure 26.2. In recent years, real-time PCR is being increasingly employed for this purpose (German et al., 2003; Chapter 30, this volume). However, resource and technical demands of real-time PCR limit its routine use by many laboratories.

***Dot Blot Hybridization for the Estimation of Copy Number in a Transformant***

To estimate copy number of transgene , one, two, three, five and ten copy reconstitutions of the probe fragment can be used as a reference. The process is explained here with the help of an example where copy number of *gus* gene is being determined in some chickpea transformants with the help of dot blot hybridization (Figure 26.2).

| | |
|---|---|
| Size of the *gus* fragment in the T-DNA | = 2.56 kb |
| Mass of *gus* gene | $= 2.56 \times 1.08 \times 10^{-6}$ pg |
| | $= 2.77 \times 10^{-6}$ pg |
| Size of haploid genome of chickpea | = 1.53 pg |
| Number of haploid genome/5 µg chickpea DNA | $= 5 \times 10^{-6}/1.53 \times 10^{-12}$ |
| | $= 3.27 \times 10^{6}$ |
| Amount of *gus* DNA equivalent to one copy of *gus* gene per haploid genome | $= 3.27 \times 10^{6} \times 2.77 \times 10^{-6}$ pg |
| | = 9.06 pg |

### *Site of Integration of Transgene*

1. Since each integration is unique with respect to its flanking host DNA, finding the exact site of integration of transgenes is important. It can also serve as a fingerprint for identifying the transgenic event.
2. Transgene integration could disrupt a host gene, resulting in mutation of the host gene. In that event, in lines homozygous for the transgene, loss of function may lead to a mutant phenotype. Such T-DNA insertional mutagenesis is extensively used for isolating genes, promoters, and enhancers.
3. The information could be used to distinguish between homozygous and heterozygous insertion lines using PCR. The sequences flanking the transgene could be used to design primers and using just three suitable primers in a multiplex reaction, the untransformed and the transformed line homozygous and heterozygous for the insertion can be unambiguously identified (Figure 26.3).

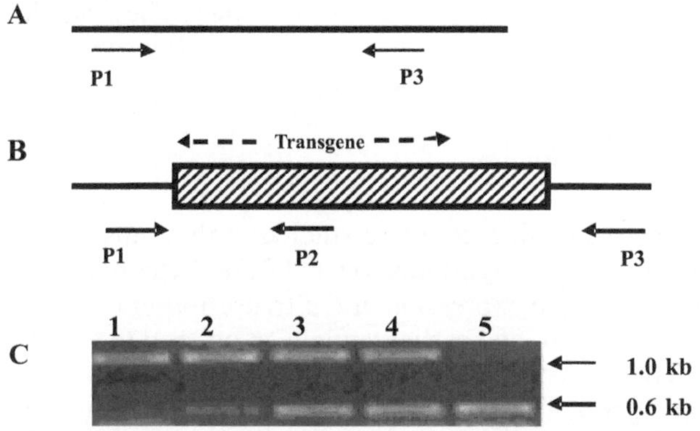


FIGURE 26.3. Multiplex PCR analysis of a transgenic line. A. A schematic representation of the untransformed plant genomic DNA. The primers P1 and P3 are used to amplify a specific fragment from the DNA of untransformed plant. B. A schematic representation of transgenic plant carrying an insertion between P1 and P3 region in the plant genomic DNA. PCR with the primers P1 and P2 (primer specific to transgene) would yield an amplicon specific to the site of insertion. C. Multiplex PCR analysis with all the three primers P1, P2, and P3 of the genomic DNA samples of plants can be used to differentiate between homozygous and heterozygous transgenic plants. The presence of both the amplicons indicates heterozyogosity, whereas the presence of a single amplicon (0.6 kb) indicates homozygosity for the T-DNA insertion in transgenic plants. Lane 1 is homozygous whereas lanes 2, 3, and 4 are heterozygous for the insertion. Lane 5 contains DNA from untransformed plant.

There are several approaches to identifying insertion sites, but all of them make use of the sequence information of the inserted DNA. For example, Southern (RFLP) information can be used for preparing a subgenomic library and to identify the clone containing the T-DNA. Alternatively, inverse PCR technique may be employed to amplify the host genome regions flanking the transgene (Radhamony et al., 2005). In some T-DNA tagging approaches, bacterial origin of replication (ori) sequences are included within the T-DNA to facilitate cloning of genomic DNA fragments containing T-DNA as plasmids (plasmid rescue vectors; Radhamony et al., 2005). Another PCR approach uses T-DNA-specific and random primers for amplifying and cloning of the fragments flanking the transgene (Liu et al., 1995). Antal et al. (2004) devised single oligonucleotide PCR technique to amplify 5' or 3' flanking regions of genes. The clones obtained by any of the above methods can be sequenced and the exact site of integration can be determined.

Methods such as fluorescent *in situ* hybridization (FISH; Pedersen et al,. 1997) and fiber-FISH (Jackson et al., 2001) are being employed for physical identification of transgenes on chromosomes. These studies are particularly useful in understanding the dispersion of transgene integration into chromosomes of the recipient host.

### *Gene Expression Studies*

Transgene expression could be studied at the transcript level or at the protein level. Northern hybridization and RT-PCR techniques are employed for determining gene expression at the transcription level. The use of RT-PCR has been mentioned earlier, and it can be used to study transgene expression. The tissue to be sampled for RNA is dictated by the promoter used to drive the gene of interest. RNA blots probed with sequences from the transgene will reveal whether the gene is transcribed or not. Relative quantification of transcripts can also be performed using northern blots. For this purpose, an RNA blot is probed simultaneously with two probes, the transgene and a constitutively expressed host gene. By comparing the intensity of signals of the transgene with the constitutively expressed host gene, relative expression of the transgene can be determined.

Transgene expression leads to either production of a new protein or alteration in the expression of native proteins (e.g., as in the case of antisense or RNAi constructs). Western blotting and ELISA techniques are commonly employed to examine these changes in transgenics. However, these techniques are dependent on the availability of specific antibodies that react with the protein in question and hence are not routinely used for studying transgene expression. More often, researchers are aware of the likely effect of transgene expression on some biochemical or phenotypic traits. In many instances, such associated changes can be measured (e.g., enzyme activity, fatty acid, protein, or starch profile) to get an indirect estimate of transgene activity. For example, McKersie et al. (1993) used isozyme analysis to study the expression of superoxide dismutase (Mn-SOD) in *Medicago* transgenics engineered for cold tolerance. It was found that the transgene for Mn-SOD gave a distinct band and thus could be used to unambiguously identify transgenics. Similarly, bioassays can also be devised to obtain indirect estimates of transgene expression. For example, Schroeder et al. (1993, 1995) and Morton et al. (2000) used insect bioassay to show that transgenic expression of bean alpha amylase inhibitor confers bruchid resistance in pea.

*In situ* hybridization can also be performed to find the exact location within the tissue or cell where the transgene product gets accumulated. Both RNA-DNA hybridization and antibody-based approaches are employed for this purpose.

## *Genetic Analysis*

Barring *Arabidopsis thaliana* and *Medicago truncatula,* transgenic plants in all other species are derived through tissue culture. This involves either de novo differentiation of shoot buds or somatic embryos from transformed cells, or multiple-shoot regeneration from shoot buds (transformation of shoot meristem). Thus, one or more shoots may arise from a single point, which may originate from the same event or from different events. Schroeder et al. (1991) observed variations in Southern patterns among plants regenerated from the same callus. Such variations may arise from somaclonal variation or due to several independent transformation events occurring in neighboring cells that give rise to a callus. Therefore, it is desirable to maintain a precise record of the lineage of each shoot during in vitro multiplication and rooting stages. Only after establishing their identity based on the Southern pattern, independent events can be taken further.

Primary transgenics ($T_0$ generation) are hemizygous for the transgene. Each $T_0$ plant is unique with respect to site of transgene integration and number of copies of the transgene. Further, somaclonal effects are also variable among different transgenics. Hence, $T_0$ transgenics are not considered ideal for detailed analysis. In general, primary transgenics are selfed to obtain a $T_1$ generation. The $T_1$ population comprises plants that are homozygous, hemizygous, or null for the transgene. Although transgenes at different loci can be regarded as duplicate loci, a significant position effect prevents such a treatment. Thus, a $T_1$ population derived from multicopy primary transgenics will be highly variable with regard to dose and composition of the transgene (Table 26.2). Plants homozygous for the transgene will also exhibit insertional mutagenesis, if the transgene is located within a functional host gene. Thus, the variability will be enlarged in the $T_1$ generation.

For proper comparison, only single-copy $T_1$ transgenics should be chosen. Most researchers prefer to pick single-copy primary transgenics for further study. However, such an approach leads to rejection of a large proportion of $T_0$ plants. Since production of transgenics is time consuming, resource demanding, and often difficult in many pulses, it is more prudent to make the best use of available transgenics. Single-copy transgenics can be isolated in subsequent generations of primary transgenics. Southern analysis can help identify single-copy transgenics in the $T_1$ generation. However, it cannot distinguish between homozygous and heterozygous individuals. Moreover, the frequency of single-copy transgenics will be low in the $T_1$ generation. If primary transgenics are crossed with untransformed control plants, a large proportion of single-copy $BC_1$ transgenics can be isolated (Table 26.2). Since all $BC_1$ transgenics are hemizygous, independent single-copy transgenics are readily comparable. Furthermore, insertional mutagenesis will not seriously affect the results. The null segregants are also invaluable in resolving transgene effects from somaclonal effects. The proportion of such null segregants is relatively high in the $BC_1$ generation as compared to the $T_1$ generation.

Therefore, we strongly suggest production of $BC_1$ rather than $T_1$ generations. For inheritance studies too, a $BC_1$ population is better than $T_1$, because the number of possible genotypes is fewer in a $BC_1$ generation. Hence, fewer $BC_1$ individuals will need to be studied to get the correct information about number of loci and linkage relationships. Selected $BC_1$

TABLE 26.2. Genetic composition of $T_1$ and $BC_1$ populations of transgenics.

| | $T_1$ generation | | | | $BC_1$ generation | | |
|---|---|---|---|---|---|---|---|
| | | Proportion of genotypes | | | | Proportion of genotypes | |
| No. of loci | No. of possible genotypes | Homozygous | Hemizygous | Null | No. of possible genotypes | Hemizygous | Null |
| 1 | 3 | 1/4 | 1/2 | 1/4 | 2 | 1/2 | 1/2 |
| 2 | 9 | 3/16 | 3/4 | 1/16 | 4 | 3/4 | 1/4 |
| 3 | 27 | 7/64 | 7/8 | 1/64 | 8 | 7/8 | 1/8 |
| 4 | 81 | 15/256 | 240/256 | 1/256 | 16 | 15/16 | 1/16 |

individuals can be selfed to obtain homozygous individuals for further analysis.

Since the majority of the pulses are not readily amenable to transformation, only a limited number of studies on genetic analysis of transgenics are available. In almost all cases, both Mendelian and non-Mendelian inheritance of transgenes have been reported (Samis et al., 2002, alfalfa; Aragao et al., 1996, bean; Saalbach et al., 1994, Vicia; Choffnes et al., 2001, soybean; Scott et al., 1998, white clover). A variety of explanations have been offered to account for distorted segregation ratios in transgenics (Yin et al., 2004). In most cases, inheritance studies are conducted in selfed generations. Micaleff et al. (1995) crossed primary transgenics of alfalfa to an agronomically superior cultivar to improve the performance of transgenics. A backcross breeding approach was also followed by McKersie et al. (1993) in *Medicago* to avoid inbreeding depression resulting from selfing. Based on expression of foreign protein, they found a 1:1 segregation for the transgene. However, PCR with *npt*II- or transgene-specific primers indicated the presence of three independent insertions, of which only one was functional. These results demonstrate the significance of multiple tests that may be necessary for correct interpretation of results.

Sharma et al. (1998) analyzed the back cross progeny of clover transgenics to understand the inheritance of transgenes. In general, intercross among primary transgenics is not recommended, as the analysis of results becomes complicated and presence of multiple copies of transgenes may lead to gene silencing. However, Sharma et al. (1998) studied progeny of intercross among different transgenics and found results matching with the backcross data. Scott et al. (1998) found 2:1 segregation for GUS in later generations of white clover transgenics. The distorted segregation ratio was attributed to either insertional mutagenesis or insertion of the transgene close to a recessive lethal gene. Christou et al. (1989) explained 1:1 segregation in soybean transgenics to nonviability of pollen carrying the transgene. In *Phaseolus,* insertional mutagenesis of an essential gene involved in female gamete development was attributed to distorted segregation of *gus, neo, AC123,* and *BC1* transgenes (Aragão et al., 1996). Distorted segregation may also result from chimerism in $T_0$ transgenics (Hiei et al., 1994).

The transgenic approach is gradually becoming indispensable for various studies in plant biology. Careful and judicious use of appropriate techniques and strategies is crucial for arriving at meaningful conclusions.

# REFERENCES

Alonso, J.M., A.N. Stepanova, T.J. Leisse, C.J. Kim, H. Chen, P. Shinn, D.K. Stevenson, J. Zimmerman, P. Barajas, R. Cheuk, et al. (2003). Genome wide insertional mutagenesis of *Arabidopsis thaliana. Science* 301:653-657.

Antal, Z., C. Rascle, M. Fevre, and C. Bruel (2004). Single oligonucleotide nested PCR: A rapid method for the isolation of genes and their flanking regions from expressed sequence tags. *Current Genetics* 46:240-246.

Aragão, F.J.L., L.M.G. Barros., A.C.M. Brasileiro, S.G. Ribeiro, F.D. Smith, J.C. Sanford, J.C. Faria, and E.L. Rech (1996). Inheritance of foreign genes in transgenic bean (*Phaseolus vulgaris* L.) co-transformed via particle bombardment. *Theoretical and Applied Genetics* 93:142-150.

Aragão, F.J.L., L.M.G. Barros, M.V. Sousa, M.F. Grossi-de-Sã, E.R.P. Almeida, E.S. Gander, and E.L. Rech (1999). Expression of a methionine-rich storage albumin from Brazil nut (*Bertholletia excelsa* H.B.K., Lecythidaceae) in transgenic bean plants (*Phaseolus vulgaris* L., Fabaceae). *Genetics and Molecular Biology* 22:445-449.

Barakat, A., P. Gallois, M. Raynal, D. Mestre-Ortega, C. Sallaud, E. Guiderdoni, M. Delseny, and G. Bernardi (2000). The distribution of T-DNA in the genomes of *Arabidopsis* and rice. *FEBS Letters* 471:1612-1614.

Bhat, S.R. and S. Srinivasan (2003). Molecular and genetic analyses of transgenic plants: Considerations and approaches. *Plant Science* 163:673-681.

Choffnes, D.S., R. Philip, and L.O. Vodkin (2001). A transgenic locus in soybean exhibits a high level of recombination. *In Vitro Cellular and Developmental Biology–Plant* 37:756-762.

Christou, P. (1990). Morphological description of transgenic soybean chimeras created by the delivery, integration and expression of foreign DNA using electric discharge particle acceleration. *Annals of Botany* 66:379-386.

Christou, P., W.F. Swain, N. Yang, and D.E. McCabe (1989). Inheritance and expression of foreign genes in transgenic soybean plants. *Proceedings of the National Academy of Sciences USA* 86:7500-7504.

Dean, C., J. Jones, M. Favreau, P. Dunsmuir, and J. Bedbrook (1988). Influence of flanking sequences on the variability in expression levels of an introduced gene in transgenic tobacco plants. *Nucleic Acids Research* 16:9267-9283.

De Block, M. (1993). The cell biology of plant transformation: Current state, problems, prospects and implications for plant breeding. *Euphytica* 71:1-14.

De Buck, S., A. Jacobs, M. van Montagu, and A. Depicker (1999). The DNA sequences of T-DNA junctions suggest that complex T-DNA loci are formed by a recombination process resembling T-DNA integration. *Plant Journal* 20: 295-304.

D'Halluin, K., J. Botterman, and W. De Greef (1990). Engineering of herbicide-resistant alfalfa and evaluation under field conditions. *Crop Science* 30:866-871.

Feldmann, K.A. (1991). T-DNA insertion mutagenesis in *Arabidopsis:* Mutational spectrum. *Plant Journal* 1:71-82.

German, M.A., M. Kandel-Kifr, D. Swarzberg, T. Matsevitz, and D. Granot (2003). A rapid method for the analysis of zygosity in transgenic plants. *Plant Science* 164:183-187.

Hiei, Y., S. Ohta, T. Komari, and T. Kumashiro (1994). Efficient transformation of rice (*Oryza sativa* L.) mediated by *Agrobacterium* and sequence analysis of the boundaries of the T-DNA. *Plant Journal* 6:271-282.

Jackson, S.A., P. Zhang, W.P. Chen, R.L. Phillips, B. Friebe, S. Muthukrishnan, and B.S. Gill (2001). High-resolution structural analysis of biolistic transgene integration into the genome of wheat. *Theoretical and Applied Genetics* 103:56-62.

Kohli, A., R.M. Twyman, R. Abranches, E. Wegel, E. Stoger, and P. Christou (2003). Transgene integration, organization and interaction in plants. *Plant Molecular Biology* 52:247-258.

Kononov, M.E., B. Bassuner, and S.B. Gelvin (1997). Integration of T-DNA binary vector backbone sequences into the tobacco genome: Evidence for multiple complex patterns of integration. *Plant Journal* 11:945-957.

Imai, T., R. Aida, and T. Ishige (1993). High frequency of tetraploidy in *Agrobacterium*-mediated transformants regenerated from tuber discs of diploid potato lines. *Plant Cell Reports* 12:299-302.

Labra, M., C. Savini, M. Bracale, N. Pelucchi, L. Colombo, M. Bardini, and F. Sala (2001). Genomic changes in transgenic rice (*Oryza sativa* L.) plants produced by infecting calli with *Agrobacterium tumefaciens*. *Plant Cell Reports* 20:325-330.

Labra, M., C. Vannini, F. Grassi, M. Bracale, M. Balsemin, B. Basso, and F. Sala (2004). Genomic stability in *Arabidopsis thaliana* transgenic plants obtained by floral dip. *Theoretical and Applied Genetics* 109:1512-1518.

Larkin, P.J. and W.R. Scowcroft (1981). Somaclonal variation: A novel source of variability from cell cultures for plant improvement. *Theoretical and Applied Genetics* 60:197-214.

Liu, Y., N. Mitsukawa, T. Oosumi, and R. Whittier (1995). Efficient isolation and mapping of *Arabidopsis thaliana* T-DNA insert junctions by thermal asymmetric interlaced PCR. *Plant Journal* 8:457-463.

McKersie, B.D., S.R. Bowley, and K.S. Jones (1999). Winter survival of transgenic alfalfa overexpressing superoxide dismutase. *Plant Physiology* 119:839-847.

McKersie, B.D., Y. Chen, M. de Beus, S.R. Bowley, C. Bowler, D. Inzé, K. D'Halluin, and J. Botterman (1993). Superoxide dismutase enhances tolerance of freezing stress in transgenic alfalfa (*Medicago sativa*). *Plant Physiology* 103:1155-1163.

Micaleff, M.C., S. Austin, and E.T. Bingham (1995). Improvement of transgenic alfalfa by backcrossing. *In Vitro Cellular and Developmental Biology–Plant* 31:187-192.

Molvig, L., M.L. Tabe, B.O. Eggum, A.E. Moore, S. Craig, D. Spencer, and T.J.V. Higgins (1997). Enhanced methionine levels and increased nutritive value of seeds of transgenic lupins (*Lupinus angustifolius* L.) expressing a sunflower seed albumin gene. *Proceedings of the National Academy of Sciences USA* 94:8393-8398.

Morton, R.L., H.E. Schroeder, K.S. Bateman, M.J. Chrispeels, E. Armstrong, and T.J.V. Higgins (2000). Bean alpha amylase inhibitor 1 in transgenic peas (*Pisum sativum*) provides complete protection from pea weevil (*Bruchus pisorum*) under field conditions. *Proceedings of the National Academy of Sciences USA* 97:3820-3825.

Nain, V., R. Jaiswal, M. Dalal, B. Ramesh, and P.A. Kumar (2005). Polymerase chain reaction analysis of transgenic plants contaminated by *Agrobacterium. Plant Molecular Biology Reporter* 23:59-65.

Nauerby, B., K. Billing, and R. Wyndaele (1997). Influence of the antibiotic timentin on plant regeneration compared to carbenicillin and cefotaxime in concentrations suitable for elimination of *Agrobacterium tumefaciens. Plant Science* 123:169-177.

Oger, P. and Y. Dessaux (2003). *Agrobacterium* mediated transformation of *Lotus* species. In *Applied genetics of Leguminosae biotechnology*, P.K. Jaiwal and R.P. Singh (eds.). Dordrecht: Kluwer Academic Publishers, pp. 255-281.

Pedersen, C., J. Zimny, D. Becker, A. Janhne-Gartner, and H. Lorz (1997). Localization of introduced genes on the chromosomes of transgenic barley, wheat and triticale by fluorescence *in situ* hybridization. *Theoretical and Applied Genetics* 94:749-757.

Perrin, Y., C. Vaquero, I. Gerrard, M. Sack, J. Drossard, E. Stoger, P. Christou, and R. Fischer (2000). Transgenic pea seeds as bioreactors for the production of a single-chain Fv fragment (scFV) antibody used in cancer diagnosis and therapy. *Molecular Breeding* 6:345-352.

Pickardt, T., I. Saalbach, D. Waddell, M. Meixner, K. Müntz, and O. Schieder (1995). Seed specific expression of the 2S albumin gene from Brazil nut (*Bertholletia excelsa*) in transgenic *Vicia narbonensis. Molecular Breeding* 1:295-301.

Pigeaire, A., D. Abernathy, P.M.C. Smith, K. Simpson, N. Fletcher, C.Y. Lu, C.A. Atkins, and E. Cornish (1997). Transformation of a grain legume lupin (*Lupinus angustifolius* L.) via *Agrobacterium tumefaciens* mediated gene transfer to shoot apices. *Molecular Breeding* 3:341-349.

Porsch, P., A. Jahnke, and K. During (1998). A plant transformation vector with minimal T-DNA II. Irregular integration patterns of T-DNA in the plant genome. *Plant Molecular Biology* 37:581-585.

Puonti-Kaerlas, J., T. Eriksson, and P. Engström (1992). Inheritance of a bacterial hygromycin phosphotransferase gene in the progeny of primary transgenic pea plants. *Theoretical and Applied Genetics* 84:443-450.

Radhamony, R.N., A.M. Prasad, and R. Srinivasan (2005). T-DNA insertional mutagenesis in *Arabidopsis:* A tool for functional genomics. *Electronic Journal of Biotechnology* 8:82-106.

Reddy, M.S., R.D. Dinkins, and G.B. Collins (2003). Gene silencing in transgenic soybean plants transformed via particle bombardment. *Plant Cell Reports* 21: 676-683.

Saalbach, I., T. Pickardt, F. Machemehl, G. Saalbach, O. Schieder, and K. Müntz (1994). A chimeric gene encoding the methionine-rich albumin of the Brazil nut

(*Bertholletia excelsa* H.B.K.) is stably expressed and inherited in transgenic grain legumes. *Molecular and General Genetics* 242:226-236.

Sahoo, P.D. (1996). Genetic Transformation studies in chickpea (*Cicer arietinum* L.). PhD thesis, Indian Agricultural Research Institute, New Delhi, India.

Sala, F., A. Arencibia, S. Castiglione, H. Yifan, M. Labra, C. Savini, M. Bracale, and N. Peluchi (2000). Somaclonal variations in transgenic plants. *Acta Horticulturae* 530:411-419.

Samis, K., S. Bowley, and B. McKersie (2002). Pyramiding Mn-superoxide dismutase transgenes to improve persistence and biomass production in alfalfa. *Journal of Experimental Botany* 53:1343-1350.

Schroeder, H.E., S. Gollasch, L. Tabe, A. Moore, D. Spencer, and T. Higgins (1995). The expression and stability of transgenes in peas (*Pisum sativum* L.). In *Proceedings of the 2nd European Conference on Grain Legumes,* Copenhagen, Denmark, p. 422.

Schroeder, H.E., M.R.I. Khan, W.R. Knibb, D. Spencer, and T.J.V. Higgins (1991). Expression of a chicken ovalbumin gene in three Lucerne cultivars. *Australian Journal of Plant Physiology* 18:495-505.

Schroeder, H.E., A.H. Schotz, T. Wardley-Richardson, D. Spencer, and T.J.V. Higgins (1993). Transformation and regeneration of two cultivars of pea (*Pisum sativum* L.). *Plant Physiology* 101:751-757.

Scott, A., D. Woodfield, and D.W.R. White (1998). Allelic composition and genetic background effects on transgene expression and inheritance in white clover. *Molecular Breeding* 4:479-490.

Sharma, S.B., K.R. Hancock, P.M. Ealing, and D.W.R. White (1998). Expression of a sulphur-rich maize seed storage protein, delta zein, in white clover (*Trifolium repens*) to improve forage quality. *Molecular Breeding* 4:435-448.

Tinland, B. (1996). The integration of T-DNA into plant genomes. *Trends in Plant Sciences* 1:178-184.

Todd, R. and B.W. Tague (2001). Phosphomannose isomerase: A versatile selectable marker for *Arabidopsis thaliana* germ-line transformation. *Plant Molecular Biology Reporter* 19:307-319.

Yin, Z., W. Plader, and S. Malepszy (2004). Transgene inheritance in plants. *Journal of Applied Genetics* 45:127-144.

Chapter 27

# Tagging Regulatory Elements in Plants

Raju Datla
Vivijan Babic
P. B. Kirti
Gopalan Selvaraj

## *INTRODUCTION*

The basal genetic potential of diverse cell and tissue types in eukaryotes is similar if not identical. However, to carry out diverse functions, various cells and tissues express only certain genes of the genome. Such precise spatial and temporal regulation of gene expression is important for maintaining highly complex processes of growth and development in a coordinated and orderly manner. Transcription is the first major committed step in gene expression and thus provides a primary and predominant control point for regulation in biological systems. Eukaryotic transcription is regulated in the majority of cases by complex interactions between *cis*-acting DNA motifs present in the 5' end of the protein-coding regions and *trans*-acting protein factors. In some cases, the regulatory elements found in the 5' untranslated region and 3' end of the coding region, and even in some of the introns are required for accurate and coordinated gene expression (Larkin et al., 1993; Seiburth and Meyerowitz, 1997; Yu et al., 2001). The DNA sequences present upstream to the transcription start site are collectively referred to as the promoter region and are involved in transcriptional initiation and regulation. Sequences downstream of the coding region are essential for transcription termination and polyadenylation, and in some cases are also necessary for transcription initiation as well (Waugh and Brown, 1991).

The promoter of eukaryotic protein coding genes can be divided into proximal core promoter (within a few hundred nucleotides upstream to the transcriptional start site) and distal promoter regions that can include 1 to 3 kb sequences beyond the core promoter area. The distal promoter region contains several key regulatory elements that confer positive or negative activities to work as activators or repressors respectively. In some cases, the activity of a gene or sets of genes can be regulated by common enhancer elements located quite far from the core promoter or at different genomic locations. In a majority of eukaryotic genes, including plants, the core promoter region consists of a signature element called a TATA box, usually found 25 to 30 bp upstream of the transcriptional start site. General factors required for transcription are recruited to the TATA box to form an initiation complex through interactions with gene specific DNA-binding factors. DNA-binding proteins regulate the general transcription factors thereby activating or repressing their expression (Roeder, 1991). Promoters and enhancers are usually composed of several discrete *cis* elements that may be specifically recognized by one or more *trans*-acting proteins. Spatial and temporal control has been shown in a number of genes controlled by the interactions of regulatory elements present in promoters with cell, tissue, developmental, and environmental specific *trans*-acting DNA-binding factors. Advances in the identification of key components of the transcriptional apparatus, along with the functional dissection of the other key associated factors, contributed to the development of a general conceptual framework for transcriptional regulation in eukaryotes (Kornberg, 1999, 2005). In general, plant genes share several of these key features and considerable progress has been made during the past few decades in the understanding of transcriptional regulation in plants (reviewed in Reichmann, 2002).

## *PLANT PROMOTERS*

Advances in gene cloning tools in the early 1980s contributed to the isolation of promoter sequences active in plant cells. Most of these correspond to highly expressed genes, and they were usually isolated based on cDNAs and their corresponding genomic sequences. In addition, plant active promoters of pathogenic origin (e.g., *Agrobacterium* and cauliflower mosaic virus) were also isolated, studied, and used extensively. The development of gene transfer technologies for several plant species in the 1980s and 1990s led to many new opportunities for applying plant genetic engineering tools to crop improvement. During the early years of development of transgenic plant technology, the CaMV35S promoter, *Agrobacterium* promoters from Ti plasmids, and a few limited promoters of plant genome origin were primarily used to express foreign genes in genetically modified (GM) crops. Although these promoters worked well for proof-of-concept experiments, they displayed less than desirable phenotypes, especially for transgenes that required precise regulation for producing optimal expression for targeted trait modifications. As plant genetic engineering technologies advanced into commercial production of GM plants, targeted gene regulation became a necessity. Accordingly, the search for suitable plant promoters became as important as molecular genetic modification. Importantly, promoters offering spatial and temporal regulatory features would also be utilized in basic studies on plant form and function (reviewed in Potenza et al., 2004).

To discern promoter activity, various reporter genes have been developed. These offer desirable properties with respect to biochemical assays and also cytological localization. The *gus, luciferase,* and *GFP* genes have been the most widely used for plant promoter studies (Jefferson et al., 1987; Kertbundit et al., 1991; Haseloff et al., 1997; Alvarado et al., 2004). In addition to their use in characterizing promoters that had already been isolated, these reporters have also been immensely useful in de novo identification of promoters. A number of plant promoters have been shown to retain their specificity in heterologous species. However, it has also been shown that many promoters differ in their expression patterns and specificities when transferred to heterologous species, suggesting that there are no universal promoters that function in an identical manner in all plant hosts (e.g., 35S). These observations raise the possibility that for certain genes in certain hosts the ideal promoter may have to come from the target plant species itself. Accordingly, it is essential to generate a collection of promoters from which suitable ones can be used as required. A broadly applicable promoter tagging system may be especially useful in this regard if the method can be extended to agronomically important crop species.

## *IN VIVO PROMOTER TAGGING FOR PLANTS*

The cDNA-directed approach for isolation of plant promoters has been successful only to a limited extent until recently and hence, alternative approaches became imperative. The efficient delivery and integration properties of T-DNA from *Agrobacterium* encouraged the development of the first in vivo tagging system in plants (Andre et al., 1986; Koncz et al., 1989). The underlying expectation was that T-DNA insertion was random enough to place a promoterless *npt*II open reading frame (ORF) at various places in the plant genome. Furthermore, when the *npt*II ORF is juxtaposed next to the right border sequence (RB) of the T-DNA, any T-DNA insertions downstream of a native promoter would result in transcriptional fusion of the promoter to the *npt*II ORF and therefore kanamycin (Km) resistance of the cells. The putative promoter region upstream of the *npt*II sequence in Km-resistant tobacco cells, when recovered and reintroduced into tobacco cells again, conferred Km resistance. This demonstrated the feasibility of in vivo tagging of plant promoters. Though promoters can be tagged using this strategy, defining the specificity of the tagged line was not possible with this reporter system. This limited its broader utility.

To address this limitation, tagging vectors were designed in which a promoterless reporter gene was placed next to RB and transgenic cells were selected with the use of a constitutively expressed positive selection marker—mostly *npt*II. In this method, only a subset of the selected transgenic lines would show reporter expression. In such a system, the *gus* reporter gene's ideal and sensitive *in situ* histochemical staining and localization features were exploited (Fobert et al., 1991). Jeoung et al. (2002) used a promoterless *gus* gene next to the RB and multiple copies of the enhancer from 35S promoter at the left border (LB). More than 13,450 promoter trap lines in rice were obtained. Plants histochemically analyzed for GUS expression revealed that the use of the enhancer doubled the frequency of GUS positive lines, suggesting that any basal expression was rendered evident by this approach. In general, screening for Gus expression identified a range of specific expression patterns in the transgenic lines. However, as there was no positive selection for successful transcriptional fusions with these reporter systems, only a small portion of the transgenic lines showed Gus expression. Thus, a larger collection of transgenics is necessary.

Webb et al. (2000) used the promoterless *gus* (*uid*A) gene for isolating promoters of the genes involved in the process of symbiotic nitrogen fixation process in association with rhizobium in the model legume, *Lotus japonicus*. They have produced 284 primary transformants, of which 7 confirmed transformants showed different GUS expression patterns. Of these, 3 expressed GUS in nodules and roots before and after inoculation with *Mesorhizobium loti,* whereas 4 lines had GUS expression only after inoculation. This study has lead to the characterization of a gene *LjCbp1* for a calcium binding protein. In a similar study, Buzas et al. (2005) introduced the promoterless *uid*A into *L. japonicus* and developed lines that expressed GUS in a stable and developmentally regulated manner. They observed 2 to 5 percent functional GUS fusion frequencies out of the total transgenic lines developed. They utilized promoter GUS fusions to investigate the genetic regulation of LjENOD40–2 and FATA MORGANA GUS. The LjENOD40–2 promoter defined a novel expression domain and the FATA MORGANA nodule expression was reiterated by a 2.0 kb sequence upstream of the T-DNA insertion. It has also been postulated that the elements that regulate LjENOD40–2 also regulate FONTANA GUS (Buzas et al., 2005).

In a separate study, 20,000 transgenic lines were obtained in *Arabidopis* using promoterless *luciferase* reporter gene in a T-DNA construct (Alvarado et al., 2004). It was apparent that a reporter that offers both positive selection and histological screenability would be more robust in promoter tagging. Initially, we developed a selectable version of *gus* by fusing its C-terminus to the N-terminus of *npt*II (Datla et al., 1991). The versatile features of *gus:npt*II fusion were demonstrated in diverse plant species both for efficient selection to produce transgenic lines and for sensitive enzymatic assays for GUS biochemical quantitative and histochemical activities. These results encouraged us to develop a new in vivo promoter tagging strategy exploiting the ideal features of the *gus:npt*II fusion gene.

### *gus:nptII-Based Promoter Tagging System*

In designing this system, a synthetic T-DNA of nopaline plasmid type flanked by 25 bp of RB and LB with multiple restriction sites was constructed in the vector backbone of Bin19 (Bevan, 1984) from which the original T-DNA had been removed. Since T-DNA insertions show more precise integration at RB than at LB sequences, the-promoterless *gus:npt*II cassette was inserted very close to the RB of the synthetic T-DNA. Since the primary focus of this construct was to generate transcriptional fusions in frame, stop codons were inserted between the *gus:npt*II initiation codon and RB so that translational fusions would be selected against. It was expected that a significant number of insertion events would likely involve fusion to low-expressing regulatory sequences. To increase the possibility of recovering these sequences, translational enhancer sequences based on AMV RNA4 were inserted between the RB and the *gus* initiation codon (Datla et al., 1993). Though this would not enhance transcription, the transcripts produced at low levels were expected to translate more efficiently, and to confer Km resistance.

### *Transformation: Transgenic Line Production*

The T-DNA-based promoter-tagging system using *gus:npt*II depends on the efficiency of the genetic transformation system. As tobacco can be efficiently transformed, we initially tested the promoter trap system in tobacco. The well-established leaf disc method was used to introduce the promoter-tagging construct and primary selection for potential transformed cells in the resulting callus was selected on kanamycin. The rationale behind this was that all cells are deregulated at the callus stage for developmental specificities and therefore even a highly tissue-specific gene would be expressed to some extent in a callus. If this were true, in some plants regenerated from a Km-resistant callus, there should be developmental regulation of the reporter gene activity. We found that the growth characteristics of the calli were different; some were much slower in growth than the others, likely a reflection of the promoter strength. In order to enhance the recovery of cells where developmentally regulated promoters are tagged, we optimized the Km selection protocols to include periods of nonselection. Approximately 1,000 independently recovered plants were analyzed for Gus activity. This revealed a wide range of expression patterns suggesting that various tissue-specific promoters could be tagged by this approach. The promoter tags included those expressing in pollen, guard cells, xylem, and phloem. Among the tagged lines, vascular expression is one of the predominant patterns. These promising results in tobacco encouraged us to apply this system in other plant species. Though the recovery frequency of transgenic lines varied, a large number of putative promoter-tagged lines were produced in *Arabidopsis, Brassica napus, Brassica carinata,* and *Nicotiana plumbaginifolia*. Gus analysis revealed a broad range of expression patterns, as observed in tobacco, further confirming the applicability of this system in these plant species. These results further suggest that any legume species for which genetic transformation protocols are available would be amenable to promoter tagging by this approach. Indeed, this has been applied to *Arachis hypogaea* (peanut), an economically important legume crop, following improvements to the transformation efficiency (Swathi Anuradha et al., 2006). Tagging with *gus:npt*II yielded several lines with a range of expression patterns, which suggests that this system can be applied to other legumes such as chickpea and pea that can be transformed readily now.

## *ISOLATION OF PROMOTERS FROM TAGGED LINES*

While histochemical analysis of Gus expression in tagged lines provides initial information for identifying lines that display unique or desirable features, the molecular properties of the integration events have to be defined before the flanking putative regulatory sequences are isolated. About 50 percent of the promoter-tagged lines in tobacco show single copy insertions, 20 percent with two or three copies, and the rest with more than three copies. In some of the multiple-copy lines, the integrations are tandem copies; these lines can also be used if a restriction map of the insertion is developed. In the case of tagged lines with insertions at different genomic locations, it is important to associate the Gus expression patterns with specific insertions. Where possible, it is preferable to genetically segregate and identify single copy lines before undertaking promoter isolation. Alternatively, the 5' flanking sequences in the insertions can be isolated and each of these can be reintroduced along with a reporter fusion in order to determine their functional properties. In some cases, complex rearrangements at the insertion sites, as reported in another study (Tax and Vernon, 2001) render promoter isolation very difficult.

In general, various approaches can be applied to isolate the tagged promoter sequence. Several plant promoters have been successfully isolated using promoter tagging techniques. These include embryo, seed coat, tapetum and vascular tissue, constitutive, guard cell, young leaves, and meristematic cell-specific promoters (reviewed in Radhamony et al., 2005). Polymerase chain reaction (PCR)-based strategies such as inverse PCR, TAIL PCR, adapter PCR, and chromosome walking have all been generally successful in isolating the 5' flanking sequences of the promoter tags (Radhamony et al., 2005). In some cases, however, construction of subgenomic libraries encompassing a given restriction pattern of tagged DNA has been necessary. This approach requires construction of a restriction map that includes approximately 4-6 kb of the region flanking the insertion site. Then appropriate restriction enzymes are used to cut the genomic DNA outside this region and size-selected DNA is cloned into a plasmid or bacteriophage λ vector. The reporter sequence is then used for identifying the clones containing the putative promoter region. The putative promoter is assessed for function by introducing it as a new fusion to a reporter in transient or stably transformed plant lines (Guilfoyle, 1997).

## *PERSPECTIVES*

With rapid advances in many areas of applied plant biology research, many new opportunities are now available for genetic improvement of crop species. The model plant *Arabidopsis* continues to be immensely helpful while plants such as rice (now fully sequenced) offer rich opportunities for trait improvement as a model and also as a target. The explosive increase in expressed sequence databases for various plant species and the increasing number of expression analyses that use microarrays have changed the landscape of promoter search. It is evident in various ways. For example, comparative genomics is being tapped to identify multispecies conserved sequences that provide a target for discovery of promoters (Hughes et al. 2005). In addition, prior knowledge of *cis*-acting elements, in association with comparative genomics, has been used to identify promoters pertaining to stress responsive gene expression (Zhang et al., 2005).

New bioinformatics tools have already been developed and applied to predict and identify regulatory elements in annotated genes (Rombauts et al., 2003). It is also now possible to search for known *cis* elements in the PlantCare database (Lescot et al., 2002) or for transcription factors that bind to these motifs using the TransFAC database (Kennedy and Wilson, 2004). Clearly, promoter tagging in vivo, although a functional screen, is no longer the only means for promoter discovery; nor has it been superseded by in silico approaches. As genomics technologies and genetic transformation protocols advance and spread to crop plants, computational approaches will become a major tool complemented by functional analysis.

## REFERENCES

Alvarado, M.C., L.M. Zsigmond, I. Kovacs, A. Cseplo, C. Koncz, and L.M. Szabados (2004). Gene trapping with firefly luciferase in *Arabidopsis:* Tagging of stress responsive genes. *Plant Physiology* 134:18-27.

Andre, D., D. Colau, J. Schell, V.M. Montagu, and J. Hernalsteens (1986). Gene tagging in plants by a T-DNA insertion mutagen that generates *APH (3')* II-plant gene fusions. *Molecular and General Genetics* 204:512-518.

Bevan, M. (1984). Binary *Agrobacterium* vectors for plant transformation. *Nucleic Acids Research* 12:8711-8721.

Buzas, D.M., D. Lohar, S. Sato, Y. Nakamura, S. Tabata, C.E. Vickers, J. Stiller, and P.M. Greshoff (2005). Promoter trapping reveals novel root and nodule GUS expression domains. *Plant and Cell Physiology* 46:1202-1212.

Datla, R.S.S., F. Bekkaoui, J.K. Hammerlindl, L.E. Pelcher, G. Pilate, D.I. Dunstan, and W.L. Crosby (1993). Improving high level constitutive foreign gene expression in plants using an AMV RNA4 untranslated leader sequence. *Plant Science* 94:139-149.

Datla, R.S.S., J.K. Hammerlindl, L.E. Pelcher, W.L. Crosby, and G. Selvaraj (1991). A bifunctional fusion between beta-glucuronidase and neomycin phosphotransferase: A broad-spectrum marker enzyme for plants. *Gene* 101:239-246.

Fobert, P.R., B.L. Miki, and V.N. Iyer (1991). Detection of gene regulatory signals in plants revealed by T-DNA mediated fusions. *Plant Molecular Biology* 17: 837-851.

Guilfoyle, J.T. (1997). The structure of plant gene promoters. In *Genetic Engineering,* Volume 19, J.K. Setlow (ed.). New York: Plenum Press, pp. 15-47.

Haseloff, J., K.R. Siemering, D.C. Prasher, and S. Hodge (1997). Removal of a cryptic intron and subcellular localization of green fluorescent protein are required to mark transgenic *Arabidopsis* plants brightly. *Proceedings of the National Academy of Sciences USA* 94:2122-2127.

Hughes J.R., J.F. Cheng, N. Ventress, S. Prabhakar, K. Clark, E. Anguita, M. De Gobbi, P. de Jong, E. Rubin, and D.R. Higgs (2005). Annotation of *cis* regulatory elements by identification, sub classification, and functional assessments of multispecies conserved sequences. *Proceedings of the National Academy of Sciences USA* 102:9830-9835.

Jefferson, R.A., T.A. Kavangh, and M.W. Bevan (1987). GUS fusions: β-glucuronidase as a sensitive and versatile gene fusion marker in higher plants. *EMBO Journal* 6:3901-3907.

Jeong, H.D., S. An, G.-H. Kang, S. Moon, J.-J. Han, S. Park, H.S. Lee, K. An, and G. An (2002). T-DNA insertional mutagenesis for activation tagging in rice. *Plant Physiology* 130:1636-1644.

Kennedy, C.G. and I.W. Wilson (2004). Plant functional genomics: Opportunities in microarray databases and data mining. *Functional Plant Biology* 31:295-314.

Kertbundit, S., H. De Greve, F. De Boeck, M. Van Montagu, and J.P. Hernalsteens (1991). In-vivo random β-glucuronidase gene fusions in *Arabidopsis thaliana. Proceedings of the National Academy of Sciences USA* 88:5212-5216.

Koncz C., N. Martini, R. Mayerhofer, Z. Koncz-Kalman, H. Korber, G.P. Redei, and J. Schell (1989). High-frequency of T-DNA mediated gene tagging in plants. *Proceedings of the National Academy of Sciences USA* 86:8467-8471.

Kornberg, D.R. (1999). Eukaryotic transcriptional control. *Trends in Biochemical Sciences* 24:46-49.

Kornberg, D.R. (2005). Mediator and the mechanism of transcriptional activation. *Trends in Biochemical Sciences* 30:235-239.

Larkin, J.C., D.G. Oppenheimer, S. Pollock, and M.D. Marks (1993). *Arabidopsis GLABROUS1* gene requires downstream sequences for function. *Plant Cell* 5:1739-1748.

Lescot, M., P. Dehais, G. Thijs, K. Marchal, Y. Moreau, Y. Van de Peer, P. Rouze, and P. Rombauts (2002). PlantCARE: A database of plant *cis*-acting regulatory elements and a portal to tools for *in silico* analysis of promoter sequences. *Nucleic Acids Research* 30:325-327.

Potenza, C., L. Aleman, and C. Sengupta–Gopalan (2004). Targeting transgene expression in research, agricultural, and environmental applications: Promoters used in transformation. *In Vitro Cellular and Developmental Biology–Plant* 40:1-22.

Radhamony, R.N., A.M. Prasad, and R. Srinivasan (2005). T-DNA insertional mutagenesis in *Arabidopsis:* A tool for functional genomics. *Electronic Journal of Biotechnology* 8:82-106.

Reichmann, L.J. (2002). Transcriptional regulation: A genomic overview. In *The Arabidopsis book,* C.R. Somerville and E.M. Meyerowitz (eds.). Rockville, MD: American Society of Plant Biologists.

Roeder, R.G. (1991). The complexities of eukaryotic transcription initiations: Regulation of pre-initiation complex assembly. *Trends in Genetics* 7:402-407.

Rombauts, S., K. Florquin, M. Lescot., K. Marchal, P. Rouze, and Y. Van de Peer (2003). Computational approaches to identify promoters and *cis*-regulatory elements in plant genomes. *Plant Physiology* 132:1161-1176.

Seiburth, L.E. and E.M. Meyerowitz (1997). Molecular dissection of the *AGAMOUS* control region shows that *cis*-elements for spatial regulation are located intragenically. *Plant Cell* 9:355-365.

Swathi Anuradha, T., S.K. Jami, R.S. Datla, and P.B. Kirti (2006). Genetic transformation of peanut (*Arachis hypogaea* L.) using cotyledonary node as explant and a promoterless *gus::npt*II fusion vector. *Journal of Biosciences* 31:1-12.

Tax, F.E. and D.M. Vernon (2001). T-DNA associated duplication/translocation in *Arabidopsis*: Implications for mutant analysis and functional genomics. *Plant Physiology* 126:1527-1538.

Waugh, R. and J.W. Brown (1991). Plant gene structure and expression. In *Plant genetic engineering,* D. Grierson (ed.). Glasgow: Blakie & Son, pp. 1-31.

Webb, K.J., L. Skot, M.N. Nicholson, B. Jorgensen, S. Mizen (2000). *Mesorhizobium loti* increases a root specific expression of a calcium binding homologue identified by promoter tagging in *Lotus japonicus. Molecular Plant-Microbe Interactions* 13:606-616.

Yu, F., C. Chen, and Z. Chen (2001). Evidence for an important role of WRKY DNA binding proteins in the regulation of *NPR1* gene expression. *Plant Cell* 13:1527-1539.

Zhang, W., J. Ruan, T.-H. David Ho, Y. You, T. Yu, and R.S. Quatrano (2005). *Cis*-regulatory element based targeted gene finding: Genome-wide identification of abscisic acid and abiotic stress-responsive genes in *Arabidopsis thaliana. Bioinformatics* 21:3074-3081.

## Chapter 28

# cDNA-AFLP Transcriptome Analysis in Legumes

Márcia Margis-Pinheiro
Jean Luis Simões-Araújo

### *DIFFERENTIALLY EXPRESSED GENES*

The expression pattern analysis of genes on a large scale is an essential aspect of functional genomics. Information about the spatial and temporal regulation of gene expression has been acquired in recent years, which frequently helps in determining the possible function of a specific gene. A multiplicity of sophisticated techniques, useful to compare steady-state mRNA concentrations, has been developed allowing access to differentially expressed genes. However, some genes are rarely expressed or are alternatively expressed in a thin cell layer or in a group of specialized cells. When the tissue is crushed, the transcripts from genes of interest can represent only a very small fraction of the overall sample. The difficulty of isolating differentially expressed genes, particularly rare ones, can be realized by taking into account that a eukaryotic cell contains ~15,000-30,000 different mRNAs with an occurrence ranging from one to several thousands in a total mass of ~100,000 mRNAs. Approximately 50 percent of the total transcript population consists of abundant mRNA, representing only 1 percent of different mRNA species, while the other half contains "rare" transcripts (Wan et al., 1996). Currently, there are fundamentally three high-throughput strategies for large-scale analysis of gene expression: serial analysis of gene expression (SAGE), hybridization-based methods such as differential hybridization and microarray, and gel-based RNA fingerprinting techniques such as differential display and cDNA-AFLP (cDNA-amplified fragment length polymorphism). Although DNA microarrays represent an appropriate tool for genome-wide expression analysis, their use is limited to organisms for which the complete genome sequence or a large cDNA collection is available (Breyne and Zabeau, 2001) and may miss transcripts expressed at very low levels. SAGE and cDNA-AFLP are alternative approaches to explore quantitative gene expression patterns without prior availability of transcript information. Here we discuss the use of cDNA-AFLP technology to identify transcript-derived fragments that correspond to differentially expressed genes in plants and describe how to employ this technique to discover new genes in legumes.

## *The cDNA-AFLP Technique*

cDNA-AFLP is an extremely efficient RNA fingerprinting method for the isolation of differentially expressed genes (Bachem et al., 1996, 1998). It is a genome-wide expression analysis technique that does not require previous sequence information, and makes it an excellent tool for gene discovery (Ditt et al., 2001). The traditional technique consists of four main steps. One, preparation of cDNAs and cDNA double restriction enzyme digestion; two, ligation of oligonucleotide adapters to the resulting restriction fragments to generate template DNA; three, two steps of polymerase chain reaction (PCR) representing a preamplification and a selective amplification. PCR primers, complementary to the adapter sequences with additional selective nucleotides at the 3' end, are used in selective amplification, which allows the amplification of a limited number of cDNA fragments, reducing the complexity of amplified samples (Figure 28.1). Four, separation and visualization of the selective amplified product by using a $^{33}P$ labeled primer.

In contrast to the differential display methods, which use small random primers (Liang and Pardee, 1992), relatively higher annealing temperatures can be reached during PCR reactions in cDNA-AFLP, providing a much more reproducible method. In relation to hybridization-based techniques, such as macro-and microarrays, cDNA-AFLP can discriminate between homologous genes belonging to gene families very common in plants. Besides, the sensitivity of the technique is very high, resulting in an excellent detection of low-abundance genes. As in the differential display method, both induced and repressed genes can be detected and more than two samples can be compared. Once a candidate band has been eluted from the gel (Figure 28.2), gene expression analysis tools such RT-PCR, reverse northern, and northern blot can be applied to confirm the expression pattern.

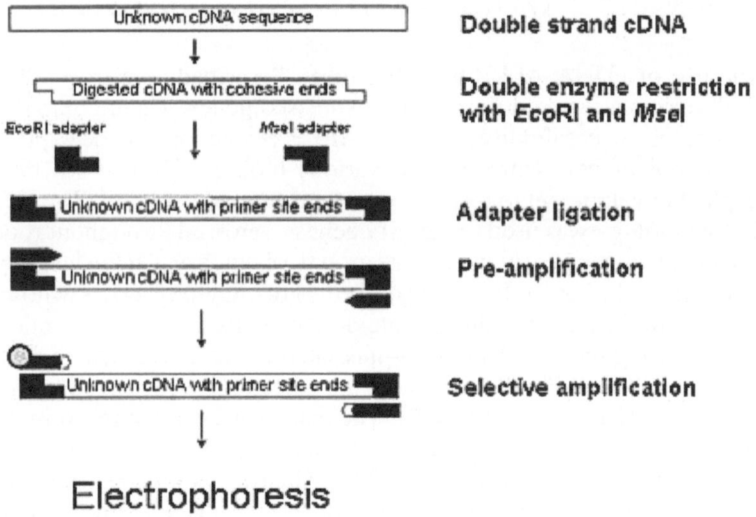


FIGURE 28.1. Schematic representation of the main steps in cDNA-AFLP.

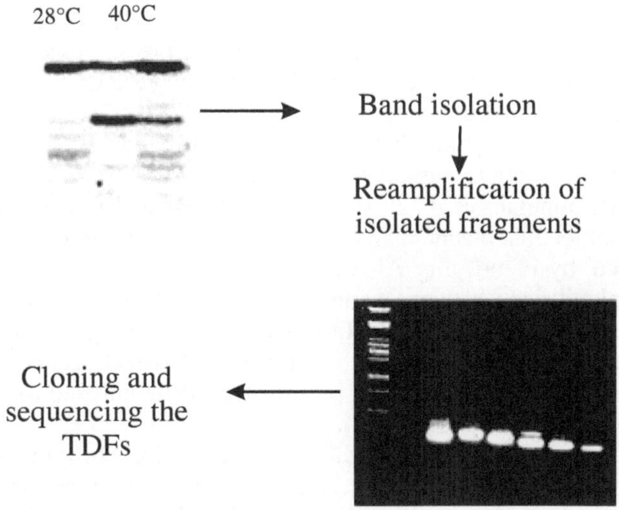


FIGURE 28.2. Visualization and isolation of differentially expressed transcript-derived fragments (TDFs) identified after the cDNA-AFLP selective amplification.

## *cDNA-AFLP SYSTEMS USED FOR GENE DISCOVERY*

The use of cDNA-AFLP has often been described as a truly efficient method for isolation of differentially expressed genes. There are many examples of the successful use of cDNA-AFLP as a genome-wide expression analysis tool of genes involved in various biological processes ranging from plant development to responses to environmental stimuli. The cDNA-AFLP technology was used to identify genes expressed throughout root enlargement during the early vegetative growth of chicory, *Cichorium intybus* L. (Goupil et al., 2003). It was also used to display transcripts whose expression is rapidly altered during interactions between plant and microorganisms: during the *Avr-9* and *Cf-9*-mediated defense response in tobacco cell cultures, 290 of 30,000 fragments inspected showed altered transcript accumulation (Durrant et al., 2000). The nature of the plant response to infection and transformation by *Agrobacterium tumefaciens* was also investigated by cDNA-AFLP (Ditt et al., 2001). Gabriels et al. (2006) identified new genes required for the hypersensitive response (HR) in tomato plants by performing cDNA-AFLP experiments followed by functional analysis of differentially expressed genes.

Studies on the morphological, physiological, or molecular variability of the root system to adapt its growth to environmental stress were also conducted using cDNA-AFLP (Dubos and Plomion, 2003). In almond [*Prunus amygdalus* (L.) Batsch], several genes strongly expressed in response to dehydration were analyzed as a means of determining the genetic basis of the mechanisms involved in drought tolerance. In this case, cDNA-AFLP was used to identify transcripts that accumulated in mature embryos and in in vitro-cultured plantlets subjected to desiccation or abscisic acid (ABA) treatment (Campalans et al., 2001). cDNA-AFLP was also carried out to inspect the molecular mechanisms of aluminum (Al) toxicity and Al tolerance of rice, by identifying Al-regulated genes in roots of an Al-tolerant tropical upland rice (Azucena), and an Al-sensitive lowland rice (IR1552). Nineteen genes with known function were found among 34 transcript-derived fragments (TDFs) regulated by Al stress (Mao et al., 2004). The investigation of plant response to high salinity was also analyzed using cDNA-AFLP (Hmida-Sayari et al., 2005).

In addition to the examples mentioned, the cDNA-AFLP method is also being employed successfully to investigate how different genomes are coordinated in the same nucleus of hexaploid wheat (*Triticum aestivum*), since it contains three different genomes derived from three distinct species. The authors compared global gene expression of a synthetic hexaploid wheat with its diploid (*Aegilops tauschii*) and tetraploid (*T. turgidum*) parents by cDNA-AFLP display. The results suggested that the expression of a significant fraction of genes was altered in the synthetic hexaploid and a majority of them appeared to be diminished or activated (He et al., 2003).

There are only a few examples of the use of this technique for isolating genes in legumes. The cDNA-AFLP approach was used to identify markers of germination in *Phaseolus vulgaris*. Among changes observed throughout the germination process, a cDNA encoding a germinin-like protein, which accumulates in the embryo axis before radicle emergence, was identified (Aubry et al., 2003). Wang et al., (2005) employed cDNA-AFLP to analyze the transcriptome of a recombinant inbred line population of soybean that was derived from soybean mosaic virus (SMV) resistant and susceptible cultivars. An *HDZip I* gene *GmHZ1,* whose expression was different between resistant and susceptible cultivars infected by SMV, was identified in this study. In addition, using the model bean common mosaic virus (BCMV) infection in *I* gene-containing *Phaseolus vulgaris,* Cadle-Davidson and Jahn (2006) carried out a cDNA-AFLP screen to discover transcriptional variation that occurs when either resistance or susceptibility develops. Scippa et al. (2006) analyzed the response of *Spartium junceum* roots to slope. The cDNA-AFLP approach revealed changes in the expression of several genes in root systems of slope-grown plants. Some differentially expressed genes are homologues of genes induced by environmental stresses in other plant species or are involved in the production of strengthening materials.

Our group has carried out cDNA-AFLP experiments to analyze genes related to thermal tolerance during cowpea-*Rhizobium* symbiosis. This symbiotic interaction has special importance since cowpea is highly tolerant to heat stress, the main stress factor that limits biological nitrogen fixation in tropical countries. We have used cDNA-AFLP to study the genes expressed in nodule during heat stress (Figure 28.3).

In our experiments, the number and length of observed TDFs were dependent on the primer combinations. Many TDFs displayed an altered expression pattern in response to heat stress and were selected for further analysis. In that work, cDNA-AFLP allowed the identification of several TDFs from cowpea nodules subjected to high temperature stress. All the characterized TDFs have homology to genes related to stress defense, suggesting that they might participate in the thermal tolerance mechanism. We consider that the data obtained will provide the first clues for guiding further functional studies of biological nitrogen fixation (Simões-Araújo et al., 2002).

The cDNA-AFLP can be combined with other high-throughput strategies for large-scale analysis of a transcriptome, such as EST (expressed se-

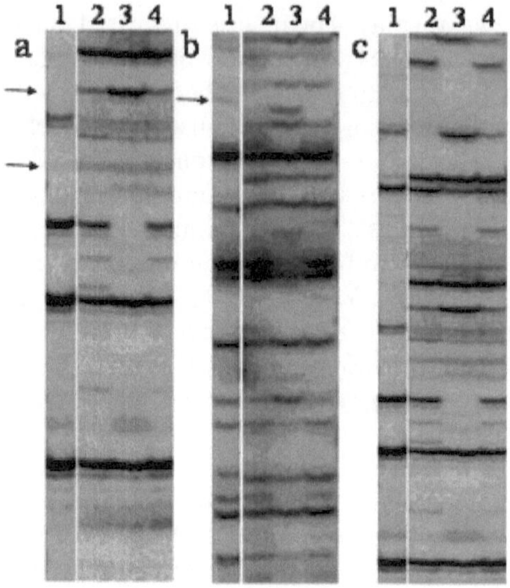


FIGURE 28.3. Differentially expressed genes revealed by cDNA-AFLP during legume-*Rhizobium* interactions under heat stress conditions. cDNA-AFLP reactions were performed with three different primer pair combinations (a, b, and c). Cowpea plants were subjected to different conditions: (1) control (28°C), (2) 30 minutes (3) 1 hour, and (4) 2 hours at 40°C. Arrows indicate examples of differentially expressed TDFs.

quence tag) sequencing. Qin et al. (2001) developed a computer program called GenEST, which predicts the size of TDFs of in silico digestion of cDNA sequences retrieved from databases. The resulting virtual TDFs can be traced back among the TDFs displayed on cDNA-AFLP gels. As a consequence, a cDNA sequence database can be screened very efficiently to identify genes with relevant expression profiles. On the other hand, the program permits quick extension of TDFs by searching for the corresponding ESTs. Using this strategy, the authors revealed various novel pathogenicity factors from the plant parasitic nematode *Globodera rostochiensis* by combining EST from infective stage juveniles with an expression profile of 4,000 genes in five developmental stages produced by cDNA-AFLP.

Finally, Breyne et al. (2003) have developed an improved cDNA-AFLP method for genome-wide expression analysis, which is efficient for quantitative transcript profiling representing an alternative to microarrays. In this technique, unique transcript tags, generated from reverse-transcribed messenger RNA by restriction enzymes, were screened through a series of selective PCR amplifications. Supported by *in silico* analysis, an enzyme combination was chosen to ensure that at least 60 percent of all the mRNAs were represented in an informative sequence tag. The sensitivity and specificity of the method allows the detection of poorly expressed genes and to distinguish between homologous sequences. Precise gene expression profiles can be determined by quantitative analysis of band intensities allowing the detection of even subtle differences in transcriptional activity. The authors demonstrated the efficacy of the method for genome-wide expression analysis by a detailed screen for cell cycle-modulated genes in tobacco (Breyne et al., 2003).

## *cDNA-AFLP PROTOCOLS*

### *Preparation of Poly (A)+ RNA and cDNA Synthesis*

- Plant tissue sample is harvested and immediately frozen in liquid $N_2$. The plant material can be stored at –80°C until RNA extraction.
- To extract RNA, the frozen material (0.5 g) is ground to a fine powder with a mortar and pestle (precooled) under liquid $N_2$. Different RNA extraction protocols (e.g., TRIzol reagent, GIBCO-BRL) can be used for this purpose. Obtaining high-quality and intact RNA is very important to obtain the best cDNA synthesis and cDNA-AFLP performance.
- Resuspend total RNA in 40 µl of RNase-free water and use 2 µl of this sample to check the RNA yield/quality on a 1 percent agarose gel. The gel and run buffer must also be prepared with RNase-free water. (All material and solutions that come into contact with RNA should be RNase free. Autoclaving usually gives enough protection; however, DEPC or RNase inhibitors treatments may be necessary.)
- Purify the poly-A+ mRNA from total RNA (20 µg) using oligo(dT) coupled to paramagnetic beads (Dynal A.S., Oslo, Norway) or Poly-ATtract mRNA Isolation System (Promega). Poly-A+ RNA isolation is not required, especially when isolating RNA from small amounts of tissue. However, when total RNA is used, DNaseI treatment is necessary.
- Synthesize double-stranded cDNA from poly-A+ mRNA according to SuperScript Double-Stranded cDNA Synthesis Kit (Invitrogen).
- Resuspend the double-stranded cDNA in sterile water (20 µl).

***Template Preparation, AFLP Reaction, and PAGE Analysis of Amplification Products***

- The template is prepared according to Bachem et al. (1996) with modifications as described here.
- Digest 10 µl of the cDNA prepared as described above, first with *Eco*RI (10U) and then with *Mse*I[1](10U) for 2 hours at 37°C.
- Ligate the *Eco*RI and *Mse*I double-strand adapters, according to the procedure described in the AFLP Core Reagent kit (Gibco-BRL).
- Perform the preamplification reaction using 28 cycles (94°C, 30 seconds; 60°C, 1 minute; 72°C, 1 minute) using primers corresponding to *Eco*RI and *Mse*I adapters without extension and 1/10 of template volume.
- Check the products of preamplification on an agarose gel (the product of amplification should appear as a smear between 100-700 bp). At this stage, products of abundant transcripts may be visible as important bands in the smear.
- Dilute the preamplification product (10× with TE (Tris-EDTA) buffer).

Use 5 µl of the diluted mix for selective amplification using 42 cycles including 14 touchdown cycles comprising a reduction of the annealing temperature from 65°C to 56°C in 0.7°C steps, which is then maintained for 28 cycles, as indicated below.

*Touchdown cycles*

(94°C, 30 seconds; 65-56°C [decrease 0.7°C each cycle], 30 seconds; 72°C, 60 seconds) × 14.

Continue with:

(94°C, 30 seconds; 56°C, 30 seconds; 72°C, 60 seconds) × 28
Primer labeled mix for each reaction:

---

[1]This is a critical step for cDNA-AFLP resolution. Although we have used the *Eco*RI and *Mse*I combination for ds-cDNA digestion, other enzyme combinations have been used by different authors, for example: *Taq*I and *Ase*I in potato (Bachem et al., 1996); *Apa*I and *Mse*I in rice (Durrant et al., 2000), *Ava*II and *Taq*I in soybean (Umezawa et al., 2002).

0.1 μl (*Eco*RI primer + NNN)[2] (50 ng·μl$^{-1}$)
0.2 μl [(γ$^{33}$P)dATP] (250 μCi/μl)
0.01 μl T4 polynucleotide kinase (Amersham Biosciences)
0.05 μl 10× T4 polynucleotide kinase buffer
0.24 μl distilled water

Incubate for 60 minutes at 37°C. Terminate the reaction by heating at 70°C for 10 minutes.

- Denature the selective amplification products in formamide (50 percent) at 95°C, and separate by electrophoresis in polyacrylamide gel (4, 5 percent) containing urea and TBE according to Sambrook et al. (1989).
- Dry the gel on 3MM Whatman paper (Whatman, Maidstone, UK) and then expose to Kodak Biomax (Sigma) for at least 24 hours at 37°C.

TDF Isolation from Polyacrylamide Gels

- Place the film on the gel (films must be aligned with markings) and select the bands to be isolated. Cut out the band from the gel using a scalpel and monitor the activity.
- Soak the excised band in 100 μl TE buffer (10 mM Tris, pH 7.5, and 1 mM EDTA, pH 8.0) and incubate overnight at 37°C.
- Heat the DNA solution 15 minutes at 95°C and then cool immediately on ice.
- Spin briefly and purify the DNA by precipitation as described by Reuber and Ausubel (1995). Add 10 μl of 3 M sodium acetate and 5 μl of glycogen (10 mg/ml) as a carrier and precipitate the DNA with 450 μl of absolute ethanol.
- We strongly recommend special attention to TDF recovery. The precipitation with glycogen as a carrier can also help DNA recovery from the excised band.
- Allow the solution to stand—80°C for at least 30 minutes (overnight is recommended) and spin 20 minutes at 4°C, wash the DNA pellet with ice cold 80 percent ethanol, and resuspend it in 10 μl of sterile water.

---

[2]Addition of selective nucleotides to the 3' end of one primer (for example, *Eco*RI+AAG or *Eco*RI+AGC combined with *Mse*I primer) is of fundamental importance to enhance the cDNA-AFLP gel resolution (Umezawa et al., 2002). However, use of primers with only two added selective nucleotides reduces the power of resolution.

- Reamplify the TDFs using 2 μl and the same primers and conditions as the selective amplification. In some cases, when the amount of DNA recovered is too small, a second round of amplification may be necessary. Under such circumstances, the PCR product of the first amplification should be diluted 10× and 2 μl used in a second PCR reaction (see the main steps of the protocol illustrated in Figures 28.1 and 28.2).

### *TDF Analysis*

- The reamplified TDFs must be cloned into PCRII plasmid using the TA Cloning Kit (Invitrogen) or plasmid pGEM-T Easy Vector System I (Promega). Use 2 μl of fresh PCR product to the vector ligation. After selection, the recombinant clones must be sequenced and analyzed. Direct TDF sequencing without cloning into the vector is possible. But a limitation at this point is that the TDFs are a mixture of PCR products (Durrant et al., 2000).
- Database searches can be performed using the BLAST Network Service (National Center for Biotechnology Information, http://www.ncbi.nlm.nih.gov/BLAST). The sequence of each TDF can be searched against all sequences in the nonredundant databases using the BLASTN, BLASTX, and TBLASTX algorithm, and in the EST database using the BLASTN program in turn. Sequences that returned with no significant homology must be compared against genomic sequence databases using the BLASTN program or Genomic BLAST pages. The function of genes may be classified according to the putative function using BLASTX and TBLASTX (E values less than 10e-5; Ditt et al., 2001).

### *Reverse Northern Blot*

- Digest plasmid DNA from each cloned TDF with appropriate restriction enzyme or amplify the cloned TDF by PCR reaction. Use enough plasmid to get 1 μg of insert.
- Separate the total volume of each digestion in a 1 percent TAE agarose gel.
- Blot it into a Hybond-$N^+$ (Amersham) nylon membrane.
- Hybridize using a cDNA complex probe derived from RNA extracted from control and treated plants. The cDNA probes are generated with superscript reverse transcriptase (Life Technologies) with 25 μg total

RNA as described by Jimérnez-Zurdo et al. (2000) prepared as follows:

1. Denature 25 µl of total RNA (1 µg/µl) at 65°C for 3 minutes;
2. Prepare the following reaction mixture:
   10 µl superscript buffer 5×
   0.5 µl RNAsin
   5 µl DTT 0,1M
   5 µl d (A, G, T) TP 10 mM
   5 µl [$^{32}$P] dCTP (50 µCi)
   1 µl superscript reverse transcriptase (Invitrogen)
   Incubate for 1 hour at 37°C
3. Add 2 µl of cold dCTP 25 mM and incubate 45 minutes at 37°C.
4. Adjust the volume to 100 µl and purify the probe using a G50-STE column.
5. Degrade the RNA with alkaline hydrolyze. Add 100 µl of STE and 60 µl of NaOH 1M; incubate for 10 minutes at room temperature. Then add 60 µl of HCl 1M and 60 µl of SSPE 20×.
6. Prehybridize the membranes for at least 1 hour at 65°C, add denatured probes (1 minute at 95°C), and hybridize overnight at 65°C, one set for each probe, in 0.5 M Na-phosphate, pH 7.2, 7 percent sodium dodecyl sulfate (SDS) buffer at 65°C (Church and Gilbert, 1984).
7. After hybridization, wash the membranes twice for 30 minutes each with 2XSSC/0.1 percent and 1XSSC/0.1 percent SDS.

## *CONCLUSION*

The use of cDNA-AFLP as a tool to explore plant genomes presents several advantages in relation to other RNA fingerprinting techniques and DNA chip-based approaches. The main benefit of a cDNA-AFLP approach over the existing EST databases is its ability to set up the precise chronology of gene expression events in the process under study. The cDNA-AFLP display gives a sensible indication of the temporal expression profiles of relevant genes. The amplification of fragments from constitutively expressed genes can provide internal control bands with each primer pair combination. Besides, the cDNA-AFLP detects reliably differential patterns of gene expression. In contrast to differential display, in which random primers are used for amplification of cDNAs, cDNA-AFLP permits a relatively easy confirmation of the displayed profiles with control genes with known expression patterns. In contrast to microarray analysis, cDNA-AFLP can be performed in organisms where transcript data are limited, allowing the exploitation of the genomes of nonmodel organisms. For these organisms, cDNA-AFLP may initially be the most suitable method of identifying new differentially expressed genes. Additionally, cDNA-AFLP analysis requires smaller amounts of RNA compared with the quantities needed for microarrays. It also provides the opportunity for automation, with the use of fluorescent dyes in a manner similar to microarray technology (Cho et al., 2001).

In conclusion, this relatively general approach can be applied to any comparable condition and combination as long as distinct mRNA populations can be isolated from cell populations representative of these conditions. The cDNA-AFLP expression profiles might reveal coordinated expression patterns for genes, which are involved in the same process. For the genome totally sequenced, *in silico* comparisons of the upstream regions of these genes might reveal common motifs controlling the expression specificity of them. The data obtained by cDNA-AFLP represent the first clues for guiding further functional studies. Functional genomics combines genome sequence analysis, genomewide mRNA or protein expression monitoring, mutant isolation, and genetic screens. All these strategies together allow the study of gene function on a genomic scale.

## REFERENCES

Aubry, C., M.C. Pave, A.L. Chateigner-Boutin, B. Teulat-Merah, C. Ricoult, D. Peltier, R. Jalouzot, and A.M. Limami (2003). A gene encoding a germin-like protein, identified by a cDNA-AFLP approach, is specifically expressed during germination of *Phaseolus vulgaris. Planta* 217:466-475.

Bachem, C.W.B., R.J.F.J. Oomen, and R.G.F. Visser (1998). Transcript imaging with cDNA-AFLP: A step-by-step protocol. *Plant Molecular Biology Reporter* 16:157-173.

Bachem, C.W.B., R.S. van der Hoeven, S.M. de Bruijn, D. Vreugdenhil, M. Zabeau, and R.G.F. Visser (1996). Visualization of differential gene expression using a novel method of RNA fingerprinting based on AFLP analysis of gene expression during potato tuber development. *Plant Journal* 9:745-753.

Breyne, P., R. Dreesen, B. Cannoot, D. Rombaut, K. Vandepoele, S. Rombauts, R. Vanderhaeghen, D. Inze, and M. Zabeau (2003). Quantitative cDNA-AFLP analysis for genome-wide expression studies. *Molecular Genetics and Genomics* 269:173-179.

Breyne, P. and M. Zabeau (2001). Genome-wide expression analysis of plant cell cycle modulated genes. *Current Opinion in Plant Biol*ogy 4:136-142.

Cadle-Davidson, M. and M.M. Jahn (2006). Differential gene expression in *Phaseolus vulgaris* I locus NILs challenged with bean common mosaic virus. *Theoretical and Applied Genetics* 112:1452-1457.

Campalans, A., M. Pagés, and R. Messeguer (2001). Identification of differentially expressed genes by the cDNA-AFLP technique during dehydration of almond (*Prunus amygdalus*). *Tree Physiology* 21:633-643.

Cho, Y.-J., J.D. Meade, J.C. Walden, X. Chen, Z. Guo, and P. Liang (2001). Multicolor fluorescent differential display. *BioTechniques* 30:562-572.

Church, G.M. and W. Gilbert (1984). Genomic sequencing. *Proceedings of the National Academy of Sciences USA* 81:1991-1995.

Ditt, R.F., E.W. Nester, and L. Comai (2001). Plant gene expression response to *Agrobacterium tumefaciens. Proceedings of the National Academy of Sciences USA* 98:10954-10959.

Dubos, C. and C. Plomion (2003). Identification of water-deficit responsive genes in maritime pine (*Pinus pinaster* Ait.) roots. *Plant Molecular Biology* 51:249-262.

Durrant, W.E., O. Rowland, P. Piedras, K.E. Hammond-Kosack, and J.D.G. Jones (2000). cDNA-AFLP reveals a striking overlap in race-specific resistance and wound response gene expression profiles. *Plant Cell* 12:963-977.

Gabriels, S.H., F.L. Takken, J.H. Vossen, C.F. de Jong, Q. Liu, S.C. Turk, L.K. Wachowski, J. Peters, H.M. Witsenboer, et al. (2006). cDNA-AFLP combined with functional analysis reveals novel genes involved in the hypersensitive response. *Molecular Plant-Microbe Interactions* 19:567-576.

Goupil, P., Y.S. Mahamoud, J. Poulain, D. Windels, P. Crete, B. Huss, and S. Rambour (2003). cDNA-AFLP display for the isolation of differentially expressed genes during chicory root development. *Journal of Plant Physiology* 160:303-309.

He, P., B.R. Friebe, B.S. Gill, and J.-M. Zhou (2003). Allopolyploidy alters gene expression in the highly stable hexaploid wheat. *Plant Molecular Biology* 52:401-414.

Hmida-Sayari, A., A. Costa, A. Leone, S. Jaoua, and R. Gargouri-Bouzid (2005). Identification of salt stress-induced transcripts in potato leaves by cDNA-AFLP. *Molecular Biotechnology* 30:31-40.

Jimérnez-Zurdo, J.I., F. Frugier, M.D. Crespi, and A. Kondorosi (2000). Expression profiles of 22 novel molecular markers for organogenetic pathways acting in alfalfa nodule development. *Molecular Plant-Microbe Interactions* 13:96-106.

Liang, P. and A.B. Pardee (1992). Differential display of eukaryotic messenger RNA by means of the polymerase chain reaction. *Science* 257:967-971.

Mao, C., K. Yi, L. Yang, B. Zheng, Y. Wu, F. Liu, and P. Wu (2004). Identification of aluminium-regulated genes by cDNA-AFLP in rice (*Oryza sativa* L.): Aluminium-regulated genes for the metabolism of cell wall components *Journal of Experimental Botany* 55:137-143.

Qin, L., P. Prins, J.T. Jones, H. Popeijus, G. Smant, J. Bakker, and J. Helder (2001). GenEST, a powerful bi-directional link between cDNA sequence data and gene expression profiles generated by cDNA-AFLP. *Nucleic Acids Research* 29: 1616-1622.

Reuber, T.L. and F.M. Ausubel (1995). Differential mRNA display. In *Methods in plant cell biology, Part A,* D.W. Galbraith, H.J. Bohnert, and D.P. Bourque (eds.). Boca Raton, FL: Academic Press, pp. 431-441.

Sambrook, J., E. Fritsch, and T. Maniatis (1989). Molecular cloning: A laboratory manual. Cold Spring Harbor, NY: Cold Spring Harbor Laboratory Press.

Scippa, G.S., M. Di Michele, A. Di Iorio, A. Costa, B. Lasserre, and D. Chiatante (2006). The response of *Spartium junceum* roots to slope: Anchorage and gene factors. *Annals of Botany (London)* 97:857-866.

Simões-Araujo, J.L., R. Rodrigues, L.B.A. Gerhardt, J.M.C. Mondego, M. Alves-Ferreira, N.G. Rumjanek, and M. Margis-Pinheiro (2002). Identification of differentially expressed genes by cDNA-AFLP technique during heat stress in cowpea nodules. *FEBS Letters* 515:44-50.

Umezawa, T., K. Mizuno, and T. Fujimura (2002). Discrimination of genes expressed in response to the ionic or osmotic effect of salt stress in soybean with cDNA-AFLP. *Plant Cell and Environment* 25:1617-1625.

Wan, J.S., S.J. Sharp, G.M.C. Poirier, P.C. Wagaman, J. Chambers, J. Pyati, Y.I. Hom, J.E. Galindo, A. Huvar, P.A. Peterson, et al. (1996). Cloning differentially expressed mRNA. *Nature Bio/Technology* 14:1685-1691.

Wang, Y.J., Li Y.D., Luo G.Z., Tian A.G., Wang H.W., Zhang J.S., and Chen S.Y. (2005). Cloning and characterization of an *HDZip I* gene *GmHZ1* from soybean. *Planta* 221:831-843.

Chapter 29

# Suppression Subtractive Hybridization to Identify Novel Genes in Legumes

Katrien Schroeyers
Sofie Goormachtig
Marcelle Holsters

## *INTRODUCTION*

Following the orchids (Orchidaceae) and asters (Asteraceae), the leguminous family (Fabaceae) is the third largest family of flowering plants. The legume family is highly diverse, with very variable types of fruits and flowers. Ecologically, legumes are found all over the world, from deserts to alpine and aquatic habitats (Doyle and Luckow, 2003).

One of the special features of legumes is their nodulation capacity. A group of soil bacteria, commonly designated as rhizobia, are capable of establishing a nitrogen-fixing symbiosis with specific legume hosts by inducing the formation of root nodules. In these root nodules, the differentiated bacteria (bacteroids) fix atmospheric nitrogen, which is delivered to the host. In this symbiotic process, a large number of plant genes are triggered. These genes, the nodulin genes (van Kammen, 1984), code for early nodulin proteins that are correlated with the infection process and nodule ontogeny, and late nodulins that participate in various aspects of nodule functioning, such as nitrogen assimilation, oxygen transport, and carbon metabolism (Nap and Bisseling, 1990; Schultze and Kondorosi, 1998).

---

This work was supported by the Interuniversity Poles of Attraction Programme, Belgian Science Policy (P5/13), and by the Research Foundation-Flanders (project no. G027601 and Krediet aan Navorsers no. 1.5.088.99N and 1.5.192.01N). K. S. is indebted to the Instituut voor de aanmoediging van Innovatie door Wetenschap en Technologie in Vlaanderen for a predoctoral fellowship.

Unraveling the complex interactions and organ morphogenesis throughout nodulation has been initiated by analyzing gene expression in both legumes and rhizobia. Because of technical difficulties and the molecular complexities faced in proteomics and metabolomics, the spotlights were first on the genes (Smith, 2000; Colebatch et al., 2002; Donson et al., 2002). However, gene expression analysis should be interpreted with care. Causative correlations to specific cellular responses in certain tissues may not always be reliable because of posttranslational control of protein location and numbers (Gygi et al., 1999). Nevertheless, in most cases, the primary control of regulation of cellular functions is performed on a transcriptional base (Donson et al., 2002) and a differential expression of a gene is a good indication for its involvement in a cellular process. Many techniques are available for high-throughput analysis of gene expression, but they differ in convenience, expense, number of transcripts assayed, and sensitivity (Kuhn, 2001).

As in many other scientific disciplines, the generation of expressed sequence tag (EST) libraries played a pivotal role in the legume research field. The initial aim of library construction was the discovery of new nodulins. More recently, since high-throughput DNA sequencing centers were established, large-scale EST sequencing provides a quantitative estimation (electronic or digital RNA gel blots) of transcripts in specific organs, such as nodules (database of the Institute for Genome Research; Fedorova et al., 2002; Weidner et al., 2003). In June 2004, more than 189,000 ESTs of *Medicago truncatula* were present in cDNA libraries consisting of almost 29,000 ESTs from roots infected with rhizobia and root nodules. Because aberrant mRNA structures, sampling size of the libraries, and input of normalized cDNA libraries compromise the evaluation of gene expression levels, this type of *in silico* quantification should be handled with caution (Ohlrogge and Benning, 2000).

As EST libraries became saturated, it was realized that large numbers of cell type-specific genes were not represented within these libraries (Martin and Pardee, 2000). Therefore, in the late 1980s and early 1990s, a variety of methods were introduced based on differential screening, normalization, or subtractive hybridizations. These techniques were indispensable for the further analysis of specific mRNA populations, especially the low abundant transcripts.

Nodules have been compared with uninoculated roots by means of general subtractive hybridizations and differential screening procedures, such as differential hybridization, differential display (Liang and Pardee, 1992), cold-plaque screening, and macro- and microarray hybridization (Lockhart and Winzeler, 2000; Küster et al., 2004; Mitra et al., 2004). For an overview of the different techniques used in the plant field, we refer to Baldwin et al. (1999), Ohlrogge and Benning (2000), Kuhn (2001), Lievens et al. (2001), and Yamazaki and Saito (2002); in the last two articles, the differential display (DD) technique of Liang and Pardee (1992) was discussed in more detail. Although DD was very popular in the 1990s, several drawbacks have been recognized, such as isolation of false positives, low sensitivity, heterogeneous ESTs present in single isolated bands, and lack of quantification and reproducibility. Hence, improved new techniques have been explored (Kuhn, 2001; Lievens et al., 2001), such as cDNA-amplified fragment length polymorphism (AFLP; Bachem et al., 1996; Breyne et al., 2003), selective amplification of differentially expressed mRNAs via biotin- and restriction-mediated enrichment (SABRE; Lavery et al., 1997), representational difference analysis (RDA; Hubank and Schatz, 1994), and suppression subtractive hybridization (SSH; Diatchenko et al., 1996). All these procedures include a cDNA synthesis via poly-dT primers, a restriction digest, ligation of linkers (adaptors), and the selective amplification by polymerase chain reaction (PCR) of differentially expressed genes. In the cDNA-AFLP and DD protocols, the PCR is performed with selective primers and the amplified bands are analyzed on a polyacrylamide gel. Compared to DD, one of the main advantages of cDNA-AFLP is the reduced variability thanks to stringent primer hybridization to adaptors. In contrast to cDNA-AFLP and DD, the protocols of SABRE, RDA, and SSH include a subtraction hybridization step.

All these techniques were designed to analyze low abundant messages (< 5 copies/cell) that represent approximately 50 to 35 percent of the mRNA population in cells and correspond to approximately 15,000 transcripts. The remaining mRNA mass (50-65 percent) included medium (15-500 copies/cell) and high (more than 500 copies/cell) abundant transcripts that consisted of approximately 2,000 copies (Martin and Pardee, 2000). In the cDNA-AFLP and DD techniques, low abundant, differentially expressed genes are visualized by using primer combinations with more selective nucleotides. A different strategy is used in SABRE, RDA, and SSH. One of these alternative techniques, the SSH, is the topic of this chapter.

## *THE SSH TECHNIQUE*

The SSH technique has been used extensively to identify differentially expressed genes in many organisms. A key feature of SSH is simultaneous subtraction and normalization, allowing equalization of abundant target cDNAs in the subtracted population and enrichment of rare differential transcripts.

Classic problems associated with subtraction methods are the requirements of large amounts of start material [>20 μg poly(A) mRNA], the time-consuming and labor-intensive nature of the technique, the necessity of multiple subtraction steps, the loss of low abundant genes after several rounds of physical purification of the double-stranded (ds) cDNAs and single-stranded (ss) cDNAs, the high rate of false positives, and the irreproducibility of the results. These drawbacks make it almost impossible to apply this strategy on a large scale. However, in the SSH technique these problems are solved. The SSH (Figure 29.1) has to be performed only once and less than 2 μg poly(A) mRNA starting material is needed. It incorporates two hybridization steps leading to both an efficient normalization and subtraction of the ss-cDNAs and this fraction is not separated on a physical basis but is amplified by PCR. The protocol is based on RDA, a technique originally designed for the study of dissimilarities between two complex genomes (Lisitsyn et al., 1993). The SSH procedure adds a hybridization step in the subtraction to equalize (normalize) the mRNA levels of the two populations, thus overcoming the problem of variations in mRNA abundance. The amplification of unwanted transcripts is circumvented by the use of PCR suppression (see below). The SSH procedure can easily be combined with macro- or microarray hybridizations for high-throughput transcriptome analysis. Consequently, the SSH has become a very attractive method for gene expression studies.

In the SSH protocol, two mRNA populations, driver (control) and tester (target of interest), are compared (Figure 29.1). From these populations, poly(A) mRNA is prepared, ds-cDNA is synthesized, and a restriction digest is performed with a 4-base pair (bp) cutter (for example, *Rsa*I). This restriction digest has two advantages: it facilitates the hybridization procedures because long DNA fragments may form complex networks and it generates small fragments for a better representation of the individual genes (Diatchenko et al., 1996). The formation of small fragments from a related gene family can compensate for cross-hybridization and elimination of particular members during the subtraction. Additionally, the smaller fragments have different characteristics (hybridization and amplification) and, while some fragments will be eliminated during the procedure, others will be picked up. Unfortunately, the restriction digestion sometimes leads to undersized fragments and to the identification of different clones derived from a single gene. For example, in a search for new genes within a normalized data set of *Arabidopsis thaliana,* the 5,722 isolated ESTs corresponded to 5,000 independent genes (Ando et al., 2004).

After the restriction digestion of the tester and driver ds-cDNA, the tester population is divided into two portions (tester1 and tester2) that are ligated

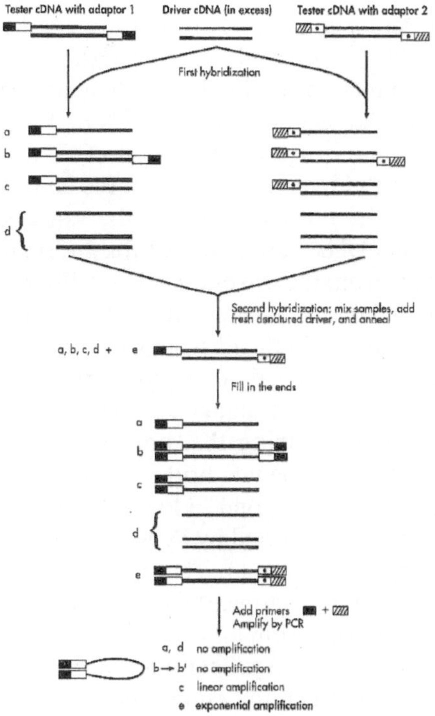


FIGURE 29.1. SSH technique. The *Rsa*I-digested tester and driver cDNAs are represented by solid lines. A solid or a shaded box symbolizes the outer part of adaptor1 and adaptor2 and corresponds to the PCR primers P1 and P2, respectively. Clear boxes indicate the inner part of the adaptors and corresponding nested PCR primers. Note that after filling in the recessed 39 ends with DNA polymerase, molecules of types a, b, and c with adaptor2 are also present, but are not shown. d-Type molecules consist of driver or driver homoduplexes and a-type molecules are the single-stranded tester molecules that all lack a primer-binding site and, as a consequence, cannot be amplified. c-Type molecules correspond to the tester-driver heteroduplexes that can be linearly amplified, due to one primer-binding site. The exponential amplification of the tester homoduplexes (b-type molecules; two primer-binding sites) is suppressed by the formation of stem-loop structures (suppression effect). Only e-type molecules can be exponentially amplified. *Source:* Reprinted from Diatchenko, L., Y.-F.C. Lau, A.P. Campbell, A. Chenchik, F. Moqadam, B. Huang, S. Lukyanov, K. Lukyanov, N. Gurskaya, et al. (1996). Suppression subtractive hybridization: A method for generating differentially regulated or tissue-specific cDNA probes and libraries. *Proceedings of the National Academy of Sciences USA* 93:6025-6030, with permission. ©1996, National Academy of Sciences, U.S.A..

at their 5' side to different adaptors (Ad1 and Ad2, or Ad2R). Next, the first hybridization is performed (tester1 + driver, tester2 + driver), with an excess of driver (no adaptors). The samples are denatured by heat and allowed to anneal. Because of the second-order kinetics in the hybridization procedure, high abundant molecules hybridize faster than low abundant ones, forming ds-cDNAs with identical adaptors at the ends, tester homoduplex. This feature leads to the normalization of the ss-cDNAs within the two tester pools. These ss-cDNAs are enriched for differentially expressed sequences as transcripts from the driver associate with their counterpart present in the tester populations, forming ds-cDNAs with only one adaptor (tester-driver heteroduplex). In the next step, the two hybridized tester populations are combined with freshly denatured driver cDNAs to further enrich for differentially expressed sequences. This second hybridization occurs without denaturing, meaning that only homologous ss-cDNAs can reassociate. During this step, new hybrids are formed with different adaptors on each strand; they correspond to low differentially expressed genes and will be exponentially amplified by PCR with primers designed to bind to adaptor sequences (Figure 29.1). The driver homoduplexes cannot be amplified (no primer binding site) and the tester-driver heteroduplexes are linearly amplified (one primer binding site). The exponential amplification of the tester homoduplexes (two primer binding sites) is suppressed by the formation of stem-loop structures (suppression effect). Stem-loop-DNA constructs are created by the ligation of long identical GC-rich adaptors to ds-cDNA ends. Hence, each ss-cDNA fragment is flanked by terminal inverted repeats. These self-complementary ends will form large hairpin structures during PCR amplifications and suppress the annealing of primers and amplification. A second (nested) PCR amplification is used to reduce the background and enrich the differential sequences further. These PCR products can be cloned and transformed to produce a subtracted and normalized cDNA library.

In an improved version of the SSH technique (Diatchenko et al., 1999), the undesirable selection of short fragments is minimized by using different adaptors (Ad2R instead of Ad2). The incorporation of 22 identical nucleotides (to Ad1) at the 5' end of AdR2 diminished the selection of small sequences because the complementary inverted terminal repeats will suppress the PCR amplification of undersized fragments more severely than for the larger fragments. In addition, the identical ends allow the amplification of the target sequences in the primary PCR by a single primer, thereby reducing the amount of nonspecific annealing and amplification (Diatchenko et al., 1999).

Often two libraries are created: a forward and a reverse library. In a forward library, the population of interest is the tester and the control population is the driver, as mentioned above, and in a reverse library, these two populations are switched (tester = control, driver = target). The forward and reverse subtracted libraries are frequently used in differential screening approaches (see below) to remove the false positives and detect the low abundant differentially expressed genes. Screening of a reverse library allows the characterization of specific down-regulated genes (Diatchenko et al., 1996).

## *APPLICATIONS*

Literature searches indicate a growing interest in SSH in the plant field (Table 29.1). Before 1999, published data on the applications of the SSH technique were only available in the animal field. Slowly, the technique spread within the plant world and in 2003, 16 papers were listed in the Web of Science, showing its increasing popularity. The studied topics varied from organ development, stress responses, and secondary metabolism to plant-pathogen or plant-symbiont interactions. The method has been applied in several plants, including model species such as *A. thaliana, M. truncatula,* and *Oryza sativa* (rice), important crop species such as *Pisum sativum* (pea), *Camellia sinensis* (tea), *Solanum tuberosum* (potato), *Lycopersicon esculentum* (tomato), *Triticum aestivum* (wheat), *Glycine max* (soybean), citrus, and other plant species with limited availability of sequence data such as *Sesbania rostrata, Eucalyptus globules,* and *Davidia involucrata* (dovetree). Although cDNA microarrays and extended EST collections are being developed for the model species and some important crop species, the SSH technique is widely used, because of its discovery (novel sequences) potential and ability to recover abundant, as well as low-copy-number mRNA transcripts.

We have reported on the use of the SSH technique to study the symbiotic association between the legume *S. rostrata* and the microsymbiont *Azorhizobium caulinodans* (Schroeyers et al., 2004). A special feature of this tropical, flooding-adapted legume species is the presence of dormant adventitious rootlets along the stem. Upon submergence, the dormant root meristems are activated and adventitious roots grow out (Duhoux and Dreyfus, 1982; Rinaudo et al., 1983). Inoculation of the adventitious rootlets with *A. caulinodans* results in the formation of stem nodules (Dreyfus and Dommergues, 1981; Goormachtig et al., 1998). The colonization of adventitious root primordia starts at the base, in the circular fissure created by the protrusion of the rootlets through the stem cortex and epidermis. The

TABLE 29.1. Applications of SSH in the plant field.

| Topic | Isolated SSH clones | Confirmation method | Confirmed clones | Confirmation method | Confirmed clones | Confirmation method | Confirmed clones | References |
|---|---|---|---|---|---|---|---|---|
| S-locus region in *Brassica napus* | NA | Genomic DNA screened with FS | NA | NA | NA | NA | NA | Cui et al. (1999) |
| Flower maturation in *Dianthus caryophyllus* | 150 | RNA gel blot on 10 sequences | 5 | One chosen (phospholipase A) | NA | NA | NA | Kim et al. (1999) |
| Potato resistance to *Phytophthora infestans* | 768 | Colony hybridization, FS-D | 100 | Sequencing | 61 | cDNA-AFLP | NA | Birch et al. (1999) |
| Salt stress in *Mesembryanthemum crystallinum L.* | 54 | RNA slot blot, T-D | 46 | Sequencing of weak expressed | 3 | NA | NA | Yen et al. (2000) |
| Ozone stress in *Pisum sativum* | 48 | Slot blot, D | 25 | Sequencing | 25 | RNA gel blot on 9 clones | 7 | Sävenstrand et al. (2000) |
| Potato interactions with *Erwinia carotovora* subsp. *atroseptica* | NM | One chosen (WRKY) | NA | NA | NA | NA | NA | Dellagi et al. (2000) |
| Leaf senescence in *Arabidopsis thaliana* | NM | One chosen (WRKY) | NA | NA | NA | NA | NA | Hinderhofer and Zentgraf (2001) |
| Low-temperature induced genes from cold-treated *Hordeum vulgare* | 160 | NM | NM | NA | NA | NA | NA | Bahn et al. (2001) |

| | | | | | | | | |
|---|---|---|---|---|---|---|---|---|
| Defense-related rice (*Oryza sativa*) genes | 768 | Colony hybridization, T-D | 140 | Cross hybridization to remove redundant sequences | 38 | RNA gel blot | 34 | Xiong et al. (2001) |
| Potato resistance to *Phytophthora infestans* | 288 | Dot blot, T-D | 43 | Sequencing of 11 clones | 10 | NA | NA | Beyer et al. (2001) |
| *Cuscuta reflexa*–incompatible interaction on tomato | 200 | Colony hybridization, T-D | 18 | Two chosen (aquaporin) | NA | NA | NA | Werner et al. (2001) |
| Adventitious root growth in *Sesbania rostrata* | 192 | Sequencing | 66 | Dot blot, FS-RS-T-D | 25 | NA | NA | Caturla et al. (2002) |
| Dovetree *(Davidia involucrata)* bracts | 250 | Dot blot, FS-RS | 167 | Sequencing of 16 clones | 10 | NA | NA | Li et al. (2002) |
| Calcium oxalate crystal formation in *Medicago truncatula* | NA | SSH, SH, dot blot, FS-RS | 7 | NA | NA | NA | NA | Nakata and McConn (2002) |
| Nitrate treatment in *O. sativa* roots | 10,000 | Colony hybridization, FS-RS | 864 | Reverse RNA gel blot | 400 | Sequencing | 92 | Wang et al. (2002) |
| Triacontanol-regulated genes in *O. sativa* | 1,600 | DNA array, FS-RS | 33 | Sequencing | 15 | RNA gel blot | 12 | Chen et al. (2002) |
| Effect of low levels of ultraviolet-B radiation in *P. sativum* | 48 | Slot blot, D | 12 | Sequencing | 9 | RNA gel blot | 7 | Sävenstrand et al. (2002) |
| *Fusarium solani* f.sp. *glycines* interaction with *Glycine max* | 384 | Sequencing | 135 | Dot blot, T-D | 28 | NA | NA | Iqbal et al. (2002) |
| OPB3 overexpression lines in *A. thaliana* | 198 | Dot blot, FS-RS | Focus on 8 up-regulated clones | Sequencing | 5 | NA | NA | Kang et al. (2003) |

TABLE 29.1 *(continued)*

| Topic | Isolated SSH clones | Confirmation method | Confirmed clones | Confirmation method | Confirmed clones | Confirmation method | Confirmed clones | References |
|---|---|---|---|---|---|---|---|---|
| Overexpression *pto* gene in tomato | NA | EST library screened with FS | NM | Microarray | NM | NA | NA | Mysore et al. (2003) |
| Phosphorus starvation in rice | NM | Colony hybridization, FS-RS sequencing | 65 | RNA gel blot on 15 clones | NA | NA | NA | Xia et al. (2003) |
| Flower development in *D. caryophyllus* | 274 | Reverse RNA gel blot, T-D | 25 | Reverse RNA gel blot, FS-RS | 112 | Sequencing of middle and high induced clones | 60 | Ok et al. (2003) |
| Kernel development in *H. vulgare* | 780 | Reverse RNA gel blot, T-D | 109 | Sequencing of some clones | NA | NA | NA | Jang et al. (2003) |
| Biosynthesis of ginsenoside in *Panax ginseng* | 40 | Sequencing | 6 novel ESTs | RNA gel blot, RT-PCR | 6 | NA | NA | Luo et al. (2003) |
| Arbuscular mycorrhiza development in *M. truncatula* | 2,000 | Sequencing | 890 | Reverse RNA gel blot on 34 clones | 22 | NA | NA | Wulf et al. (2003) |
| Submergence-induced genes in maize (*Zea mays*) | 408 | Dot blot, FS-RS-T-D | 184 | Sequencing | 95 | NA | NA | Gao et al. (2003) |

| | | | | | | | | |
|---|---|---|---|---|---|---|---|---|
| NaCl stress in *G. max* | NA | PCR on FL | NA | NA | NA | NA | NA | Liao et al. (2003) |
| Stress defense in *A. thaliana* | 6,000 | cDNA blotting, FS-RS-T-D | 1,461 | Sequencing | 1,058 | Digital RNA gel blots | 577 rare | Mahalingam et al. (2003) |
| Leaf senescence in *A. thaliana* | 350 | Sequencing | 175 | RNA gel blot | 127 | NA | NA | Gepstein et al. (2003) |
| Oxidative stress in sugarcane | 288 | Reverse RNA gel blot, T-D | 182 | Sequencing of 50 most up-regulated clones | 50 | NA | NA | Watt (2003) |
| *Puccinia triticina* nonhost interaction in *H. vulgare* | 75 | Sequencing | 34 | RNA gel blot on 6 clones | 6 | NA | NA | Neu et al. (2003) |
| *Orobanche ramosa* infection of *A. thaliana* | 96 | Dot blot, T-D sequencing | 13 | cDNA blotting, T-D | 11 | NA | NA | Vieira Dos Santos et al. (2003) |
| *M. truncatula* infected with *Aphanomyces euteiches* | 2,000 | Reverse RNA gel blot, T-D | 560 | Sequencing | 404 | NA | NA | Nyamsuren et al. (2003) |
| *A. thaliana*, self subtraction | 5,396 | Sequencing | 4,375 | NA | NA | NA | NA | Ando et al. (2004) |
| Secondary metabolism in *Camellia sinensis* (tea) | 588 | Sequencing | 149 | Quantitative RT-PCR on 5 clones | 5 | NA | NA | Park et al. (2004) |
| Adventitious root nodulation *S. rostrata–Azorhizobium caulinodans* | 300 | Sequencing | 102 | cDNA array, T-D | 26 | NA | NA | Schroeyers et al. (2004) |
| Heavy metal treatment of *P. sativum* roots | 48 | Slot blot, D, sequencing | 9 | RNA gel blot | 6 | NA | NA | Sävenstrand et al. (2004) |

TABLE 29.1 *(continued)*

| Topic | Isolated SSH clones | Confirmation method | Confirmed clones | Confirmation method | Confirmed clones | Confirmation method | Confirmed clones | References |
|---|---|---|---|---|---|---|---|---|
| Senescence in *Alstroemeria* petals | 991 | Sequencing | 500 | Microarray T-D | 251 | NA | NA | Breeze et al. (2004) |
| Potato resistance to *P. infestans* | 384-NM | Dot blot, FS-D | 119 | Sequencing | 10 | RNA gel blot | 8 | Avrova et al. (2004) |
| *O. minuta* defense on *Magnaporthe grisea* | 960 | Microarray | 377 | Sequencing | 180 | RNA gel blot on 8 clones | 8 | Shim et al. (2004) |
| Rice defense on *M. grisea* | 7,680 | Colony hybridization, FS-RS | 658 | Reverse RNA gel blot, FS-RS | 47 | RNA gel blot | 35 | Lu et al. (2004) |
| Powdery mildew infection of *H. vulgare* | 192 | Dot blot, FS-RS, sequencing | 21 | NA | NA | NA | NA | Hein et al. (2004) |

*Note:* FS, forward subtracted probe; RS, reverse subtracted probe; FL, forward library; D, driver probe; T, tester probe.

bacteria invade this crack and induce the formation of infection pockets in the cortex. Infection pockets are intercellular spaces occupied by dividing rhizobia. This type of invasion is called crack entry. A similar invasion procedure takes place during lateral root base nodulation on hydroponic roots of *S. rostrata*. An alternative invasion process occurs on well-aerated roots, namely the "classical" root hair invasion. This invasion path starts in root zone I of developing root hairs via an active colonization of root hairs. The subsequent colonization and formation of the root nodules occurs in a more or less similar fashion. Infection threads guide the bacteria to the deeper cortical cells, where they are internalized into cells of the newly formed nodule primordia. Subsequently, the bacteria differentiate into bacteroids and fix atmospheric nitrogen (Ndoye et al., 1994; Goormachtig et al., 1997). In submerged conditions, root hair invasion cannot take place because of the inhibitory effect of the plant hormone ethylene (Goormachtig et al., 2004). The versatile nodulation features of *S. rostrata* are related to the adaptive nature of the plant that permits survival under changing environmental conditions.

We have applied SSH assays to study adventitious root nodulation of *S. rostrata*. The predisposition of the adventitious roots (sites for the stem nodule formation) on the stem provides the opportunity to collect material at extremely early time points in a very specific manner. In addition, large quantities of material can be harvested, because the adventitious root primordia are abundantly present on the stem. Thus, *S. rostrata* can be considered a good model plant to follow the alternative invasion pathway of crack entry infection.

Previously, the DD technique had been implemented on adventitious root nodulation and 25 up-regulated mostly abundantly transcribed genes have been identified (Goormachtig et al., 1995; Lievens et al., 2001). In parallel, adventitious root outgrowth of *S. rostrata* has been investigated at the transcription level with the SSH technique (Caturla et al., 2002). Because of the available experience, we were interested in using the SSH in an ensuing study on adventitious root nodulation. The hope and aim were to generate low differentially expressed genes during adventitious root nodulation (Table 29.2).

## *EXPERIMENTAL SETUP*

As start material for the tester population, different time points after inoculation (4, 6, and 12 hours; 1, 2, 3, 4, and 6 days) were pooled in equal proportions and untreated adventitious root primordia were used as a driver population. Next, 300 inserts were randomly selected and sequenced, re-

TABLE 29.2. SSH clones, corresponding arbitrary intensity units (AIU), and first homologue after BLASTN or BLASTX (x) analyses against the plant databases available on NCBI or the database of TIGR for *Medicago truncatula*.

| Name | Accession | Nuc (bp) | AIU Primordia | AIU Nodule | First homologue BLAST hit | Accession number | E value |
|---|---|---|---|---|---|---|---|
| **Up-regulated clones** | | | | | | | |
| **Metabolism** | | | | | | | |
| sshnod1-Srchr1 | CK233208 | 445 | 69 | 1267 | *Sesbania rostrata* mRNA for putative chalcone reductase | AJ223291 | 0 |
| sshnod2 (*) | CK233209 | 284 | 169 | 326 | *Medicago truncatula* similar to ferredoxin | TC87861 | 5E-21 |
| sshnod3 | CK233210 | 309 | 1163 | 2215 | *Cicer arietinum* mRNA for cytosolic fructose-1,6-bisphosphate aldolase | AJ005041 | 1E-58 |
| sshnod4 | CK233211 | 530 | 88 | 556 | *M. truncatula* mRNA for sucrose synthase | AJ131943 | 0 |
| sshnod5 | CK233212 | 147 | 49 | 375 | *Glycine max* acyl-CoA oxidase (ACX1;2) mRNA | AF404404 | 2E-40 |
| sshnod6 (x) | CK233213 | 373 | 125 | 285 | *Lotus corniculatus* pyrophosphate-fructose-phophotransferase α subunit | AP006425 | 7E-24 |
| sshnod7 (x) | CK233214 | 382 | 129 | 262 | *Petunia integrifolia* S1 self-incompatibility locus-linked pollen 3.15 protein | AAQ09997 | 1E-23 |

| | | | | | | | |
|---|---|---|---|---|---|---|---|
| **Protein-protein interaction and signal transduction** | | | | | | | |
| sshnod8-SrRING | CK233215 | 539 | 418 | 1613 | *Lotus japonicus* mRNA for RING finger protein (Gifu B-129) | Z36750 | 1E-149 |
| sshnod9 (x) | CK233216 | 388 | 106 | 351 | *Arabidopsis thaliana* ankyrin-like protein | AAM62711 | 1E-23 |
| sshnod10-Sr14-3-3 | CK233217 | 498 | 371 | 2259 | *G. max* mRNA for 14-3-3-like protein | AJ004898 | 6E-25 |
| sshnod11 | CK233218 | 187 | 36 | 278 | *G. max* 14-3-3 related protein (SGF14B) mRNA | U70534 | 3E-21 |
| **Cell wall functioning and synthesis** | | | | | | | |
| sshnod12 | CK233219 | 166 | 275 | 755 | *G. max SbPRP1* gene encoding a proline-rich protein | J02746 | 8E-31 |
| **Hormone perception and synthesis** | | | | | | | |
| sshnod14-SrSAMS | CK233221 | 348 | 300 | 603 | *Lycopersicon esculentum* S-adenosyl-L-methionine synthetase mRNA | Z24742 | 4E-93 |
| **Transcription-translation** | | | | | | | |
| sshnod15 | CK233222 | 332 | 164 | 666 | *P. sativum* GA mRNA (60S ribosomal protein L9) | X65155 | 2E-45 |
| **Stress** | | | | | | | |
| sshnod16 | CK233223 | 524 | 787 | 1964 | *Gossypioides kirkii* putative SAH7 (allergen) protein gene | AY117069 | 0.008 |

TABLE 29.2 *(continued)*

| Name | Accession | Nuc (bp) | AIU Primordia | AIU Nodule | First homologue BLAST hit | Accession number | E value |
|---|---|---|---|---|---|---|---|
| **Unknown-hypothetical** | | | | | | | |
| sshnod17 (*) | CK233224 | 203 | 306 | 683 | *M. truncatula* similar to unknown protein AAM62719 (*A. thaliana*) | TC80537 | 5.3E-23 |
| sshnod18 (x) | CK233225 | 351 | 209 | 525 | *A. thaliana* expressed protein (At5g18920) | NP197393 | 1E-15 |
| sshnod19 (x) | CK233226 | 241 | 201 | 401 | *A. thaliana* unknown protein | AAG51032 | 1E-14 |
| **No significant hit** | | | | | | | |
| sshnod20 | CK233227 | 288 | 648 | 1872 | No significant hit | | |
| sshnod21 | CK233228 | 279 | 572 | 1109 | No significant hit | | |
| sshnod22 | CK233229 | 170 | 247 | 483 | No significant hit | | |
| sshnod23 | CK233230 | 200 | 211 | 428 | No significant hit | | |
| sshnod24 | CK233231 | 223 | 164 | 337 | No significant hit | | |
| **Down-regulated clones Transcription-translation** | | | | | | | |
| sshnod101 | CK233308 | 202 | 7681 | 3392 | *M. sativa* mRNA for translationally controlled tumor protein | X63872 | 7E-60 |
| sshnod102 (*) | CK233309 | 309 | 1657 | 861 | *M. truncatula* homologue to protein translation factor (SUI1) | TC76587 | 1.5E-17 |

(*) AIU is calculated by averaging the values of the six spots (belonging to a clone) on two separate filters and by correcting it with the local background. The background was calculated by averaging the intensities of the pixels on the border and multiplying it by the total pixels of each cell. Table adapted from Schroeyers et al. (2004).

sulting in 102 unique clones (sshnod clones). Some redundancy of the sshnod clones was noticed, which is not surprising because of the PCR amplification, the last step in the technique. Occasionally, some of these clones corresponded to severely enriched differentially expressed transcripts and some matched false positives. At times, redundancies can have serious consequences. For example, in a screening for genes of potato plants infected with *Phytophthora infestans,* 119 differentially expressed sequences corresponded to only 10 unique ESTs (Avrova et al., 2004). In a SSH library of rice infected by *Magnaporthe grisea,* 70 percent of the sequenced clones matched one EST, encoding a putative alcohol dehydrogenase (Lu et al., 2004), whereas in a library of *Hordeum vulgare* (barley) infected with *Puccinia triticina,* 45 percent of the identified clones corresponded to the very abundant transcript of the small subunit of ribulose-1,5-bisphosphate carboxylase/oxygenase (Neu et al., 2003). In a kernel development library of barley, three clones (homologous to sequences encoding a hordein, toxin α-hordothionin, and α-amylase inhibitor) were very well represented in the differential gene pool (Jang et al., 2003). In a report on petal senescence in *Alstroemeria,* the interference of a highly abundant metallothionein transcript was noted.

In the adventitious root nodulation population, most clones were one, two, or three times present, indicating enrichment but no overrepresentation. Nevertheless, several clones corresponding to abundant genes, such as genes encoding chlorophyll-related proteins and parts of the ribulose-1,5-bisphosphate carboxylase/oxidase complex, were identified (Schroeyers et al., 2004). In the subsequent analysis of the sshnod clones, the above-mentioned genes were not reconfirmed as differentially expressed (see below).

In the SSH approach, the quality of the starting mRNA is crucial; sometimes impurities in the samples such as rRNAs interfere with the subtraction. Birch et al. (1999) and Sävenstrand et al. (2000) reported the presence of several rRNA clones in the subtracted EST library. These clones came through the subtraction procedure and gave a strong hybridization signal after differential hybridizations with a subtracted cDNA probe. To prevent the interference of rRNAs in subsequent studies, rRNAs were added together with the driver cDNAs in the subtraction (Sävestrand et al., 2002; Sävestrand and Strid, 2004). We too observed interference of rRNAs in the SSH technique, but most rRNA transcripts could be excluded in a differential screening with driver and tester population (Schroeyers et al., 2004). Other impurities in the biological test samples, such as dilution of nonreacting (noninfection, nondeveloping) tissue, could also be the cause of false positives.

In addition to these contaminations, problems can also occur in the subtraction and PCR amplification steps (Rebrikov et al., 2000). Nonligated adaptor sequences can anneal during the subtractive hybridization and serve as a template during PCR amplification. Also, PCR primers can generate some background due to nonspecific annealing. Another type of background amplification is caused by the redundancy of some cDNA molecules. This background can be high, especially when the nondifferentially expressed redundant cDNAs exceed the number of target clones in a subtracted library. By hybridization to driver sequences, they can be amplified in the subsequent PCR.

To reveal the false positive SSH clone, a subsequent analysis is necessary. Except for false positives generated by the nonspecific annealing of adaptors and primers, most types of contamination cannot be detected by hybridization with the subtracted probes. Alternative screening with tester and driver cDNA probes can be used for such differential screenings. However, frequently several clones that represent rare transcripts give no signal (see below). To avoid the amplification of redundant clones, Rebrikov et al. (2000) developed a technique called mirror orientation selection. The method is based on the orientation of the adaptor sequences to the DNA sequence. For the background molecules only, one orientation corresponding to the progenitor (original sequence in the tester population) was observed after the PCR amplification compared to a bidirectional orientation for the target molecules, because of enrichment of several progenitor molecules. In the plant field, this technique was successfully used by Shim et al. (2004) to reduce the amount of background with decreased occurrence of false positives.

## *LARGE-SCALE PRIMARY CONFIRMATION*

SSH clones can be analyzed further in a high-throughput manner by differential screening. Large-scale screening of spotted colonies, PCR products, or reverse-transcribed RNA samples (reverse RNA gel blots) is used for initial confirmation. The sensitivity of these techniques diverges a lot (colony versus DNA hybridization). To circumvent sensitivity problems in the hybridization procedure, the commonly used nitrocellulose filters were replaced by (expensive) glass slides (microarray) in a search for genes in senescent petals of *Alstroemeria* (Breeze et al., 2004) and in the interaction of *M. grisea* with *Oryza minuta* (Shim et al., 2004). These confirmation methods are really necessary (Table 29.1) because results from a first screening fluctuate from 9 to 85 percent positive clones (Yen et al., 2000; Werner et al., 2001; Wang et al., 2002; Lu et al., 2004).

We used macroarray (spotted PCR product) hybridization as the initial screening procedure. Two filters were prepared carrying the 102 sshnod clones and several controls corresponding to previously characterized nodulins of *S. rostrata*. One filter was hybridized with a probe derived from RNA extracted from adventitious root primordia and the other with a probe derived from RNA of adventitious root primordia inoculated with *A. caulinodans* (RNA preparations at the same time points as those of the SSH experiment were mixed). By taking a standard twofold variation into account as a cutoff value, 24 up-regulated and 2 down-regulated genes were identified (25 percent; Table 29.2). On the filter hybridized with a developing nodule-derived probe, 98 percent of the clones could be detected, indicating that most genes were detected with this unmodified probe. If the sshnod clones were divided into groups based on their abundancy, in correlation with the positive controls (based on RNA gel blot analysis, RT-PCR, and *in situ* hybridization), we could clearly see that we were not very successful in the identification of low abundant expressed genes. Only three genes fell into the low abundant group, 67 in the medium abundant group, and the remaining 32 genes in the high abundant group. In the same type of experiments, only faint signals had been detected after screening with unmodified probes, showing the low abundance of the identified clones (Diatchenko et al., 1999; Caturla et al., 2002; Ok et al., 2003). Subsequently, forward or reverse subtracted probes are used instead of total mRNA probes. In comparison to tester and driver analysis, the latter type of hybridization revealed the lower (enriched) expressed sequences identified by the SSH. However, these results should be treated with care, as this approach is not quantitative and is sometimes difficult to interpret because the degree of subtraction and the level of normalization within the probe and subsequent screened library can vary considerably (Birch et al., 1999; Caturla et al., 2002; Chen et al., 2002; Wang et al., 2002; Kang et al., 2003; Mahalingam et al., 2003; Ok et al., 2003; Hein et al., 2004; Lu et al., 2004).

One of the reasons why we were not very successful in detecting the low abundant genes is pooling of several time points in the tester population. By combining different samples, the low abundant genes could be diluted from a particular time point into a bigger pool and smaller expression variations could disappear. To counter this problem, Nyamsuren et al. (2003) used the cDNA-AFLP technique before SSH as a primary screening approach to score for the samples with the highest percentage of differential transcripts.

The sensitivity of macroarrays could also disable the detection of small differences in expression levels. These problems are difficult to circumvent, because large-scale screenings are necessary. A differential screening with subtracted probes (forward-reverse) may be a solution. Nevertheless, the outcome of this screening is not quantitative because high and low abundant sequences are equalized in the subtracted probes and the sequences are enriched for low abundant differentially expressed genes. In addition, false positives cannot be distinguished (see above). In contrast, small-scale analyses, such as RNA gel blots, RT-PCR, or *in situ* hybridizations, could reveal these small variations in induction, but their application as an initial approach after the subtraction is not feasible due to their labor-intensive nature and high costs.

An additional explanation is that SSH is not as sensitive as reported previously. Diatchenko et al. (1996) first described an enrichment of target sequences (φX174 in a skeletal muscle library) of up to 1,000-fold in tester populations containing 0.1 percent, 0.01 percent, or 0.001 percent of viral DNA. The original abundance of a target molecule, the ratio of concentrations in tester to driver, and the number of differentially expressed genes within a sample could jeopardize this level of enrichment. Ji et al. (2002) investigated these limitations at both theoretical and practical levels. In an experiment analogous to that of Diatchenko et al. (1996), two identical cDNA samples from fibroblasts were compared. Different concentrations of φX174 DNA were artificially added in the tester sample. The φX174 DNA could be visualized on an agarose gel at a concentration of 0.01 percent, whereas Diatchenko et al. (1996) could detect it at a concentration of 0.001 percent. When the concentration of the φX174 DNA is less than 0.01 percent of tester cDNA, no clearly visible bands were apparent, but a smear of fragments was seen indicating that most fragments were not the differentially expressed targets (φX174), but were predominantly randomly amplified cDNAs. In conclusion, the SSH could not exclude all nondifferentially expressed genes. A reason for the failure in amplification of the target at a concentration of 0.01 percent may be the necessary minimal concentration of the target to drive (complete) hybridizations (Ji et al., 2002).

The efficiency of the SSH technique for enrichment of different ratios of an expressed gene was also tested. Different amounts of φX174 DNAs were added to the tester and driver populations. The φX174 DNAs could only be detected as differentially expressed if a ratio of the target between the tester and driver was 5. As a consequence, Ji et al. (2002) suggested that the SSH approach could only be used in a process with a dramatic alteration of gene expression and for effective enrichment of the targets; transcripts must at least represent 0.1 percent of the total mRNA. Hence, low abundance genes would not be detected (Ji et al., 2002).

Analysis of the sshnod clones revealed that out of the 26 differentially expressed clones, 5 had no significant hit (E value higher than 0.01), 3 were similar to hypothetical or unknown genes, and the others were identical or similar to known genes (Table 29.2). This last group can be divided into different categories. The largest category (7) is involved in primary metabolism, including transcripts involved in sugar metabolism (sshnod3, a putative cytosolic fructose-1,6-bisphosphate aldolase; sshnod4, a putative sucrose synthase; and sshnod6, a putative pyrophosphate-fructose 1-phosphotransferase), fatty acid metabolism (sshnod5, a putative acyl-CoA oxidase) and electron transport (sshnod2, a putative ferredoxin), or in secondary metabolism (sshnod1, a putative chalcone reductase). One of these genes encodes a putative chalcone reductase that corresponds to a formerly identified nodulin gene (Srchr1-sshnod1) of *S. rostrata* (Goormachtig et al., 1995, 1999). The other genes could be divided into groups linked to protein-protein interaction and signal transduction (sshnod8, a putative RING finger protein; sshnod9, a putative ankyrin-like protein; and sshnod10 and sshnod11, two putative 14-3-3 proteins), transcription and translation (sshnod15, a putative ribosomal protein; sshnod101, a putative translationally controlled tumor protein; and sshnod102, a putative translation factor, SUI1), hormone perception and synthesis (sshnod13, a putative GA protein and sshnod14, a putative *S*-adenosyl methionine synthetase), cell wall proteins (sshnod12, a putative proline-rich protein), and stress responses (sshnod15, a putative allergen protein).

Besides clones without significant hit and unknown proteins, the other ESTs did not correspond to new nodulins. Analogous genes were obtained in other nodule libraries. We can conclude that approximately 25 percent of the genes corresponded to putative novel nodulins. Further analysis of the related expression pattern or protein function is necessary to understand their involvement in nodulation. As already mentioned, the SSH is a powerful technique for identification of novel genes. In addition to the above-mentioned results, other experimental setups were even more successful. For example, 327 novel ESTs were identified in three normalized SSH libraries of *A. thaliana* (Ando et al., 2004). In a study of stress-responsive genes in the same model plant, 180 new ESTs, previously not identified in *A. thaliana* EST databases, were retrieved from 8 differentially screened SSH libraries and more than two thirds of the 1,058 differential expressed genes had never been identified in earlier reports on stress (Mahalingam et al., 2003).

## *SMALL-SCALE SECONDARY CONFIRMATION*

After sequencing and a large-scale analysis of clones, a more detailed analysis of putative genes of interest is needed as a confirmation, because all methods have their drawbacks. For example, often not all genes that are confirmed in the primary screening represent true differentially expressed genes. In the interaction between rice and *M. grisea,* only 35 of the 47 primary confirmed clones could be reconfirmed in a secondary RNA gel blot hybridization (Lu et al., 2004). The opposite can also be true: valid differentially expressed genes are not found in the primary screening. For instance, a metallothionein transcript was extremely enriched in a subtracted senescence petal library but could not be detected via microarray hybridization, although its induction was clearly shown by RNA gel blot analyses. We noticed similar problems in our gene pools. Two genes, *SrERF1* and *SrRab1c,* gave no significant signal increase after hybridization with the nodule-derived probe compared to the noninoculated probe. However, *in situ* hybridization with the same clone showed clear up-regulation in nodule tissue. Two causative justifications for this contradictory result could be the insensitivity of the macroarray hybridization and the interference of multigene families during the hybridizations.

Generally, RNA gel blot analysis is used as a confirmation method, but the sensitivity of this hybridization is sometimes too low. Instead of adding more RNA (up to 40 μg), use of poly(A) as material, exposure for weeks, RT-PCR, or cDNA-AFLP analysis could give better results (Birch et al., 1999; Luo et al., 2003; Park et al., 2004). We used the very sensitive *in situ* hybridization method to obtain an immediate overview of the expression pattern within the nodules. When the development of a whole organ is studied, the tissue- or cell-specific expression patterns could provide hints about gene functions.

After the analysis of all *in situ* hybridization patterns, we ended up with a list of several interesting clones. Two of them (*SrERF1* and *SrSAMS*) gave additional information on the presence of the gaseous hormone ethylene during nodule initiation at lateral root bases of *S. rostrata*. Ethylene plays an important role throughout nodulation, but its role in *S. rostrata* is rather exceptional because this plant is modified specifically to resist water-rich environments and has different ethylene responses compared to terrestrial plants (Goormachtig et al., 2004).

In the plant field, the use of the SSH technique has identified various interesting genes. Besides many new genes, putative transcription factors such as WRKY, and genes encoding signaling proteins, such as putative receptor-like kinases, MAP kinases, phosphatases, G proteins, GTP proteins, and stress- and pathogen-related genes have been characterized (Birch et al., 1999; Kim et al., 1999; Dellagi et al., 2000; Sävenstrand et al., 2000, 2002; Beyer et al., 2001; Hinderhofer and Zentgraf, 2001; Xiong et al., 2001; Caturla et al., 2002; Iqbal et al., 2002; Li et al., 2002; Wang et al., 2002; Vieira Dos Santos et al., 2003; Gao et al., 2003; Gepstein et al., 2003;

Kang et al., 2003; Liao et al., 2003; Mahalingam et al., 2003; Neu et al., 2003; Nyamsuren et al., 2003; Ok et al., 2003; Sanchez-Ballesta et al., 2003; Watt, 2003; Xia et al., 2003; Ando et al., 2004; Avrova et al., 2004; Breeze et al., 2004; Hein et al., 2004; Lu et al., 2004; Park et al., 2004; Sävenstrand and Strid, 2004; Schroeyers et al., 2004; Shim et al., 2004).

## *SSH AND LEGUMES*

Besides our own experience with SSH to study adventitious root nodulation and root development in *S. rostrata,* the SSH technique has been implemented to investigate different aspects of the legume field. In a quest for genes for stress resistance from *P. sativum,* 13 genes have been identified in ozone and UV-B stress and 6 have been found to be involved in heavy metal stress (Sävenstrand et al., 2000, 2002; Sävenstrand and Strid, 2004). Several interesting stress-induced genes, such as genes encoding pathogenesis-related (PR) proteins, extensin, the Toll-interleukin-1 receptor, *S*-adenosyl-L-methionine, salicylic acid carboxyl methyl transferase, and *S*-adenosyl-L-methionine synthase were characterized further within this gene pool (Sävenstrand et al., 2000, 2002; Sävenstrand and Strid, 2004).

The SSH technique was applied in *G. max* also. One gene, a novel purple acid phosphatase characterized from a salt stress library, was analyzed in more detail (Liao et al., 2003) and six quantitative trait loci, each conferring partial resistance to the sudden death syndrome caused by *Fusarium solani* f.sp. *glycines* were investigated (Iqbal et al., 2002). In the latter investigation, a unigene set of 135 ESTs was established and analyzed by macroarray analysis in several soybean varieties. Twenty-eight cDNA fragments were increased in the inoculated roots of RIL 23 (Essex x Forrest recombinant inbred line, six resistance alleles) compared to uninoculated roots. In a similar (inoculated-uninoculated) differential screening experiment with the soybean varieties Forrest (four resistance alleles) and Essex (two resistance alleles), only one gene was upregulated (Iqbal et al., 2002). Most of the analyzed mRNAs (61/135) had reduced abundance in the variety Essex. The identified genes were similar to signal recognition proteins, such as phospholipase D, defense-like proteins involved in the flavonoid and isoflavonoid biosynthesis (phenylalanine ammonia lyase, cinnamic acid 4-hydroxylase, and chalcone synthase), cell wall proteins (proline-rich protein), and proteins associated with general metabolism (vacuolar ATP synthase, ATP/ADP transporter, and alcohol dehydrogenase).

A more elaborate application of the SSH technique was performed on the symbiotic interaction between *M. truncatula* and *Glomus intraradices* mycorrhiza (Wulf et al., 2003). A subtracted cDNA library enriched for mycorrhiza-specific transcripts was produced with a driver consisting of four mixtures of diverse plant material. Besides uninoculated roots, roots supplemented with phosphate, *Sinorhizobium meliloti*-inoculated roots, and *A. euteiches*-infected roots were exploited. This pooling was performed to avoid cloning of genes induced by the improved phosphate nutrition, general symbiosis, and pathogenesis, respectively, in the subtracted library. From 2,000 clones initially sequenced, 1,805 ESTs clustered into 600 singletons and 290 tentative consensus (TC) sequences. Reverse RNA gel blot hybridizations with tester and driver probes confirmed 22 of the 34 amplified clones. From this pool of genes, 11 genes revealed significant similarities with several plant genes and were confirmed to be differentially expressed in mycorrhiza-treated roots by quantitative RT-PCR. Within this gene pool, three genes showed a similarity to lectins and genes encoding a phosphate transporter, a nitrate transporter, a germin-like protein, a miraculin-like protein, and a glutathione *S*-transferase. This last gene was confirmed by promoter-*gus* fusions to be highly expressed in response to root colonization by *G. intraradices* (Wulf et al., 2003).

In another extended study, the interaction of *M. truncatula* with the oomycete *A. euteiches* was investigated (Nyamsuren et al., 2003). As mentioned above, before using the SSH technique, the cDNA-AFLP technique was utilized to determine the time point with the highest percentage of differentially expressed genes during the disease development (Nyamsuren et al., 2003). From the 2,000 initially screened clones, 560 were confirmed to be differentially expressed, of which 74 TC and 330 singletons were revealed. Several stress-responsive genes that encode PR proteins, an ABA-responsive protein, and senescence-associated proteins were identified. Furthermore, 8 percent of the 560 characterized genes could be classified as new *M. truncatula* sequences.

A whole different approach was followed by Nakata and McConn (2002), who combined techniques to characterize genes involved in calcium oxalate crystal formation in *M. truncatula*. First, the SSH technique was applied to ensure the retention and amplification of low abundant differentially expressed genes, and this library was subjected to several rounds of subtraction to avoid identification of unwanted common transcripts. A differential screening was performed subsequently to reduce the level of false positives. Sequence analysis of the clones resulted in seven gene-encoding proteins mostly involved in lipid or cell wall metabolism (outer membrane lipoprotein, two xylosidases, xylose isomerase, xylosidase, and UDP galactose epimerase; Nakata and McConn, 2002).

## *CONCLUSION*

SSH can be considered a profitable approach for gene-hunting strategies in model species as well as varieties with few molecular investigations. This technique can be adapted efficiently for the characterization of specific target sequences by pooling tester samples, adding rRNA, or combining diverse samples in the driver and regulating the degree of subtraction or normalization, and the level of amplification. The power of the SSH lies to a substantial extent in the easy combination with large-scale methodologies, such as colony hybridizations, reverse RNA gel blots or dot blot (macroarray), and microarray analyses. These large screening approaches in combination with sequencing are indispensable for the further analysis of the subtracted target population as a consequence of the sometimes high quantity of false positives and the degree of redundancies. Nevertheless, these drawbacks are common to almost all high-throughput transcriptome analysis. Broadly speaking, SSH is very powerful for identifying novel and low abundant differentially expressed genes, although rare transcripts are not always detected and the accumulation of false positives is sometimes high.

## REFERENCES

Ando, K., S. Yamakawa, K. Miyashita, K. Yoshida, A. Yokota, A. Shinmyo, and T. Kohchi (2004). Efficient construction of cDNA microarrays utilizing normalized cDNA libraries of *Arabidopsis thaliana. Journal of Bioscience and Bioengineering* 97:85-88.

Avrova, A.O., N. Taleb, V.-M. Rokka, J. Heilbronn, E. Campbell, I. Hein, E.M. Gilroy, L. Cardle, J.E. Bradshaw, H.E. Stewart, et al. (2004). Potato oxysterol binding protein and cathepsin B are rapidly up-regulated in independent defense pathways that distinguish *R* gene-mediated and field resistances to *Phytophthora infestans. Molecular Plant Pathology* 5:45-56.

Bachem, C.W.B., R.S. van der Hoeven, S.M. de Bruijn, D. Vreugdenhil, M. Zabeau, and R.G.F. Visser (1996). Visualization of differential gene expression using a novel method of RNA fingerprinting based on AFLP: Analysis of gene expression during potato tuber development. *Plant Journal* 9:745-753.

Bahn, S.C., M.S. Bae, Y.B. Park, S.-I. Oh, J.U. Jeung, J.M. Bae, Y.S. Chung, and J.S. Shin (2001). Molecular cloning and characterization of a novel low temperature-induced gene, *blti2,* from barley (*Hordeum vulgare* L). *Biochimica et Biophysica Acta* 1522:134-137.

Baldwin, D., V. Crane, and D. Rice (1999). A comparison of gel-based, nylon filter and microarray techniques to detect differential RNA expression in plants. *Current Opinion in Plant Biology* 2:96-103.

Beyer, K., A. Binder, T. Boller, and M. Collinge (2001). Identification of potato genes induced during colonization by *Phytophthora infestans. Molecular Plant Pathology* 2:125-134.

Birch, P.R.J., A.O. Avrova, J.M. Duncan, G.D. Lyon, and R.L. Toth (1999). Isolation of potato genes that are induced during an early stage of the hypersensitive response to *Phytophthora infestans. Molecular Plant-Microbe Interactions* 12: 356-361.

Breeze, E., C. Wagstaff, E. Harrison, I. Bramke, H. Rogers, A. Stead, B. Thomas, and V. Buchanan-Wollaston (2004). Gene expression patterns to define stages of post-harvest senescence in *Alstroemeria* petals. *Plant Biotechnology Journal* 2:155-168.

Breyne, P., R. Dreesen, B. Cannoot, D. Rombaut, K. Vandepoele, S. Rombauts, R. Vanderhaeghen, D. Inzé, and M. Zabeau (2003). Quantitative cDNA-AFLP analysis for genome-wide expression studies. *Molecular Genetics and Genomics* 269:173-179.

Caturla, M., C. Chaparro, K. Schroeyers, and M. Holsters (2002). Suppression subtractive hybridization to enrich low-abundance and submergence-enhanced transcripts of adventitious root primordia of *Sesbania rostrata. Plant Science* 162:915-921.

Chen, X., H. Yuan, R. Chen, L. Zhu, B. Du, Q. Weng, and G. He (2002). Isolation and characterization of triacontanol-regulated genes in rice (*Oryza sativa* L.): Possible role of triacontanol as a plant growth stimulator. *Plant and Cell Physiology* 43:869-876.

Colebatch, G., B. Trevaskis, and M. Udvardi (2002). Functional genomics: Tools of the trade. *New Phytologist* 153:27-36.

Cui, Y., N. Brugière, L. Jackman, Y.-M. Bi, and S.J. Rothstein (1999). Structural and transcriptional comparative analysis of the *S* locus regions in two self-incompatible *Brassica napus* lines. *Plant Cell* 11:2217-2231.

Dellagi, A., J. Heilbronn, A.O. Avrova, M. Montesano, E.T. Palva, H.E. Stewart, I.K. Toth, D.E.L. Cooke, G.D. Lyon, and P.R.J. Birch (2000). A potato gene encoding a WRKY-like transcription factor is induced in interactions with *Erwinia carotovora* subsp. *atroseptica* and *Phytophthora infestans* and is coregulated with class I endochitinase expression. *Molecular Plant-Microbe Interactions* 13:1092-1101.

Diatchenko, L., Y.-F.C. Lau, A.P. Campbell, A. Chenchik, F. Moqadam, B. Huang, S. Lukyanov, K. Lukyanov, N. Gurskaya, et al. (1996). Suppression subtractive hybridization: A method for generating differentially regulated or tissue-specific cDNA probes and libraries. *Proceedings of the National Academy of Sciences USA* 93:6025-6030.

Diatchenko, L., S. Lukyanov, Y.-F.C. Lau, and P.D. Siebert (1999). Suppression subtractive hybridization: A versatile method for identifying differentially expressed genes. In *cDNA preparation and characterization,* Methods In Enzymology, Vol. 303, S.M. Weissman (ed.). San Diego, CA: Academic Press, pp. 349-380.

Donson, J., Y. Fang, G. Espiritu-Santo, W. Xing, A. Salazar, S. Miyamoto, V. Armendarez, and W. Volkmuth (2002). Comprehensive gene expression analysis by transcript profiling. *Plant Molecular Biology* 48:75-97.

Doyle, J.J. and M.A. Lucknow (2003). The rest of the iceberg. Legume diversity and evolution in a phylogenetic context. *Plant Physiology* 131:900-910.

Dreyfus, B. and Y.R. Dommergues (1981). Nitrogen-fixing nodules induced by *Rhizobium* on the stem of the tropical legume *Sesbania rostrata. FEMS Microbiology Letters* 10:313-317.

Duhoux, E. and B.L. Dreyfus (1982). Nature des sites d'infection par le *Rhizobium* de la tige de la légumineuse *Sesbania rostrata* Brem. *Comptes Rendus Hebdomadaires des Séances de l'Académie des Sciences Paris* 294:407-411.

Fedorova, M., J. van de Mortel, P.A. Matsumoto, J. Cho, C.D. Town, K.A. VandenBosch, J.S. Gantt, and C.P. Vance (2002). Genome-wide identification of nodule-specific transcripts in the model legume *Medicago truncatula. Plant Physiology* 130:519-537.

Gao, P., G.-Y. Wang, H.-J. Zhao, L. Fan, and Y.-Z. Tao (2003). Isolation and identification of submergence-induced genes in maize *Zea mays* seedlings by suppression subtractive hybridization. *Acta Botanica Sinica* 45:479-483.

Gepstein, S., G. Sabehi, M.-J. Carp, T. Hajouj, M.F.O. Nesher, I. Yariv, C. Dor, and M. Bassani (2003). Large-scale identification of leaf senescence-associated genes. *Plant Journal* 36:629-642.

Goormachtig, S., M. Alves-Ferreira, M. Van Montagu, G. Engler, and M. Holsters (1997). Expression of cell cycle genes during *Sesbania rostrata* stem nodule development. *Molecular Plant-Microbe Interactions* 10:316-325.

Goormachtig, S., W. Capoen, E.K. James, and M. Holsters (2004). Switch from intracellular to intercellular invasion during water stress-tolerant legume nodulation. *Proceedings of the National Academy of Sciences USA* 101:6303-6308.

Goormachtig, S., S. Lievens, S. Herman, M. Van Montagu, and M. Holsters (1999). Chalcone reductase-homologous transcripts accumulate during development of stem-borne nodules on the tropical legume *Sesbania rostrata. Planta* 209:45-52.

Goormachtig, S., P. Mergaert, M. Van Montagu, and M. Holsters (1998). The symbiotic interaction between *Azorhizobium caulinodans* and *Sesbania rostrata.* Molecular cross-talk in a beneficial plant-bacterium interaction. In *Plant-microbe interactions,* Subcellular Biochemistry, Vol. 29, B.B. Biswas and H.K. Das (eds.). New York: Plenum Press, pp. 117-164.

Goormachtig, S., M. Valerio-Lepiniec, K. Szczyglowski, M. Van Montagu, M. Holsters, and F.J. de Bruijn (1995). Use of differential display to identify novel *Sesbania rostrata* genes enhanced by *Azorhizobium caulinodans* infection. *Molecular Plant-Microbe Interactions* 8:816-824.

Gygi, S.P., Y. Rochon, B.R. Franza, and R. Aebersold (1999). Correlation between protein and mRNA abundance in yeast. *Molecular and Cellular Biology* 19: 1720-1730.

Hein, I., E.I. Campbell, M. Woodhead, P.E. Hedley, V. Young, W.L. Morris, L. Ramsay, J. Stockhaus, G.D. Lyon, et al. (2004). Characterisation of early transcriptional changes involving multiple signaling pathways in the *Mla13* barley interaction with powdery mildew (*Blumeria graminis* f.sp. *hordei*). *Planta* 218:803-813.

Hinderhofer, K. and U. Zentgraf (2001). Identification of a transcription factor specifically expressed at the onset of leaf senescence. *Planta* 213:469-473.

Hubank, M. and D.G. Schatz (1994). Identifying differences in mRNA expression by representational difference analysis of cDNA. *Nucleic Acids Research* 22: 5640-5648.

Iqbal, M.J., S. Yaegashi, V.N. Njiti, R. Ahsan, K.L. Cryder, and D.A. Lightfoot (2002). Resistance locus pyramids alter transcript abundance in soybean roots inoculated with *Fusarium solani* f.sp. *glycines*. *Molecular Genetics and Genomics* 268:407-417.

Jang, C.S., M.S. Lee, J.Y. Kim, D.S. Kim, and Y.W. Seo (2003). Molecular characterization of a cDNA encoding putative calcium binding protein, HvCaBP1, induced during kernel development in barley (*Hordeum vulgare* L.). *Plant Cell Reports* 22:64-70.

Ji, W., M.B. Wright, L. Cai, A. Flament, and K. Lindpaintner (2002). Efficacy of SSH PCR in isolating differentially expressed genes. *BMC Genomics* 3:12.1-12.7 (http://www.biomedcentral.com/1471-2144/3/12).

Kang, H.-G., R.C. Foley, L. Oñate-Sánchez, C. Lin, and K.B. Singh (2003). Target genes for OBP3, a *Dof* transcription factor, include novel basic helix-loop-helix domain proteins inducible by salicylic acid. *Plant Journal* 35:362-372.

Kim, J.Y., Y.S. Chung, S.H. Ok, S.G. Lee, W.I. Chung, I.Y. Kim, and J.S. Shin (1999). Characterization of the full-length sequences of phospholipase $A_2$ induced during flower development. *Biochimica et Biophysica Acta* 1489:389-392.

Kuhn, E. (2001). From library screening to microarray technology: Strategies to determine gene expression profiles and to identify differentially regulated genes in plants. *Annals of Botany* 87:139-155.

Küster, H., N. Hohnjec, F. Krajinski, F. El Yahyaoui, K. Manthey, J. Gouzy, M. Dondrup, F. Meyer, J. Kalinowski, et al. (2004). Construction and validation of cDNA-based Mt6k-RIT macro- and microarrays to explore root endosymbioses in the model legume *Medicago truncatula*. *Journal of Biotechnology* 108:95-113.

Lavery, D.J., L. Lopez-Molina, F. Fleury-Olela, and U. Schibler (1997). Selective amplification via biotin- and restriction-mediated enrichment (SABRE), a novel selective amplification procedure for detection of differentially expressed mRNAs. *Proceedings of the National Academy of Sciences USA* 94:6831-6836.

Li, Y.-X., L. Chen, L. Juan, Y. Li, and F. Chen (2002). Suppression subtractive hybridization cloning of cDNAs of differentially expressed genes in dovetree (*Davidia involucrata*) bracts. *Plant Molecular Biology Reporter* 20:231-238.

Liang, P. and A.B. Pardee (1992). Differential display of eukaryotic messenger RNA by means of the polymerase chain reaction. *Science* 257:967-971.

Liao, H., F.-L. Wong, T.-H. Phang, M.-Y. Cheung, W.-Y.F. Li, G. Shao, X. Yan, and H.-M. Lam (2003). *GmPAP3,* a novel purple acid phosphatase-like gene in soybean induced by NaCl stress but not phosphorus deficiency. *Gene* 318:103-111.

Lievens, S., S. Goormachtig, and M. Holsters (2001). A critical evaluation of differential display as a tool to identify genes involved in legume nodulation: Looking back and looking forward. *Nucleic Acids Research* 17:3459-3468.

Lisitsyn, N., N. Lisitsyn, and M. Wigler (1993). Cloning the differences between two complex genomes. *Science* 259:946-951.

Lockhart, D.J. and E.A. Winzeler (2000). Genomics, gene expression and DNA arrays. *Nature* 405:827-836.

Lu, G., C. Jantasuriyarat, B. Zhou, and G.-L. Wang (2004). Isolation and characterization of novel defense response genes involved in compatible and incompatible interactions between rice and *Magnaporthe grisea. Theoretical and Applied Genetics* 108:525-534.

Luo, Z.-Y., Q.-H. Lu, S.-P. Liu, X.-H. Chen, J.-Q. Luo, L.-J. Tan, and W.-X. Hu (2003). Screening and identification of novel genes involved in biosynthesis of ginsenoside in *Panax ginseng* plant. *Acta Biochimica et Biophysica Sinica* 35:554-567.

Mahalingam, R., A.M. Gomez-Buitrago, N. Eckardt, N. Shah, A. Guevara-Garcia, P. Day, R. Raina, and N.V. Fedoroff (2003). Characterizing the stress/defense transcriptome of *Arabidopsis. Genome Biology* 4:R20.1-R20.14.

Martin, K.J., and A.B. Pardee (2000). Identifying expressed genes. *Proceedings of the National Academy of Sciences USA* 97:3789-3791.

Mitra, R.M., C.A. Gleason, A. Edwards, J. Hadfield, J.A. Downie, G.E.D. Oldroyd, and S.R. Long (2004). A $Ca^{2+}$/calmodulin-dependent protein kinase required for symbiotic nodule development: Gene identification by transcript-based cloning. *Proceedings of the National Academy of Sciences USA* 101:4701-4705.

Mysore, K., M.D. D'Ascenzo, X. He, and G.B. Martin (2003). Overexpression of the disease resistance gene *Pto* in tomato induces gene expression changes similar to immune responses in human and fruit fly. *Plant Physiology* 132:1901-1912.

Nakata, P.A. and M.M. McConn (2002). Sequential subtractive approach facilitates identification of differentially expressed genes. *Plant Physiology and Biochemistry* 40:307-312.

Nap, J.-P. and T. Bisseling (1990). Developmental biology of a plant-prokaryote symbiosis: The legume root nodule. *Science* 250:948-954.

Ndoye, I., F. de Billy, J. Vasse, B. Dreyfus, and G. Truchet (1994). Root nodulation of *Sesbania rostrata. Journal of Bacteriology* 176:1060-1068.

Neu, C., B. Keller, and C. Feuillet (2003). Cytological and molecular analysis of the *Hordeum vulgare-Puccinia triticina* nonhost interaction. *Molecular Plant-Microbe Interactions* 16:626-633.

Nyamsuren, O., F. Colditz, S. Rosendahl, M. Tamasloukht, T. Bekel, F. Meyer, H. Kuester, P. Franken, and F. Krajinski (2003). Transcriptional profiling of *Medicago truncatula* roots after infection with *Aphanomyces euteiches* (Oomycota) identifies novel genes upregulated during this pathogenic interaction. *Physiological and Molecular Plant Pathology* 63:17-26.

Ohlrogge, J. and C. Benning (2000). Unraveling plant metabolism by EST analysis. *Current Opinion in Plant Biology* 3:224-228.

Ok, S.H., H.M. Park, J.Y. Kim, S.C. Bahn, J.M. Bae, M.C. Suh, J.-U. Jeung, K.-N. Kim, and J.S. Shin (2003). Identification of differentially expressed genes during flower development in carnation (*Dianthus caryophyllus*). *Plant Science* 165:291-297.

Park, J.-S., J.-B. Kim, B.-S. Hahn, K.-H. Kim, S.-H. Ha, J.-B. Kim, and Y.-H. Kim (2004). EST analysis of genes involved in secondary metabolism in *Camellia sinensis* (tea) using suppression subtractive hybridization. *Plant Science* 166: 953-961.

Rebrikov, D.V., O.V. Britanova, N.G. Gurskaya, K.A. Lukyanov, V.S. Tarabykin, and S.A. Lukyanov (2000). Mirror orientation selection (MOS): A method for eliminating false positive clones from libraries generated by suppression subtractive hybridization. *Nucleic Acids Research* 28:e90.

Rinaudo, G., B. Dreyfus, and Y.R. Dommergues (1983). *Sesbania rostrata* green manure and the nitrogen content of rice crop and soil. *Soil Biology and Biochemistry* 15:111-113.

Sanchez-Ballesta, M.T., Y. Lluch, M.J. Gosalbes, L. Zacarias, A. Granell, and M.T. Lafuente (2003). A survey of genes differentially expressed during long-term heat-induced chilling tolerance in citrus fruit. *Planta* 218:65-70.

Sävenstrand, H., M. Brosché, M. Ängehagen, and Å. Strid (2000). Molecular markers for ozone stress isolated by suppression subtractive hybridization: Specificity of gene expression and identification of a novel stress-regulated gene. *Plant, Cell and Environment* 23:689-700.

Sävenstrand, H., M. Brosché, and Å. Strid (2002). Regulation of gene expression by low levels of ultraviolet-B radiation in *Pisum sativum*: Isolation of novel genes by suppression subtractive hybridization. *Plant and Cell Physiology* 43:402-410.

Sävenstrand, H. and Å. Strid (2004). Six genes strongly regulated by mercury in *Pisum sativum* roots. *Plant Physiology and Biochemistry* 42:135-142.

Schroeyers, K., C. Chaparro, S. Goormachtig, and M. Holsters (2004). Nodulation-enhanced sequences from the water stress-tolerant tropical legume *Sesbania rostrata. Plant Science* 167:207-216.

Schultze, M. and A. Kondorosi (1998). Regulation of symbiotic root nodule development. *Annual Reviews of Genetics* 32:33-57.

Shim, K.S., S.K. Cho, J.U. Jeung, K.W. Jung, M.K. You, S.H. Ok, Y.S. Chung, K.H. Kang, H.G. Hwang, H.C. Choi, et al. (2004). Identification of fungal (*Magnaporthe grisea*) stress-induced genes in wild rice (*Oryza minuta*). *Plant Cell Reports* 22:599-607.

Smith, R.D. (2000). Probing proteomes—seeing the whole picture? *Nature Biotechnology* 18:1041-1042.

van Kammen, A. (1984). Suggested nomenclature for plant genes involved in nodulation and symbiosis. *Plant Molecular Biology Reporter* 2:43-45.

Vieira Dos Santos, C., P. Delavault, P. Letousey, and P. Thalouarn (2003). Identification by suppression subtractive hybridization and expression analysis of *Ara-*

*bidopsis thaliana* putative defence genes during *Orobanche ramosa* infection. *Physiological and Molecular Plant Pathology* 62:297-303.

Wang, X., P. Wu, M. Xia, Z. Wu, Q. Chen, and F. Liu (2002). Identification of genes enriched in rice roots of the local nitrate treatment and their expression patterns in split-root treatment. *Gene* 29:93-102.

Watt, D.A. (2003). Aluminium-responsive genes in sugarcane: Identification and analysis of expression under oxidative stress. *Journal of Experimental Botany* 54:1163-1174.

Weidner, S., A. Pühler, and H. Küster (2003). Genomics insights into symbiotic nitrogen fixation. *Current Opinion in Biotechnology* 14:200-205.

Werner, M., N. Uehlein, P. Proksch, and R. Kaldenhoff (2001). Characterization of two tomato aquaporins and expression during the incompatible interaction of tomato with the plant parasite *Cuscuta reflexa. Planta* 213:550-555.

Wulf, A., K. Manthey, J. Doll, A.M. Perlick, B. Linke, T. Bekel, F. Meyer, P. Franken, H. Küster, and F. Krajinski (2003). Transcriptional changes in response to arbuscular mycorrhiza development in the model plant *Medicago truncatula. Molecular Plant-Microbe Interactions* 16:306-314.

Xia, M., S.-F. Wang, X.-B. Wang, H.-B. Li, and P. Wu (2003). Identification of phosphorus starvation inducible genes in rice by suppression subtractive hybridization. *Acta Botanica Sinica* 45:736-741.

Xiong, L., M.-W. Lee, M. Qi, and Y. Yang (2001). Identification of defense-related rice genes by suppression subtractive hybridization and differential screening. *Molecular Plant-Microbe Interactions* 14:685-691.

Yamazaki, M. and K. Saito (2002). Differential display analysis of gene expression in plants. *Cellular and Molecular Life Sciences* 59:1246-1255.

Yen, H.E., S.-M. Wu, Y.-H. Hung, and S.-K. Yen (2000). Isolation of 3 salt-induced low-abundance cDNAs from light-grown callus of *Mesembryanthemum crystallinum* by suppression subtractive hybridization. *Physiologia Plantarum* 110: 402-409.

Chapter 30

# Quantitative PCR in the Analysis of Transgenic Plants

Hiroyuki Koyama
Tomonori Kihara
Eiji Takita
Munechika Honda

## *INTRODUCTION*

Polymerase chain reaction (PCR) technology has advanced during the last decade, becoming the most popular experimental tool in molecular plant sciences. There is a wide variety of PCR applications covering both basic and applied research. For example, PCR-detectable genetic markers, namely SSLP (simple sequence length polymorphism) and AFLP (amplified fragment length polymorphism), are frequently used for genotype mapping of plant populations. In addition, analyzing contamination by genetically modified crops (Germini et al., 2004) and detecting pathogenic bacteria (e.g., *Escherichia coli* O157; Maurer et al., 1999) are commonly performed using PCR during inspection of food safety. PCR coupled with reverse transcription (RT-PCR) is also a key technology in plant molecular biology, including transcript analysis and other derivative procedures for identifying specific gene expression such as cDNA subtraction and differential display techniques. The PCR procedures can also be applicable in the quantification of target DNA and transcripts in biological samples. Moreover, quantitative PCR (Q-PCR) procedures, because of their extremely high sensitivity and specificity (Gachon et al., 2004; Ingham et al., 2001)

A part of this study was supported by Grants-in-Aid from the JSPS (Nos. 60168903 and 90234912 to H.K.) and the RITE Research Program.

have also become extremely common, replacing conventional blotting technologies such as northern and Southern blotting. For example, the quantification of integration of gene copies into transgenic *Arabidopsis* genome through Southern blotting requires >200 ng of genomic DNA, whereas Q-PCR procedures require only 10 ng (Honda et al., 2002). Further, quantitative RT-PCR allows accurate and specific determination of target transcripts in the presence of highly homologous subfamilies (Ohno et al., 2004), which is usually difficult using northern blot analysis. In the present chapter, we focus on the application of Q-PCR in the analysis of transgenic plants; for example, in isolation of single-copy transgenic plants using quantitative genomic PCR and quantification of ectopic expression of integrated genes using quantitative RT-PCR.

## *QUANTITATIVE PCR PROCEDURES*

There exist several Q-PCR procedures, each employing different amplicon detection processes during quantification (Freeman et al., 1999). For example, competitive PCR was established in the early 1990s for quantification of initial amounts of target DNA at the end point of PCR reactions in the presence of competitor DNA. Competitor DNA is a kind of target DNA mimic, similar in size and carrying identical end sequences. In this situation, the extent of amplicons derived from the target DNA and competitive DNA will be the same, if the initial amounts were the same. The amplicons can be detected with a moderately sensitive fluorescent probe, namely, ethidium bromide after standard agarose gel electrophoresis, and because of this simplicity, competitive PCR is easily applicable in smaller laboratories.

The limitations of this procedure are related to the need for construction of competitor DNA for each target. However, it is also a useful procedure if the integrated target gene number is limited; for example, in selection of transgenic plants carrying single copies of the integrated gene (see next section). On the other hand, real-time PCR procedures detect the target DNA by monitoring fluorescence generated by the reaction between the amplicon and fluorescent probe; for example, nonspecific double strand DNA binding dye (SYBR Green) and specific hydrolysis probes (TaqMan probes). In this PCR reaction, a significant increase in fluorescence is observed after appropriate PCR cycles and correlated with the initial amount of target DNA (Figure 30.1A, B). Hence, by using a known amount of control DNA as a standard in the reactions, the initial amount of target DNA can be quantified. The advantage of real-time PCR compared to competitive

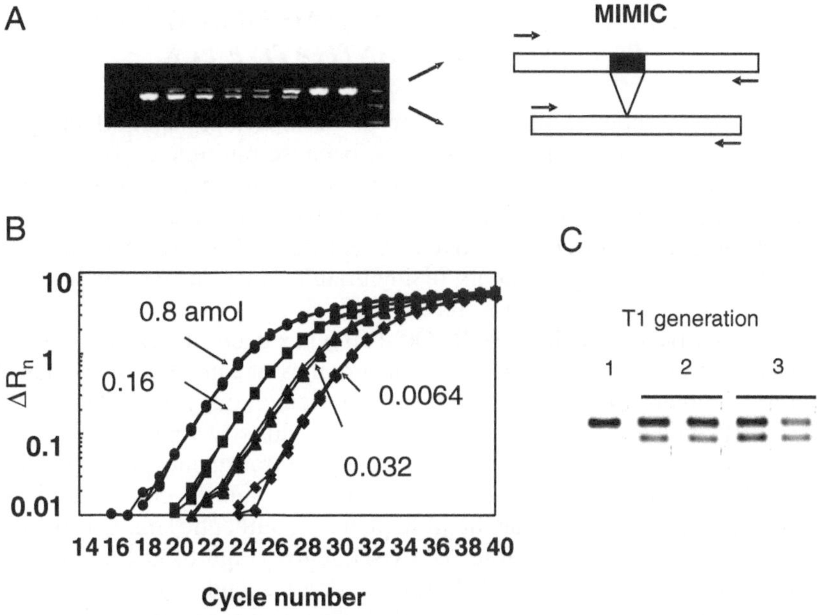


FIGURE 30.1. Competitive PCR and real-time PCR. (A) A typical band pattern obtained with competitive PCR and the structure of mimic DNA. In a previous study, mimic DNA was derived by inserting an intron between authentic sequences of the target DNA (see Honda et al., 2002). (B) Changes in fluorescence during real-time PCR reactions with various amounts of initial template. The amount of initial target DNA is correlated with a significant increase in intensity ($\Delta R_n$). (C) Typical genotyping of $T_1$ transgenic plants using dual-target PCR (Kihara et al., 2006). 1: Null segregants, 2: heterozygous, and 3: homozygous transgenic plants carrying a single copy of the integrated gene.

PCR is short-term detection without the need for post-PCR processing of products, but a potential limitation is its high running cost and the price of the apparatus required to monitor fluorescence.

Both competitive and real-time PCR possess a wide dynamic range ($>10^3$) but cannot be used to identify small differences in the initial amount of target DNA. In such cases, the initial amount of target DNA can be quantified using agarose electrophoresis with sensitive fluorescent probes such as SYBR green, because the amount of amplicons is directly correlated with the initial quantity of DNA at early stages of PCR (i.e., the exponential phase). Thus, this variety of Q-PCR procedures enables accurate quantification of initial amounts of target DNA within a wide range.

## *DETERMINATION OF THE COPY NUMBER IN TRANSGENIC PLANTS USING QUANTITATIVE PCR*

The copy number of integrated genes is an important factor affecting the stability of transgenic plant phenotypes, because multiple copies enhance gene silencing. Also, isolation of homozygous transgenic plants with a single copy of the transferred gene is important for transgenic studies and molecular breeding. Both competitive and real-time PCR have been used for determination of copy numbers of integrated genes in transgenic plants (Ingham et al., 2001; Honda et al., 2002).

Quantitative competitive PCR (QC-PCR) has been successfully used to determine the copy number of genes in genomes of a number of species including humans (Deng et al., 1993) and the filamentous fungus *Aspergillus oryzae* (Honda et al., 2000). This procedure is directly applicable in determining copy numbers of integrated genes in transgenic plants such as *Arabidopsis* and rice (Honda et al., 2002). In these studies, competitor DNA was prepared by adding an artificial sequence or authentic small intron; that is, mimic DNA, into the target DNA sequence (Figure 30.1C). Because both target DNA and mimic DNA possess identical end sequences, amplification efficiency is only affected by their size. For example, when the difference in size is less than 20 percent, amplification efficiency should be very similar between the targets and mimic DNA (Figure 30.1B).

In QC-PCR, the accuracy of genomic DNA quantification is the most important factor affecting the reliability of copy number determination. To increase reliability, a two-step QC-PCR protocol is useful (Honda et al., 2002). In the first step, the amount of genomic DNA is accurately quantified using known single-copy genes as a standard, while in the second step, the copy number of the integrated gene is quantified (Honda et al., 2002). Two-step QC-PCR can determine up to four copies of integrated DNA per haploid genome in *Arabidopsis* but can also efficiently isolate single-copy transgenic plants from primary transgenic plants. A similar procedure has been reported for detection of integrated genes in maize (Song et al., 2002).

Dual-target PCR (a kind of multiplex PCR), which could be categorized as quantitative PCR, is also very useful for determining copy numbers of integrated genes in transgenic plants (Kihara et al., 2006). In this method, the PCR reaction is carried out using two sets of PCR primers with the same template, which amplify the integrated gene and internal single-copy gene in the same PCR reaction. When primers are designed to have identical Tm values in such a way that they amplify similar-sized amplicons (less than 20 percent difference), the amplification efficiencies of both target genes in the PCR reaction are very similar. Accordingly, copy number can be determined by comparing amounts of amplicons visualized after electrophoresis. Homozygous transgenic plants carrying single copies of an integrated gene possess a 1:1 ratio of amplicons derived from the integrated gene and internal control gene.

## *QUANTITATIVE RT-PCR FOR DETERMINATION OF TRANSCRIPT LEVELS*

In quantifying transcript levels, quantitative RT-PCR possesses several advantages compared to conventional northern blotting. For example, its high sensitivity of detection allows the sample size to be reduced during isolation of RNA, which also makes it possible to detect small amounts of transcripts in tiny tissue samples such as a piece of hair root (Jones and Grierson, 2003). In addition, its high specificity of PCR amplification enables determination of target gene transcript levels in the presence of a series of isoforms, which cannot be distinguished by northern blotting. In the analysis of transgenic plants, high specificity also contributes to accurate determination of ectopic gene expression levels in the presence of authentic orthologues.

Both competitive and real-time PCR can be applied to the quantification of transcript levels as well as determining changes in transcript levels, for example, during characterization of inducible genes. However, if the difference of amount of the target amplicons between samples is not large (i.e., less than five times), direct comparison of amplicons at the early stage of PCR is useful because, as mentioned above, the amount of amplicon at this stage is correlated with the initial amount of target cDNA.

Since the extent of amplicon amplification is very small at early stages of PCR, very high sensitivity is required for detection. Canel et al. (1996) employed Southern blot analysis to detect amplicons at low intensities and others successfully performed very accurate determination (e.g., Takita et al., 1999). However, the sensitive fluorescent DNA probe SYBR Green enables detection of amplicons at the early stage of PCR, directly after agarose electrophoresis (Kihara, Ohno, et al., 2003; Kihara, Wada, et al., 2003). Using SYBR Green, transcript levels of a series of isoforms of NADP-specific isocitrate dehydrogenase in carrots (Kihara, Ohno, et al., 2003) and white lupin (Kihara, Wada, et al., 2003), and $H^+$ATPase in carrots (Ohno et al., 2004) were quantified with high efficiency. In this approach, the specificity of amplification is very important in accurate determination. In some studies, specific primers were designed for use after isolation of full-sized cDNA using degenerated PCR followed by 3' and 5' RACE. This approach can be used to compare transcript levels between transgenic plants carrying

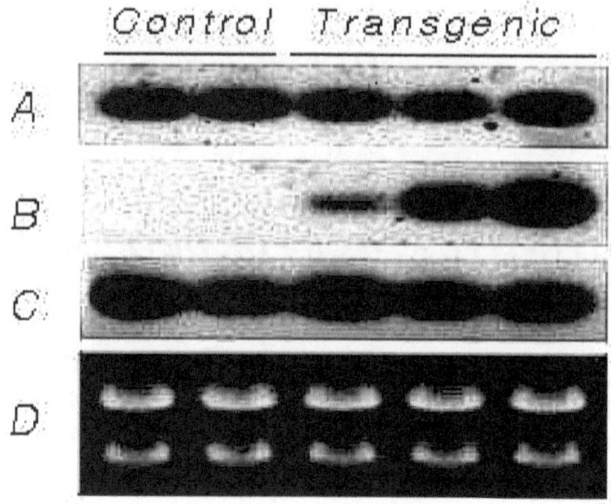


FIGURE 30.2. Quantitative RT-PCR for determination of ectopic gene expression in a transgenic carrot cell line carrying mitochondrial citrate synthase from *Arabidopsis.* Both carrot (A) and *Arabidopsis* (B) citrate synthase gene expression were determined by RT-PCR. After 13 cycles with specific primers, amplicons were detected by Southern blotting. An internal control gene (carrot actin 1 gene: C) and ribosomal RNA (D) were determined to test the quality of RNA and uniformity of the RT-PCR reactions. The figure is modified from Koyama et al. (1999).

foreign genes (Koyama et al., 1999; Jones and Grierson, 2003). For example, ectopic gene expression (citrate synthase) in a carrot cell line was distinguished from authentic gene expression using quantitative RT-PCR with specific primers (Koyama et al., 1999; Figure 30.2).

In all RT-PCR procedures, the efficiency of reverse transcription and the quality of RNA are very important in obtaining accurate results. Also, normalization of RNA samples using internal standard genes (e.g., ubiquitin and actin) is necessary when comparing transcript levels between different samples. Such factors therefore need to be carefully examined in each plant species before applying such approaches.

## REFERENCES

Bubner, B. and I.T. Baldwin (2004). Use of real-time PCR for determining copy number and zygosity in transgenic plants. *Plant Cell Reports* 23:263-271.

Canel, C., J.N. Bailey-Serres, M.L. Roose, and L. Mikeal (1996). Molecular characterization of the mitochondrial citrate synthase gene of an acid less pummelo (*Citrus maxima*). *Plant Molecular Biology* 31:143-147.

Deng G, M. Yu, and H.S. Smith (1993) An improved method of competitive PCR for quantitation of gene copy number. *Nucleic Acid Research* 21:4848-4849.

Freeman W.M., S.J. Walker, and K.E. Vrana (1999). Quantitative RT-PCR: Pitfalls and potential. *Biotechniques* 26:112-122.

Gachon, C., A. Mingam, and B. Charrier (2004). Real-time PCR: What relevance to plant studies? *Journal of Experimental Botany* 55:1445-1454.

Germini, A., A. Zanetti, C. Salati, S. Rossi, C. Forre, S. Schmid, R. Marchelli, and C. Fogher (2004). Development of a seven-target multiplex PCR for the simultaneous detection of transgenic soybean and maize in feeds and foods. *Journal of Agricultural and Food Chemistry* 52:275-280.

Hobbs, S. L. A., P. Kpodar, and C. M. O. DeLong (1990). The effect of T-DNA copy number and methylation on reporter gene expression in tobacco transformants. *Plant Molecular Biology* 15:851-864.

Honda, M., T. Minetoki, K. Gomi, and K. Machida (2000). Rapid detection of homologously integrated DNA fragments and accurate quantitation of their copy number in transgenic *Aspergillus oryzae* by PCR. *Journal of Bioscience and Bioengineering* 90:577-579.

Honda, M., Y. Muramoto, T. Kuzuguchi, S. Sawano, M. Machida, and H. Koyama (2002). Determination of gene copy number and genotype of transgenic *Arabidopsis thaliana* by competitive PCR. *Journal of Experimental Botany* 53: 1515-1520.

Ingham, D.J., S.S. Beer, S. Money, and G. Hansen (2001). Quantitative real-time PCR assay for determining transgene copy number in transformed plants. *BioTechniques* 31:132-140.

Jones, M.A. and C.S. Grierson (2003). A simple method for obtaining cell-specific cDNA from small numbers of growing root-hair cells in *Arabidopsis thaliana. Journal of Experimental Botany* 54:1373-1378.

Kihara, T., T. Ohno, H. Koyama, T. Sawafuji, and T. Hara (2003). Characterization of NADP-isocitrate dehydrogenase expression in a carrot mutant cell line with enhanced citrate excretion. *Plant and Soil* 248:145-153.

Kihara, T., T. Wada, Y. Suzuki, T. Hara, and H. Koyama (2003). Alteration of citrate metabolism in cluster roots of white lupin. *Plant and Cell Physiology* 44:901-908.

Kihara, T., C-R. Zhao, Y. Kobayashi, E.Takita, T. Kawazu, and H. Koyama, (2006). Simple identification of transgenic *Arabidopsis* plants carrying a single copy of the integrated gene. *Bioscience Biotechnology and Biochemistry* 70:1780-1783.

Koyama, H., E. Takita, A. Kawamura, T. Hara, and D. Shibata (1999). Over expression of mitochondrial citrate synthase gene improves the growth of carrot cells in Al-phosphate medium. *Plant and Cell Physiology* 40:482-488.

Maurer, J.J., D. Schmidt, P. Petrosko, S. Sanchez, L. Bolton, and M.D. Lee (1999). Development of primers to O-antigen biosynthesis genes for specific detection of *Escherichia coli* O157 by PCR. *Applied and Environmental Microbiology* 65:2954-2960.

Ohno, T., S. Nakahira, Y. Suzuki, T. Kani, T. Hara, and H. Koyama (2004). Molecular characterization of plasma membrane $H^+$-ATPase in a carrot mutant cell line with enhanced citrate excretion. *Physiologia Plantarum* 122:265-274.

Song, P., C.Q. Cai, M. Skokut, B.D. Kosegi, and J.F. Petolino (2002). Quantitative real-time PCR as a screening tool for estimating transgene copy number in WHISKERS-derived transgenic maize. *Plant Cell Reports* 20:948-954.

Takita, E., H. Koyama, Y. Shirano, D. Shibata, and T. Hara (1999). Structure and expression of the mitochondrial citrate synthase gene in carrot cells utilizing Al-phosphate. *Soil Science and Plant Nutrition.* 45:197-205.

Chapter 31

# Gene Expression Analysis with DNA Array Technology in Legumes

Masao Watanabe
Makoto Endo
Hirokazu Hakozaki
Hiromi Masuko
Jong-In Park
Makiko Kawagishi-Kobayashi
Tohru Tsuchiya

## *INTRODUCTION*

The family Fabaceae is highly divergent at the species level and comprises several important species, which are used for food and forage. In addition, symbiosis with *Rhizobium* and other related bacteria for nitrogen fixation is an important characteristic in the legume root hairs (Brewin, 1991). Furthermore, Gregor Mendel used the garden pea (*Pisum sativum*) in breeding experiments in establishing the laws of inheritance (Mendel, 1866). To date, two of Mendel's genes have been isolated and characterized using molecular techniques (Bhattacharya et al., 1990; Lester et al., 1997). In addition to these genetic studies, legume plants have been used in many physiological studies on environmental responses, metabolism, develop-

This work was supported in part by a grant from the Ministry of Agriculture, Forestry and Fisheries of Japan (Rice Genome Project MA-2208 and MA-2211) to T. T. and M. W., respectively, by a Grant-in-Aid for the 21st Century Center of Excellence Program from the Ministry of Education, Culture, Sports, Science and Technology of Japan to M. W., and by a grant from the Promotion of Basic Research Activities for Innovative Biosciences from the BRAIN to M. K.-K., M. E., H. H., and J. I. P. who are recipients of Research Fellowships of the Japan Society for the Promotion of Science for Young Scientists.

ment, biotic interactions, and so on. Recently, two plant species, *Lotus japonicus* and *Medicago truncatula,* have emerged as model legumes, because of the advantageous features like small genome size, short generation time, and self-compatibility for whole-genome analysis (Handberg and Stougaard, 1992; Cook, 1999). Using these two legumes, several physiological phenomena have been dissected out with newer molecular technology tools (VandenBosch and Stacey, 2003).

Microarrays, which are made from cDNA or genomic DNA spotted onto glass slides at a high density, are widely recognized as a significant technological advance, which provide genome-scale information about gene expression patterns. Generally, microarrays are hybridized simultaneously with two sets of color fluorescently labeled cDNAs (Cy3 and Cy5 fluorescent dyes), which are prepared from RNA samples of different cells or organs treated with different stresses (Schena et al., 1995; Eisen and Brown, 1999; Schaffer et al., 2000). Furthermore, macroarrays, which are made from cDNA or genomic DNA spotted onto a nylon filter at a low density, have also been developed, and are widely used in analysis of gene expression (Baldwin et al., 1999). In plant sciences, this array technology has been applied to identify the genes specific to several physiological phenomena (Reymond et al., 2000; Schenk et al., 2000; Kawasaki et al., 2001; Seki et al., 2001). In a majority of cases involving microarrays, expressed sequence tags (ESTs) of cDNA clones have been spotted on the array. In some cases, genomic clones have also been spotted (Hanano et al., 2002). Genome sequence projects have begun in several plant species of Fabaceae (Sato et al., 2001).

This chapter reviews the advances in array experiments and related technology in legumes.

## *ANALYSIS OF EXPRESSED SEQUENCE TAGS AND DETERMINATION OF GENOME SEQUENCE*

Expressed sequence tags (ESTs) have provided a large volume of information on gene identification for genome research. A 300 to 500 bp sequence obtained from EST analysis is sufficient to identify an encoding gene by similarity analysis against the public databases. Both the expression level and the specificity of expression of genes can be estimated from the comparison of ESTs from different treatments. Furthermore, EST analysis is necessary for precise assignment of genes on a genomic sequence.

In the case of the model legume *L. japonicus,* more than 98,000 ESTs from several libraries (whole plants, flower buds) have been isolated and characterized (Asamizu et al., 2000, 2004; Endo et al., 2000; Endo, Hakozaki, et al., 2002; Endo, Matsubara, et al., 2002). These data are available on the Web site of Kazusa DNA Research Institute (http://www.kazusa.or.jp). In the other model legume, *M. truncatula,* more than 140,000 ESTs from 30 different cDNA libraries derived from various vegetative and reproductive tissues have also been isolated (Covitz et al., 1998; Journet et al., 2002). In the related species, *M. sativa,*, 400 ESTs have also been developed (Hays and Skinner, 2001). More than 250,000 ESTs from several libraries of soybean have been deposited into dbEST soybean (Shoemaker et al., 2002).

Determination of the genome sequence of *L. japonicus* has been initiated at Kazusa Research Institute and partial sequences are available on the Internet (Sato et al., 2001; Nakamura et al., 2002; http://www.kazusa.or.jp).

## *GENE EXPRESSION ANALYSIS WITH MICRO- AND MACROARRAYS*

The analysis of gene expression with microarray and macroarray was applied to several legume species. For identification of repetitive genomic sequences, a modified DNA microarray-based technique was used in *Vicia* species. From this experiment, rDNA genes, fragments of chloroplast genome, and retroelement sequences were identified and characterized (Nouzova et al., 2000).

To understand the molecular mechanisms intrinsic to reproductive organ development, a cDNA microarray fabricated from flower bud cDNA clones was used to isolated genes, which were specifically expressed during the development of anther and pistil in *L. japonicus*. The genes, which were abundantly expressed in anthers and pistil tissues, were defined as clones with a minimum of sixfold up-regulation in their transcript level against the control leaf tissues. After cluster analysis with microarray data, Endo and co-workers found 22 independent cDNA clones, which were specifically or predominantly expressed in immature anther. Similarly, 111 independent cDNA clones were identified to be specifically or predominantly expressed in mature anther; the authors identified clones homologous to genes related to cell wall reorganization (pectin methyl esterase, ascorbate oxidase, and β-galactosidase), cytoskeleton (actin and actin depolymerization factor), and sugar metabolism (sucrose transporter, glucose transporter, and cell wall invertase). Thus, cDNA microarray technology is a powerful tool for identification of novel reproductive organ-specific genes (Endo, Matsubara, et al., 2002). Furthermore, an *in situ* hybridization experiment was also performed in order to determine the spatial and temporal expression pattern of these specific genes. In the case of immature anther-specific clones, about half of them were specifically expressed in tapetum, which supplies nutrients to the developing microspores and precursors for the synthesis of pollen exine. In contrast, most of the mature anther-specific clones were specifically expressed in mature pollen grain (Figure 31.1; Hakozaki et al., 2004; Masuko et al., 2006). Similar research on anther development, pollination, pollen-tube elongation, fertilization, and embryogenesis was performed with microarrays of *japonica* rice (Endo et al., 2004; Yoshida et al., 2005). Somatic embryogenesis in tissue culture is a model phenomenon occurring in the ovule after fertilization. To dissect this phenomenon in soybean, one interesting culture system was established, in which somatic

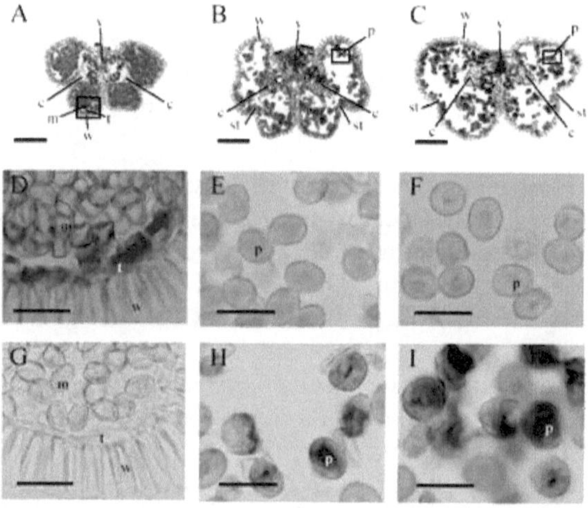


FIGURE 31.1. Cross-section of the anther at different developmental stages and *in situ* localization of anther-specific genes during anther development of *L. japonicus* is shown in (A) to (C). Each cross-section of anther was stained with toluidine blue. (A) Anther at stage 1 contained uninucleate microspores and tapetal cells. (B) Anther at stages 1-2 contained binucleate pollen grains and tapetal cells. Tapetal cells started degenerating in this stage. (C) Anther at stage 2 contained mature pollen grains. Tapetal cells disappeared during this stage just before dehiscence. c, connective; m, microspore; p, pollen grain; st, stomium; v, vascular bundle; w, anther wall. Dig-labeled antisense RNA probe was hybridized to a cross-section of the anther tissues at different developmental stages. A cross-section of the anther locule at stages 1, 1-2, and 2 are shown in (D) to (I). The gene-encoding lipid transfer protein (LjImfb-R39) was specifically expressed in immature anther tapetum (D), and not in mature pollen grains (E, F). In the case of gene-encoding late embryogenesis abundant protein (LjMfb-U92), a hybridization signal was specifically detected in pollen grains (H, I), but not in anther tapetum (G). Bar = 100 μm. (See also color gallery.)

embryos could be induced from the adaxial side of the immature cotyledon, but not from the abaxial side.

By the combination of this tissue culture system and cDNA microarray fabricated from 9,280 cDNA clones, gene expression was surveyed according to developmental stages of embryos. The transcripts related to cell proliferation indicated that cell division was induced early in both sides of the cotyledon and persisted until 28 days on the adaxial side. The genes related to a possible oxidative burst concomitant with cell division reached a peak at 14 days. Transcripts corresponding to storage proteins also accumulated in the developing somatic embryos along with the up-regulation of genes for gibberellic acid (GA) biosynthesis on the adaxial side of the cotyledon (Thibaud-Nissen et al., 2003).

In root hairs of legumes, symbiotic responses to *Rhizobium* and related bacteria for nitrogen fixation are important characteristics (Brewin, 1991). Since the *M. truncatula* EST database contains over 140,000 sequences from 30 different cDNA libraries, it could be used for identification of the tissue-specific genes *in silico*. In order to investigate the genes specific to nodule development and function in legumes, *in silico* identification of nodule-specific clones was first performed. From this analysis, 340 putative genes that expressed exclusively in root nodules were identified. Highly abundant nodule-specific genes from this analysis included clones encoding plantacyanin, agglutinin, embryo-specific protein, purine permease, calmodulin-like proteins, and small cysteine-rich proteins. Subsequent to this, the nodule-enhanced expression of the specific genes was confirmed experimentally by macroarray and RNA gel blot hybridization (Fedorova et al., 2002).

Plants have developed a multitude of inducible defense mechanisms against aggressive biotic and abiotic agents. They can emit specific blends of volatiles, which attract natural carnivorous enemies of herbivorous insect pests. A tritrophic system consisting of lima bean plants (*Phaseolus lunatus*), spider mites (*Tetranychus urticae*), and predatory mites (*Phytoseiulus persimilis*) has been well characterized (Takabayashi and Dickeb, 1996). In order to comprehensively understand the volatiles released from the neighboring leaves infected with spider mites at the gene expression level, a cDNA microarray constructed from 2,032 cDNA clones of leaves of lima bean was used. The expression of about 100 genes was enhanced by infestation with spider mites. These genes were related to such broad functions as responses to pathogenesis, wounding, hormones, ethylene biosynthesis, flavonoid biosynthesis, transcriptional modifications, translation, chaperones, secondary signaling messengers, membrane transport, protein degradation, and photosynthesis. These results indicate that herbivorous damage elicits dramatic changes in metabolism of lima bean leaves (Arimura et al., 2000).

Understanding plant-microbe interactions is important for breeding resistant lines and improving agricultural productivity (Jackson and Taylor, 1996). In order to understand plant-microbe interactions, a macroarray constructed from 135 EST clones was applied to soybean sudden death syndrome (SDS), which was caused by the fungal pathogen *Fusarium solani* f.sp. *glycines*. Plant defense-related genes encoding phospholipase D, chalcone synthase, *myo*-inositol 1-phosphate synthase, isoflavone *O*-methyl transferase, phenylalanine ammonia lyase, calmodulin-like protein, and cinnamic acid 4-hydroxylase were up-regulated in inoculated roots. In contrast, genes encoding vacuolar ATP synthase, ATP/ADP transporter, and alcohol dehydrogenase were down-regulated in inoculated roots, indicating a decrease in metabolic activity in infected roots. Taking these results together, the resistance of soybeans to SDS suggested the involvement of several metabolic pathways and responses in the process of resistance to fungal infection (Iqbal et al., 2002).

The Soybean Genomics and Microarray Database has been established, and it provides analytical tools for data mining (Alkharouf and Matthews, 2004). The *Medicago* Genome Initiative was also established (Bell et al., 2001). These databases are useful for surveying the genes specifically expressed in particular tissues or organs *in silico*.

## CONCLUSION

Array technology has been applied to several physiological phenomena with legume plant species. Furthermore, a related technology, EST and the genome projects have also been established in model legume and in other important crops as well. Array technology may be used in dissecting out different complicated phenomena. For example, genes involved in maturation of fruits of strawberry were identified using this technology (Aharoni et al., 2000). By combining other new technologies such as proteome, metabolome, interactome, and ionome, physiological phenomena should be characterized completely (Fiehn, 2002; Canovas et al., 2004).

In another model plant, *Arabidopsis thaliana*, gene expression profiles in several physiological conditions have been deposited in the public databases and they could be used to identify specific genes by *in silico* analysis (Finkelstein et al., 2002; Hilson et al., 2003). In the future, this *in silico* analysis could be performed in model legumes. Finally, mutant screens should be used in confirming the function of specific genes identified by the array technology.

## REFERENCES

Aharoni, A., L.C.P. Keizer, H.J. Bouwmeister, Z. Sun, M. Alvarez-Huerta, H.A. Verhoeven, J. Blaas, A.M.M.L. van Houwelingen, R.C.H. De Vos, H. van der Voet, et al. (2000). Identification of the *SAAT* gene involved in strawberry flavor biogenesis by use of DNA microarrays. *Plant Cell* 12:647-662.

Alkharouf, N.W. and B.F. Matthews (2004). SGMD: The soybean genomics and microarray database. *Nucleic Acids Research* 32:D398-D400.

Arimura, G., K. Tashiro, S. Kuhara, T. Nishioka, R. Ozawa, and J. Takabayashi (2000). Gene responses in bean leaves induced by herbivory and by herbivore-induced volatiles. *Biochemical and Biophysical Research Communications* 277: 305-310.

Asamizu, E., Y. Nakamura, S. Sato, and S. Tabata (2000). Generation of 7137 non-redundant expressed sequence tags from legume, *Lotus japonicus*. *DNA Research* 7:127-130.

Asamizu, E., Y. Nakamura, S. Sato, and S. Tabata (2004). Characteristics of the *Lotus japonicus* gene repertoire deduced from large-scale expressed sequence tag (EST) analysis. *Plant Molecular Biology* 54:405-414.

Baldwin, D., V. Crane, and D. Rice (1999). A comparison of gel-based, nylon filter and microarray techniques to detect differential RNA expression in plants. *Current Opinion in Plant Biology* 2:96-103.

Bell, C.J., R.A. Dixon, A.D. Farmer, R. Flores, J. Inman, R.A. Gonzales, M.J. Harrison, N.L. Paiva, A.D. Scott, J.W. Weller, et al. (2001). The *Medicago* Genome Initiative: A model legume database. *Nucleic Acids Research* 29:114-117.

Bhattacharyya, M.K., A.M. Smith, T.H.N. Ellis, C. Hedley, and C. Martin (1990). The wrinkled-seed character of pea described by Mendel is caused by a transposon-like insertion in a gene encoding starch-branching enzyme. *Cell* 60:115-122.

Brewin, N.J. (1991). Development of the legume root nodule. *Annual Review of Cell Biology* 7:191-226.

Canovas, F.M., E. Dumas-Gaudot, G. Recorbet, J. Jorrin, H.P. Mock, and M. Rossignol (2004). Plant proteome analysis. *Proteomics* 4:285-298.

Cook, D.R. (1999). *Medicago truncatula*: A model in the making! *Current Opinion of Plant Biology* 2:301-304.

Covitz, P.A., L.S. Smith, and S.R. Long (1998). Expressed sequence tags from a root-hair-enriched *Medicago truncatula* cDNA library. *Plant Physiology* 117: 1325-1332.

Eisen, M.B. and P.O. Brown (1999). DNA arrays for analysis of gene expression. *Methods in Enzymology* 303:179-205.

Endo, M., T. Kokubun, Y. Takahata, A. Higashitani, S. Tabata, and M. Watanabe (2000). Analysis of expressed sequence tags of flower buds in *Lotus japonicus*. *DNA Research* 7:213-216.

Endo, M., H. Hakozaki, T. Kokubun, H. Masuko, Y. Takahata, T. Tsuchiya, A. Higashitani, S. Tabata, and M. Watanabe (2002). Generation of 919 expressed sequence tags from immature flower buds and gene expression analysis using expressed sequence tags in the model plant *Lotus japonicus. Genes and Genetic Systems* 77:277-282.

Endo, M., H. Matsubara, T. Kokubun, H. Masuko, Y. Takahata, T. Tsuchiya, H. Fukuda, T. Demura, and M. Watanabe (2002). The advantages of cDNA microarray as an effective tool for identification of reproductive organ-specific genes in a model legume, *Lotus japonicus. FEBS Letters* 514:229-237.

Endo, M., T. Tsuchiya, H. Saito, H. Matsubara, H. Hakozaki, H. Masuko, M. Kamada, A. Higashitani, H. Takahashi, H. Fukuda, T. Demura, and M. Watanabe (2004). Identification and molecular characterization of novel anther-specific genes in *japonica* rice, *Oryza sativa* L. by using cDNA microarray. *Genes and Genetic Systems* 79:213-226.

Fedorova, M., J. van de Mortel, P.A. Matsumoto, J. Cho, C.D. Town, K.A. Vanden Bosch, J.S. Gantt, and C.P. Vance (2002). Genome-wide identification of nodule-specific transcripts in the model legume *Medicago truncatula. Plant Physiology* 130:519-537.

Fienh, O. (2002). Metabolomics-the link between genotypes and phenotypes. *Plant Molecular Biology* 48:155-171.

Finkelstein, D., R. Ewing, J. Gollub, F. Sterky, J.M. Cherry, and S. Somerville (2002). Microarray data quality analysis: Lessons from the AFGC project. Arabidopsis Functional Genomics Consortium. *Plant Molecular Biology* 48:119-131.

Hakozaki, H., M. Endo, H. Masuko, J.-I. Park, H. Ito, M. Uchida, M. Kamada, H. Takahashi, A. Higashitani, and M. Watanabe (2004). Cloning and expression pattern of a novel microspore-specific gene encoding hypersensitive-induced response protein (LjHIR1) from the model legume, *Lotus japonicus. Genes and Genetic Systems* 79:307-310.

Hanano, S., M. Amagai, T. Kaneko, C. Kuwata, T. Tsugane, N. Sakurai, Y. Nakamura, D. Shibata, and S. Tabata (2002). Analysis of gene expression in *Arabidopsis thaliana* by array hybridization with genomic fragments aligned along chromosomal regions. *Plant Journal* 30:247-255.

Handberg, K. and J. Stougaard (1992). *Lotus japonicus,* an autogamous, diploid legume species for classical and molecular genetics. *Plant Journal* 2:487-496.

Hays, D.B. and D.Z. Skinner (2001). Development of an expressed sequence tag (EST) library for *Medicago sativa. Plant Science* 161:517-526.

Hilson, P., I. Small, and M. Kuiper (2003). European consortia building integrated resources for *Arabidopsis* functional genomics. *Current Opinion of Plant Biology* 6:426-429.

Iqbal, M.J., S. Yaegashi, V.N. Njiti, R. Ahsan, K.L. Cryder, and D.A. Lightfoot (2002). Resistance locus pyramids alter transcript abundance in soybean roots inoculated with *Fusarium solani* f. sp. *glycines. Molecular Genetics and Genomics* 268:407-417.

Jackson, A.O. and C.B. Taylor (1996). Plant-microbe interactions: Life and death at the interface. *Plant Cell* 8:1651-1668.

Journet, E.-P., D. van Tuinen, J. Gouzy, H. Crespeau, V. Carreau, M.-J. Farmer, A. Niebel, T. Schiex, O. Jaillon, O. Chatagnier, et al. (2002). Exploring root symbiotic programs in the model legume *Medicago truncatula* using EST analysis. *Nucleic Acids Research* 30:5579-5592.

Kawasaki, S., C. Borchert, M. Deyholos, H. Wang, S. Brazille, K. Kawai, D. Galbraith, and H.J. Bohnert (2001). Gene expression profiles during the initial phase of salt stress in rice. *Plant Cell* 13:889-906.

Lester, D.R., J.J. Ross, P.J. Davies, and J.B. Reid (1997). Mendel's stem length gene (*Le*) encodes a gibberellin 3 β-hydroxylase. *Plant Cell* 9:1435-1443.

Masuko, H., M. Endo, H. Saito, H. Hakozaki, P.-J. Park, M. Kawagishi-Kobayashi, Y. Takada, M. Kamada, H. Takahashi, A. Higashitani, et al. (2006). Anther-specific genes, which expressed through microsporogenesis, are temporally and spatially regulated in model legume, *Lotus japonicus. Genes and Genetic Systems* 81:57-62.

Mendel, G.J. (1866). Versuche uber Pflanzen-Hybriden. *Verh. Naturforsch. Ver. Brunn.* 4:3-47.

Nakamura, Y., T. Kaneko, E. Asamizu, T. Kato, S. Sato, and S. Tabata (2002). Structural analysis of a *Lotus japonicus* genome. II. Sequence features and mapping of sixty-five TAC clones which cover the 6.5-Mb regions of the genome. *DNA Research* 9:63-70.

Nouzava, M., P. Neumann, A. Navratilova, D.W. Galbraith, and J. Macas (2000). Microarray-based survey of repetitive genomic sequences in *Vicia* spp. *Plant Molecular Biology* 45:229-244.

Reymond, P., H. Weber, M. Damond, and E.E. Farmer (2000). Differential gene expression in response to mechanical wounding and insect feeding in *Arabidopsis. Plant Cell* 12:707-720.

Sato, S., T. Kaneko, Y. Nakamura, E. Asamizu, T. Kato, and S. Tabata (2001). Structural analysis of a *Lotus japonicus* genome. I. Sequence features and mapping of fifty-six TAC clones which cover the 5.4 Mb region of the genome. *DNA Research* 8:311-318.

Schaffer, R., J. Landgraf, M. Perez-Amador, and E. Wisman (2000). Monitoring genome-wide expression in plants. *Current Opinion in Biotechnology* 11:162-167.

Schena, M., D. Shalon, R.W. Davis, and P.O. Brown (1995). Quantitative monitoring of gene expression patterns with a complimentary DNA microarray. *Science* 270:467-470.

Schenk, P.M., K. Kazan, I. Wilson, J.P. Anderson, T. Richmond, S.C. Somerville, and J.M. Manners (2000). Coordinated plant defense responses in *Arabidopsis* revealed by microarray analysis. *Proceedings of the National Academy of Sciences USA* 97:11655-11660.

Seki, M., M. Narusaka, H. Abe, M. Kasuga, K. Yamaguchi-Shinozaki, P. Carninci, Y. Hayashizaki, and K. Shinozaki (2001). Monitoring the expression pattern of 1300 *Arabidopsis* genes under drought and cold stresses by using a full-length cDNA microarray. *Plant Cell* 13:61-72.

Shoemaker, R., P. Keim, L. Vodkin, E. Retzel, S.W. Clifton, R. Waterston, D. Smoller, V. Coryell, A. Khanna, J. Erpelding, et al. (2002). A compilation of soybean ESTs: Generation and analysis. *Genome* 45:329-338.

Takabayashi, J. and M. Dickeb (1996). Plant-carnivore mutualism through herbivore-induced carnivore attractants. *Trends in Plant Science* 1:109-113.

Thibaud-Nissen, F., T. Shealy, A. Khanna, and L.O. Vodkin (2003). Clustering of microarray data reveals transcript patterns associated with somatic embryogenesis in soybean. *Plant Physiology* 132:118-136.

VandenBosch, K.A. and G. Stacey (2003). Summaries of legume genomics projects from around the globe. Community resources for crops and models. *Plant Physiology* 131:840-865.

Yoshida, K.T., M. Endo, M. Nakazono, H. Fukuda, T. Demura, T. Tsuchiya, and M. Watanabe (2005). cDNA microarray analysis of gene expression changes during pollination, pollen-tube elongation, fertilization, and early embryogenesis in rice pistils. *Sexual Plant Reproduction* 17:269-275.

Chapter 32

# Advent of Diverse DNA Markers to Decipher Genome Sequence Polymorphism

Deepmala Sehgal
Vishnu Bhat
Soom Nath Raina

## *INTRODUCTION*

Over time, genetic markers have evolved from morphological markers to isozyme markers to the most advanced DNA markers. The DNA-based markers, which screen both nuclear and organelle genomes with precision, have acted as versatile tools for fast and unambiguous genetic analysis of plant species for a multitude of benefits, some of which were impossible to comprehend before the advent of the aforementioned technology. They are developmentally stable, detectable in all tissues, unaffected by environmental conditions, insensitive to epistatic or pleiotropic effects, and provide a choice of codominant or dominant markers. With the development of huge sequence databases (NCBI, EMBL, and DDBJ), the use of sequence-targeted functional markers has gained tremendous importance over random DNA markers owing to their linkage with trait locus alleles. This review provides a consolidated account of the methodologies and basic principles underlying different hybridization-based and polymerase chain reaction (PCR) based DNA markers reported so far.

The first author (D.S.) is grateful to the Council of Scientific and Industrial Research, Government of India, for financial support. This work was supported, in part, by a grant-in-aid for scientific research from Defense Research Development Organization and Department of Biotechnology, Government of India.

## *NON-PCR-BASED MARKERS*

### *Restriction Fragment Length Polymorphism (RFLP)*

The discovery and isolation of restriction endonucleases in the 1960s set the stage for the detection of polymorphism directly at the DNA sequence level. This approach, called RFLP (Botstein et al., 1980), involves digestion of total genomic DNA with a restriction endonuclease, separation of the resultant DNA fragments by gel electrophoresis, blotting the fragments onto a membrane, and hybridizing the membrane with the probe of choice. The probes could be low copy sequences derived from the genomic DNA clones, specific gene probes of nuclear, mitochondrial, and chloroplast genome origin, and characterized or uncharacterized cDNAs. The usage of microsatellites (oligofingerprinting) and multigene families as probes is most common in deciphering polymorphism in nuclear genomes. Codominant RFLP markers are highly reproducible. The low molecular weight of chloroplast genomes (120-217 kb) enables analytical visualization of restriction site variation directly from the gel.

### *Restriction Landmark Genomic Scanning (RLGS)*

The method uses restriction sites as landmarks (Hatada et al., 1991). Genomic DNA is radioactively labeled at cleavage sites specific for a rare cleaving restriction enzyme before being size fractionated in one dimension. The fractionated DNA is further digested with another more frequently cutting enzyme and separated in a second dimension. This gives a two-dimensional pattern with thousands of scattered spots corresponding to sites for the first enzyme. The position and intensity of a spot reflect its locus and the copy number of the corresponding restriction site.

## *PCR-BASED MARKERS*

PCR-based markers can be subdivided into two categories based on the type of primer used.

### *Multiple Arbitrary Amplicon Profiling (MAAP) Markers*

In this category, many amplicons are generated with the use of arbitrary sequence primers. Depending on the length of the primer, amplification stringency, and the procedure used, the MAAP markers may be categorized as follows.

### *Random Amplified Polymorphic DNA (RAPD)*

RAPD uses primers that are 10 nucleotides long, and amplification products are usually resolved on agarose gels (Williams et al., 1990). Visualization of the bands by polyacrylamide gel electrophoresis (PAGE) or denaturing gradient gel electrophoresis (DGGE) is not uncommon. The RAPD technique generates dominant markers. The results may not always be reproducible (Jones et al., 1997).

### *Universally Primed PCR (UP-PCR)*

An analogue of RAPD, UP-PCR has been widely used to reveal polymorphism in fungi and bacterial species (Bulat et al., 1996). The technique consists of DNA amplification with a single universal random primer that is 16 to 20 nucleotides long. In comparison to RAPD (34°-38°C), the annealing temperature is higher (55°C) in this marker system. The universal primers that have been utilized are 45, 3-2, 0.3.1, and AS15, whose sequences are described elsewhere (Bulat et al., 1996).

### *Arbitrarily Primed PCR (AP-PCR)*

Random primers 10 to 15 nucleotides in length are used to amplify target DNA under low stringency conditions for two amplification cycles (Welsh and McClelland, 1990). Increased stringency of annealing at later cycles of amplification generates reproducible products, which are resolvable by PAGE and detected by autoradiography.

### *DNA Amplification Fingerprinting (DAF)*

Numerous amplicons are generated using arbitrary primers merely five-nucleotides in length (Caetano-Anollés et al., 1991). The amplified fragments are resolved with PAGE only.

### *Randomly Amplified DNA Fingerprinting (RAF)*

Among MAAP markers, RAF is the most effective technique developed by Waldron et al. (2002). RAF is a modification of DAF, which utilizes 10 mer primers in the hot start (85°C) PCR followed by touchdown PCR in subsequent cycles. The radiolabeled PCR products are separated on PAGE and detected by autoradiography. Use of fluorescein-tagged oligonucleotide (fluorescent RAF) can be an alternative to radiolabeling. The major advantage of RAF over other markers described above is that it has the highest marker index, equivalent to that of the AFLP marker system.

### *Direct Amplification of Length Polymorphism (DALP)*

The technique developed by Desmarais et al. (1998) has the advantage of a high-resolution fingerprint technique with the possibility of characterizing the polymorphism more effectively. The technique uses modified AP-PCR to produce genomic fingerprints and enables sequencing of DNA polymorphisms in virtually any species. Two primers are utilized; the selective primer is composed of universal M13 sequencing primer as the core sequence followed by an arbitrary sequence; the reverse primer is universal M13 sequencing primer alone. Such pairs are efficient in producing multiband patterns where interindividual length variations, including microsatellite polymorphism, can be detected. The amplification products are detected on denaturing polyacrylamide gels. Because of the special design of the selective primers, each band of different patterns can be sequenced with the universal sequencing primers, no matter which selective primer was used in AP-PCR.

### *Multiple Amplicon Profiling (MAP) Markers*

Barring a few, markers in this class generate multiple amplification products either by employing semispecific primers or sequence-specific primers, thus targeting partially or completely known genomic regions, respectively.

### *MAP Markers Employing Semispecific Primers*

In this category, markers do not require prior sequence information for their development but still target partially known genomic regions. The markers, such as AFLP, SRAP, and microsatellite-based markers, target restriction sites, introns and exons, and microsatellite regions, respectively.

### *Amplified Fragment Length Polymorphism (AFLP)*

The AFLP technique (Vos et al., 1995) combines the efficiency of PCR with the specificity and reliability of RFLP. It involves total genomic DNA digestion with restriction endonucleases followed by ligation of double-stranded DNA adaptors to the digested genomic DNA. Pairs of oligonucleotide primers complementary to the adaptor sequences and with one to four additional 3' nucleotides are utilized for PCR. For large and complex genomes, two PCR reactions are performed utilizing primers that are composed of the AFLP adaptor sequences plus one additional 3' nucleotide in the preselective amplification and three or four additional 3' nucleotides in selective amplification. The products are separated on large polyacrylamide sequencing gel. Labeling of the primer can be performed using either radioactive or fluorescent tags. Visualization of bands can be achieved by silver staining of gels as well. By use of only a single AFLP primer combination, up to 100 bands can be amplified in a single experiment. The polymorphism in the DNA fingerprints arises from alteration in the DNA sequences including mutations abolishing or creating a restriction site and insertions, deletions, or inversions between two restriction sites (Rani and Raina, 2003).

### *Infrequent Restriction Site-PCR (IRS-PCR)*

This AFLP-based technique is widely used for bacterial typing (Handley and Regnery, 2000).

### *Sequence-Related Amplification Polymorphism (SRAP)*

Li and Quiros (2001) developed this technique for amplification of open reading frames (ORFs) using pairs of primers with AT- and GC-rich cores. The forward and reverse primers are 17 or 18 nucleotides long. The core sequence or filler sequence is 13 to 14 bases long in both the primers, followed by sequence CCGG in the forward primer and AATT in the reverse primer. The core is followed by three selective nucleotides at the 3' end. The purpose of using CCGG and AATT sequences in the core sequence of forward and reverse primer, respectively, is to target exons and introns (exons are normally rich in GC-rich regions while introns and promoters are rich in AT-rich regions). The amplified DNA fragments are separated on denaturing acrylamide gels and detected by autoradiography. The technique combines simplicity, reliability, and moderate throughput ratio and allows sequencing of selected bands. It targets coding sequences in the genome and results in a moderate number of codominant markers. SRAP results from either fragment size changes due to insertions and deletions leading to codominant markers or nucleotide changes leading to dominant markers.

## *MICROSATELLITE-BASED MARKERS*

### *Microsatellite-Primed PCR (MP-PCR)*

A specific simple sequence repeat (SSR) motif, usually tri-, tetra-, and pentanucleotide repeat, is used as a primer for amplification of sequence between two inversely oriented microsatellites (Meyer et al., 1993). The technique has also been termed single primer amplification reaction by Gupta et al. (1994).

### *Inter-Simple Sequence Repeat (ISSR)*

The technique is similar to MP-PCR but utilizes anchored (5' or 3' anchored) SSR primers (Zietkiewicz et al., 1994). Different scientists have variously called this technique anchored microsatellite primed PCR (AMP-PCR), inter-microsatellite PCR (IM-PCR), inter-SSR amplification (ISA), or anchored simple sequence repeats (ASSR) (Gupta and Varshney, 2000).

### *Random Amplified Microsatellite Polymorphisms (RAMPs)*

Amplification is performed using two primers, a 5'-anchored SSR primer and a RAPD primer (Wu et al., 1994). Amplified products resolve length differences due to the SSR target itself or the sequence between two primers. The digestion of amplification products with restriction enzymes produces digested RAMPs (dRAMPs; Becker and Heun, 1995).

### *Random Amplified Microsatellite Polymorphism (RAMPO)*

Genomic DNA is amplified with a single arbitrary primer (usually 10-mer as in RAPD). The gel containing separated products is blotted. The membrane is subsequently hybridized with microsatellite probes to produce fingerprints that are microsatellite based (Richardson et al., 1995). Cifarelli et al. (1995) called this technique random amplified hybridization microsatellites (RAHM) while Ender et al. (1996) termed it randomly amplified microsatellite sequences (RAMS).

The annealing temperature for all the microsatellite-based markers can be calculated using Thein and Wallace's rule (1986). According to this rule, 2°C and 4°C are to be added for the addition of either A or T and G or C in the primer sequence, respectively.

### *Selective Amplification of Microsatellite Polymorphic Loci (SAMPL)*

The SAMPL technique, introduced by Morgante and Vogel (1994), is a microsatellite-based modification of AFLP. It involves construction of a preamplified library in the same way as in AFLP. Selective amplification, however, utilizes an AFLP primer in combination with a microsatellite-based SAMPL primer. Since one of the primers utilized in this technique involves a microsatellite primer and the steps involved are the same as in AFLP, the technique is also referred to as microsatellite fragment length polymorphism (MFLP). The SAMPL primer commonly utilized is a compound SSR, 18 to 20 nucleotides in length. Primers with a single SSR anchored at the 5' end with a nonmicrosatellite sequence have allowed the amplification of different kinds of tri-, tetra-, and pentanucleotide repeats in addition to compound microsatellites (Rakoczy-Trojanowska and Bolibok, 2004).

### *MAP Markers Employing Sequence-Specific Primers*

Enormous amounts of sequence data available in the GenBank databases (NCBI, EMBL, and DDBJ) have led to the generation of a large number of sequence-targeted markers hitting both coding and noncoding regions of the genome. The markers included in this class require prior sequence information.

## *EXPRESSED SEQUENCE TAG (EST)-BASED MARKERS*

ESTs are typically unedited, single-read sequences produced from cDNAs. Currently, they are the most widely sequenced nucleotide commodity from plant genomes in terms of the number of sequences and the total nucleotide count. Swelling EST databases archive all the available ESTs and provide methods to search for individual sequences on the basis of species, clone, or homology attributes. The molecular markers that have been developed by exploiting EST databases include EST-SSRs, EST-SNPs, ESTP, COS, and TRAP.

### *Expressed Sequence Tag–Simple Sequence Repeat (EST- SSR)*

With the availability of large numbers of expressed sequence tags, development of SSR markers through data mining has become a fast, efficient, and low-cost option. Since EST-SSR markers are derived from transcribed regions of the genome, the markers are more conserved and have a higher rate of transferability than the genomic SSRs. A modest 1 to 5 percent of the ESTs in various plant species have been found to have SSRs of suitable length for marker development (Eujayl et al., 2004). EST-SSRs also have a high probability of being associated with gene expression and gene function and thus are capable of assaying polymorphism carried by coding regions of the genome.

### *Expressed Sequence Tag—Single Nucleotide Polymorphism (EST-SNP)*

An SNP marker is just a single base change in a DNA sequence, with a usual alternative of two possible nucleotides at a given position. SNPs are widely distributed throughout the genomes. Within the coding regions, an SNP is either nonsynonymous and results in an amino acid sequence change, or it is synonymous and does not alter the amino acid sequence (neutral SNP; Soleimani et al., 2003). Development of SNP markers from EST databases relies upon the underlying redundancy within EST collections. SNPs can be searched, either in the individual specific ESTs from different accessions of the same species or in the contigs. In the former case, EST sequences derived from different genotypes can be aligned for the identification of SNPs. Cleaved amplified polymorphic sequence (CAPS) analysis is the most widely used approach for the detection of single nucleotide polymorphisms. However, this technique is limited to mutations, which create or disrupt a restriction enzyme recognition site. A modification of CAPS, termed derived CAPS (dCAPS), eliminates the need for SNP to fall within a restriction site. Rather, a restriction enzyme recognition site, which includes the SNP, is introduced into the PCR product by a primer containing one or more mismatches to template DNA. The PCR product thus modified is subjected to restriction digestion, and the presence or absence of SNP is determined by the restriction pattern (Neff et al., 1998). A host of restriction enzyme-independent SNP assays have been developed such as allele- specific PCR, SSCP, primer-directed nucleotide incorporation assays, dideoxy fingerprinting, and oligonucleotide fluorescence-quenching assays. To date, a large number of SNP mining tools are available to automate the process of SNP discovery.

### *Expressed Sequence Tag Polymorphism (ESTP)*

The technique involves designing primers separated by an amplifiable EST segment, and using these primers for PCR amplification of genomic DNA. Detection of length and sequence polymorphism is done by agarose gel electrophoresis, and DGGE or SSCP, respectively (Gupta and Rustogi, 2004).

### *Conserved Orthologue Set (COS)*

COS markers or anchored reference loci are gene sequence markers that anchor genes to common syntenic islands between related genomes (Fulton et al., 2002). COS markers require sequencing from distinct but somehow related genomes (Rudd, 2003). A conserved orthologue set of markers, approximately 1,025 in number, developed by comparison of the *Arabidopsis* genomic sequence with an EST database of tomato is shown to be conserved over a wide range of dicotyledonous plants. These markers are highly useful in comparative mapping studies.

### *Target Region Amplification Polymorphism (TRAP)*

The technique developed by Hu and Vick (2003) uses two primers of 18 nucleotides in length to generate polymorphic markers around targeted genes. One of the primers, the fixed primer, is designed from the targeted EST sequence in the database (partial sequence of a candidate gene); the second primer is an arbitrary sequence with either an AT- or GC-rich core to anneal with an intron or exon, respectively. For different plant species, each PCR reaction can generate as many as 50 scorable fragments on a polyacrylamide sequencing gel. The technique is useful in tagging genes governing desirable agronomic traits of crop plants. Like RAPDs, TRAP markers are quick and easy to set up, and each reaction can produce a profile comparable to AFLP or RAF marker systems.

## *RETROTRANSPOSON/MINIATURE-INVERTED REPEAT TRANSPOSABLE ELEMENT (MITE)-BASED MARKERS*

Transposons are mobile segments of DNA that can move to different positions in the genome of a single cell. These mobile segments of DNA are of three distinct types: class I transposon or retrotransposon, class II transposon, and class III transposon or MITE. More recently, different marker techniques based on transposable elements (primarily retrotransposons and MITEs) have started emerging.

Retrotransposons are major dispersed components of most eukaryotic genomes. They replicate by a cycle of transcription, reverse transcription, and integration of new copies, without excising from the genome in the process. They fall into two groups: elements containing long terminal repeats (LTRs; LTR retrotransposons) and those lacking LTRs. Because they represent a major share of the genome, cause easily detectable genetic changes having known ancestral and derived states, and contain conserved regions for which PCR primers may be designed, retrotransposons (primarily LTR retrotransposons) have been exploited to produce powerful DNA marker systems (Schulman et al., 2004). MITEs are similar to class II transposons in mechanism of DNA transposition and presence of terminal inverted repeats (~15 bp) but are smaller in size (< 500 bp) and have a target site preference for either TAA or TA (Chang et al., 2001). MITEs insert near or within genes and are present in moderate to high copy numbers in many plant genomes and thus have been exploited to generate DNA markers.

### *Sequence-Specific Amplified Polymorphism (SSAP)*

The first retrotransposon-based method to be developed (Waugh et al., 1997) is a modification of AFLP. The amplified products are between a retrotransposon integration site and a restriction site. Restriction digestion of genomic DNA, adaptor ligation, and preamplification PCR are steps similar to AFLP. Two primers are used in selective amplification; one of them is a normal AFLP primer and the other is an LTR primer based on the sequence of LTR or sequence of polypurine tract (if LTR is very small; Schulman et al., 2004) region adjacent to LTR.

### *Inter-Retrotransposon Amplified Polymorphism (IRAP)*

The technique detects two retrotransposons or LTRs sufficiently close to one another in the genome to permit PCR amplification of the intervening region utilizing outward-facing LTR primers (Kalendar et al., 1999). The amplification products are resolved generally by agarose gel electrophoresis. In case of labeled primers, a sequencing gel system may be employed.

### *Retrotransposon Microsatellite Amplified Polymorphism (REMAP)*

This method uses an outward-facing LTR primer in conjunction with an SSR primer containing a set of repeats and one or more nonrepeat bases at the 3' end to serve as an anchor. The primer allows amplification if retrotransposon or LTR derivatives are sufficiently close to SSR (Kalendar et al., 1999).

### *Exon Retrotransposon Amplification Polymorphism (ERAP)*

This is similar to REMAP except that it utilizes an LTR primer in conjunction with a gene-specific primer. The retrotransposons sufficiently close to gene regions are thus amplified.

### *Retrotransposon Based Insertion Polymorphism (RBIP)*

This is a single-locus, codominant marker technique, that detects retrotransposon insertions using primers flanking the insertion site and primers from the insertion (LTR) itself (Flavell et al., 1998). The size of the PCR product indicates which allele (occupied by retrotransposon or unoccupied) has been amplified. Because retrotransposon insertions are thousands of bases in length, the unoccupied-site PCR (complementary PCR reaction using host-specific primers) produces no product from an occupied site.

### *Inter-MITE Polymorphism (IMP)*

The technique developed by Chang et al. (2001) involves amplification between two adjacent MITEs. To generate IMPs, primers are designed from consensus sequences of terminal inverted repeats of MITEs. Many families of MITEs such as *Stowaway, Tourist,* and *Barfly* have generally been utilized to design outwardly directed MITE primers. A single primer or a combination of different MITE primers can be used. PCR amplification is done in two steps. The first step is preamplification with nonlabeled primers. An aliquot of the preamplification product is used for the second step, amplification with labeled primers. IMPs can be detected using both agarose and fluorescence-based detection approaches.

### *Transposon Display (TD)*

The technique introduced by Van den Broeck et al. (1998) is also known as MITE-transposon display or MITE-AFLP if MITEs are utilized to design primers. The technique is a modification of AFLP. Unlike AFLP, restriction digestion is performed with only one frequent cutter enzyme (generally *Mse*I) followed by ligation of adaptors based on the sequence of the frequent cutter. PCR amplification is performed with a primer anchored in a transposon (MITE in case of MITE-TD or MITE-AFLP) and primer based on the sequence of frequent cutter enzyme used.

## MICROSATELLITE-AND MINISATELLITE-BASED MARKERS

Micro- and minisatellites are classes of repetitive DNA sequences present in all organisms, both prokaryotes and eukaryotes. They consist of tandemly arranged repeats of several nucleotides (microsatellites, 1-5 bp; and minisatellites, 10-60 bp) that vary in number of repeat units between genotypes and thus are referred to as variable number of tandem repeats (VNTRs) or hypervariable regions (HVRs; Somers et al., 1996). Both micro- and minisatellites have been exploited to generate molecular markers given below.

### Simple Sequence Repeats

Development of SSRs or microsatellites is a technically demanding, expensive, and time-consuming procedure that involves the creation of a small insert genomic library, library screening by hybridization with microsatellite probes, DNA sequencing of positive clones and primer designing. A new technology named sequence tagged microsatellite profiling (STMP) has been developed (Hayden and Sharp, 2001) to rapidly generate large numbers of SSR markers from genomic or cDNA. This eliminates the need for library screening. STMP generates short but characteristic nucleotide sequence tags for fragments present within a pool of SSR amplicons. These tags are then ligated together to form concatemers for cloning and sequencing. The sequence tags can be used as primers to directly amplify corresponding SSR loci from genomic DNA, thereby reducing the cost of developing a microsatellite marker. Primer sequences can also be accessed through public databases. All the nucleotide sequences from databases (NCBI, http://www.ncbi.nlm.nih.gov, EMBL, http://www.embl-heidelberg.de) can be screened using BLASTX, BLASTN, and TBLASTX (homology search programs) and downloaded in FASTA format. The repeat patterns in the sequences can be identified using software such as SSRIT (SSR Identification Tool, http://www.gramene.org/gramene/searches/ssrtool) or Tandem Repeat Finder Program (http://c3.biomath.mssm.edu/trf.html). Primers can then be designed using software such as Primer 3 (http://frodo.wi.mit.edu/cgi-bin/primer3/primer3-www.cgi) or DNA Star Primer Select (DNAStar, Madison, WI). Primer pairs so designed can be used for amplification of SSR alleles. The polymorphism results from copy number variation at a corresponding SSR locus. The results can be visualized on agarose or metaphor gels.

### *Sequence Tagged Microsatellite Sites (STMS)*

The STMS technique is similar to SSRs with regard to isolation of microsatellite sequences from the genome but differs with regard to two primers that are designed from sequences flanking a specific locus. The STMS markers reveal polymorphisms due to variations in the length of microsatellite at specific individual loci. Thus, STMS are locus-specific and codominant markers. Both denaturing and nondenaturing PAGE can be used to resolve allele size differences.

### *Directed Amplification of Minisatellite DNA PCR (DAMD-PCR)*

The technique developed by Heath et al. (1993) requires sequence information from a highly conserved core sequence from minisatellite DNA for designing the primers. It is based on the principle that events such as local inversions or other chromosomal rearrangements move repeats so that single-copy DNA is present between them, and then PCR using a single VNTR core primer may amplify sequences historically adjacent to highly variable loci (Heath et al., 1993).

### *Intron-Exon Splice Junction Markers*

Intron-exon splice junctions are highly conserved. Consensus sequences of the junctions in plants (Brown, 1986) can be exploited for construction of intron splice junction (ISJ) primers. Since introns are subjected to weak selective pressure in comparison to exons, they are highly variable in sequence and length (Weining and Langridge, 1991), producing highly variable amplification patterns. Predigestion of the template DNA with a restriction endonuclease is required in some cases when ISJ primers are used alone or in conjunction with gene primers to reduce the complexity of bands. Also, predigestion may allow improved denaturation of template DNA fragments as a result of reduced length and better access for the *Taq* polymerase to the DNA.

### *Markers Based on Selective Genic/Intergenic Amplifications*

The structural organization of many genes, such as ribosomal RNA gene (18S-5.8S-26S) clusters originating from nuclear as well as organelle genomes, genes meant for synthesis of large subunits of rbcL and placed in chloroplast genomes, gene clusters meant for the synthesis of 5S ribosomal RNA, genes meant for synthesis of ATP subunits placed in mitochondrial genomes, maturase K genes placed in chloroplast genomes, and so on, has led to the design of many universal primers. Most of the genes listed above, more specifically rDNA gene families (18S-5.8S-26S, 45S, and 5S rRNA), comprise conservative (transcribed) and nonconservative (partly transcribed) regions. The latter region shows the degree of variation sometimes most suitable for constructing phylogenies. These genes, therefore, have been very widely used in genetic analysis. The methodology is broadly concerned with amplification using universal primers and cloning followed by sequencing of the cloned fragments. In case of longer PCR products, PCR is sometimes followed by restriction analysis. This approach, called PCR-RFLP or cleaved amplified polymorphic sequences (CAPS), has been widely used for reconstructing phylogenies at lower taxonomic levels (Demesure et al., 1995; Duminil et al., 2002). PCR-RFLP provides a highly effective method for screening organelle genome sequence variations. Sequences with relatively rapid evolutionary rates such as noncoding intergenic spacer regions have proven to be most appropriate for analyzing relationships between species within a genus. Simplicity and reproducibility provide this approach with an edge over conventional RFLP.

Numerous genes for resistance or R genes involved in gene-for-gene resistance have been cloned from a variety of plant species. The cloned R genes fall into several classes defined by the shared structural motifs. The largest class is defined by the presence of a central nucleotide-binding site (NBS) and C-terminal leucine-rich repeats (LRRs). Conserved sequences of the NBS domain of R genes have been used to design primers for resistance gene analog polymorphism (RGAP) or nucleotide binding site (NBS) profiling to sample genetic variation in and around resistance genes. Various resistance gene analogs (RGAs) have also been cloned using these primers. Designing of primers is important for amplification of TIR (toll and interleukin receptor) and non-TIRNBS-LRR families of R genes from the genome (Zhu et al., 2002).

## *OTHER MARKERS*

### *Sequence Characterized Amplified Regions (SCARs)*

Michelmore et al. (1991) introduced this technique, wherein RAPD marker termini are sequenced and longer primers (22-24 nucleotides) are designed for specific amplification of a particular locus. The technique involves purification of a specific RAPD amplification product, ligation of the purified DNA fragment to a linearized plasmid vector, transformation of bacterial cells with the recombinant DNA molecule, screening of bacterial colonies, sequence analysis of the cloned DNA fragment, and design of pairs of primers. Initially, only RAPDs have been used to obtain SCARs but polymorphic bands in AFLP, DAF, and SAMPL can also be converted to locus-specific SCARs (Gupta and Varshney, 2000). SCARs are usually dominant markers but can be converted into codominant markers by digesting them with four cutter restriction enzymes, and polymorphism can be deduced by either denaturing gradient gel electrophoresis (DGGE) or single-stranded conformational polymorphism (SSCP). Compared to arbitrary primers, SCARs exhibit several advantages in mapping studies and map-based cloning.

### *Sequence-Tagged Sites (STS)*

STSs are similar to SCARs in construction but primers are designed to match the nucleotide sequence of an RFLP probe. Using this technique, tedious hybridization procedures involved in RFLP analysis can be bypassed.

### *Allele-Specific Associated Primers (ASAP)*

The PCR primers are designed from the sequence of specific alleles, either in homozygous or heterozygous states in an RFLP profile. These markers tag specific alleles in the genome.

In addition to these, there are some markers that do not require prior sequence information but demand infrastructure and are expensive.

### *cDNA-AFLP*

This technique involves cDNA instead of genomic DNA. The technique was developed by Bachem et al. (1996) and involves four steps: 1) synthesis of cDNA using a poly-dT oligonucleotide, (2) production of primary template by restriction digestion with two restriction enzymes and ligation of anchors to their termini, (3) pre-amplification with primers corresponding to anchors from a secondary template, and (4) selective restriction fragment amplification with primers extended with one or more specific bases. The bands present in cDNA-AFLP profiles are called transcript-derived fragments (TDFs). Unique TDFs identified in cDNA-AFLP profiles can be excised, amplified by PCR, and cloned in an appropriate cloning vector and sequenced. Both single-gene approaches such as northern blot analysis and multiple-gene approaches such as microarrays can be utilized to unravel the function of a gene.

### *Diversity Array Technology (DArT)*

The DArT genotyping method, introduced by Jaccoud et al. (2001), involves a new use of microarrays that does not require sequence knowledge. Restriction fragments representing the diversity of a gene pool are cloned. Polymorphic clones in the library are identified by arraying inserts (six replicates per fragment) from a random set of clones and hybridizing the array to different samples. The inserts from polymorphic clones are immobilized on a chip. The "diversity panel" thus created is used for genetic fingerprinting of any organism or a group of organisms belonging to the gene pool from which the panel was developed. The sample to be genotyped is fluorescently labeled and hybridized against the array. For each array spot, the amount of hybridization signal is measured and the presence or absence of the corresponding fragment is scored.

## REFERENCES

Bachem, C.W.B., R.S. van der Hoeven, S.M. de Bruijn, D. Vreugdenhil, M. Zabeau, and R.G.F. Visser (1996). Visualization of differential gene expression using a novel method of RNA fingerprinting based on AFLP: Analysis of gene expression during potato tuber development. *Plant Journal* 9:745-753.

Becker, J. and M. Heun (1995). Mapping of digested and undigested random amplified microsatellite polymorphisms in barley. *Genome* 38:991-998.

Botstein, D., R.L. White, M. Skolnick, and R.W. Davis (1980). Construction of a genetic linkage map in man using restriction fragment length polymorphisms. *American Journal of Human Genetics* 32:314-331.

Brown, J.W. (1986). A catalogue of splice junction and putative branch point sequences from plant introns. *Nucleic Acids Research* 14:9549-9559.

Bulat, S.A., N.V. Mironenko, M.N. Lapteva, and P.P. Strelchenko (1996). Polymerase chain reaction with universal primers (UP PCR) and its applications to plant genome analysis. In *Conservation of plant genes II: Utilization of ancient and modern DNA,* R.P. Adams, J.S. Miller, E.M. Goldenberg, and J.E. Adams (eds.). vol. 48, St. Louis: Missouri Botanical Garden, pp. 113-129.

Caetano-Anollés, G., B.J. Bassam, and P.M. Gresshoff (1991). DNA amplification fingerprinting: A strategy for genome analysis. *Plant Molecular Biology Reporter* 9:292-305.

Chang, R.Y., L.S. O'Donoughue, and T.E. Bureau (2001). Inter-MITE polymorphism (IMP): A high throughput transposon-based genome mapping and finger printing approach. *Theoretical and Applied Genetics* 102:773-781.

Cifarelli, R.A., M. Gallitelli, and F. Celliini (1995). Random amplified hybridization microsatellites (RAHM): Isolation of a new class of microsatellite-containing DNA clones. *Nucleic Acids Research* 23:3802-3803.

Demesure, B., N. Sodzi, and R.J. Petit (1995). A set of universal primers for amplification of polymorphic non-coding regions of mitochondrial and chloroplast DNA in plants. *Molecular Ecology* 4:129-131.

Desmarais, E., I. Lanneluc, and J. Lagnel (1998). Direct amplification of length polymorphisms (DALP), or how to get and characterize new markers in many species. *Nucleic Acids Research* 26:1458-1465.

Duminil, J., M.H. Pemonge, and R.J. Petit (2002). A set of 35 consensus primer pairs amplifying genes and introns of plant mitochondrial DNA. *Molecular Ecology Notes* 2:428-430.

Ender, A., K. Schwenk, T. Stadler, B. Streit, and B. Schierwater (1996). RAPD identification of microsatellites in *Daphina. Molecular Ecology* 5:437-441.

Eujayl, I., M.K. Sledge, L. Wang, G.D. May, K. Chekhovskiy, J.C. Zwonitzer, and M.A.R. Mian (2004). *Medicago truncatula* EST-SSRs reveal cross-species genetic markers for *Medicago* spp. *Theoretical and Applied Genetics* 108:414-422.

Flavell, A.J., M.R. Knox, S.R. Pearce, and T.H.N. Ellis (1998). Retrotransposon-based insertion polymorphisms (RBIP) for high throughput marker analysis. *Plant Journal* 16:643-650.

Fulton, T.M., R. Van der Hoeven, N.T. Eannetta, and S.D. Tanksley (2002). Identification, analysis and utilization of conserved ortholog set markers for comparative genomics in higher plants. *Plant Cell* 14:1457-1467.

Gupta, M., J. Chyi, J. Romero-Severson, and J.L. Owen (1994). Amplification of DNA markers from evolutionary diverse genomes using primers of simple-sequence repeats. *Theoretical and Applied Genetics* 89:998-1006.

Gupta, P.K. and S. Rustogi (2004). Molecular markers from the transcribed/expressed region of the genome in higher plants. *Functional and Integrative Genomics* 4:139-162.

Gupta, P.K. and R.K. Varshney (2000). The development and use of microsatellite markers for genetic analysis and plant breeding with emphasis on bread wheat. *Euphytica* 113:163-185.

Handley, S.A. and R.L. Regnery (2000). Differentiation of pathogenic *Bartonella* species by infrequent restriction site PCR. *Journal of Clinical Microbiology* 38:3010-3015.

Hatada, I., Y. Hayashizaki, S. Hirotsune, and H. Komatsubara (1991). A genomic scanning method for higher organisms using restriction sites as landmarks. *Proceedings of the National Academy of Sciences USA* 88:9523-9527.

Hayden, M.J. and P.J. Sharp (2001). Sequence-tagged microsatellite profiling (STMP): A rapid technique for developing SSR markers. *Nucleic Acids Research* 29: e43.

Heath, D.D., G.K. Iwama, and R.H. Devlin (1993). PCR primed with VNTR core sequences yields species specific patterns and hypervariable probes. *Nucleic Acids Research* 21:5782-5785.

Hu, J. and B.A. Vick (2003). Target region amplification polymorphism: A novel marker technique for plant genotyping. *Plant Molecular Biology Reporter* 21: 289-294.

Jaccoud, D., K. Peng, D. Feinstein, and A. Kilian (2001). Diversity arrays: A solid state technology for sequence information independent genotyping. *Nucleic Acids Research* 49:425.
Jones, C.J., K.J. Edwards, S. Castaglioni, M.O. Winfield, F. Sala, C. van de Wiel, G. Bredemeijer, B. Vosman, M. Matthes, A. Daly, et al. (1997). Reproducibility testing of RAPD, AFLP and SSR markers in plants by a network of European laboratories. *Molecular Breeding* 3:381-390.
Kalendar, R., T. Grob, M. Regina, A. Suoniemi, and A.H. Schulman (1999). IRAP and REMAP: Two new retrotransposon-based DNA fingerprinting techniques. *Theoretical and Applied Genetics* 98:704-711.
Li, G. and C.F. Quiros (2001). Sequence-related amplified polymorphism (SRAP), a new marker system based on a simple PCR reaction: Its application to mapping and gene tagging in *Brassica. Theoretical and Applied Genetics* 103:455-461.
Meyer, W., T.G. Michell, E.Z. Freedman, and R. Vilgalys (1993). Hybridization probes for conventional DNA fingerprinting used as single primers in the polymerase chain reaction to distinguish strains of *Cryptococcus neoformans. Journal of Clinical Biology* 31:2274-2280.
Michelmore, R.W., I. Paran, and R.V. Kesseli (1991). Identification of markers linked to disease-resistance genes by bulked segregant analysis: A rapid method to detect markers in specific genomic regions by using segregating populations. *Proceedings of the National Academy of Sciences USA* 88:9828-9832.
Morgante, M. and J. Vogel (1994). Compound microsatellite primers for the detection of genetic polymorphisms. U.S. Patent Application No. 08/326456.
Neff, M.M., J.D. Neff, J. Chory, and A.E. Pepper (1998). dCAPS, a simple technique for the genetic analysis of single nucleotide polymorphisms: Experimental applications in *Arabidopsis thaliana* genetics. *Plant Journal* 14:387-392.
Rakoczy-Trojanowska, M. and H. Bolibok (2004). Characterization and comparison of three classes of microsatellite-based markers and their application in plants. *Cellular and Molecular Biology Letters* 9:221-238.
Rani, V. and S.N. Raina (2003). Molecular DNA marker analysis to assess the genetic fidelity of micropropagated woody plants. In *Micropropagation of woody trees and fruits,* S.M. Jain and K. Ishii (eds.). Dordrecht, Netherlands: Kluwer Academic Publishers, pp. 75-101.
Richardson, T., S. Cato, J. Ramser, G. Kahl, and K. Weising (1995). Hybridization of microsatellites to RAPD: A new source of polymorphic markers. *Nucleic Acids Research* 23:3798-3799.
Rudd, S. (2003). Expressed sequence tags: Alternative or complement to whole genome sequences? *Trends in Plant Science* 8:321-329.
Schulman, A.H., A.J. Flavell, and T.H.N. Ellis (2004). The application of LTR retrotransposons as molecular markers in plants. In: Mobile genetic elements: Protocols and genomic applications, W.J. Miller and P. Capy (eds.). Totowa, NJ: Humana Press, pp. 145-171.
Soleimani, V.D., B.R. Baum, and D.A. Johnson (2003). Efficient validation of single nucleotide polymorphisms in plants by allele-specific PCR with an example from barley. *Plant Molecular Biology Reporter* 21:281-288.

Somers, D.J., Z. Zhou, P.J. Bebeli, and J.P. Gustafson (1996). Repetitive, genome-specific probes in wheat (*Triticum aestivum* L. em Thell) amplified with minisatellite core sequences. *Theoretical and Applied Genetics* 93:982-989.

Thein, S.I. and R.R. Wallace (1986). The use of synthetic oligonucleotides as specific hybridization probes in the diagnosis of genetic disorders. In *Human genetic diseases: A practical approach,* K.E. Davis (ed). Oxford: IRL, pp. 33-50.

Van den Broeck, D., T. Maes, M. Sauer, J. Zethof, P. De Keukeleire, M. D'Hauw, M. Van Montagu, and T. Gerats (1998). Transposon display identifies individual transposable elements in high copy number lines. *Plant Journal* 13:121-129.

Vos, P., R. Hogers, M. Bleeker, M. Reijans, V. de Lee, T. Hornes, M. Freijters, J. Pot, J. Peleman, et al. (1995). AFLP: A new technique for DNA fingerprinting. *Nucleic Acids Research* 23:4407-4414.

Waldron, J., C.P. Peace, I.R. Searle, A. Furtado, N. Wade, I. Findlay, M.W. Graham, and B.J. Carroll (2002). Randomly amplified DNA fingerprinting: A culmination of DNA marker technologies based on arbitrarily-primed PCR amplification. *Journal of Biomedicine and Biotechnology* 2:141-150.

Waugh, R., K. McClean, A.J.. Flavell, S.R. Pearce, A. Kumar, and B.B.T. Thomas (1997). Genetic distribution of BARE-1-like retrotransposable elements in the barley genome revealed by sequence-specific amplification polymorphisms (SSAP). *Molecular and General Genetics* 253:687-694.

Weining, S. and P. Langridge (1991). Identification and mapping of polymorphisms in cereals based on the polymerase chain reaction. *Theoretical and Applied Genetics* 82:209-216.

Welsh, J. and M. McClelland (1990). Finger printing genomes using PCR with arbitrary primers. *Nucleic Acids Research* 18:7213-7218.

Williams, J.G.K., A.R. Kubelik, K.J. Livak, J.A. Rafalski, and S.V. Tingey (1990). DNA polymorphisms amplified by arbitrary primers are useful genetic markers. *Nucleic Acids Research* 18:6531-6535.

Wu, K., R. Jones, L. Dannaeberger, and P.A. Scolnik (1994). Detection of microsatellite polymorphisms without cloning. *Nucleic Acids Research* 22:3257-3258.

Zhu, H., S.B. Cannon, N.D. Young, and D.R. Cook (2002). Phylogeny and genomic organization of the TIR and non-TIR NBS-LRR resistance gene family in *Medicago truncatula. Molecular Plant-Microbe Interactions* 15:529-539.

Zietkiewicz, E., A. Rafalski, and D. Labuda (1994). Genome fingerprinting by simple sequence repeat (SSR)–anchored polymerase chain reaction amplification. *Genomics* 20:176-183.

Chapter 33

# Applicability of DNA Markers for Genome Diagnostics of Grain Legumes

Deepmala Sehgal
Vishnu Bhat
Soom Nath Raina

## *INTRODUCTION*

Legumes provide the largest single source of vegetable protein in human diets and livestock feed. The approximately 750 genera constituted by around 19,000 species include important grain, pasture, and agroforestry taxa. These taxa are cultivated on 180 million hectares (~12 percent of earth's arable land) thus accounting for 27 percent of the world's primary crop production system (Graham and Vance, 2003) with grain legumes alone contributing to 33 percent of the dietary protein needs of humans (Vance et al., 2000).

The advent of DNA-based markers has facilitated fast and unambiguous genetic analysis of plant species including that of grain legumes of both diploid and polyploid origin. The methodologies and underlying principles of the various DNA markers have been reviewed in chapter 32, this volume. The effective utilization of DNA markers to (1) examine the level of genetic diversity within the species, (2) highlight any observed trends in the diversification and differentiation of taxons at the infraspecific level, (3) fingerprint the released cultivars, (4) reconstruct phylogenetic relationships between cultivated and its close and distant relatives, (5) estimate the level and nature of introgression, and (6) construct linkage maps facilitating trait tag-

The first author (D.S.) is grateful to the Council of Scientific and Industrial Research, Government of India, for financial support. This work was supported, in part, by a grant-in-aid for scientific research from the Defense Research Development Organization and Department of Biotechnology, Government of India.

ging for marker-assisted breeding programs, map-based cloning, comparative mapping, and QTL analysis of most of the grain legumes of interest are highlighted in this review. It also provides source material for the reader who seeks information on specific areas of interest.

## *DNA MARKERS IN GRAIN LEGUMES*

### *Assessment of Genetic Diversity Including Analysis of Population Genetic Structure*

Knowledge of genetic diversity in a crop species is fundamental to its improvement. There is a need to identify germplasm with distinct DNA profiles for genetic enhancement as they are likely to contain the largest number of unique and agronomically useful alleles. DNA markers have been employed to assess genetic diversity in many crop legumes. With regard to this, random amplified polymorphic DNA (RAPD), restriction fragment length polymorphism (RFLP), and amplified fragment length polymorphism (AFLP) have been the most commonly used markers (Link et al., 1995; Fatokun et al., 1997; Kisha et al., 1998; Thompson et al., 1998; Beebe et al., 2000; Eichenberger et al., 2000; Mienie et al., 2000; Singh et al., 2000; Galván et al., 2001; Gimenes, Lopes, Galgaro, et al., 2002; Nowosielski et al., 2002; Papa and Gepts, 2003; Simioniuc et al., 2002; Ude et al., 2003) whereas microsatellite/microsatellite-derived and other gene-targeted markers have been utilized in a few grain legumes (Banerjee et al., 1998; Garber et al., 1999; Sant et al., 1999; Udupa et al., 1999; Métais et al., 2000, 2002; Burstin et al., 2001; Ford et al., 2002; Abe et al., 2003; Hudcovicová and Kraic, 2003; Galván et al., 2003; Baranger et al., 2004; McClean et al., 2004).

In chickpea (*Cicer arietinum*), genetic diversity studies with RFLP and RAPD markers detected a very narrow genetic base (Banerjee et al., 1999). RAPD and inter simple sequence repeats (ISSR) markers, however, successfully discriminated Kabuli and Desi chickpea cultivars (Iruela et al. 2002). RFLP studies using microsatellite probes (Sharma et al., 1995; Udupa et al., 1999), and random amplified microsatellite polymorphism (RAMPO; Banerjee et al., 1998) markers revealed considerable variation at the intraspecific level. Sant et al. (1999) studied genetic diversity in Indian elite chickpea cultivars using RAPD and microsatellite markers. Microsatellite markers in this study revealed genetic diversity in the range of 39 to 82 percent. Hence, microsatellites and microsatellite-based markers have been proposed as an ideal marker system for characterization of the germplasm resources in chickpea.

Among the cultivated types of pea (*Pisum sativum*), distinction has been made between fodder and feed peas by RAPD and AFLP (Samec and Nasinec, 1995; Simioniuc et al., 2002) markers. Investigations using RAPD, ISSR, simple sequence repeat (SSR), and sequence tagged sites (STS; Baranger et al., 2004), and RAPD, SSR, and ISSR (Tar'an et al., 2005) markers structured the diversity in pea according to end uses (fodder, food, and feed peas), sowing types (spring sown or winter sown), and geographical regions, thereby suggesting a divergent selection in as many gene pools.

In soybean (*Glycine max*), primarily North American soybean cultivars and their ancestral forms were the focus of genetic diversity studies using DNA markers (Diwan and Cregan, 1997; Kisha et al., 1998; Thompson et al., 1998; Song et al., 1999; Brown-Guedira et al., 2000; Narvel et al., 2000; Li et al., 2001; Ude et al., 2003). Asian soybean containing important germplasm has been analyzed using RAPD (Li and Nelson, 2001) and nuclear SSR markers (Abe et al., 2003). A relatively high genetic diversity was observed to occur among Asian germplasm pools as compared to the North American improved cultivars and ancestral forms (Ude et al., 2003). The results also indicated the existence of exclusive Japanese and Chinese gene pools, which can be used as exotic genetic resources to enlarge the genetic base of Asian soybean as such. The Japanese cultivars were found to be more distant from North American soybean ancestors and cultivars than the Chinese cultivars, which indicated that the Japanese gene pool may constitute a genetically distinct source of useful genes for yield improvement of North American germplasm.

*Phaseolus* beans include five cultivated species viz., common bean (*P. vulgaris*), runner bean (*P. coccineus*), year bean (*P. polyanthus*), lima bean (*P. lunatus*), and tepary bean (*P. acutifolius*). Of these, common bean is the most widely grown throughout the world. Two gene pools have been reported to occur in cultivated common bean (*P. vulgaris*), a Mesoamerican and an Andean one, by RAPD (Briand et al., 1998; Galván et al., 2001; Maciel et al., 2001), RFLP (Nodari et al., 1992; Becerra Velasquez and Gepts, 1994), and ISSR markers (Galván et al., 2003). This confirms the pattern of genetic diversity identified previously with biochemical and morphological markers. The two gene pools hold the pivotal position for the genetic improvement of common bean. A narrow genetic base of the Andean gene pool was revealed in a study by AFLP markers with no differentiation into races N (Nueva Granad), C (Chile) and P (Peru) as identified previously on morphological and ecogeographical criteria (Beebe et al., 2001). The Mesoamerican gene pool, on the other hand, was found to be relatively more diverse and structured into races M (Mesoamerica), D (Durango) and

J (Jalisco) (Beebe et al., 2000; Serna et al., 2005). An additional race 'Guatemela' has also been reported. A well-defined genetic structuration in the Mesoamerican gene pool and very narrow differences in Andean gene pool races were also previously reported with RFLP markers (Becerra Velasquez and Gepts, 1994). Recombination between Andean (race N) and Mesoamerican (races M, D, and J) gene pools is supposed to be the causal factor for the genetic continuity among races M, D, J, and N in Mexico (Serna et al., 2005). AFLP markers could validate the classification of Brazilian accessions (Maciel et al., 2003) according to phaseolin types (S type in Mesoamerican beans and T type in Andean beans). Investigations using AFLP (Papa and Gepts, 2003; Papa et al., 2005) and ISSR (González et al., 2005) markers have proven useful to infer the predominant direction of gene flow between domesticated and wild beans as well as to compare the spatial differentiation of domesticated and wild beans. Broadening the genetic base of the Andean gene pool becomes a priority and would require crosses with either more diverse Mesoamerican beans or landraces with introgressed Mesoamerican genes. When the genetic diversity of *P. vulgaris* was compared with that of *P. coccineus* by using RAPD and AFLP (Nowosielski et al., 2002) markers, and ISSR and nuclear and chloroplast SSRs (Sicard et al., 2005), *P. vulgaris* was found to be more polymorphic than *P. coccineus*. A stronger founder effect for *P. coccineus* and selection practiced by farmers for *P. vulgaris* is responsible for the observed trends in diversity. Similar results were obtained when comparisons were made between *P. vulgaris, P. lunatus,* and related wild species from South America (Caicedo et al., 1999). Heterogeneity was greater in *P. vulgaris* than in either *P. lunatus* or the highly cohesive *P. augusti–P. pachyrrhizoides–P. bolivianus* group. Within *P. lunatus,* diversity in the Colombian subpopulation was of the same magnitude as total diversity for the whole *P. lunatus* population, thus pointing to the existence of a third minor gene pool along with two major gene pools identified previously by RAPD (Fofana et al., 1997; Nienhuis et al., 1995) markers. Chloroplast DNA RFLP (cpDNA-RFLP; Fofana et al., 2001) analysis has also identified a secondary diversification spot (northern dry forest of Argentina) for wild lima bean. In tepary bean, genetic diversity was assessed using orthologous SSR primers derived from common bean (Christensen et al., 2005). It indicated a high degree of monomorphism.

Screening by RAPD and RFLP markers documented a low level of genetic diversity in peanut (*Arachis hypogaea;* Halward et al., 1991, 1992; Kochert et al., 1991; Paik-Ro et al., 1992; Singh et al., 1998). This was attributed to the barriers to gene flow from related diploid species to domesticated peanut as a consequence of polyploidization events or recent polyploidization events combined with self-pollination. Later studies, however, using AFLP (Gimenes, Lopes and Valls, 2002), AFLP and DNA amplification fingerprinting (DAF; He and Prakash, 1997), RAPD and ISSR (Raina, Rani, et al., 2001), and in particular SSR markers (Hopkins et al., 1999; He et al., 2003; Krishna et al., 2004; Ferguson, Bramel, et al., 2004; Ferguson, Burow, et al., 2004) detected substantial genetic variation in cultivated peanut, which opened up the possibilities of its genetic improvement through marker-assisted breeding programs.. Six botanical varieties of the cultivated peanut (varieties *hypogaea* and *hirsuta* belonging to the subspecies *hypogaea* and varieties *fastigiata, peruviana, aequatoriana,* and *vulgaris* belonging to the subspecies *fastigiata*) have also been investigated by DNA markers (Ferguson, Bramel, et al., 2004; He et al., 2005). Using sequence tagged microsatellite sites (STMS), Ferguson, Bramel, et al. (2004) suggested a limited introgression of *A. hypogaea* subsp. *fastigiata* var. *peruviana* with three other varieties of subspecies *fastigiata.* Also it was shown that *A. hypogaea* subsp. *hypogaea* var. *hypogaea* and *A. hypogaea* subsp. *hypogaea* var. *hirsuta* are distantly related. SSR markers have been used to generate variety-specific DNA markers (He et al., 2005).

In the genus *Vigna,* several species are widely cultivated for their multipurpose values. These among other members include cowpea (*V. unguiculata*), mung bean (*V. radiata*), azuki bean (*V. angularis*), black gram (*V. mungo*), and the Bambara peanut (*V. subterranea*). Cowpea cultivars have been variously classified as groups Unguiculata, Biflora, Textilis, Sesquipedalis and Melanophthalmus under either subspecies *unguiculata* or have been included in *V. unguiculata* ssp. *catjang, V. unguiculata* ssp. *sesquipedalis,* and *V. unguiculata* ssp. *cylindrica*. The genetic diversity in cultivated cowpea has been screened by nuclear genome-based AFLP (Coulibaly et al., 2002; Polegri et al., 2004), AFLP and SSR (Gillespie et al., 2005), RAPD (Mignouna et al., 1998; Ba et al., 2004), SSR (Li et al., 2001), RAPD, AFLP, and selective amplification of microsatellite polymorphic loci (SAMPL; Tosti and Negri, 2002), and AFLP and SAMPL (Tosti and Negri, 2005) markers and chloroplast DNA polymorphism (Vaillancourt and Weeden, 1992). Studies involving both domesticated and wild forms have indicated that wild cowpea is more diverse than domesticated cowpea (Coulibaly et al., 2002; Ba et al., 2004). The gene flow between wild and domesticated forms had considerable impact on the organization of overall genetic diversity. This genetic exchange has blurred to a large extent the distinction between the ancestral and postdomestication state. According to Ba et al. (2004) and Gillespie et al. (2005), no differentiation could be made between so-called cultivar groups or subspecies. The old cultivar groups (Biflora and Textilis) are more diverse than the recent cultivar groups (Melanophthalmus, Sesquipedalis, and Unguiculata). In a comparative analysis of efficiency of RAPD, AFLP, and SAMPL markers in assessing genetic variation among cowpea landraces, AFLP and SAMPL markers have proven to be more useful in genetic diversity diagnostics among cowpea landraces (Tosti and Negri, 2002, 2005).

Azuki bean consists of two varieties, *V. angularis* var. *angularis* (the cultigen) and *V. angularis* var. *nipponensis* (the presumed wild type of azuki bean), along with a weedy or semiwild azuki. The three types constitute the crop species complex. The genetic diversity within the azuki bean complex has been studied using RAPD (Xu, Tomooka, Vaughan, and Doi, 2000), SSR (Wang et al., 2004), AFLP (Xu, Tomooka and Vaughan, 2000; Zong et al., 2003), and RAPD and AFLP (Yee et al., 1999) markers. Diversity screened across Asian populations has revealed a high degree of variation, more particularly in the Himalayan and North Eastern Asian accessions (Zong et al., 2003; Mimura et al., 2000). The results suggest that the domestication of azuki bean occurred at least twice, once in the Himalayan region of southern Asia and once in northeast Asia. The subtropical highlands of Asia have been suggested as a potential genetic resource for azuki improvement (Mimura et al., 2000). A marked reduction in genetic diversity from wild species to the cultivated azuki bean complex was shown to occur in Japan (Xu, Tomooka, and Vaughan, 2000; Xu, Tomooka, Vaughan, and Doi, 2000) by AFLP markers. AFLP markers further provided molecular evidence to support the existence of three types, the cultigen, the wild type, and a weedy type, in the complex. SSR markers (Wang et al., 2004) utilized to study the genetic structure of the azuki bean complex in Japan revealed that genetic structure differed more between north and south than between east and west regions.

Genetic diversity analysis in black gram by AFLP (Sivaprakash et al., 2004) and RAPD and ISSR markers (Souframanien and Gopalakrishna, 2004) revealed high genetic polymorphism in the landraces (Sivaprakash et al., 2004), while a moderate level of diversity was detected in elite genotypes (Souframanien and Gopalakrishna, 2004). RAPD marker analysis of mung bean cultivars revealed a narrow genetic base, thus suggesting a need to exploit the large germplasm collections having diverse morpho-agronomic traits or collection of diverse germplasm from centers of diversity (Lakhanpal et al., 2000; Afzal et al., 2004; Betal et al., 2004). In contrast, considerable genetic diversity was found among accessions and landraces of Bambara peanut by RAPD (Amadou et al., 2001; Massawe et al., 2003) and AFLP (Massawe et al., 2002; Ntundu et al., 2004) markers. AFLP markers not only revealed high polymorphism in Bambara peanut landraces

(84 percent polymorphism), but landrace-specific markers were also identified (Massawe et al., 2002).

In faba bean (*Vicia faba*), evaluation of European and Mediterranean germplasm resources by RAPD markers (Link et al., 1995) revealed that European small-seeded lines and Mediterranean lines form distinct heterotic groups with European large-seeded lines dispersed between two heterotic groups. AFLP markers were employed to analyze genetic diversity among 22 faba bean inbred lines derived from elite cultivars (Zeid et al., 2001) and also the faba bean material used for breeding in Asia, Europe, and North America (Zeid et al., 2003). The latter study revealed Asian germplasm to be quite distinct from European and North African germplasm, indicating that it can be a potential source for introgression of traits into European germplasm.

Genetic diversity studies carried out in cultivated lentil (*Lens culinaris* ssp. *culinaris;* Abo-elwafa et al., 1995; Sharma et al., 1995; Mayer and Soltis, 1994) by RAPD and cpDNA RFLP markers have revealed a low degree of intraspecific genetic variation. The AFLP marker system (Sharma et al., 1996) proved useful to differentiate micro- and macrosperma varietal groups of the cultivated lentil.

### *Phylogenetic Reconstruction between the Cultivated and Related Wild Species, and Other Related Taxa*

An understanding of the phylogenetic relationships between a cultivated species and its wild relatives constituting primary, secondary, and tertiary gene pools is essential for crop improvement and developing gene transfer strategies. A total of 33 annual and 9 perennial species constitute the genus *Cicer,* in which *C. arietinum* is the only cultivated species. Investigations into the phylogeny of *Cicer* using RAPD (Sudupak et al., 2002; Javadi and Yamaguchi, 2004b), ISSR (Sudupak, 2004; Rajesh, Sant, et al., 2002), STMS (Choumane et al., 2000), RAPD and ISSR (Iruela et al., 2002), RFLP (Patil et al., 1995; Serret et al., 1997), and AFLP (Nguyen et al., 2004; Sudupak et al., 2004) markers revealed that *C. reticulatum* is the most likely candidate to be the wild progenitor of chickpea, *C. arietinum. Cicer echinospermum* is also closely related to *C. arietinum*. A noteworthy and significant observation was the close relationship between *C. anatolicum, C. reticulatum, C. echinospermum,* and *C. arietinum* (Choumane et al., 2000). The close genetic relationships between these wild species and the cultigen can be utilized for transfer of desirable genes. An investigation of nine perennial and eight annual species by AFLP markers revealed three main species groups in *Cicer* (Nguyen et al., 2004). Group I included *C.*

*arietinum, C. reticulatum,* and *C. echinospermum*. The annual species *C. bijugum, C. judaicum,* and *C. pinnatifidum* formed group II. Group III contained all nine perennial species and two annual species, *C. yamashitae* and *C. cuneatum*. The genetic distance detected between group I and group III was equivalent to the genetic distance detected between group I and group II. These results are significant since they indicated that perennial tertiary species (group III) are as valuable as annual tertiary species in *Cicer* for incorporating novel genes into the cultigen. The results obtained by both ISSR (Rajesh, Sant, et al., 2002) and chloroplast DNA sequence data (Javadi and Yamaguchi, 2004a) revealed multiple origins of annual species in the genus *Cicer*.

There are about 55 species in the genus *Phaseolus*. Of these, five species including *P. vulgaris* are cultivated. Based on chloroplast DNA and internal transcribed sequence (ITS) sequence data (Delgado-Salinas et al., 1993, 1999), the 55 species investigated formed a monophyletic clade with *P. microcarpus* as the basal species. Barring *P. lunatus,* the remaining four cultivated species—*P. vulgaris, P. coccineum, P. acutifolium,* and *P. polyanthus*—formed one group. In a later study, Goel et al. (2002) also supported the monophyletic origin of the genus with *P. microcarpus* as the basal species. That three species, *P. augusti, P. bolivianus,* and *P. pachyrrhizoides* do not merit a separate taxonomic ranking at species level was revealed by AFLP (Caicedo et al., 1999) markers.

The genus *Pisum* contains *P. fulvum* and a sativum species complex variously assigned from one species (*P. sativum*) to three species (*P. humile, P. elatius,* and *P. sativum*). The monophyletic origin of this complex has been revealed based on chloroplast RFLP (Palmer et al., 1985) and RAPD (Hoey et al., 1996) markers. *P. humile* has been proposed as the wild progenitor of cultivated pea. *P. fulvum* is the most divergent species.

The genus *Glycine* comprises the subgenera *Soja* and *Glycine*. Subgenus *Soja* includes the cultivated soybean (*G. max*), *G. soja,* and *G. gracilis*. The subgenus *Glycine* comprises 23 wild perennial herbaceous species. The species relationships within the subgenus *Glycine* have been extensively investigated by RFLP (Menancio et al., 1990; Doyle et al., 1990; Doyle and Beachy, 1985) and RAPD (Taylor-Grant and Soliman, 1999) markers, and comparative nuclear ITS (Kollipara et al., 1997) and chloroplast DNA sequence data (Xu, Abe, et al., 2000). The results of Xu, Abe et al. (2000) showed that *G. microphylla* diverged from other species in the subgenus *Glycine* at an early stage of evolution. The study by Kollipara et al. (1997) not only remained useful in resolving genomic groups in *Glycine,* but it also gave new genome symbols to *G. arenaria* (HH), *G. hirticaulis* (H1H1), *G. pindanica* (H2H2), *G. albicans* (II), and *G. lactovirens* (I1I1). Chloroplast

RFLP markers provided insight into relationships between the species of subgenus *Soja*. The results showed that *G. soja* is the putative wild progenitor of *G. max* and *G. gracilis* (Shoemaker et al., 1986).

The genus *Arachis* comprises 80 species, 69 of which have been described and distributed among 9 taxonomic sections. Section *Arachis* contains the largest number of species including cultivated peanut. The phylogenetic relationships between the *Arachis* species have been investigated extensively by RAPD (Halward et al., 1992; Hilu and Stalker, 1995), RAPD and ISSR (Raina, Rani, et al., 2001), RFLP (Kochert et al., 1991, 1996; Gimenes, Lopes, Galgaro, et al., 2002; Paik-Ro et al., 1992), RFLP and RAPD (Galgaro et al., 1998), AFLP (Gimenes, Lopes, and Valls, 2002) markers, and comparative sequence analysis of genes such as stearoyl-ACP desaturase and oleoyl-PC desaturase (Jung et al., 2003). The identification of diploid progenitors of tetraploid *A. hypogaea* involved comprehensive investigations at the DNA sequence level, but controversy still persists. Six diploid species, *A. villosa, A. duranensis, A. cardenasii, A. ipaensis, A. batizocoi,* and *A. correntina,* have been proposed as diploid progenitors of *A. hypogaea. A. batizocoi,* earlier reported to be involved in the evolution of *A. hypogaea* as a B genome donor species, has been found to be distantly related to domesticated peanut by DNA markers. Similarly, tetraploid *A. monticola,* reported to be closely related to *A. hypogaea* by DNA markers (Paik-Ro et al., 1992; Hilu and Stalker, 1995; Raina, Rani, et al., 2001; Moretzsohn et al., 2004), has been excluded from the possibility of being the tetraploid progenitor of the cultivated peanut on the basis of sequences of stearoyl-ACP desaturase and oleoyl-PC desaturase genes (Jung et al., 2003). Studies carried out by Gimenes, Lopes, Galgaro, et al. (2002) using RFLP markers suggested that B genome designation should be restricted to *A. ipaensis* alone. The unresolved controversy with respect to the identification of wild progenitors of *A. hypogaea* was resolved to a large extent by the study based on molecular cytogenetics (Raina and Mukai, 1999a, 1999b). They reported that *A. villosa* and *A. ipaensis* are A and B genome donors to *A. hypogaea,* respectively, and that *A. monticola* is very closely related to the cultivated species. DNA marker-based studies on sections *Extranervosae, Caulorhizae, Heterenthae, Triseminatae* (Galgaro et al., 1998), and section *Rhizomatosae* (Nübile et al., 2004) have been very useful in resolving various phylogenetic controversies in the genus, more particularly regarding the origin of tetraploid species of section Rhizomatosae and intersectional relationships.

The genus *Vigna* includes seven subgenera, of which three subgenera, *Ceratotropis, Plectotropis,* and *Vigna* include cultivated species. Subgenus *Plectotropis* includes the cultivated species *V. vexillata* (zombie bean). Subgenus *Vigna* includes *V. unguiculata* (cowpea) and *V. subterranea* (Bambara peanut). The subgenus *Ceratotropis* includes eight cultivated species, of which mung bean (*V. radiata*), black gram (*V. mungo*), and azuki bean (*V. angularis*) are most important. The phylogenetic relationships in *Vigna* have been studied by RFLP (Fatokun, Danesh, Young, et al., 1993), AFLP (Coulibaly et al., 2002; Tomooka et al., 2002), AFLP and RAPD (Saravanakumar et al., 2004), RAPD (Kaga et al., 1996) and ISSR markers (Ajibade et al., 2000) and nuclear rDNA ITS (Goel et al., 2002), and *atpB-rbcL* intergenic spacer of cpDNA sequences (Doi et al., 2002). Wild *V. unguiculata* has been suggested as the wild progenitor of cultivated *V. unguiculata* (Ajibade et al., 2000; Goel et al., 2002). *V. umbellata* and *V. angularis* have been suggested as diploid progenitors of the allotetraploid species *V. glabrescens* (Goel et al., 2002). Studies have revealed two (Fatokun, Danesh, Young, et al., 1993; Kaga et al., 1996; Ajibade et al., 2000; Goel et al., 2002) or three (Doi et al., 2002; Tomooka et al., 2002) groups in the subgenus *Ceratotropis*. The proponents of three groups have suggested three sections, *Aconitifoliae, Angulares,* and *Ceratotropis* in the subgenus *Ceratotropis*. Sections *Angulares* and *Aconitifoliae* in the subgenus *Ceratotropis* are considered the most recent and most ancient sections, respectively. The species *V. trinervia,* occupying a central position among the three sections, may be a useful species to facilitate gene transfer among sections of the subgenus *Ceratotropis* (Doi et al., 2002). Tomooka et al. (2002) have analyzed the relationship between the newly described *Vigna* species, *V. aridicola, V. exilis, V. nepalensis,* and *V. tenuicaulis,* and other species in the subgenus *Ceratotropis* using AFLP markers.

The genus *Vicia* comprises approximately 180 species under two subgenera, *Vicia* and *Vicilla*. Subgenus *Vicia* contains the agriculturally important species of the genus including faba bean (*V. faba*), common vetch (*V. sativa*), narbon vetch (*V. narbonensis*), and their close relatives. The taxonomic history of the genus is extensive and contentious. The phylogeny, more particularly between *V. faba* and its close relatives, has been investigated using multiple DNA markers such as RFLP of both nuclear and chloroplast genomes (Shiran and Raina, 2001; Raina and Ogihara, 1994, 1995), gene-specific PCR (Van de Ven et al., 1993), and chloroplast PCR RFLP and RAPD (Potokina et al., 1999) markers. There is overwhelming evidence to suggest that none of the seven species in the *narbonensis* species complex (morphologically very close to *V. faba*) can be considered as putative allies of faba bean. The evidence based on nuclear and organellar RFLPs and PCR analysis suggests that the seven taxa within the *V. sativa* species complex should be relegated to a rank, perhaps of subspecies within the complex. The data on physical mapping of rDNA genes on chromosomes suggest four distinct species within the *V. villosa* species complex (Raina, Mukai, et al., 2001). The quantitative change in nuclear DNA amounts between *Vicia* species was achieved by changes in amounts of both repetitive and nonrepetitive DNA sequences (Raina and Narayan, 1984).

The genus *Lens* has seven taxa, including the cultivated species *Lens culinaris* ssp. *culinaris* and its wild relatives. The nuclear DNA RFLP (Havey and Muehlbauer, 1989; Patil et al., 1995), cpDNA RFLP (Mayer and Soltis, 1994; van Oss et al., 1997), mitochondrial DNA RFLP (mt DNA RFLP; Rajora and Mohan, 1997), RAPD (Abo-elwafa et al., 1995; Sharma et al., 1995, Ahmad and McNeil, 1996), AFLP (Sharma et al., 1996) markers, and ITS (Sonnante et al., 2003) and lectin nucleotide sequence (Galasso et al., 2004) analysis have unambiguously revealed that *Lens culinaris* ssp. *orientalis* is the wild progenitor of the cultivated *Lens*. *Lens nigricans* is genetically the most diverged species (Havey and Muehlbauer, 1989; van Oss et al., 1997; Galasso, 2003; Sonnante et al., 2003; Galasso et al., 2004). Two more recently recognized species, *L. lamottei* and *L. tomentosus,* have been found to be two independent entities (van Oss et al., 1997; Sonnante et al., 2003).

The cultivated pigeon pea (*Cajanus cajan*) has many wild relatives (~32 species) that belong to 11 genera grouped under the subtribe Cajaninae. In addition to wild *Cajanus* species that constitute primary, secondary, and tertiary gene pools, other genera constituting important gene pools for its improvement include *Dunberia, Rhyncosia, Paracalyx, and Flemingia*. Nuclear RFLP, RAPD, and chloroplast PCR-RFLP (Ratnaparkhe et al., 1995; Lakshmi et al., 2000) markers revealed that *C. cajanifolius* is the wild progenitor of *C. cajan*. Detailed investigations including the closely related genera (Lakshmi et al., 2000; Sivaramakrishnan et al., 2002) have proven to some extent useful in resolving systematic problems that the genus poses but controversy still persists. For instance, while chloroplast PCR-RFLP (Lakshmi et al., 2000) markers have shown genus *Rhyncosia* to be a connecting link between the *Dunberria-Cajanus* complex and the *Flemingia* and *Paracalyx* complex, mt RFLP (Sivaramakrishnan et al., 2002) markers, on the other hand, have shown *Cajanus* and *Rhyncosia* to be quite distinct from each other.

## *Genome Analysis of Hybrids and Introgressed Lines*

DNA markers have been used to study interspecific and intergeneric hybrids and to characterize gene introgression. Detection of the *Glycine tomentella* genome in *G. max* × *G. tomentella* derived amphidiploid intersubgeneric line and monosomic alien addition lines (MAALs) was successfully made out using SSR and ITS-cleaved amplified polymorphic sequences (ITS-CAPS) markers (Zou et al., 2004). The amphidiploid line displayed a combination of parental DNA bands confirming the hybrid nature. Absence of donor genome patterns in an amphidiploid line, in several cases, might be due to the loss of donor genome during the meiotic process. It was also observed that in a *G. max* cv. Clark 63 × (*G. max* cv. Altona × *G. tomentella* PI 483218) cross, Altona genome was replaced by Clark 63 genome after backcrossing four times. One ITS-CAPS marker (ITS-*Hinc*II digest) could clearly identify bands from *G. tomentella* in the amphidiploid and four MAALs. These MAALs contained 40 chromosomes from *G. max* plus one chromosome from *G. tomentella*. They are excellent bridge materials to transfer useful genes from wild to cultivated species.

RAPD markers were utilized to study interspecific hybrids between *Cajanus acutifolius* and *C. cajan* (Mallikarjuna and Saxena, 2002). cDNA-RFLP markers were used to study interspecific hybrids between *Phaseolus vulgaris* and *P. coccineus* (Guo et al., 1994). Preferred transmission of *P. vulgaris* alleles was observed for 24 of the 28 loci. Congruity and recurrent backcross interspecific hybrids between common bean (*P. vulgaris*) and tepary bean (*P. acutifolius*) were compared for the amount of introgression occurring between genomes by AFLP markers (Muñoz et al., 2004). The level of introgression of tepary bean marker bands into the common bean genome background was higher in the congruity backcross lines than in the recurrent backcross-derived genotypes. Congruity backcrossing was suggested to increase introgression rates between the species and to transfer oligogenic traits from tepary bean to common bean.

ITS-RFLP marker analysis of putative intergeneric *Pisum sativum* × *Lathyrus sativus* and interspecific *P. sativum* cv. Baccara × *P. fulvum* D 76 (1) hybrids demonstrated unequivocally the true hybrid nature of interspecific hybrids (Ochatt et al., 2004). The so-called intergeneric hybrids were not hybrids at all. RAPD markers have been utilized successfully in characterizing *Arachis batizocoi* × *A. chacoense* synthetic amphidiploids, and hybrids between this amphidiploid and *A. hypogaea* (Lanham et al., 1992). Forty-six introgression lines from a cross between *Arachis hypogaea* and *A. cardenasii* were analyzed by 73 RFLP probes and 70 RAPD primers for the introgression of *A. cardenasii* chromosome segments (Garcia et al., 1995). Thirty-four RFLP probes and 45 RAPD primers identified putative *A. cardenasii* introgressed chromosome segments in a few lines. The smallest introgressed fragment was detected by a single RFLP marker and the largest was detected by three or four adjacent markers representing introgressed segments of 30 to 40 cM.

## *DNA Marker Diagnostics for Cultivars and Breeding Lines*

For numerous beneficial reasons, for instance, protection of cultivars from pilferage, protection of intellectual property rights, and so on, cultivar diagnostics is a prerequisite for their commercialization. Such a technique was long needed, but there was no foolproof technology to achieve the desired goals to perfection. Now DNA marker diagnostics of the cultivars, breeding lines, and any other resource material of importance is an ultimate technology for achieving the desired goals. For example, microsatellites (Rongwen et al., 1995), RAPD (Chowdhury et al., 2002), and arbitrarily primed PCR (AP-PCR), SSR, and ISSR (Brick and Sivolap, 2001) markers have been employed for cultivar identification in soybean. Of these, the investigation by Brick and Sivolap (2001) deserves special mention as not only individual marker techniques could differentiate all 19 cultivars but also the cultivar formula (the formula of the allelic state of fixed loci that reflects the specificity of the given genotype and permits its identification and differentiation from others), also called cultivar certificate, was assigned to each cultivar by the markers individually. Similarly, randomly amplified microsatellite sequences (RAMS) and STMS markers have been employed in genotyping pea cultivars that represented the genetic base for the Australian field pea-breeding program (Ford et al., 2002). Twenty-four commercial lines of *Phaseolus vulgaris* have been evaluated by RFLP, directed amplification of minisatellite DNA- PCR (DAMD-PCR), ISSR, and RAPD marker analysis (Métais et al., 2000). A repeat sequence of 130 bp containing seven repeats of a 15 bp minisatellite sequence isolated from the common bean genome (Métais et al., 1998) was utilized as a probe in RFLP as well as for designing primers in DAMD-PCR. Genotyping of cowpea varieties by RAPD markers revealed some DNA bands specific to higher or lower nitrogen-fixing varieties, which suggested that some genes could govern the higher nitrogen fixation characters in this crop (Fall et al., 2003).

## *Construction of Linkage Maps*

With the advent of DNA markers, high-resolution linkage maps have been developed in many crop legumes. Table 33.1 lists molecular DNA marker-based maps in legumes. Prior to the availability of DNA markers, maps based, for example, on macromuations, morphological and cytogenetical criteria, or biochemical markers provided fewer opportunities to geneticists and breeders to identify the source of genetic variation in the germplasm resources or traits. DNA marker-based maps have many advan-

TABLE 33.1. DNA marker-based linkage maps in grain legumes.

| Species | Marker (s) employed | Reference |
|---|---|---|
| *Cicer arietinum* | RAPD, RFLP | Simon and Muehlbauer, 1997 |
| | STMS | Winter et al., 1999 |
| | STMS, DAF, AFLP, ISSR, RAPD, cDNA, SCAR | Winter et al., 2000 |
| | STMS, RGA | Tekeoglu et al., 2002 |
| | STMS, AFLP, DAF, ERAP | Pfaff and Kahl, 2003 |
| | STMS, ISSR, RGA | Flandez-Galvez, Ford, et al., 2003 |
| *Glycine max* | RFLP | Keim et al., 1990; Shoemaker and Olson, 1993 |
| | AFLP | Keim et al., 1997 |
| | RAPD, RFLP | Ferreira et al., 2000 |
| | RFLP, SSR, RAPD, AFLP, SCAR | Liu et al., 2000 |
| | AFLP, SCAR | Yang et al., 2000 |
| *Phaseolus vulgaris* | RFLP | Vallejos et al., 1992; Nodari, Tsai, Gilbertson, et al.,1993; Gepts et al., 1993; Boutin et al., 1995; Yu et al., 1998 |
| | RAPD | Jung et al., 1996, 1997; Skroch et al., 1996 |
| | RAPD and RFLP | Freyre et al., 1998 |
| | SSR | Yu et al., 2000 |
| | SSR | Blair et al., 2003 |
| *Pisum sativum* | RFLP | Ellis et al., 1992 |
| | RBIP | Ellis et al.,1998 |
| | cDNA-RFLP, RAPD, AFLP, EST | Gilpin et al., 1997 |
| | RAPD, SSR | Laucou et al., 1998; Loridon et al., 2005 |
| *Arachis hypogaea* | RFLP | Burrow et al., 2001 |
| | SSR | Moretzsohn et al., 2005 |
| *Vigna unguiculata* | cDNA-RFLP, RAPD, AFLP | Menendez et al., 1997 |
| | DAF, AFLP | Li et al., 1999 |
| | AFLP, RAPD, RFLP | Ouedraogo, Gowda, et al., 2002 |

| | | |
|---|---|---|
| *Vicia faba* | RFLP, RAPD<br>RAPD | Van de Ven et al., 1991<br>Vaz Patto et al., 1999 |
| | RAPD, SSR | Roman et al., 2004 |
| *Lens culinaris* | RFLP | Havey and Muehlbauer, 1989; Tahir et al.,1993 |
| | RAPD, AFLP | Eujayl et al., 1998 |
| | RAPD, ISSR, RGA | Rubeena et al., 2003 |
| | RAPD, ISSR, AFLP, SSR | Durán et al., 2004 |
| *Medicago truncatula* | RAPD, AFLP, gene markers<br>EST, BAC, RGA | Thoquet et al., 2002<br>Choi et al., 2004 |
| *Lotus japonicus* | AFLP, SSRP, dCAPS<br>AFLP, RAPD, SSR, RFLP, gene markers | Hayashi et al., 2001<br>Sandal et al., 2002 |

tages in identifying markers linked to genes of agronomic or horticultural importance and bringing out the relationship between physical and genetic distances, and determining linkage drag.

In *Phaseolus vulgaris* (common bean, $2n = 22$), RFLPs were principally used as framework markers to develop linkage maps (Vallejos et al., 1992; Gepts et al., 1993; Nodari, Tsai, Gilbertson, et al., 1993; Nodari, Tsai, Guzmán, et al., 1993; Boutin et al., 1995; Yu et al., 1998). RAPD marker-based maps (Jung et al., 1996, 1997; Skroch et al., 1996) were aligned with these RFLP maps to produce a core map based on recombinant inbred population BAT 93 × lalo EEP 558 (Freyre et al., 1998). The core map had a total length of 1226 cM and comprised 563 markers including 120 RFLP and 430 RAPD markers. This consensus map has been subsequently saturated with microsatellites (Yu et al., 2000; Blair et al., 2003). Fifteen gene-based microsatellite markers were placed on an integrated linkage map on seven linkage groups (Yu et al., 2000). The map of Blair et al. (2003) was based on two RIL populations, 'DOR 364 × G198333' and 'BAT93 × JaloEEP558' and had genomic as well as gene-based microsatellites.

In *Glycine max* (soybean, $2n = 40$) the first map was published in 1990 (Keim et al., 1990) using RFLP markers, which was then expanded with the addition of more than 350 RFLP loci (Shoemaker and Olson, 1993). Shoemaker and Specht (1995) achieved partial integration of the various marker types into a common linkage map. A total of 606 SSR loci were mapped in one or more of the three mapping populations (Cregan et al., 1999), which included 544 new loci that were not previously reported by either Akkaya et al. (1995) or Mansur et al. (1996). Later, a high-density genetic map was developed based on AFLP markers (Keim et al., 1997). More recently, a linkage map of *Glycine max* carrying 356 markers was established by anchoring 106 RAPD markers to an existing RFLP map (Ferreira et al., 2000). The RAPD markers showed similar distribution throughout the genome and identified a level of polymorphism similar to the RFLP marker. By using a subset population to anchor the RAPD markers, it was possible to enhance the throughput of selecting and adding reliable marker loci to the existing map. Currently, over 1,000 SSR markers have been positioned in the soybean genetic map (Song et al., 2004). The recent single nucleotide polymorphism (SNP) marker development (Van et al., 2004) from ESTs will facilitate construction of an SNP-based linkage map in soybean, thus fulfilling the gaps in existing maps.

RAPD markers were also used to construct a genetic linkage map of *Pisum sativum* (pea, $2n = 14$) in a population of 139 recombinant inbred lines (Laucou et al., 1998). This map included nine linkage groups covering 1,139 cM and was aligned with the RFLP map (Ellis et al., 1992), followed by further enrichment with anchored PCR markers corresponding to MADS-box genes and *Ty1*-copia-like insertions (Ellis et al., 1998). Another map based on $F_2$ plants consisted of 209 markers covering 1,330 cM (Gilpin et al., 1997); 29 loci representing genes of known function have been incorporated onto this map. Linkage maps based on linkage relationships from multiple maps and linkage studies (Ellis and Poyser, 2002) and microsatellites (Loridon et al., 2005) have been reported.

Havey and Muehlbauer (1989) were the first to include RFLP in genetic linkage mapping of *Lens culinaris* ssp. *culinaris* (lentil, $2n = 14$). Weeden et al. (1992) published the 560 cM linkage map of lentil and showed that about 40 percent of the linkage maps of lentil and pea were conserved. A compiled map of 10 arbitrary linkage groups was developed by Tahir et al. (1993). RAPD and AFLP markers enabled construction of saturated maps in this species (Eujayl et al., 1997, 1998) followed by the localization of genes of agronomic interest for faster improvement in breeding programs (Tullu et al., 2003). The use of RILs as mapping populations helped overcome the limitations such as the dominant mode of inheritance of RAPD and AFLP markers for genetic mapping, thus paving the way for construction of an extensive genetic linkage map spanning 1,073 cM with 177 markers (Eujayl et al., 1998). To explore the utility of ISSR and resistant gene analog (RGA) markers in mapping of the lentil genome, a linkage map was developed with 114 markers (100 RAPD, 11 ISSR, and 3RGA) using an $F_2$ population (Rubeena et al., 2003). Gene loci governing desirable traits such as ascochyta blight resistance were also located. In order to maximize polymorphism for map construction in lentil, intersubspecific hybrid populations were also used. Durán et al. (2004) have developed a map based on the segregation analysis of five kinds of molecular and morphological markers in 113 $F_2$ plants obtained from a single hybrid of *L. culinaris* ssp. *culinaris* x *L. culinaris* ssp. *orientalis*.

Simon and Muehlbauer (1997) developed the first genetic linkage map of *Cicer arietinum* (chickpea, $2n = 16$) genome using RFLP and RAPD markers. The map also had morphological and isozyme markers. Since microsatellite markers successfully revealed intraspecific polymorphism in *C. arietinum* in comparison to other markers, STMS markers became the principal framework marker for generation of high-density marker maps for chickpea genome. One hundred and seventy four STMS loci characterized from the chickpea genome have proven to be polymorphic in chickpea at an intraspecific level (Flandez-Galvez, Ford, et al., 2003). The map constructed by Winter et al. (2000) utilizing STMS, DAF, AFLP, ISSR, RAPD, cDNA, and sequence characterized amplified region (SCAR) markers is the most extensive linkage map currently available for *C. arietinum* × *C. reticulatum* hybrid genomes. A total of 118 STMS, 96 DAF, 70 AFLP, 37 ISSR, 17 RAPD, 3 cDNA, and 2 SCAR markers were mapped in the RILs. In addition, three loci that confer resistance against different races of fusarium wilt were also mapped. The maps developed thereafter involved integration of (1) gene-specific markers onto existing maps based on STMS, AFLP, and DAF markers (Pfaff and Kahl, 2003); and (2) resistance gene analogs (RGAs) on the maps (Flandez-Galvez, Ford, et al., 2003; Tekeoglu et al., 2002). The most recent map (Cobos et al., 2005) based on two RILs derived from intraspecific crosses with a common parental line (JG62) was developed by employing RAPD, STMS, morphological (flower color, seed coat thickness, single/double podding locus), and resistant gene (FOC-0, gene for resistance to fusarium wilt race 0) markers. Flower color (B/b) and seed coat thickness (Tt/tt) appear to be linked to STMS. The single/double podding locus was located on LG 9 (total 10 LGs obtained) jointly with two RAPD markers and STMS marker TA 80. LG 3 included a gene for resistance to FOC-0 (Foc 01/foc 01) flanked by RAPD marker $OPJ20_{600}$ and STMS marker TR 59. This is the first map to assign the resistance gene FOC-0 to any linkage group.

In *Arachis hypogaea* (peanut, $2n = 40$), the only tetraploid genetic linkage map currently available is based on an interspecific cross (Burow et al., 2001). A $BC_1$ population consisting of 78 progeny from the cross *A. hypogaea* cv. Florunner ($2n = 4x = 40$) × [(*A. batizocoi* × (*A. cardenasii* × *A. diogoi*)) 4*x*] was used for mapping. Approximately 370 RFLP loci have been mapped onto 23 linkage groups comprising a total map length of approximately 2,210 cM. Analysis of the origin of markers on each linkage group indicated disomic pairing, with the largest number of markers derived from *A. batizocoi* (56 percent of the total) and the remainder from *A. cardenasii* (20 percent) and *A. diogoi* (16 percent). To compare the tetraploid and diploid peanut maps, results from the tetraploid map of *A. hypogaea* were compared to a 115 marker RFLP map made from a cross between two wild diploid peanut species (*A. cardenasii* × *A. stenosperma;* Halward et al., 1993). This was extended to *A. hypogaea* by mapping the $F_{10}$ $C_9$ progeny of *A. cardenasii* × *A. hypogaea* cross (Garcia et al., 1995). In this cross, 11 linkage groups representing only the A genome of *A. hypogaea* were identified. Then 23 cDNAs mapped as RFLP markers on the diploid map (*A. cardenasii* × *A. stenosperma*; Halward et al., 1993) were hybridized to the tetraploid mapping population, and five sets of corresponding linkage groups were identified. More recently, a linkage map based on microsatellite markers was constructed using an $F_2$ population obtained from a cross between two diploid wild species with AA genome (*A. duranensis* and *A. stenosperma;* Moretzsohn et al., 2005). This map consists of 11 linkage groups covering 1,230.89 cM of total map distance.

Van de Ven et al. (1991) used RFLP and RAPD markers in establishing the first linkage groups in *Vicia faba* (faba bean, $2n = 12$). Since then, specific genomic regions have been saturated with RAPD and RFLP markers (Torres et al. 1993; Satovic et al., 1996; Vaz Patto et al., 1999). More recently, Roman et al. (2004) developed a composite map based on RAPDs, seed protein genes, and microsatellite markers in combination with morphological markers and isozymes. The map incorporates data from 11 $F_2$ families for a total of 654 individuals all sharing the common female parent Vf6. This integrated map is arranged in 14 linkage groups, 5 of which were located on specific chromosomes. These linkage groups include 192 loci covering 1,559 cM with an overall average marker interval of 8 cM.

In *Vigna unguiculata* (cowpea, $2n = 22$), Fatokun, Danesh, Menancio-Hautea, et al., (1993) made the first attempt to generate a comprehensive linkage map. They used polymorphism detected by 87 random genomic DNA fragments, 5 cDNAs, and RAPD markers that generated a map consisting of 10 linkage groups spanning 680 cM. Further improvement on this initial map was made by Menéndez et al. (1997), who developed a linkage map for cowpea consisting of 181 loci falling into 12 linkage groups. Thereafter, linkage maps were constructed (Li et al., 1999; Ouédraogo, Gowda, et al., 2002) by using DAF, AFLP, RFLP, and RAPD markers in combination with biochemical markers and biological resistance traits in a recombinant inbred population derived from a cross between IT 84S-2049 and 5L4B.

The impetus toward the construction of linkage maps in so-called model legumes, *Medicago truncatula* and *Lotus japonicus,* had arisen primarily from difficulties in genetic analysis of large and complicated legume genomes of interest. Construction of linkage maps in model legumes with small genomes could boost the studies on genetics, genomics, and breeding of grain legumes such as pea, chickpea, and beans. Various applications of model legume maps are discussed in the section Comparative Mapping and Map Based Cloning later in this chapter.

## *Trait Tagging and Marker-Assisted Selection*

Considerable progress has been made in tagging and mapping many agriculturally important genes with DNA markers in grain legumes (Table 33.2). The fundamental advantage of tightly linked DNA markers to these traits, for instance, disease resistance, will not only provide the opportunity to efficiently screen resistant genotypes at early stages of development but also facilitate the cloning of underlying genes. In common bean, RAPD markers have been used extensively in tagging important disease-resistant traits such as rust, anthracnose, and angular leaf spot resistance. Bean rust is a disease that results in reduced bean yield and lowers the quality of beans as well. The pathogenic variability of the fungus is broad, with over 300 races recognized. Bulked segregant analysis was used to detect different rust resistance genes in the germplasm of Mesoamerican (*Ur-3, Ur-5,* and *Ur-11*) and Andean (*Ur-4, Ur-6,* and *Ur-9*) origin by means of RAPD markers (Park et al., 2003). RAPD markers were detected in coupling phase linkage with the *Ur-7* gene (Park et al., 2003). RAPD marker (OBC06.300) linked in coupling phase with *Ur-6* gene (Park et al., 2004) and AFLP marker (EAAC/MACC. 405) linked to *Ur-13* gene (Mienie et al., 2005) were converted to SCAR markers. These markers linked to rust resistance genes are useful in pyramiding resistance genes into a single cultivar for durable rust resistance. Park, Coyne, Bokesi, et al. (1999) reported close linkage between markers tagged to a rust resistance gene (*Ur-9*) and a gene for indeterminate growth habit (*fin*), an association between a quantitative and a qualitative trait. In addition, three independent RAPD markers tagged to anthracnose resistance genes *Co-1, Co-5,* and *Co-6* have been reported (Young and Kelly, 1997). Angular leaf spot (ALS), caused by the fungus *Phaeoisariopsis griseola,* is one of the most serious diseases of common bean in tropical and subtropical bean-producing regions. Investigations have shown RAPD, SSR, and AFLP markers linked to angular leaf spot resistant alleles in line 'ESAL550' and common bean accession G10474 (Da

TABLE 33.2. Genes tagged with DNA markers.

| Species | Trait | Genes tagged | Population | Linkage group | DNA marker linked | Reference |
|---|---|---|---|---|---|---|
| *Pisum sativum* | Fusarium wilt resistance | *Fw* | $F_2$ 'Erygel × 661' | | RFLP | Dirlewanger et al., 1994 |
| | Powdery mildew resistance | *er* | $F_2$ 'Erygel × 661' | | | |
| | Pea common mosaic virus resistance | *Mo* | $F_2$ 'Erygel × 661' | | | |
| | Plant architecture | *det* | $F_2$ 'JI2121 × Térèse' | LG V | RAPD, SCAR | Rameau et al., 1998 |
| | | *fa* | $F_2$ 'JI814 × Térèse' | LG IV | | |
| | | *sn* | $F_2$ 'HL59 × Térèse' | LG VII | | |
| | | *Dne* | $F_2$ 'K218 × Térèse' | LG III | | |
| | | *rms2* | $F_2$ 'K524 × Térèse' | LG I | | |
| | | *rms4* | $F_2$ 'M3T-946 × Torsdag' | LG VII | | |
| | | *Rms3* | $F_2$ 6 parents in cross | LG II | | |
| | Bacteriod and symbiosome differentiation | *Sym31* | $F_2$ 'sprint-Fix-2 × NGB101238' | LG III | DAF | Men et al., 1999 |

| | | | | | | |
|---|---|---|---|---|---|---|
| *Pisum sativum* | Flower color | *A* | $F_2$ 'L-111 × Chi-15' | | RAPD | Cheghamirza et al., 2002 |
| | | | $F_2$ 'L-1238 × Vio' | | | |
| | Seed color | *I* | $F_2$ 'L-1238 × Chi-15' | | | |
| | Pod color | *Gp* | $F_2$ 'L-1238 × Chi-15' | | | |
| | | | $F_2$ 'L-1238 × Vio' | | | |
| | Seed form | *R* | $F_2$ 'L-1238 × Xa-18' | | AFLP | Von Stackelberg et al., 2003 |
| | Seeds linkage | *S* | $F_2$ 'L-1238 × Chi-15' | | | |
| | Leaf color | *Chi-15* | $F_2$ 'L-111 × Chi-15' | | | |
| | Pod shattering | *Xa-18* | $F_2$ 'L-1238 × Xa-18' | | | |
| | | | $F_2$ 'L-84 × Xa-18' | | | |
| | | *Def* | $F_2$ 'DGV × PF' | | | |
| | Powdery mildew resistance | er | $F_2$ 'DMR11$^{er}$ × DMR11$^{Er}$' | LG VI | RAPD SCAR | Janila and Sharma, 2004 |
| | Plant color | *chi115* | $F_2$ 'WL1238 × Chi115' | LG III | RAPD, ISSR | Cheghamirza et al., 2004 |
| | Rust resistance | *Ruf* | $BC_1F_1$ 'HUVP1 × (HUVP1 × FC1)' | | RAPD | Vijayalakshmi et al., 2005 |
| *Cicer arietinum* | Fusarium wilt resistance | | RIL '*C. arietinum* (ICC-4958) × *C. reticulatum* (PI489777)' | LG VI | ISSR | Ratnaparkhe et al., 1998 |
| | Fusarium wilt resistance | Resistance gene cluster | RIL '*C. arietinum* (ICC-4958) x *C. reticulatum* (PI489777)' | LGII, LGIII, LGV, LGVI | RGA | Huttel et al., 2002 |

TABLE 33.2 *(Continued)*

| Species | Trait | Genes tagged | Population | Linkage group | DNA marker linked | Reference |
|---|---|---|---|---|---|---|
| | Double podding | *s* | NIL 'CA-2156 × JG-62'<br>RIL 'surutato-77 × JG-62' | LG VI | STMS | Rajesh, Tullu, et al., 2002 |
| *Glycine max* | Phytophthora root rot resistance | *Rps1*<br>*Rps2*<br>*Rps3*<br>*Rps4* | NIL 'Clark × Harosoy'<br>NIL 'Williams × Harosoy' | LGN, LGJ, LGF, LGG | SSR | Demirabas et al., 2001 |
| | Frogeye leaf spot resistance | | $F_2$ 'Parana × Bossier'<br>$F_2$ 'Cristalina × Bossier'<br>$F_2$ 'Uberaba × Bossier' | LG J | RAPD, SCAR | Filho et al., 2002 |
| | Grassy-beany and bitter flavors | *$Lx_2$* | RIL 'Pureunkong × Jinpumkong' | LG F | SSR, SNP | Kim et al., 2004 |
| | Soybean mosaic virus resistance | *Rsv1* and *Rsv3* | $F_2$ 'Myeongjunamulkong × Lee68' | LG F and LG B2 | RAPD RFLP, SNP | Jeong and Saghai Maroof, 2004 |
| *Phaseolus vulgaris* | Rust resistance | *Rust resistance gene block* | $BC_6F_2$ 'BBL-47'<br>$BC_5F_2$ 'slenderette' | | RAPD | Haley et al., 1993 |
| | Anthracnose resistance | *Co-1, Co-5, and Co-6* | Heterogeneous inbred populations | | RAPD | Young and Kelly, 1997 |

| | | | | | | |
|---|---|---|---|---|---|---|
| | Rust resistance | *Rust resistance gene block* | $F_2$ 'Ouro Negro × US pinto 111' | | RAPD | Faleiro et al., 2000 |
| | Rust resistance | *Ur-9* | $F_2$ 'PC-50 × Chichara 83-109' | | RAPD | Park et al., 1999 |
| | Indeterminate growth habit | *Fin* | | | | |
| *Phaseolus vulgaris* | Seed coat color | *C* | RIL 'BAT 93 × Jalo EEP558' | B8 | RAPD | McClean et al., 2002 |
| | | *G* | | B4 | | |
| | | *V* | | B6 | | |
| | Flower, seed coat color | *P* | RIL 'BAT 93 × Jalo EEP558' | B7 | RAPD | Erdmann et al., 2002 |
| | Rust resistance | *Ur-7* | $F_2$ 'GN1140 × GN Nebr #1' | LG 11 | RAPD | Park, Coyne, Bokesi, et al., 2003 |
| | | | RIL 'GN Bel Neb-RR-1 × A55' | | | |
| | Angular leaf spot | | $F_{2:3}$ 'ESAL 550 × Carioca MG' | | RAPD, SSR | Da Silva et al., 2003 |
| | Cooking time | | RIL 'Bayo Macentral × Bayo Victoria' | | RAPD | Jacinto-Hernandez, et al., 2003 |
| | Rust resistance | *Ur-6* | $F_2$ 'Olathe × Nebr # 1 Sel. 27' | LG 11 | RAPD, SCAR | Park et al., 2004 |
| | Angular leaf spot | *ALS* | $F_2$ 'G10474 × Sprite' | | AFLP, SCAR | Mahuku et al., 2004 |
| | Rust resistance | *Ur-13* | RIL 'BAT 93 × Jalo EEP558' | LG B8 | AFLP, SCAR | Mienie et al., 2005 |

TABLE 33.2 *(Continued)*

| Species | Trait | Genes tagged | Population | Linkage group | DNA marker linked | Reference |
|---|---|---|---|---|---|---|
| *Arachis hypogaea* | Root knot nematode resistance | *Mae, Mag* | $F_2$ 'GA6 × PI 261942' | LG 1 | RAPD | Garcia et al., 1996 |
| | | | $BC_4F_{42}$ 'TxAG-7 × 5 *A. hypogaea* varieties' | | | Burow et al., 1996 |
| | Resistance to aphid vector | Single recessive gene | $F_2$ 'ICG 12991 × ICGV-SM 93541' | LG 1 | AFLP | Herselman et al., 2004 |
| *Vigna unguiculata* | Resistance to *Striga gesnerioides* race 1 | *Rsg2-1* | $F_2$ 'Tvx 3236 × IT 82D-849' | LG 1 | AFLP | Ouédroago et al., 2001 |
| | race 3 | *Rsg4-3* | $F_2$ 'IT84S-2246-4 × Tvu 14676' | | | |
| | Resistance to *Striga* race 1 | *Rsg3* | $F_2$ 'Gorom × Tvx 3236' | LG VI | AFLP | Ouédroago, |
| | | 994-Rsg | $F_2$ 'Tvx 3236 × IT81D-994' | LG VI | | Tignegre, et al., 2002 |
| | Resistance to *Striga* race 3 | *Rsg1* | $F_2$ 'IT93K-693-2 × IARI696' | | SCAR | Boukar et al., 2004 |

| | | | | | |
|---|---|---|---|---|---|
| *Vicia faba* | Resistance to *Uromyces viciae-fabae* race 1 | *Uvf-1* | $F_2$ '2N52 x VF-176' | RAPD | Avila et al., 2003 |
| *Lens culinaris* | Radiation frost tolerance | *Frt* | $F_2$ 'ILL5588 × L692-16-1' | RAPD | Eujayl et al., 1999 |
| | *Ascochyta* blight resistance | *ral2* | $F_2$ 'Eston × Indianhead' | RAPD, SCAR | Chowdhury et al., 2001 |
| | Resistance to *Colletotrichum truncatum* | *LCt-2* | RIL 'Eston × PI320937' | RAPD, AFLP | Tullu et al., 2003 |
| | Resistance to *Ascochyta lentis* | *ral1* *AbR1* | RIL 'CDC Robin × 964a-46' | SCAR, RAPD | Tar'an, Buchwaldt, et al., 2003 |
| | Resistance to *Colletotrichum truncatum* | a major gene | RIL 'CDC Robin × 964a-46' | | |

TABLE 33.3. SCAR markers linked with disease resistance in *Phaseolus vulgaris.*

| Disease | *Tagged locus* | Linkage group | Reference |
|---|---|---|---|
| Bean common mosaic virus (BCMV) | I | B2 | Melotto et al., 1996 |
| | *bc-3* | B6 | Johnson et al., 1997 |
| | *bc-1²* | B3 | Miklas et al., 2000 |
| | *bgm-1* | | Urrea et al. ,1996 |
| Bean curly top virus (BCTV) | *Ctv-1* | B7 | Larsen and Miklas, 2004 |
| Rust | *Ur-3* | B1 | Nemchin and Stavely, 1998 |
| | *Ur-5* | B4 | Melotto et al., 1998 |
| | *Ur-11* | B11 | Boone et al., 1999 |
| | *Ur-5* | B4 | Miklas et al., 2000 |
| | Ouro Negro | B4 | Correa et al., 2000 |
| | *Ur-3, Ur-11* | B11 | Miklas et al., 2002 |
| | Ouro Negro | B4 | |
| | *Ur-6* | B11 | Park et al., 2003 |
| Anthracnose | *Co-2* | B11 | Geffroy et al., 1998 |
| | *Co-4²* | B8 | Young et al., 1998 |
| | *Co-4²* | B8 | Awale and Kelly, 2001 |
| | *Co-9* | B4 | Mendez de Vigo et al., 2002 |
| | *Co-5* | | Vallejo and Kelly, 2001 |
| | *Co-4²* | B8 | Kelly et al., 2003 |
| Angular leaf spot | *Phg-2* | B8 | Nietsche et al., 2000 |
| | *Phg-2* | B8 | Miklas, 2002 |

Silva et al., 2003; Mahuku et al., 2004). A big list of SCAR markers (Table 33.3) tagged to disease resistance genes in common bean has been compiled by Phil Miklas (Web site: http://www.usda.prosser.wsu.edu/miklas/Scartable3.pdf). In common bean, RAPD markers have also been used to tag genes controlling flower and seed coat color (Erdmann et al., 2002; McClean et al., 2002) and cooking time (Jacinto-Hernandez et al., 2003).

Fusarium wilt caused by *Fusarium oxysporum* is a devastating disease in chickpea. The existence of pathogenic races of *F. oxysporum* is well established, with races 1, 2, 3, and 4 reported from India and races 0, 5, and 6 from Spain. Identification of DNA markers closely linked to fusarium wilt-resistant genes would have considerable potential for marker-assisted selection (MAS) in chickpea. RAPD markers linked to genes for resistance to race 1 (Mayer et al., 1997) and race 4 (Tullu et al., 1998, 1999) in RILs developed from the 'C-104 × WR-315' cross have been identified. An ISSR marker linked to a gene for resistance to fusarium wilt race 4 is also linked to a gene for fusarium wilt race 1 (Ratnaparkhe et al., 1998). This indicated that genes for resistance to races 1 and 4 are closely linked. Such closely linked markers and a complete genetic map can be used to investigate additional genes, which can account for different classes of resistance reported in chickpea. The mapping of resistance gene analogs on linkage maps has been used as a candidate gene approach to identify genes for resistance to various pathogens. This approach has been utilized successfully in chickpea to tag a fusarium resistance gene cluster (Huettel et al., 2002). Double podding trait is one of the important traits governing yield improvement in chickpea. An STMS marker 'TA-80' linked to an s gene, conferring the double podding trait, was identified (Rajesh, Tullu, et al., 2002). The marker could be useful in infering information about the allelic state of the 's' locus and also be effective in eliminating linkage drag during marker-assisted backcrossing.

The peanut root knot nematode (*Meloidogyne arenaria* Chidwood race 1) is responsible for significant economic losses throughout peanut-producing areas. Identifying markers closely linked to nematode resistance genes would help in peanut breeding by MAS and backcross introgression. Two dominant genes that conferred resistance to the root knot nematode in a segregating $F_2$ population derived from a cross between wild *Arachis cardenasii* and *A. hypogaea* were mapped using RAPD and SCAR markers (Garcia et al., 1996). In addition, Burow et al. (1996) reported three RAPD markers linked to a single dominant *M. arenaria* resistance gene. Stalker and Mozingo (2001) identified RAPD markers for early and late leaf spot resistance in a peanut population, which contained the wild species *A. cardenasii* in its pedigree. Herselman et al. (2004) identified 20 putative AFLP markers, of which 12 mapped to five linkage groups, and a single recessive gene was mapped on linkage group 1 for resistance to peanut rosette disease.

In pea, there are important agronomic advantages of controlling plant architecture. Five *Rms* (ramosus) genes and several alleles are known to have direct effects on apical dominance. RAPD markers linked to two morphological markers (fa and det), three ramosus genes (*rms2, rms3,* and *rms4*), and two genes conferring flowering response to photoperiod (*sn, dne*) were identified by bulked segregant analysis of $F_2$ populations (Rameau et al., 1998). In addition, markers linked to genes of other important morphological traits have been identified using RAPD and ISSR markers (Cheghamirza et al., 2002, 2004). RAPD markers linked to *rms3, dne,* and *Chi15* genes have also been converted to SCAR markers. Several mutants defective in the nodulation process and rhizobial endosymbiosis of pea have been identified. Four DAF markers linked to the *Sym31* gene responsible for bacteriod and symbiosome differentiation have been identified using BSA (Men et al., 1999). Two markers flanking the *Sym31* gene cover a 4 to 5 cM interval of pea linkage group 3. Schneider et al. (2002) integrated the map positions of two such nodulation mutations, *Sym9* and *Sym10,* into the molecular map of pea by AFLP-BSA analysis and subsequently mapped by RFLP. Ascochyta blight, fusarium wilt, powdery mildew, and pea common mosaic virus are serious diseases of pea. Single genes govern resistance to these diseases. Molecular markers linked to each resistance gene were found using RFLP, RAPD, and microsatellite markers (Dirlewanger et al., 1994). Screening of powdery midew-resistant cultivar DMR 11 and its susceptible near isogenic line for polymorphism revealed two RAPD markers linked to the *er* gene (Janila and Sharma, 2004). Earlier, SCAR markers linked to *er* were developed in Canadian germplasm (Timmerman-Vaughen et al., 1994; Tiwari et al., 1998). Two RAPD markers flanking a rust resistance gene (*Ruf*) with a distance of 10.8 cM and 24.5 cM were also identified (Vijayalakshmi et al., 2005).

Many resistance or *R* genes have been mapped in soybean (Ashfield et al., 2003). They revealed a complex locus containing several tightly linked families of nucleotide binding site-lucine rich repeats (NBS-LRR) genes. *R* genes present included *Rpg1-b* and *Rpg1-r* (effective against *Pseudomonas syringae;* Ashfield et al., 1998), *Rps3* (effective against *Phytophthora soja;* Diers et al., 1992), *Rsv1* (effective against soybean mosaic virus; Yu et al., 1994; Gore et al., 2002), and *Rpv 1* (effective against peanut mottle virus; Gore et al., 2002; Roane et al., 1983). All these genes mapped within 3 cM of the RFLP marker pR-45. Another RFLP clone, A519, mapped close to *Rsv3* (soybean mosaic virus resistance gene; Jeong et al., 2002). By using a modified allele-specific PCR procedure, Jeong and Saghai Maroof (2004) reported detection and genotyping of SNPs in the genomic DNA fragments tightly linked to two soybean mosaic virus resistance genes, *Rsv1* and *Rsv3*. Filho et al. (2002) reported a SCAR marker linked to soybean frogeye leaf spot caused by *Cercospora sojina.* By using a bulked segregant analysis approach, Demirabas et al. (2001) have identified SSR markers linked to *Rps1, Rps2, Rps3,* and *Rps4,* which are dominant alleles causing monogenic race specific resistance to phytophthora root rot.

The parasitic weed *Striga gesnerioides* is one of the most important constraints to cowpea production. Several host resistance genes have been identified that are effective against specific races of *S. gesnerioides*. Ouédraogo et al. (2001) identified three AFLP markers tightly linked to the resistance gene *Rsg 2-1,* effective against *S. gesnerioides* race 1, and six AFLP markers associated with resistance gene *Rsg 4-3,* effective against race 3. A SCAR marker linked to *Rsg 1* that gives resistance to race 3 of *S. gesnerioides* has been developed from a polymorphic AFLP marker (Boukar et al., 2004). Many DNA markers associated with various biological resistance traits, resistance genes, and RGAs were placed on the genetic map (Ouédraogo, Tignegre, et al., 2002) including markers for resistance to *S. gesnerioides* races 1 and 3, cowpea mosaic virus (CPMV), cowpea severe mosaic virus (CPSMV), southern bean mosaic virus (SBMV), fusarium wilt, and root knot nematodes. Toward a better understanding of the structure and organization of *R* genes in cowpea, Gowda et al. (2002) have identified several RGAs in cowpea that were associated with RFLP markers, which allowed them to be placed on the cowpea genomic map.

In *Vicia faba,* bulked segregant analysis was used to identify RAPD markers linked to a gene determining hypersensitive resistance against race 1 of the rust *Uromyces viciae-fabae* (Avila et al., 2003), which can cause up to 70 percent loss in yield. Three RAPD markers were mapped in the coupling phase while two additional markers were found linked to a gene in the repulsion phase. The RAPD markers linked in repulsion phase to the resistance gene would allow discrimination of homozygous susceptible individuals from heterozygous ones.

Two major diseases, ascochyta blight and anthracnose, threaten lentil production. A major dominant gene *AbR1* and a recessive gene *ral2* are reported for resistance to *Ascochyta lentis*. Two RAPD markers each were found linked to *ral2* and *AbR1* genes (Chowdhury et al., 2001; Ford et al., 1999). For resistance to anthracnose disease, a major dominant gene, namely *LCt-2,* was identified and two RAPD and three AFLP markers were found linked to this locus (Tullu et al., 2003). Employing RAPD and SCAR markers, Tar'an, Buchwaldt, et al. (2003) distinguished between individuals that have the same disease phenotype, but carry different genes so as to identify single genes in pyramided plants. Radiation frost injury is an important abiotic constraint to lentil production in west Asia. A single gene, *Frt,* controls this trait and a RAPD marker has been found linked to the *Frt* locus in an RIL population (Eujayl et al., 1999).

### *Quantitative Trait Loci Analysis*

Most of the agronomic characters like yield and yield components, plant height, or days to flowering are controlled by several genes. Such traits are called quantitative traits, and the loci governing these traits are called quantitative trait loci (QTL). Through the use of molecular linkage maps and QTL-DNA marker association, characterization of quantitatively inherited traits has been facilitated including identifying the genomic regions containing contributing loci, postulating the types of gene action involved, and determining the role of epistatic effects in specifying phenotype (Gilpin et al., 1997). The ability to find an association between a QTL and a DNA marker depends on the magnitude of the QTL's effect on the trait, the size of the population studied, and the recombination frequency between the marker and the QTL. DNA marker analysis for identification of QTLs has mainly been performed using analysis of variance or interval mapping. By combining interval mapping with multiple regression, a further improved method of QTL mapping has been developed by Zeng (1993, 1994). Many QTLs have been identified and mapped by using DNA markers in crop legumes.

In pea and chickpea, QTL analyses using DNA markers have been carried out mainly to detect genomic regions conferring resistance to ascochyta blight. In chickpea, resistance to this disease, caused by the ascomycete *Ascochyta rabiei,* is encoded by two or three QTLs, QTL1, QTL2, and QTL3 (Rakshit et al., 2003). Two RAPD markers, one ISSR and one RAPD marker, and two ISSR markers were found flanked to QTL1, QTL2, and QTL3, respectively (Santra et al., 2000). In addition, Rakshit et al. (2003) reported DAF and STMS markers tightly linked to QTL1 on linkage group 4. Collard et al. (2003) identified RAPD, ISSR, STMS, and RGA markers flanking QTLs for ascochyta blight resistance. While single-locus composite interval mapping (CIM) helped identify six QTLs for ascochyta blight resistance, multiple interval mapping (MIM) resolved only two QTLs (Flandez-Galvez, Ades, et al., 2003). All the QTLs were mapped near an RGA marker. The identification of DNA markers flanking these QTLs will not only help to elucidate the complex resistance against different *Ascochyta* pathotypes in the future, but the markers may also assist in monitoring introgressions, accelerating the recovery of a recurrent parent, and minimizing donor genetic material. Table 33.4 shows that LG 4 is an important region of *Cicer* genome for resistance to ascochyta blight.

In pea, 13 QTLs flanked by RAPD, RFLP, AFLP, SCAR, and STS markers were detected for ascochyta blight resistance, caused by *Mycosphaerella pinodes,* on seven linkage groups. Eight of these QTLs were detected in multiple environments or by multiple traits scores (Timmerman-Vaughan et al., 2002). Similarly, Tar'an, Warkentin, et al. (2003) identified two QTLs for lodging resistance, and three QTLs each for plant height and resistance to *Mycosphaerella* blight using AFLP, RAPD, and STS markers. Recently,

TABLE 33.4. QTL-marker associations.

| Species | Trait | QTLs | Population | Linkage groups covered | Marker linked to QTL | Reference |
|---|---|---|---|---|---|---|
| *Pisum sativum* | Partial resistance to field epidemics of ascochyta blight | 13 | $F_{2:3}$ and $F_{2:4}$ '3148-A88 × Rovar' | 7 | RAPD, RFLP, AFLP, SCAR, STS | Timmerman-Vaughan et al., 2002 |
| | Lodging resistance | 2 | RIL 'Carneval × MP1401' | 10 | AFLP, RAPD, STS | Tar'an, Warkentin, et al., 2003 |
| | Plant height | 3 | RIL 'Carneval × MP1401' | | | |
| | Resistance to *Mycosphaerella* | 3 | RIL 'Carneval × MP1401' | | | |
| | Partial resistance to ascochyta blight at seedling stage | 6 | RIL 'DP × JI 296' | 8 | RAPD, SSR, STS | Prioul et al., 2004 |
| | Adult stage | 10 | RIL 'DP × JI 296' | | | |
| | Common to both stages | 4 | RIL 'DP × JI 296' | | | |
| | Grain yield, seed protein concentration, and early maturity | 11 | RIL 'Carneval × MP1401' | 5 | AFLP, RAPD, STS | Tar'an et al. 2004 |

TABLE 33.4 *(Continued)*

| Species | Trait | QTLs | Population | Linkage groups covered | Marker linked to QTL | Reference |
|---|---|---|---|---|---|---|
| *Cicer arietinum* | Ascochyta blight resistance | QTL-1<br>QTL-2<br>QTL-3 | RIL '*C. arietinum* (FLIP84-92C) × *C. reticulatum* (PI599072)' | LG 6, LG 1, LG 4 | RAPD, ISSR | Santra et al., 2000 |
| | Seedling resistance to ascochyta blight | 2 | $F_2$ 'Lasseter × *C. echinospermum* (PI527930)' | LG 4 | RAPD, ISSR, STMS, RGA | Collard et al., 2003 |
| | Ascochyta blight resistance | Major QTL | $F_2$ 'ICC1 2004 × Lasseter' | LG 3 | RGA | Flandez-Galvez, Ades, et al., 2003 |
| | Ascochyta blight resistance | QTL-1 | RIL '*C. arietinum* × *C. reticulatum*' | LG 4 | DAF, STMS | Rakshit et al., 2003 |
| *Glycine max* | Resistance to corn earworm | 1 major and 2 minor QTLs | $F_2$ 'Cobb × PI 229358' | LG M, LG H, LG D1 | RFLP | Rector et al., 1998 |
| | Brown stem rot resistance | 2 linked QTLs | RIL 'BSR 101 × PI 437654' | LG J | RFLP, AFLP, RGA | Lewers et al., 1999 |
| | Field resistance to soybean cyst nematode race 3 | 2 | RIL and NIL 'Forrest × Essex' | LG G, LG A2 | AFLP | Meksem et al., 2001 |
| | Salt tolerance | Major QTL | $F_2$ 'S-100 × Tokyo' | LG N | RFLP, SSR | Lee et al., 2004 |
| | Ten agronomic traits | 64 | RIL 'Kefeng No. 1 × Nannong 1138-2' | 12 | RFLP, SSR, EST | Zhang et al., 2004 |

| | | | | | | |
|---|---|---|---|---|---|---|
| | Phosphorous deficiency tolerance | 7 | RIL 'Kefeng No. 1 × Nannong 1138-2' | LG F1, LG F2 | RAPD, SSR, EST, and resistance gene loci | Li et al., 2005 |
| *Phaseolus vulgaris* | Seed yield, yield components and plant architecture | 20 | $F_{2:4}$ 'OAC Seaforth × OAC 95-4' | 12 | RAPD, AFLP, SCAR, SSR | Tar'an et al., 2002 |
| | White mold resistance | Major QTL and other significant QTLs | RIL 'Bunsi × Newport'<br>RIL 'Huron × Newport' | LG B2 | RAPD, AFLP | Kolkman and Kelly, 2003 |
| | Seed mass, Ca, Fe, Zn, and tannin content in bean seed | 14 | $F_{2:3}$ 'Bayo Baranda × G-22837' | 5 | AFLP | Guzmán-Maldonado et al., 2003 |
| | Agronomical and architectural traits | 21 | RIL 'WO3391 × OAC Speedvale' | 7 | RAPD, SSR, STS | Beattie et al., 2003 |
| *Vicia faba* | Ascochyta blight resistance | 6 | $F_2$ '29H × VF136'<br>$F_2$ 'IT2246-4 × TVNI' | LG VIII | RAPD | Avila et al., 2004 |
| *Vigna unguiculata* | Seed weight | Major QTL | $F_2$ 'IT2246-4 × TVN1 963'<br>$F_2$ 'VC 3890 × TC 1966' | LG II, LG VI | RFLP | Fatokun et al., 1992 |
| *Lens culinaris* | Winter hardiness | 7 | RIL 'WA8649090 × Precoz.' | LG I, LG IV | RAPD, ISSR, AFLP | Kahraman et al., 2004 |

four stable QTLs flanked by RAPD, SSR, and STS markers were identified for resistance to *Mycosphaerella pinodes* common at seedling and adult stages (Prioul et al., 2004). Selection for high yield, protein-rich seeds, and early maturity has been extensively practiced by pea breeders to develop suitable cultivars with superior performance. Putative QTLs for grain yield, seed protein concentration, and early maturity have been identified utilizing a linkage map consisting of 193 AFLP, 13 RAPD, and one STS markers (Tar'an et al., 2004). Four QTLs each for grain yield and days to maturity, and three QTLs for seed protein concentration were detected.

Based on the construction of genetic linkage maps, QTL mapping has been reported for a number of agronomic traits in soybean (Zhang et al., 2004). Mansur et al. (1996) used an RIL population to investigate the genetic basis of 15 agronomic traits and found that most of the QTLs were clustered. For most traits, major loci appeared to act independently and additively. Lee et al. (1996) detected a major locus of *Dt1* associated with plant height on linkage group L. Using two populations, Mian et al. (1996) identified two QTLs for seed weight on linkage groups F and K, respectively. They were located at similar genomic regions in the two populations. Several ESTs have been found to be linked with QTLs for flowering time, leaflet shape, and leaf area (Yamanaka et al., 2001; Mathews et al., 2001). In addition, many QTLs conferring resistance to corn ear worm (Rector et al., 1998), soybean cyst nematode (Guo et al., 2005), sclerotinia stem rot (Arahana et al., 2001), and salt tolerance (Lee et al., 2004) have been identified. Recently, 184 RILs, developed from soybean varieties Kefeng No. 1 and Nanong 1138-2 were used to identify QTLs associated with phosphorous deficiency tolerance (Li et al., 2005). Seven QTLs were detected and mapped on two linkage groups. Five QTLs were mapped on linkage group F2 and two on linkage group F1.

In cowpea, major quantitative trait loci were located for seed weight (Fatokun et al., 1992) covering two unlinked genomic regions. A significant interaction was detected between the QTLs in the conserved region and an unlinked marker locus.

In faba bean, previously reported linkage maps have allowed the mapping of QTLs associated with seed weight, *Orobanche crenata* resistance, and *Ascochyta fabae* resistance (Vaz Patto et al., 1999; Roman et al., 2002, 2003). Several QTLs for seed weight were identified, the most important of which, located on chromosome 6, explained approximately 30 percent of the total phenotypic variation. By simple interval mapping, Roman et al. (2003) detected two putative QTLs, Af 1 and Af 2, for *Ascochyta fabae* resistance on linkage groups VIII (chromosome 3) and IVa (chromosome 2), respectively. Af 1 displayed pure additive gene action and Af2 showed considerable dominant effect. Avila et al. (2004) used two pathogenically distinct *Ascochyta* isolates, CO99-01 and LO98-01, and identified six QTLs, Af 3, Af 4, Af 5, Af 6, Af 7, and Af 8. The QTLs Af 3 and Af 4 were effective against both isolates; Af 5 was effective against only CO99-01 while Af 6, Af 7, and Af 8 were effective against only LO98-01.

QTL analysis has so far been used in common bean to study disease resistance traits (Kelly and Miklas, 1998), canning quality (Walters et al., 1997; Posa-Macalincag et al., 2000), $N_2$ fixation (Nodari, Tsai, Guzmán, et al., 1993), drought resistance (Schneider et al., 1997), common bacterial blight (Jung et al., 1999; Park, Coyne, Mutlu, et al., 1999), white mold resistance (Miklas et al., 2001; Park et al., 2001; Kolkman and Kelly, 2003), seed mass, Ca, Fe, Zn, and tannin content in the seed (Guzman-Maldonado et al., 2003), and fusarium rot resistance (Schneider et al., 2001). In addition, 21 QTLs were identified over three environments for eight agronomic and architectural traits (Beattie et al., 2003). Some of these QTLs will be useful for early generation selection of tall, upright, high-yielding lines in a breeding program.

Susceptibility to cold temperature limits the use of lentil as a fall-sown winter annual crop in temperate highland areas of the world. Genetics of winter hardiness in lentil were analyzed by QTL analysis (Kahraman et al., 2004). Seven QTLs flanked by RAPD, ISSR, and AFLP markers were detected from three different environments, although only one of them was common to all environments.

### *Comparative Mapping and Map-Based Cloning*

The complex (polyploid nature) and large sized genome, for example, 12,000 Mb in faba bean, 1,115 Mb in soybean, and 4,000 Mb in pea, have not only precluded their in-depth characterization but have also hindered the isolation of agronomically important genes from crop species. Model legumes such as *Lotus japonicus* and *Medicago truncatula* with their small genome size (~500 Mb) can be subjected to the most advanced genome analysis methods. The development of comparative maps between model legumes and crop legumes will facilitate the cloning of agriculturally important genes from the legume crops. This approach, called map-based cloning, relies on the generation of DNA markers closely linked to the target gene. Using markers such as AFLP, RFLP, and gene specific markers, the genetic map of *Lotus japonicus* (Sandal et al., 2002) has been developed into a very effective tool for map-based cloning of symbiotic genes and other genes of interest. Similarly, the genetic linkage map of *Medicago truncatula* by Thoquet et al. (2002) will be useful in cloning *Mtsym6* (gene for symbiosis) and *SPC* (gene for anticlockwise pod coiling) genes. *Medicago truncatula* LG 5 has been shown to have a high degree of macrosynteny and microsynteny with pea LG 1. LG 1 of pea is populated with many symbiosis and resistance-related genes, in particular a non-toll and interleukin receptor RGA (non-TIR RGA) cluster (AW 123698). The close syntenic relatedness between these genomes opens up the opportunities for cloning of resistance genes and also movement of genes across species boundaries by transgenic approaches. The study by Stracke et al. (2004) has served as an excellent example for the feasibility of model legume strategy to clone genes from more complex genomes. In this study, AFLP marker screening and comparative mapping led to successful cloning of *Lotus LjSYM 2* and pea *PsSYM 19* genes. Other examples of the successful application of model legume genomes for the identification of crop genes include the *NIN* orthologue from pea and *HAR1* orthologue from soybean and pea (Stracke et al., 2004). DNA markers linked tightly to genes and local colinearity among crop legume genomes and model legume genomes may facilitate the identification and cloning of more and more genes from crop legumes.

In addition to the above, the information generated by comparative mapping studies has proven useful in predicting the genome organization of species under study, and also linkage relationships in closely related or distant taxa and cross-incompatible taxa. Using this strategy, it has been shown that pea linkage groups I, II, III, IV, and V are largely syntenic to *Medicago* chromosomes 5, 1, 3, 8, and 7, respectively. Pea linkage group VI appears to include genes present on two *Medicago* chromosomes (2 and 6) and in turn one of these chromosomes (6) appears to have sequences present on two pea linkage groups (VI and VII). One small part of pea linkage group III corresponds to a part of *Medicago* chromosome 2. This suggests that a fragment/fusion event, a translocation, and a large duplication in pea distinguish these genomes (http://medicago.toulouse.inra.fr/EU/documents/medicagoleaf.pdf). The comparative mapping studies between *Medicago sativa* and pea (Kaló et al., 2004), and between *Medicago truncatula* and soybean (Yan et al., 2003) have shown that the genome of soybean has evolved by duplications (an ancient polyploidization event), and that of pea by multiple retrotransposition events. *Vigna radiata* and *Phaseolus vulgaris* represent closely related members of the Phaseoleae. The genetic analysis of 22 gene-specific markers and 16 PCR markers revealed a combination of marker colinearity, inferred translocations or duplications, and nonsyntenic loci. Also, 29 of the 38 markers tested revealed evidence of conserved gene order between *M. truncatula* and *V. radiata* (Choi et al., 2004). Comparative mapping analysis of mung bean, common bean, and soybean by Boutin et al. (1995) has revealed that mung bean and common bean have a high degree of conserved linkage groups and marker order. These two genomes have as much as or more in common than the commonality between mung bean (*V. radiata*) and cowpea (*V. unguiculata*). Extensive conservation of linkage relationships has also been reported between pea and lentil (Weeden et al., 1992) and between chickpea, lentil, and pea (Simon and Muelbauer, 1997).

## *CONCLUSIONS*

The utilization of DNA markers in crop legumes has given unambiguous measures of estimation of relationships within the hierarchy of individuals and populations of particular species, which is of fundamental importance to the understanding of evolutionary patterns and processes. The amount, dynamics, and spatial partitioning of genetic variability largely determine the evolutionary and genetic enhancement potential of a species. DNA-based studies, for example, in *Arachis hypogaea* and *Cicer arietinum,* have shown that substantial genetic variation can be unveiled through the use of proper DNA marker technology. In comparison to RAPD and RFLP markers, SSR and AFLP markers have shown more genetic variation in both peanut and chickpea. In legume species in which the genetic diversity has been reported to be low, and genetic diversity assessment projects are freshly initiated, the use of PCR-based markers, which either amplify multiple loci (AFLP, randomly amplified DNA fingerprinting (RAF), and SAMPL) or multiple alleles per locus (SSRs) should be preferred for revealing polymorphisms. The chip-based technologies, albeit the most cost intensive, such as diversity array technology (DArT) also remain promising. At higher taxonomic levels, DNA markers have provided unequivocal understanding of the relationships between species within a genus and provided a retrospective window on how they have evolved.

The expression-based markers are superior to all random markers owing to their linkage with trait locus alleles. However, deployment of such markers is feasible in only those legume species where abundant sequence data is available. The sequence data at present are not available in most legume crops. Data mining of public databases for marker development for legume crops with scarce sequence databases can be an alternative approach. Particularly, other related crop specialists can mine the databases of the model systems for designing markers for marker-assisted selection. For grain legumes such as chickpea, pigeon pea, cowpea, and peanut in which availability of genomic sequences and DNA markers is limited, venturing into such approaches may provide better alternatives than waiting for their genomic information. Bioinformatics, in this regard, can play an important role. For instance, it can be used to screen model species databases for EST-SSRs, which can be used to cross-reference genes between species for enhancing the resolution of comparative genomics studies and identifying conserved genomic regions among species. The significant transferability of *M. truncatula* EST-SSRs (twofold higher than that of genomic SSRs) across three pulse crops, faba bean, chickpea, and pea has been accomplished (Gutierrez et al., 2005). In addition, the information on conservation of chickpea STMS loci between accessions (Choumane et al., 2000) and transferability of STMS across major pulses (Pandian et al., 2000; Choumane et al., 2004) will facilitate overcoming laborious cloning, sequencing, and screening procedures employed in identifying potential species-specific STMS markers. Especially for saturation of less dense genetic linkage maps of legumes of semiarid tropics, such as chickpea, peanut, and pigeon pea, developing markers through the mining of public databases available for model legumes would be advantageous. Using soybean UNIGENE data, 23 primer sets were designed as universal primers for SNP markers, which could serve as useful genetic markers for the study of comparative legume genomics (Web site: http://www22.sede.embrapa.br/labex/perry_cregan_group_poster.pdf).

The major advantage of most miniature inverted repeat transposable element (MITE) based markers is a preference for insertion in or near transcriptionally active genomic regions. This feature may be especially valuable in studying the large genomes of agriculturally important legume species such as peanut, faba bean, soybean, pea, and so on. Having a class of markers that are enriched in genic regions coupled with the ease of isolating MITE markers could expedite chromosome walks and map-based cloning protocols in these species. Mapping studies require a large number of markers distributed over the genome. For such studies, choice of markers such as AFLP, RAF, and SAMPL, each with a high marker index, would be advantageous.

## REFERENCES

Abe, J., D.H. Xu, Y. Suzuki, A. Kanazawa, and Y. Shimmed (2003). Soybean germplasm pool in Asia revealed by nuclear SSRs. *Theoretical and Applied Genetics* 106:445-453.

Abo-elwafa, A., K. Murai, and T. Shimada (1995). Intra- and inter-specific variations in *Lens* revealed by RAPD markers. *Theoretical and Applied Genetics* 90: 335-340.

Afzal, M.A., M.M. Haque, and S. Shanmugasundaram (2004). Random amplified polymorphic (RAPD) analysis of selected mungbean [*Vigna radiata* (L.) Wilczek] cultivars. *African Journal of Plant Sciences* 3:20-24.

Ahmad, M. and D.L. McNeil (1996). Comparison of crossability, RAPD, SDS-PAGE and morphological markers for revealing genetic relationships within and among *Lens* species. *Theoretical and Applied Genetics* 93:788-793.

Ajibade, S.R., N.F. Weeden, and S.M. Chite (2000). Inter simple sequence repeat analysis of genetic relationships in the genus *Vigna. Euphytica* 111:47-55.

Akkaya, M.S., R.C. Shoemaker, J.E. Specht, A.A. Bhagwat, and P.B. Cregan (1995). Integration of simple sequence repeat DNA markers into a soybean linkage map. *Crop Science* 35:1439-1445.

Amadou, H.I., P.J. Bebeli, and P.J. Kaltsikes (2001). Genetic diversity in Bambara groundnut (*Vigna subterranea* L.) germplasm revealed by RAPD markers. *Genome* 44:995-999.

Arahana, V.S., G.L. Graef, J.E. Specht, and J.R. Steadman (2001). Identification of QTLs for *Sclerotinia* stem rot resistance in soybean using SSR markers. *Crop Science* 41:180-188.

Ashfield, T., A. Bocian, D. Held, A.D. Henk, L.F. Marek, D. Danesh, S. Peñuela, K. Meksem, D.A. Lightfoot, N.D. Young, et al. (2003). Genetic and physical localization of the soybean *Rpg1-b* disease resistance gene reveals a complex locus containing several tightly linked families of NBS-LRR genes. *Molecular Plant-Microbe Interactions* 16:817-826.

Ashfield, T., J.R. Danzer, D. Held, K. Clayton, and P. Keim (1998). *Rpg1*, a soybean gene effective against races of bacterial blight, maps to a cluster of previously identified disease resistance genes. *Theoretical and Applied Genetics* 96: 1013-1021.

Avila, C.M., Z. Satovic, J.C. Sillero, D. Rubiales, M.T. Moreno, and A.M. Torres (2004). Isolate and organ-specific QTLs for ascochyta blight resistance in faba bean (*Vicia faba* L.). *Theoretical and Applied Genetics* 108:1071-1078.

Avila, C.M., J.C. Sillero, D. Rubiales, M.T. Moreno, and A.M. Torres (2003). Identification of RAPD markers linked to the *Uvf-1* gene conferring hypersensitive resistance against rust (*Uromyces viciae-fabae*) in *Vicia faba* L. *Theoretical and Applied Genetics* 107:353-358.

Ba, F.S., R.S. Pasquet, and P. Gepts (2004). Genetic diversity in cowpea [*Vigna unguiculata* (L.) Walp.] as revealed by RAPD markers. *Genetic Resources and Crop Evolution* 51:539-550.

Banerjee, H., A. Leele, and N. Seetharama (1998). Random amplified microsatellite polymorphism (RAMPO) analysis of chickpea accessions. *International Chickpea Newsletter* 5:16-18.

Banerjee, H., R.A. Pai, and R.P. Sharma (1999). Restriction fragment length polymorphism and random amplified polymorphic DNA analysis of chickpea accessions. *Biologia Plantarum* 42:197-208.

Baranger, A., G. Aubert, G. Arnau, A.L. Lainé, G. Deniot, J. Potier, C. Weinachter, I. Lejeune-Hénaut, J. Lallemand, and J. Burstin (2004). Genetic diversity within

*Pisum sativum* using protein- and PCR-based markers. *Theoretical and Applied Genetics* 108:1309-1321.

Beattie, A.D., J. Larsen, T.E. Michaels, and K.P. Pauls (2003). Mapping quantitative trait loci for a common bean (*Phaseolus vulgaris* L.). *Genome* 46:411-422.

Becerra Velasquez, L. and P. Gepts (1994). RFLP diversity of common bean (*Phaseolus vulgaris*) in its centre of origin. *Genome* 37:256-263.

Beebe, S., J. Rengifo, E. Gaitan, M.C. Duque, and J. Tohme (2001). Diversity and origin of Andean landraces of common bean. *Crop Science* 41:854-862.

Beebe, S.E., P.W. Skroch, J. Tohme, M.C. Duque, F. Pedraza, and J. Nienhuis (2000). Structure of genetic diversity among common bean landraces of Middle American origin based on correspondence analysis of RAPD. *Crop Science* 40:264-273.

Betal, S., P.R. Chowdhury, S. Kundu, and S.S. Raychaudhuri (2004). Estimation of genetic variability of *Vigna radiata* cultivars by RAPD analysis. *Biologia Plantarum* 48:205-209.

Blair, M.W., F. Pedraza, H.F. Buendia, E. Gaitan-Solis, S.E. Beebe, and P. Gepts (2003). Development of a genome-wide anchored microsatellite map for common bean (*Phaseolus vulgaris* L.). *Theoretical and Applied Genetics* 107:1362-1374.

Boukar, O., L. Kong, B.B. Singh, L. Murdock, and H.W. Ohm (2004). AFLP and AFLP-derived SCAR markers associated with *Striga gesnerioides* resistance in cowpea. *Crop Science* 44:1259-1264.

Boutin, S.R., N.D. Young, T.C. Olson, Z.-H. Yu, R.C. Shoemaker, and C.E. Vallejos (1995). Genome conservation among three legume genera detected with DNA markers. *Genome* 38:928-937.

Briand, L., A.E. Brown, J.M. Lenné, and D.M. Teverson (1998). Random amplified polymorphic DNA variation within and among bean landrace mixtures (*Phaseolus vulgaris* L.) from Tanzania. *Euphytica* 102:371-377.

Brick, A.F. and Y. M. Sivolap (2001). Molecular genetic identification and certification of soybean (*Glycine max* L.) cultivars. *Russian Journal of Genetics* 37:1061-1067.

Brown-Guedira, G.L., J.A. Thompson, R.L. Nelson, and M.L. Warburton (2000). Evaluation of genetic diversity of soybean introductions and North American ancestors using RAPD and SSR markers. *Crop Science* 40: 815-823.

Burow, M.D., C.E. Simpson, J.L. Starr, and A.H. Paterson (2001). Transmission genetics of chromatin from a synthetic amphidiploid to cultivated peanut (*Arachis hypogaea* L.): broadening the gene pool of a monophyletic polyploid species. *Genetics* 159:823-837.

Burow, M.D., J.L. Starr, C.E. Simpson, and A.H. Paterson (1996). Identification of RAPD markers in peanut (*Arachis hypogaea*) associated with root-knot nematode resistance derived from *A. cardenasii*. *Molecular Breeding* 2:307-319.

Burstin, J., G. Deniot, J. Potier, C. Weinachter, G. Aubert, and A. Baranger (2001). Microsatellite polymorphism in *Pisum sativum*. *Plant Breeding* 120:311-317.

Caicedo, A.L., E. Gaitán, M.C. Duque, O.T. Chica, D.G. Debouck, and J. Tohme (1999). AFLP fingerprinting of *Phaseolus lunatus* L. and related wild species from South America. *Crop Science* 39: 1497-1507.

Cheghamirza, K., O. Koveza, F. Konovalov, and S. Gostimsky (2002). Identification of RAPD markers and their use for molecular mapping in pea (*Pisum sativum* L.). *Cellular and Molecular Biology Letters* 7 (2B):649-655.

Cheghamirza, K., O.V. Koveza, F.A. Konovalov, and S.A. Gostimsky (2004). Identification and mapping of *chi115* gene and DNA markers linked to it in pea (*Pisum sativum* L.). *Russian Journal of Genetics* 40:737-742.

Choi, H., D. Kim, T. Uhm, E. Limpens, H. Lim, J. Mun, P. Kalo, R.V. Penmetsa, A. Seres, O. Kulikova, et al. (2004). A sequence-based genetic map of *Medicago truncatula* and comparison of marker colinearity with *M. sativa. Genetics* 166:1463-1502.

Choumane, W., P. Winter, M. Baum, and G. Kahl (2004). Conservation of microsatellite flanking sequences in different taxa of Leguminosae. *Euphytica* 138: 239-245.

Choumane, W., P. Winter, F. Weigand, and G. Kahl (2000). Conservation and variability of sequence-tagged microsatellite sites (STMSs) from chickpea (*Cicer arietinum* L.) within the genus *Cicer. Theoretical and Applied Genetics* 101: 269-278.

Chowdhury, A.K., P. Srinives, P. Tongpamnak, P. Saksoong, and P. Chatwachirawong (2002). Genetic relationship among exotic soybean introductions in Thailand: Consequence for varietal registration. *Science Asia* 28:227-239.

Chowdhury, M.A., C.P. Andrahennadi. A.E. Slinkard, and A. Vandenberg (2001). RAPD and SCAR markers for resistance to ascochyta blight in lentil. *Euphytica* 118:331-337.

Christensen, S., B.W. Rampton, D.C. Butler, R. Papa, E.N. Jellen, P.J. Maughan, and M.R. Stevens (2005). Assessing genetic diversity in tepary bean using orthologous SSR primers derived from common bean. Paper presented at Plant and Animal Genome Conference, San Diego, January 15-19.

Cobos, M.J., M.J. Fernández, J. Rubio, M. Kharrat, M.T. Morreno, J. Gill, and T. Millán (2005). A linkage map of chickpea (*Cicer arietinum* L.) based on populations from Kabuli × Desi crosses: Location of genes for resistance to fusarium wilt race 0. *Theoretical and Applied Genetics* 110:347-1353.

Collard, B.C.Y., E.C.K. Pang, P.K. Ades, and P.W.J. Taylor (2003). Preliminary investigation of QTLs associated with seedling resistance to ascochyta blight from *Cicer echinospermum,* a wild relative of chickpea. *Theoretical and Applied Genetics* 107:719-729.

Coulibaly, S., R.S. Pasquet, R. Papa, and P. Gepts (2002). AFLP analysis of the phenetic organization and genetic diversity of *Vigna unguiculata* L. Walp. reveals extensive gene flow between wild and domesticated types. *Theoretical and Applied Genetics* 104:358-366.

Cregan, P.B., T. Jarvik, A.L. Bush, R.C. Shoemaker, K.G. Lark, A.L. Kahler, N. Kaya, T.T. Van Toai, D.G. Lohnes, J. Chung, and J.E. Specht (1999). An integrated genetic linkage map of the soybean genome. *Crop Science* 39:1464-1490.

Da Silva, G.F., J.B. dos Santos, and M.A.P. Ramalho (2003). Identification of SSR and RAPD markers linked to a resistance allele for angular leaf spot in the common bean (*Phaseolus vulgaris*) line ESAL 550. *Genetics and Molecular Biology* 26:459-463.

Delgado-Salinas, A., A. Bruneau, and J.J. Doyle (1993). Chloroplast phylogenetic studies in New World Phaseolinae (Leguminosae: Papilionoideae: Phaseoleae). *Systematic Botany* 18:6-17.

Delgado-Salinas, A., T. Turley, A. Richman, and M. Lavin (1999). Phylogenetic analysis of the cultivated and wild species of *Phaseolus* (Fabaceae). *Systematic Botany* 24:438-460.

Demirabas, A., B.G. Rector, D.G. Lohnes, R.J. Fioritto, G.L. Graef, P.B. Cregan, R.C. Shoemaker, and J.E. Specht (2001). Simple sequence repeat markers linked to the soybean *Rps* genes for Phytophthora resistance. *Crop Science* 41:1220-1227.

Diers, B.W., L. Mansur, J. Imsande, and R.C. Shoemaker (1992). Mapping phytophthora resistance loci in soybean with restriction fragment length polymorphism markers. *Crop Sciences* 32:377-383.

Dirlewanger, E., P.G. Isaac, S. Ranade, M. Belajouza, R. Cousin, and D. de Vienne (1994). Restriction fragment length polymorphism analysis of loci associated with disease resistance genes and developmental traits in *Pisum sativum* L. *Theoretical and Applied Genetics* 88:17-27.

Diwan, N. and P.B. Cregan (1997). Automated sizing of fluorescent labeled simple sequence repeat (SSR) markers to assay genetic variation in soybean. *Theoretical and Applied Genetics* 95:723-733.

Doi, K., A. Kaga, N. Tomooka, and D.A. Vaughan (2002). Molecular phylogeny of genus *Vigna* subgenus *Ceratotropis* based on rDNA ITS and *atpB-rbcL* intergenic spacer of cpDNA sequences. *Genetica* 114:129-145.

Doyle, J.J. and R.N. Beachy (1985). Ribosomal gene variation in soybean (*Glycine*) and its relatives. *Theoretical and Applied Genetics* 70:369-376.

Doyle, J.J., J.L. Doyle, and A.H.D. Brown (1990). A chloroplast-DNA phylogeny of the wild perennial relatives of soybean (*Glycine* subgenus *Glycine*): Congruence with morphological and crossing groups. *Evolution* 44:371-389.

Durán, Y., R. Fratini, P. Garcia, and M. Pérez de la Vega (2004). An intersubspecific genetic map of *Lens*. *Theoretical and Applied Genetics* 108:1265-1273.

Eichenberger,K., F. Gugerli, and J.J. Schneller (2000). Morphological and molecular diversity of Swiss common bean cultivars (*Phaseolus vulgaris* L., Fabaceae) and their origin. *Botanica Helvetica* 110:61-77.

Ellis, T.H.N. and S.J. Poyser (2002). An integrated and comparative view of pea genetic and cytogenetic maps. *New Phytologist* 153:17-25.

Ellis, T.H.N., S.J. Poyser, M.R. Knox, A.V. Vershinin, and M.J. Ambrose (1998). Polymorphism of insertion sites of Ty1-copia class retrotransposons and its use for linkage and diversity analysis in pea. *Molecular and General Genetics* 260: 9-19.

Ellis, T.H.N., L. Turner, R.P. Hellens, D. Lee, C.L. Harker, C. Enard, C. Domoney, and D.R. Davis (1992). Linkage maps in pea. *Genetics* 130:649-663.

Erdmann, P.M., K.L. Rian, J.B. Mark, and P.E. McClean (2002). A molecular marker tightly linked to *P,* a gene required for flower and seed coat color in common bean. *Genome* 45(4):728-736.

Eujayl, I., M. Baum, W. Erskine, E. Pehu, and F.J. Muehlbauer (1997). The use of RAPD markers for lentil genetic mapping and the evaluation of distorted $F_2$ segregation. *Euphytica* 96:405-412.

Eujayl, I., M. Baum, W. Powell, W. Erskine, and E. Pehu (1998). A genetic linkage map of lentil (*Lens* sp.) based on RAPD and AFLP markers using recombinant inbred lines. *Theoretical and Applied Genetics* 97(1-2):83-89.

Eujayl, I., W. Erskine, M. Baum, and E. Pehu (1999). Inhertance and linkage analysis of frost injury in lentil. *Crop Science* 39:639-642.

Faleiro, F.G., W.S. Vinhadelli, V.A. Ragagnin, R.X. Corrêa, M.A. Moreira, and E.G. de Barros (2000). RAPD markers linked to a block of genes conferring rust resistance to the common bean. *Genetics and Molecular Biology* 23:399-402.

Fall, L., D. Diouf, M.A. Fall-Ndiaye, F.A. Badiane, and M. Gueye (2003). Genetic diversity in cowpea [*Vigna unguiculata* (L.) Walp] varieties determined by ARA and RAPD techniques. *African Journal of Biotechnology* 2:48-50.

Fatokun, C.A., D. Danesh, D.I. Menancio-Hautea, and N.D. Young (1993). A linkage map for cowpea (*Vigna unguiculata* [L.] Walp.) based on DNA markers. In *A compilation of linkage and restriction maps of genetically studied organisms. Genetic maps 1992*. J.S. O'Brien (Ed.). Cold Spring Harbor: Cold Spring Harbor Laboratory Press, pp. 6256-6258.

Fatokun, C.A., D. Danesh, N.D. Young, and E.L. Stewart (1993). Molecular taxonomic relationships in the genus *Vigna* based on RFLP analysis. *Theoretical and Applied Genetics* 86:97-104.

Fatokun, C.A., D.I. Menancio-Hautea, D. Danesh, and N.D. Young (1992). Evidence for orthologous seed weight genes in cowpea and mungbean based on RFLP mapping. *Genetics* 132:841-846.

Fatokun, C.A., H.D. Mignouna, M.R. Knox, and T.H.N. Ellis (1997). AFLP variation among cowpea varieties. In *Agronomy Abstracts.* Madison, WI: ASA, p. 156.

Ferguson, M.E., P.J. Bramel, and S. Chandra (2004). Gene diversity among botanical varieties in peanut (*Arachis hypogaea* L.). *Crop Science* 44:1847-1854.

Ferguson, M.E., M.D. Burow, S.R. Schulze, P.J. Bramel, A.H. Paterson, S. Kresovich, and S. Mitchell (2004). Microsatellite identification and characterization in peanut (*Arachis hypogaea* L.). *Theoretical and Applied Genetics* 108:1064-1070.

Ferreira, A.R., K.R. Foutz, and P. Keim (2000). Soybean genetic map of RAPD markers assigned to an existing scaffold RFLP map. *Journal of Heredity* 91:392-396.

Filho, S.M., C.S. Sediyama, M.A. Moreira, and E.G. de Barros (2002). RAPD and SCAR markers linked to resistance to frogeye leaf spot in soybean. *Genetics and Molecular Biology* 25 (3):317-321.

Flandez-Galvez, H., P.K. Ades, R. Ford, E.C.K. Pang, and P.W.J. Taylor (2003). QTL analysis for ascochyta blight resistance in an intraspecific population of chickpea (*Cicer arietinum* L.). *Theoretical and Applied Genetics* 107:1257-1265.

Flandez-Galvez, H., R. Ford, E.C.K. Pang, and P.W.J. Taylor (2003). An intraspecific linkage map of the chickpea (*Cicer arietinum* L.) genome based on sequence tagged microsatellite site and resistance gene analog markers. *Theoretical and Applied Genetics* 106:1447-1456.

Fofana, B., P. du Jardin and J.P. Baudoin (2001). Genetic diversity in the Lima bean (*Phaseolus lunatus* L.) as revealed by chloroplast DNA (cpDNA) variations. *Genetic Resources and Crop Evolution* 48:437-445.

Fofana, B., X. Vekemans, P. du Jardin, and J.P. Baudoin (1997). Genetic diversity in Lima bean (*Phaseolus lunatus L.*) as revealed by RAPD markers. *Euphytica* 95:157-165.

Ford, R., E.C.K. Pang and P.W.J. Taylor (1999). Genetics of resistance to ascochyta blight (*Ascochyta lentis*) of lentil and the identification of closely linked RAPD markers. *Theoretical and Applied Genetics* 98:93-98.

Ford, R., K.L. Roux, C. Itman, J.B. Brouwer, and P.W.J. Taylor (2002). Diversity analysis and genotyping in *Pisum* with sequence tagged microsatellite site (STMS) primers. *Euphytica* 124:397-405.

Freyre, R., P.W. Skroch, V. Geffroy, A.-F. Adam-Blondon, A. Shirmohamadali, W.C. Johnson, V. Llaca, R.O. Nodari, P.A. Pereira, S.-M. Tsai, et al. (1998). Towards an integrated linkage map of common bean. 4. Development of a core linkage and alignment of RFLP maps. *Theoretical and Applied Genetics* 97: 847-856.

Galasso, I. (2003). Distribution of highly repeated DNA sequences in species of the genus *Lens* Miller. *Genome* 46:1118-1124.

Galasso, I., L. Lioi, C. Lanave, R. Bollini, and F. Sparvoli (2004). Identification and isolation of lectin nucleotide sequences and species relationships in the genus *Lens* (Miller). *Theoretical and Applied Genetics* 108:1098-1102.

Galgaro, L., C.R. Lopes, M. Gimenes, J.F.M. Valls, and G. Kochert (1998). Genetic variation between several species of sections *Extranervosae, Caulorrhizae, Heteranthae,* and *Trisominatae* (genus *Arachis*) estimated by DNA polymorphism. *Genome* 41:445-454.

Galván, M., M. Aulicino, S. García Medina, and P.A. Balatti (2001). Genetic diversity among northwestern Argentinian cultivars of common bean (*Phaseolus vulgaris* L.) as revealed by RAPD markers. *Genetic Resources and Crop Evolution* 48:251-260.

Galván, M.Z., B. Bornet, P.A. Balatti, and M. Branchard (2003). Inter simple sequence repeat (ISSR) markers as a tool for the assessment of both genetic diversity and gene pool origin in common bean (*Phaseolus vulgaris* L.). *Euphytica* 132:297-301.

Garber, K., I. Bilic, O. Pusch, J. Tohme, A. Bachmair, D. Schweizer, and V. Jantsch (1999). The Tpv2 family of retrotransposons of *Phaseolus vulgaris*: Structure, integration characteristics and use for genotype classification. *Plant Molecular Biology* 39:797-807.

Garcia, G.M., H.T. Stalker, and G. Kochert (1995). Introgression analysis of an interspecific hybrid population in peanuts (*Arachis hypogeae* L.) using RFLP and RAPD markers. *Genome* 38:166-176.

Garcia, G.M., H.T. Stalker, E. Shroeder, and G. Kochert (1996). Identification of RAPD, SCAR and RFLP markers tightly linked to nematode resistance genes introgressed from *Arachis cardensii* into *Arachis hypogeae. Genome* 39:836-845.

Gepts, P., R. Nodari, R. Tsai, E.M.K. Koinange, V. Llaca, R. Gilbertson, and P. Guzman (1993). Linkage mapping in common bean. Annual Report of the Bean Improvement Cooperative. 36:24-38.

Gillespie Jr., A.G., M.S. Hopkins, and R.E. Dean (2005). Determining genetic diversity between lines of *Vigna unguiculata* subspecies by AFLP and SSR markers. *Genetic Resources and Crop Evolution* 52:245-247.

Gilpin, B.J., J.A. McCallum, T.J. Frew, and G.M. Timmerman-Vaughan (1997). A linkage map of the pea (*Pisum sativum* L.) genome containing cloned sequences of known function and expressed sequence tags (ESTs). *Theoretical and Applied Genetics* 95:1289-1299.

Gimenes, M.A., C.R. Lopes, M.L. Galgaro, J.F.M. Valls, and G. Kochert (2002). RFLP analysis of genetic variation in species of section Arachis, genus *Arachis* (Leguminoseae). *Euphytica* 123:421-429.

Gimenes, M.A., C.R. Lopes, and J.F.M. Valls (2002). Genetic relationship among *Arachis* species based on AFLP. *Genetics and Molecular Biology* 25 (3):349-353.

Goel, S., S.N. Raina, and Y. Ogihara (2002). Molecular evolution and phylogenetic implications of internal transcribed spacer sequences of nuclear ribosomal DNA in the *Phaseolus-Vigna* complex. *Molecular Phylogenetics and Evolution* 22:1-19.

González, A., A. Wong, A. Delagado-Salinas, R. Papa, and P. Gepts (2005). Assessment of inter simple sequence repeat markers to differentiate sympatric wild and domesticated populations of common bean. *Crop Science* 45:606-615.

Gore, M.A., A.J. Hayes, S.C. Jeong, Y.G. Yue, and G.R. Buss (2002). Mapping tightly linked genes controlling polyvirus infection at the *Rsv1* and *Rpv1* region in soybean. *Genome* 45:592-599.

Gowda, B.S., J.L. Miller, S.S. Rubin, D.R. Sharma, and M.P. Timko (2002). Isolation, sequencing and linkage mapping of resistance–gene analogs in cowpea (*Vigna unguiculata* L.). *Euphytica* 126:365-377.

Graham, P.H. and C.P. Vance (2003). Legumes: Importance and constraints to greater use. *Plant Physiology* 131:872-877.

Guo, B., D.A. Sleper, P.R. Arelli, J.G. Shannon, and H.T. Nguyen (2005). Identification of QTLs associated with resistance to soybean cyst nematode races 2, 3 and 5 in soybean PI 90763. *Theoretical and Applied Genetics* 111:965-971.

Guo, M., M.C. Mok, and D.W. Mok (1994). RFLP analysis of preferential transmission in interspecific hybrids of *Phaseolus vulgaris* and *P. coccineus*. *Journal of Heredity* 85:174-178.

Gutierrez, M.V., M.C. Vaz Patto, T. Huguet, J.I. Cubero, M.T. Moreno, and A.M. Torres (2005). Cross-species amplification of *Medicago truncatula* microsatellites across three major pulse crops. *Theoretical and Applied Genetics* 110:1210-1217.

Guzman-Maldonado, S.H., O. Martínez, J.A.A. Gallegos, F.G. Lara, and O.P. López (2003). Putative quantitative trait loci for physical and chemical components of common bean. *Crop Science* 43:1029-1035.

Haley, S.D., P.N. Miklas, J.R. Stavely, J. Byrum, and J.D. Kelly (1993). Identification of RAPD markers linked to a major rust resistance gene block in common bean. *Theoretical and Applied Genetics* 86:505-512.

Halward, T., H.T. Stalker, and G. Kochert (1993). Development of an RFLP linkage map in peanut species. *Theoretical and Applied Genetics* 87:379-394.

Halward, T.M., H.T. Stalker, E.A. Larue, and G. Kochert (1991). Genetic variation detectable with molecular markers among unadapted germ-plasm resources of cultivated peanut and related wild species. *Genome* 34:1013-1020.

Halward, T., H.T. Stalker, E. LaRue, and G. Kochert (1992). Use of single-primer DNA amplification in genetic studies of peanut (*Arachis hypogeae* L.). *Plant Molecular Biology* 18:315-325.

Havey, M.J. and F.J. Muehlbauer (1989). Variability of restriction fragment lengths and phylogenies in lentil. *Theoretical and Applied Genetics* 77:839-843.

Hayashi, M., A. Miyahara, S. Sato, T. Kato, M.Yoshikawa, M. Taketa, M. Hayashi, A. Pedrosa, R. Onda, H. Imaizumi-Anraku, et al. (2001). Construction of a genetic linkage map of the model legume *Lotus japonicus* using an intraspecific F2 population. *DNA Research* 8:301-310.

He, G., R. Meng, H. Gao, B. Guo, G. Gao, M. Newman, R.N. Pittman, and C.S. Prakash (2005). Simple sequence repeat markers for botanical varieties of cultivated peanut (*Arachis hypogaea* L.). *Euphytica* 142:131-136.

He, G., R. Meng, M. Newman, G. Gao, R.N. Pittman, and C.S. Prakash (2003). Microsatellites as DNA markers in cultivated peanut (*Arachis hypogaea* L.). *BMC Plant Biology* 3:3.

He, G. and C.S. Prakash (1997). Identificaion of polymorphic DNA markers in cultivated peanuts (*Arachis hypogeae* L.). *Euphytica* 97:143-149.

Herselman, L., R. Thwaites, F.M. Kimmins, B. Courtois, P.J.A.. van der Merwe, and S.E. Seal (2004). Identification and mapping of AFLP markers linked to peanut (*Arachis hypogaea* L.) resistance to the aphid vector of groundnut rosette disease. *Theoretical and Applied Genetics* 109:1426-1433.

Hilu, K.W. and H.T. Stalker (1995). Genetic relationships between peanut and wild species of *Arachis* (Fabaceae): Evidence from RAPDs. *Plant Systematics and Evolution* 198:167-178.

Hoey, B.K., K.R. Crowe, V.M. Jones, and N.O. Polans (1996). A phylogenetic analysis of *Pisum* based on morphological characters, allozyme and RAPD patterns. *Theoretical and Applied Genetics* 92:92-100.

Hopkins, M.S., A.M. Casa, T. Wang, S.E. Mitchell, R.E. Dean, G.D. Kochert, and S. Kresovich (1999). Discovery and characterization of polymorphic simple sequence repeats (SSRs) in peanut. *Crop Science* 39:1243-1247.

Hudcovicová, M. and J. Kraic (2003). Utilization of SSRs for characterization of the soybean [*Glycine max* (L.) Merr.] genetic resources. *Czech Journal of Genetics and Plant Breeding* 39:120-126.

Huettel, B., D. Santra, F.J. Muehlbauer, and G. Kahl (2002). Resistance gene analogues of chickpea (*Cicer arietinum* L.): Isolation, genetic mapping and association with a fusarium resistance gene cluster. *Theoretical and Applied Genetics* 105:479-490.

Iruela, M., J. Rubio, J.I. Cubero, J.Gil and T. Millán (2002). Phylogenetic analysis in the genus *Cicer* and cultivated chickpea using RAPD and ISSR markers. *Theoretical and Applied Genetics* 104:643-651.

Jacinto-Hernandez, C., S. Azpiroz-Rivero, J.A. Acosta-Gallegos, H. Hernandez-Sanchez, and I. Bernal-Lugo (2003). Genetic analysis and random amplified polymorphic DNA markers associated with cooking time in common bean. *Crop Science* 43:329-332.

Janila, P. and B. Sharma (2004). RAPD and SCAR markers for powdery mildew resistance gene *er* in pea. *Plant Breeding* 123:271-274.

Javadi, F. and H. Yamaguchi (2004a). Interspecific relationships of the genus *Cicer* L. (Fabaceae) based on *trnT-F* sequences. *Theoretical and Applied Genetics* 109:317-322.

Javadi, F. and H. Yamaguchi (2004b). RAPD and seed coat morphology variation in annual and perennial species of the genus *Cicer* L. *Genetic Resources and Crop Evolution* 51:783-794.

Jeong, S.C., S. Krishtipati, A.J. Hayes, P.J. Maughan, S.L. Noffsinger, I. Gunduz, G.R. Buss, and M.A. Saghai Maroof (2002). Genetics and sequence analysis of markers tightly linked to the soybean mosaic virus resistance gene, *Rsv3*. *Crop Science* 42:265-270.

Jeong, S.C. and M.A. Saghai Maroof (2004). Detection and genotyping of SNPs tightly linked to two disease resistant loci, *Rsv1* and *Rsv3* of soybean. *Plant Breeding* 123:305-310.

Jung, G., D. Coyne, P. Skroch, J. Nienhuis, E. Arnaud-Santana, J. Bokosi, H. Ariyarathne, J. Steadman, J. Beaver, and S. Kaeppler (1996). Molecular markers associated with plant architecture and resistance to common blight, web blight and rust in common beans. *Journal of American Society of Horticultural Science* 121:794-803.

Jung, G., P.W. Kroch, J. Nienhuis, D.P. Coyne, H.M. Ariyarathne, and E. Arnaud-Santana (1999). Confirmation of QTL associated with common bacterial blight resistance in four different genetic backgrounds in common bean. *Crop Science* 39:1448-1455.

Jung, G., P. Skroch, D. Coyne, J. Nienhuis, H. Ariyarathne, S. Kaeppler, and M. Bassett (1997). Molecular-marker-based genetic analysis of tepary-bean-derived common bacterial blight resistance in different developmental stages of common bean. *Journal of American Society of Horticultural Science* 122:329-337.

Jung, S., P.L. Tate, R. Horn, G. Kochert, K. Moore, and A.G. Abbott (2003). The phylogenetic relationship of possible progenitors of cultivated peanut. *Journal of Heredity* 94:334-340.

Kaga, A., N. Tomooka, Y. Egawa, K. Hosaka, and O. Kamijima (1996). Species relationships in the subgenus *Ceratotropis* (genus *Vigna*) as revealed by RAPD analysis. *Euphytica* 88:17-24.

Kahraman, A., I. Kusmenoglu, N. Aydin, A. Aydogan, W. Erskine, and F.J. Muehlbauer (2004). QTL mapping of winter hardiness genes in lentil. *Crop Science* 44:13-22.

Kaló, P., A. Seres, S.A. Taylor, J. Jakab, Z. Kevei, A. Kereszt, G. Endre, T.H.N. Ellis, and G.B. Kiss (2004). Comparative mapping between *Medicago sativa* and *Pisum sativum*. *Molecular Genetics and Genomics* 272:235-246.

Keim, P., B.W. Diers, T.C. Olson, and R.C. Shoemaker (1990). RFLP mapping in soybean: Association between marker loci and variation in quantitative traits. *Genetics* 126:735-742.

Keim, P., J.M. Schupp, S.E. Travis, K. Clayton, T. Zhu, L. Shi, A. Ferriera, and D.M. Webb (1997). A high-density soybean genetic map based on AFLP markers. *Crop Science* 37:537-543.

Kelly, J.D. and J.R. Miklas (1998). The role of RAPD markers in breeding for disease resistance in common bean. *Molecular Breeding* 4:1-11.

Kim, M.Y., B.K. Ha, T.H. Jun, E.Y. Hwang, K. Van, Y.I. Kuk, and S.H. Lee (2004). Single nucleotide polymorphism discovery and linkage mapping of lipoxygenase-2 gene (*Lx2*) in soybean. *Euphytica* 135:169-177.

Kisha, T.J., B.W. Diers, J.M. Hoyt, and C.H. Sneller (1998). Genetic diversity among soybean plant introductions and North American germplasm. *Crop Science* 38:1669-1680.

Kochert, G., T. Halward, W.D. Branch, and C.E. Simpson (1991). RFLP variability in peanut (*Arachis hypogeae*) cultivars and wild species. *Theoretical and Applied Genetics* 81:565-570.

Kochert, G., H. T. Stalker, M. Gimenes, L. Galgaro, C.R. Lopes, and K. Moore (1996). RFLP and cytogenetic evidence on the origin and evolution of allotetraploid domesticated peanut, *Arachis hypogeae* (Leguminosae). *American Journal of Botany* 83 (10):1282-1291.

Kolkman, J.M. and J.D. Kelly (2003). QTL conferring resistance and avoidance to white mold in common bean. *Crop Science* 43:539-548.

Kollipara, K.P., R.J. Singh, and T. Hymowitz (1997). Phylogenetic and genomic relationships in the genus *Glycine* Willd. based on sequences from the ITS region of nuclear rDNA. *Genome* 40:57-68.

Krishna, G.K., J. Zhang, M. Burow, R.N. Pittman, S.G. Delikostadinov, Y. Lu, and N. Puppala (2004). Genetic diversity analysis in Valencia peanut (*Arachis hypogaea* L.) using microsatellite markers. *Cellular and Molecular Biology Letters* 9:685-697.

Lakhanpal, S., S. Chadha, and K.V. Bhat (2000). Random amplified polymorphic DNA (RAPD) analysis in Indian mung bean [*Vigna raiata* (L.) Wilczek] cultivars. *Genetica* 109:227-234.

Lakshmi, N., P. Senthilkumar, M. Parani, M. N. Jithesh, and A.K. Parida (2000). PCR-RFLP analysis of chloroplast gene region in *Cajanus* (Leguminoseae) and allied genera. *Euphytica* 116:245-250.

Lanham, P.G., S. Fennell, J.P. Moss, and W. Powell (1992). Detection of polymorphic loci in *Arachis* germplasm using random amplified polymorphic DNAs. *Genome* 35:885-889.

Laucou, V., K. Haurogné, N. Ellis, and C. Rameau (1998). Genetic mapping in pea. 1. RAPD-based genetic linkage map of *Pisum sativum. Theoretical and Applied Genetics* 97:905-915.

Lee, G.J., H.R. Boerma, M.R. Villagarcia, X. Zhou, T.E. Carter Jr., Z. Li, and M.O. Gibbs (2004). A major QTL conditioning salt tolerance in S-100 soybean and descendent cultivars. *Theoretical and Applied Genetics* 109:1610-1619.

Lee, S.H., M.A. Bailey, M.A.R. Mian, E.R. Shipe, D.A. Ashley, W.A. Parrott, R.S. Hussey, and H.R. Boerma (1996). Identification of quantitative trait loci for plant height, lodging, and maturity in a soybean population segregating for growth habit. *Theoretical and Applied Genetics* 92:516-523.

Lewers, K.S., E.H. Crane, C.R. Bronson, J.M. Schupp, P. Keim, and R.C. Shoemaker (1999). Detection of linked QTL for soybean brown stem rot resistance in 'BSR 101' as expressed in a growth chamber environment. *Molecular Breeding* 5:33-42.

Li, C., C.A. Fatokun, B. Ubi, B.B. Singh, and G.J. Scoles (2001). Determining genetic similarities and relationships among cowpea breeding lines and cultivars by microsatellite markers. *Crop Science* 41:189-197.

Li, J., G. He, P. Gepts, and C.S. Prakesh (1999). Development of a genetic map for cowpea (*Vigna unguiculata*) using DNA markers. *Plant and Animal Genome VII Conference,* January 17-21, San Diego, CA, p. 237.

Li, Y.D., Y.J. Wang, Y.P. Tong, J.G. Gao, and J.S. Zhang (2005). QTL mapping of phosphorous deficiency tolerance in soybean [*Glycine max* (L.) Merr.]. *Euphytica* 42:137-142.

Li, Z. and R.L. Nelson (2001). Genetic diversity among soybean accessions from three countries measured by RAPDs. *Crop Science* 41:1337-1347.

Li, Z., L. Qiu, J.A. Thompson, M.M. Welsh, and R.L. Nelson (2001). Molecular genetic analysis of U.S. and Chinese soybean ancestral lines. *Crop Science* 41: 1330-1336.

Link, W., C. Dixkens, M. Singh, M. Schwall, and A.E. Melchinger (1995). Genetic diversity in European and Mediterrean faba bean germ plasm revealed by RAPD markers. *Theoretical and Applied Genetics* 90:27-32.

Liu, F., B.C. Zhuang, J.S. Zhang, and S.Y. Chen (2000). Construction and analysis of soybean genetic map. *Yi Chuan Xue Bao* 27:1018-1026.

Loridon, K., K. McPhee, J. Morin, P. Dubreuil, M.L. Pilet-Nayel, G. Aubert, C. Rameau, A. Baranger, C. Coyne, I. Lejeune-Hènaut, et al. (2005). Microsatellite marker polymorphism and mapping in pea (*Pisum sativum* L.). *Theoretical and Applied Genetics* 111:1022-1031.

Maciel, F.L., S. Echeverrigaray, L.T.S. Gerald, and F.G. Grazziotin (2003). Genetic relationships and diversity among Brazilian cultivars and landraces of common beans (*Phaseolus vulgaris* L.) revealed by AFLP markers. *Genetic Resources and Crop Evolution* 50:887-893.

Maciel, F.L., L.T.S. Gerald, and S. Echeverrigaray (2001). Random amplified polymorphic DNA (RAPD) markers variability among cultivars and landraces of common beans (*Phaseolus vulgaris* L.) of South-Brazil. *Euphytica* 120:257-263.

Mahuku, G., C. Montoya, M.A. Henriquez, C. Jara, H. Teran, and S. Beebe (2004). Inheritance and characterization of angular leaf spot resistance gene present in common bean accession G 10474 and identification of an AFLP marker linked to the resistance gene. *Crop Science* 44:1817-1824.

Mallikarjuna, N. and K.B. Saxena (2002). Production of hybrids between *Cajanus acutifolius* and *C. cajan*. *Euphytica* 124:107-110.

Mansur, L.M., J.H. Orf, K. Chase, T. Jarvik, P.B. Cregan, and K.G. Lark (1996). Genetic mapping of agronomic traits using recombinant inbred lines of soybean. *Crop Science* 36:1327-1336.

Massawe, F.J., M. Dickinson, J.A. Roberts, and S.N. Azam-Ali (2002). Genetic diversity in bambara groundnut [*Vigna subterranea* (L.) Verde] landraces revealed by AFLP markers. *Genome* 45:1175-1180.

Massawe, F.J., J.A. Roberts, S.N. Azam-Ali, and M.R. Davey (2003). Genetic diversity in bambara groundnut [*Vigna subterranea* (L.) Verde] landraces assessed by random amplified DNA (RAPD) markers. *Genetic Resources and Crop Evolution* 50:737-741.

Mathews, L.M., T.E. Devine, J.M. Weisemann, H.S. Beard, K.S. Lewers, M.H. McDonald, Y.-B. Park, R. Maiti, J.-J. Lin, J. Kuo, et al. (2001). Incorporation of sequenced cDNA and genomic markers into the soybean genetic map. *Crop Science* 41:516-521.

Mayer, M.S., A. Tullu, C.J. Simon, J. Kumar, W.J. Kaiser, J.M. Kraft, and F.J. Muehlbauer (1997). Development of a DNA marker for fusarium wilt resistance in chickpea. *Crop Science* 37:1625-1629.

Mayer, M.S. and P.S. Soltis (1994). Chloroplast DNA phylogeny of *Lens* (Leguminosae): Origin and diversity of the cultivated lentil. *Theoretical and Applied Genetics* 87:773-781.

McClean, P.E., R.K. Lee, and P.N. Miklas (2004). Sequence diversity analysis of dihydroflavanol reductase intron I in common bean. *Genome* 47:266-280.

McClean, P.E., R.K. Lee, C. Otto, P. Gepts, and M.J. Bassett (2002). Molecular and phenotypic mapping of genes controlling seed coat patterns and color in common bean (*Phaseolus vulgaris* L.). *Journal of Heredity* 93:148-152.

Meksem, K., T. Pantazopoulos, V.N. Njiti, L.D. Hyten, P.R. Arelli, and D.A. Lightfoot (2001). 'Forrest' resistance to the soybean cyst nematode is bigenic: Saturation mapping of the *Rhg* 1 and *Rhg* 4 loci. *Theoretical and Applied Genetics* 103:710-717.

Men, A.E., A.Y. Borisov, S.M. Rozov, K.V. Ushakov, V.E. Tsyganoe, I.A. Tikhonovich, and P.M. Gresshoff (1999). Identification of DNA amplification fingerprinting (DAF) markers close to the symbiosis-ineffective sym31 mutation of pea (*Pisum sativum* L.). *Theoretical and Applied Genetics* 98:929-936.

Menancio, D.I., A.G. Hepburn, and T. Hymowitz (1990). Restriction fragment length polymorphism (RFLP) of wild perennial relatives of soybean. *Theoretical and Applied Genetics* 79:235-240.

Menéndez, C.M., A.E. Hall, and P. Gepts (1997). A genetic linkage map of cowpea (*Vigna unguiculata*) developed from a cross between two inbred domesticated lines. *Theoretical and Applied Genetics* 95:1210-1217.

Métais, I., C. Aubry, B. Hamon, and R. Jalouzot (1998). Cloning, quantification and characterization of a minisatellite DNA sequence from common bean *Phaseolus vulgaris* L. *Theoretical and Applied Genetics* 97:232-237.

Métais, I., C. Aubry, B. Hamon, R. Jalouzot, and D. Peltier (2000). Description and analysis of genetic diversity between commercial bean lines (*Phaseolus vulgaris* L.). *Theoretical and Applied Genetics* 101:1207-1214.

Métais, I., B. Hamon, R. Jalouzot, and D. Peltier (2002). Structure and level of genetic diversity in various bean types evidenced with microsatellite markers isolated from a genomic enriched library. *Theoretical and Applied Genetics* 104: 1346-1352.

Mian, M.A.R., M.A. Bailey, J.P. Tamulonis, E.R. Shipe, T.E. Carter, J.W.A. Parrott, D.A. Ashley, R.S. Hussey, and H.R. Boerma (1996). Molecular markers associated with seed weight in two soybean populations. *Theoretical and Applied Genetics* 93:1011-1016.

Mienie, C.M.S., M.M. Liebenberg, Z.A. Pretorius, and P.N. Miklas (2005). SCAR markers linked to the common bean rust resistance gene *Ur-13*. *Theoretical and Applied Genetics* 111:972-979.

Mienie, C.M.S., L. Herselman, and R.E. Terhlanche (2000). RAPD analysis of selected common bean (*Phaseolus vulgaris* L.) cultivars from three African countries. *South African Journal of Plant and Soil* 17:93-94.

Mignouna, H.D., N.Q. Ng, J. Ikca, and G. Thottapilly (1998). Genetic diversity in cowpea as revealed by random amplified polymorphic DNA. *Journal of Genetics and Breeding* 52:151-159.

Miklas, P.N., W.C. Johnson, R. Delorme, and P. Gepts (2001). QTL conditioning physiological resistance and avoidance to white mold in dry bean. *Crop Science* 41:309-315.

Mimura, M., K. Yasuda, and H. Yamaguchi (2000). RAPD variation in wild, weedy and cultivated azuki beans in Asia. *Genetic Resources and Crop Evolution* 47:603-610.

Moretzsohn, M.C., M.S. Hopkins, S.E. Mitchell, S. Kresovich, J.F.M. Valls, and M.E. Ferreira (2004). Genetic diversity of peanut (*Arachis hypogaea* L.) and its wild relatives based on the analysis of hypervariable regions of the genome. *BMC Plant Biology* 4:11.

Moretzsohn, M.C., L. Leoi, K. Proite, P.M. Guimarães, S.C.M. Leal-Bertioli, M.A. Gimenes, W.S. Martins, J.F.M. Valls, D. Grattapaglia, and D.J. Bertioli (2005).

A microsatellite-based, gene-rich linkage map for the AA genome of *Arachis* (Fabaceae). *Theoretical and Applied Genetics* 111:1060-1071.

Muñoz, L.C., M.W. Blair, M.C. Duque, J. Tohme, and W. Roca (2004). Introgression in common bean x tepary bean interspecific congruity-backcross lines as measured by AFLP markers. *Crop Science* 44:637-645.

Narvel, J.M., W.R. Fehr, W. Chu, D. Grant, and R.C. Shoemaker (2000). Simple sequence repeat diversity among soybean plant introductions and elite genotypes. *Crop Science* 40:1452-1458.

Nienhuis, J., J. Tivang, P. Skroch, and J.B. Santos (1995). Genetic relationships among cultivars and landraces of lima bean (*Phaseolus lunatus* L.) as measured by RAPD markers. *Journal of American Society of Horticultural Science* 120: 300-306.

Nguyen, T.T., P.W.J. Taylor, R.J. Redden, and R. Ford (2004). Genetic diversity estimates in *Cicer* using AFLP analysis. *Plant Breeding* 123:173-179.

Nübile, P.M., M.A. Gimenes, J.F.M. Valls, and C.R. Lopes (2004). Genetic variation within and among species of genus *Arachis,* section Rhizomatosae. *Genetic Resources and Crop Evolution* 51:299-307.

Nodari, R.O., E.M.K. Koinange, J.D. Kelly, and P. Gepts (1992). Towards an integrated linkage map of common bean. 1. Development of genomic DNA probes and levels of restriction fragment length polymorphism. *Theoretical and Applied Genetics* 84:186-192.

Nodari, R.O., S. M. Tsai, R.L. Gilbertson, and P. Gepts (1993). Towards an integrated linkage map of common bean. 2. Development of an RFLP-based linkage map. *Theoretical and Applied Genetics* 85:513-520.

Nodari, R.O., S.M. Tsai, P. Guzmán, R.L. Gilbertson, and P. Gepts (1993). Towards an integrated linkage map of common bean. III. Mapping genetic factors controlling host-bacteria interactions. *Genetics* 134:341-350.

Nowosielski, J., W. Podyma, and D. Nowosielski (2002). Molecular research on the genetic diversity of polish varieties and landraces of *Phaseolus coccineus* L. and *Phaseolus vulgaris* L. using RAPD and AFLP methods. *Cellular and Molecular Biology Letters* 7:753-762.

Ntundu, W.H., I.C. Bach, J.L. Christiansen, and S.B. Andersen (2004). Analysis of genetic diversity in bambara groundnut [*Vigna subterranea* (L.) Verde] landraces using amplified fragment length polymorphism (AFLP) markers. *African Journal of Biotechnology* 3:220-225.

Ochatt, S.J., A. Benabdelmouna, P. Marget, G. Aubert, F. Moussy, C. Pontécaille, and L. Jacas (2004). Overcoming hybridization barriers between pea and some of its wild relatives. *Euphytica* 137:353-359.

Ouédraogo, J.T., B.S. Gowda, M. Jean, T.J. Close, J.D. Ehlers, A.E. Hall, A.G. Gillaspie, P.A. Roberts, A.M. Ismail, G. Bruening, et al. (2002). An improved genetic linkage map for cowpea (*Vigna unguiculata* L.) combining AFLP, RFLP, RAPD, biochemical markers, and biological resistance traits. *Genome* 45:175-188.

Ouédraogo, J.T., V. Maheshwari, D.K. Berner, C.-A. St-Pierre, F. Belzile, and M.P. Timko (2001). Identification of AFLP markers linked to resistance of cowpea

(*Vigna unguiculata* L.) to parasitism by *Striga gesnerioides*. *Theoretical and Applied Genetics* 102:1029-1036.

Ouédraogo, J.T., J.B. Tignegre, M.P. Timko, and F.J. Belzile (2002). AFLP markers linked to resistance against *Striga gesnerioides* race 1 in cowpea (*Vigna unguiculata*). *Genome* 45(5):787-793.

Paik-Ro, O.G., R.L. Smith, and D.A. Knauft (1992). Restriction fragment length polymorphism evaluation of six peanut species within the *Arachis* section. *Theoretical and Applied Genetics* 84:201-208.

Palmer, J.D., R.A. Jorgensen, and W.F. Thompson (1985). Chloroplast DNA variation and evolution in *Pisum:* Patterns of change and phylogenetic analysis. *Genetics* 109:195-213.

Pandian, A., R. Ford, and P.W.J. Taylor (2000). Transferability of sequence tagged microsatellite site (STMS) primers across four major pulses. *Plant Molecular Biology Reporter* 18:395a-395h.

Papa, R., J. Acosta, A. Delgado-Salinas, and P. Gepts (2005). A genome-wide analysis of differentiation between wild and domesticated *Phaseolus vulgaris* from Mesoamerica. *Theoretical and Applied Genetics* 111:1147-1158.

Papa, R. and P. Gepts (2003). Asymmetry of gene flow and differential geographical structure of molecular diversity in wild and domesticated common bean (*Phaseolus vulgaris* L.) from Mesoamerica. *Theoretical and Applied Genetics* 106:239-250.

Park, S.O., D.P. Coyne, J.M. Bokosi, and J.R. Steadman (1999). Molecular markers linked to genes for specific rust resistance and indeterminate growth habit in common bean. *Euphytica* 105:133-141.

Park, S.O., D.P. Coyne, N. Mutlu, G. Jung, and J.R. Steadman (1999). Confirmation of molecular markers and flower color associated with QTL for resistance to common bacterial blight in common beans. *Journal of American Society of Horticultural Science* 124:519-526.

Park, S.O., D.P. Coyne, J.R. Steadman, K.M. Crosby, and M.A. Brick (2004). RAPD and SCAR markers linked to the *Ur-6* Andean gene controlling specific rust resistance in common bean. *Crop Science* 44:1799-1807.

Park, S.O., D.P. Coyne, J.R. Steadman, and P.W. Skroch (2001). Mapping of QTL resistance for whitemold disease in common bean. *Crop Science* 41:1253-1262.

Park, S.O., D.P. Coyne, J.R. Steadman, and P.W. Skroch (2003). Mapping of the *Ur-7* gene for specific resistance to rust in common bean. *Crop Science* 43:1470-1476.

Patil, P., P.L. Vrinten, G.J. Scoles, and A.E. Slinkard (1995). Variation in the ribosomal RNA units of the genera *Lens* and *Cicer*. *Euphytica* 83:33-42.

Pfaff, T. and G. Kahl (2003). Mapping of gene specific markers on the genetic map of chickpea (*Cicer arietinum* L.). *Molecular Genetics and Genomics* 269:243-251.

Polegri, L., P. Taviani, B. Tiranti, and V. Negri (2004). Molecular markers to protect cowpea landraces. Proceedings of the XLVIII Italian Society of Agricultural Genetics-SIFV-SIGA Joint Meeting, September 15-18, Leece, Italy.

Posa-Macalincag, M.-C. T., G.L. Hosfield, K.F. Grafton, M.A. Uebersax, and J.D. Kelly (2000). Quantitative trait loci (QTL) analysis of canning quality traits in kidney bean (*Phaseolus vulgaris* L.). *Journal of American Society of Horticultural Science* 127:608-615.

Potokina, E., N. Tomooka, D.A. Vaughan, T. Alexandrova, and R. Xu (1999). Phylogeny of *Vicia* subgenus *Vicia* (Fabaceae) based on analysis of RAPDs and RFLP of PCR-amplified chloroplast genes. *Genetic Resources and Crop Evolution* 46:149-161.

Prioul, S., A. Frankewitz, G. Deniot, G. Morin, and A. Baranger (2004). Mapping of quantitative trait loci for partial resistance to *Mycosphaerella pinodes* in pea (*Pisum sativum* L.), at the seedling and adult plant stages. *Theoretical and Applied Genetics* 108:1322-1334.

Raina, S.N. and Y. Mukai (1999a). Genomic *in situ* hybridization in *Arachis* (Fabaceae) identifies the diploid progenitors of cultivated (*A. hypogeae* and wild *A. monticola*) peanut species. *Plant Systematics and Evolution* 214:251-262.

Raina, S.N. and Y. Mukai (1999b). Detection of variable number of 18S-5.8S-26S and 5S ribosomal DNA loci by fluorescent *in situ* hybridization in diploid and tetraploid *Arachis* species. *Genome* 42:52-59.

Raina, S.N., Y. Mukai, K. Kawaguchi, S. Goel, and A. Jain (2001). Physical mapping of 18S-5.8S-26S and 5S ribosomal RNA gene families in three important vetches (*Vicia* species) and their allied taxa constituting three species complexes. *Theoretical and Applied Genetics* 103:839-845.

Raina, S.N. and R.K.G. Narayan (1984). Changes in DNA composition in the evolution of *Vicia* species. *Theoretical and Applied Genetics* 68:187-192.

Raina, S.N. and Y. Ogihara (1994). Chloroplast DNA diversity in *Vicia faba* and its close wild relatives: Implications for reassessment. *Theoretical and Applied Genetics* 88:261-266.

Raina, S.N., and Y. Ogihara (1995). Ribosomal DNA repeat unit polymorphism in 49 *Vicia* species. *Theoretical and Applied Genetics* 90:477-486.

Raina, S.N., V. Rani, T. Kojima, Y. Ogihara, K.P. Singh, and R.M. Devarumath (2001). RAPD and ISSR fingerprints as useful genetic markers for analysis of genetic diversity, varietal identification and phylogenetic relationships in peanut (*Arachis hypogaea*) cultivars and wild species. *Genome* 44:1-10.

Rajesh, P.N., V.J. Sant, V.S. Gupta, F.J. Muehlbauer, and P.K. Ranjekar (2002). Genetic relationships among annual and perennial wild species of *Cicer* using inter simple sequence repeat (ISSR) polymorphism. *Euphytica* 129:15-23.

Rajesh, P.N., A. Tullu, J. Gill, V.S. Gupta, P.K. Ranjekar, and F.J. Muehlbauer (2002). Identification of an STMS marker for the double-podding gene in chickpea. *Theoretical and Applied Genetics* 105:604-607.

Rajora, O.P. and J.D. Mohan (1997). Mitochondrial and nuclear DNA variation, genotype fingerprinting and genetic relationship in lentil (*Lens culinaris* Medik.). *Canadian Journal of Plant Science* 77:515-521.

Rakshit, S., P. Winter, M. Tekeoglu, J. Juarez Muñoz, T. Pfaff, A.M. Benko-Iseppon, F.J. Muehlbauer, and G. Kahl (2003). DAF marker tightly linked to a major locus for ascochyta blight resistance in chickpea (*Cicer arietinum* L.). *Euphytica* 132:23-30.

Rameau, C., D. Dénoue, F. Fravell, K. Haurogné, J. Josserand, V. Laucou, S. Batge, and I.C. Murfet (1998). Genetic mapping in pea. 2. Identification of RAPD and SCAR markers linked to genes affecting plant architecture. *Theoretical and Applied Genetics* 97:916-928.

Ratnaparkhe, M.B., V.S. Gupta, and P.K. Ranjekar (1995). Genetic fingerprinting of pigeonpea [*Cajanus cajan* (L.) Millsp.] and its wild relatives using RAPD markers. *Theoretical and Applied Genetics* 91:893-898.

Ratnaparkhe, M.B., D.K. Santra, A. Tullu, and F.J. Muehlbauer (1998). Inheritance of inter-simple-sequence-repeat polymorphisms and linkage with a fusarium wilt resistance gene in chickpea. *Theoretical and Applied Genetics* 96:348-353.

Rector, B.G., J.N. All, W.A. Parrott, and H.R. Boerma (1998). Identification of molecular markers linked to quantitative trait loci for soybean resistance to corn earworm. *Theoretical and Applied Genetics* 96:786-790.

Roane, C.W., S.A. Tolin and G.R. Buss (1983). Inheritance of reaction to two viruses in the soybean cross 'York' x 'Lee 68.' *Journal of Heredity* 74:289-291.

Roman, B., Z. Satovic, C.M. Avila, D. Rubiales, M.T. Moreno, and A.M. Torres (2003). Locating genes associated with *Ascochyta fabae* resistance in *Vicia faba*. *Australian Journal of Agricultural Research* 54:85-90.

Roman, B., Z. Satovic, D. Pozarkova, J. Macas, J. Dolezal, J.I. Cubero, and A.M. Torres (2004). Development of a composite map in *Vicia faba*, breeding applications and future prospects. *Theoretical and Applied Genetics* 108:1079-1088.

Roman, B., A.M. Torres, D. Rubiales, J.I. Cubero, and Z. Satovic (2002). Mapping of quantitative trait loci controlling broomrape (*Orobanche crenata* Forsk.) resistance in faba bean (*Vicia faba* L.). *Genome* 45:1057-1063.

Rongwen, J., M.S. Akkaya, A.A. Bhagwat, U. Lavi, and P.B. Cregan (1995). The use of microsatellite DNA markers for soybean genotype identification. *Theoretical and Applied Genetics* 90:43-48.

Rubeena, R. Ford, and P.W.J. Taylor (2003). Construction of intraspecific linkage map of lentil (*Lens culinaris* ssp. *culinaris*). *Theoretical and Applied Genetics* 107:910-916.

Samec, P. and V. Nasinec (1995). Detection of DNA polymorphism among pea cultivars using RAPD technique. *Biologia Plantarum* 37:321-327.

Sandal, N., L. Krusell, S. Radutoiu, M. Olbryt, A. Pedrosa, S. Stracke, S. Sato, T. Kato, S. Tabata, M. Parniske, et al. (2002). A genetic linkage map of the model legume *Lotus japonicus* and strategies for fast mapping of new loci. *Genetics* 161:1673-1683.

Sant, V.J., A.G. Patankar, N.D. Sarode, L.B. Mhase, M.N. Sainani, R.B. Deshmukh, P.K. Ranjekar, and V.S. Gupta (1999). Potential of DNA markers in detecting divergence and in analyzing heterosis in Indian elite chickpea cultivars. *Theoretical and Applied Genetics* 98:1217-1225.

Santra, D.K., M. Tekeoglu, M. Ratnaparkhe, W.J. Kaiser, and F.J. Muehlbauer (2000). Identification and mapping of QTLs conferring resistance to ascochyta blight in chickpea. *Crop Science* 40:1606-1612.

Saravanakumar, P., A. Kaga, N. Tomooka, and D.A. Vaughan (2004). AFLP and RAPD analyses of intra- and interspecific variation in some *Vigna* subgenus *Ceratotropis* (Leguminosae) species. *Australian Journal of Botany* 52:417-424.

Satovic, Z., A.M. Torres, and J.I. Cubero (1996). Genetic mapping of new morphological, isozyme and RAPD markers in *Vicia faba* L. using trisomics. *Theoretical and Applied Genetics* 93:1130-1138.

Schneider, A., S.A. Walker, M. Sagan, G. Due, T.H.N. Ellis, and J.A. Downie (2002). Mapping of the nodulation loci *sym9* and *sym10* of pea (*Pisum sativum* L.). *Theoretical and Applied Genetics* 104:1312-1316.

Schneider, K.A., M.E. Brothers, and J.D. Kelly (1997). Marker-assisted election to improve drought resistance in common bean. *Crop Science* 37:51-60.

Schneider, K.A., K.F. Grafton, and J.D. Kelly (2001). QTL analysis of resistance to fusarium root rot in bean. *Crop Science* 41:535-542.

Serna, R., S.H. Delgado, M.G. Paz, J.A.A. Gallegos, and N.M. Pérez (2005). Genetic relationships and diversity revealed by AFLP markers in Mexican common bean bred cultivars. *Crop Science* 45:1951-1957.

Serret, M.D., S.M. Udupa, and F. Weigand (1997). Assessment of genetic diversity of cultivated chickpea using microsatellate-derived RFLP markers: Implications for origin. *Plant Breeding* 116:573-578.

Sharma, P.C., P. Winter, T. Bünger, B. Hüttel, F. Weigand, K. Weising, and G. Kahl (1995). Abundance and polymorphism of di-, tri- and tetra-nucleotide tandem repeats in chickpea (*Cicer arietinum* L.). *Theoretical and Applied Genetics* 90:90-96.

Sharma, S.K., M.R. Knox, and T.H.N. Ellis (1996). AFLP analysis of the diversity and phylogeny of *Lens* and its comparison with RAPD analysis. *Theoretical and Applied Genetics* 93:751-758.

Shiran, B. and S.N. Raina (2001). Evidence of rapid evolution and incipient speciation in *Vicia sativa* species complex based on nuclear and organellar RFLPs and PCR analysis. *Genetic Resources and Crop Evolution* 48:519-532.

Shoemaker, R.C., P.M. Hatfield, R.G. Palmer, and A.G. Atherly (1986). Chloroplast DNA variation in the genus *Glycine* subgenus *Soja*. *Journal of Heredity* 77:26-30.

Shoemaker, R. and T. Olson (1993). Molecular linkage map of soybean. In *Genetic maps: Locus maps of complex genomes* (6th ed.). S. O'Brien (Ed.). Cold Spring Harbor, NY: Cold Spring Harbor Laboratory Press.

Shoemaker R.C. and J.E. Specht (1995). Integration of the soybean molecular and classical genetic linkage groups. *Crop Science* 35:436-446.

Sicard, D., L. Nanni, O. Porfiri, D. Bulfon, and R. Papa (2005). Genetic diversity of *Phaseolus vulgaris* L. and *P. coccineus* L. landraces in central Italy. *Plant Breeding* 124:464-472.

Simioniuc, D., R. Uptmoor, W. Friedt, and F. Ordon (2002). Genetic diversity and relationships among pea cultivars revealed by RAPDs and AFLPs. *Plant Breeding* 121:429-435.

Simon, C.J., and F.J. Muehlbauer (1997). Construction of a chickpea linkage map and its comparison with maps of pea and lentil. *Journal of Heredity* 88:115-119.

Singh, A.K., J. Smartt, C.E. Simpson, and S.N. Raina (1998). Genetic variation vis-à-vis molecular polymorphism in groundnut, *Arachis hypogaea* L. *Genetic Resources and Crop Evolution* 45:119-126.

Singh, S.P., F.I. Morales, P.N. Miklas, and H. Teran (2000). Selection for bean golden mosaic resistance in intra- and interracial bean populations. *Crop Science* 40:1565-1572.

Sivaprakash, K.R., S.R. Prashanth, B.P. Mohanty, and A. Parida (2004). Genetic diversity of black gram (*Vigna mungo*) landraces as evaluated by amplified fragment length polymorphism markers. *Current Science* 86:1411-1416.

Sivaramakrishnan, S., S. Kannan, and L.J. Reddy (2002). Diversity in selected wild and cultivated species of pigeonpea using RFLP of mtDNA. *Euphytica* 125: 21-28.

Skroch, P., G. Jung, J. Nienhuis, and D. Coyne (1996). Integration of RAPD marker linkage maps and comparative mapping of OTL for disease resistance in common bean. *Annual Report of the Bean Improvement Cooperation* 39:48-49.

Song, Q.J., L.F. Marek, R.C. Shoemaker, K.G. Lark, V.C. Concibido, X. Delannay, J.E. Specht, and P.B. Cregan (2004). A new integrated genetic linkage map of the soybean. *Theoretical and Applied Genetics* 109:122-128.

Song, Q.J., C.V. Quigley, R.L. Nelson, T.E. Carter, H.R. Boerma, J.L. Strachan, and P.B. Cregan (1999). A selected set of trinucleotide simple sequence repeat markers for soybean cultivar identification. *Plant Varieties and Seeds* 12:207-220.

Sonnante, G., I. Galasso, and D. Pignone (2003). ITS sequence analysis and phylogenetic inference in the genus *Lens* Mill. *Annals of Botany* 91:49-54.

Souframanien, J. and T. Gopalakrishna (2004). A comparative analysis of genetic diversity in black gram genotypes using RAPD and ISSR markers. *Theoretical and Applied Genetics* 109:1687-1693.

Stalker, H.T. and L.G. Mozingo (2001). Molecular markers of *Arachis* and marker-assisted selection. *Peanut Science* 28:117-123.

Stracke, S., S. Sato, N. Sandal, M. Koyama, T. Kaneko, S. Tabata, and M. Parniske (2004). Exploitation of colinear relationships between the genomes of *Lotus japonicus, Pisum sativum* and *Arabidopsis thaliana,* for positional cloning of a legume symbiosis gene. *Theoretical and Applied Genetics* 108:442-449.

Sudupak, M.A. (2004). Inter and intra-species inter simple sequence repeat (ISSR) variation in the genus *Cicer. Euphytica* 135:229-238.

Sudupak, M.A., M.S. Akkaya, and A. Kence (2002). Analysis of genetic relationships among perennial and annual *Cicer* species growing in Turkey using RAPD markers. *Theoretical and Applied Genetics* 105:1220-1228.

Sudupak, M.A., M.S. Akkaya, and A. Kence (2004). Genetic relationships among perennial and annual *Cicer* species growing in Turkey assessed by AFLP finger printing. *Theoretical and Applied Genetics* 108:937-944.

Tahir, M., C.J. Simon and F.J. Muehlbauer (1993). Gene map of lentil: A review. *Lens Newsletters* 20:3-10.

Tar'an, B., L. Buchwaldt, A. Tullu, S. Banniza, T.D. Warkentin, and A. Vandenberg (2003). Using molecular markers to pyramid genes for resistance to ascochyta blight and anthracnose in lentil (*Lens culinaris* Medik). *Euphytica* 134(2): 223-230.

Tar'an, B., T.E. Michaels, and K.P. Pauls (2002). Genetic mapping of agronomic traits in common bean. *Crop Science* 42:544-556.

Tar'an, B., T. Warkentin, D.J. Somers, D. Miranda, A. Vandenberg, S. Blade, and D. Bing (2004). Identification of quantitative trait loci for grain, seed protein concentration and maturity in field pea (*Pisum sativum* L.). *Euphytica* 136:297-306.

Tar'an, B., T. Warkentin, D.J. Somers, D. Miranda, A. Vandenberg, S. Blade, S. Woods, D. Bing, A. Xue, D. Dekoeyer, et al. (2003). Quantitative trait loci for lodging resistance, plant height and partial resistance to mycosphaerella blight in field peas (*Pisum savitum* L.). *Theoretical and Applied Genetics* 107(8):1482-1491.

Tar'an, B., C. Zhang, T. Warkentin, A. Tullu, and A. Vandenberg (2005). Genetic diversity among varieties and wild species accessions of pea (*Pisum sativum* L.) based on molecular markers, and morphological and physiological characters. *Genome* 48:257-272.

Taylor-Grant, N. and K.M. Soliman (1999). Detection of polymorphic DNA and taxonomic relationships among ten wild perennial soybean species using specific and arbitrary nucleotide primers. *Biologia Plantarum* 42:25-37.

Tekeoglu, M., P.N. Rajesh, and F.J. Muehlbauer (2002). Integration of sequence tagged microsatellite sites to the chickpea genetic map. *Theoretical and Applied Genetics* 105:847-854.

Thompson, J.A., R.L. Nelson, and L.O. Vodkin (1998). Identification of diverse germplasm using RAPD markers. *Crop Science* 38:1348-1355.

Thoquet, P., M. Ghérardi, E. Journet, A. Kereszt, J. Ané, J. Prosperi, and T. Huguet (2002). The molecular genetic linkage map of the model legume *Medicago truncatula:* An essential tool for comparative legume genomics and the isolation of agronomically important genes. *BMC Plant Biology* 2:1.

Timmerman-Vaughen, G.M., T.J. Frew, A.C. Russell, T. Khan, R. Butler, M. Gilpin, S. Murray and K. Falloon (2002). QTL mapping of partial resistance to field epidemics of ascochyta blight of pea. *Crop Science* 42:2100-2111.

Timmerman-Vaughen, G.M., T.J. Frew, N.F. Weeden, A.L. Miller, and D.S. Goulden (1994). Linkage analyis of *er-1*, a recessive *Pisum sativum* gene for resistance to powdery mildew fungus (*Erysiphe pisi* D.C.). *Theoretical and Applied Genetics* 88:1050-1055.

Tiwari, K.R., G.A. Penner, and T. D. Warkentin (1998). Identification of coupling and repulsion phase RAPD markers for powdery mildew resistance gene *er-1* in pea. *Genome* 41:440-444.

Tomooka, N., M.S. Yoon, K. Doi, A. Kaga, and D. Vaughan (2002). AFLP analysis of diploid species in the genus *Vigna* subgenus *Ceratotropis. Genetic Resources and Crop Evolution* 49:521-530.

Torres, A.M., N.F. Weeden, and A. Martin (1993). Linkage among isozyme, RFLP and RAPD markers in *Vicia faba. Theoretical and Applied Genetics* 85:937-945.

Tosti, N. and V. Negri (2002). Efficiency of three PCR-based markers in assessing genetic variation among cowpea (*Vigna unguiculata* subsp. *unguiculata*) landraces. *Genome* 45:268-275.

Tosti, N. and V. Negri (2005). On-going on-farm micro evolutionary processes in neighboring cowpea landraces revealed by molecular markers. *Theoretical and Applied Genetics* 110:1275-1283.

Tullu, A., L. Buchwaldt, T. Warkentin, B. Tar'an, and A. Vandenberg (2003). Genetics of resistance to anthracnose and identification of AFLP and RAPD markers linked to the resistance gene in PI 320937 germplasm of lentil (*Lens culinaris* Medikus). *Theoretical and Applied Genetics* 106:428-434.

Tullu, A., W.J. Kaiser, J.M. Kraft, and F.J. Muehlbauer (1999). A second gene for resistance to race 4 of fusarium wilt in chickpea and linkage with a RAPD marker. *Euphytica* 109:43-50.

Tullu, A., F.J. Muehlbauer, W.J. Kaiser, C.J. Simon, M.S. Mayer, J. Kumar, and J.M. Kraft (1998). Inheritance and linkage of a gene for resistance to race 4 of fusarium wilt and RAPD markers in chickpea. *Euphytica* 102:227-232.

Ude, G.N., W.J. Kenworthy, J.M. Costa, P.B. Cregan, and J. Alvernaz (2003). Genetic diversity of soybean cultivars from China, Japan, North America, and North American ancestral lines determined by amplified fragment length polymorphism. *Crop Science* 43:1858-1867.

Udupa, S.M., L.D. Robertson, F. Weigand, M. Baum and G. Kahl (1999). Allelic variation at $(TAA)_n$ microsatellite loci in a world collection of chickpea (*Cicer arietinum* L.) germplasm. *Molecular and General Genetics* 261:354-363.

Vaillancourt, R.E. and N.F. Weeden (1992). Chloroplast DNA polymorphism suggests a Nigerian centre of domestication for the cowpea *Vigna unguiculata* (Leguminosae). *American Journal of Botany* 79:1194-1199.

Vallejos, E.C, N.S. Sakiyama, and C.D. Chase (1992). A molecular marker-based linkage map of *Phaseolus vulgaris* L. *Genetics* 131:733-740.

van Oss, H., Y. Aron, and G. Ladizinsky (1997). Chloroplast DNA variation and evolution in the genus *Lens* Mill. *Theoretical and Applied Genetics* 94:452-457.

Van, K., E.-Y. Hwang, M.Y. Kim, Y.-H. Kim, Y.-I. Cho, P.B. Cregan, and S.-H. Lee (2004). Discovery of single nucleotide polymorphisms in soybean using primers designed from ESTs. *Euphytica* 139:147-157.

Vance, C.P., P.H. Graham, and D.L. Allan (2000). Biological nitrogen fixation. Phosphorous: A critical future need. In *Nitrogen fixation: From molecules to crop productivity,* F.O. Pedrosa, M. Hungaria, M.G. Yates, and W.E. Newton (eds.). Dordrecht, Netherlands: Kluwer Academic Publishers, pp 506-514.

Van de Ven, W.T.G., N. Duncan, G. Ramsay, M. Phillips, W. Powell, and R. Waugh (1993). Taxonomic relationships between *V. faba* and its relatives based on nuclear and mitochondrial RFLPs and PCR analysis. *Theoretical and Applied Genetics* 86:71-80.

Van de Ven, W.T.G., R. Waugh, N. Duncan, G. Ramsay, N. Dow, and W. Powell (1991). Development of a genetic linkage map in *Vicia faba* using molecular and biochemical techniques. *Aspects of Applied Biology* 27:49-54.

Vaz Patto, M.C., A.M. Torres, A. Koblizkova, J. Macas, and J.I. Cubero (1999). Development of a genetic composite map of *Vicia faba* using $F_2$ populations derived from trisomic plants. *Theoretical and Applied Genetics* 98:736-743.

Vijayalakshmi, S., K. Yadav, C. Kushwaha, S.B. Sarode, C.P. Srivastava, R. Chand, and B.D. Singh (2005). Identification of RAPD markers linked to the rust (*Uromyces fabae*) resistance gene in pea (*Pisum sativum*). *Euphytica* 144:265-274.

Von Stackelberg, M., S. Lindemann, M. Menke, S. Riesselmann, and H.-J. Jacobsen (2003). Identification of AFLP and STS markers closely linked to *def* locus in pea. *Theoretical and Applied Genetics* 106:1293-1299.

Walters, K.L., G.L. Hosfield, M.A. Uebersax, and J.D. Kelly (1997). Navy bean canning quality: Correlations, heritability estimates, and randomly amplified polymorphic DNA markers associated with component traits. *Journal of the American Society of Horticultural Science* 122:338-343.

Wang, X.W., A. Kaga, N. Tomooka, and D.A. Vaughan (2004). The development of SSR markers by a new method in plants and their application to gene flow studies in azuki bean [*Vigna angularis* (Wild.) Ohwi & Ohashi]. *Theoretical and Applied Genetics* 109:352-360.

Weeden, N.R., F.J. Muehlbauer, and G. Ladizinsky (1992). Extensive conservation of linkage relationships between pea and lentil genetic maps. *Journal of Heredity* 83:123-129.

Winter, P., A.M. Benko-Iseppon, B. Huettel, M. Ratnaparkhe, A. Tullu, G. Sonnante, T. Pfaff, M. Tekeoglu, D. Santra, V.J. Sant, et al. (2000). A linkage map of the chickpea (*Cicer arietinum* L.) genome based on recombinant inbred lines from a *C. areitinum* x *C. reticulatum* cross: Localization of resistance genes for fusarium wilt races 4 and 5. *Theoretical and Applied Genetics* 101:1155-1163.

Winter, P., T. Pfaff, S.M. Udupa, B. Huettel, P.C. Sharma, S. Sahi, R. Arreguin-Espinoza, F. Weigand, F.J. Muehlbauer, and G. Kahl (1999). Characterization and mapping of sequence-tagged microsatellite sites in the chickpea (*Cicer arietinum* L.) genome. *Molecular and General Genetics* 262:90-101.

Xu, D.H., J. Abe, M. Sakai, A. Kanazawa, and Y. Shimamoto (2000). Sequence variation of non-coding regions of chloroplast DNA of soybean and related wild species and its implications for the evolution of different chloroplast haplotypes. *Theoretical and Applied Genetics* 101:724-732.

Xu, R.Q., N. Tomooka, and D.A. Vaughan (2000). AFLP markers for characterizing the azuki bean complex. *Crop Science* 40:808-815.

Xu, R.Q., N. Tomooka, D.A. Vaughan, and K. Doi (2000). The *Vigna angularis* complex: Genetic variation and relationships revealed by RAPD analysis, and their implication for *in situ* conservation and domestication. *Genetic Resources and Crop Evolution* 47:123-134.

Yamanaka, N., S. Ninomiya, M. Hoshi, Y. Tsubokura, M. Yano, Y. Nagamura, T. Sasaki and K. Harada (2001). An informative linkage map of soybean reveals QTLs for flowering time, leaflet morphology and regions of segregation distortion. *DNA Research* 8:61-72.

Yan, H.H., J. Mudge, D.J. Kim, D. Larsen, R.C. Shoemaker, D.R. Cook, and N.D. Young (2003). Estimates of conserved microsynteny among the genomes of *Glycine max, Medicago truncatula* and *Arabidopsis thaliana. Theoretical and Applied Genetics* 106:1256-1265.

Yang, W., D.B. Weaver, B.L. Nelson, and J.Qiu (2000). A preliminary genetic linkage map of soybean using an intraspecific cross of two cultivars: 'Peking' and 'Lee'. *Soybean Genetics Newsletter* 27.

Yee, E., K.K. Kidwell, G.R. Sills, and T.A. Lumpkin (1999), Diversity Among selected *Vigna angularis* (azuki) accessions on the basis of RAPD and AFLP markers. *Crop Science* 39:268-275.

Young, R.A. and J.D. Kelly (1997). RAPD markers linked to three major anthracnose resistance genes in common bean. *Crop Science* 37:940-946.

Yu, K., S.J. Park, V. Poysa, and P. Gepts (2000). Integration of simple sequence repeat (SSR) markers into a molecular linkage map of common bean (*Phaseolus vulgaris* L.). *Journal of Heredity* 91:429-434.

Yu, Y.G., M.A. Saghai-Maroof, G.R. Buss, P.J. Maughan, and S.A. Tolin (1994). RFLP and microsatellite mapping of a gene for soybean mosaic virus resistance. *Phytopathology* 84:60-64.

Yu, Z.H., R.E. Stall, and C.E. Vallejos (1998). Detection of genes for resistance to common bacterial blight of beans. *Crop Science* 38:1290-1296.

Zeid, M., C.C. Schön, and W. Link (2001). Genetic diversity in a group of recent elite faba bean lines. *Czech Journal of Genetics and Plant Breeding* 37:34-40.

Zeid, M., C.C. Schön, and W. Link (2003). Genetic diversity in recent elite faba bean lines using AFLP markers. *Theoretical and Applied Genetics* 107:1304-1314.

Zeng, Z.-B. (1993). Theoretical basis of separation of multiple linked gene effects on mapping quantitative trait loci. *Proceedings of National Academy of Sciences, USA* 90:10972-10976.

Zeng, Z.-B. (1994). Precision mapping of quantitative trait loci. *Genetics* 136:1457-1468.

Zhang, W.K., Y.J. Wang, G.Z. Luo, J.S. Zhang, C.Y. He, X.L. Wu, J.Y. Gai, and S.Y. Chen (2004). QTL mapping of ten agronomic traits on the soybean (*Glycine max* L. Merr.) genetic map and their association with EST markers. *Theoretical and Applied Genetics* 108:1131-1139.

Zong, X.X., A. Kaga, N. Tomooka, X.W. Wang, O.K. Han, and D. Vaughan (2003). The genetic diversity of the *Vigna angularis* complex in Asia. *Genome* 46:647-648.

Zou, J.J., R.J. Singh, and T. Hymowitz (2004). SSR marker and ITS cleaved amplified polymorphic sequence analysis of soybean × *Glycine tomentella* intersubgeneric derived lines. *Theoretical and Applied Genetics* 109:769-774.

Chapter 34

# Genetic Maps and Molecular Markers in the Genus *Medicago*

Bernadette Julier
Thierry Huguet

## *INTRODUCTION*

The genus *Medicago* comprises more than 60 species, two-thirds being annuals and one-third being perennials (Quiros and Bauchan, 1988). Its primary center of diversity is located in the Caucasian region, and it is endemic to the Mediterranean region. The genus is present across the world in varying climates and soils. The history of this diversification is described (review by Michaud et al., 1988; Musial et al., 2005) and completed with new elements obtained through molecular data (Bena, Jubier, et al., 1998; Bena, Lejeune et al., 1998; Bena, Prosperi, et al.,1998; Muller et al., 2003). Most species have a chromosome number of eight; they are diploid or tetraploid, autogamous (annual species) or allogamous (perennial species). A small number of species are cultivated. Among the perennial species, the *Medicago sativa* complex, which was formed from intercrosses between the subspecies *sativa, falcata,* and *glutinosa,* contains cultivated alfalfa. Several annual species (*M. truncatula, M. polymorpha,* and others) are cultivated in rangelands in Mediterranean climates.

The cultivated populations of alfalfa are autotetraploid (Stanford, 1951) with $2n = 4x = 32$, perennial, and allogamous, making their genetic study very complicated. However, breeding programs aimed at improving forage production and quality, resistance to pests and diseases, persistence, and seed production are being developed worldwide. The development of genomic tools for this species is hampered by the complexity of the species at the genetic level (tetraploidy and allogamy) as well as the physiological level (perenniality).

The need for a model species devoted to legume crops was noted in the late 1980s. Such a species had to be phylogenetically related to the main temperate legume crops (alfalfa, clover, and pea); it needed to be diploid, autogamous, and annual, and have a small genome. *Medicago truncatula* was selected as this model species, first to analyze the molecular aspects of the symbiosis (Barker et al., 1990; Cook, 1999) and subsequently to analyze all other components related to agronomic traits in cultivated species (Huguet et al., 2004). *M. truncatula* belongs to the same genus as alfalfa, thus leading to possible information exchange between scientists in charge of the model and those working on the cultivated species.

Genetic maps are essential tools for the analysis of genomes and quantitative traits of agronomic interest. They require the use of specific populations, studied with molecular markers. In this chapter, an overview of the advances in mapping programs on both the model species *M. truncatula* and the cultivated species *M. sativa* is given, first giving a rapid overview on the available markers in *Medicago* spp., second describing the progress in genetic mapping in the model and in the cultivated species, and third, showing the synteny between the two species. Prospects on the use of these maps for research programs are examined.

## *MARKERS IN* MEDICAGO *SPP.*

An overview of the markers available in *Medicago* species is given in Table 34.1. As in any species, "anonymous" markers are mass-revealed and usually scored as dominant markers: AFLPs are the most reproducible but RAPDs are also usable. They were useful for the early mapping programs, when the number of informative markers was limited (Echt et al., 1993; Julier et al., 2003; Kaló et al., 2000; Thoquet et al., 2002). Nowadays, they can help in saturating a specific zone of the genome in a fine mapping program or in a bulk-segregant analysis. A few morphological traits are also available, like pod coiling (Thoquet et al., 2002) or foliar spot (T. Huguet, unpublished results) in *M. truncatula,* and flower color (Barnes, 1972; Demarly, 1954), dwarfism, or sticky leaf mutation (Kiss et al., 1993) in *M. sativa*.

Many informative markers (i.e., developed from DNA sequences), codominant, reproducible, and polymorphic, are now available. They include RFLPs, SSR, isozymes, STS, CAPs, dCAPs, or SNPs (Choi, Kim, et al., 2004; Eujayl et al., 2004; Julier et al., 2003; Kaló et al., 2000; Mun et al., 2006; Sledge et al., 2005; Thoquet et al., 2002). Many of these markers were developed from EST or BAC sequences of *M. truncatula*. The transferability of SSR markers from *M. truncatula* to *M. sativa* was tested: for

TABLE 34.1. Markers available in *Medicago*.

| Types of markers | Nature | Advantage | Disadvantage | Reference |
|---|---|---|---|---|
| Isozymes | Codominant | Cheap<br>Simple to reveal | Low number of loci Low polymorphism | Birouk and Dattée, 1989; Brunel, 1982; De Vienne, 1978; Jenczewski et al., 1999; Kaló et al., 2000; Quiros, 1981, 1983; Quiros and Morgan, 1981; Small et al., 1999; Thoquet et al., 2002 |
| RFLP | Codominant | Locus specific<br>Portability | Expensive<br>Time consuming | Brummer et al., 1993; Echt et al., 1993; Kaló et al., 2000; Tavoletti et al., 1996; Thoquet et al., 2002 |
| SSR | Codominant | Simple to reveal<br>Low cost<br>Many loci available<br>High polymorphism | | Baquerizo-Audiot et al., 2001; Diwan et al., 1997; Eujayl et al., 2004; Julier et al., 2003; Mun et al., 2006; Sledge et al., 2005; T. Huguet, unpublished data, http://www.medicago.org/ |
| STS | Codominant | Simple to reveal | Low number of loci<br>Moderate polymor-phism | Choi, Kim, et al., 2004; Choi, Mun, et al., 2004 |
| CAPS, dCAPS | Codominant | Locus specific | Expensive | Choi, Kim, et al., 2004; Choi, Mun, et al., 2004 |
| SNP | Codominant | Locus specific | Expensive | Choi, Kim, et al., 2004; Choi, Mun, et al., 2004 |
| RAPD | Dominant | Mass-revealed<br>Genome coverage<br>Cheap | Low reproducibility No portability | Echt et al., 1993; Kaló et al., 2000; Musial et al., 2005; Thoquet et al., 2002 |
| AFLP | Dominant | Mass-revealed<br>Genome coverage | Low portability | Julier et al., 2003; Musial et al., 2005; Thoquet et al., 2002 |

SSR developed from *M. truncatula* EST, 80 percent gave amplification products in *M. sativa,* and 50 percent gave polymorphism among two genotypes (Julier et al., 2003). Similar levels of cross-amplification and polymorphism were described by Eujayl et al. (2004) and Sledge et al. (2005).

Isozymes were also used in a number of studies (Birouk and Dattée, 1989; Brunel, 1982; De Vienne, 1978; Jenczewski et al., 1999; Kaló et al., 2000; Quiros, 1981; Quiros and Morgan, 1981; Small et al., 1999). Similarly, seed protein markers obtained in two-dimensional electrophoresis have also been generated (Kaló et al., 2000).

## *GENETIC MAPS IN* MEDICAGO TRUNCATULA

Three maps are currently available, two from $F_2$ populations and one from recombinant inbred line (RIL) populations (Figure 34.1). The first saturated map published on *M. truncatula* (Thoquet et al., 2002) was based on an $F_2$ mapping population, obtained from the cross between two lines, an Australian line 'Jemalong' and an Algerian genotype 'DZA315-16' showing polymorphism for phenotypic traits (flowering date, morphology) and molecular markers. The population comprised 124 $F_2$ individuals that were genotyped with 289 RAPDs, AFLPs, genes, and isoenzymes. The map covered 1,225 cM arranged in eight linkage groups with an average distance of 470 kb per cM.

Another $F_2$ population with 93 individuals was also obtained from the cross of two Australian accessions, 'Jemalong' and 'A20,' two lines that proved to be phenotypically different (Penmetsa and Cook, 2000). A genetic map was constructed (Choi et al., 2004) integrating markers developed from ESTs, BAC-end sequences, and resistance gene analogs (RGAs), anchored with an *M. truncatula* cytogenetic map (Kulikova et al., 2001) through FISH mapping of BAC clones. This map comprises 286 markers covering 513 cM. A set of 346 new SSR markers developed from BAC sequences was added to integrate genetic and physical maps (Mun et al., 2006). Detailed information on this map is available at http://www.medicago.org/genome.

However, $F_2$ mapping populations are difficult (or impossible) to maintain, and each individual cannot be correctly replicated to proceed to a robust phenotypic evaluation. For these reasons, therefore, RILs as mapping populations are more adequate. The $F_6$ lines resulting from selfing of $F_2$ plants retain only 3.1 percent of residual heterozygosity. Hence, they can be multiplied by selfing, stored, exchanged, and analyzed for any traits in replicated designs. An $F_6$ RIL population of 199 lines was obtained on the

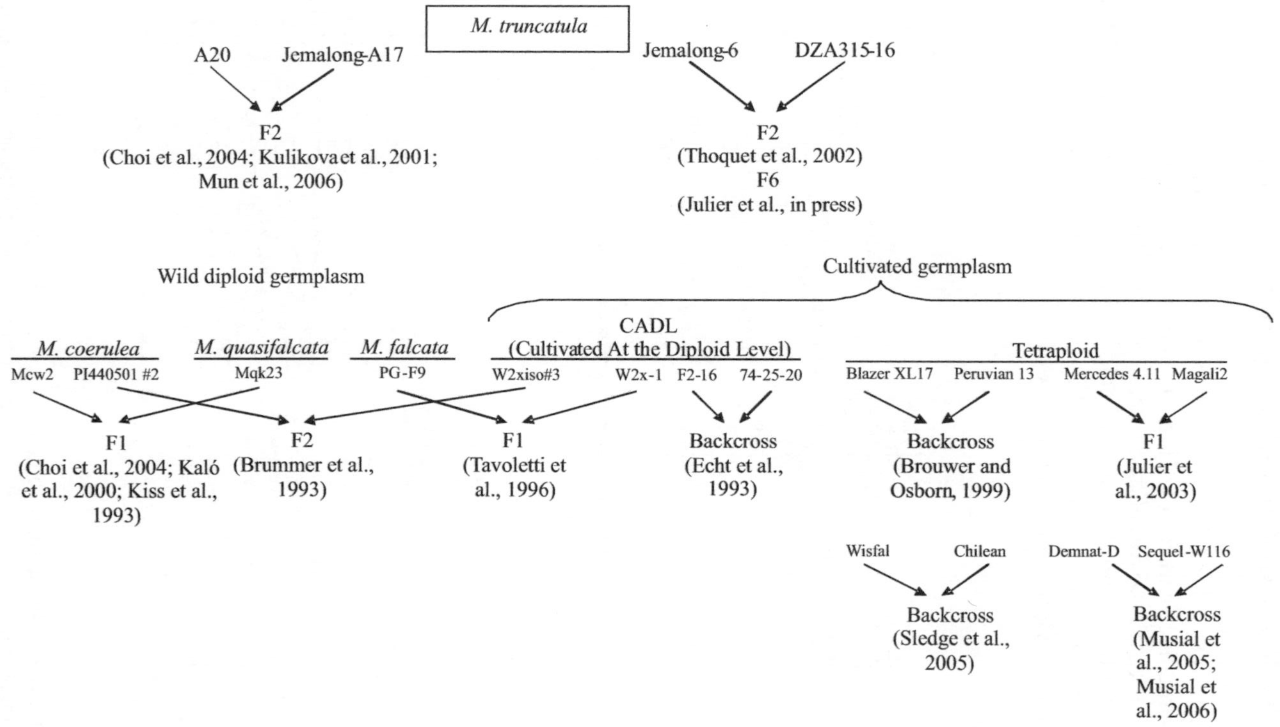


FIGURE 34.1. General view of the mapping populations in *Medicago* spp.

cross Jemalong × DZA315-16 and typed with more than 600 anonymous and informative markers (T. Huguet, unpublished data). Typing data for 246 AFLP, RAPD, ISSR, and SSR markers is available on a core-RIL population of 93 lines (http://medicago.toulouse.inra.fr/Mt/GeneticMAP/LR4_MAP.html). A map, presented in Figure 34.2, was built with JoinMap software (Van Ooijen and Voorrips, 2001).

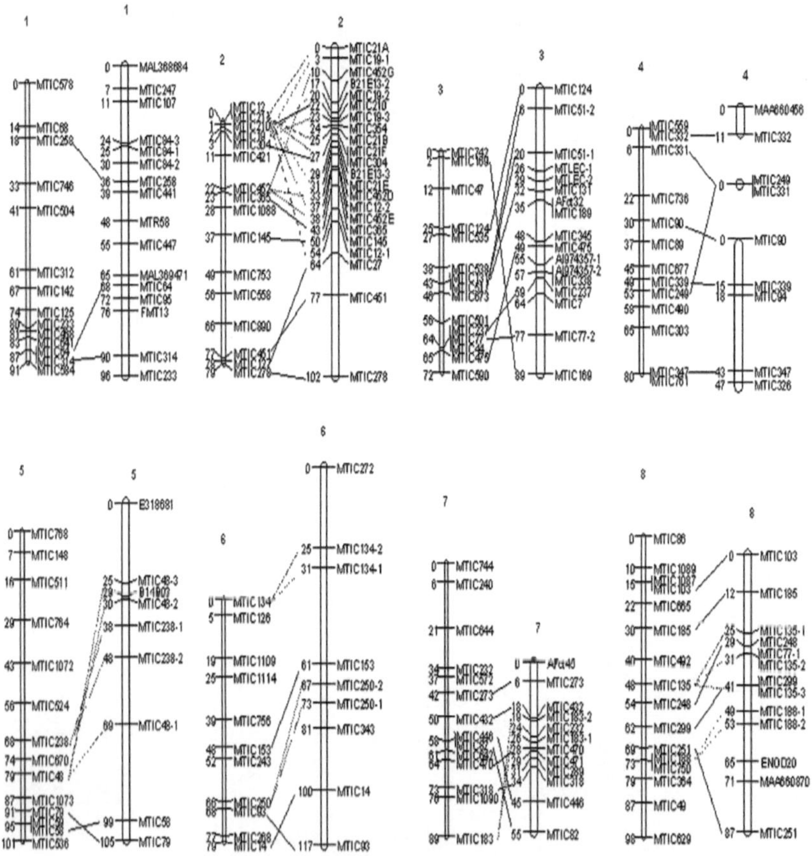

FIGURE 34.2. Genetic maps obtained on an $F_6$ RIL population of *M. truncatula* (Jemalong-6 × DZA315-16), and on an $F_1$ population of tetraploid alfalfa (at the right-hand side) with SSR markers, showing the synteny between the two species. Some markers identified amplified several loci in *M. sativa*.

These three genetic maps are oriented similarly and the linkage groups are numbered in accordance with a diploid *M. sativa* map (Kaló et al., 2000). They show a similar degree of segregation distortion at the $F_2$ level: 27 percent of the markers of the cross Jemalong × DZA315.16 (Thoquet et al., 2002) and 20 percent in the cross Jemalong × A20 (Choi, Kim, et al., 2004). However, these distorted markers are not randomly distributed over the genetic maps since only three linkage groups show clusters of distorted markers. Interestingly, these linkage groups are not the same in each cross, suggesting the existence of specific interactions, which favor some alleles.

## *GENETIC MAPS IN* MEDICAGO SATIVA *AT THE DIPLOID LEVEL*

Until the 1990s, autotetraploidy was an obstacle to the development of a genetic map of alfalfa. Hence, several groups focused on diploid alfalfas that belong to the *M. sativa* complex (Brummer et al., 1993; Echt et al., 1993; Kiss et al., 1993; Tavoletti et al., 1996). These species (*M. coerulea, M. quasifalcata*) are allogamous, so the plants are heterozygous. Mapping populations were either $F_1$ between *M. quasifalcata* (a diploid *falcata* wild type) and *M. coerulea* (a diploid *sativa* wild type; Kiss et al., 1993), or a backcross with two parents coming from a series of cultivated alfalfa established at the diploid level (CADL; Echt et al., 1993), or $F_2$ between a CADL plant and a wild *M. coerulea* plant (Brummer et al., 1993). The markers were RFLPs, RAPDs, and isozymes. In 1993, all three maps identified at least eight linkage groups, but the marker density was poor. The $F_1$ map was further improved by the addition of more anonymous markers, seed proteins, and RFLPs (Kaló et al., 2000), producing a very dense map. The segregation distortion in these four maps is very variable, ranging from 9 percent to 62 percent. The $F_1$ map from Kaló et al. (2000) is the reference map for chromosome numbering and orientation. Later this map was anchored to an *M. truncatula* map by common markers (Choi et al., 2004).

This approach, based on diploid plants from *M. sativa* complex species, has simplified mapping procedures. It is easier to analyze marker segregation and recombination rate calculation, and marker ordering on chromosomes is more efficient than in autotetraploid species (Hackett et al., 1998). However, the plant material is wild or suffers from diploidization (CADL), which induces a strong inbreeding depression. In both cases, the allelic variation is possibly different from that observed in cultivated populations. Furthermore, the use of maps to locate quantitative trait loci (QTL) is greatly hampered by the wild phenotypes of the mapping populations. It is therefore difficult to score some of the traits of agronomic interest. Another constraint is related to the type of mapping populations: due to the allogamy of the species, the $F_1$, $F_2$, or backcross plants cannot be multiplied except through cuttings. The maintenance or exchange of these mapping populations is therefore very difficult.

## *GENETIC MAPS AT THE TETRAPLOID LEVEL*

The autotetraploid species usually express a strong inbreeding depression. For this reason, inbreeding generations should be avoided to produce mapping populations. $F_1$ populations are the most common, but $F_2$ or backcross can also be used. In this case, attention must be paid to a possible bias induced by the loss of gametes or individuals carrying lethal alleles.

The theory of linkage mapping in tetraploid species first relied on the segregation of single-dose (simplex) markers or SDRF for single-dose restriction fragments (Wu et al., 1992; Yu and Pauls, 1993). The alleles showing single-dose segregation (i.e., present in one parent, absent in the other and segregating in 1:1 in an $F_1$ population) can be used for calculating the linkage between pairs of markers. Indeed, the calculations are the same in this case as in diploid species. However, this procedure induces some loss of information when alleles originate from codominant markers, because the pattern of segregation of all alleles is not used. In these early steps, the linkage analysis was also reduced to the coupling phase (i.e., markers that are located on the same linkage group). But in autotetraploid organisms, it is important to identify the four chromosomes that belong to a homology group so as to take into account the repulsion linkage phase.

In the late 1990s, several groups developed theories of linkage analysis in autopolyploid organisms (Hackett et al., 1998; Ripol et al., 1999; Wu et al., 2001). The dominant markers segregating as double dose (duplex) or simplex markers present in both parents can be used for linkage analysis, even if the variance of the estimate of recombination rate is higher than that with simplex markers. Major progress was made possible using all the information provided by codominant markers (Luo et al., 2000, 2001). A simplex marker can be mapped on only one chromosome (over eight if we count both parents), but a codominant marker can be mapped on each of the eight linkage groups, if it carries eight different alleles. A software specially developed for mapping $F_1$ populations of autotetraploids, TetraploidMap, was developed (Hackett and Luo, 2003) and released on the Web site, http://www.bioss.sari.ac.uk/knowledge/tetraploidmap/. This software analyzes the segregation of each marker, either dominant or codominant. For codominant markers, it evaluates the frequency of double-reduction events. Map construction is performed in four steps. First, TetraploidMap infers the most likely parental genotypes for each locus and tests to determine if there is a significant amount of double reduction. Second, it calculates a $\chi^2$ test of independence between markers, which leads to the partition of the markers into homology groups for each parent. Third, within each homology group, TetraploidMap determines the linkage phase, recombination rate, and LOD score between markers. Linkage groups (i.e., chromosomes) within homology groups are identified using the linkage phase information. Last, the markers are ordered and the distances between markers are calculated.

The first linkage study in tetraploid alfalfa was conducted on an $F_1$ population with RAPD markers (Yu and Pauls, 1993). A similar study was conducted with SSR markers (Diwan et al., 2000). However, the number of markers was the limiting factor in both cases. In two backcross populations with RFLP markers, a map containing seven linkage groups (instead of eight) covering 453 cM was obtained (Brouwer and Osborn, 1999). The linkage mapping was based on single-dose markers that showed less than 10 percent of segregation distortion.

From an $F_1$ population, two dense linkage maps were obtained with AFLP and SSR markers developed from an *M. truncatula* EST database (Julier et al., 2003). Eight homology groups were found for each parent, each containing four linkage groups. For the female parent, the map length was 3,045 cM with 339 marker positions and an average distance between markers of 9.0 cM. The map of the male parent was 2,649 cM, with 350 marker positions and an average distance between markers of 6.9 cM. In both maps, the distorted markers (25 percent of the SSRs and 34 percent of the AFLPs) were spread across the genome. Using only SSR markers, a composite map of eight linkage groups was built, covering 709 cM (Figure 34.2). Similar linkage studies were realized in an $F_1$ population (Campbell et al., 2003) and backcrosses (Musial et al., 2006; Sledge et al., 2005). For linkage mapping in autotetraploids, it seems that the codominant markers, even if they are not mass-revealed as are AFLPs or RAPDs, give rise to very useful information for building homology groups (Hackett et al., 2003). For this reason, efforts must be made to obtain SSR or RFLP markers.

The analysis of mapping populations in tetraploid alfalfa is also a way for (1) estimating the double reduction frequency, (2) testing the autotetraploidy, and (3) evaluating the synteny with the model species *M. truncatula*. Since the discovery of the autotetraploidy of alfalfa and the observation of the formation of quadrivalents during meiosis, it was stressed that if crossing-over occurred, with the recombinant chromatids migrating to the same pole at anaphase I, then gametes with copies of the same gene are formed. This phenomenon is described as double reduction, and meiosis is said to be pseudo-equational (Demarly, 1963; Gallais, 2003; Mather, 1935, 1936; Ronfort et al., 1998). The quantification of double reduction events requires codominant markers and adequate statistics (Luo et al., 2000, 2004). The SSR loci often give a large number of alleles useful for such calculation. Out of 104 SSR loci in alfalfa, 26 statistically showed significant double reduction (Julier et al., 2003) in an $F_1$ population. In the same $F_1$ population extended to more than 500 individuals, the analysis of two SSR loci indicates that the double reduction frequency is about 2 to 3 percent (Ayadi et al., 2005). This low level of double reduction is in accordance with the low frequency of tetravalent formation at meiosis (Armstrong, 1954).

The autotetraploid species were often described as progressively evolving to allotetraploidy, in which chromosomes pair preferentially (Soltis and Soltis, 1993). Mapping populations are useful to test if chromosomes show a preferential pairing or if they pair randomly. Indeed, a random pairing was observed (Julier et al., 2003), with repulsion coupling occurring between each pair of chromosomes from a homology group. Another group (Ma et al., 2002) described preferential pairing, which, however, could not be deduced after a corrected statistical analysis (Cao et al., 2004).

## *SYNTENY*

Starting from the earliest dense map obtained on diploid *M. sativa* (Kaló et al., 2000), chromosomes in all successive maps of the *Medicago* genus were numbered accordingly when informative markers were mapped. Indeed, the number of chromosomes is conserved across species. The synteny, defined by the conservation of marker order along the chromosomes, was first shown between tetraploid alfalfa and *M. truncatula* (Julier et al., 2003) and then between diploid alfalfa and *M. truncatula* (Choi, Kim, et al., 2004; Choi, Mun, et al., 2004). This absence of rearrangements in the genomes of the two species probably reflects a recent divergence. It is clearly an advantage for the genetic analysis of traits in alfalfa, a species with a complicated autotetraploid genome, whose cultivated varieties are synthetic populations. It offers the possibility to analyze traits in the model species *M. truncatula,* before a transfer of information toward the cultivated one, *M. sativa* (Delseny, 2004). But even if the regulation of traits is likely to be similar, it is not known if genes displaying variability are the same.

## *PROSPECTS*

### *Mapping*

Studies on molecular markers and mapping populations have produced significant results both for the model legume *M. truncatula* and for cultivated alfalfa. The number of codominant markers available for mapping is no longer a constraint. Furthermore, current progress in physical mapping in *M. truncatula* through sequencing and the anchorage of the physical map to the genetic map has given good access to much information about gene location and genome structure. The synteny between *Medicago* species is very high. The alignment of genomes of different legume species would help in the positional cloning of genes (Stracke et al., 2004).

### *QTL Identification*

The mapping populations are useful in further identification of QTLs of agronomic traits. The number of published studies on tetraploid alfalfa is still limited, but some groups have shown interest in mapping winter hardiness, autumn growth, and freezing injury (Alarcon-Zuniga et al., 2004; Brouwer et al., 2000), drought resistance (Sledge et al., 2004), forage yield (Musial et al., 2006; Robins et al., 2003), stem growth and morphogenesis (Julier, Ecalle, et al., 2002; Julier, Guines, et al., 2002), and disease resistance (Musial et al., 2005). For such studies, the choice of parents for the mapping populations is of major importance. To extend its genetic diversity to a maximum, a very nondormant *M. sativa* cultivated plant was crossed with a very dormant plant of the *M. falcata* germplasm (Alarcon-Zuniga et al., 2004; Campbell et al., 2003; Sledge et al., 2004). As the two parents belong to two different species or subspecies, there is a risk of the genomes showing preferential pairing. An alternative is to work within the cultivated germplasm by crossing two extreme types for dormancy range (Brouwer and Osborn, 1999) or reducing the range of variation for dormancy to cultivated material in Europe (Julier et al., 2003) or in Australia (Musial et al., 2006). From available results, variation was observed in all crosses. As seen in other species, QTL identification will rely on this choice of material. The comparison of QTL position between studies would require that the linkage groups are numbered as in the reference maps of *M. truncatula* or diploid *M. sativa*. Eventually, marker-assisted selection can be carried out within the mapping population on the basis of both the phenotypes and the QTLs. However, the development and mapping of populations are extremely time consuming, making the efficiency of such a procedure probably low. In the future, a marker-assisted selection would be more efficient in a tetraploid and perennial species like alfalfa than in a diploid and allogamous species, because a selection of plants carrying four doses of positive alleles made at an early stage (plantlets) would improve the efficiency of breeding schemes and reduce their duration.

In the case of *M. truncatula,* RIL populations are ideal tools for QTL identification. Several RIL populations are being developed currently (T. Huguet, J.M. Prosperi, personal communication) involving plant lines of contrasting phenotypes like phenology, morphology, or pest resistance. To date, the number of published QTL analyses on model legumes is restricted (Julier et al., 2007) but should increase rapidly. In this study, several regions involved in growth and morphogenesis were identified. A major QTL related to flowering date was found on chromosome 7. QTL mapping is also possible in diploid alfalfa (Tavoletti et al., 2000). QTL location on *M. truncatula* offers the prospect of identifying the underlying genes with either a candidate gene approach or positional cloning strategy. QTL identification and gene cloning would help in the analysis of genetic bases in alfalfa if the genes responsible for variation in a trait in the model species are also involved in the variation in the cultivated species.

### *Analysis of Genetic Diversity*

The codominant microsatellite markers mapped in *M. truncatula* can be used for the analysis of genetic structure in *M. truncatula* or alfalfa. In *M. truncatula*, SSR markers were used to build nested core collections, useful for the study of natural variation (Ronfort, 2006). In alfalfa, the SSR markers showed that the within-variety variation is extremely large but were unable to clearly distinguish between varieties coming from a single breeder (Flajoulot et al., 2005). The availability in SSR markers will probably assist the analysis of genetic resources in alfalfa.

## REFERENCES

Alarcon-Zuniga, B., P. Scott, K.J. Moore, D. Luth, and E.C. Brummer (2004). Quantitative trait locus mapping of winter hardiness metabolites in autotetraploid alfalfa (*M. sativa*). In *Molecular breeding of forage and turf,* A. Hopkins, Z.Y. Wang, R. Mian, M. Sledge, and R.E. Barker (eds.). Dordrecht: Kluwer Academic Publishers, pp. 97-104.

Armstrong, J.M. (1954). Cytological studies in alfalfa polyploids. *Canadian Journal of Botany* 32:531-542.

Ayadi, R., P. Barre, C. Huyghe, and B. Julier (2005). Estimation of the coefficient of double-reduction in autotetraploid lucerne. In *XX International Grassland Congress, Proceedings of the 4th International Symposium on the Molecular Breeding of Forage and Turf,* M. O. Humphreys (ed.). Wageningen:Academic Publishers, p. 138.

Baquerizo-Audiot, E., B. Desplanque, J.M. Prosperi, and S. Santoni (2001). Characterization of microsatellite loci in the diploid legume *Medicago truncatula* (barrel medic). *Molecular Ecology Notes* 1:1-3.

Barker, D., S. Bianchi, F. Blondon, Y. Dattée, G. Duc, S. Essad, P. Flament, P. Gallusci, G. Génier, P. Guy, X. Muel, J. Tourneur, J. Denarie, and T. Huguet (1990). *Medicago truncatula,* a model plant for studying the molecular genetics of the *Rhizobium*-legume symbiosis. *Plant Molecular Biology Reporter* 8:40-49.

Barnes, D.K. (1972). A system for visually classifying alfalfa flower color. *Agriculture handbook.* Washington, D. C. Agricultural Research Service, U.S. Department of Agriculture.

Bena, G., M.F. Jubier, I. Olivieri, and B. Lejeune (1998). Ribosomal external and internal transcribed spacers: Combined use in the phylogenetic analysis of *Medicago* (Leguminosae). *Journal of Molecular Evolution* 46:299-306.

Bena, G., B. Lejeune, J.M. Prosperi, and I. Olivieri (1998). Molecular phylogenetic approach for studying life-history evolution: The ambiguous example of the genus *Medicago* L. *Proceedings of the Royal Society of London, Series B* 265: 1141-1151.

Bena, G., J.M. Prosperi, B. Lejeune, and I. Olivieri (1998). Evolution of annual species of the genus *Medicago:* A molecular phylogenetic approach. *Molecular Phylogenetics and Evolution* 9:552-559.

Birouk, A. and Y. Dattée (1989). Organisation de la variabilité enzymatique des populations marocaines de luzerne (*Medicago sativa* L.): Structures géniques et génotypiques. *Genome* 32:120-128.

Brouwer, D.J., S.H. Duke, and T.C. Osborn (2000). Mapping genetic factors associated with winter hardiness, fall growth, and freezing injury in autotetraploid alfalfa. *Crop Science* 40:1387-1398.

Brouwer, D.J. and T.C. Osborn (1999). A molecular marker linkage map of tetraploid alfalfa. *Theoretical and Applied Genetics* 99:1194-1200.

Brummer, E.C., J.H. Bouton, and G. Kochert (1993). Development of an RFLP map in diploid alfalfa. *Theoretical and Applied Genetics* 86:329-332.

Brunel, D. (1982). Mise en évidence de 4 locus enzymatiques chez la luzerne (*Medicago sativa* L.) di- et tétraploïde. *Agronomie* 2:133-148.

Campbell, T.A., E.C. Brummer, D. Luth, G.R. Bauchan, Z.L. Xia, and C. He (2003). Screening autotetraploid alfalfa and *Medicago truncatula* SSR markers for inclusion in a cultivated alfalfa linkage map. In *Molecular breeding of forage and turf: Third international symposium,* A. Hopkins and R.E. Barker (eds.). Dallas, Texas: Samuel Roberts Noble Foundation, p. 153.

Cao, D.C., T.C. Osborn, and R.W. Doerge (2004). Correct estimation of preferential chromosome pairing in autotetraploids. *Genome Research* 14:459-462.

Choi, H.K., D. Kim, T. Uhm, E. Limpens, H. Lim, J.H. Mun, P. Kaló, R.V. Penmetsa, A. Seres, O. Kulikova, et al. (2004). A sequence-based genetic map of *Medicago truncatula* and comparison of marker colinearity with *Medicago sativa. Genetics* 166:1463-1502.

Choi, H.K., J.H. Mun, D.J. Kim, H. Zhu, J.M. Baek, J. Mudge, B. Roe, N. Ellis, J. Doyle, G.B. Kiss, et al. (2004). Estimating genetic conservation between crop and model legume species. *Proceedings of the National Academy of Sciences USA* 101:15289-15294.

Cook, D.R. (1999). *Medicago truncatula*—a model in the making! *Current Opinion in Plant Biology* 2:301-304.

Delseny, M. (2004). Re-evaluating the relevance of ancestral shared synteny as a tool for crop improvement. *Current Opinion in Plant Biology* 7:126-131.

Demarly, Y. (1954). Etude de l'hérédité de la bigarrure de la fleur chez la luzerne. *Annales de l'Amélioration des Plantes* 4:5-20.

Demarly, Y. (1963). Génétique quantitative des tétraploïdes et amélioration des plantes. *Annales de l'Amélioration des Plantes* 13:307-400.

De Vienne, D. (1978). Variabilité chez une espèce tétraploïde: Analyse isoenzymatique et biométrique du pollen de quelques familles apparentées de luzerne. *Annales de l'Amélioration des Plantes* 28:289-307.

Diwan, N., A.A. Bhagwat, G.R. Bauchan, and P.B. Cregan (1997). Simple sequence repeat DNA markers in alfalfa and perennial and annual *Medicago* species. *Genome* 40:887-895.

Diwan, N., J.H. Bouton, G. Kochert, and P.B. Cregan (2000). Mapping of simple sequence repeat (SSR) DNA markers in diploid and tetraploid alfalfa. *Theoretical and Applied Genetics* 101:165-172.

Echt, C.S., K.K. Kidwell, S.J. Knapp, T.C. Osborn, and T.J. McCoy (1993). Linkage mapping in diploid alfalfa (*Medicago sativa*). *Genome* 37:61-71.

Eujayl, I., M.K. Sledge, L. Wang, G.D. May, K. Chekhowskiy, J.C.Z. Zwonitzer, and M.A.R. Mian (2004). *Medicago truncatula* EST-SSRs reveal cross-species genetic markers for *Medicago* spp. *Theoretical and Applied Genetics* 108:414-422.

Flajoulot, S., J. Ronfort, P. Baudouin, P. Barre, T. Huguet, C. Huyghe, and B. Julier (2005). Genetic diversity among alfalfa (*Medicago sativa*) cultivars coming from a single breeding program, using SSR markers. *Theoretical and Applied Genetics* 111:1420-1429.

Gallais, A. (2003). Quantitative genetics and breeding methods in autopolyploid plants. Paris: INRA Editions.

Hackett, C.A., J.E. Bradshaw, R.C. Meyer, J.W. McNicol, D. Milbourne, and R. Waugh (1998). Linkage analysis in tetraploid species: A simulation study. *Genetical Research Cambridge* 71:143-154.

Hackett, C.A. and Z.W. Luo (2003). TetraploidMap: Construction of a linkage map in autotetraploid species. *Journal of Heredity* 94:358-359.

Hackett, C.A., B. Pande, and G.J. Bryan (2003). Constructing linkage maps in autotetraploid species using simulated annealing. *Theoretical and Applied Genetics* 106:1107-1115.

Huguet, T., P. Thoquet, M. Ghérardi, G. Cardinet, S. Prioul, F. Lazrek, M.E. Aouani, M. Laouar, A. Abdelguerfi, O. Kurchak, et al. (2004). A post-genomic approach of the natural variations of the model-legume *Medicago truncatula*. In *Legumes for the benefit of agriculture, nutrition and the environment: Their genomics, their products, and their improvement.* Dijon:AEP, pp. 169-170.

Jenczewski, E., J.M. Prosperi, and J. Ronfort (1999). Evidence for gene flow between wild and cultivated *Medicago sativa* (Leguminosae) based on allozyme markers and quantitative traits. *American Journal of Botany* 86:677-687.

Julier, B., C. Ecalle, and C. Huyghe (2002). Identification of molecular markers associated to morphological traits in alfalfa. In *Multifunction grasslands: Quality forages, animal products and landscape,* J.L. Durand, J.C. Emile, C. Huyghe, and G. Lemaire (eds.). Versailles:AFPF, 7:318-319.

Julier, B., S. Flajoulot, P. Barre, G. Cardinet, S. Santoni, T. Huguet, and C. Huyghe (2003). Construction of two genetic linkage maps in cultivated tetraploid alfalfa (*Medicago sativa*) using microsatellite and AFLP markers. *BMC Plant Biology* 3:9.

Julier, B., F. Guines, C.A. Hackett, T. Huguet, and C. Huyghe (2002). Genetic mapping and identification of QTL for stem morphogenesis in tetraploid alfalfa. Paper presented at North American Alfalfa Improvement Conference, Sacramento, CA.

Julier, B., T. Huguet, F. Chardon, R. Ayadi, J.B. Pierre, J.M. Prosperi, P. Barre, and C. Huyghe (2007). Identification of quantitative trait loci influencing aerial morphogenesis in the model legume *Medicago truncatula. Theoretical and Applied Genetics* 114:1391-1406.

Kaló, P., G. Endre, L. Zimányi, G. Csanádi, and G.B. Kiss (2000). Construction of an improved linkage map of diploid alfalfa (*Medicago sativa*). *Theoretical and Applied Genetics* 100:641-657.

Kiss, G.B., G. Csanadi, K. Kalman, P. Kaló, and L. Ökrész (1993). Construction of a basic genetic map for alfalfa using RFLP, RAPD, isozyme and morphological markers. *Molecular and General Genetics* 238:129-137.

Kulikova, O., G. Gualrieri, R. Geurts, D.J. Kim, D. Cook, T. Huguet, J. Hans de Jong, P.F. Fransz, and T. Bisseling (2001). Integration of the FISH pachytene and genetic maps of *Medicago truncatula. Plant Journal* 27:49-58.

Luo, Z.W., C.A. Hackett, J.E. Bradshaw, J.W. McNicol, and D. Milbourne (2000). Predicting parental genotypes and gene segregation for tetrasomic inheritance. *Theoretical and Applied Genetics* 100:1067-1073.

Luo, Z.W., C.A. Hackett, J.E. Bradshaw, J.W. McNicol, and D. Milbourne (2001). Construction of a genetic linkage map in tetraploid species using molecular markers. *Genetics* 157:1369-1385.

Luo, Z.W., R.M. Zhang, and M.J. Kearsey (2004). Theoretical basis for genetic linkage analysis in autotetraploid species. *Proceedings of the National Academy of Sciences USA* 101:7040-7045.

Ma, C.X., G. Casella, Z.J. Shen, T.C. Osborn, and R. Wu (2002). A unified framework for mapping quantitative trait loci in bivalent tetraploids using single-dose restriction fragments: A case study from alfalfa. *Genome Research* 12:1974-1981.

Mather, K. (1935). Reductional and equational separation of the chromosomes in bivalents and multivalents. *Journal of Genetics* 30:53-78.

Mather, K. (1936). Segregation and linkage in autotetraploids. *Journal of Genetics* 32:287-314.

Michaud, R., W.F. Lehman, and M.D. Rumbauch (1988). World distribution and historical development. In *Agronomy, series of monograph 2,* A.A. Hanson, D.K. Barnes, and R.R. Hill (eds.). Madison, WI: ASA, CSSA, SSSA, pp. 25-91.

Muller, M.H., J.M. Prosperi, S. Santoni, and J. Ronfort (2003). Inferences from mitochondrial DNA patterns on the domestication history of alfalfa (*Medicago sativa*). *Molecular Ecology* 12:2187-2199.

Mun, J.H., D.J. Kim, H.K. Choi, J. Gish, F. Debelle, J. Mudge, R. Denny, G. Endre, O. Saurat, A.M. Dudez, et al. (2006). Distribution of microsatellites in the genome of *Medicago truncatula:* A resource of genetic markers that integrate genetic and physical maps. *Genetics* 172:2541-2555.

Musial, J.M., K.S. Aitken, J.M. Mackie, and J.A.G. Irwin (2005). A genetic linkage map in autotetraploid lucerne adapted to northern Australia, and use of the map to identify DNA markers linked to resistance to *Phytophthora medicaginis. Australian Journal of Agricultural Research* 56:333-344.

Musial, J.M., K.F. Lowe, J.M. Mackie, K.S. Aitken, and J.A.G. Irwin (2006). DNA markers linked to yield, yield components, and morphological traits in autotetraploid lucerne (*Medicago sativa* L.). *Australian Journal of Agricultural Research* 57:801-810.

Penmetsa, R.V. and D.R. Cook (2000). Production and characterization of diverse development mutants of *Medicago truncatula. Plant Physiology* 123:1387-1397.

Quiros, C. (1981). Starch gel electrophoresis technique used with alfalfa and other *Medicago* species. *Canadian Journal of Plant Science* 61:745-749.

Quiros, C. (1983). Alfalfa, lucerne (*Medicago sativa* L.). In *Isozymes in plant genetics and breeding. Part B,* S.D. Tanksley and T.J. Orton (eds). Amsterdam: Elsevier Science Publishers, pp. 253-294.

Quiros, C.F. and G.R. Bauchan (1988). The genus *Medicago* and the origin of the *Medicago sativa* complex. In *Agronomy, series of monographs, 29,* A.A. Hanson, D.K. Barnes, and R.R. Hill (eds.). Madison, WI:ASA, CSSA, SSSA, pp. 93-124.

Quiros, C. and K. Morgan (1981). Peroxidase and leucine-aminopeptidase in diploid *Medicago* species closely related to alfalfa: Multiple gene loci, multiple allelism, and linkage. *Theoretical and Applied Genetics* 60:221-228.

Ripol, M.I., G.A. Churchill, J.A.G. Da Sylva, and M. Sorrells (1999). Statistical aspects of genetic mapping in autopolyploids. *Gene* 235:31-41.

Robins, J.G., D.R. Viands, T.A. Campbell, D. Luth, and E.C. Brummer (2003). Mapping biomass yield in an intersubspecific cross of alfalfa. In *Molecular breeding of forage and turf, third international symposium,* A. Hopkins and R.E. Barker (eds.). Dallas: Samuel Roberts Noble Foundation, p. 133.

Ronfort, J. (2006). Microsatellite diversity and broad scale geographic structure in a model legume: Building a set of nested core collections for studying naturally occurring variation in *Medicago truncatula. BMC Plant Biology* 6:28.

Ronfort, J., E. Jenczewski, T. Bataillon, and F. Rousset (1998). Analysis of population structure in autotetraploid species. *Genetics* 150:921-930.

Sledge, M.K., I.M. Ray, and G. Jiang (2004). *Medicago truncatula* EST-SSRs for genetic mapping in alfalfa. In *5th European conference on grain legumes—ICLGG 2004, conference handbook.* Dijon:AEP-INRA, p. 182.

Sledge, M., I. Ray, and G. Jiang (2005). An expressed sequence tag SSR map of tetraploid alfalfa (*Medicago sativa* L.). *Theoretical and Applied Genetics* 111: 980-992.

Small, E., S.I. Warwick, and B. Brookes (1999). Allozyme variation in relation to morphology and taxonomy in *Medicago* sect. *Spirocarpos* subsect. *Intertextae* (Fabaceae). *Plant Systematics and Evolution* 214:29-47.

Soltis, D.E. and P.S. Soltis (1993). Molecular data and the dynamic nature of polyploidy. *Critical Reviews in Plant Sciences* 12:243-273.

Stanford, E.H. (1951). Tetrasomic inheritance in alfalfa. *Agronomy Journal* 43: 222-225.

Stracke, S., S. Sato, N. Sandal, M. Koyama, T. Kaneko, S. Tabata, and M. Parniske (2004). Exploitation of colinear relationships between the genomes of *Lotus japonicus, Pisum sativum* and *Arabidopsis thaliana,* for positional cloning of a legume symbiosis gene. *Theoretical and Applied Genetics* 108:442-449.

Tavoletti, S., P. Pesaresi, G. Barcaccia, E. Albertini, and F. Veronesi (2000). Mapping the jp (jumbo pollen) gene and QTLs involved in multinucleate microspore formation in diploid alfalfa. *Theoretical and Applied Genetics* 101:372-378.

Tavoletti, S., F. Veronesi, and T. Osborn (1996). RFLP linkage map of an alfalfa meiotic mutant based on an F1 population. *Journal of Heredity* 87:167-170.

Thoquet, P., M. Ghérardi, E.P. Journet, A. Kereszt, J.M. Ané, J.M. Prosperi, and T. Huguet (2002). The molecular genetic linkage map of the model legume *Medicago truncatula:* An essential tool for comparative legume genomics and the isolation of agronomically important genes. *BMC Plant Biology* 2:1.

Van Ooijen, J.W., and R.E. Voorrips (2001). JoinMap Version 3.0, software for the calculation of genetic linkage maps. Wageningen, Netherlands: Plant Research International.

Wu, K.K., W. Burnquist, M.E. Sorrels, T.L. Tew, P.H. Moore, and S.D. Tanksley (1992). The detection and estimation of linkage in polyploids using single-dose restriction fragments. *Theoretical and Applied Genetics* 83:294-300.

Wu, R., M. Gallo-Meagher, R.C. Littell, and Z.B. Zeng (2001). A general polyploid model for analyzing gene segregation in outcrossing tetraploid species. *Genetics* 159:869-882.

Yu, K.F. and K.P. Pauls (1993). Segregation of random amplified polymorphic DNA markers and strategies for molecular mapping in tetraploid alfalfa. *Genome* 36:844-851.

Chapter 35

# Management of Abiotic Stresses in Grain Legumes Through Manipulation of Genes for Compatible Solutes

P. Sharmila
Firoz Anwar
K. R. Sharma
P. Pardha Saradhi

## *INTRODUCTION*

Grain legumes are the second most important group of crops after cereals that are grown to meet human food demand. Being rich in protein (often referred to as "poor man's meat"), they constitute an important dietary component, especially for millions of Asians. The occurrence of bovine spongiform encephalopathy (BSE or mad cow disease), bird flu, and so on has posed the challenge of providing an alternative to animal-derived protein sources for both human and animal consumption. Grain legumes, being a rich source of protein, will prove to be a good substitute, and demand for grain legumes will continue to grow. Food legumes, besides ensuring nutritional security (grain for human consumption and residue as nutritious feed for livestock), play a vital role in sustainable agriculture by improving soil fertility through biological nitrogen fixation and increasing soil organic content. Pulse grains possess 25 to 30 percent protein on a dry weight basis, which is about three times more than that of cereals. In spite of being so important, food legume farmers are presently faced with enormous challenges such as very low productivity, due to increased incidence of biotic and abiotic factors.

Legumes are predominantly exposed to abiotic stresses such as drought, salinity, cold or frost, high temperature, anoxia or waterlogging, mineral nutrient deficiency or imbalance, and high or low light intensity, individually or in combination. These adverse environmental factors prevent grain legumes from realizing their full genetic potential. For instance, although the potential seed yield of chickpea exceeds 6 t/ha, the realized yield is only about 850 kg/ha as the crop is mostly cultivated in marginal lands that generally experience receding soil moisture, forcing the crop to experience terminal drought (Economic Survey, 2005). As exposure of grain legumes to these stresses (which decrease their productivity to a significant extent) is usually unpredictable, farmers complain of yield inconsistency. Rain-fed legumes are often exposed to terminal drought due to inadequate rains to meet water requirements during pod filling. On the other hand, cultivation of legumes under irrigation for prolonged periods or use of ground water for irrigation invariably leads to higher sodicity or salinity in the agricultural fields. Besides terminal drought and salinity, high or low temperature drastically affects grain filling in legumes (Croser et al., 2003; Singla and Garg, 2005). In their quest to feed the ever-increasing human population, scientists have to contend with these adverse environmental factors and redesign the genetic architecture of concerned crops, including grain legumes, through modern genetic manipulation technology to cope better with abiotic stresses so that agricultural production, which is otherwise expected to decrease, can be increased dramatically. Advances in understanding plant abiotic stress tolerance mechanisms and the advent of molecular genetic technology allow us to address these issues much more efficiently than in the past (Chinnusamy et al., 2005).

In this chapter, we review the most significant achievements of the genetic engineering approach to improve plant abiotic stress tolerance, paying special attention to compatible solutes, antioxidant systems, and ion homeostasis. Improved tolerance of plants to various abiotic stresses has been observed in transgenic plants that express or overexpress genes regulating cell osmoticum, antioxidation, ion homeostasis, and functional three-dimensional structure of proteins, and so on that either directly or indirectly ensure protection of various cell components (membranes, proteins, electron transport complexes, transcriptional and translational machinery, etc.) and overall cell metabolism against toxic aspects of stress. Although the results are not always consistent, these studies collectively foretell a scenario where biotechnology will arm our future crops with new tactics to survive in hostile environments.

Environmental factors such as drought, salinity, low or high temperature, freezing, anoxia, high light intensity, and nutrient imbalances that cause abnormal growth and development of living systems are collectively termed abiotic stresses. Abiotic stresses to which major grain legumes are sensitive are listed in Table 35.1.

TABLE 35.1. Key abiotic stresses influencing yield of major grain legumes.

| Grain legume | Abiotic stress |
|---|---|
| Chickpea (*Cicer arietinum* L.) | Low temperature, frost, terminal drought, salt, waterlogging, mineral deficiency (Fe, B, Zn, Mn) |
| Pigeon pea (*Cajanus cajan* L.) | Cold, waterlogging, terminal drought |
| Black gram (*Phaseolus mungo* L.) | Drought, temperature, preharvest sprouting |
| Mung bean (*Vigna radiata* L.) | Drought, temperature |
| Pea (*Pisum sativum* L.) | Temperature, moisture |
| Lentil (*Lens ensculenta* L.) | Cold, terminal drought |

## *STRATEGIES FOR ENHANCING ABIOTIC STRESS TOLERANCE*

Plants have a built-in potential to monitor and adjust to adverse environmental stresses, but the degree of capacity to withstand (i.e., tolerance potential) varies from species to species and in fact even from genotype to genotype within the same species. Adaptive strategies, which are under genetic control, may vary significantly from plant system to plant system. Accordingly, adaptation entails changes in the gene expression pattern, that is, it is up- or down-regulated under abiotic stress. Tolerance to environmental stresses requires concerted action of several gene products and is often associated with several sequential events involving stress perception, signal transduction, transcription regulation, and expression or suppression of structural genes followed by adaptation for better survival. Although significant success has been obtained in the production and commercialization of insect-and herbicide-resistant plants and to some extent in biotic stress-resistant plants using transgenic technology, success in the production of abiotic stress-tolerant plants has been limited, largely because of the complex genetic mechanisms that govern abiotic stress tolerance.

In fact, owing to the extreme complexity of the cellular response of plants to abiotic stresses, it was believed for a long time that enhancing tolerance of plants to abiotic stresses is an impossible task. However, rapid advancement in technology and modern molecular tools paved the way for identifying the factors or genes that could be used for imparting abiotic stress tolerance. Transcriptome analyses using microarray technology have revealed the regulation (induction or suppression) of several genes in plant species by abiotic stresses. Depending on their overall functional role, the stress-induced genes can be broadly classified into two major categories: (1) the first group encodes products that directly protect cells against stress;

and (2) the second group encodes products that regulate gene expression and signal transduction (Shinozaki et al., 2003). The first category are structural genes that code for enzymes that are involved either directly for detoxification processes or indirectly by catalyzing biochemical reactions or pathways involved in synthesis of metabolites that are known to impart stress tolerance. The second category, on the other hand, includes regulatory genes that govern expression of structural genes at hierarchically upstream positions, such as genes that control expression of transcription factors, signal transduction components, or receptor-related proteins. Modern molecular tools have revealed the presence of several different transcriptional regulatory systems that are involved in stress-responsive gene induction. Several different sets of *cis*- and *trans*-acting factors are involved in stress-responsive transcription. One of the plant growth regulators, abscisic acid (ABA, whose levels increase significantly in plants exposed to abiotic stresses, especially drought and salinity) plays a critical role in imparting stress tolerance (Pardha Saradhi et al., 2000). However, stress-induced gene expression may or may not depend on ABA and, accordingly, both ABA-dependent and -independent regulatory systems exist for stress-responsive gene expression (Shinozaki et al., 2003). Many genes are induced when plants are exposed to two or more types of abiotic stresses such as drought, salinity, and cold stress, suggesting the existence of probable crosstalk between various kinds of abiotic stress-signaling pathways (Shinozaki et al., 2003; Li et al., 2005; Mahajan and Tuteja, 2005). Such findings have unequivocally revealed the existence of extreme degrees of complexity in stress signaling and adaptive strategies that are operational in living systems to counter abiotic stresses. A full elucidation of abiotic stress tolerance mechanisms is essential to provide fact-based answers to a number of vital questions, so that an intelligent breeding strategy for stress tolerance can be planned effectively. Due to space constraints, we have restricted ourselves to review the work carried out so far in enhancing tolerance of plants to abiotic stresses by manipulating the level of compatible solutes. Most of the genes for regulating the production of compatible solutes that have so far been reported to be involved in imparting abiotic stress tolerance are listed in Table 35.2.

## COMPATIBLE SOLUTES

The accumulation of compatible solutes is one of the key strategies evolved by plants to cope with various abiotic stresses (Arora and Pardha Saradhi, 1995, 2002; Sharmila and Pardha Saradhi, 2002; Pardha Saradhi and Sharmila, 2003; Waditee et al., 2003, 2005). Compatible solutes are

TABLE 35.2. Genes involved in synthesis of compatible solutes reported to enhance abiotic stress tolerance in plants.

| Gene/source | Resulting translation product (enzyme)/end product | Enhanced tolerance to | References |
|---|---|---|---|
| *mtlD/Escherichia coli* | Mannitol-1-phosphate dehydrogenase/mannitol | Salt, drought, and oxidative stress | Tarczynski et al., 1992, 1993; Thomas et al., 1995; Karakas et al., 1997; Shen et al., 1997; Abebe et al., 2003 |
| *IMT1/Mesembryanthemum crystallium* | Myo-inositol-o-methyltransferase/D-ononitol | Better $CO_2$ fixation under salinity stress and better recovery after drought stress | Sheveleva et al., 1997 |
| *PIN01/Porteresia coarctata* | L-myo-inositol-1-phosphate synthase/meso-inositol | Salt | Majee et al., 2004 |
| *P5CS/Vigna aconitifolia* | Pyrroline carboxylate synthetase/proline | Salt, cold, and drought | Kishor et al., 1995; Zhu et al., 1998; Hong et al., 2000; Sawahel and Hassan, 2002; Anoop and Gupta, 2003; Su and Wu, 2004; Gleeson et al., 2005 |
| *P5CS/Arabidopsis thaliana* | Pyrroline carboxylate synthetase/proline | Salt, drought, and heat | de Ronde et al., 2001; Hmida-Sayari et al., 2005 |
| *AtproDH /Arabidopsis* | Proline dehydrogenase/proline | Freezing and salt | Nanjo et al., 1999 |
| *AtP5CS/Arabidopsis thaliana* | Pyrroline carboxylate synthetase/proline | Drought | Yamada et al., 2005 |
| *OsP5CS/Oryza sativa* | Pyrroline carboxylate synthetase/proline | Drought | Yamada et al., 2005 |

TABLE 35.2 *(Continued)*

| Gene/source | Resulting translation product (enzyme)/end product | Enhanced tolerance to | References |
|---|---|---|---|
| *OsP5CS2/Oryza sativa* | Pyrroline carboxylate synthetase/proline | Salt and cold | Hur et al., 2004 |
| δ-*OAT/Arabidopsis thaliana* | Ornithine-δ aminotransferase/proline | Salt and drought | Roosens et al., 2002; Wu et al., 2003 |
| *P5CR/Arabidopsis thaliana* | Pyrroline carboxylate reductase/proline | Heat and osmotic | de Ronde et al., 2004 |
| *S6PDH/Prunus* | Sorbitol,6-phosphate dehydrogenase | Salt | Gao et al., 2001 |
| *BADH/Escherichia coli* | Betaine aldehyde dehydrogenase/ glycinebetaine | Osmotic and high temperature | Holmstrom et al., 1994; Yang et al., 2005 |
| *BADH/Hordeum vulgare* | Betaine aldehyde dehydrogenase/ glycinebetaine | Salt, cold, and heat | Ishitani et al., 1995; Nakamura et al., 1997; Kishitani et al., 2000 |
| *BADH/Atriplex hortensis* | Betaine aldehyde dehydrogenase/ glycinebetaine | Salt | Guo et al., 1997, 2000; Jia et al., 2002 |
| *BADH/Spinacea oleracea* | Betaine aldehyde dehydrogenase/ glycinebetaine | Salt | Li et al., 2000 |
| *BADH/Spinacea oleracea; Beta vulgaris* | Betaine aldehyde dehydrogenase/ glycinebetaine | Not given | Rathinasabapathi et al., 1994 |

| | | | |
|---|---|---|---|
| *betB/Daucus carota* | Betaine aldehyde dehydrogenase/ glycinebetaine | Salt | Kumar et al., 2004 |
| *betA/Escherichia coli* | Choline dehydrogenase/betaine aldehyde + glycinebetaine | Salt, drought, and cold | Lilius et al., 1996; Takabe et al., 1998; Holmstrom et al., 2000; Bhattacharya et al., 2004; Quan et al., 2004 |
| *CMO/Spinacea oleracea* | Choline monooxygenase/betaine aldehyde + glycinebetaine | — | Nuccio et al., 1998 |
| *codA/Arthrobacter globiformis* | Choline oxidase/glycinebetaine | High light intensity, salt, cold/chilling/freezing, drought and high temperature | Hayashi et al., 1997; Sakamoto et al., 1998, 2000; Alia et al., 1998a,b; Prasad et al., 2000a,b; Gao et al., 2000; Mohanty et al., 2002; Sawahel, 2003; Sulpice et al., 2003; Park et al., 2004; Prasad and Pardha Saradhi, 2004 |
| *COX/Arthrobacter pascens* | Choline oxidase/ glycinebetaine | Salt, drought, and freezing | Huang et al., 2000 |
| *otsA and ostB/Escherichia coli* | Trehalose-6-phosphate synthase and trehalose-6-phosphate phosphatase/trehalose | Salt, drought, and low temperature | Pilon-Smits et al. 1995; Garg et al., 2002 |
| *TPS/TPP/Escherichia coli* | Trehalose-6-phosphate synthase and trehalose-6-phosphate phosphatase/ trehalose | Salt and drought | Jang et al., 2003 |

TABLE 35.2 *(Continued)*

| **Gene/source** | **Resulting translation product (enzyme)/end product** | **Enhanced tolerance to** | **References** |
|---|---|---|---|
| TPS1/Saccharomyces cerevisiae | Trehalose-6-phosphate synthase/trehalose | Drought | Holmström et al., 1996; Zhao et al., 2000; Yeo et al., 2000 |
| *GSMT and DMT/Aphanothece halophytica* | Glycine sarcosine methyltransferase and dimethyl sarcosine methyltransferase/ glycinebetaine | Salt and cold | Waditee et al., 2005 |
| *SAMDC/Tritodeum* | S-adenosylmethionine decarboxylase/polyamine | Salt | Malabika and Wu, 2002 |
| *SAMDC/Homo sapiens* | S-adenosylmethionine decarboxylase/polyamine | Drought and salt | Waie and Rajam, 2003 |
| *SPS/Arabidopsis thaliana* | Sucrose phosphate synthase/sucrose | Freezing tolerance | Strand et al., 2003 |
| *SacB/Bacillus subtilis* | Levansucrase/fructan | Drought | Ebskamp et al., 1994; Pilon-Smits et al., 1995, 1999 |
| *adc/Avena sativa* | Arginine decarboxylase/ putrescine | Drought and salt | Capell et al., 1998; Roy and Wu, 2001; Sivamani et al., 2001 |
| *adc/Datura stramonium* | Arginine decarboxylase/ putrescine | Drought | Capell et al., 2004 |
| *apo-Inv /Saccharomyces cerevisiae* | Apoplastic invertase/ apo-inv protein | Osmotic and salt | Fukushima et al., 2001 |

small organic molecules that are accumulated by living systems as one of the common adaptive strategies to tackle stresses such as salinity, drought, heavy metals, mineral deficiency, UV radiation, high light intensity, and gaseous pollutants. Compatible solutes do not perturb any of the essential cellular metabolic events, even when present at very high levels. The type of compatible solutes that are accumulated under stress differ among plant species and include amino or imino acids such as proline, betaines such as glycinebetaine, sugars such as trehalose, sugar alcohols like mannitol, myo-inositol, and sorbitol, and polyamines such as putrescine, spermidine, and spermine (Tarczynski et al., 1993; Holmstrom et al.,1994; Kishor et al., 1995; Hayashi et al., 1997; Sheveleva et al., 1997; Nuccio et al., 1998; Rajam et al., 1998; Sharmila and Pardha Saradhi, 2002; Shen et al., 2002; Pardha Saradhi and Sharmila, 2003). Compatible solutes are believed to protect plants against abiotic stresses by:

1. Maintaining an osmotic balance with the environment (Sharmila and Pardha Saradhi, 2002; Molinari, et al., 2004; Roosens et al., 2004);
2. Stabilizing several functional units such as oxygen-evolving photosystem II complex and complex II of mitochondria (Alia et al., 1991; Papageorgiou and Murata, 1995; Hamilton and Heckathron, 2001);
3. Acting as a reservoir of carbon and nitrogen;
4. Protecting the structure of proteins by facilitating hydration and thereby retaining its native conformational state including that of rubisco (Sivakumar, Yadav, et al., 2001; Pardha Saradhi and Sharmila, 2003);
5. Maintaining structural integrity of membranes (Papageorgiou and Murata, 1995; Alia et al., 1997; Pardha Saradhi and Sharmila, 2003; Quan et al., 2004);
6. Scavenging toxic oxygen species such as hydroxyl radicals, singlet oxygen, and superoxide radicals (Alia et al., 1991, 1993, 1995, 1997; Sharmila and Pardha Saradhi, 2002; Pardha Saradhi and Sharmila, 2003);
7. Controlling cytosolic pH (Venekamp, 1989);
8. Regulating $NAD(P)^+/NAD(P)H$ ratio (Alia and Pardha Saradhi, 1991, 1993; Pardha Saradhi et al., 1993, 1996; Sharmila and Pardha Saradhi, 2002);
9. Protecting the transcriptional and translational machinery under stress (Kadpal and Rao, 1985; Allard et al., 1998; Houry et al., 1999);

10. Acting as molecular chaperones in refolding of enzymes (Sivakumar, Yadav, et al., 2001);
11. Regulating structure and function of DNA by promoting transcription and replication under high-salt conditions (Rajendrakumar et al., 1997).

Realizing the significance of compatible solutes in enhancing the tolerance of plants to abiotic stress, successful attempts have been made to (1) overexpress the genes; (2) introduce recombinant or fusion genes; and (3) introduce genes for the synthesis of compatible solutes whose biosynthetic pathways do not exist in desired species (Table 35.2).

Since the first report on the accumulation of compatible solutes like proline in *Lolium perenne* under drought conditions, innumerable reports have indicated the existence of a positive correlation between accumulation of compatible solutes and adaptation of various plant species to osmotic stresses such as drought and salinity. Therefore, in general it is presumed that their accumulation plays a key role in osmoregulation in plants under various environmental stresses. Moreover, compatible solutes are extremely soluble in water, that is, have very high affinity for water. Owing to their high affinity for water, it is possible to prepare solutions of compatible solutes of very high molar strength. For instance, one can prepare 14 M proline and 11.6 M glycinebetaine without much difficulty (Dawson et al., 1969; Sharmila and Pardha Saradhi, 2002). In fact, compatible solutes can significantly contribute to colligative properties of the cytoplasm by neutralizing differences in osmotic potential.

Numerous studies have demonstrated the potential of compatible solutes in ameliorating the deleterious effects of environmental stresses on the structural and functional integrity of enzymes and proteins (Sharmila and Pardha Saradhi, 2002; Pardha Saradhi and Sharmila, 2003). It has been shown that compatible solutes exert protective effects on various enzymes against the effects of salt, heat, cold, and PEG-induced dehydration or precipitation (Sivakumar et al., 1998, 2000; Flowers, 2004). However, exceptions to this rule exist in the literature reporting the inhibition of activities of certain enzymes by the compatible solute proline (Sivakumar et al., 1998, 2000, 2002; Sivakumar, Sharmila, et al., 2001; Sivakumar, Yadav, et al., 2001). Rubisco is a bifunctional enzyme with the capacity to competitively use $CO_2$ or $O_2$, and plants exposed to abiotic stresses show enhanced oxygenation at the cost of carboxylation (Sivakumar et al., 2000). Figure 35.1 depicts salt-stress promoted suppression in carboxylase accompanied by an increase in oxygenase activity of rubisco from *Sesbania sesban*.

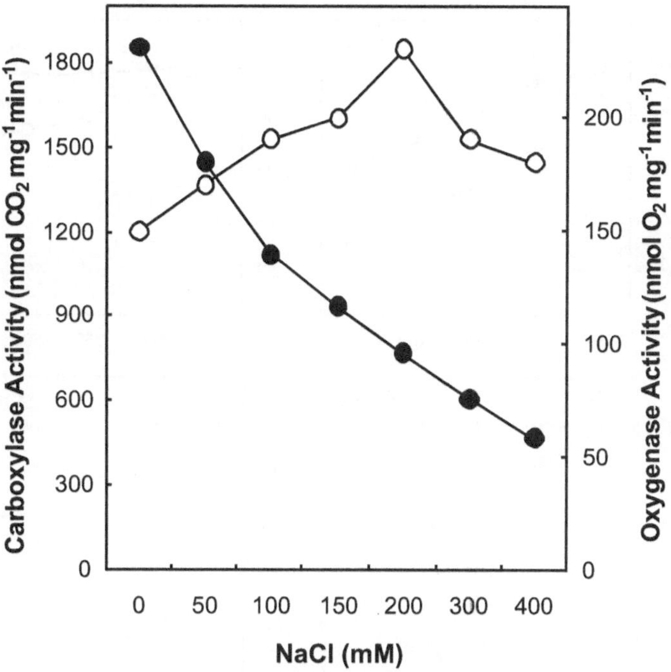


FIGURE 35.1. Effect of salt stress on carboxylase and oxygenase activity of rubisco. Please note rapid decline in carboxylase activity (•—•) with the stimulation of oxygenase (o—o) activity by NaCl.

Earlier, research findings in our laboratory clearly demonstrated that proline suppresses carboxylase as well as oxygenase activities of rubisco from higher plants even when present at concentrations as low as 100 mM. However, proline-induced suppression in oxygenase activity was significantly higher than that of carboxylase activity (Sivakumar et al., 1998, 2000). These studies further demonstrated that proline suppresses rubisco activity by dissociating small subunits from the octamer core of large subunits, probably by weakening hydrophobic interactions between them (Sivakumar, Sharmila, et al., 2001). These findings also caution the negative aspects of overexpressing genes for proline accumulation in chloroplasts. However, other compatible solutes like sugars and sugar alcohols suppress oxygenase activity of rubisco without exhibiting any significant inhibitory effect on its carboxylase activity (Sivakumar et al., 2002; Figure 35.2). These findings clearly indicated that stress-induced accumulations of com-

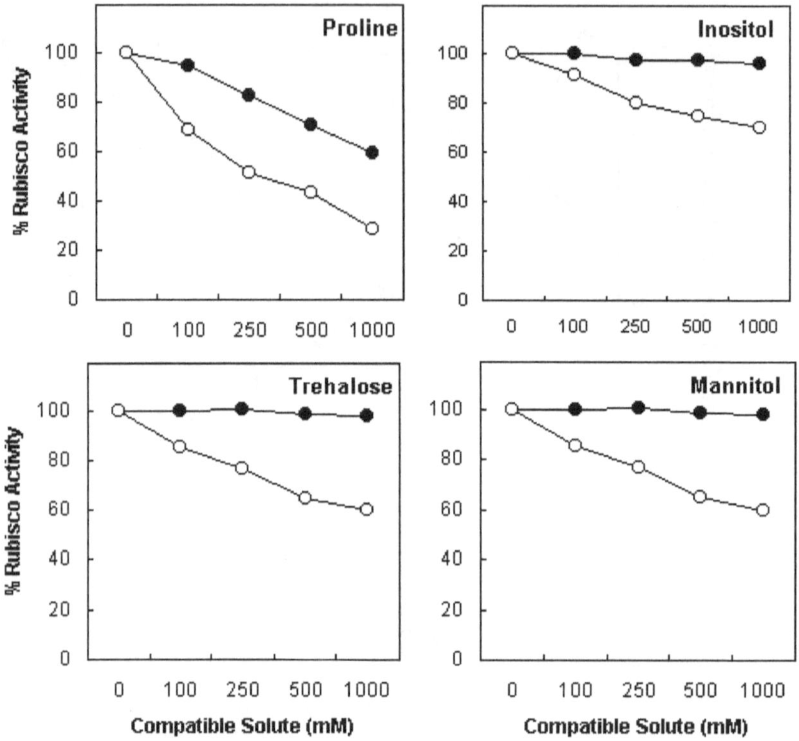


FIGURE 35.2. Percentage alteration in carboxylase (•—•) and oxygenase (o—o) activities of rubisco exposed to different levels of compatible solutes in the presence of 200 mM NaCl. Please note potential of (1) all the compatible solutes to suppress oxygenase activity; and (2) proline to suppress even carboxylase.

patible solutes have a critical role in lowering the loss of fixed carbon due to salt stress-promoted oxygenase activity of rubisco. Chloroplasts, which harbor the most vital metabolic pathway in all autotrophic plants, are extremely sensitive to all kinds of abiotic stresses. Therefore they are the chief site for biosynthesis of compatible solutes such as proline, glycinebetaine, and myo-inositol in plants under stress (Hanson et al., 1994; Sharmila and Pardha Saradhi, 2002; Majee et al., 2004). Earlier investigations demonstrated that the chloroplast from both control and salt-stressed plants of *Vigna radiata* are sensitive to strong white light with high photon flux density and exhibited significant reduction in water oxidation capacity in a time-dependent manner. The loss in activity was more pronounced in chloroplasts from NaCl-stressed plants than the controls (Pardha Saradhi, unpublished data). Salt stress substantially enhances the susceptibility of thylakoids to photoinhibition (Alia et al., 1991, 1993). Reduction in photochemical activities in thylakoids exposed to high light intensity could be due to the damage to photosystems, PS II pigment-protein complex in particular (Alia et al., 1991, 1993, 1997). Interestingly, the presence of proline in the incubation medium brought about significant reduction in time-dependent loss in photochemical activity of thylakoids (from both control and salt-stressed plants) exposed to strong light with concomitant increase in lipid peroxidation measured in terms of malondialdehyde.

Most likely, proline protects the photochemical activities, especially those mediated by the PS II complex of chloroplasts, by arresting the photoinhibitory damage, through thylakoid lipid peroxidation and cross-linking of proteins due to high light intensity–induced generation of reactive oxygen species (Alia et al., 1991, 1993, 1997). Similarly, Murata's research team (Papageorgiou and Murata, 1995) demonstrated the potential of glycinebetaine to protect the PS II complex.

Transgenic plants expressing or overexpressing the genes for biosynthesis of proline, glycinebetaine, trehalose, mannitol, or myoinositol showed superior PS II-mediated photoreactions under abiotic stresses such as salinity, drought, low temperature, or high light intensity individually or in combination in comparison to the respective wild-type plants (Prasad, Sharmila, and Pardha Saradhi, 2000; Prasad, Sharmila, Kumar, et al., 2000; Garg et al., 2002; Pardha Saradhi and Sharmila, 2003; Eung-Jun et al., 2004; Prasad and Pardha Saradhi, 2004; Chiang et al., 2005) suggesting the involvement of compatible solutes in protecting photosynthetic machinery against photo damage under stress.

Significant enhancement in potential to withstand abiotic stress-accelerated photoinhibitory damage was recorded in the transgenic plants expressing choline oxidase in chloroplasts (Alia, Hayashi, Chen, et al., 1998; Alia, Hayashi, Sakamoto et al., 1998; Pardha Saradhi and Sharmila, 2003; Prasad and Pardha Saradhi, 2004; Eung-Jun et al., 2004). Shen et al. (1997a, 1997b) observed significant enhancement in tolerance of transgenic plants to oxidative stress by targeting the *mtl*D enzyme into chloroplasts, due to an increased capacity to scavenge hydroxyl radicals by mannitol.

Transgenic plants expressing the *P5CS* gene for overproduction of proline also exhibited marked reduction in osmotic stress-induced oxidative damage (Lie et al., 2000). Similarly, Garg et al. (2002) reported that the transgenic rice plants expressing trehalose biosynthetic genes (*otsA* and *otsB*) as a fusion gene exhibited sustained plant growth, less photo-oxidative damage, and a more favorable mineral balance under salt, drought, and low-temperature stress conditions compared to wild-type, that is, nontransgenic plants. Transgenic plants expressing *IMT1* encoding myo-inositol-o-methyltransferase for synthesis of ononitol and the *sacB* gene encoding levansucrase for synthesis of fructan from fructose also exhibited significantly better drought and freezing tolerance than did wild-type plants (Sheveleva et al., 1997; Parvanova et al., 2004).

Earlier, de Ronde et al. (2001, 2004) noted transgenic lines of soybean expressing *Arabidopsis P5CS* or *P5CR* genes possessed enhanced tolerance to salt, drought, osmotic and high-temperature stresses. Transgenic soybean plants expressing the *P5CR* gene coding for L-$\Delta^1$-pyrroline-5-carboxylate reductase in the sense direction exhibited higher tolerance to drought and high temperature with lower $H_2O_2$ and lipid hydroperoxide concentrations in comparison to wild-type plants. In contrast, the transgenics expressing this gene in an antisense direction were more susceptible to these abiotic stresses as they generated high levels of $H_2O_2$ and lipid hydroperoxides in comparison to wild type. These investigations demonstrated the antioxidant role of proline (Kocsy et al., 2005). Kocsy and co-workers (2005) recorded significantly higher levels of (1) proline and ascorbate in the sense transformants; and (2) AA/dehydroascorbate and GSH/GSSG ratios, and ascorbate peroxidase activity in the antisense transgenic genotypes, suggesting that the manipulation of proline synthesis affects not only the GSH concentrations but also the levels of other antioxidants.

Grain legumes exposed to various environmental stresses show a marked decline in the mitochondrial electron transport activity associated with several-fold increase in proline levels (Table 35.3) as reported earlier (Alia and Pardha Saradhi, 1993; Pardha Saradhi et al., 1996). Amazingly, *Vigna radiata* seedlings exposed to various mitochondrial electron transport inhibitors—potassium cyanide (KCN), antimycin A, and rotenone—showed a significantly high accumulation of proline, similar to that noted in plants under heavy metal stress (Figure 35.3) as shown earlier in rice (Alia and Pardha Saradhi, 1993). These results gave a clue that the accumulation of compatible solutes such as proline in plants under stress might be related to the suppression in electron transport associated with mitochondria and chloroplasts. An increase in the level of NADH is expected due to suppression of mitochondrial electron transport activity (Alia and Pardha Saradhi, 1993; Pardha Saradhi et al., 1996). The seedlings of *Vigna mungo* exposed to 15 percent PEG (drought stress) and 200 mM NaCl (salt stress) as well as various inhibitors of mitochondrial transport activity showed a significant increase in the level of NADH. Another prominent observation made in

TABLE 35.3. Electron transport activity of mitochondria isolated from the shoots of etiolated seedlings of plant species raised in mineral growth medium without (control) and supplemented with 200 mM sodium chloride.

| | nmol $O_2$/min/mg protein | | | |
|---|---|---|---|---|
| | NADH | | Succinate | |
| **Plant species** | **Control** | **Stressed** | **Control** | **Stressed** |
| *Vigna radiata* | 242.4 ± 23.1 | 129.5 ± 15.5 | 116.1 ± 11.3 | 59.5 ± 6.7 |
| *Cajanus cajan* | 183.0 ± 17.1 | 63.5 ± 8.3 | 77.0 ± 6.1 | 36.4 ± 5.5 |
| *Cicer arietinum* | 215.7 ± 30.0 | 72.9 ± 7.7 | 95.6 ± 9.3 | 54.8 ± 9.4 |

*Notes:* Electron transport activity was measured in terms of $O_2$ consumed due to the oxidation of NADH and succinate. Experiments were carried out at least 5 times. Data represent mean ± standard error.

plants exposed to abiotic stresses is the significant decline in the carboxylase activity of rubisco, resulting in curtailment of the usual progress in the Calvin cycle (Sivakumar et al., 1998, 2000; Sivakumar, Sharmila, et al. 2001). This would obviously result in increased levels of NAD(P)H. Accordingly, an increase in NAD(P)H to NAD(P)$^+$ ratio was recorded invariably in plants subjected to various abiotic stresses such as salt, drought, and heavy metal stress (Alia and Pardha Saradhi, 1993; Pardha Saradhi et al., 1996). Such an increase in NAD(P)H might even affect substrate-level phosphorylation indirectly, besides inhibiting important metabolic reactions that need NAD(P)$^+$. For example, the conversion of glyceraldehyde-3-phosphate to 1,3-diphosphoglyceric acid mediated by the key enzyme of glycolysis, glyceraldehyde-3-phosphate dehydrogenase, depends on the availability of NAD$^+$, the rate-limiting step of the pentose phosphate pathway catalyzed by glucose-6-phosphate dehydrogenase, which in turn is dependent on the availability of NADP$^+$ and is inhibited by NADPH and some enzymes of Kreb's cycle depend on the availability of NAD$^+$ (see Sharmila and Pardha Saradhi, 2002).

To let such essential reactions continue, living systems have evolved a number of strategies to readily maintain the ratio of NAD(P)$^+$ to NAD(P)H. This includes activation of certain reactions that can readily oxidize NAD(P)H. The most pronounced response shown by plants under stress is a significant increase in the activity of various dehydrogenases that can readily maintain the NAD(P)$^+$/NAD(P)H ratio. The majority of stress-induced or accelerated dehydrogenase-catalyzed reactions lead to accumula-

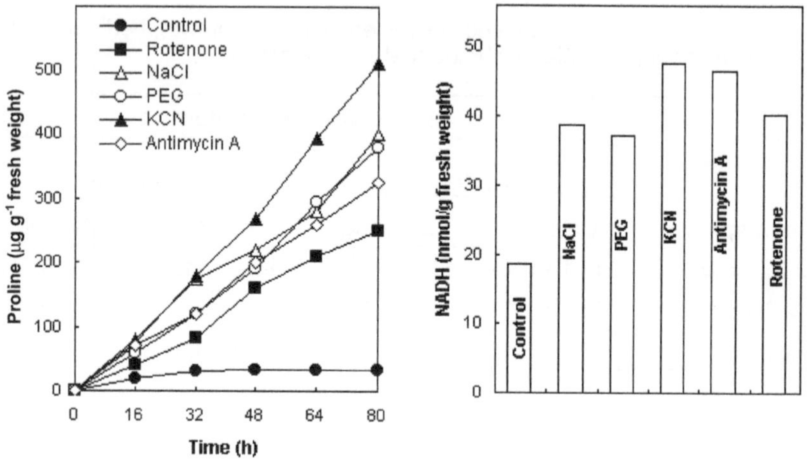


FIGURE 35.3. (left) Time-dependent change in the level of proline; and (right) NADH levels in shoots in *Vigna mungo* seedlings exposed to mineral growth medium (Arora and Pardha Saradhi, 1995) without (control) and supplemented with NaCl (200 mM), PEG (15 percent), rotenone (50 mM), antimycin A (10 µg/ml), or Potassium cyanide (KCN; 5 mM).

tion of organic acids such as malate, lactate, and glycollate. An increase in the level of these organic acids is recorded in plants exposed to abiotic stress (see Venekamp, 1989). The accumulation of yet another organic acid, citrate could result due to limited availability of $NAD^+$ needed for its metabolism. The accumulation of organic acids would, however, bring about a disturbance in several metabolic processes in cells by lowering the pH. Interestingly, the synthesis of the majority of compatible solutes is linked to the maintenance of the cellular reducing power, in particular $NAD(P)^+$/NAD(P)H and/or $Fd_{ox}/Fd_{red}$.

Unlike the organic acids produced in other reactions, associated with the oxidation of NAD(P)H or $Fd_{red}$, the compatible solutes do not possess any negative influence on the cellular environment (Venekamp, 1989). In fact, it has been proposed that compatible solutes like proline and mannitol play a role of redox buffer by storage of excess reductants in a nontoxic form (Alia and Pardha Saradhi, 1993; Pardha Saradhi et al., 1993, 1996) without perturbing reactions that are most vital for maintenance of living cells and organisms.

Many legumes such as *Vigna* species, *Cajanus Cajan,* and *Microptilium* species have been reported to accumulate cyclic polyols such as D-pinitol (1-D-3-o-methyl-chiro-inositol) or D-ononitol (1-D-4-o-methyl-myo-inositol) in response to water deficit (Wanek and Richter, 1997). Inositols (cyclohexane hexitols) play a vital role in growth regulation, membrane biogenesis, and osmotolerance, besides acting as a second messenger in signal transduction pathways (Majee et al., 2004). L-myo-inositol-1-phosphate synthase exists in the chloroplasts of a few legumes such as *Pisum sativum, Vigna radiata,* and *Phaseolus vulgaris* (Lackey et al., 2003). Majee et al. (2004) isolated the gene *PINO1* for a novel salt-tolerant L-myo-inositol-1-phosphate synthase from the wild halophytic rice, *Porteresia coarctata.* Introgression of *PINO1* rendered transgenic tobacco plants capable of growth in 200 to 300 mM NaCl with retention of ~40 to 80 percent of photosynthetic competence (i.e., maintenance of photosynthetic potential) with concomitant increased inositol production compared with unstressed controls. MIPS protein isolated from *PINO1* transgenics showed salt-tolerant properties in vitro, confirming functional expression of the *PINO1* gene. Use of genes encoding functional units such as enzymes that have the potential to withstand extreme degrees of uncontrollable or unpredictable stressful factors such as excess of ions or water deficit would be of utmost significance. Such key findings cast doubt on the validity of using the genes for transcription factors, which might switch on a number of cellular events involved in stress tolerance, but many of the enzymes and other cellular components that are produced might be sensitive to extreme stress factors.

Plant polyamine content has been modulated by the overexpression or down-regulation of arginine decarboxylase (*adc*), ornithine decarboxylase (*odc*), and *S*-adenosyl methionine decarboxylase (*samdc;* Lepri et al., 2001; Mehta et al., 2002; Thu-Hang et al., 2002; Trung-Nghia et al., 2003; Waie and Rajam, 2003). Overexpression of heterologous *adc* or *odc* cDNAs in plants generally results in the production of high levels of putrescine (Capell et al., 1998). Engineering of the plant polyamine biosynthetic pathway leads to generation of transgenics with significantly altered polyamine content and enhanced tolerance to drought and salinity (Malabika and Wu, 2002; Waie and Rajam, 2003; Capell and Christou, 2004; Capell et al., 2004). Roy and Wu (2001, 2002) found enhanced salt tolerance in transgenic plants expressing oat *adc* or *Tritordeum Samdc* genes with superior seedling growth and higher biomass production in comparison to wild type. Polyamines, particularly spermidine and spermine, are involved in regulation of gene expression by enhancing DNA-binding activities, in particular transcription factors. Polyamines are believed to have an osmoprotectant function in plant cells under water deficit (Waie and Rajam, 2003). Sivamani and co-workers (2001) were successful in expressing a heterologous oat arginine decarboxylase gene in *Cajanus cajan* and the transgenic cell lines showed an increase in putrescine levels.

## *EPILOGUE*

As there has been a growing realization of evaluating the metabolic pathways in the context of the whole plant or cell rather than at the single pathway level, attention is shifting from single-gene engineering strategies toward more complex approaches involving the simultaneous expression or overexpression, or suppression of multiple genes. Two basic approaches are gaining popularity in this direction: first is the use of regulatory factors controlling several genes at a time, and second is the pyramiding of vital genes. However, both have their limitations. At this juncture, the best option for achieving the goal of developing abiotic stress-tolerant genotypes of crop plants, especially grain legumes, is to pyramid one gene each from the categories of (1) compatible solutes (preferably one whose biosynthetic pathway is not available in the candidate plant species); (2) antioxidant systems (preferably monodehydroascorbate reductase); and (3) ion homeostasis (preferably one maintaining high levels of intracellular $K^+$ to handle water deficit or vacuolar $Na^+/H^+$ antiporter in case of salt stress). We wish to conclude by emphasizing that in spite of our best efforts, we remain far behind in unraveling the basic mechanisms evolved by living systems to counteract environmental stresses.

## REFERENCES

Abebe, T., A.C. Guenzi, B. Martin, and J.C. Cushman (2003). Tolerance of mannitol-accumulating transgenic wheat to water stress and salinity. *Plant Physiology* 131:1748-1755.

Alia, H. Hayashi, T.H.H. Chen, and N. Murata (1998). Transformation with gene for choline oxidase enhances the cold tolerance of *Arabidopsis* during germination and early growth. *Plant Cell and Environment* 21:232-239.

Alia, H. Hayashi, A. Sakamoto, and N. Murata (1998). Enhancement of the tolerance of *Arabidopsis* to high temperatures by genetic engineering of the synthesis of glycine betaine. *Plant Journal* 16:155-161.

Alia, and P. Pardha Saradhi (1991). Proline accumulation under heavy metal stress. *Journal of Plant Physiology* 138:554-558.

Alia, and P. Pardha Saradhi (1993). Suppression in mitochondrial electron transport is the prime cause behind stress induced proline accumulation. *Biochemical and Biophysical Research Communications* 193:54-58.

Alia, P. Pardha Saradhi, and P. Mohanty (1991). Proline enhances primary photochemical activities in isolated thylakoid membranes of *Brassica juncea* by arresting photo inhibitory damage. *Biochemical and Biophysical Research Communications* 181:1238-1244.

Alia, P. Pardha Saradhi, and P. Mohanty. (1993). Proline in relation to free radical production in seedlings of *Brassica juncea* raised under sodium chloride stress. *Plant and Soil* 156:497-500.

Alia, P. Pardha Saradhi, and P. Mohanty (1997). Involvement of proline in protecting thylakoid membranes against free radical-induced photo damage. *Journal of Photochemistry and Photobiology B: Biology* 38:253-257.

Alia, K.V.S.K. Prasad, and P. Pardha Saradhi (1995). Effect of zinc on free radicals and proline in *Brassica juncea* and *Cajanus cajan. Phytochemistry* 39:45-47.

Allard, F., M. Houde, M. Krol, A. Ivanov, N.P.A. Hunter, and F. Sarhan (1998). Betaine improves freezing tolerance in wheat. *Plant and Cell Physiology* 39: 1194-1202.

Anoop, N. and A.K. Gupta (2003). Transgenic *indica* rice cv IR-50 overexpressing *Vigna aconitifolia,* 1-pyrroline-5-carboxylate synthetase cDNA shows tolerance to high salt. *Journal of Plant Biochemistry and Biotechnology* 12:109-116.

Arora, S. and P. Pardha Saradhi (1995). Light induced enhancement in proline levels in *Vigna radiata* exposed to environmental stresses. *Australian Journal of Plant Physiology* 22:383-386.

Arora, S. and P. Pardha Saradhi (2002). Light induced enhancement in proline levels under stress is regulated by non-photosynthetic events. *Biologia Plantarum* 45:629-632.

Bhattacharya, R.C., M. Maheshwari, V. Dineshkumar, P.B. Kirti, S.R. Bhat, and V.L. Chopra (2004). Transformation of *Brassica oleracea* cv. *capitata* with bacterial *betA* gene enhances tolerance to salt stress. *Scientific Horticulture* 100: 215-227.

Capell, T., L. Bassie, and P. Christou (2004). Modulation of the polyamine biosynthetic pathway in transgenic rice confers tolerance to drought stress. *Proceedings of the National Academy of Sciences USA* 101:9909-9914.

Capell, T. and P. Christou (2004). Progress in plant metabolic engineering. *Current Opinion in Biotechnology* 15:148-154.

Capell, T., C. Escobar, H. Liu, D. Burtin, O. Lepri, and P. Christou (1998). Overexpression of the oat arginine decarboxylase cDNA in transgenic rice (*Oryza sativa* L.) affects normal development patterns in vitro and results in putrescine accumulation in transgenic plants. *Theoretical and Applied Genetics* 97:246-254.

Chiang, Y.J., C. Stushnoff, A.E. McSay, M.L. Jones, and H.J. Bohnert (2005). Overexpression of mannitol-1-phosphate dehydrogenase increases mannitol accumulation and adds protection against chilling injury in *Petunia. Journal of the American Society for Horticultural Science* 130:605-610.

Chinnusamy, V., A. Jagendorf, and J. K. Zhu (2005). Understanding and improving salt tolerance in plants. *Crop Science* 45:437-448.

Croser, J.S., H.J. Clarke, K.H.M. Siddique, and T.N. Khan (2003). Low-temperature stress: Implications for chickpea (*Cicer arietinum* L.) improvement. *Critical Reviews in Plant Science* 22:185-219.

Dawson, R.W.C., D.C. Elliot, W.H. Elliot, and K.M. Jones (1969). *Data for biochemical research.* 2nd ed. Oxford:Clarendon Press, p. 12.

de Ronde, J.A., W.A. Cress, G.H.J. Kruger, R.J. Strasser, and J. Van Staden (2004). Photosynthetic response of transgenic soybean plants, containing an *Arabidopsis P5CR* gene, during heat and drought stress. *Journal of Plant Physiology* 161:1211-1224.

de Ronde, J.A., W.A. Cress, and J. van Staden (2001). Interaction of osmotic and temperature stress on transgenic soybean. *South African Journal of Botany* 67:655-660.

Ebskamp, M.J., I.M. van der Meer, B.A. Spronk, P.J. Weisbeek, and S.C. Smeekens (1994). Accumulation of fructose polymers in transgenic tobacco. *Nature Biotechnology* 12:272-275.

*Economic survey, 2004-05* (2005). Ministry of Finance, Govt. of India.

Eung-Jun, P., J. Zoran, A. Sakamoto, J. DeNoma, Y. Raweewan, N. Murata, and T.H.H. Chen (2004). Genetic engineering of glycine betaine synthesis in tomato protects seeds, plants, and flowers from chilling damage. *Plant Journal* 40:474-487.

Flowers, T.J. (2004). Improving crop salt tolerance. *Journal of Experimental Botany* 55:307-319.

Fukushima, E., Y. Arata, T. Endo, U. Sonnewald, and F. Sato (2001). Improved salt tolerance of transgenic rice expressing apoplastic yeast derived inverstase. *Plant Cell Physiology* 42: 245-249.

Gao, M., A. Sakamoto, K. Miura, N. Murata, A. Sugiura, and R. Tao (2000). Transformation of Japanese persimmon (*Diospyros kaki* Thunb.) with a bacterial gene for choline oxidase. *Molecular Breeding* 6:501-510.

Gao, M., R. Tao, K. Miura, A.M. Dandekar, and A. Sugiura (2001). Transformation of Japanese persimmon (*Diospyros kaki* Thunb.) with apple cDNA encoding NADP-dependent sorbitol-6-phosphate dehydrogenase. *Plant Science* 160:837-845.

Garg, A.K., J.K. Kim, T.G. Owens, A.P. Ranwala, Y.D. Choi, L.V. Kochian, and R.J. Wu (2002). Trehalose accumulation in rice plants confers high tolerance levels to different abiotic stresses. *Proceedings of the National Academy of Sciences USA* 99:15898-15903.

Gleeson, D., L. Walter, M. Anne, and P. Michael (2005). Overproduction of proline in transgenic hybrid larch [*Larix* × *leptoeuropaea* (Dengler)] cultures renders them tolerant to cold, salt and frost. *Molecular Breeding* 9:21-29.

Guo, B.H., Y.M. Zhang, H.J. Li, L.Q. Du, Y.X. Li, J.S. Zhang, S.Y. Chen, and Z.Q. Zhu (2000). Transformation of wheat with a gene encoding for the betaine aldehyde dehydrogenase (*BADH*). *Acta Botanica Sinica* 42:279-283.

Guo, Y., L. Zhang, G. Xiao, S.Y. Cao, D.M. Gu, W.Z. Tian, and S.Y. Chen (1997). Expression of betaine aldehyde dehydrogenase gene and salinity tolerance in rice transgenic plants. *Science China* 40:496-501.

Hamilton III, E.W. and S.A. Heckathron (2001). Mitochondrial adaptations to NaCl. Complex I is protected by anti-oxidants and small heat shock proteins, whereas complex II is protected by proline and betaine. *Plant Physiology* 126:1266-1274.

Hanson, A.D., B. Rathinasabapathi, J. Revoal, M. Burnet, M.O. Dillon, and D.A. Gage (1994). Osmoprotective compounds in Plumbaginaceae: A natural experiment in metabolic engineering of stress tolerance. *Proceedings of the National Academy of Science USA* 91:306-310.

Hayashi, H., Alia, L. Mustardy, P. Deshnium, M. Ida, and N. Murata (1997). Transformation of *Arabidopsis thaliana* with the *codA* gene for choline oxidase; accumulation of glycine betaine and enhanced tolerance to salt and cold stress. *Plant Journal* 12:133-142.

Hmida-Sayari, A., R. Gargouri-Bouzid, A. Bidani, L. Jaoua, A. Savouré, and S. Jaoua (2005). Overexpression of $\Delta^1$-pyrroline-5-carboxylate synthetase increases proline production and confers salt tolerance in transgenic potato plants. *Plant Science* 169:746-752.

Holmström, K.O., E. Mäntylä, B. Welin, A. Mandal, E.T. Palva, O.E. Tunnela, and J. Londesborough (1996). Drought tolerance in tobacco. *Nature* 379:683-684.

Holmstrom, K.O., S. Somersalo, A. Mandal, T.E. Palva, and B. Welin (2000). Improved tolerance to salinity and low temperature in transgenic tobacco producing glycine betaine. *Journal of Experimental Botany* 51:177-185.

Holmstrom, K.O., B. Weilin, A. Mandal, I. Kristiansdottir, T.H. Teeri, T. Lamark, A.R. Strom, and E.T. Palva (1994). Production of the *Escherichia coli* betaine-aldehyde dehydrogenase, an enzyme required for the synthesis of osmoprotectant glycinebetaine in transgenic plants. *Plant Journal* 6:749-758.

Hong, H., K. Lakkineni, Z. Zhang, and D.P.S. Verma (2000). Removal of feedback inhibition of $\Delta^1$-pyrroline-5-carboxylate synthetase results in increased proline accumulation and protection of plants from osmotic stress. *Plant Physiology* 122:1129-1136.

Houry, W.A., D. Frishman, C. Eckerskorn, F. Lottspeicch, and F. Hartl (1999). Identification of in vivo substrates of the chaperonin GroEL. *Nature 402*:147-154.

Huang, J., R. Hirji, L. Adam, K.L. Rozwadowski, J.K. Hammerlindi, W.A. Keller, and G. Selvaraj (2000). Genetic engineering of glycinebetaine production toward enhancing stress tolerance in plants: Metabolic limitations. *Plant Physiology* 122:747-756.

Hur, J., J.K. Hong, C.H. Lee, and G. An (2004). Stress-inducible *OsP5CS2* gene is essential for salt and cold tolerance in rice. *Plant Science* 167:417-426.

Ishitani, M, T. Nakamura, S.Y. Han, and T. Takebe (1995) Expression of betaine aldehyde dehydrogenase gene in barley in response to osmotic stress and abscisic acid. *Plant Molecular Biology* 27:307-315.

Jang, I.C., S.J. Oh, J.S. Seo, W.B. Choi, I.K.S. Sang, C.H. Kim, Y.S. Kim, H.S. Seo, Y.D. Choi, B.H. Nahm, et al. (2003). Expression of a bifunctional fusion of the *Escherichia coli* genes for trehalose-6-phosphate synthase and trehalose-6-phosphate phosphatase in transgenic rice plants increases trehalose accumulation and abiotic stress tolerance without stunting growth. *Plant Physiology* 131:516-524.

Jia, G.-X, Z.-Q. Zhu, F.-Q. Chang, and Y.-X. Li (2002). Transformation of tomato with the *BADH* gene from Atriplex improves salt tolerance. *Plant Cell Reports* 21:141-146.

Kadpal, R.P. and N.A. Rao (1985). Alterations in the biosynthesis of proteins and nucleic acids in finger millet (*Eleucine coracana*) seedlings during water stress and the effect of proline on protein biosynthesis. *Plant Science* 40:73-79.

Karakas, B., P. Ozias Akins, C. Stushnoff, M. Suefferheld, and M. Rieger (1997). Salinity and drought tolerance of mannitol accumulating transgenic tobacco. *Plant Cell and Environment* 20:609-616.

Kishitani, S., T. Takanami, M. Suzuki, M. Oikawa, S. Yokoi, M. Ishitani, A.M. Alvarez-Nakase, T. Takabe, and T. Takabe (2000). Compatibility of glycine-betaine in rice plants: Evaluation using transgenic rice plants with a gene for peroxisomal betaine aldehyde dehydrogenase from barley. *Plant Cell and Environment* 23:107-114.

Kishor, P.B., Z.L. Hong, G.H. Miao, C.A.A. Hu, and D.P.S. Verma (1995). Over-expression of delta-pyrroline-5-carboxylate synthetase increases proline production and confers osmotolerance in transgenic plants. *Plant Physiology* 108: 1387-1394.

Kocsy, G., R. Laurie, G. Szalai, V. Szilágyi, L. Simon-Sarkadi, G. Galiba, and J.A. de Ronde (2005). Genetic manipulation of proline levels affects antioxidants in soybean subjected to simultaneous drought and heat stresses. *Physiologia Plantarum* 124:227-235.

Kumar, S., A. Dhingra, and H. Daniell (2004). Plastid-expressed *betaine aldehyde dehydrogenase* gene in carrot cultured cells, roots and leaves confers enhanced salt tolerance. *Plant Physiology* 136:2843-2854.

Lackey, K.H., P.M. Pope, and M.D. Johnson (2003). Expression of 1L-myoinositol-1-phosphate synthase in organelles. *Plant Physiology* 132:2240-2247.

Lepri, O., L. Bassie, G. Safwat, P. Thu-Hang, P. Trung-Nghia, E. Hölttä, P. Christou, and T. Capell (2001). Over-expression of the human ornithine decarboxylase cDNA in transgenic rice plants alters the polyamine pool in a tissue-specific manner. *Molecular General Genetics* 266:303-312.

Li, X.P., A.G. Tian, G.Z. Luo, Z.Z. Gong, J.S. Zhang, and S.Y. Chen (2005). Soybean DRE-binding transcription factors that is responsive to abiotic stresses. *Theoretical and Applied Genetics* 110:1355-1362.

Li, Y.X., F.Q. Chang, L.Q. Du, B.H. Guo, H.J. Li, J.S. Zhang, S.Y. Chen, and Z.Q. Zhu (2000). Genetic transformation of watercress with a gene encoding for betaine aldehyde dehydrogenase (BADH). *Acta Botanica Sinica* 42:480-484.

Lie, H.J., K. Lakkineni, Z.M. Zhang, D.P.S. Verma, Z.L. Hong, and Z.M. Zhang (2000). Removal of feedback inhibition of $\Delta^1$-pyrroline-5-carboxylate synthetase results in increased proline accumulation and protection of plants from osmotic stress. *Plant Physiology* 122:1129-1136.

Lilius, G., N. Holmberg, and L. Bülow (1996). Enhanced NaCl stress tolerance in transgenic tobacco expressing bacterial choline dehydrogenase. *Nature Bio/ Technology* 14:177-180.

Mahajan, S. and N. Tuteja (2005). Cold, salinity and drought stresses: An overview. *Archives of Biochemistry and Biophysics* 444:139-158.

Majee, M., S. Maitra, K.G. Dastidar, S. Pattnaik, A. Chatterjee, N.C. Hait, K.P. Das, and A.L. Majumder (2004). A novel salt-tolerant L-*myo*-inositol-1-phosphate synthase from *Porteresia coarctata* (Roxb.) Tateoka, a halophytic wild rice. *Journal of Biological Chemistry* 279:28539-28552.

Malabika, R. and R. Wu (2002). Overexpression of *S*-adenosyl methionine decarboxylase gene in rice increases polyamine level and enhances sodium chloride-stress tolerance. *Plant Science* 163:987-992.

Mehta, R.A., T. Cassol, N. Li, N. Ali, A.K. Handa, and A.K. Mattoo (2002). Engineered polyamine accumulation in tomato enhances phytonutrient content, juice quality, and vine life. *Nature Bio/Technology* 20:613-618.

Mohanty, A., H. Kathuria, A. Ferjani, A. Sakamoto, P. Mohanty, N. Murata, and A.K. Tyagi (2002). Transgenics of an elite *indica* rice variety Pusa Basmati 1 harboring the *codA* gene are highly tolerant to salt stress. *Theoretical and Applied Genetics* 106:51-57.

Molinari, H.B.C., C.J. Marur, J.C.B. Hilho, A.K. Kobayashi, R. Meeley, M. Muszynski, and J. Habben (2004). Osmotic adjustment in transgenic citrus rootstock Carrizo citrange (*Citrus sinensis* Osb. × *Poncirus trifoliate* L. Raf.) overproducing proline. *Plant Science* 167:1375-1381.

Nakamura, T., S. Yokota, Y. Muramoto, K. Tsutsui, Y. Oguri, K. Fukui, and T. Takabe (1997). Expression of a betaine aldehyde dehydrogenase gene in rice, a glycinebetaine nonaccumulator, and possible localization of its protein in peroxisomes. *Plant Journal* 11:1115-1120.

Nanjo, T., M. Kobayashi, Y. Yoshiba, Y. Kakubari, K. Yamaguchi-Shinozaki, and K. Shinozaki (1999). Antisense suppression of proline degradation improves tolerance to freezing and salinity in *Arabidopsis thaliana. FEBS Letters* 461:205-210.

Nuccio, M.L., B.L. Russell, K.D. Nolte, B. Rathinasabapthi, D.A. Gage, and A.D. Hanson (1998). The endogenous choline supply limits glycine betaine synthesis in transgenic tobacco expressing choline monooxygenase. *Plant Journal* 16: 487-496.

Papageorgiou, G.C. and N. Murata (1995). The unusually strong stabilizing effects of glycine betaine on the structure and function in the oxygen-evolving photosystem-II complex. *Photosynthetic Research* 44:243-252.

Pardha Saradhi, P., Alia, and B. Vani (1993). Inhibition of mitochondrial electron transport is the prime cause behind proline accumulation during mineral deficiency in *Oryza sativa. Plant and Soil* 155/156:465-468.

Pardha Saradhi, P., S. Arora, B. Vani, and J.T. Puthur (1996). Alteration in $NAD^+$/NADH ratio regulates salt stress induced proline accumulation in *Vigna radiata. Proceedings of the National Academy of Sciences India* 66:89-100.

Pardha Saradhi, P. and P. Sharmila (2003). Improvement of crop plants for abiotic stress tolerance through introduction of glycinebetaine pathway. In *Plant genetic engineering,* R.P. Singh and P.K. Jaiwal (eds.). Houston, USA: SCI Tech Publishing, pp. 147-177.

Pardha Saradhi, P., I. Suzuki, A. Katoh, A. Sakamoto, P. Sharmila, D.J. Shi, and N. Murata (2000). Protection against the photo-induced inactivation of the photosystem II complex by abscisic acid. *Plant Cell and Environment* 23:711-718.

Park, E.J., Z. Jeknić, A. Sakamoto, J. DeNoma, R. Yuwansiri, N. Murata, and T.H.H. Chen (2004). Genetic engineering of glycinebetaine synthesis in tomato protects seeds, plants, and flowers from chilling damage. *Plant Journal* 40:474-487.

Parvanova, D., A. Popova, I. Zaharieva, P. Lambrev, Konstantinova, S. Taneva, A. Atanassov, V. Goltsev, and D. Djilianov (2004). Low temperature tolerance of tobacco plants transformed to accumulate proline, fructans, or glycine betaine. Variable chlorophyll fluorescence evidence. *Photosynthetica* 42:179-185.

Pilon-Smits, E.A.H., M.J.M. Ebstamp, M.J. Paul, and M.J.W. Jeuken (1995). Improved performance of transgenic fructan-accumulating tobacco under drought stress. *Plant Physiology* 107:125-130.

Pilon-Smits, E.A.H., N. Terry, T. Sears, and K. van Dun (1999). Enhanced drought resistance in fructan-producing sugar beet. *Plant Physiology and Biochemistry* 37:313-317.

Prasad, K.V.S.K. and P. Pardha Saradhi (2004). Enhanced tolerance to photoinhibition in transgenic plants through targeting of glycine betaine biosynthesis into the chloroplasts. *Plant Science* 166:1197-1212.

Prasad, K.V.S.K., P. Sharmila, P.A. Kumar, and P. Pardha Saradhi (2000). Transformation of *Brassica juncea* (L.) Czern with bacterial *codA* gene enhances its tolerance to salt stress. *Molecular Breeding* 6:489-499.

Prasad, K.V.S.K., P. Sharmila, and P. Pardha Saradhi (2000). Enhanced tolerance of transgenic *Brassica juncea* to choline confirms successful expression of the bacterial *codA* gene. *Plant Science* 159:233-242.

Quan, R., M. Shang, H. Zhang, Y. Zhao, and J. Zhang (2004). Improved chilling tolerance by transformation with *betA* gene for the enhancement of glycine betaine synthesis in maize. *Plant Science* 166:141-149.

Rajam, M.V., S. Dagar, B. Waie, J.S. Yadav, P.A. Kumar, F. Shoeb, and R. Kumria (1998). Genetic engineering of polyamine and carbohydrate metabolism for osmotic stress tolerance in higher plants. *Journal of Biosciences* 23:473-482.

Rajendrakumar, C.S.V., T. Suryanarayana, and A.R. Reddy (1997). DNA helix destabilization by proline and betaine: Possible role in the salinity tolerance process. *FEBS Letters* 410:201-205.

Rathinasabapathi, B., K.F. McCue, D.A. Gage, and A.D. Hanson (1994). Metabolic engineering of glycine betaine synthesis: Plant betaine aldehyde dehydrogenase lacking transit peptide are targeted to tobacco chloroplast where they confer betaine aldehyde resistance. *Planta* 193:155-162.

Roosens, N.H., F.A. Bitar, K. Loenders, G. Angenon, and M. Jacobs (2004). Overexpression of ornthine-*d*-aminotransferase increases proline biosynthesis and confers osmotolerance in transgenic plants. *Molecular Breeding* 9:73-80.

Roy, M. and R. Wu (2001). Arginine decarboxylase transgene expression and analysis of environmental stress tolerance in transgenic rice. *Plant Science* 160:869-875.

Roy, M. and R. Wu (2002). Overexpression of *S*-adenosyl methionine decarboxylase gene in rice increases polyamine level and enhances sodium chloride-stress tolerance. *Plant Science* 163:987-992.

Sakamoto, A., Alia, and N. Murata (1998). Metabolic engineering of rice leading to biosynthesis of glycinebetaine and tolerance to salt and cold. *Plant Molecular Biology* 38:1011-1019.

Sakamoto, A., R. Valverde, Alia, T.H.H. Chen, and N. Murata (2000). Transformation of *Arabidopsis* with the *codA* gene for choline oxidase enhances freezing tolerance of plants. *Plant Journal* 22:449-453.

Sawahel, W. (2003). Improved performance of transgenic glycinebetaine-accumulating rice plants under drought stress. *Biologia Plantarum* 47:39-44.

Sawahel, W.A. and A.H. Hassan (2002). Generation of transgenic wheat plants producing high levels of the osmoprotectant proline. *Biotechnology Letters* 24:721-725.

Sharmila, P. and P. Pardha Saradhi (2002). Proline accumulation in heavy metal stressed plants. An adaptive strategy. In *Physiology and biochemistry of metal toxicity and tolerance in plants,* M.N.V. Prasad and K. Strzalka (eds.). Dordrecht: Kluwer Academic Publishers, pp. 179-199.

Shen, B., R.G. Jensen, and H.J. Bohnert (1997a). Increased resistance to oxidative stress in transgenic plants by targeting mannitol biosynthesis to chloroplasts. *Plant Physiology* 113:1177-1183.

Shen, B., R.G. Jensen, and H.J. Bohnert (1997b). Mannitol protects against oxidation by hydroxyl radicals. *Plant Physiology* 115:527-532.

Shen, Y.G., B.X. Du, W.K. Zhang, J.S. Zhang, and S.Y. Chen (2002). *AhCMO,* regulated by stresses in *Atriplex hortensis,* can improve drought tolerance in transgenic tobacco. *Theoretical and Applied Genetics* 105:815-821.

Sheveleva, E., W. Chmara, H.J. Bohnert, and R.G. Jensen (1997). Increased salt and drought tolerance by D-ononitol production in transgenic *Nicotiana tabacum* L. *Plant Physiology* 115:1211-1219.

Shinozaki, K, K. Yamaguchi-Shinozaki, and M. Seki (2003). Regulatory network of gene expression in drought and cold stress response. *Current Opinion in Plant Biology* 6:410-417.

Singla, R. and N. Garg (2005). Influencing of salinity on growth and yield attributes in chickpea cultivars. *Turkish Journal of Agriculture Forestry* 29:231-235.

Sivakumar, P., P. Sharmila, and P. Pardha Saradhi (1998). Proline suppresses rubisco activity in higher plants. *Biochemical and Biophysical Research Communications* 252:428-432.

Sivakumar, P., P. Sharmila, and P. Pardha Saradhi (2000). Proline alleviates salt stress induced enhancement in ribulose 1,5-bisphosphate oxygenase activity. *Biochemical and Biophysical Research Communications* 279:512-515.

Sivakumar, P., P. Sharmila, and P. Pardha Saradhi (2001). Proline suppresses rubisco activity by dissociating small subunits from holoenzyme. *Biochemical and Biophysical Research Communications 282*:236-241.

Sivakumar, P., P. Sharmila, and P. Pardha Saradhi (2002). Sugars have potential to curtail oxygenase activity of rubisco. *Biochemical and Biophysical Research Communications* 298:247-250.

Sivakumar, P., S. Yadav, and P. Pardha Saradhi (2001). Rubisco folding—A model system for the functional investigation of molecular chaperones. In *Biophysical processes in living systems*, P. Pardha Saradhi (ed.). Enfield, NH: Science Publishers, pp. 91-116.

Sivamani, E., L. Bassie, P. Christou, and T. Capell (2001). Development of a novel gene transfer system for *Cajanus cajan* and expression of monocot arginine decarboxylase cDNA in transformed cell lines. *Plant Physiology and Biochemistry* 39:575-582.

Strand, Å., C.H. Foyer, P. Gustafsson, P. Gardeström, and V. Hurry (2003). Altering flux through the sucrose biosynthesis pathway in transgenic *Arabidopsis thaliana* modifies photosynthetic acclimation at low temperatures and the development of freezing tolerance. *Plant Cell and Environment* 26:523-535.

Su, J. and R. Wu (2004). Stress-inducible synthesis of proline in transgenic rice confers faster growth under stress conditions than that with constitutive synthesis. *Plant Science* 166:941-948.

Sulpice, R., H. Tsukaya, H. Nonaka, L. Mustardy, T.H.H. Chen, and N. Murata (2003). Enhanced formation of flowers in salt-stressed *Arabidopsis* after genetic engineering of the synthesis of glycine betaine. *Plant Journal* 36:165-176.

Tarczynski, M.C., R.G. Jensen, and H.J. Bohnert (1992). Expression of a bacterial *mtlD* gene in transgenic tobacco leads to production and accumulation of mannitol. *Proceedings of the National Academy of Sciences USA* 89:2600-2604.

Tarczynski, M.C., R.G. Jensen, and H.J. Bohnert (1993). Stress protection of transgenic tobacco by production of the osmolyte mannitol. *Science* 259:508-510.

Thomas, J.C., M. Sepahi, B. Arendall, and H.J. Bohnert (1995). Enhancement of seed germination in high salinity by engineering mannitol expression in *Arabidopsis thaliana. Plant Cell and Environment* 18:801-806.

Thu-Hang, P., L. Bassie, G. Safwat, P. Trung-Nghia, P. Christou, and T. Capell (2002). Expression of a heterologous *S-adenosyl methionine decarboxylase* cDNA in plants demonstrates that changes in *S*-adenosyl-L-methionine decarboxylase activity determine levels of the higher polyamines spermidine and spermine. *Plant Physiology* 129:1744-1754.

Trung-Nghia, P., L. Bassie, G. Safwat, P. Thu-Hang, O. Lepri, P. Rocha, P. Christou, and T. Capell (2003). Reduction in the endogenous arginine decarboxylase transcript levels in rice leads to depletion of the putrescine and spermidine pools with no concommitant changes in the expression of downstream genes in the polyamine biosynthetic pathway. *Planta* 218:125-134.

Venekamp, J.H. (1989). Regulation of cytosolic acidity in plants under conditions of drought. *Physiologia Plantarum* 76:112-117.

Waditee, R., M.N.H. Bhuiyan, V. Rai, K. Aoki, Y. Tanaka, T. Hibino, S. Suzuki, J. Takano, A.T. Jagendorf, and T. Takabe (2005). Genes for direct methylation of glycine provide high levels of glycine betaine and abiotic-stress tolerance in *Synechococcus* and *Arabidopsis. Proceedings of the National Academy of Sciences USA* 102:1318-1323.

Waditee, R., Y. Tanaka, K. Aoki, T. Hibino, H. Jikuya, J. Takano, and T. Takabe (2003). Isolation and functional characterization of *N*-methyl transferases that catalyze betaine synthesis from glycine in a halotolerant photosynthetic organism *Aphanothece halophytica. Journal of Biological Chemistry* 278:4932-4942.

Waie, B. and M.V. Rajam (2003). Effect of increased polyamine biosynthesis on stress responses in transgenic tobacco by introduction of human *S*-adenosyl methionine gene. *Plant Science* 164:727-734.

Wanek, W. and A. Richter (1997). Biosynthesis and accumulation of d-ononitol in *Vigna umbellata* in response to drought stress. *Physiologia Plantarum* 101:416-424.

Wu, L.-Q, Z.-H. Fan, L. Guo, Y.-Q. Li, W.-J. Zhang, L.-J. Qi, and Z.-E. Chen (2005). Overexpression of δ-*OAT* gene enhances salt and drought tolerance in transgenic rice. *Chinese Science Bulletin* 48:2050-2056.

Yamada, M., H. Morishita, K. Urano, N. Shiozaki, K. Yamaguchi-Shinozaki, K. Shinozaki, and Y. Yoshiba (2005). Effects of free proline accumulation in petunias under drought stress. *Journal of Experimental Botany* 56:1975-1981.

Yang, X., Z. Liang, and C. Lu (2005). Genetic engineering of the biosynthesis of glycine betaine enhances photosynthesis against high temperature stress in transgenic tobacco plants. *Plant Physiology* 138:2299-2309.

Yeo, E.T., H.B. Kwon, S.E. Han, J.T. Lee, J.C. Ryu, and M.O. Byun (2000). Genetic engineering of drought-resistant potato plants by introduction of the trehalose-6-phosphate synthase (TPS1) gene from *Saccharomyces cerevisiae. Molecular Cell* 10:263-268.

Zhao, H.W., Y.J. Chen, Y.L. Hu, Y. Gao, and Z.P. Lin (2000). Construction of a trehalose-6-phosphate synthase gene driven by drought responsive promoter and expression of drought-resistance in transgenic tobacco. *Acta Botanica Sinica* 42:616-619.

Zhu, B.C., J. Su, M.C. Chan, D.P.S. Verma, Y.L. Fan, and R. Wu (1998). Overexpression of a delta (1)-pyrroline-5-carboxylate synthetase gene and analysis of tolerance to water- and salt-stress in transgenic rice. *Plant Science* 139:41-48.

## Chapter 36

# Legume-Nematode Interactions with Reference to Model Legumes

Dasharath P. Lohar

## *INTRODUCTION*

Nematodes, unsegmented roundworms, are among the most abundant creatures on earth, representing 80 to 90 percent of all animals (Boucher and Lambshead, 1994; Blaxter et al., 1998). Nematodes probably establish parasitic relationships with all other complex animals and plants (Blaxter and Bird, 1997). Plant-parasitic nematodes are serious crop pests, causing an average yield loss of 12.3 percent annually, and as high as 20 percent in crops such as banana (Sasser and Freckman, 1987; Koening et al., 1999). In monetary terms, the crop loss is over U.S. $100 billion annually worldwide. In addition to direct crop loss, the control of nematodes requires the use of nematicides in many circumstances, costing a large amount of money and causing environmental pollution. For example, over 49,000 metric tons of nematicidal active ingredient was applied to crops in the United States alone at a cost exceeding U.S. $1 billion in 1982 (Landels, 1989).

Legumes are most important to humans for food, pasture, and agroforestry, second only to the plant family Gramineae (Graham and Vance, 2003). Moreover, legumes fix atmospheric nitrogen in the soil in symbiosis with rhizobia for use by crop plants, reducing the cost of crop production directly and environmental damage caused by nitrogenous fertilizer application indirectly. Like other crop plants, legumes are also susceptible to attack by parasitic nematodes. For example, soybean cyst nematode (SCN,

---

The data presented here on *L. japonicus* as a new model of plant-nematode interactions and its resistance to SCN were generated in David Bird's laboratory at North Carolina State University, Raleigh, with funding from the U.S. Department of Agriculture and the National Science Foundation.

*Heterodera glycine*) is the principal pathogen of soybean in the United States, causing economic damage of U.S. $500-1,000 million annually. Durable host resistance is the best approach for managing plant-parasitic nematodes in crops. However, naturally occurring nematode resistance is difficult to obtain and it can take many years to transfer resistance to a desirable crop variety. Another approach would be to identify plant genes involved in response to parasitic nematodes at the molecular level, and engineer such genes for resistance. Unfortunately, crop legumes possess larger genomes with resistance to genetic manipulation, less developed molecular tools, and are often polyploids, making them unsuitable for routine molecular biological studies. In recent years, model legumes with small genomes, easy transformation protocols, diploid inheritance, and well-developed molecular tools have been promoted for studying legume biology, particularly legume-rhizobia symbiosis. Two such model legumes include *Lotus japonicus* (Stougaard, 2001) and *Medicago truncatula* (Cook, 1999). Attempts have been made to use these models to study plant-nematode interactions (Lohar and Bird, 2003) and to compare them with legume-rhizobia symbiosis (Koltai et al., 2001; Mathesius, 2003; Lohar et al., 2004). In this chapter, we present some of the recent findings in legume-nematode interaction with reference to model legumes. The nematodes discussed are root-knot nematodes (RKN, *Meloidogyne* spp.) and SCN, the two most important parasitic nematodes of crop plants.

## *DEVELOPMENT OF LEGUME MODELS IN PLANT-NEMATODE INTERACTIONS*

Two legume models, *L. japonicus* and *M. truncatula*, have been developed for nematode studies in legumes. Table 36.1 lists some of the features and tools available in both model legumes. *L. japonicus* is diploid and self-fertile with six chromosomes in the haploid genome (Jiang and Gresshoff, 1997). It has 3 to 4 months generation time (seed to seed), profuse flowering and pod set, and produces small primary plants that are suitable for in vitro and high-density handling (Handberg and Stougaard, 1992). Currently, the genome of *L. japonicus,* which is about 450 megabases (Mb), is being sequenced by the Kazusa DNA Research Institute (Young et al., 2005). Similarly, *M. truncatula* is also diploid and self-fertile with eight chromosomes in the haploid genome (Cook, 1999; Young et al., 2005). The *M. truncatula* genome, which is about 500 Mb, is being sequenced with funding from the U.S. National Science Foundation and the European Union 6th Framework Program (Young et al., 2005). Both legume models are used widely for legume-rhizobia symbiosis and legume biology research.

TABLE 36.1. Some important features or tools of model legumes, *L. japonicus* and *M. truncatula.*

| | **Status** | |
|---|---|---|
| **Features/tools** | ***L. japonicus*** | ***M. truncatula*** |
| Generation time | 3-4 months | 3-4 months |
| Haploid chromosome number | 6 | 8 |
| Haploid genome size | ~450 Mbp | ~500 Mbp |
| Chloroplast genome | Sequenced | Sequenced |
| Nuclear genome sequencing (expected completion) | 2006 | 2006 |
| Expressed sequence tag data | Medium | Large |
| Genome mapping | Good | Good |
| Positional cloning of genes | Routine | Routine |
| RNA interference | Possible | Possible |
| Hairy root transformation | Routine | Routine |
| Whole plant transformation | Routine | Routine |
| Ecotype diversity | Available | Available |
| Promoter trapping | Available | Available |
| Insertional mutagenesis | Possible | Possible |
| TILLING* | Possible | Possible |
| Expression profiling | Possible | Possible |
| Proteomics | Possible | Possible |
| Metabolomics | Possible | Possible |
| Microsymbiont | *Mesorhizobium loti* | *Sinorhizobium meliloti* |
| Microsymbiont genome | Sequenced | Sequenced |
| Symbiotic mutants | Present | Present |
| Nematode work | Possible | Possible |
| Nematode resistance | Available | Available |

*TILLING = Targeting-induced local lesions in genomes.

Lohar and Bird (2003) optimized the conditions for studying plant-nematode interactions in *L. japonicus*. One-week-old plants grown in an equal sand-peat mixture at 26.0°C had normal root development and robust RKN infection. Galls were visible as bumps one week after inoculation with RKN eggs. Due to the thin and translucent roots of *L. japonicus,* the authors reported that the study of the nematode infection process was easy since they could visualize the invading nematode larvae in roots. Freshly hatched

RKN second-stage larvae (L2) penetrate the root in the zone of elongation, and migrate intercellularly in the cortex to locate the root meristem (Figure 36.1). Upon finding the root meristem, the L2 turns around and migrates basipetally to the differentiated vascular bundle, where it induces giant cells in the parenchyma cells. The L2 develops into a pear-shaped female and the cells surrounding the giant cells proliferate, giving rise to a gall. The adult females feed on giant cells and lay eggs in a gelatinous matrix on the surface of the gall. Sometimes the development of a gall stops the growth of the root, resulting in a terminal gall. The infection process of *L. japonicus* roots by RKN is similar to the infection process of some other species such as *Arabidopsis thaliana* (Wyss et al., 1992). The galls induced by *Meloidogyne hapla* are usually smaller in size than those induced by *M. incognita,* probably due to more cell proliferation in the latter (Figure 36.1; Lohar and Bird, 2003). The lateral roots originate from both *M. incognita*- and *M. hapla*-induced galls. More lateral roots are formed from *M. hapla* galls than from *M. incognita* galls, giving rise to witches' broom structures. The RKN infection process and morphological changes in *Medicago truncatula* roots are the same as in *L. japonicus*. The infection of *Lotus* roots by SCN is optimum in an equal mix of sand and peat at 26.0°C constant temperature (Lohar and Bird, 2003). The hatched L2s enter the root above the root tip, orient along the vascular bundle, and migrate in the cortex basipetally to find a

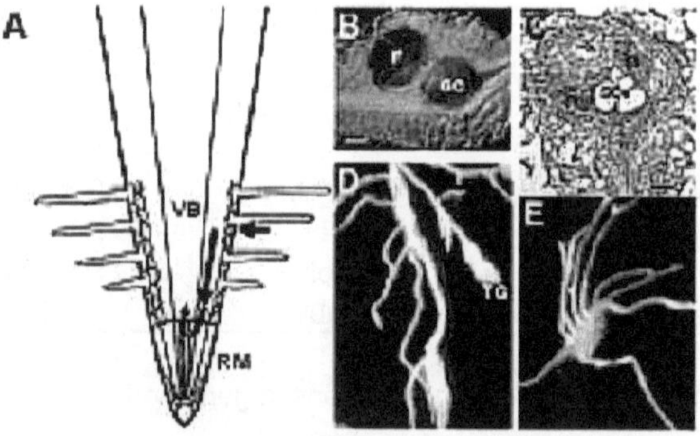


FIGURE 36.1. Infection of *L. japonicus* roots by RKN. (A) A diagram showing penetration and migration of RKN second-stage larvae (L2) into the root of *L. japonicus.* Arrows represent the direction of larval movement. (B) A fully developed RKN female (F) feeding on a giant cell (GC) enclosed in a gall. (C) A cross-section through a gall with giant cells surrounded by small proliferating cells. (D) A part of the root with galls. (E) An *M. hapla*-induced gall with multiple lateral roots. Bars in B and C = 100 μm. VB = vascular bundle; RM = root meristem; TG = a terminal gall.

feeding site in the vascular bundle (Figure 36.2). The larvae have been observed to enter the root in the mature part of the root also. In this case, the movement is both basipetal and acropetal. Similarly, some larvae also enter the root at the site of lateral root eruption. The penetration of *L. japonicus* root by SCN L2 appears to be different from some other species such as soybean and *A. thaliana*. In these species, larvae penetrate the root predominantly in the region above the root tip (Wyss, 1997; Kinloch, 1998).

## *RESISTANCE TO ROOT PARASITIC NEMATODES IN MODEL LEGUMES*

Mutant plants defective in establishing interaction with nematodes are useful tools in studying plant-nematode interactions. In recent years, both *L. japonicus* and *M. truncatula* ecotypes have been identified that are resistant to RKN and SCN. *L. japonicus* ecotype Gifu is resistant to the MO line of SCN (Lohar and Bird, 2003). As described above, SCN can penetrate and migrate freely into *L. japonicus* roots without inducing any microscopic response. However, when the larva starts feeding in the vascular bundle, a strong hypersensitive response is induced. Subsequently, the larva either dies or migrates out of the root. In rare instances, a syncytium can also be formed devoid of a living worm (Figure 36.2).

Lohar and Bird (2003) tested the response of two *L. japonicus* lines that produced more nodules than the wild type in symbiosis with *Mesorhizobium loti*. One such mutant, *har1*, which is mutated for *CLAVATA1*-like receptor kinase (Wopereis et al., 2000; Krussel et al., 2002; Nishimura et al., 2002), is also hyperinfected by RKN in that it makes more galls than the wild type. An *A. thaliana* mutant for the *CLAVATA1* gene is also hyperinfected by nematodes (David Bird, personal communication). However, another *L. japonicus* hypernodulating line, which is resistant to ethylene due to the transgenic overexpression of *A. thaliana etr1-1* allele, has the same RKN infection level as the wild type (Lohar and Bird, 2003). Since *har1* makes more lateral roots, the hypergalling observed with RKN could just be due to more root tips providing potential infection sites.

Screening of about 100 ecotypes of *M. truncatula* identified a large variation for resistance to RKN (*M. incognita, M. arenaria, and M. hapla*) and clover cyst nematode (*Heterodera trifolii;* Opperman and Bird, 2002; Dhandaydham et al., 2002). They identified a large variation for resistance in the screened accessions ranging from complete resistance to all three RKN species to complete susceptibility to all. To identify the inheritance of resistance and map the resistance locus, Dhandaydham et al. (2002) analyzed 212 recombinant inbred lines (RILs) from an accession resistant to all

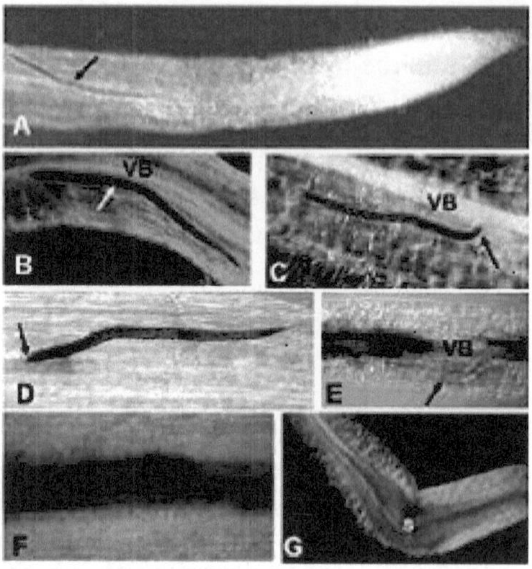


FIGURE 36.2. Infection of *L. japonicus* roots by SCN. Roots were inoculated with 1,000 SCN eggs per plant, and imaged 6 days (A-D) and 10 days (E-G) after inoculation. (A) A second-stage larva (arrow) entered the root above the tip. (B) A larva (arrow) migrating along the vascular bundle. (C) A larva trying to feed in the vascular bundle (arrow). (D) Induction of hypersensitive response (dark coloration) at the site of feeding (arrow). (E) A dead larva (arrow) and hypersensitive response as seen by dark coloration in the vascular bundle. (F) A hypersensitive response seen as dark coloration. The larva is either dead or migrated out of the root. (G) A rare syncytium induced on *L. japonicus* root by SCN. The nematode is either dead or migrated out of the root. VB = vascular bundle; S = syncytium.

three RKN species (DZA045) and an accession susceptible to all three (F83005). RIL analysis demonstrated a role for two independently segregating loci conferring resistance in DZA045. The effects of genes are additive and the presence of only one of the two results in an intermediate phenotype. In spite of the resistance in DZA045 being temperature sensitive, Dhandaydham et al. (2002) did not observe a hypersensitive response to RKN infection, which is commonly seen in Solanaceous hosts. Instead, most of the RKN larvae that entered the root either died as juveniles or developed as males and migrated out of the root, with only 3 percent becoming females. Development of females in infected roots is an index of susceptibility, since egg-laying adult females provide a continuous source of inoculum in the field. Opperman and Bird (2002) also reported a strain of *H. glycine* (SCN) that could infect and reproduce on *M. truncatula*.

## *ADVANCES IN UNDERSTANDING LEGUME-NEMATODE INTERACTIONS USING MODEL LEGUMES*

One useful aspect of model legumes in plant-nematode interaction studies is the ability to compare legume-rhizobia symbiosis with nematode parasitism. Since both legume-rhizobia symbiosis and legume-nematode parasitism induce a similar plant developmental program (Mathesius, 2003), any finding in one system can be tested in the other. Therefore, attempts have been made to do such comparative studies (such as Lohar et al., 2004). One of the requirements for both legume-rhizobia symbiosis and nematode parasitism is the induction of the plant cell cycle. The formation of giant cells and galls by RKN and syncytia by SCN requires cell cycle activation (de Almeida Engler et al., 1999). Nodule formation also requires cell division (Mathesius et al., 2000; Kondorosi et al., 2005). Koltai et al. (2001) compared the expression of *CCS52* (which encodes a mitotic inhibitor) and *ENOD40* (encodes an early nodulation mitogen) during rhizobia-induced nodule formation and RKN-induced giant cell and gall formation in *M. truncatula.* Transcripts of both genes were detected in dividing cells of nodules and galls, and in giant cells. *PHAN* and *KNOX* genes are required for the establishment of meristems (Waites et al., 1998: Scheeberger et al., 1998). Not surprisingly, transcripts for *M. truncatula* orthologues of *PHAN* (*MtPHAN*) and *KNOX* (*MtKNOX*) accumulate in dividing cells of nodules as well as in giant or dividing cells of galls (Koltai et al., 2001).

The plant hormones auxin and cytokinin have been implicated in cell division and nodule formation (Hirsch, 1992; Hirsch et al., 1995; Kondorosi et al., 2005). Using auxin-responsive promoter fused to a *gus* reporter gene (*GH3:gus*) in white clover roots, Mathesius, Bayliss, et al. (1998) and Mathesius, Schlaman, et al. (1998) have shown that auxin accumulates in cortical cells targeted for division during nodule formation. However, auxin accumulation is absent in already divided cells. A similar pattern of *GH3:gus* expression has been reported in white clover during root parasitism by RKN (*M. javanica*; Hutangura et al., 1999; Mathesius, 2003). These authors observed a high *GH3:gus* activity in the gall 48 to 72 hours postinoculation and in a ring around the giant cells later. High activity was also observed in giant cell precursor cells initially (48-72 hours postinoculation), which later subsided. Therefore, Hutangura et al. (1999) suggested that auxin is required for giant cell initiation but not for later enlargement.

Another important plant hormone for cell cycle regulation is cytokinin, which affects nodulation positively (Fang and Hirsch, 1998). Lohar et al. (2004) compared the roles of cytokinin in nodulation and RKN gall formation in *L. japonicus*. The study was done using a cytokinin-inducible promoter from *A. thaliana* fused to a *gus* reporter gene (*ARR5:gus*), and cytokinin oxidase genes from *A. thaliana* (*CKX3*) and maize (*CKX1*) in transgenic hairy roots. The background level of *ARR5* induction was in root tips (in meristem plus columella root cap in some roots), newly formed lateral root meristems, and patches in the vascular bundle. After inoculation with *Mesorhizobium loti,* the cytokinin response was seen in deforming root hairs and in dividing cortical cells in the nodule primordium but not in the mature nodules. RKN (*M. incognita*) induced a cytokinin response in and around the vascular bundle where the gall or giant cell would be formed. The response continued in the dividing cells around the giant cells but not in developed giant cells. In this regard, the cytokinin response is similar to the auxin response in gall or giant cell formation as reported by Hutangura et al. (1999). The transgenic overexpression of *CKX* genes made transgenic hairy roots resistant to exogenous cytokinin treatment (Lohar et al., 2004). Such transgenic hairy roots had significantly more lateral roots but less nodules and RKN galls than the control hairy roots. It indicated a similar effect of cytokinin on nodulation and RKN gall formation.

A parallel between the initial response of roots to rhizobia and nematodes has been reported in terms of root hair deformation. Nod factor treatment or rhizobia inoculation causes root hair deformation such as wavy growth, tip swelling, and branching on a compatible legume root. An identical root hair deformation response in *L. japonicus* roots to RKN inoculation or RKN extract application has been reported (Weerasinghe et al., 2005). Wild-type root hairs inoculated with RKN for 4 hours produced wavy root hairs with swollen tips, and approximately 40 percent of the root hairs displayed branching. Additionally, the authors observed a similar microtubulin and actin disintegration and repolymerization after Nod factor treatment and RKN inoculation. A nematode larvae-free supernatant of the nematode suspension also produced wavy root hairs, suggesting the generation of a diffusible signal from nematodes akin to Nod factors from rhizobia. Mutations in putative Nod factor receptor genes, *nfr1* and *nfr5,* abolish interaction of *L. japonicus* roots with their microsymbiont including root hair deformation responses (Madsen et al., 2003; Radutoiu et al., 2003). Strikingly, Weerasinghe et al. (2005) observed almost an absence of root hair branching in these mutants after RKN inoculation. The RKN-induced gall formation and the number of adult egg-laying females were significantly reduced in these mutants compared to wild-type controls. This finding further strengthens the view that plant-nematode interaction shares genetic pathways with legume-rhizobia symbiosis.

## *CONCLUSION*

The interaction between plant roots and nematodes is beginning to be understood at the molecular level. The use of model legumes and comparative studies with legume-rhizobia symbiosis have aided in the identification of several signaling components in plant root-nematode parasitism. However, a lot still remains unknown. For example, most of the plant genes involved in the formation of giant cells, syncytia, and galls are still unknown. It is expected that the genomic sequence data combined with tissue-specific gene expression profiling experiments and reverse genetics approaches such as RNA interference, TILLING (targeting-induced local lesions in genomes), and gene overexpression will lead to a better understanding of the biological processes of nematode parasitism at the molecular level. This in turn will lead to better control of root parasitic nematodes in plants.

## NOTE

Recently, Ithal et al. (2007) have analyzed the genome wide expression patterns in the host (soybean, *Glycine max*) and the nematode (*Heterodera glycines,* soybean cyst nematode) during the course of infection in a compatible interaction using the Affymetrix GeneChip carrying soybean genome array. Among 35,611 soybean transcripts monitored, they identified 429 genes that showed differential expression between uninfected and nematode-infected root tissues. These included genes encoding enzymes involved in primary metabolism; biosynthesis of phenolic compounds, lignin, and flavonoids; genes related to stress and defense responses; cell wall modification; cellular signaling and transcriptional regulation giving new insights into legume nematode interaction. They have also analyzed the global gene expression patterns in the nematode. Transcript profiling on the nematode side indicated changes occurring in it as it infects and establishes a permanent feeding site within a host plant root.

## REFERENCES

Blaxter, M. and D.M. Bird (1997). Parasitic nematodes. In *Coenorhabditis elegans II,* D.L. Riddle, T. Blumenthal, B.J. Meyer, and J.R. Priess (eds.). Cold Spring Harbor, NY: Cold Spring Harbor Laboratory Press, pp. 851-878.

Blaxter, M.L., P. De Ley, J.R. Garey, L.X. Liu, P. Scheldeman, A. Vierstraete, J.R. Vanfleteren, L.Y. Mackey, M. Dorris, L.M. Frisse, et al. (1998). A molecular evolutionary framework for the phylum Nematoda. *Nature* 392:71-75.

Boucher, G. and P.J.D. Lambshead (1994). Ecological biodiversity of marine nematodes in samples from temperate, tropical and deep sea regions. *Conservation Biology* 9:1594-1604.

Cook, D.R. (1999). *Medicago truncatula*—a model in the making! *Current Opinion in Plant Biology* 2:301-304.

de Almeida Engler, J., V. De Vleesschauwer, S. Burssens, J.L. Celenza Jr., D. Inzé, M. Van Montagu, G. Engler, and G. Gheysen (1999). Molecular markers and cell cycle inhibitors show the importance of cell cycle progression in nematode-induced galls and syncitia. *Plant Cell* 11:793-807.

Dhandaydham, M., L. Charles, H. Zhu, J. Starr, T. Huguet, D. Cook, J.M. Prosperi, and C. Opperman (2002). Two genes control root knot nematode resistance in *Medicago truncatula.* Poster presented at the First International Conference on Legume Genomics and Genetics: Translation to Crop Improvement, June 2-6, Minneapolis-St. Paul, MN.

Fang, Y. and A.M. Hirsch (1998). Studying early nodulin gene *ENOD40* expression and induction by nodulation factor and cytokinin in transgenic alfalfa. *Plant Physiology* 116:53-68.

Graham, P.H. and C.P. Vance (2003). Legumes: Importance and constraints to greater use. *Plant Physiology* 131:872-877.

Handberg, K. and J. Stougaard (1992). *Lotus japonicus*, an autogamous, diploid legume species for classical and molecular genetics. *Plant Journal* 2:487-496.

Hirsch, A.M. (1992). Developmental biology of legume nodulation. *New Phytologist* 122:211-237.

Hirsch, A.M., Y. Fang, L.M. Brill, K.L. Wycoff, B.M. Niner, J.P. Bredt, and P. van Rhyn (1995). Nodule development in legumes: The early stages, involvement of early nodulins, lectins and other proteins. In *Nitrogen fixation: Fundamentals and applications,* I.A. Tikhonovich, N.A. Provorov, V.I. Romanov, and W.E. Newton (eds.). St. Petersburg, Russia: Proceedings of the 10th International Congress on Nitrogen Fixation, pp. 299-304.

Hutangura, P., U. Mathesius, M.G.K. Jones, and B.G. Rolfe (1999). Auxin induction is a trigger for root gall formation caused by root-knot nematodes in white clover and is associated with the activation of the flavonoid pathway. *Australian Journal of Plant Physiology* 26:221-231.

Ithal, N., J. Recknor, D. Nettleton, L. Hearne, T. Maier, T.J. Baum, and M.G. Mitchum (2007). Parallel genome-wide expression profiling of host and pathogen during soybean cyst nematode infection of soybean. Molecular Plant-Microbe Interactions 20:393-405.

Jiang, Q. and P.M. Gresshoff (1997). Classical and molecular genetics of the model legume *Lotus japonicus. Molecular Plant-Microbe Interactions* 10:559-568.

Kinloch, R.A. (1998). Soybean. In *Plant and nematode interactions,* K.R. Barker, G.A. Pederson, and G.L. Windham (eds.). Madison, WI: American Society of Agronomy, Crop Science Society of America, Soil Science Society of America, pp. 317-334.

Koenning, S.R., C. Overstreet, J.W. Noling, P.A. Donald, J.O. Becker, and B.A. Fortnum (1999). Survey of crop losses in response to phytoparasitic nematodes in the United States for 1994. *Journal of Nematology* 31:587-618.

Koltai, H., M. Dhandaydham, C. Oppeman, J. Thomas, and D.M. Bird (2001). Overlapping plant signal transduction pathways induced by a parasitic-nematode and a rhizobial endosymbiont. *Molecular Plant-Microbe Interactions* 14: 1168-1177.

Kondorosi, E., M. Redondo-Neito, and A. Kondorosi (2005). Ubiquitin-mediated proteolysis: To be in the right place at the right moment during nodule development. *Plant Physiology* 137:1197-1204.

Krussel, L., L.H. Madsen, S. Sato, G. Aubert, A. Genua, K. Szczyglowski, G. Duc, T. Kaneko, S. Tabata, F. de Bruijn, et al. (2002). Shoot control of root development and nodulation is mediated by a receptor-like kinase. *Nature* 420:422-426.

Landels, S. (1989). Fumigants and nematicides. In *Chemical economics handbook.* Stanford, CA: Stanford Research Institute International.

Lohar, D.P. and D.M. Bird (2003). *Lotus japonicus:* A new model to study root-parasitic nematodes. *Plant Cell Physiology* 44:1176-184.

Lohar, D.P., J.E. Schaff, J.G. Laskey, J.J. Kieber, K. Bilyeu, and D.M. Bird (2004). Cytokinins play opposite roles in lateral root formation, and nematode and rhizobial symbioses. *Plant Journal* 38:203-214.

Madsen, E.B., L.H. Madsen, S. Radutoiu, M. Olbryt, M. Rakwaslska, K. Szczyglowski, S. Sato, T. Kaneko, S. Tabata, N. Sandal, et al. (2003). A receptor kinase gene of the LysM type is involved in legume perception of rhizobial signals. *Nature* 425:637-640.

Mathesisus, U. (2003). Conservation and divergence of signaling pathways between roots and soil microbes—the *Rhizobium*-legume symbiosis compared to the development of lateral roots, mycorrhizal interactions and nematode-induced galls. *Plant and Soil* 255:105-119.

Mathesius, U., C. Bayliss, J.J. Weinman, H.R.M. Schlaman, H.P. Spaink, B.G. Rolfe, M.E. McCully, and M.A. Djordjevic (1998). Flavonoids synthesized in cortical cells during nodule initiation are early developmental markers in white clover. *Molecular Plant-Microbe Interactions* 11:1223-1232.

Mathesius, U., C. Charon, B.G. Rolfe, A. Kondorosi, and M. Crespi (2000). Temporal and spatial order of events during the induction of cortical cell divisions in white clover by *Rhizobium leguminosarum* bv. *trifolii* inoculation or localized cytokinin addition. *Molecular Plant-Microbe Interactions* 13:617-628.

Mathesius, U., H.R.M. Schlaman, H.P. Spaink, C. Sautter, B.G. Rolfe, and M.A. Djordjevic (1998). Auxin transport inhibition precedes nodule formation in white clover roots and is regulated by flavonoids and derivatives of chitin oligosaccharides. *Plant Journal* 14:23-34.

Nishimura, R., M. Hayashi, G.-J. Wu, H. Kouchi, H. Imaizumi-Anraku, Y. Murakami, S. Kawasaki, S. Akao, M. Ohmori, M. Nagasawa, et al. (2002). *HAR1* mediates systemic regulation of symbiotic organ development. *Nature* 420:426-429.

Opperman, C.H. and D.M. Bird (2002). The *Medicago truncatula*-plant-parasitic nematode interaction: Susceptibility, shared signal transduction pathways, and host resistance. Paper presented at the First International Conference on Legume Genomics and Genetics: Translation to Crop Improvement, June 2-6, Minneapolis-St. Paul, MN.

Radutoiu, S., L.H. Madsen, E.B. Madsen, H.H. Felle, Y. Umehara, M. Grønlund, S. Sato, Y. Nakamura, S. Tabata, N. Sandal, et al. (2003). Plant recognition of symbiotic bacteria requires two LysM receptor-like kinases. *Nature* 425:585-592.

Sasser, J.N. and D.W. Freckman (1987). A world perspective on nematology: The role of the society. In *Vistas on nematology,* J.A. Veech and D.W. Dickson (eds.). Hyattsville: Society of Nematology, pp. 7-14.

Schneeberger, R., M. Tsiantis, M. Freeling, and J.A. Langdale (1998). The *rough sheath2* gene negatively regulates homeobox gene expression during maize leaf development. *Development* 125:2857-2865.

Stougaard, J. (2001). Genetics and genomics of root symbiosis. *Current Opinion in Plant Biology* 4:328-335.

Waites, R., H.R.N. Selvandurai, I.R. Oliver, and A. Hudson (1998). The *PHANTASTICA* gene encodes a *MYB* transcription factor involved in growth and dorsoventrality of lateral organs in *Antirrhinum. Cell* 93:779-789.

Weerasinghe, R.R., D.M. Bird, and N.S. Allen (2005). Root-knot nematodes and bacterial Nod factors elicit common signal transduction events in *Lotus japonicus. Proceedings of the National Academy of Sciences USA,* 102:3147-3152.

Wopereis, J., E. Pajuelo, F.B. Dazzo, Q. Jiang, P.M. Gresshoff, F.J. de Bruijn, J. Stougaard, and K. Szczyglowski (2000). Short root mutant of *Lotus japonicus* with a dramatically altered symbiotic phenotype. *Plant Journal* 23:97-114.

Wyss, U. (1997). Root parasitic nematodes: An overview. In *Cellular and molecular aspects of plant-nematode interactions,* C. Fenoll, F.M.W. Grundler, and S.A. Ohl (eds.). Dordrecht: Kluwer Academic Publishers, pp. 5-24.

Wyss, U., F.M.W. Grundler, and A. Munch (1992). The parasitic behaviors of second-stage juveniles of *Meloidogyne incognita* in *Arabidopsis thaliana. Nematology* 38:98-111.

Young, N.D., S.B. Cannon, S. Sato, D. Kim, D.R. Cook, C.D. Town, B.A. Roe, and S. Tabata (2005). Sequencing the gene spaces of *Medicago truncatula* and *Lotus japonicus. Plant Physiology* 137:1174-1181.

Chapter 37

# Defensins in Legumes

Jack H. Wong
T.B. Ng

## *INTRODUCTION*

Every living organism, ranging from microorganisms to plants or animals, is continuously exposed to assault by pathogens from the environment. Humans have an immune system for protection against pathogens, which recruits antibodies and killer cells to recognize and annihilate specific invaders, respectively. But these adaptive immune responses act only in higher vertebrates (Matsunaga and Rahman, 1998). In plants, innate immunity is a ubiquitous defense mechanism involving production of antimicrobial proteins and peptides in the host that does not possess the antigen-recognition specificity of antibodies (Boman, 1995). Nevertheless, since these peptides and proteins are products of transcription and translation of a single gene, they can be delivered soon after infection, with only a small input of energy and biomass. Although the innate immune response of plants has been considered as a primordial defense system, it can exhibit activity against a variety of microorganisms (Thomma et al., 1998).

To combat pathogen attack, legumes produce some antimicrobial proteins and peptides. They include chitinases, ribosome-inactivating proteins, cyclophilin-like proteins, protease inhibitors, lectins, peroxidases, and embryo-abundant proteins (Ng, 2004). In this chapter, emphasis is placed on defensins in legumes, which exert potent antimicrobial activity.

The skilled secretarial assistance of Miss Fion Yung is much appreciated.

## *DEFINITION AND STRUCTURE OF DEFENSIN*

Defensin is derived from the Latin word defendo, which means to repel. In 1985, defensin became the name of a family of antimicrobial peptides with a characteristic motif with many β sheets and a framework of six disulphide-linked cysteines. Ganz et al. (1985) isolated a group of structurally related peptides, which were capable of destroying microorganisms such as *Staphylococcus aureus, Pseudomonas aeruginosa,* and *Escherichia coli*. The peptides were designated as defensins in view of their role in host defense. Subsequently, defensins were also detected in insects (Lambert et al., 1989; Dimarq et al., 1990) and plants (Colilla et al., 1990).

Plant defensins are 45 to 54 amino acids long and are highly basic cysteine-rich peptides that are apparently ubiquitously distributed among plants. Colilla et al. (1990) and Mendez et al. (1990) purified the first plant defensins from wheat and barley grains, respectively. These proteins were then called γ-thionins because their size and cysteine content resembled previously reported thionins (Carrasco et al., 1981). Structural analysis has subsequently shown, however, that γ-thionins are unrelated to thionins (Bruix et al., 1993). Due to their structural resemblance to mammalian and insect defensins, γ-thionins were renamed plant defensins (Carrasco et al., 1981).

Plant defensins have some structural properties in common with mammalian and insect defensins. All plant defensins hitherto identified possess eight cysteines forming four structure-stabilizing disulfide bridges. Investigation of the three-dimensional structure of plant defensins has disclosed that the structure consists of a triple-stranded β-sheet with a parallel α-helix (Bruix et al., 1993, 1995; Bloch et al., 1998; Fant et al., 1998, 1999; Almeida et al., 2002). The core of the global fold of plant and invertebrate defensins includes a cysteine-stabilized α-helix β-sheet (CS) motif (Cornct et al., 1995). In this motif, two cysteine residues, located one turn apart in the α-helix, form two disulfide bridges with two cysteine residues separated by a single amino acid in the last strand of the β-sheet. As pointed out by Bruix et al. (1995), the core global three-dimensional structure of plant defensins is very similar to that of insect defensins, except that insect defensins are devoid of the domain corresponding to the amino terminal β-sheet of plant defensins.

Defensins in legumes have eight cysteine residues although only several beans have been reported to have defensins or defensin-like peptides. They are 40 to 50 amino acids in length and carry a net positive charge. The structure of legume defensins is mainly a triple-stranded β-sheet and a single α-helix lying in parallel with the β-sheet to form a cysteine-stabilized α-helix β-sheet (CSαβ) motif. Table 37.1 presents a comparison of the amino acid sequences of legume defensins isolated from different kinds of beans. Defensins and defensin-like peptides have been isolated from *Trigonella foenum-graecum* (Olli and Kirti, 2006; Olli et al., 2007), *Medicago truncatula* (Hanks et al., 2005), *Pachyrrhizus erosus* (Song et al., 2004, 2005), *Arabidopsis* species (Silverstein et al., 2005), azuki bean (Chen et al., 2005), mung bean (Liu et al., 2006), *Pisum sativum* (Cabral et al., 2003), large and small scarlet runner beans (Ngai and Ng, 2004, 2005), shelf bean, haricot bean, white cloud bean, and ground bean (Wong and Ng, 2005a,b, 2006a,b; Wong et al., 2006).

TABLE 37.1. Amino acid sequence homology of legume defensins (results using basic local alignment search tool, BLAST, available at www.ncbi.nlm.nih.gov).

| **Species name** | **N-terminal sequences** |
|---|---|
| *Trigonella foenum-graecum* | ktcenladkyrgpcfsgcdthcttkehavsgrcrddfrcwctkr |
| *Medicago truncatula* | ntcenlagtfrgpcfgnsncdfhcrtkehlvsgrcrddfrcwctrnc |
| *Pachyrhizus erosus* | ktcenladtfrgpcftdgscddhcknkehlikgrcrddfrcwctrnc |
| *Arabidopsis thaliana* | ricerrsktwtgfcgntrgcdsqckrwerashgachaqfpgfacfcyfnc |
| *Arabidopsis halleri* | klcerasgtwsgvcgnnnacknqcirlekarhgscnyvfpahkcicyfpc |
| *Vigna radiata* cv. mung bean | rtcmikkegwgkclidttcahscknrgyiggnckgmtrtcyclvnc |
| *Vigna radiata* cv. azuki bean | rtcmikkegwgkclidttcahscknrgyiggnckgmtrtcyclvnc |
| *Pisum sativum* | ntcehladtyrgvcftdascddhcknkahlisgtchnfkc |
| *Phaseolus coccineus* cv. Major | kqtenlanty |
| *Phaseolus coccineus* cv. Minor | kctenladtykgpppffttg |
| *Phaseolus limensis* cv. shelf bean | ktcenladtykgpcfttggcddhcknkehllsgrcrddfrcwctrnc |
| *Phaseolus vulgaris* cv. white cloud bean | ktcenladtfrgpcfatsnc |
| *Phaseolus vulgaris* cv. haricot bean | ktcenladtykgpcftsggd |
| *Vigna sesquipedalis* cv. ground bean | ktcenladty |

## *MECHANISMS OF ACTION OF DEFENSINS*

Mammalian and insect defensins form channels in microbial, mammalian, and artificial lipid membranes, increasing membrane permeability in a charge- or voltage-dependent manner. Mammalian defensins can permeabilize membranes of target mammalian cells and increase permeability of the target cell to the dye trypan blue (Lichtenstein et al.,1988). In an artificial phospholipid membrane study, Kagan et al. (1990) suggesed that the cationic defensin molecule is inserted into the bilayer. The insect defensins act on *Micrococcus luteus* with an analogous mechanism, forming membrane channels that leak potassium and induce limited membrane depolarization (Cociancich et al., 1993).

Unlike mammalian and insect defensins, plant defensins have not been demonstrated to induce ion-permeable pores in artificial membrane, nor alter the electrical properties of the lipid bilayer of plasma membrane (Thevissen et al., 1996; Caaveiro et al., 1997). The exact mode of action of plant defensins has not yet been resolved. For the majority of plant defensins, the molecular components implicated in signaling and putative intracellular targets remain to be elucidated. Only for defensins from alfalfa (*Medicago sativa*) MsDef1, dahlia plant (*Dahlia merckii*) DmAMP1, and radish (*Raphanus sativus*) RsAFP2 is the molecular basis of their inhibitory activity partly understood.

## *M(IP)$_2$C AS TARGET FOR DAHLIA PLANT DEFENSIN DMAMP1*

The *IPT1* gene regulates the sensitivity of *Saccharomyces cerevisiae* to DmAMP1 (Thevissen, Cammue, et al., 2000). The gene encodes inositol phosphotransferase (Ipt1p), which catalyzes the final step of the pathway of mannose-(inositol-phosphate)$_2$-ceramide [M(IP)$_2$C] synthesis (Dickson et al., 1997). *S. cerevisiae* strains with a nonfunctional *IPT1* allele are devoid of M(IP)$_2$C in plasma membrane. They have an attenuated DmAMP1 binding capacity, and are recalcitrant to DmAMP1-induced membrane permeabilization and growth suppression (Thevissen, Cammue, et al., 2000). The aforementioned phenotypic abnormalities can be rectified by reinsertion of a functional *IPT1* gene. Under conditions of nutrient shortage, an *ipt1*-null mutant of *S. cerevisiae* (*ipt1*) does not differ in sensitivity toward DmAMP1 from the corresponding wild-type yeast strain (Im et al., 2003). This mutant is capable of producing meager quantities of M(IP)$_2$C via an alternative biosynthetic pathway. It appears that the presence of M(IP)$_2$C, rather than that of a functional *IPTI* gene, determines DmAMP1-sensitivity. Thus

$M(IP)_2C$, rather than Ipt1p, is a component of the DmAMP1 binding site on the fungal plasma membrane. Furthermore, DmAMP1 interacts directly with *S. cerevisiae* sphingolipids and this interaction is facilitated by equimolar concentrations of ergosterol (Thevissen et al., 2003).

The highly homologous defensins Ah-AMP1 and Ct-AMP1 (Thevissen, Osborn, et al., 2000), from *Aesculus hippocastanum* and *Clitoria ternatea,* respectively, compete for the binding site of DmAMP1 (Osborn et al., 1995; Thomma et al., 2002). Elimination of *IPT1* brings about resistance to DmAMP1, Ah-AMP1, and Ct-AMP1. A remotely related plant defensin, such as *Heuchera sanguinea* HsAFP1, fails to compete for the DmAMP1 binding site (Thevissen, Osborn, et al., 2000). Although it is active on wild-type *S. cerevisiae,* ablation of *IPT1* from *S. cerevisiae* does not result in enhanced resistance to HsAFP1. Apparently, HsAFP1 binding sites on the plasma membrane are separate from those of DmAMP1, Ah-AMP1, and Ct-AMP1.

## *FUNGAL GLUCOSYL CERAMIDES AS TARGET FOR THE RADISH PLANT DEFENSIN RSAFP2*

The *GCS* gene determines the sensitivity of the yeasts *P. pastoris* and *C. albicans* to RsAFP2. It encodes glucosyl ceramide (GlcCer) synthase, which regulates glucosyl ceramide synthesis. *P. pastoris* and *C. albicans gcs* null mutants without the RsAFP2 sensitivity gene are much more resistant to RsAFP2 and the plant defensin-like insect antifungal peptide heliomicin than the corresponding wild-type strains. Defensin peptides may have evolved from a single precursor and are conserved across the eukaroytes. Plant and insect antifungal peptides alike target similar structures in the fungal plasma membrane, lending credence to evolutionary conservation of defensins as defense peptides in the plant and animal kingdoms. RsAFP2 interacts with fungal GlcCer, but not with human GlcCer, soybean GlcCer, or structurally different complex lipid components such as soybean MGDG, due to structural variations in ceramide among fungal, plant, and human GlcCer. Various fungal and yeast species share the same GlcCer (*N*-2'-hydroxyoctadecanoyl-1-*O*-β-d-glucopyranosyl-(4*E,* 8*E*)-9-methyl-sphingadienine). Fungal GlcCer displays some structural characteristics that differentiate it from plant GlcCer such as the 9-methyl group branching of the sphingoid base and variable levels of unsaturation and fatty acid chain length. These structural features of fungal GlcCer might be essential for determining the interaction with antifungal defensins and ensuing growth inhibition (Thevissen et al., 2003).

## *L-TYPE CALCIUM CHANNEL AS TARGET FOR ALFALFA DEFENSIN MSDEF1*

Spelbrink et al. (2004) first demonstrated that alfalfa defensin MsDef1 could block the L-type $Ca^{2+}$ channel in mammalian cells, and Jackson and Heath (1993) reported that interference with fungal $Ca^{2+}$ gradients would result in hyperbranching (Jackson and Heath, 1993).

Both MsDef1 and the $Ca^{2+}$ channel blocker, 1,2-bis [(2-aminophenoxy) ethane-*N, N, N', N'*-tetra acetate] EGTA, suppressed the growth of fungal hyphae and induced hyperbranching of fungal hyphae (Spelbrink et al., 2004). Another experiment also showed that growth in *U. maydis* was inhibited by EGTA. These findings indicate that defensins retard hyphal growth by adversely affecting $Ca^{2+}$ transport.

The $Ca^{2+}$ ion is a common signaling molecule with crucial roles in fungi and controls hyphal tip elongation during fungal growth (Jackson and Heath, 1993), which is a dynamic and complex process entailing control of localized synthesis and expansion of the growing tip and is under the control of a cytosolic $Ca^{2+}$ gradient generated by $Ca^{2+}$ channels located at the tip (Tsien and Tsien, 1990). Disruption of this tip gradient induces hyperbranching in growing hyphae like that noted in the case of MsDef1.

MsDef1 and Rs-AFP2 resemble each other in three-dimensional structure (Lay et al., 2003) and yet are distinct in primary amino acid sequence (Fant et al., 1998). Rs-AFP2 appears to act without blocking effects on any of the three $Ca^{2+}$ channels (Spelbrink et al., 2004). Both of them possess antifungal activity. This indicates dissimilar modes of action for these two structurally related defensins.

## *APPLICATIONS OF PLANT DEFENSINS*

Legumes can be used as bioreactors for producing antibodies used in cancer diagnosis and therapy (Perrin et al., 2000). We could also easily farm the defensins, the antifungal agents in the field just by transforming and overexpressing the defensin gene in legumes.

## *EXPRESSION OF PLANT DEFENSINS IN TRANSGENIC PLANTS*

Defensins are small peptides with antimicrobial activity. Obviously, they can be employed in transgenic crops to increase protection against pathogens. Terras et al. (1995) produced the first transgenic plants expressing another plant defensin. They transferred radish defensin Rs-AFP2 to tobacco plants, which as a result acquired reinforced antifungal activity against the tobacco foliar pathogen *A. longipes*. Gao et al. (2000) showed that constitutive expression of alfalfa antifungal peptide MsAFP conferred in potato a resistance against the agronomically important fungus *Verticillium dahliae* under field conditions. This was the first demonstration that agronomically useful levels of fungal control can be accomplished by expressing a single transgene in agricultural crops.

Transgenic plants carrying foreign genes of pathogen resistance have many advantageous features such as reduced pesticide usage, environmentally friendly nature, decreased input costs to farmers, and season-long protection independent of weather conditions.

A limited number of antifungal drugs are currently available for the treatment of an expanding range of pathogenic fungi. They can be grouped into three classes based on their targets, namely azoles (Walsh et al., 2000), polyenes (Bolard, 1986), and fluorinated pyrimidines (Waldorf and Polak, 1983). However, owing to the development of resistance, it is imperative to develop new chemical classes of antifungal therapeutics directed toward new fungal targets. Antifungal agents should target molecules that are ubiquitous in fungal cells but absent in mammalian cells such as components of the cell membrane or cell wall, and virulence factors. In microbes, developing drug resistance would mean that they have to change fundamental elements of their cells like membrane components. Sphingolipids are essential membrane components in eukaryotic cells that are structurally different in mammalian and fungal cells. Sphingolipid $M(IP)_2C$ on fungal plasma membrane is involved in the binding site for DmAMP1. Binding with the sphingolipid is the first step of killing. That means DmAMP1 is a natural antifungal agent, and this needs to be confirmed. The structure of DmAMP1 should be analyzed to elucidate the mechanism of the specific binding, which could pave the way for generating new antifungal agents.

Since the mode of action of only a small number of plant defensins has been examined extensively, they might form a family of promising membrane-targeting compounds for development into antifungal drugs.

## *ISOLATION OF LEGUME DEFENSINS*

The methodology used in the isolation of legume defensins from legume seeds is straightforward. Commonly, the first step involves homogenization and extraction of the beans. The homogenization step physically breaks up the beans into small pieces. To continue the extraction, extraction buffer is added to the homogenized slurry. The buffer contains chemicals to maintain pH and ion strength. Sometimes protease inhibitors are added. The mixture is then centrifuged. The supernatant containing the defensin is collected. In the next step, the supernatant is subjected to different types of column chromatography to purify the defensin, including ion exchange chromatography on DEAE-cellulose, CM- or Q-Sepharose, Mono S and Mono Q, size exclusion chromatography on Superdex 75, Sephadex G-75 and Superdex Peptide, reversed-phase chromatography on an analytical C8 column, or affinity chromatography on an Affi-gel blue gel column. Gel electrophoresis such as tricine gel electrophoresis and mass spectrometry are used to assess the homogeneity and molecular mass of defensins (Perrin et al., 2000; Almeida et al., 2000; Wong and Ng, 2005a,b).

## REFERENCES

Almeida, M.S., K.M. Cabral, E. Kurtenbach, F.C. Almeida, and A.P. Valente (2002). Solution structure of *Pisum sativum* defensin 1 by high resolution NMR: Plant defensins, identical backbone with different mechanisms of action. *Journal of Molecular Biology* 315:749-757.

Almeida, M.S., K.M. Cabral, R.B. Zingali, and E. Kurtenbach (2000). Characterization of two novel defense peptides from pea (*Pisum sativum*) seeds. *Archives of Biochemistry and Biophysics* 378:278-286.

Bloch, C. Jr., S.U. Patel, F. Baud, M.J. Zvelebil, M.D. Carr, P.J. Sadler, and J.M. Thornton (1998). $^1$H NMR structure of an antifungal gamma-thionin protein SIalpha1: Similarity to scorpion toxins. *Proteins* 32:334-349.

Bolard, J. (1986). How do the polyene macrolide antibiotics affect the cellular membrane properties? *Biochimica et Biophysica Acta* 22:257-304.

Boman, H.G. (1995). Peptide antibiotics and their role in innate immunity. *Annual Review of Immunology* 13:61-92.

Bruix, M., C. Gonzalez, J. Santoro, F. Soriano, A. Rocher, E. Mendez, and M. Rico (1995). $^1$H-NMR studies on the structure of a new thionin from barley endosperm. *Biopolymers* 36:751-763.

Bruix, M., M.A. Jimenez, J. Santoro, C. Gonzalez, F.J. Colilla, E. Mendez, and M. Rico (1993). Solution structure of gamma 1-H and gamma 1-P thionins from barley and wheat endosperm determined by $^1$H-NMR: A structural motif common to toxic arthropod proteins. *Biochemistry* 32:715-724.

Caaveiro, J.M., A. Molina, J.M. Gonzalez-Manas, P. Rodriguez-Palenzuela, F. Garcia-Olmedo, and F.M. Goni (1997). Differential effects of five types of antipathogenic plant peptides on model membranes. *FEBS Letters* 410:338-342.

Cabral, K.M., M.S. Almeida, A.P. Valente, F.C. Almedia, and E. Kurtenbach (2003). Production of the active antifungal *Pisum sativum* defensin 1 (Psd1) in *Pichia pastoris* overcoming the inefficiency of the STE13 protease. *Protein Expression and Purification* 31:115-122.

Carrasco, L., D. Vazquez, C. Hernandez-Lucas, P. Carbonero, and F. Garcia-Olmedo (1981). Thionins: Plant peptides that modify membrane permeability in cultured mammalian cells. *European Journal of Biochemistry.* 116:185-189.

Chen, G.H., M.P. Hsu, C.H. Tan, H.Y. Sung, C.G. Kuo, M.J. Fan, H.M. Chen, S. Chen, and C.S. Chen (2005). Cloning and characterization of a plant defensin VaD1 from azuki bean. *Journal of Agricultural and Food Chemistry* 53:982-982.

Cociancich, S., A. Ghazi, C. Hetru, J.A. Hoffmann, and L. Letellier (1993). Insect defensin, an inducible antibacterial peptide, forms voltage-dependent channels in *Micrococcus luteus. Journal of Biological Chemistry* 268:19239-19245.

Colilla, F.J., A. Rocher, and E. Mendez (1990). Gamma-purothionins: Amino acid sequence of two polypeptides of a new family of thionins from wheat endosperm. *FEBS Letters* 270:191-194.

Cornet, B., J.M. Bonmatin, C. Hetru, J.A. Hoffmann, M. Ptak, and F. Vovelle (1995). Refined three-dimensional solution structure of insect defensin A. *Structure* 3:435-448.

Dickson, R.C., E.E. Nagiec, G.B. Wells, M.M. Nagiec, and R.L. Lester (1997). Synthesis of mannose-(inositol-P)2-ceramide, the major sphingolipid in *Saccharomyces cerevisiae,* requires the *IPT1* (*YDR072c*) gene. *Journal of Biological Chemistry* 272:29620-29625.

Dimarcq, J.L., D. Zachary, J.A. Hoffmann, D. Hoffmann, and J.M. Reichhart (1990). Insect immunity: Expression of the two major inducible antibacterial peptides, defensin and diptericin, in *Phormia terranovae. EMBO Journal* 9:2507-2515.

Fant, F., W.F. Vranken, and F.A. Borremans (1999). The three-dimensional solution structure of *Aesculus hippocastanum* antimicrobial protein 1 determined by 1H nuclear magnetic resonance. *Proteins* 37:388-403.

Fant, F., W. Vranken, W. Broekaert, and F. Borremans (1998). Determination of the three-dimensional solution structure of *Raphanus sativus* antifungal protein 1 by $^{1}$H NMR. *Journal of Molecular Biology* 279:257-270.

Ganz, T., M.E. Selsted, D. Szklarek, S.S.L. Harwing, K. Daher, D.F. Bainton, and R.I. Lehrer (1985). Defensins. Natural peptide antibiotics of human neutrophils. *Journal of Clinical Investigations* 76:1427-1435.

Gao, A.G., S.M. Hakimi, C.A. Mittanck, Y. Wu, B.M. Woerner, D.M. Stark, D.M. Shah, J. Liang, and C.M. Rommens (2000). Fungal pathogen protection in potato by expression of a plant defensin peptide. *Nature Biotechnology* 18:1307-1310.

Hanks, J.N., A.K. Snyder, M.A. Graham, R.K. Shah, L.A. Blaylock, M.J. Harrison, and D.M. Shah (2005). Defensin gene family in *Medicago truncatula:* Structure, expression and induction by signal molecules. *Plant Molecular Biology* 58:385-399.

Im, Y.J., J. Idkowiak-Baldys, K. Thevissen, B.P.A. Cammue, and J.Y. Takemoto (2003). *IPT1*-independent sphingolipid biosynthesis and yeast inhibition by syringomycin E and plant defensin DmAMP1. *FEMS Microbiology Letters* 223:199-203.

Jackson, S.L. and I.B. Heath (1993). Roles of calcium ion in hyphal tip growth. *Microbiology Reviews* 57:367-382.

Kagan, B.L., M.E. Selsted, T. Ganz, and R.I. Lehrer (1990). Antimicrobial defensin peptides form voltage-dependent ion-permeable channels in planar lipid bilayer membranes. *Proceedings of the National Academy of Sciences USA* 87:210-214.

Lambert, J., E. Keppi, J.L. Dimarcq, C. Wicker, J.M. Reichhart, B. Dunbar, P. Lepage, A. Van Dorsselaer, J. Hoffmann, and J. Fothergill (1989). Insect immunity: Isolation from immune blood of the dipteran *Phormia terranovae* of two insect antibacterial peptides with sequence homology to rabbit lung macrophage bactericidal peptides. *Proceedings of the National Academy of Sciences USA* 86:262-266.

Lay, F.T., H.J. Schirra, M.J. Scanlon, M.A. Anderson, and D.J. Craik (2003). The three-dimensional solution structure of NaD1, a new floral defensin from *Nicotiana alata* and its application to a homology model of crop defense protein alfAFP. *Journal of Molecular Biology* 325:175-188.

Lichtenstein, A.K., T. Ganz, T.M. Nguyen, M.E. Selsted, and R.I. Lehrer (1988). Mechanism of target cytolysis by peptide defensins. Target cell metabolic activities, possibly involving endocytosis, are crucial for expression of cytotoxicity. *Journal of Immunology* 140:2686-2694.

Liu, Y.J., C.S. Cheng, S.M. Lai, M.P. Hsu, C.S. Chen, and P.C. Lyu (2006). Solution structure of the plant defensin VrD1 from mung bean and its possible role in insecticidal activity against bruchids. *Proteins* 63:777-786.

Matsunaga, T. and A. Rahman (1998). What brought the adaptive immune system to vertebrates? The jaw hypothesis and the seahorse. *Immunology Reviews* 166:177-186.

Mendez, E., A. Moreno, F. Colilla, F. Pelaez, G.G. Limas, R. Mendez, F. Soriano, M. Salinas, and C. de Haro (1990). Primary structure and inhibition of protein synthesis in eukaryotic cell-free system of a novel thionin, gamma-hordothionin, from barley endosperm. *European Journal of Biochemistry* 194:533-539.

Ng, T.B. (2004). Antifungal proteins and peptides of leguminous and non-leguminous origins. *Peptides* 25:1215-1222.

Ngai, P.H. and T.B. Ng (2004). Coccin, an antifungal peptide with antiproliferative and HIV-1 reverse transcriptase inhibitory activities from large scarlet runner beans. *Peptides* 25:2063-2068.

Ngai, P.H. and T.B. Ng (2005). Phaseococcin, an antifungal protein with anti-proliferative and anti-HIV-1 reverse transcriptase activities from small scarlet runner beans. *Biochemistry and Cell Biology* 83:212-220.

Olli, S. and P.B. Kirti (2006). Cloning, characterization and antifungal activity of defensin *Tfgd1* from *Trigonella foenum-graecum* L. *Journal of Biochemistry and Molecular Biology* 39:278-283.

Olli, S., L. Guruprasasd, and P.B. Kirti (2007). Characterization of defensin (*Tfgd2*) from *Trigonella foenum graecum. Current Science* 93:365-369.

Osborn, R.W., G.W. De Samblanx, K. Thevissen, I. Goderis, S. Torrekens, and F. Van Leuven (1995). Isolation and characterization of plant defensins from seeds of Asteraceae, Fabaceae, Hippocastanaceae and Saxifragaceae. *FEBS Letters* 368:257-262.

Perrin, Y., C. Vaguero, I. Gerrard, M. Sack, J. Drossard, E. Stoger, P. Christou, and R. Fischer (2000). Transgenic pea seeds as bioreactors for the production of a single chain Fv fragment (scFV) antibody used in cancer diagnosis and therapy. *Molecular Breeding* 6:345-352.

Silverstein, K.A., M.A. Graham, T.D. Pappe, and K.A. VandenBosch (2005). Genome organization of more than 300 defensin-like genes in *Arabidopsis. Plant Physiology* 138:600-610.

Song, X., J. Wang, F. Wu, X. Li, M. Teng, and W. Gong (2005). cDNA cloning, functional expression and antifungal activities of dimeric plant defensin SPE10 from *Pachyrrhizus erosus* seeds. *Plant Molecular Biology* 57:13-20.

Song, X., Z. Zhou, J. Wang, F. Wu, and W. Gong (2004). Purification, characterization and preliminary crystallographic studies of a novel plant defensin from *Pachyrrhizus erosus* seeds. *Acta Crystallographia Section D: Biological Crystallography* 60:1121-1124.

Spelbrink, R.G., N. Dilmac, A. Allen, T.J. Smith, D.M. Shah, and G.H. Hockerman (2004). Differential antifungal and calcium channel-blocking activity among structurally related plant defensins. *Plant Physiology* 135:2055-3067.

Terras, F.R., K. Eggermont, V. Kovaleva, N.V. Raikhel, R.W. Osborn, A. Kester, S.B. Rees, S. Torrekens, F. Van Leuven, and J. Vanderleyden (1995). Small cysteine-rich antifungal proteins from radish: Their role in host defense. *Plant Cell* 7:573-588.

Thevissen, K., B.P. Cammue, K. Lemaire, J. Winderickx, R.C. Dickson, and R.L. Lester (2000). A gene encoding a sphingolipid biosynthesis enzyme determines the sensitivity of *Saccharomyces cerevisiae* to an antifungal plant defensin from dahlia (*Dahlia merckii*). *Proceedings of the National Academy of Science USA* 97:9531-9536.

Thevissen, K., K.K.A. Ferket, I.E.J.A. Francois, and B.P.A. Cammue (2003). Interactions of antifungal plant defensins with fungal membrane components. *Peptides* 24:1705-1712.

Thevissen, K., A. Ghazi, G.W. De Samblanx, C. Brownlee, R.W. Osborn, and W.F. Broekaert (1996). Fungal membrane responses induced by plant defensins and thionins. *Journal of Biological Chemistry* 271:15018-15025.

Thevissen, K., R.W. Osborn, D.P. Acland, and W.F. Broekaert (2000). Specific binding sites for an antifungal plant defensin from Dahlia (*Dahlia merckii*) on fungal cells are required for antifungal activity. *Molecular Plant-Microbe Interactions* 13:54-61.

Thomma, B.P.H.J., B.P.A. Cammue, and K. Thevissen (2002). Plant defensins. *Planta* 216:193-202.

Thomma, B.P.H.J., K. Eggermont, I.A.M.A. Penninckx, B. Mauch-Mani, R. Vogelsang, B.P.A. Cammue, and W.F. Broekaert (1998). Separate jasmonate-dependent and salicylate-dependent defense-response pathways in *Arabidopsis* are essential for resistance to distinct microbial pathogens. *Proceedings of the National Academy of Sciences USA* 95:15107-15111.

Tsien, R.W. and R.Y. Tsien (1990). Calcium channels, stores and isolations. *Annual Review of Cell Biology* 6:715-760.

Waldorf, A.R. and A. Polak (1983). Mechanisms of action of 5-fluorocytosine. *Antimicrobial Agents and Chemotherapeutics* 23:79-85.

Walsh, T.J., M.A. Viviani, E. Arathoon, C. Chiou, M. Ghannoum, A.H. Groll, and F.C. Odds (2000). New targets and delivery systems for antifungal therapy. *Medical Mycology* 38:335-347.

Wong, J.H. and T.B. Ng (2005a). Lunatusin, a trypsin-stable antimicrobial peptide from lima beans (*Phaseolus lunatus* L.). *Peptides* 26:2086-2092.

Wong, J.H. and T.B. Ng (2005b). Sesquin, a potent defensin-like antimicrobial peptide from ground beans with inhibitory activities toward tumor cells and HIV-1 reverse transcriptase. *Peptides* 26:1120-1126.

Wong, J.H. and T.B. Ng (2006a). Limenin, a defensin-like peptide with multiple exploitable activities from shelf beans. *Journal of Peptide Science* 12:341-346.

Wong, J.H. and T.B. Ng (2006b). Vulgurinin, a broad-spectrum antifungal peptide from haricot beans (*Phaseolus vulgaris*). *International Journal of Biochemistry and Cell Biology* 37:1626-1632.

Wong, J.H., X.Q. Zhang, H.X. Wang, and T.B. Ng (2006). A mitogenic defensin from white cloud bean (*Phaseolus vulgaris*). *Peptides* 27:2075-2081.

# Index

('i' indicates an illustration; 'n' indicates a note; 't' indicates a table)


2,4-D (dichlorophenoxyacetic acid) medium, 30, 32, 94, 95, 104

Abiotic stresses, 577-578, 579t, 585
  compatible solutes, 580, 581t-584t, 585-586
  tolerance strategies, 579-580
*AdhI* (alcohol dehydrogenase) marker gene, 205
Adventitious root nodulation, 439, 440t-442t
Agricultural Biotechnology Center (ABC), Hungary, 207
*Agrobacterium rhizogenes*, 57-58, 161-162
  *L. japonicus*, 310
  strains, 58, 59t, 153-154
*Agrobacterium tumefaciens*, 29, 33-34, 45, 111, 336, 343
  A209 wild-type strain, 181-182
  EHA101 strain, 259, 333
  LBA4404 strain, 135, 190
  MP90 C58, 320
  overgrowth control, 95
  strains, 49, 50t, 51t, 52, 73, 189-190, 235t-236t, 258t, 291, 309, 333t
  T-DNA instability, 83
*Agrobacterium*-mediated transformation
  alfalfa, 31-37
  *Arabidopsis*, 275-277
  azuki bean, 171-173
  chickpea, 149-152
  faba bean, 288, 297-298
*Agrobacterium*-mediated transformation *(continued)*
  Lotus species, 303-310
  low-ODAP grass pea, 181-183
  *Medicago truncatula*, 21-24
  meristematic cell regeneration, 91-94, 97
  organogenesis, 55-57, 92
  pea, 190-194, 191t
  peanut, 233, 235t-236t, 237-239, 254
  *phaseolus acutifolius*, 161
  pigeon pea, 135-139
  root endosymbiosis, 58-63
  somatic embryogenesis, 46-54, 94-96
  sonication-assisted (SAAT), 103, 113-119
  soybean, 91-94, 111-112
  transgene copies, 389-390
  white clover, 25-26
Agronomic characters, DNA marker use, 525
Agronomic field performance, 84
Agronomic limitations, 241
Albatross, faba bean cultivar, 290, 297
Alcohol dehydrogenase selection agent. *See AdhI*
Alfalfa (*Medicago sativa*), 12n-13n, 559
  *Agrobacterium rhizogenes*-mediated transformation, 57-63
  *Agrobacterium tumefaciens*-mediated transformation, 29-30, 34-37, 45-57
  digestibility improvement, 8-11

Alfalfa (*Medicago sativa) (continued)*
as forage,1, 3, 29
genetic maps, 565-566
HCT levels, 12n-13n
lignin synthesis, 9, 10-11
Met protein increase, 6
nutritional improvement, 3-8
protein utilization improvement, 11-12
Alfalfa callus, embryos, 36i
Alfalfa mosaic virus (AMV), 20
Allele-specific associated primers (ASAP), 491
Alpha tocopherol, 78i, 78
Amino acids, 70-71, 141
alfalfa, 3
defensins sequence, 619t
legumes, 163
*Lotus* spp., 307
soybean, 71
Amplified fragment length polymorphism (AFLP), 459, 480-481, 498, 533, 561t
Analysis techniques, transgenic plants, 387t
Angular leaf spot (ALS) resistance
common bean, 515
SCAR marker, 522t
Annual species, *Medicago*, 559
Anthracnose resistance
lentils, 525
SCAR marker, 522t
Anthranilate synthase (AS), soybean, 73
Antibiotic selection agents, 34, 53, 94
rooting, 138
Antioxidant treatment, SAAT coculture, 116-117, 117t
Antisense technology, 75, 178, 391
Aphids, cowpea, 162
*Aph*II marker gene, soybean, 94
*Arabidopsis* CγS (*At*CγS), 6, 7, 8
*Arabidopsis thaliana*, xxi, 340
*Agrobacterium in planta* transformation, 274, 276-277
*in planta* transformation, 276-277
Met synthesis, 6, 71, 72
napin promotion, 83-84
PA synthesis, 12
RNAi-induced silencing, 357

*Arabidopsis thaliana (continued)*
root infection, 608
vitamin E production, 79, 80, 82
Arbitrarily primed PCR (AP-PCR), 479
Array technology, 472. *See also* DNA Microarray analysis, Macroarray analysis, Microarray analysis
Ascochyta blight, 287
resistance, 525, 526
Aseptic seedlings, 134
Aspartate (Asp), alfalfa, 6, 8
*Aspergillus flavus*, peanut disease, 228, 242
Autotetraploidy, 565
Auxin (2,4-D) induction, 30, 32, 94, 95, 104
Auxins, 133
*Avena sativa*, electroporation, 218
Axillary meristem
legumes, 148
soybean, 92
Azuki bean (*Vigna angularis*)
*Agrobacterium* transformation, 171-173, 173i
epicotyl explants, 170-171
features, 169-170
genetic diversity analysis, 502
traditional crop, 169
weevils, 195

*Baby Boom* gene, 164
*Bacillus thuringiensis*, 139, 141, 149, 154, 155i
Bacterial suspension, SAAT, 117
Bacterial wilt, peanut, 243
*Bar* (phosphinothricin acetyl transferase) marker gene, 389
alfalfa, 53, 56, 61, 94
pea, 193, 194, 195
Barrel medic. *See Medicago truncatula*
BASTA solution, 293-294
Bath sonicators, 114
$BC_1$ transgenes, 396, 396t, 397
Bean bruchid, 163
Bean common mosaic virus (BCMV) resistance, SCAR marker, 522t

Bean curly top virus (BCTV) resistance, SCAR marker, 522t
Benzyl aminopurine (BAP) growth regulator, 92, 128, 129, 131, 132
β-N-oxalyl L-amino alanine (BOAA), grass pea toxicity, 177
β-N-oxalyl-L-α, β-diamino propionic acid (ODAP)
  degradation plasmids, 181
  grass pea toxicity, 177, 178
Binary vectors, chickpea transformation,149, 151i, 155i
Biofortification, peanut, 243-244
Bioinformatics tools, 346, 534
Biolistic transformation
  pea, 203-204
  pigeon pea, 139-140
Black-eyed pea. *See* Cowpea
Black gram bean (*Phaseolus mungo*)
  abiotic stresses, 579t
  cotyledonary nodes, 263
  genetic diversity, 502
Boisson-Dernier protocol, 61

*Caenorhabditis elegans*, RNAi technology, 353
Caffeic acid 3-O-methyl transferase (COMT), alfalfa, 9, 10, 10i, 39n
Caffeoyl CoA 3-O-methyl transferase (CCOMT), alfalfa, 9, 10, 10i
Callogenesis, 47i, 50t, 51t, 52
CaMV 35S promoter, 2, 4, 7, 84, 205, 404
Cauliflower mosaic virus (CaMV) 35S promoter, 2, 4, 7, 84, 151i, 182
cDNA-AFLP (amplified fragment length polymorphism), 413, 415i, 423-424, 429, 491
  protocols, 419-423
  systems, 416-419
  technique, 414
Cell selection, 53
Center for *Medicago* Genomics Research, Nobel Foundation, 345
Chickpea (*Cicer arietinum*)
  abiotic stresses, 579t
  cotyledonary node explant procedure, 148-152, 150i
  cotyledonary node explant results, 152-155
  cultivated/wild species relationship, 503-504
  embryo axes, 260, 261t
  genetic diversity analysis, 498
  linkage map, 510t, 513
  pulse crop, 147
  Pusa 362 genotype, 152t
  QTL analysis, 526, 528t
  trait tagging, 517t-518t
Chimeras, 56, 61, 108, 357
*Chlamydomonas reinhardtii*, electroporation, 218
Chloroplasts, abiotic stress sensitivity, 588-589
Chocolate spot disease, 287
*Cis* elements, trans-acting protein recognition, 404, 410
Class I transposable elements, 360, 363t
Class II transposable elements, 359-360, 363t
Cleaved amplified polymorphic sequences (CAPS), 490, 561t
"Clip 'n' squirt" transformation, 276
Clone recovery, particle bombardment, 108-109
Clover yellow vein virus (CYVV), 20
Cocultivation, 52, 91
  azuki bean transformation, 172
  chickpea transformation, 149-152
  faba bean transformation, 290
  SAAT, 116
  temperature, 61-62
  thiol, 93
  timing, 154
Color
  pigeon pea shoots, 137
  soybean tissue, 106
Common bean (*Phaseolus vulgaris*)
  abiotic stresses, 593
  comparative mapping analysis, 532-533
  cultivar identification, 509

Common bean (*Phaseolus vulgaris*) *(continued)*
- cultivated/wild species relationship, 504
- food use, 159
- genetic diversity analysis, 499-500
- insect resistance, 163
- linkage map, 510t, 511
- methionine increase, 163-164
- parasitic plants, 163
- QTL analysis, 529t, 531
- trait tagging, 519t
- virus resistance, 163

Comparative mapping analysis, 531-533
Composite plant, 47i, 57-58, 162
Compression technology, 75-76
Condensed tannins (CT), 9, 11, 20
- synthesis, 10i

Conserved orthologue set (COS), 485
Container, soybean sonication, 114-115, 115t
Copy number, QC-PCR, 462
Copy number, transgenic plants, 389-392
Cornstalk borer, 239
- resistance in peanut, 240

Cotyledonary node (CN)
- efficient transformation, 275
- protocol, 91
- regenerative cells, 148

Cotyledonary node (CN) explants, 23, 25, 27
- chickpea, 148-149, 155-156
- peanut, 254-257, 258t
- peanut preparation method, 264-266
- pigeon pea, 137
- soybean, 92-93
- tree legume, 319

Cowpea (*Vigna unguiculata*)
- comparative mapping, 533
- cultivated/wild species relationship, 505-506
- drought tolerance, 159
- genetic diversity analysis, 501
- insect resistance, 162
- linkage map, 510t, 514
- methionine increase, 163-164
- parasitic plants, 163
- QTL analysis, 529t, 531
- trait tagging, 520t, 524-525
- virus resistance, 163

Cowpea leaf beetle, 162
Cowpea mosaic virus (CPMV), 525
Cowpea weevils, 195
CP4 marker gene, soybean, 94
Cultivar protection, DNA marker use, 508
Cultivated and wild species relationship, 503-507
Culture media. *See also* Media
- alfalfa regeneration, 32-33, 50t, 51t, 52-53, 60
- embryogenic soybean, 106-107
- peanut regeneration, 229, 230t-232t

*Current Opinion in Plant Biology*, 343
Cysteine (Cys)
- legumes, 141
- Met synthesis, 6, 8, 71
- pea, 187, 203
- soybean, 71

Cystathionine γ-synthase (CγS), 6, 38
Cytokinins, shoot induction, 130

D20 cultures, soybean, 109, 110t, 113
DDBJ sequence database, 477, 483
De novo regeneration protocol
- faba bean, 288, 291-292
- preparation methodology, 289-291
- results, 294-296

Dedifferentiation, 52
Defensins, xxiii, 618
- amino acid sequence, 619t
- application in legumes, 622
- expression in transgenic plants, 622-623
- isolation in legumes, 623-624
- mechanisms of action, 620
- structure, 618-619

Differential display (DD) technique, 429, 439
Diploid alfalfa mapping, 565-566
Direct amplification of length polymorphism (DALP), 480
Directed amplification of minisatellite DNA PCR (DAMD-PCR), 489
Disease resistance
- DNA marker use, 515
- SCAR markers, 522t

Distal promoter region, 404

Diversity array technology (DArT), 492, 533
DNA amplification fingerprinting (DAF), 479
DNA-based markers, 477, 533-534
  comparative mapping analysis, 531-533
  cultivar/wild species relationship, 503-507
  diversity assessment, 498-503
  genome analysis, 507-508
  introgression characterization, 507-508
  linkage map construction, 508, 510t-511t, 511-515
  map-based cloning, 531-533
  quantitative trait loci (QTL) analysis, 525, 527t-529t, 526-531
  trait tagging, 515, 523-525
  uses, 497-498
DNA microarray analysis, 413-414, 468
DNA uptake, electric charge, 217
Dneasy Plant Mini Kit (Qiagen), 294
Dot blot hybridization, 391-392, 392i
*Drosophila* system, 346
DsRed fluorescent marker, 61
Dual-target PCR, 462

Eicosapentaenoic acid (EPA), 77
ELAK Ltd., Hungary, 207
Electrical discharge particle bombardment device, 107-108
Electron transport activity, abiotic stress reduction, 590-591, 591t
Electropermeability, plant cells, 218
Electroporation
  pea, 219-223
  peanut, 235t-236t, 241
  technique, 217-219
ELISA techniques, 387t, 394
EMBL sequence database, 477, 483
Embryo axis explants, 257-259, 260, 261t-262t, 263
Embryo origin, soybean, 105
Embryogenesis, 46, 275. *See also* Somatic embryogenesis
Embryonal axes, pigeon pea, 137
Endoplasmic reticulum (ER) protein retention, 3
Ensiling, protein loss, 12
Environmental stresses, xxii, 159, 287
Epicotyl explants, azuki bean, 170-171
Erucic acid, 76
Escapes, 56
Exon retrotransposon amplification polymorphism (ERAP), 487
Explants
  cotyledonary node (CN), 23, 25, 27, 92-93, 148-149, 155-156, 254-257, 258t, 264-266
  embryo axis, 257-259, 260, 261t-262t, 263
  high-cytokinin medium, 91
  pea, 191t
  peanut, 230t-232t
  pigeon pea, 128, 137
  *tumefaciens*-mediated transformation, 50t-51t
Expressed sequence tag (EST) libraries
  legumes, 428, 468-469
  *M. truncatula*, 471
Expressed sequence tag (EST)-based markers, 344, 483-485
Expressed sequence tag nucleotide polymorphism (ESTP), 484-485
Expressed sequence tag-simple nucleotide polymorphism (EST-SNP), 484
Expressed sequence tag-simple sequence repeat (EST-SSR), 483-484
Expression-based markers, 533
Expression pattern analysis, 413

Faba bean (*Vicia faba*)
  cultivars, 290, 297
  cultivated/wild species relationship, 506-507
  *de novo* regeneration protocol, 288, 291-292
  disease and stresses, 287
  embryo axes transformation system, 289, 292-294
  genetic diversity analysis, 503

Faba bean (*Vicia faba*) *(continued)*
genome size, 531
linkage map, 511t, 514
preparation methodology, 289-291
QTL analysis, 529t, 530
regeneration results, 294-296
trait tagging, 521t, 525
transformation difficulties, 288
Fast neutron technology, 359
Fatty acid desaturase gene expression, 77
Field trials, soybean, 70, 84
Finer Laboratory web site, 109
Flavonoid pathway, CT manipulation, 11-12
Floral branches, 49
Floral dip method, 277
Fluorescence *in situ* hybridization (FISH), 387t, 394
DNA sequences, 371
hybridization and probe detection, 378-381
imaging and analysis, 381-382
labeling probes, 378
*M. truncatula*, 372-375
preparations, 375-378
two-color detection, 381t
Fluorescent markers, 61
Flux control, 83
Folioles, 48, 50t-51t
Food fortification, 142
Forages
ensiling, 12
*Lotus* spp. as, 301, 302
ruminant diet, 8-9
traits, 19
Fusarium wilt resistance, chickpea, 522-523

GenBank databases, 483
Gene expression studies, transgenic plants, 394-395
Gene-silencing technologies, 178, 188, 357-358
GENEBOOSTER, 204, 207
cost, 211
optimal nitrogen pressure, 210t, 212
shooting distance, 210-211, 212
Genetic analysis, transgenic plants, 395-397
Genetic diversity
alfalfa, 570
azuki bean, 502
black gram, 502
chickpea, 498
common bean, 499-500
cowpea, 501
faba bean, 503
lentil, 503
markers, 498
pea, 499
peanut, 500-501
soybean, 499
Genetic engineering, alfalfa, 1
Genetic transformation, alfalfa, 30, 37-39
Genetics
history, 340-342
maps, 343
molecular tools, 343
Genome analysis, DNA marker use, 507-508
Genome sequencing, 344
Geranylgeranyldiphosphate (GGDP), 78i, 79, 80, 81, 82
Ghent University, Department of Molecular Genetics, 161
Gibberellic acid (GA), 131
Giza 2, faba bean cultivar, 290, 297
Global vegetable oil production, 76
Glucuronidase selection agent. *See* *Gus, uid*A
Glycinin, soybean, 75
Grass pea (*Lathyrus sativus*)
*Agrobacterium*-mediated transformation, 181-183
leaf explants, 179-180
low-ODAP clones, 181
low-toxin somaclones, 184
somatic embryogenesis, 183-184
toxicity, 177-178
winter pulse crop, 177
Green fluorescent protein (GFP) marker, 62, 115, 388
Groundnut rosette assistor virus (GRAV), 243

Growth regulators, 2,4-D, 30, 32, 94, 95, 104
BAP, 92, 128, 129, 150, 325
NAA, 32, 130, 131
peanut transformation, 230t-232t
TDZ, 130, 132, 189
*Gus* (glucuronidase) marker gene, 149, 153, 154, 155, 356, 405, 406
pea, 195, 205, 208
peanut, 257, 258t, 388
tree legume, 320, 321
*Gus:npt*II-based promoter tagging system, 407

Hairy root fragment regeneration, 62, 161
Hairy root transformation
*Lotus* spp., 304
T-DNA, 355
*Heliothis armigera* resistance, chickpea, 260
Herbicide markers, soybean, 94
HN (hemagglutinin neuraminidase) gene, 135, 141
Homogentisate geranylgeranyl transferase (HGGT), 78i, 79, 80, 81, 82
Homogentisic acid (HGA), 78i, 79, 80, 81, 82
Hormone-free media, 134
House dust mite, 308
*Hpt* marker gene
alfalfa transformation, 53
pea transformation, 193, 195
Hybridization and probe detection, *M. truncatula* for FISH, 377-381, 381t
Hydroxycinnamoyl transferase (HCT), alfalfa, 12n-13n
Hygromycin, 34, 53, 94, 193
*Robinia pseudoacacia*, 335-336

Imaging and analysis, *M. truncatula* for FISH, 381-382
Imbibition time, 93
*In planta* transformation
alfalfa, 20, 39n
*Arabidopsis*, 276
electroporation, 218-219
large seed legume embryo, 278-282
methods, 20, 356
*In silico* identification, 471
*In situ* hybridization, xxiii, 387t, 395
*L. japonicus*, 469-470, 470i
In vitro regeneration
approaches, 128, 275. *See also* *Agrobacterium*-mediated transformation, Embryogenesis, Organogenesis
avoiding, 45
difficulties of, xxii
tree legume, 317-319
In vivo tagging systems, 405-407
Indole-3-butyric acid (IBA), 91
Indole acetic acid (IAA), 130, 131, 132, 133, 134
Infrequent restriction site-PCR (IRS-PCR), 481
Insertional mutagenesis, 276
Integration, transgenic plants, 389-392
Integration site, transgenic plants, 393-394
Inter-MITE polymorphism (IMP), 487
Inter-retrotransposon amplified polymorphism (IRAP), 486
Inter-simple sequence repeat (ISSR), 482C
Introgression characterization, DNA marker use, 507-508
Intron-exon splice junction markers, 489
Isogenic transformation, white clover, 25-26, 26i
Isozyme markers, 477, 560, 561t

Jack, soybean cultivar, 106
Jemalong A 17, 372, 373i
Jemalong 2HA, *M. truncatula* genotype, 46-47, 47i, 50t
Jemalong, *M. truncatula* genotype, 46, 48, 59t

Kanamycin, 34, 53, 94, 135, 193, 324, 407, 408
Khesari. *See* Grass pea
Kinetin, 129

Labelling probes, *M. truncatula* for FISH, 378
Lathyrism, 177
Lauric oils, 76
LB (Luria-Bertani) medium
- alfalfa, 21
- azuki bean, 172
- faba bean, 291
- peanut, 264
- pigeon pea, 136

Leaf spot, peanut disease, 228, 267n
Legume-*rhizobium* interaction, xxiv, 19, 273, 427, 606
Legumes (Fabaceae)
- abiotic stresses, 577-578, 579t, 585
- amino acid composition, 70-71, 141, 163
- defensin, 617, 623-624
- divergence, 467
- DNA-based markers, 497-498
- EST libraries, 428
- high-quality protein, 19
- importance of, xxi, 605
- large-seed embryo *in planta* transformation, 278-282
- low productivity, 162
- nodulation capacity, 427
- parasitic nematodes, 605-606
- plant-nematode interaction, 606, 611-612
- SSH novel gene identification, legumes, 449-450
- storage fractions, 141
- transformation difficulties, 274-275
- "tropical tribe," 159

Lentil bean (*Lens culinaris*)
- abiotic stresses, 579t
- cultivated/wild species relationship, 507
- genetic diversity, 503
- linkage map, 511t, 512
- QTL analysis, 529t, 531
- trait tagging, 521t

Leucoanthocyanidin dioxygenase (ANS), 11, 12
Leucoanthocyanidin reductase (LAR), 11, 12
Lignin
- methylation, 9, 10i

Linkage map construction, DNA marker use, 508, 510t-511t, 511-515
Liquid medium protocols, 95
Long chain saturated fatty acids (LSFAs), peanut, 244
Long terminal repeats (LTRs), 486
*Lotus corniculatus*, 302, 304, 306, 308-309
*Lotus glaber*, 302, 304, 309
*Lotus japonicus*, 302, 303, 305, 309-310, 406
- comparative mapping, 531
- features/tools, 607i
- legume-rhizobia symbiosis, 606, 611-612
- linkage map, 511t, 515
- as model, 273, 354, 468
- parasitic nematode resistance, 609-610, 610i
- root infection, 608

*Lotus species*
- *Agrobacterium*-mediated transformation, 303-308
- agronomic use, 302
- in vitro manipulation, 302-303
- as model system, 302, 310
- protocols, 308-310
- tannins, 302, 306-307

Lucerne. *See Medicago truncatula*
Lysine (Lys), soybean, 71

Macroarray analysis, 469-472
Macroprojectiles, 207
Map-based cloning, 531-533
MAP markers employing sequence-specific primers, 483
Marker-free alfalfa transgenic plants, 39n
Markers, 343

Media. *See also* LB medium, MS medium, RIM medium, SIM medium, YEB medium
chickpea coculture, 149-152
embryonic soybean, 110t
faba bean transformation, 289-290
high-quality factors, 106
protocols, 95-96
regeneration, 32-33, 50t, 51t, 52-53, 60
Medicago Inc., 39
*Medicago* species, 559
mapping populations, 563i
markers, 560, 561t, 562
*Medicago truncatula* (lucerne, barrel medic), xxi, 3, 275
*Agrobacterium*-mediated transformation, 21-24
chromosome mapping, 372, 373i, 374i
comparative mapping, 531-532
EST libraries, 428, 471
explants, 48-49, 50t-51t
features/tools, 607i
fluorescence *in situ* hybridization (FISH), 372-375
genetic diversity, 570
genetic maps, 562, 564, 564i, 563i, 565
genome, 343, 344, 372
genotypes, 46-48, 59t
legume-rhizobia symbiosis, 606-609
linkage map, 511t, 515, 567
as model system, 19, 273, 339-346, 468, 560
morphological and molecular markers, 341i
references in literature, 347i, 347-348
researchers, 346-348
reverse genetics, 359-362, 363t
root infection, 608
*Tnt1* transposition, 361, 362, 364
transformation methods, 20, 46
Medicago.org, 347
Meiotic pachytene chromosome preparations, *M. truncatula* for FISH, 376-377
Mendel, Gregor, 340, 354
Meristematic cell proliferation, 96-97
faba bean, 289
Metabolomics, 345-346
Methionine (Met) protein
alfalfa sink source model, 7i
legumes, 141, 164
pea, 187, 203
peanut, 243
soybean, 71-73
synthesis, 4, 6, 71-72
Microarray analysis, 469-472
Microprojectile delivery
pea, 204-212
pigeon pea, 140-141
Microprojectiles, size, 209-210, 210t, 212
Microsatellite-based markers, 482-483, 488
Microsatellite-primed PCR (MP-PCR), 482
Microwounding, sonication, 113
Minisatellite-based markers, 488
Mitotic chromosome preparations, *M. truncatula* for FISH, 375-376
Model system
*Cajanus cajan* as, 142
definition, 339
*Lotus* species, as, 302, 310
*Medicago truncatula* as, 19, 273, 339-340
*Phaseolus acutifolius* as for common bean, 161
Molecular biology, alfalfa, 1
*Molecular Biology of the Gene*, 339
Monsanto and Forge Genetics International, 38
Morphological markers, 477
MS (Murashige and Skoog) medium
alfalfa, 21, 37
chickpea, 151
faba bean, 289
grass pea, 180
pea, 219
peanut, 230t-232t
pigeon pea, 131, 136
*Robinia*, 331
Multiple amplicon profiling (MAP) markers, 480
Multiple arbitrary amplicon profiling (MAAP) markers, 478-479
Multiple transgene control, 83-84

Mung bean (*Vigna radiata*)
abiotic stresses, 579t, 588-589, 590, 593
comparative mapping, 533
Mythos, faba bean cultivar, 290, 297

Naphthalene acetic acid (NAA) medium, 32, 130, 131, 132, 133, 134
Napin promoter, 83
NCBI sequence database, 477, 483
Neomycin phosphotransferase. *See* *Npt*II
*Nicotiana tabacum*, electroporation, 218
Nitrogen fixation
legumes, 187, 427, 467
pea, 203
research, xxi-xxii
Nitrogen gas pressure, microprojectile bombardment pea, 209, 210t, 212
Nitrogen seed sink, 196
Nodal explants, 49
Nodule-specific promoters, 2
Nodulin genes, 427
Non-PCR-based markers, 478
Nonstorage fraction, legume seed, 141
Northern hybridization, 387t, 394
*Npt*II (neomycin phosphotransferase) marker gene
alfalfa transformation, 53, 61
chickpea transformation, 151i
grass pea transformation, 182
pea, 205, 389
pigeon pea transformation, 135, 138, 140, 405, 406, 407

O-methyltransferase (OMT), 326
Oil crops, 76, 77
Optical density
chickpea transformation, 151
SAAT bacterial suspension, 117-119, 118t
Organogenesis, 46, 275
direct-shoot faba bean, 289
pigeon pea 128-134
*Orobanche*, 163

Parasitic nematodes, 605-606
Particle bombardment transformation, 103, 107-109
peanut, 235t-236t, 239-241
transgene integration, 390
Pasture bloat, 11, 20
PCR (polymerase chain reaction) analysis, 135, 138, 140, 220, 359, 362, 387t, 388, 409, 459-460
multiplex, 393i
PCR-based markers, 478-481
PCR restriction fragment length polymorphism (PCR-RFLP), 490
Pea (*Pisum sativum*)
abiotic stresses, 579t, 593
*Agrobacterium*-mediated transformation, 190-195
biolistic transformation, 203-240
cultivated/wild species relationship, 504
electroporation, 219-223
embryo axes, 261t, 263
genetic diversity analysis, 499
genome organization, 532
genome size, 531
genotypes/cultivars, 189, 190, 191t
linkage map, 510t, 512
pest resistance, 195-196
protein content, 187, 203
QTL analysis, 526, 527t, 530
trait tagging, 516t-517t, 523-524
Pea enation mosaic virus (PEMV), 220, 222
Peanut (*Arachis hypogaea*)
*Agrobacterium*-mediated transformation, 233, 235t-236t, 237-239
agronomic limitations, 241
bacterial disease resistance, 242-243
biofortification, 243-244
CN preparation method, 264-266
cotyledonary node explants, 254-257, 258t
cultivated/wild species relationship, 505
cultivation area, 227
efficient transformation, 233-234
electroporation, 235t-236t, 239

Peanut *(Arachis hypogaea*) *(continued)*
embryo axes, 257-259, 260, 261t-262t, 263
embryo *in planta* transformation, 278-282
explants, 230t-232t, 235t-236t
genetic diversity analysis, 500-501
genotypes/cultivars, 230t-232t, 235t-236t
linkage map, 510t, 513-514
oil quality, 244
particle bombardment, 235t-236t, 239-241
pest resistance, 242, 267n
stresses, 228
tissue culture methods, 229, 230t-232t
trait tagging, 520t, 523
value-added traits, 242
virus resistance, 243
Peanut allergenicity reduction, 266n
Peanut bud necrosis virus (IPCV), 243
Peanut mottle virus (PMV), 243
Peanut root knot nematode resistance, 523
Peanut stripe virus (PStV), 240, 243
Perennial species, *Medicago*, 559
Peste des petits ruminant virus (PPRV), 139
Petiolar leaf region explant, 139
Petioles, 48, 49
*Phaseolus acutifolius*, as model for common bean, 161
Phaseomics program, 160
Phosphinothricin selection agent. *See Bar*
Phytol kinase, 82
Phytyldiphosphate (PDP), 78i, 79, 80, 81, 82
Pigeon pea (*Cajanus cajan*)
abiotic stresses, 579t, 593
*Agrobacterium tumefaciens*-mediated transformation, 135-139
biolistic transformation, 139-140
crop constraints, 125-126
crop improvement, 126-128
cultivated/wild species relationship, 507
embryo axes, 259-260, 262t
Pigeon pea (*Cajanus cajan*) *(continued)*
microprojectile delivery, 140-141
as model system, 142
organogenesis, 128-134
protein content, 125, 141
somatic embryogenesis, 134
Plant-nematode interaction, 606, 611-612
Plantlet conversion, 47i, 52
Pleiotropic effects, 84
Plumules, pea, 206
Pod borer, cowpea, 162
Pod-sucking insect, cowpea, 162
Polygalacturonase inhibiting protein (PGIP), 196
Polyphenol oxidase (PPO), alfalfa, 12, 38
Pot-grown plants, 48
Primary transgenics (T0), 395, 396t
Proanthocyanidins (PA), 11, 20
Probe sonicators, 114
Progenetor electroporation system, 219
Proline, compatible solute, 581t, 586, 587, 589, 590, 592i
Promoters, xxiii, 404-405
alfalfa, 2-3
analysis, 408-409
CaMV 35S 2, 4, 7, 84, 205
choice, 83
napin, 83-84
tagging systems, 405-408
Proteomics, 345
Protocols
*Agrobacterium*-mediated transformation, 21-24
*Agrobacterium*-mediated transformation pea immature cotyledons, 198
*Agrobacterium*-mediated transformation pea lateral buds, 199
cDNA-AFLP (amplified fragment length polymorphism), 419-423
cotyledonary node, 91, 97, 112
*de novo* regeneration faba bean, 288, 291-296
fluorescence *in situ* hybridization (FISH), 375-382

Protocols *(continued)*
in vitro regeneration pea immature cotyledons, 197
liquid vs. solid medium, 95-96
microprojective bombardment pea, 206-208
particle bombardment, 103, 109-111
somatic embryogenesis, 94-96
sonication-assisted *Agrobacterium*-mediated transformation (SAAT), 103, 113-119
tree legume transformation, 319-323
Proximal core promoter, 403
Pusa 362, chickpea genotype, 152t

Quantitative competitive PCR (QC-PCR) analysis, 461i, 462
Quantitative PCR (Q-PCR) analysis, 459, 464i
copy number, 462-463
procedures, 460-461
transcript levels, 463-464
Quantitative trait loci (QTL) analysis
DNA marker use, 525-526, 527t-529t, 530-531
*M. sativa*, 569-570

R108 *M. truncatula* genotype, 46, 48, 50t, 51t, 59t, 372
R(everse)T(ranscription)-PCR analysis, xxiii, 138, 140, 387t, 389, 394, 414, 459-460, 463-464
Radicles, pea, 206
Random amplified microsatellite polymorphism (RAMPO), 482
Random amplified microsatellite polymorphisms (RAMPs), 482
Random amplified polymorphic DNA (RAPD), 479, 498, 533, 561t
Randomly amplified fingerprinting (RAF), 479-480
Recombinant inbred line (RIL) populations, *M. truncatula*, 562, 563i, 564i, 570
Red clover (*Trifolium pratense*)
ensiling, 12
PPO, 38
transformation protocols, 21
Regen S, alfalfa genotype, 30, 31
Regen SY alfalfa genotype, 34, 39n
Regen Y, alfalfa genotype, 30, 31
Regeneration, alfalfa, 31
*RepA* gene, cell cycle regulators, 164
Reporter genes, 405
Resistance genes (R genes), 490, 524
Restriction enzymes, 390
Restriction fragment length polymorphism (RFLP), 391, 394, 478, 498, 533, 561t
Restriction landmark genomic scanning (RLGS), 478
Retrotransposon based insertion polymorphism (RBIP), 487
Retrotransposon microsatellite amplified polymorphism (REMAP), 486
Retrotransposon/miniature-inverted repeat transposable element (MITE)-based markers, 485-487, 534
Retrotransposons, reverse genetics, 360-362, 363t
Reverse genetics
in legumes, 354-358
*Medicago truncatula*, 341-342, 361-362, 363t
in plants, 353-354
Reverse northern blot, cDNA-AFLP analysis, 422-423
*Rhizobia* bacteria
and legumes, 19, 467
and plants, 355
*Rhizobium meliloti*, 20, 48, 341
*Rhizobium*-legume symbiosis, 273, 606
RIM (root induction medium), peanut, 264
Rinderpest virus, 136, 141
RNAi (interfering RNA) technology, 357
*Robinia pseudoacacia*
assays, 335-336
honey production, 331
plantlet regeneration, 334-335
selection markers, 333

*Robinia pseudoacacia (continued)*
transformation procedure, 331-332, 336
*Rol* gene expression, 62
Root induction, media, 133-134
Root-knot nematodes (RKN), 606, 607-608, 608i
model legume resistance, 609-610, 611-612
Root selection, 61
Root-specific promoters, 2
Roundup Ready alfalfa, 38, 275
Roundup Ready soybean, 108
RPH (rinderpest hemagglutinin) gene, 135
Rubisco, abiotic stress reduction, 586, 587, 588i, 591
Rust resistance, 515
faba bean, 525
SCAR marker, 522t

S-adenosyl methionine (SAM), 71
S(ulfur)-amino acid deficiency, alfalfa, 3
S-methyl methionine (SMM), 7, 8, 71
*Saccharomyces cerevisiae*, reverse genetics, 353
Salt-stress suppression, 586, 587t
*Sclerotinia minor* resistance, peanut, 259
Secondary embryogenesis, 47i, 51t, 52
Seed specific promoters, 84
Selection agents
alfalfa, 34, 53
chickpea, 149, 153, 153i
peanut, 235t-236t, 237
*Robinia pseudoacacia*, 333
soybean, 94
Selective amplification biotin/restriction-mediated enrichment (SABRE) technique, 429
Selective amplification of microsatellite polymorphic loci (SAMPL), 483
Selective genic/intergenic amplification markers, 489-490
Selective subtractive hybridization (SSH)
applications, 433, 434t-438t, 439
clone large-scale analysis, 444-447
Selective subtractive hybridization (SSH) *(continued)*
clone small-scale analysis, 447-449
legumes, 449-450
setup, 439, 443-444
technique, 429-433, 431i
Sequence characterized amplified regions (SCARs), 490-491
Sequence-related amplification polymorphism (SRAP), 481
Sequence-specific amplified polymorphism (SSAP), 486
Sequence tagged microsatellite profiling (STMP), 488
Sequence tagged microsatellite sites (STMS), 489
Sequence-tagged sites (STS), 491, 561t
Serial analysis of gene expression (SAGE), 344-345, 413, 414
Shoot culture, *Robinia pseudoacacia*, 331-332
Shoot induction
media, 131-133
pigeon pea, 128-134
Shoot regeneration, pigeon pea, 128-129
Shooting distance, gene delivery, 210, 211t, 212
SIM (shoot induction) medium, peanut, 264
Simple sequence length polymorphism (SSLP), 459
Simple sequence repeats (SSRs), 488, 561t
Single copy T1 transgenics, 396, 396t
*Sinorhizobium meliloti*, 20, 48, 354
Slot blot hybridization, 391-392, 392i
Soil organisms, ODAP utilization, 177
Solid medium proliferation culture system, 95-96
Somatic embryo production, soybean, 94-95
Somatic embryogenesis, 46, 273-274, 356
difficulty of, 275
low-ODAP grass pea, 183-184
pigeon pea, 134
protocols, 94-96
soybean, 104-105

Sonication-assisted *Agrobacterium*-mediated transformation (SAAT), 103, 113-119
Sonication duration, 115-116
Sonicator, SAAT, 114
Southern bean mosaic virus (SBMV), 525
Southern blot hybridization, 135, 140, 221i, 322-323, 387t, 389-391, 396
Soybean (*Glycine max*)
  abiotic stresses, 590
  *Agrobacterium* interaction, 90-91
  *Agrobacterium*-mediated transformation, 91-94, 111-112
  color, 106
  composition, 74-75
  cultivated/wild species relationship, 504
  embryo explants, 261t
  embryogenic cultures, 104-107
  field trials, 70, 84
  genetic diversity analysis, 499
  genome organization, 532
  genome size, 531
  linkage map, 510t, 511-512
  Met increase, 71-73
  nematode infection, 605-606, 613n
  as oil crop, 69, 76-77
  QTL analysis, 528t-529t, 530
  SAAT, 103, 113-119
  somatic embryo production, 94-95
  trait tagging, 518t, 524
  Trp increase, 73-76
  uses, 69
  vitamin E enhancement, 77-82
Soybean cyst nematode (SCN), 605-606, 613n
Soybean sudden death syndrome (SDS), 472
Spodoptera resistance, peanut, 267n
Stacking multiple transgenes, 83
Stilbene synthase, 242
Storage fraction, legume seed, 141
Streptomycin, chickpea culture, 150
*Striga*, 163
*Striga gesnerioides* resistance, cowpea, 524-525
Subepidermal embryo origin, 105
Superoxide dismutase (Mn-SOD), 395
Suspension cultures, soybean, 109
Synteny, xxiii, 568

T(ransferred)-DNA
  *Agrobacterium*, 111, 112, 148
  binary vectors, 149, 151i
  instability, 83
  *M. truncatula*, 364
  technology, 355-358
  transkingdom, 33
T-DNA-based promoter tagging system, 407-408
Tannins, 9, 11, 20
  *Lotus* spp., 302, 306-307
Target region amplification polymorphism (TRAP), 485
TATA box, 404
Tetraploid species mapping, 566-568
Thidiazuron (TDZ), 130, 132, 189
  faba bean de novo regeneration, 288, 289, 294
Thiol use, soybean cotyledonary node transformation, 93
Threonine (Thr), soybean, 71
Thrips, cowpea, 162
TILLING (targeting induced local lesions in genomes)
  EMS, 355
  model legumes, 607t, 613
  technology, 358-359
Tissue culture, 395
Tissue culture independent transformation, 276
*Tnt1* transposition, *Medicago truncatula*, 361, 362, 364
Tobacco streak virus (TSV), 243
Tocopherols, 77, 78i
  biosynthesis, 78i, 79-80
  R1 soybean seeds, 81t
Toxicity, and antisense technology, 178
Trait tagging, DNA marker use, 515, 523-525
Transcript-derived fragments (TDFs)
  analysis, 422
  cDNA-AFLP, 415i, 418
  isolation, 421-422
Transcriptional regulation, 83
Transcriptomics, 344

Transformation, 343
Transformation efficiency
plantlets, 54
transgenic roots, 61-62
Transformation occurrence analysis, transgenic plants, 387t, 387-389
Transformed roots, 58, 61
Transgene silencing, 83
Transgenic plants
gene expression studies, 394-395
genetic analysis, 395-397
integration and copy number, 389-392
integration site, 393-394
transformation occurrence analysis, 387t, 387-389
variation, 385-386
Transkingdom DNA transfer, 33-34
Transposon display (TD), 487
Transposons, 485
Tree legume (*Leucaena leucocephala*)
explants, 318t
forestry uses, 319
in vitro regeneration, 319-317
protocol, 319-323
results, 323-326
Tryptophan (Trp), soybean, 71, 73-76
Tungsten microcarriers, 207
optimal size, 210t, 212

U. S. Animal and Plant Health Inspection Service (APHIS), 70
*Uid*A (glucuronidase) marker gene, 389
chickpea transformation, 151i
grass pea transformation, 182
pigeon pea transformation, 135, 140, 406
Universally primed PCR (UP-PCR), 479

Vaccine, rinderpest, 136
Vaccines, plant-derived, 142
Vacuum infiltration transformation, 20
Value-added traits, 242
Variation, transgenic plants, 385-386
*Vigna vexillata*, pod borer resistance, 162
*Vir* genes, T-DNA transfer, 112
Virus-induced gene silencing (VIGS), 188
Vitamin E enhancement
oil crops, 80-82
soybean, 77-79

Western blot analysis, 138, 387t, 394
White clover (*Trifolium repens*), 19
*Agrobacterium*-mediated transformation, 21-24
isogenic transformation, 25-26, 26i
transformation protocol, 20-21
virus resistance, 20
White clover mosaic virus (WCMV), 20
Winter hardiness
alfalfa, 569
lentil, 531
*Wushel* gene, 164

Yeast extract broth (YEB) medium, 34, 37, 150, 291

Zein protein
accumulation, 4-6, 5i, 8
as Met sink, 7i